TODAY'S TECHNICIAN™

T0198113

SHOP MANUAL

For Automotive Engine Performance

TODAY'S TECHNICIAN™

SHOP MANUAL

For Automotive Engine Performance

SEVENTH EDITION

Ken Pickerill
Southern Illinois University at Carbondale

CENGAGE
Learning®

Australia • Brazil • Japan • Korea • Mexico • Singapore • Spain • United Kingdom • United States

Today's Technician: Automotive Engine Performance, Seventh Edition

Ken Pickerill

SVP, GM Skills & Global Product Management: Jonathan Lau

Product Director: Matthew Seeley

Senior Product Manager: Katie McGuire

Senior Director, Development: Marah Bellegarde

Senior Product Development Manager: Larry Main

Senior Content Developer: Meaghan Tomaso

Product Assistant: Mara Ciacelli

Vice President, Marketing Services: Jennifer Ann Baker

Marketing Manager: Jonathan Sheehan

Senior Production Director: Wendy Troeger

Production Director: Andrew Crouth

Senior Content Project Manager: Cheri Plasse

Managing Art Director: Jack Pendleton

Cover image(s): Umberto Shtanzman/ Shutterstock.com

Library of Congress Control Number: 2016959776

Book Only ISBN: 978-1-305-95827-2
Package ISBN: 978-1-305-95828-9

Cengage Learning
20 Channel Center Street
Boston, MA 02210
USA

Cengage Learning is a leading provider of customized learning solutions with employees residing in nearly 40 different countries and sales in more than 125 countries around the world. Find your local representative at **www.cengage.com**

Cengage Learning products are represented in Canada by Nelson Education, Ltd.

To learn more about Cengage Learning, visit **www.cengage.com**

Purchase any of our products at your local college store or at our preferred online store **www.cengagebrain.com**

Notice to the Reader

Publisher does not warrant or guarantee any of the products described herein or perform any independent analysis in connection with any of the product information contained herein. Publisher does not assume, and expressly disclaims, any obligation to obtain and include information other than that provided to it by the manufacturer. The reader is expressly warned to consider and adopt all safety precautions that might be indicated by the activities described herein and to avoid all potential hazards. By following the instructions contained herein, the reader willingly assumes all risks in connection with such instructions. The publisher makes no representations or warranties of any kind, including but not limited to, the warranties of fitness for particular purpose or merchantability, nor are any such representations implied with respect to the material set forth herein, and the publisher takes no responsibility with respect to such material. The publisher shall not be liable for any special, consequential, or exemplary damages resulting, in whole or part, from the readers' use of, or reliance upon, this material.

Printed in the United States of America
Print Number: 09   Print Year: 2024

DEDICATION

I want to dedicate this book to the memory of my father, Miles K. Pickerill. Dad taught me love, honesty, and craftsmanship.

Also to Peggy, my wife of 40 years, and my two sons, Adam and Eric.

My thanks for their contributions to my career go out to:

The late Doug McNally, who gave me my first automotive job.

The late John Riddle, who guided me as a young apprentice many years ago.

Dennis Price, who always believed in my ability.

Ken Pickerill

CONTENTS

PHOTO SEQUENCES

JOB SHEETS

Thanks to the support the *Today's Technician™ Series* has received from those who teach automotive technology, Cengage Learning, the leader in automotive-related textbooks, is able to live up to its promise to provide new editions of the series every few years. By revising this series on a regular basis, we can respond to changes in the industry, changes in technology, changes in the certification process, and to the ever-changing needs of those who teach automotive technology.

The *Today's Technician™ Series* features textbooks and digital learning solutions that cover all mechanical and electrical systems of automobiles and light trucks. The individual titles correspond to the ASE Education Foundation and National Institute for Automotive Service Excellence (ASE) certification areas and are specifically correlated to the 2013 standards for Automotive Service Technicians (AST) and Master Automotive Service Technicians (MAST).

The mission of ASE Education Foundation is to improve the quality of automotive technician training programs. ASE Education Foundation evaluates programs against standards developed by the automotive industry and recommends qualified programs for certification (accreditation) from ASE. ASE Education Foundation national standards reflect the skills that students must master to be successful. ASE certification through ASE Education Foundation ensures that certified training programs meet or exceed industry-recognized, uniform standards of excellence. All titles in the *Today's Technician™ Series* include remedial skills and theories common to all of the certification areas, and advanced or specific subject areas that reflect the latest technological trends, such as this updated title on engine performance.

The technician of today and the future must know the underlying theory of all automotive systems and be able to service and maintain those systems. Dividing the material into two volumes, a Classroom Manual and a Shop Manual, provides the student with the information needed to begin a successful career as an automotive technician without interrupting the learning process by mixing cognitive and performance learning objectives into one volume.

The design of Delmar's *Today's Technician™ Series* was based on features that are known to promote improved student learning. The design was further enhanced by a careful study of survey results, in which the respondents were asked to value particular features. Some of these features can be found in other textbooks, while others are unique to this series.

Each Classroom Manual contains the principles of operation for each system and subsystem. The Classroom Manual also discusses design variations in key components used by the different vehicle manufacturers, and considers emerging technologies that will be standard or optional features in the near future. This volume is organized to build upon basic facts and theories. Its primary objective is to help the reader gain an understanding of how each system and subsystem operates. This understanding is necessary to diagnose the complex automobiles of today and tomorrow. Although the basics contained in the Classroom Manual provide the knowledge needed for diagnostics, diagnostic procedures appear only in the Shop Manual. An understanding of the underlying theories is also a requirement for competence in the skill areas covered in the Shop Manual.

A spiral-bound Shop Manual delivers hands-on learning experiences with step-by-step instructions for diagnostic and repair procedures. Photo Sequences are used to illustrate some of the common service procedures. Other common procedures are listed and are accompanied with fine line drawings and photos that let the reader visualize and conceptualize the finest details of the procedure. This volume explains the reasons for performing the procedures, as well as the circumstances when each particular service is appropriate.

The two volumes are designed to be used together and are arranged in corresponding chapters. Not only are the chapters in the volumes linked together, the contents of the chapters are also linked. The linked content is indicated by marginal callouts that refer the reader to the chapter and page where the same topic is addressed in the companion volume. This valuable feature saves users the time and trouble of searching the index or table of contents to locate supporting information in the other volume. Instructors will find this feature especially helpful when planning the presentation of material and when making reading assignments.

Both volumes contain clear and thoughtfully selected illustrations, many of which are original drawings or photos specially prepared for inclusion in this series. This means that the art is a vital part of each textbook and not merely inserted to increase the number of illustrations.

The page design of this series uses available margin space to deliver helpful information efficiently without interrupting the pedagogical lesson material. This information includes examples of concepts just introduced in the text, explanations or definitions of terms that are not defined in the text, examples of common trade jargon used to describe a part or operation, and unique applications of the system or service described in the text. Many textbooks also include this information but insert it in the main body of text; this tends to interrupt the reader's thought process. By placing this information to the side of the main text, students can read through the text uninterrupted and refer to the additional information when it is best for them.

Jack Erjavec,
Series Editor

HIGHLIGHTS OF THIS NEW EDITION—CLASSROOM MANUAL

We are very proud of the changes made to make the seventh edition of *Today's Technician™: Automotive Engine Performance*. The Classroom Manual has been updated to reflect current changes in automotive engine performance, while retaining some of the more pertinent information on older models that are still in wide use. Engine performance technology is often seen by students as an array of complicated subjects, all of which can be difficult to understand. This is why *Automotive Engine Performance* follows a natural progression of subjects, including basic theories, engine construction, electrical theory, and computer control, to bring all these topics together. More experienced students and working technicians can review the basic theories and then move on to more advanced areas in the latter half of the text.

This edition features an even further investigation of digital fuel injection (DFI) use and operation, pulse width modulated fuel pump control, and variable valve timing, along with expanded and updated coverage of hybrid vehicle technology. Coverage of ignition systems and fuel injection has been updated with coverage that reflects the vehicles in use today. This edition continues to present the latest in sensor technology and new developments in automotive engine performance. The Classroom Manual also has new ASE-style multiple-choice questions to help assess student understanding of the material.

HIGHLIGHTS OF THIS NEW EDITION—SHOP MANUAL

The *Automotive Engine Performance*, seventh edition, Shop Manual follows the same natural progression of the Classroom Manual but focuses on the skill set required to be a successful technician in today's shop. The Shop Manual starts with shop procedures and ASE certification, progresses through electrical testing, basic engine testing, and scan tool procedures, and finally through the most advanced diagnostics used today. The coverage includes new basic engine technology, such as variable valve timing, while retaining a thorough coverage of basic engine testing such as vacuum, compression, cylinder leakage, and engine balance that are the cornerstone of engine performance diagnosis. Students are then led through electrical troubleshooting. This coverage includes basic multimeter, test lamp, logic probe, and oscilloscope pattern reading and use. Electrical wiring and connector repair are also covered, and a review of battery starting and charging diagnostics is given. Engine performance ignition systems coverage reflects modern usage of coil-over-plug systems, while retaining some distributor ignition coverage. Fuel injection system coverage reflects port fuel and direct fuel injection. Engine control systems such as variable valve timing, fuel systems such as returnless fuel systems, pulse width modulated fuel pump control, and emission controls such as PCV, EGR, and evaporative emissions are covered in depth. A section on computer operation and networking is included, as well as a chapter on related systems to bring the operation of the whole vehicle together for the student. This broad coverage makes this edition suitable for a wide range of student capability, from beginning technician to a more seasoned veteran. Other important updates are updated photo sequences. Many job sheets have been added to provide complete coverage of MAST, AST, and MLR tasks.

CLASSROOM MANUAL

Features of the Classroom Manual include the following:

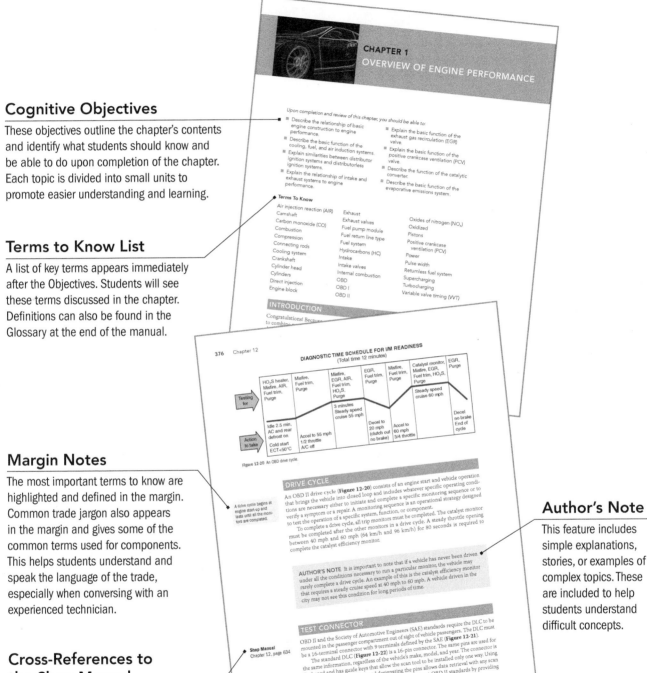

Cognitive Objectives

These objectives outline the chapter's contents and identify what students should know and be able to do upon completion of the chapter. Each topic is divided into small units to promote easier understanding and learning.

Terms to Know List

A list of key terms appears immediately after the Objectives. Students will see these terms discussed in the chapter. Definitions can also be found in the Glossary at the end of the manual.

Margin Notes

The most important terms to know are highlighted and defined in the margin. Common trade jargon also appears in the margin and gives some of the common terms used for components. This helps students understand and speak the language of the trade, especially when conversing with an experienced technician.

Cross-References to the Shop Manual

References to the appropriate page in the Shop Manual appear whenever necessary. Although the chapters of the two manuals are synchronized, material covered in other chapters of the Shop Manual may be fundamental to the topic discussed in the Classroom Manual.

Author's Note

This feature includes simple explanations, stories, or examples of complex topics. These are included to help students understand difficult concepts.

A Bit of History

This feature gives the student a sense of the evolution of the automobile. This feature not only contains nice-to-know information, but also should spark some interest in the subject matter.

Summary

Each chapter concludes with a summary of key points from the chapter. These key points help the reader review the chapter contents.

Review Questions

Short-answer essays, fill-in-the-blanks, and multiple-choice questions are found at the end of each chapter. These questions are designed to accurately assess the student's competence in the stated objectives at the beginning of the chapter.

SHOP MANUAL

To stress the importance of safe work habits, the Shop Manual also dedicates one full chapter to safety. Other important features of this manual include:

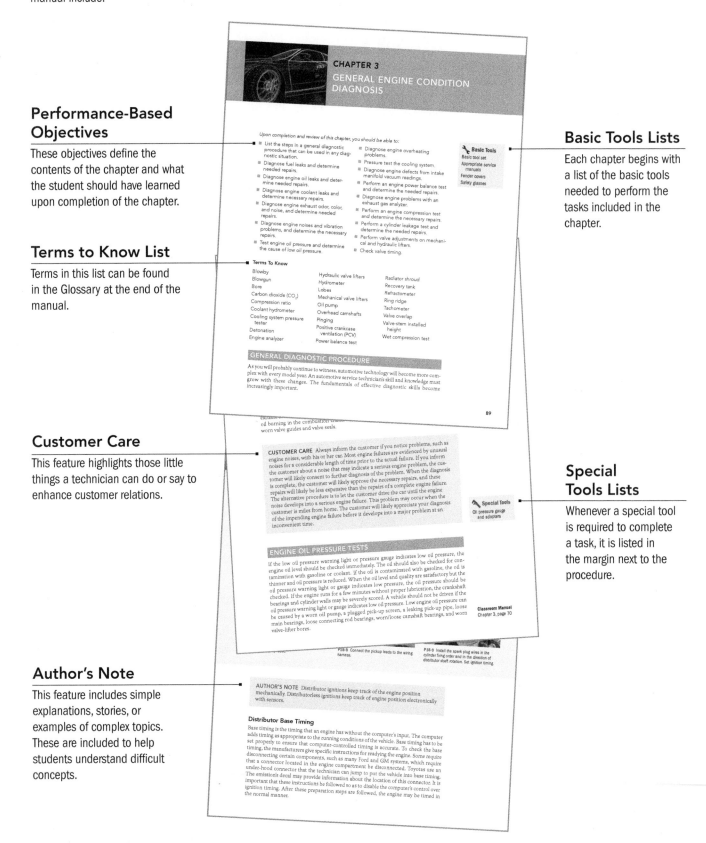

Performance-Based Objectives

These objectives define the contents of the chapter and what the student should have learned upon completion of the chapter.

Terms to Know List

Terms in this list can be found in the Glossary at the end of the manual.

Customer Care

This feature highlights those little things a technician can do or say to enhance customer relations.

Author's Note

This feature includes simple explanations, stories, or examples of complex topics. These are included to help students understand difficult concepts.

Basic Tools Lists

Each chapter begins with a list of the basic tools needed to perform the tasks included in the chapter.

Special Tools Lists

Whenever a special tool is required to complete a task, it is listed in the margin next to the procedure.

Photo Sequences

Many procedures are illustrated in detailed photo sequences. These detailed photographs show the students what to expect when they perform particular procedures. They also can provide the student a familiarity with a system or type of equipment, which the school might not have.

References to the Classroom Manual

References to the appropriate page in the Classroom Manual appear whenever necessary. Although the chapters of the two manuals are synchronized, material covered in other chapters of the Classroom Manual may be fundamental to the topic discussed in the Shop Manual.

Service Tips

Whenever a shortcut or special procedure is appropriate, it is described in the text. Generally, these tips describe common procedures used by experienced technicians.

Warnings and Cautions

Cautions appear throughout the text to alert readers to potentially hazardous materials or unsafe conditions. Warnings advise the students of things that can go wrong if instructions are not followed or if an incorrect part or tool is used.

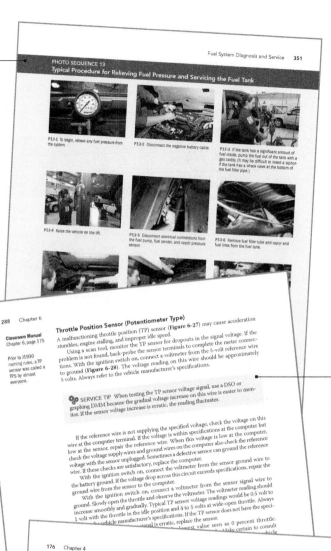

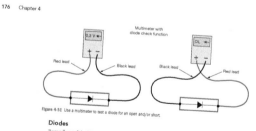

Case Studies

Each chapter ends with a Case Study describing a particular vehicle problem and the logical steps a technician might use to solve the problem. These studies focus on system diagnosis skills and help students gain familiarity with the process.

ASE-Style Review Questions

Each chapter contains ASE-Style Review Questions that reflect the performance objectives listed at the beginning of the chapter. These questions can be used to review the chapter as well as to prepare for the ASE certification exam.

Diagnostic Charts

Some chapters include detailed diagnostic charts that list common problems and most probable causes. They also list a page reference in the Classroom Manual for the student to use to gain a better understanding of the system's operation and a page reference in the Shop Manual to refer the student to details on the procedure necessary for correcting the problem.

---

650   Chapter 12

CASE STUDY

A customer brought in his OBD II–equipped car into the dealership complaining of poor fuel economy. The customer stated that the gas mileage has been declining since the car was new. This is not normal. Gas mileage normally improves slightly as the engine is broken in, the customer had no other complaints. Because this is a very difficult problem to verify, the technician began the diagnostic process with a visual inspection and found nothing out of the ordinary. The car's MIL was not lit.

He then connected a scan tool to the DLC and reviewed the data. Comparing the input data being displayed to the normal range of values listed in the service information, he discovered that the upstream O_2 sensor was biased rich. That meant the PCM was seeing a lean condition and adding fuel to correct for this problem. To verify this, the technician watched the fuel trim. Sure enough, the long-term FT had moved to add more fuel. Normally this is caused by a vacuum leak or restricted fuel injectors. The latter probable cause seemed unlikely since the O_2 sensor showed that added fuel was being delivered. Therefore, the technician began to look for a possible cause of a vacuum leak.

The process continued for quite some time as he checked all of the vacuum hoses and components. Nothing appeared to be leaking. As he leaned over to check something in the back of the engine, he heard a slight "Pfft." The noise had somewhat of a rhythm to it, and he focused his attention to it. It did not sound like a vacuum leak. As he increased the engine's speed, the noise pulses became closer together and soon became a constant noise. The noise appeared to be coming from the lower part of the engine.

Using a stethoscope, he was able to identify the source of the noise. It appeared that the gasket joining the exhaust manifold to the exhaust pipe was not seated properly. To verify this, he raised the car on a hoist and took a look. He found that the retaining bolts were loose. He took a quick look at the gasket and found it to be in reasonable shape, then tightened the bolts to specifications.

Not sure that the noise or the exhaust leak was related to the problem but suspected that it could be, he connected a lab scope to the oxygen sensor and watched its activity. The sensor's signal no longer showed a bias. It appears that the leak in the exhaust was pulling air into the exhaust between each pulse. This was adding oxygen to the exhaust stream causing the computer to think the mixture was lean.

ASE-STYLE REVIEW QUESTIONS

1. While discussing the adaptive learning of a PCM, *Technician A* says that when the battery is disconnected from the vehicle, the learning process resets. *Technician B* says that the vehicle must be driven under part throttle with moderate acceleration for the PCM to relearn the system.
   Who is correct?
   A. A only
   B. B only
   C. Both A and B
   D. Neither A nor B

2. *Technician A* says that P0 codes with definitions can be accessed through a generic scan tool.

3. *Technician A* says that if the PCM is correcting for a lean condition, the long-term FT will indicate a negative number.
   Technician B says that a rich condition causes the long-term FT to show a positive number.
   Who is correct?
   A. A only
   B. B only
   C. Both A and B
   D. Neither A nor B

Who is correct?
A. A only
B. B only
C. Both A and B
D. Neither A nor B

Onboard Diagnostic (OBD II) System Diagnosis and Service   635

DTC PO108 MAP sensor circuit high voltage

Step	Action	Values	Yes	No
1	Was the Powertrain On-Board Diagnostic (OBD) System Check performed?	—	Go to Step 2	Go to Powertrain OBD System Check 2.4 L Powertrain OBD System Check 2.2 L
2	1. Install a scan tool 2. Engine at idle. Does the scan tool display the MAP voltage specified?	4.0V	Go to Step 3	Go to Step 4
3	1. Turn the ignition switch OFF. 2. Disconnect the MAP sensor electrical connector. 3. Turn the ignition switch ON. Does the scan tool display the MAP voltage specified?	1.0V	Go to Step 5	Go to Step 6
4	1. Turn the ignition switch ON, with the engine OFF, review freeze frame data. 2. Operate the vehicle within the freeze-frame conditions. Does the scan tool display the MAP voltage specified?	4.0V	Go to Step 3	Go to diagnostic aids
5	Probe the MAP sensor signal ground circuit with a test light connected to battery voltage. Does the test light illuminate?	—	Go to Step 7	Go to Step 11
6	Check the MAP sensor signal circuit for a short to voltage and repair as necessary. Was a repair necessary?	—	Go to Step 14	Go to Step 12
7	With a DVM connected to ground, probe the 5-volt reference circuit. Does the DVM display the specified voltage?	5V	Go to Step 8	Go to Step 9
8	Check the MAP sensor vacuum source for being plugged or leaking. Was a problem found?	—	Go to Step 10	Go to Step 13
9	Check the 5-volt reference circuit for a short to voltage and repair as necessary.	—	Go to Step 14	Go to Step 12
10	Repair the vacuum hose as necessary. Is the action complete?	—	Go to Step 14	
11	Check for an open in the MAP sensor ground circuit and repair as necessary. Was a repair necessary?	—	Go to Step 14	Go to Step 12
12	Replace the PCM. Is the action complete?	—	Go to Step 14	
13	Replace the MAP sensor. Is the action complete?	—	Go to Step 14	—
14	1. Using the scan tool, clear the DTCs. 2. Start the engine and idle at normal temperature. 3. Operate the vehicle within the conditions for setting the DTC. Does the scan tool indicate that the diagnostic ran and passed?	—	Go to Step 16	Go to Step 2
15	Check for any additional DTCs. Are any DTCs displayed that have not been diagnosed?	—	Go to specific DTC charts	System OK

Figure 12-26 Diagnostic chart for a DTC–P0108.

Job Sheets

Located at the end of each chapter, the Job Sheets provide a format for students to perform procedures covered in the chapter. A reference to the ASE Task addressed by the procedure is included on the Job Sheet.

Name _____

Date _____

Diagnosing Related Systems   683

INSPECTING DRIVE BELT

Upon completion of this job sheet, you should be able to visually inspect a serpentine drive belt and check its tightness.

JOB SHEET
58

ASE Education Foundation Correlation

This job sheet addresses the following **AST/MAST** task: VIII. Engine Performance; A. General: Engine Diagnosis

Task #3   Diagnose abnormal engine noises or vibration concerns; determine needed action. **(P-3)**

Tools and Materials
- Vehicle with a serpentine belt
- Service information for the above vehicle

Describe the vehicle being worked on:

Year _____

Model _____   Make _____

VIN _____

Procedure

1. Carefully inspect the belt and describe the general condition.

_____
_____
_____

2. Check the tension of the belt. Belt tension and wear are generally assessed by looking at a wear indicator on the belt tensioner. (See Service Information.)
   You found _____

3. Based on the above, what are your recommendations?

_____

ASE Challenge Questions

Each technical chapter ends with five ASE Challenge Questions. These are not mere review questions; rather, they test the students' ability to apply general knowledge to the contents of the chapter.

680   Chapter 13

ASE CHALLENGE QUESTIONS

1. *Technician A* says that improper tire sizes can cause the speedometer to be inaccurate.
   Technician B says that the PCM can be recalibrated to correct this condition.
   Who is correct?
   A. A only
   B. B only
   C. Both A and B
   D. Neither A nor B

2. While discussing a vibration problem,
   Technician A says as a general rule, a U-joint problem is most noticeable in the 30 to 60 mph range.
   Technician B says that tire balance problems are most noticeable on acceleration and deceleration.
   Who is correct?
   A. A only
   B. B only
   C. Both A and B
   D. Neither A nor B

3. *Technician A* says that a four-wheel alignment tells the technician whether the rear axle or wheels are square with the front wheels.
   Technician B says that total toe for all four wheels must be determined and rear toe adjusted where possible to bring the rear axle or wheels into square with the chassis.

4. While discussing electronically controlled transmissions,
   Technician A says that line pressure is controlled by a pulse width modulated solenoid.
   Technician B says that the solenoid is controlled by the PCM.
   Who is correct?
   A. A only
   B. B only
   C. Both A and B
   D. Neither A nor B

5. While discussing wheel alignment angles,
   Technician A says that camber angle changes while driving.
   Technician B says that the camber angle also changes when the vehicle is loaded and sags under the weight.
   Who is correct?
   A. A only
   B. B only
   C. Both A and B
   D. Neither A nor B

SUPPLEMENTS

Instructor Resources

The *Today's Technician* series offers a robust set of instructor resources, available online at Cengage's Instructor Resource Center and on DVD. The following tools have been provided to meet any instructor's classroom preparation needs:

- An Instructor's Guide provides lecture outlines, teaching tips, and complete answers to end-of-chapter questions.
- PowerPoint presentations include images, videos, and animations that coincide with each chapter's content coverage.
- Cengage Learning Testing Powered by Cognero® delivers hundreds of test questions in a flexible, online system. You can choose to author, edit, and manage test bank content from multiple Cengage Learning solutions and deliver tests from your Learning Management System, or you can simply download editable Word documents from the DVD or Instructor Resource Center.
- An Image Gallery includes photos and illustrations from the text.
- The Job Sheets from the Shop Manual are provided in Word format.
- End-of-Chapter Review Questions are also provided in Word format, with a separate set of text rejoinders available for instructors' reference.
- To complete this powerful suite of planning tools, a pair of correlation guides map this edition's content to the ASE Education Foundation tasks and to the previous edition.

REVIEWERS

The author and publisher would like to extend a special thanks to the following reviewers for their contributions to this text:

Terry L. Enyart
University of Northwestern Ohio
Lima, OH

Tim Mulready
University of Northwestern Ohio
Lima, OH

Chris McNally
Hudson Valley Community College
Troy, NY

Christopher D. Parrott
Vatterott College
Wichita, KS

CHAPTER 1
TOOLS AND SAFETY

Upon completion and review of this chapter, you should be able to:

- Understand the importance of safety and accident prevention in an automotive shop.
- Explain the basic principles of personal safety, including protective eyewear, clothing, gloves, shoes, and hearing protection.
- Demonstrate proper lifting procedures and precautions.
- Explain the procedures and precautions for safely using tools and equipment.
- Follow safety precautions regarding the use of power tools.
- Demonstrate proper safety precautions during the use of compressed-air equipment.
- Demonstrate proper vehicle lift operating and safety procedures.
- Observe all safety precautions when hydraulic tools are used in the automotive shop.
- Follow the recommended procedure while operating hydraulic tools such as presses, floor jacks, and vehicle lifts to perform automotive service tasks.

- Follow safety precautions while using cleaning equipment in the automotive shop.
- Observe all shop rules when working in the shop.
- Operate vehicles in the shop according to shop driving rules.
- Explain what should be done to maintain a safe work area, including handling vehicles in the shop and venting carbon monoxide gases.
- Fulfill employee obligations when working in the shop.
- Describe the Automotive Service Excellence (ASE) technician testing and certification process, including the nine areas of certification.
- Follow safety precautions while handling hazardous waste materials.
- Dispose of hazardous waste materials in accordance with state and federal regulations.

Terms To Know

Abrasive cleaning
ASE blue seal of excellence
Carbon monoxide (CO)
Chemical cleaning
Closed-loop system
Corrosive
Environmental Protection Agency (EPA)
Floor jack

Hazard Communication Standard
Hydraulic press
Jack stand
Lift
Material safety data sheets (MSDS)
National Institute for Automotive Service Excellence (ASE)

Neutral
Occupational Safety and Health Act (OSHA)
Park
Particulates
Pneumatic tools
Power tools
Reactive

Resource Conservation
  and Recovery Act (RCRA)
Right-to-know laws
Safety glasses

Steam cleaning
Sulfuric acid
Thermal cleaning

Toxic
Workplace Hazardous
  Materials Information
  Systems (WHMIS)

INTRODUCTION

Safety and accident prevention must be top priorities in all automotive shops. There is great potential for serious accidents simply because of the nature of the business and the equipment used. In fact, the automotive repair industry is rated as one of the most dangerous occupations in the United States.

Vehicles, equipment, and many parts are very heavy, and parts often fit tightly together. Many components become hot during operation, and high fluid pressures can build up inside the cooling system, fuel system, or battery. Batteries contain highly corrosive and potentially explosive acids. Fuels and cleaning solvents are flammable. Exhaust fumes are poisonous. During some repairs, technicians can be exposed to harmful dust particles and vapors.

Good safety practices eliminate these potential dangers. A careless attitude and poor work habits invite disaster. Shop accidents can cause serious injury, temporary or permanent disability, and death. Therefore, safety is a very serious matter. Both the employer and employees must work together to protect the health and welfare of all who work in the shop.

This chapter contains many safety guidelines concerning personal safety, tools and equipment, and work areas, as well as some of the responsibilities you have as an employee and responsibilities your employer has to you. In addition to these rules, special warnings have been used throughout this book to alert you to situations where carelessness could result in personal injury. Finally, when working on cars, always follow the safety guidelines given in service manuals and other technical literature. They are there for your protection.

PERSONAL SAFETY

Personal safety simply involves those precautions you take to protect yourself from injury. These include wearing protective gear, dressing for safety, and handling tools and equipment correctly.

Eye Protection

Your eyes can become permanently damaged by many things in a shop. Some repair procedures, such as grinding, result in tiny particles of metal and dust that are thrown off at very high speeds. These metal and dirt particles can easily get into your eyes, causing scratches or cuts on your eyeball. Pressurized gases and liquids escaping a ruptured hose or fuel line fitting can spray a great distance. If these chemicals get into your eyes, they can cause blindness. Dirt and sharp bits of corroded metal can easily fall down into your eyes while you are working under a vehicle.

Eye protection should be worn whenever you are exposed to these risks. To be safe, you should wear **safety glasses** whenever you are working in the shop. Many types of eye protection are available (**Figure 1-1**). To provide adequate eye protection, safety glasses have lenses made of safety glass. They also offer some sort of side protection. Regular prescription glasses do not offer sufficient protection and therefore should not be worn as a substitute for safety glasses.

If you wear prescription glasses, wear safety goggles that will securely fit over your glasses.

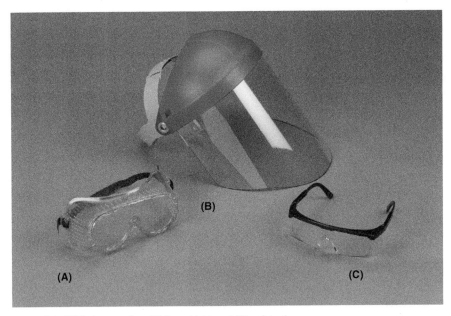

Figure 1-1 (A) Safety goggles, (B) face shield, and (C) safety glasses.

Wearing safety glasses at all times is a good habit to have. To help develop this habit, wear safety glasses that fit well and feel comfortable.

If chemicals such as battery acid, fuel, or solvents get into your eyes, flush them continuously with clean water. Have someone call a doctor and get medical help immediately.

Clothing

Your clothing should be well fitted and comfortable but made with strong material. Loose, baggy clothing can easily become caught in moving parts and machinery. Neckties should not be worn. Some technicians prefer to wear coveralls or shop coats to protect their personal clothing. Cutoffs and short pants are not appropriate for shop work.

Caustic chemicals can burn through cloth and skin.

Foot Protection

Automotive work involves handling many heavy objects that can be accidentally dropped on your feet or toes. Always wear leather or similar material shoes or boots with nonslip soles. Steel-toe safety shoes can give added protection for your feet. Jogging or basketball shoes, street shoes, and sandals are not appropriate in the shop.

Hand Protection

Good hand protection is often overlooked. A scrape, cut, or burn can limit your effectiveness at work for many days. A well-fitted pair of heavy work gloves should be worn during operations such as grinding and welding or when handling high-temperature components. Always wear approved heavy rubber gloves when handling corrosive chemicals.

It is also a very good idea to wear disposable gloves (**Figure 1-2**) when working with products such as used oil, grease, throttle body cleaners, fuel injection cleaners, brake cleaners, and so on. Many of these chemicals are either carcinogens or dangerous chemicals that can be absorbed directly into the skin and can cause serious health problems from paralysis to liver damage.

Caustic chemicals are strong and dangerous chemicals. They can easily burn your skin. Be very careful when handling this type of chemical.

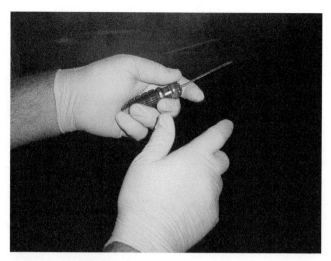

Figure 1-2 Disposable gloves.

Ear Protection

Exposure to very loud noise levels for extended periods of time can lead to a loss of hearing. Air wrenches, engines running under a load, and vehicles running in enclosed areas can all generate annoying and harmful levels of noise. Simple ear plugs or earphone-type protectors should be worn in constantly noisy environments.

Hair and Jewelry

Long hair and loose hanging jewelry can create the same type of hazard as loose-fitting clothing. They can become caught in moving engine parts and machinery. If you have long hair, tie it back or cover it with a brimless cap.

Never wear rings, watches, bracelets, or neck chains. These can easily get caught in moving parts and cause serious injury, especially when working on or near batteries and electrical systems.

Other Personal Safety Warnings

- Never smoke while working on a vehicle or while working with any machine in the shop.
- Playing around or horseplay is not fun when it sends someone to the hospital. Air nozzle fights, creeper races, practical jokes, and the like have no place in the shop.
- To prevent serious burns, keep your skin away from hot metal parts such as the radiator, exhaust manifold, tailpipe, catalytic converter, and muffler.
- When working with a hydraulic press, make sure that the pressure is applied in a safe manner. It is generally wise to stand to the side when operating the press. Always wear safety glasses.
- Properly store all parts and tools by putting them away in a place where people will not trip over them. This practice not only cuts down on injuries but also reduces time wasted looking for a misplaced part or tool.

LIFTING AND CARRYING

Knowing the proper way to lift heavy objects is important. You should also use back protection devices when you are lifting a heavy object. Always lift and work within your ability, and ask others to help when you are not sure that you can handle the size or weight of an object. Even small, compact parts can be surprisingly heavy or unbalanced. Think

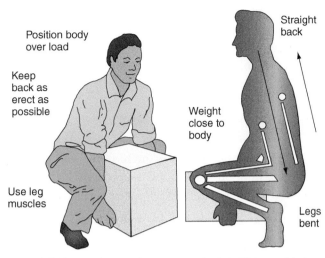

Figure 1-3 Use your leg muscles, never your back, to lift heavy objects.

about how you are going to lift something before beginning. When lifting any object, follow these steps.

1. If the object is to be carried, be sure your path is free from loose parts or tools.
2. Place your feet close to the object. Always position your feet so that you will be able to maintain a good balance.
3. Keep your back and elbows as straight as possible. Bend your knees until your hands reach the best place to get a strong grip on the object (**Figure 1-3**).
4. If the part is in a cardboard box, make sure the box is in good condition. Old, damp, or poorly sealed boxes will tear and the part will fall out.
5. Firmly grasp the object or container. Never try to change your grip as you move the load.
6. Keep the object close to your body and lift it up by straightening your legs. Use your leg muscles, not your back muscles.
7. If you must change your direction of travel, never twist your body. Turn your whole body, including your feet.
8. When placing the object on a shelf or counter, do not bend forward. Place the edge of the load on the shelf and slide it forward. Be careful not to pinch your fingers.
9. When setting down a load, bend your knees and keep your back straight. Never bend forward—this strains the back muscles.
10. Set the object onto blocks of wood to protect your fingers when lowering something heavy onto the floor.

OCCUPATIONAL SAFETY AND HEALTH ACT

The **Occupational Safety and Health Act (OSHA)** was passed by the U.S. government in 1970. The purposes of this legislation are these:

1. To assist and encourage the citizens of the United States in their efforts to assure safe and healthful working conditions by providing research, information, education, and training in the field of occupational safety and health.
2. To assure safe and healthful working conditions for working men and women by authorizing enforcement of the standards developed under the Act.

The *Occupational Safety and Health Act (OSHA)* regulates working conditions in the United States.

Because approximately 25 percent of workers are exposed to health and safety hazards on the job, OSHA is necessary to monitor, control, and educate workers regarding health and safety in the workplace. Employers and employees should be familiar with **Workplace Hazardous Materials Information Systems (WHMIS)**.

SHOP HAZARDS

Shop hazards must be recognized and avoided to prevent personal injury.

Service technicians and students encounter many hazards in an automotive shop. When these hazards are known, basic shop safety rules and procedures must be followed to avoid personal injury. Some of the hazards in an automotive shop are listed below:

1. Flammable liquids such as gasoline and paint must be handled and stored properly in special metal cabinets.
2. Flammable materials such as oily rags must be stored properly in special covered receptacles (**Figure 1-4**) to avoid a fire hazard.
3. Batteries contain a corrosive sulfuric acid solution and produce explosive hydrogen gas while charging.
4. Loose sewer and drain covers may cause foot or toe injuries.
5. Caustic liquids such as those in hot cleaning tanks are harmful to skin and eyes.
6. High-pressure air in the shop's compressed-air system can be very dangerous if it penetrates the skin and enters the bloodstream.
7. Frayed cords on electric equipment and lights may result in severe electrical shock.
8. Hazardous waste material such as batteries and the caustic cleaning solution from a hot or cold cleaning tank must be handled properly to avoid harmful effects.
9. Carbon monoxide from vehicle exhaust is poisonous.
10. Loose clothing or jewelry, or long hair, may become entangled in rotating parts on equipment or vehicles, resulting in serious injury.
11. Dust and vapors generated during some repair jobs are harmful. Asbestos dust generated during brake-lining service and clutch service is a contributor to lung cancer.
12. High noise levels from shop equipment such as an air chisel may be harmful to the ears.
13. Oil, grease, water, or parts-cleaning solutions on shop floors may cause someone to slip and fall, resulting in serious injury.

Figure 1-4 Keep combustibles in safety containers.

AUTHOR'S NOTE The old saying "Had I known I was going to live this long, I would have taken better care of myself" should serve as a friendly reminder to all technicians. We do preventive maintenance on our customers' cars, right? So why not take the same measures to prevent temporary or permanent injury to ourselves? Taking care of yourself in the shop in even the simplest way can reward you greatly later in life. This means wearing gloves, eye protection, ear protection, and dust masks when appropriate. Try to anticipate the potential risks of an injury while performing a particular service on a vehicle, then act on methods of protection that will mitigate the risks.

HAND TOOL SAFETY

Many shop accidents are caused by improper use and care of hand tools. These hand tool safety steps must be followed:

1. Keep tools clean and in good condition. Worn tools may slip and result in hand injury. If a hammer with a loose head is used, the head may fly off and cause personal injury or vehicle damage. Keep all hand tools free of grease and in good condition. Tools that slip can cause cuts and bruises. If a tool slips and falls into a moving part, it can fly out and cause serious injury.
2. Use the proper tool for the job. Make sure the tool is of professional quality. Using poorly made tools or the wrong tools can damage parts or the tool itself, or cause injury. Do not use broken or damaged tools.
3. Be careful when using sharp or pointed tools. Do not place sharp tools or other sharp objects in your pockets. They can stab or cut your skin, ruin automotive upholstery, or scratch a painted surface.
4. Tool tips that are intended to be sharp should be kept in a sharp condition. Sharp tools such as chisels will do the job faster and with less effort if they are kept sharp. Dull tools can be more dangerous than sharp tools.
5. Never use a tool for which it is not intended, such as using a screwdriver as a pry bar. The blade or blade tip can break and cause personal injury.
6. Use only impact sockets that are designed for use with an impact wrench. Ordinary sockets can break or shatter if used improperly (**Figure 1-5**).

POWER TOOL SAFETY

Power tools are operated by an outside source of power, such as electricity, compressed air, or hydraulic pressure. Safety around power tools is very important. Serious injury can result from carelessness. Always wear safety glasses when using power tools.

If the tool is electrically powered, make sure it is properly grounded. Check the wiring for cracks in the insulation and bare wires before using it. When using electrical power tools, never stand on a wet or damp floor. Disconnect the power source before performing any service on the machine or tool. Before plugging in any electric tool, make sure the switch is off to prevent serious injury. When you are through using the tool, turn it off and unplug it. Never leave a running power tool unattended. When you leave, turn it off.

When using power equipment on a small part, never hold the part in your hand. Always mount the part in a bench vise or use vise grip pliers. Never try to use a

Figure 1-5 A worn-out impact socket must be replaced.

machine or tool beyond its stated capacity or for operations requiring more than the rated power of the tool.

When working with larger power tools, such as bench or floor equipment, check the machines for signs of damage before using them. Place all safety guards into position (**Figure 1-6**). A safety guard is a protective cover over a moving part. It is designed to prevent injury. Wear safety glasses or a face shield. Make sure there are no people or parts around the machine before starting it up. Keep your hands and clothing away from the moving parts. Maintain a balanced stance while using the machine.

Figure 1-6 Safety guards on a bench grinder.

Hydraulic Press

 WARNING When operating a hydraulic press, always be sure that the components being pressed are properly supported on the press bed with steel supports.

When two components have a tight precision fit between them, a **hydraulic press** is used to separate them or press them together. The hydraulic press rests on the shop floor. An adjustable steel beam bed is retained to the lower press frame with heavy steel pins. A hydraulic cylinder and ram are mounted on the top part of the press with the ram facing downward toward the press bed (**Figure 1-7**). The component being pressed is placed on the press bed with the appropriate steel supports. A hand-operated hydraulic pump is mounted on the side of the press. When the handle is pumped, hydraulic fluid is forced into the cylinder, and the ram is extended against the component on the press bed to complete the pressing operation. A pressure gauge on the press indicates the pressure applied from the hand pump to the cylinder. The press frame is designed for a certain maximum pressure. This pressure must not be exceeded during operation.

Electrical Safety

1. Frayed cords on electrical equipment must be replaced or repaired immediately.
2. All electric cords from lights and electric equipment must have a ground connection. The ground connector is the round terminal in a three-prong electrical plug. Do not use a two-prong adapter to plug in a three-prong electrical cord. Three-prong electrical outlets should be mandatory in all shops.
3. Do not leave electrical equipment running and unattended.

On the bench grinder, always make sure the tool rest is no more than 1/8 inch away from the grinding stone.

⚠ Caution

When using a hydraulic press, never operate the pump handle when the pressure gauge exceeds the maximum pressure rating of the press. If this pressure is exceeded, some part of the press may suddenly break and cause severe personal injury.

Hybrid vehicles use high voltage and amperage for their electric motor.

Figure 1-7 A hydraulic press.

Battery Safety

Batteries give off hydrogen and oxygen gases, which are explosive! Use extreme caution when working around or handling batteries.

1. Always wear eye protection when working near a battery.
2. Do not smoke or allow any open flame around a battery.
3. Do not connect or disconnect a battery charger while the charger is in the "on" position. In addition, it is a good practice to unplug the battery charger from the 110-volt AC outlet before connecting or disconnecting the charger from the battery.
4. Observe proper polarity when installing a battery, connecting a battery charger, or jump-starting the vehicle.
5. Do not allow the battery to exceed 125°F while charging the battery.
6. When using load-test equipment, be certain the load controller is off prior to connecting or disconnecting the unit from the battery.
7. If battery acid contacts your skin, wash immediately for at least 10 minutes.
8. Battery acid can harm skin, eyes, clothing, and some paint finishes.
9. Neutralize any battery acid spills with baking soda.

> **AUTHOR'S NOTE** Hybrid vehicles have as many as three separate voltages running in the same vehicle. Many of these vehicles cannot be charged with a conventional charger (even the 12-volt battery). Some have 36- and 42-volt batteries that should *never* be charged with a conventional charger. *Always* check with the manufacturer's service information *before* connecting any charging equipment to a hybrid vehicle.

Hybrid High-Voltage Safety

Hybrid vehicles require the use of some special procedures and tools. The higher voltage and amperage can be deadly. Whenever a hybrid vehicle is in for service that might involve the electrical system in any way, the high-voltage battery must be disconnected. Hybrid vehicles have a disconnection point (**Figure 1-8**, **Figure 1-9**) that varies according to the

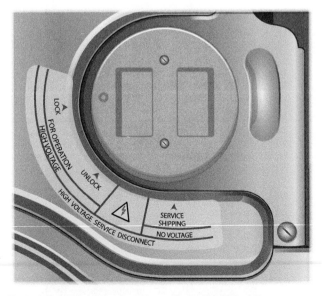

Figure 1-8 A high-voltage battery cutoff on a Hybrid Ford Escape.

Figure 1-9 A high-voltage battery cutoff on a Mini E.

Figure 1-10 Special high-voltage gloves are worn inside leather outer gloves when working on hybrid vehicles' high-voltage electrical systems.

make and model of the vehicle. Additionally, special class zero high-voltage gloves are required whenever working on a hybrid vehicle. The inner, thick rubber gloves are specially made to resist high voltage, and must be retested periodically to ensure they are still safe. Thick leather gloves are worn over the rubber gloves to help protect the rubber gloves from abrasion (**Figure 1-10**). The gloves have to be periodically sent to a lab to recertify their insulating capabilities.

Gasoline Safety

Gasoline is a very flammable liquid! One exploding gallon of gasoline has energy equal to 14 sticks of dynamite. It is the expanding vapors from gasoline that are extremely dangerous. These vapors are present even in cold temperatures. Vapors formed in gasoline tanks in cars are controlled, but vapors from a gasoline storage container may escape from the can, resulting in a hazardous situation. Therefore, gasoline storage containers must be placed in a well-ventilated space.

Approved gasoline storage cans have a flash-arresting screen at the outlet (**Figure 1-11**). These screens prevent external ignition sources from igniting the gasoline within the can

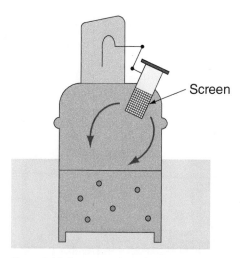

Figure 1-11 An approved gasoline container.

while the gasoline is being poured. Follow these safety precautions regarding gasoline containers:

1. Always use approved gasoline containers that are designed specifically for gasoline and properly labeled as such.
2. Do not fill gasoline containers completely full. Always leave the level of gasoline at least one inch from the top of the container. This action allows expansion of the gasoline at higher temperatures. If gasoline containers are completely full, gasoline vapors will expand when the temperature increases. This expansion forces gasoline from the can and creates a dangerous spill.
3. If gasoline containers must be stored, place them in a well-ventilated area such as a storage shed. Do not store gasoline containers in your home or in the trunk of a vehicle.
4. When a gasoline container must be transported, be sure it is secured against upsets.
5. Do not store a partially filled gasoline container for long periods of time because it may give off vapors and produce a potential danger.
6. Never leave gasoline containers open except while filling or pouring gasoline from the container.
7. Do not prime an engine with gasoline while cranking the engine.
8. Never use gasoline as a cleaning agent.
9. Avoid contact with unprotected skin.

Refrigerant Safety

It is illegal to knowingly discharge or vent air-conditioning refrigerant into the atmosphere. When servicing an air-conditioning system, recover and/or recycle the refrigerant using approved, certified equipment. These services must be performed by an **Environmental Protection Agency (EPA)**—recognized, certified technician following specific procedures. When working around refrigerant or air-conditioning service equipment, it is important to wear full eye protection. Refrigerant can cause immediate blindness.

General Shop Safety

1. All sewer covers must fit properly and be kept securely in place.
2. Always wear a face shield, protective gloves, and protective clothing when necessary. Gloves should be worn when working with solvents and caustic solutions, handling hot metal, or grinding metal. Various types of protective gloves are available. Shop coats and coveralls are the most common types of protective clothing.

3. Never direct high-pressure air from an air gun against human flesh or *any* part of the body. If this action is allowed, air may penetrate the skin and enter the bloodstream, causing serious health problems or even death. Always keep air hoses in good condition. If an end blows off of an air hose, the hose may whip around and result in personal injury. Use only OSHA-approved air gun nozzles.

4. Handle all hazardous waste materials according to state and federal regulations. (These regulations are explained later in this chapter.)

5. Always place a shop exhaust hose on the tailpipe of a vehicle if the engine is running in the shop, and be sure the shop exhaust fan is turned on.

6. Keep hands, long hair, jewelry, and tools away from rotating parts such as fan blades and belts on running engines. Remember that an electric-drive fan may start turning without warning. Avoid wearing rings, watches, or bracelets, especially when working on or around batteries or other electrical systems.

7. When servicing brakes or clutches from manual transmissions, always clean asbestos dust from these components with an approved asbestos dust parts washer, and wear a dust mask.

8. Always use the correct tool for the job. For example, never strike a hardened steel component, such as a piston pin, with a steel hammer. This type of component may shatter, and fragments may penetrate the eyes or skin.

9. Follow the car manufacturer's recommended service procedures.

10. Be sure that the shop has adequate ventilation.

11. Make sure the work area has adequate lighting.

12. Use only fluorescent or LED-style trouble lamps. Incandescent trouble lamps are very dangerous around flammable liquids due to their risk of bursting, and they cause burns to the skin if touched.

13. When servicing a vehicle, always apply the parking brake and place the transmission in **park** with an automatic transmission or **neutral** with a manual transmission if the engine is running. When the engine is stopped, place the transmission in park with an automatic transmission or either reverse or first gear with a manual transmission.

14. Avoid working on a vehicle parked on an incline.

15. Never work under a vehicle unless the vehicle chassis is supported securely on jack stands.

16. When one end of a vehicle is raised, place wheel chocks on both sides of the wheels remaining on the floor.

17. Be sure that you know the location of shop first-aid kits, eyewash fountains, and fire extinguishers.

18. Collect oil, fuel, brake fluid, and other liquids in the proper safety containers.

19. Use only approved cleaning fluids and equipment. Do not use gasoline to clean parts.

20. Obey all state and federal safety, fire, and hazardous material regulations.

21. Always operate equipment according to the equipment manufacturer's recommended procedure.

22. Do not operate equipment unless you are familiar with the correct operating procedure.

23. Do not leave running equipment unattended.

24. Be sure the safety shields are in place on rotating equipment.

25. Ensure all shop equipment has regularly scheduled maintenance and adjustment.

26. Some shops have safety lines around equipment. Always work within these lines when operating equipment.

27. Be sure that shop heating equipment is well ventilated.

28. Do not run in the shop or engage in horseplay.

29. Post emergency phone numbers near the phone. These numbers should include 911, a doctor, and a hospital, as well as direct numbers for ambulance, fire department, and police services.
30. Do not place hydraulic jack handles where someone can trip over them.
31. Keep aisles clear of debris.
32. Use caution when working near a raised lift. Be aware of the available clearance.

Fire Safety

1. Familiarize yourself with the location and operation of all shop fire extinguishers.
2. If a fire extinguisher is used, report it to management so the extinguisher can be recharged.
3. Do not use any type of open-flame heater to heat the work area.
4. Do not turn on the ignition switch or crank the engine with a gasoline line disconnected.
5. Store all combustible materials such as gasoline, paint, and oily rags in approved safety containers.
6. Clean up gasoline, oil, or grease spills immediately.
7. Always wear clean shop clothes. Do not wear oil-soaked clothes.
8. Do not allow sparks and flames near batteries.
9. Welding tanks must be securely fastened in an upright position.
10. Do not block doors, stairways, or exits.
11. Do not smoke when working on vehicles.
12. Do not smoke or create sparks near flammable materials or liquids.
13. Store combustible shop supplies such as paint in a closed steel cabinet.
14. Gasoline must be kept in approved safety containers.
15. If a gasoline tank is removed from a vehicle, do not drag the tank on the shop floor.
16. Know the approved fire escape route from your classroom or shop to the outside of the building.
17. If a fire occurs, do not open doors or windows. This action creates extra draft which makes the fire worse.
18. Do not put water on a gasoline fire because the water will make the fire worse.
19. Call the fire department as soon as a fire begins and then attempt to extinguish the fire.
20. If possible, stand 6 to 10 feet from the fire and aim the fire extinguisher nozzle at the base of the fire with a sweeping action.
21. If a fire produces a lot of smoke in the room, remain close to the floor to obtain oxygen and avoid breathing smoke.
22. If the fire is too hot or the smoke makes breathing difficult, get out of the building.
23. Do not reenter a burning building.
24. Keep solvent containers covered except when pouring from one container to another. When flammable liquids are transferred from bulk storage, the bulk container should be grounded to a permanent shop fixture such as a metal pipe. During this transfer process, the bulk container should be grounded to the portable container. These ground wires prevent the buildup of a static electric charge, which could result in a spark and a disastrous explosion. Always discard or clean empty solvent containers, because fumes in these containers are a fire hazard.
25. Familiarize yourself with different types of fires and fire extinguishers, and know the type of extinguisher to use on each fire.

COMPRESSED-AIR EQUIPMENT SAFETY

Tools that use compressed air are called **pneumatic tools**. Compressed air is used to inflate tires, apply paint, and drive tools. Compressed air can be dangerous when it is not used properly. The shop air supply contains high-pressure air in the shop compressor and air lines. Serious injury or property damage may result from careless operation of compressed-air equipment. Follow these guidelines when working with compressed air:

1. Safety glasses or a face shield should be worn for all shop tasks, including those tasks involving the use of compressed-air equipment.
2. Wear ear protection when using compressed-air equipment.
3. Always maintain air hoses and fittings in good condition. If an end suddenly blows off an air hose, the hose will whip around, and this may cause personal injury.
4. Do not direct compressed air against the skin. This air may penetrate the skin, especially through small cuts or scratches. If compressed air penetrates the skin and enters the bloodstream, it can be fatal or cause serious health complications. Use only air gun nozzles approved by OSHA.
5. Never use an air gun to blow off clothing or hair.
6. Do not clean the workbench or floor with compressed air. This action may blow very small parts against your skin or into your eyes. Small parts blown by compressed air may cause vehicle damage. For example, if the car in the next stall has the air cleaner removed, a small part may go into the throttle body. When the engine is started, this part will likely be pulled into the cylinder by engine vacuum and will penetrate through the top of a piston.
7. Never spin bearings with compressed air because the bearing will rotate at extremely high speed. Under this condition, the bearing may be damaged or it may disintegrate, causing personal injury.
8. All pneumatic tools must be operated according to the manufacturer's recommended operating procedure.
9. Follow the equipment manufacturer's recommended maintenance schedule for all compressed-air equipment.

LIFT SAFETY

A **lift** is used to raise a vehicle so a technician can work under it. The lift arms must be placed under the car at the manufacturer's recommended lift points. There are several types of vehicle lifts in use today. Categories of lifts include single, double, and four-post varieties. Typically, single-post lifts are considered an in-ground type with the hydraulic post and lines located directly under the lift. Above-ground types include twin and four posts with adjustable arms and contact pads or drive-on runways. Others consist of a scissors-type design (**Figure 1-12**). Some lifts have an electric motor that drives a hydraulic pump to create fluid pressure and force the lift upward. Other lifts use air pressure from the shop air supply to force the lift upward. If shop air pressure is used for this purpose, the air pressure is applied to fluid in the lift cylinder. A control lever or switch is placed near the lift. The control lever supplies shop air pressure to the lift cylinder, and the switch turns on the lift pump motor. Always be sure that the safety lock is engaged after the lift is raised. When the safety lock is released, a release lever is operated slowly to lower the vehicle.

Always be careful when raising a vehicle on a lift or a hoist. Adapters and hoist plates must be positioned correctly on twin-post and rail-type lifts to prevent damage to the

Figure 1-12 A scissors-type, drive-on lift.

underbody of the vehicle. There are specific lift points. These points allow the weight of the vehicle to be evenly supported by the adapters or hoist plates. The correct lift points can be found in the vehicle's service manual. **Figure 1-13** shows typical locations for frame and unibody cars. These diagrams are for illustration only, so always follow the manufacturer's instructions. Before operating any lift or hoist, carefully read the operating manual and follow the operating instructions.

⚠ **WARNING** **Never use a lift or jack to move something heavier than it is designed for. Always check the rating before using a lift or jack. If a jack is rated for 2 tons, do not attempt to use it for a job requiring 5 tons. It is dangerous for you and the vehicle.**

When guiding someone driving a vehicle onto a lift, be sure to stand off to the driver's side of the vehicle rather than in front of it. Use clear, deliberate hand signals and/or verbal

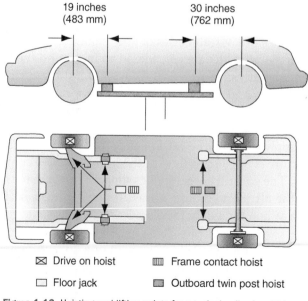

Figure 1-13 Hoisting and lifting points for a typical unibody vehicle.

instructions to the driver to indicate vehicle direction. In the event of an unexpected action by the vehicle, always leave yourself a clear path. It is a good practice to check the clearance underneath the vehicle prior to driving over the lift. Low-hanging accessories or exhaust systems may be damaged when coming in contact with the lift.

Before driving a vehicle over a lift, position the arms and supports to provide an unobstructed clearance. Do not hit or run over lift arms, adapters, or axle supports. This could damage the lift, vehicle, or tires.

Position the lift supports to contact the vehicle at its lifting points. Raise the lift until the supports contact the vehicle. Then, check the supports to make sure they are in full contact with the vehicle. Raise the lift to the desired working height.

Make sure the vehicle's doors, hood, and trunk are closed before raising the vehicle. Never raise a car with someone inside.

⚡ WARNING **Before working under a car, make sure the lift's locking device is engaged.**

After lifting a vehicle to the desired height, always lower it onto its mechanical safeties. On some vehicles, the removal (or installation) of components can cause a critical shift of the vehicle's weight, which may cause the vehicle to be unstable on the lift. Refer to the vehicle's service manual for the recommended procedures to prevent this from happening.

Make sure tool trays, stands, and other equipment are removed from under the vehicle. Release the lift's locking devices according to the instructions before attempting to lower the lift.

JACK AND JACK STAND SAFETY

An automobile can be raised off the ground by a hydraulic jack (**Figure 1-14**). **Jack stands** (**Figure 1-15**) are supports of different heights that sit on the floor. They are placed under a sturdy chassis member, such as the frame or axle housing, to support the vehicle. Like jacks, jack stands also have a capacity rating. Always use the correct rating of jack stand.

A **floor jack** is a portable unit mounted on wheels. The lifting pad on the jack is placed under the chassis of the vehicle, and the jack handle is operated with a pumping action. This jack handle operation forces fluid into a hydraulic cylinder in the jack, and the cylinder extends to force the jack lift pad upward and lift the vehicle. Always be sure that the lift pad is positioned securely under one of the car manufacturer's recommended lift points. To release the hydraulic pressure and lower the vehicle, the handle or release lever must be turned slowly.

The jack should be removed after the jack stands are set in place. This eliminates a hazard, such as a jack handle sticking out into a walkway. A jack handle that is bumped or kicked can cause a tripping accident or cause the vehicle to fall. Never use a jack by itself to support an automobile. Always use a jack stand with the jack as a safety precaution. Make sure the jack stands are properly placed under the vehicle.

Accidents involving the use of floor jacks and jack stands may be avoided if these safety precautions are followed:

1. Never work under a vehicle unless jack stands are placed securely under the vehicle chassis and the vehicle is resting on these stands.
2. Prior to lifting a vehicle with a floor jack, be sure that the jack lift pad is positioned securely under a recommended lift point on the vehicle. Lifting the front end of a vehicle with the jack placed under a radiator support may cause severe damage to the radiator and support.

Jack stands are also called "safety stands."

As the vehicle is lifted, the floor jack will roll under the load. Make sure that the wheels on the floor jack are free to roll and are not blocked by gravel or other obstructions on the shop floor. If the jack cannot roll, the lifting plate may be pulled off the vehicle lifting point.

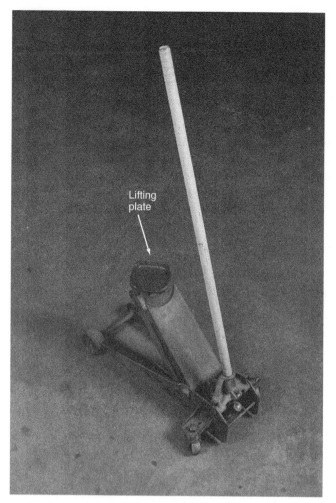

Figure 1-14 A typical hydraulic floor jack.

Figure 1-15 Support stands.

3. Position the jack stands under a strong chassis member, such as the frame or axle housing. The jack stands must contact the vehicle manufacturer's recommended lift points.
4. Because the floor jack is on wheels, the vehicle and jack tend to move as the vehicle is lowered from a floor jack onto jack stands. Always be sure the jack stands remain under the chassis member during this operation, and be sure the jack stands do not tip. All the jack stand legs must remain in contact with the shop floor.

CLEANING EQUIPMENT SAFETY

All technicians are required to clean parts during their normal work routines. Face shields and protective gloves must be worn while operating cleaning equipment. The solution in hot and cold cleaning tanks may be caustic, and contact between this solution and the skin or eyes must be avoided. Parts cleaning often creates a slippery floor, and care must be taken when walking in the parts-cleaning area. The floor in this area should be cleaned frequently. When the caustic cleaning solution in hot or cold cleaning tanks is replaced, environmental regulations require that the old solution be handled as hazardous waste. Use caution when placing aluminum or aluminum alloy parts in a cleaning solution. Caustic solutions will damage these components by reacting with the aluminum. Always follow the cleaning equipment manufacturer's recommendations. Parts cleaning is a necessary step in most repair procedures. Cleaning automotive parts can be divided into four basic categories.

Chemical cleaning relies primarily on some type of chemical action to remove dirt, grease, scale, paint, or rust. A combination of heat, agitation, mechanical scrubbing, or washing may also be used to help remove dirt. Chemical-cleaning equipment includes small-parts washers, hot and cold tanks, pressure washers, spray washers, and salt baths.

Some parts washers provide electromechanical agitation of the parts to provide improved cleaning action (**Figure 1-16**). These parts washers may be heated with gas or electricity, and various water-based hot-tank cleaning solutions are available depending on the type of metals being cleaned. For example, Kleer-Flo Greasoff Number 1 powdered detergent is available for cleaning iron and steel. Non-heated electromechanical parts washers are also available, and these washers use cold cleaning solutions such as Kleer-Flo Degreasol formulas.

Many cleaning solutions, such as Kleer-Flo Degreasol 99R, contain no ingredient listed as hazardous by the EPA's **Resource Conservation and Recovery Act (RCRA)**. Degreasol 99R is a blend of sulfur-free hydrocarbons, wetting agents, and detergents. It does not contain aromatic or chlorinated solvents, and it conforms to California's Rule 66 for clean air. Always use the cleaning solution recommended by the equipment manufacturer.

Some parts washers have an agitated immersion chamber under the shelves that provides thorough parts cleaning. Folding work shelves provide a large upper cleaning area with a constant flow of solution from the dispensing hose. This cold parts washer operates on Degreasol 99R.

An aqueous parts-cleaning tank uses a water-based, environmentally friendly cleaning solution such as Greasoff 2 rather than traditional solvents. The immersion tank is heated and agitated for effective parts cleaning (**Figure 1-17**). A sparger bar pumps a constant flow of cleaning solution across the surface to push floating oils away, and an integral skimmer removes these oils. This action prevents floating surface oils from redepositing on cleaned parts.

Thermal cleaning relies on heat, which bakes off or oxidizes the dirt. Thermal cleaning leaves an ash residue on the surface that must be removed by an additional cleaning process such as airless shot blasting or spray washing.

Abrasive cleaning relies on physical abrasion to clean the surface. This includes everything from a wire brush to glass bead blasting, airless steel shot blasting, abrasive tumbling, and vibratory cleaning. Chemical in-tank solution sonic cleaning might also be

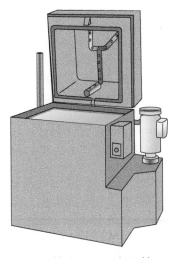

Figure 1-16 A parts washer with an electromechanical agitator.

Figure 1-17 An aqueous parts-cleaning tank.

included here because it relies on the scrubbing action of ultrasonic sound waves to loosen surface contaminants.

Steam cleaning uses hot water vapor mixed with chemical cleaning agents to clean dirt from an object. After steam cleaning, the object should be thoroughly hosed down with clean water, then air dried.

There are several reasons why the use of steam cleaning has rapidly declined in recent years. Concerns about the environment have led to mandates that a **closed-loop system** be used for steam cleaning. This means that the runoff from the cleaning process must be collected and contained within the steam-cleaning system. The runoff cannot flow into a public sewage system.

Steam cleaning is normally done in a non-congested portion of the shop or in a separate building. Care must be taken to protect all painted surfaces and exposed skin. If these were to come into contact with the steam's heat and chemicals, it could result in injury or damage. In addition, care must also be taken when working on the slippery floor that this process creates.

Before using a steam-cleaning machine, check the electrical cords. Pay special attention to the grounding connector at the plug. If the machine is not properly grounded, there is a great possibility of getting an electrical shock. Finally, steam cleaning takes a great deal of time and work. Most shops cannot justify the labor cost for using an open steam-cleaning system.

VEHICLE OPERATION

A Note on Hybrid Vehicle Safety

Even though this book is not intended to be about hybrid vehicles, we discuss them throughout this text. In the interest of safety, I have done some research to come up with a few general statements about working safely around hybrid vehicles. As I looked at the different types from several manufacturers (some of whom offered more than one type), I collected the best possible advice.

At this time, there are too many changes occurring for technicians to service a hybrid without extensive factory training in the specific models they are attempting to service. One such example is the fact that many hybrids use an automatic engine restart to recharge the battery anytime the key is in the ignition or the smart key is in the vicinity of the vehicle. Imagine what might happen if you were changing the oil or had your hands in the engine compartment when this happened!

⚡ **WARNING Hybrids have very high-voltage systems that can kill you if they are mishandled! Do not attempt to service a hybrid vehicle until you have been trained and understand the proper procedures necessary to keep you safe.**

Carbon monoxide (CO) is an odorless, poisonous gas. When it is breathed in, it can cause headaches, nausea, ringing in the ears, tiredness, and heart flutter. A large amount of CO can kill you.

When the customer brings in a vehicle for service, certain shop safety guidelines should be followed. For example, when moving a car into the shop, check the brakes before beginning. Then buckle the safety belt. Drive carefully in and around the shop. Make sure no one is near, that the way is clear, and that there are no tools or parts under the car before you start the engine.

When road testing the car, obey all traffic laws. Drive only as far as is necessary to check the automobile. Never make excessively quick starts, turn corners too quickly, or drive faster than conditions or speed limits allow.

If the engine must be running while you are working on the car, block the wheels to prevent the car from moving. Place the transmission in park for automatic transmissions or in neutral for manual transmissions. Set the emergency brake. Never stand directly in front of or behind a running vehicle.

Figure 1-18 When running an engine in a shop, make sure the vehicle's exhaust is connected to the shop's exhaust ventilation system.

Run the engine only in a well-ventilated area to avoid the danger of poisonous CO in the engine exhaust. If the shop is equipped with an exhaust ventilation system (**Figure 1-18**), use it. If not, use a hose and direct the exhaust out of the building.

⚡ WARNING **Never run the engine in a vehicle inside the shop without an exhaust hose connected to the tailpipe.**

Carbon Monoxide

Vehicle exhaust contains small amounts of CO, which is a poisonous gas. Strong concentrations of CO may be fatal for human beings. All shop personnel are responsible for air quality in the shop. Shop management is responsible for providing an adequate exhaust system to remove exhaust fumes from the maximum number of vehicles that may be running in the shop at the same time. Technicians should never run a vehicle in the shop unless a shop exhaust hose is installed on the tailpipe of the vehicle. The exhaust fan must be switched on to remove exhaust fumes.

If shop heaters or furnaces have restricted chimneys, they release CO emissions into the shop air. Therefore, chimneys should be checked periodically for restriction and proper ventilation.

Monitors are available to measure the level of CO in the shop. Some of these monitors read the amount of CO in the shop air, while other monitors provide an audible alarm if the concentration of CO exceeds the danger level. When CO is breathed in, it can cause headaches, nausea, ringing in the ears, tiredness, and heart flutter. A large amount of CO can kill you.

Diesel exhaust contains some CO, but particulates are also present in the exhaust from these engines. **Particulates** are small carbon particles that can be harmful to the lungs.

Batteries

The **sulfuric acid** solution in car batteries is a very corrosive, poisonous liquid. If a battery is charged with a fast charger at a high rate for a period of time, the battery becomes hot, and the sulfuric acid solution begins to boil. Under this condition, the battery may emit a strong sulfuric acid smell, and these fumes may be harmful to the lungs. If this condition occurs in the shop, the battery charger should be turned off or the charger rate should be reduced considerably.

When an automotive battery is charged, hydrogen and oxygen gases escape from the battery. If these gases are combined, they form water, but hydrogen gas by itself is very explosive. While a battery is charged, sparks, flames, and other sources of ignition must not be allowed near the battery.

Some modern vehicles use up to several thousand pounds per square inch (psi) fuel pressures that can puncture skin. Use external caution around fuel fitting.

Use disposable solvent-resistant gloves when working with solvents such as brake parts cleaner or carburetor cleaner. Many of these solvents can penetrate the skin and may cause serious health concerns if used over a period of time.

Personal injury, vehicle damage, and property damage must be avoided by following safety rules regarding personal protection, substance abuse, electrical safety, gasoline safety, housekeeping safety, fire safety, and general shop safety.

Air Bag Safety

Before working on the steering column or dash areas, make certain that the air bag system has been disarmed according to the manufacturer's recommended procedure. Some general points to keep in mind are as follows:

1. The air bag can deploy for several minutes even after the battery has been disconnected.
2. Never test an air bag circuit using a test lamp or self-powered test lamp.
3. Never split into the air bag harness. The harness can be identified by its bright yellow color and identification markings.
4. Always carry the air bag with the trim cover facing away from your body.
5. Do not place the air bag on a bench with the trim cover down.
6. If the steering column is removed from the vehicle, do not stand the column with the air bag facing down.
7. Do not deploy an air bag by any procedure other than the manufacturer's recommendations.
8. If it is necessary to handle a deployed air bag, make certain to take precautions against the residue left behind; it can cause skin irritation.

Personal Protection

1. Always wear safety glasses or a face shield in the shop (**Figure 1-19**).
2. Wear ear plugs or covers if high noise levels are encountered.
3. Always wear boots or shoes that provide adequate foot protection. Safety work boots or shoes with steel-toed caps are best for working in the automotive shop. Most safety shoes also have slip-resistant soles. Footwear must protect against heavy falling objects, flying sparks, and corrosive liquids. Soles on footwear must protect against punctures by sharp objects. Sneakers and street shoes are not recommended in the shop.

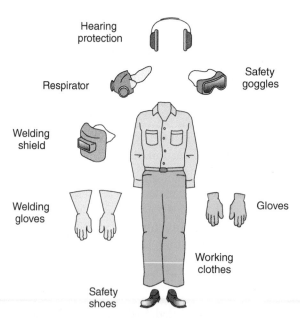

Figure 1-19 Shop safety equipment: this includes safety goggles, a respirator, a welding shield, proper work clothes, ear protection, gloves, and safety shoes.

4. Do not wear watches, jewelry, or rings when working on a vehicle. Severe burns occur when jewelry makes contact between an electric terminal and ground. Jewelry may catch on an object, resulting in painful injury.

5. Do not wear loose clothing, and keep long hair tied behind your head. Loose clothing or long hair is easily entangled in rotating parts.

6. Wear a respirator to protect your lungs when working in dusty conditions.

Smoking, Alcohol, and Drugs in the Shop

Do not smoke when working in the shop. If the shop has designated smoking areas, smoke only in these areas. Do not smoke in customers' cars; nonsmokers may not appreciate cigarette odor in their cars. A spark from a cigarette or lighter may ignite flammable materials in the workplace.

The use of drugs or alcohol must be avoided while working in the shop. Even a small amount of drugs or alcohol affects reaction time. In an emergency situation, slow reaction time may cause personal injury. If a heavy object falls off the workbench and your reaction time is slowed by drugs or alcohol, you may not get your foot out of the way in time, resulting in foot injury. When a fire starts in the workplace and you are a few seconds slower when getting a fire extinguisher into operation because of alcohol or drug use, it could make the difference between extinguishing a fire and incurring expensive fire damage.

WORK AREA SAFETY

Your work area should be kept clean and safe. The floor and bench tops should be kept clean, dry, and orderly. Any oil, coolant, or grease on the floor can make it slippery. Slips can result in serious injuries. To clean up oil, use commercial oil absorbent. Keep all water off the floor. Water is slippery on smooth floors, and electricity flows well through water. Aisles and walkways should be kept clean and wide enough to easily move through. Make sure the work areas around machines are large enough to safely operate the machine.

Proper ventilation of space heaters, which are used in some shops, is necessary to reduce the CO levels in the shop. Also, proper ventilation is very important in areas where volatile solvents and chemicals are used. A volatile liquid is one that vaporizes very quickly.

Keep an up-to-date list of emergency telephone numbers clearly posted next to the telephone. These numbers should include 911, a doctor, a hospital, and direct numbers for local fire and police departments. Also, the work area should have a first-aid kit for treating minor injuries. There should also be eye-flushing kits readily available.

Gasoline is a highly flammable volatile liquid. Always keep gasoline or diesel fuel in an approved safety can, and never use it to clean your hands or tools. Oily rags should also be stored in an approved metal container. When these oily, greasy, or paint-soaked rags are left lying about or are not stored properly, they can cause spontaneous combustion. Spontaneous combustion results in a fire that starts by itself without a match.

Make sure that all drain covers are snugly in place. Open drains or covers that are not flush to the floor can cause toe, ankle, and leg injuries.

Oil absorbent must be treated as hazardous waste.

Handle all solvents (or any liquids) with care to avoid spillage. Keep all solvent containers closed, except when pouring. Be extra careful when transferring flammable materials from bulk storage (**Figure 1-20**). Static electricity can build up enough to create a spark that could cause an explosion. Discard or clean all empty solvent containers. Solvent fumes in the bottom of these containers are very flammable. Never light matches or smoke near flammable solvents and chemicals, including battery acids. Solvent and other combustible materials must be stored in approved and designated storage cabinets or rooms (**Figure 1-21**). Storage rooms should have adequate ventilation.

Special class-zero (also called "class0") gloves and rubber-soled shoes are required for working around hybrid vehicles due to these vehicles' high voltage.

The improper or excessive use of alcoholic beverages and/or drugs may be referred to as substance abuse.

 Caution

Always know the location of all safety equipment in the shop, and be familiar with the operation of this equipment.

Shop rules, vehicle operation in the shop, and shop housekeeping are serious business. Each year a significant number of technicians are injured and vehicles are damaged by disregarding shop rules, careless vehicle operation, and sloppy housekeeping.

Figure 1-20 Flammable liquids should be stored in safety-approved containers.

Figure 1-21 Store combustible materials in approved safety cabinets.

Know where the fire extinguishers are and what types of fires they put out (**Figure 1-22**). A multipurpose dry chemical fire extinguisher will put out ordinary combustibles, flammable liquids, and electrical fires. Never put water on a gasoline fire. The water will just spread the fire. Use a fire extinguisher to smother the flames. Remember that during a fire, never open doors or windows unless it is absolutely necessary; the extra draft will only make the fire worse. A good rule is to call the fire department first and then attempt to extinguish the fire.

To extinguish a fire, stand 6 to 10 feet from the fire. Hold the extinguisher firmly in an upright position. Aim the nozzle at the base, and use a side-to-side motion, sweeping the entire width of the fire. Stay low to avoid inhaling the smoke. If it becomes too hot or too smoky, get out. Remember: never go back into a burning building for anything.

First-Aid Kits

First-aid kits should be clearly identified and conveniently located (**Figure 1-23**). These kits contain such items as bandages and ointment required for minor cuts. All shop personnel must be familiar with the location of first-aid kits. At least one of the shop personnel should have basic first-aid training, and this person should be in charge of administering first aid and keeping first-aid kits filled.

HAZARDOUS WASTE DISPOSAL

Hazardous waste materials in automotive shops are chemicals or components that the shop no longer needs and that pose a danger to people and the environment if they are disposed of in ordinary garbage cans or sewers. However, it should be noted that no material is considered hazardous waste until the shop has finished using it and is ready to dispose of it. When handling any hazardous waste, always wear proper protective clothing, which may include a respirator and gloves (**Figure 1-24**), and use equipment detailed in the right-to-know laws. The EPA publishes a list of hazardous materials that is included

	Class of Fire	Typical Fuel Involved	Type of Extinguisher
Class A Fires (green)	**For Ordinary Combustibles** Put out a Class A fire by lowering its temperature or by coating the burning combustibles.	Wood Paper Cloth Rubber Plastics Rubbish Upholstery	Water*[1] Foam* Multipurpose dry chemical[4]
Class B Fires (red)	**For Flammable Liquids** Put out a Class B fire by smothering it. Use an extinguisher that gives a blanketing, flame-interrupting effect; cover whole flaming liquid surface.	Gasoline Oil Grease Paint Lighter fluid	Foam* Carbon dioxide[5] Halogenated agent[6] Standard dry chemical[2] Purple K dry chemical[3] Multipurpose dry chemical[4]
Class C Fires (blue)	**For Electrical Equipment** Put out a Class C fire by shutting off power as quickly as possible and by always using a nonconducting extinguishing agent to prevent electric shock.	Motors Appliances Wiring Fuse boxes Switchboards	Carbon dioxide[5] Halogenated agent[6] Standard dry chemical[2] Purple K dry chemical[3] Multipurpose dry chemical[4]
Class D Fires (yellow)	**For Combustible Metals** Put out a Class D fire of metal chips, turnings, or shaving by smothering or coating with a specially designed extinguishing agent.	Aluminum Magnesium Potassium Sodium Titanium Zirconium	Dry powder extinguishers and agents only

*Cartridge-operated water, foam, and soda-acid types of extinguishers are no longer manufactured. These extinguishers should be removed from service when they become due for their next hydrostatic pressure test.

Notes:

(1) This freezes in low temperatures unless treated with antifreeze solution, usually weighs over 20 pounds, and is heavier than any other extinguisher mentioned.

(2) Also called "ordinary" or "regular" dry chemical (solution bicarbonate).

(3) Has the greatest initial fire-stopping power of the extinguishers mentioned for class B fires. Be sure to clean residue immediately after using the extinguisher so that sprayed surfaces will not be damaged (potassium bicarbonate).

(4) The only extinguishers that fight A, B, and C class fires. However, they should not be used on fires in liquified fat or oil of appreciable depth. Be sure to clean residue immediately after using the extinguisher so sprayed surfaces will not be damaged (ammonium phosphates).

(5) Use with caution in unventilated, confined spaces.

(6) May cause injury to the operator if the extinguishing agent (a gas) or the gases produced when the agent is applied to a fire is inhaled.

Figure 1-22 Guide to fire extinguisher selection.

Figure 1-24 Wear recommended safety clothing and equipment when handling hazardous materials.

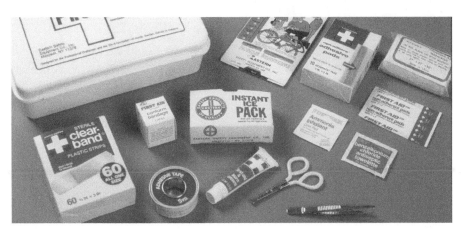

Figure 1-23 First-aid kit.

The *Environmental Protection Agency (EPA)* is a government agency that oversees activities and laws designed to protect water, air, land, and the public from harmful pollution or contamination.

in the Code of Federal Regulations. Waste is considered hazardous if it is included on the EPA list of hazardous materials or has one or more of these characteristics:

1. *Reactive.* Any material that reacts violently with water or other chemicals is considered hazardous. When exposed to low-pH acid solutions, if a material releases cyanide gas, hydrogen sulfide gas, or similar gases, it is hazardous.
2. *Corrosive.* If a material burns the skin or dissolves metals and other materials, it is considered hazardous.
3. *Toxic.* Materials are **toxic** if they leach one or more of eight heavy metals in concentrations greater than 100 times the primary drinking water standard.
4. *Ignitable.* A liquid is ignitable if it has a flashpoint below 140°F (60°C), and a solid is ignitable if it ignites spontaneously.

The automotive service industry is considered a generator of hazardous wastes. However, the vehicles it services are the real generators. Once you drain oil from an engine, you have generated the waste and now become responsible for the proper disposal of this hazardous waste. There are many other wastes that need to be handled properly after you have removed them, such as batteries, brake fluid, and transmission fluid.

A material that reacts violently with water or other chemicals is referred to as **reactive**.

Engine coolant should not be allowed to go down sewage drains. This is also true for all liquids drained from a car. Coolant should be captured and recycled or disposed of properly.

Filters for fluids (transmission, fuel, and oil filters) also need to be handled in designated ways. Used filters need to be drained and then crushed or disposed of in a special shipping barrel. Most regulations demand that oil filters be drained for at least 24 hours before they are disposed of or crushed.

Corrosive materials burn the skin or dissolve metals or other materials.

Federal and state laws control the disposal of hazardous waste materials. Every shop employee must be familiar with these laws. Hazardous waste disposal laws include the Resource Conservation and Recovery Act. This law basically states that hazardous material users are responsible for hazardous materials from the time they become a waste until the proper waste disposal is completed. Many automotive shops hire an independent hazardous waste hauler to dispose of hazardous waste material (**Figure 1-25**). The shop owner or manager should have a written contract with the hazardous waste hauler. Rather than have hazardous waste material hauled to an approved hazardous waste disposal site, a shop may choose to recycle the material in the shop. Therefore, the user must store hazardous waste material properly and safely and be responsible for the transportation of this material until it arrives at an approved hazardous waste disposal site and is processed according to the law.

Materials are toxic if they exceed levels greater than 100 times that in drinking water.

Figure 1-25 Hazardous waste hauler.

The RCRA controls these types of automotive waste:

1. Paint and body repair products waste
2. Solvents for parts and equipment cleaning
3. Batteries and battery acid
4. Mild acids used for metal cleaning and preparation
5. Waste oil, engine coolants, or antifreeze
6. Air-conditioning refrigerants
7. Engine oil filters

Never, under any circumstances, use these methods to dispose of hazardous waste material:

1. Pour hazardous wastes on weeds to kill them.
2. Pour hazardous wastes on gravel streets to prevent dust.
3. Throw hazardous wastes in a dumpster.
4. Dispose of hazardous wastes anywhere but an approved disposal site.
5. Pour hazardous wastes down sewers, toilets, sinks, or floor drains.
6. Bury hazardous wastes in the ground.

The **right-to-know laws** state that employees have a right to know when the materials they use at work are hazardous. The right-to-know laws started with the **Hazard Communication Standard** published by OSHA in 1983. This document was originally intended for chemical companies and manufacturers that required employees to handle hazardous materials in their work situation. At the present time, most states have established their own right-to-know laws. Meanwhile, the federal courts have decided to apply these laws to all companies, including automotive service shops. Under the right-to-know laws, the employer has three responsibilities regarding the handling of hazardous materials by its employees.

First, all employees must be trained about the types of hazardous materials they will encounter in the workplace. The employees must be informed about their rights under legislation regarding the handling of hazardous materials. All hazardous materials must be properly labeled, and information about each hazardous material must be posted on **material safety data sheets (MSDS)**, which are available from the manufacturer (**Figure 1-26**). In Canada, MSDS sheets are called workplace hazardous materials information systems (WHMIS).

The employer has a responsibility to place MSDS where they are easily accessible by all employees. The MSDS provide extensive information about the hazardous material such as the following:

1. Chemical name
2. Physical characteristics
3. Protective equipment required for handling
4. Explosion and fire hazards
5. Other incompatible materials
6. Health hazards such as signs and symptoms of exposure, medical conditions aggravated by exposure, and emergency and first-aid procedures
7. Safe handling precautions
8. Spill and leak procedures

Second, the employer has a responsibility to make sure that all hazardous materials are properly labeled. The label information must include health, fire, and reactivity hazards posed by the material, as well as the protective equipment necessary to handle the material. The manufacturer must supply all warning and precautionary information about

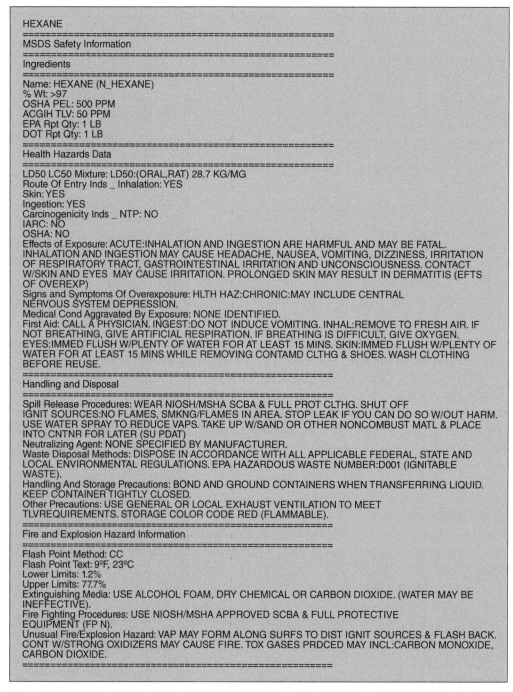

```
HEXANE
==================================================
MSDS Safety Information
==================================================
Ingredients
==================================================
Name: HEXANE (N_HEXANE)
% Wt: >97
OSHA PEL: 500 PPM
ACGIH TLV: 50 PPM
EPA Rpt Qty: 1 LB
DOT Rpt Qty: 1 LB
==================================================
Health Hazards Data
==================================================
LD50 LC50 Mixture: LD50:(ORAL,RAT) 28.7 KG/MG
Route Of Entry Inds _ Inhalation: YES
Skin: YES
Ingestion: YES
Carcinogenicity Inds _ NTP: NO
IARC: NO
OSHA: NO
Effects of Exposure: ACUTE:INHALATION AND INGESTION ARE HARMFUL AND MAY BE FATAL.
INHALATION AND INGESTION MAY CAUSE HEADACHE, NAUSEA, VOMITING, DIZZINESS, IRRITATION
OF RESPIRATORY TRACT, GASTROINTESTINAL IRRITATION AND UNCONSCIOUSNESS. CONTACT
W/SKIN AND EYES  MAY CAUSE IRRITATION. PROLONGED SKIN MAY RESULT IN DERMATITIS (EFTS
OF OVEREXP)
Signs and Symptoms Of Overexposure: HLTH HAZ:CHRONIC:MAY INCLUDE CENTRAL
NERVOUS SYSTEM DEPRESSION.
Medical Cond Aggravated By Exposure: NONE IDENTIFIED.
First Aid: CALL A PHYSICIAN. INGEST:DO NOT INDUCE VOMITING. INHAL:REMOVE TO FRESH AIR. IF
NOT BREATHING, GIVE ARTIFICIAL RESPIRATION. IF BREATHING IS DIFFICULT, GIVE OXYGEN.
EYES:IMMED FLUSH W/PLENTY OF WATER FOR AT LEAST 15 MINS. SKIN:IMMED FLUSH W/PLENTY OF
WATER FOR AT LEAST 15 MINS WHILE REMOVING CONTAMD CLTHG & SHOES. WASH CLOTHING
BEFORE REUSE.
==================================================
Handling and Disposal
==================================================
Spill Release Procedures: WEAR NIOSH/MSHA SCBA & FULL PROT CLTHG. SHUT OFF
IGNIT SOURCES:NO FLAMES, SMKNG/FLAMES IN AREA. STOP LEAK IF YOU CAN DO SO W/OUT HARM.
USE WATER SPRAY TO REDUCE VAPS. TAKE UP W/SAND OR OTHER NONCOMBUST MATL & PLACE
INTO CNTNR FOR LATER (SU PDAT)
Neutralizing Agent: NONE SPECIFIED BY MANUFACTURER.
Waste Disposal Methods: DISPOSE IN ACCORDANCE WITH ALL APPLICABLE FEDERAL, STATE AND
LOCAL ENVIRONMENTAL REGULATIONS. EPA HAZARDOUS WASTE NUMBER:D001 (IGNITABLE
WASTE).
Handling And Storage Precautions: BOND AND GROUND CONTAINERS WHEN TRANSFERRING LIQUID.
KEEP CONTAINER TIGHTLY CLOSED.
Other Precautions: USE GENERAL OR LOCAL EXHAUST VENTILATION TO MEET
TLVREQUIREMENTS. STORAGE COLOR CODE RED (FLAMMABLE).
==================================================
Fire and Explosion Hazard Information
==================================================
Flash Point Method: CC
Flash Point Text: 9ºF, 23ºC
Lower Limits: 1.2%
Upper Limits: 77.7%
Extinguishing Media: USE ALCOHOL FOAM, DRY CHEMICAL OR CARBON DIOXIDE. (WATER MAY BE
INEFFECTIVE).
Fire Fighting Procedures: USE NIOSH/MSHA APPROVED SCBA & FULL PROTECTIVE
EQUIPMENT (FP N).
Unusual Fire/Explosion Hazard: VAP MAY FORM ALONG SURFS TO DIST IGNIT SOURCES & FLASH BACK.
CONT W/STRONG OXIDIZERS MAY CAUSE FIRE. TOX GASES PRDCED MAY INCL:CARBON MONOXIDE,
CARBON DIOXIDE.
==================================================
```

Figure 1-26 Material safety data sheets (MSDS) inform employees about hazardous materials. N-hexane is commonly used in solvents and many brake-cleaning products.

hazardous materials, and this information must be read and understood by the employee before handling the material.

Third, employers are responsible for maintaining permanent files regarding hazardous materials. These files must include information on hazardous materials in the shop, proof of employee training programs, and information about accidents such as spills or leaks of hazardous materials. The employer's files must also include proof that employees' requests for hazardous material information such as MSDS have been met. A general right-to-know compliance procedure manual must be maintained by the employer.

HAND TOOLS AND DIAGNOSTIC EQUIPMENT

Automotive engine performance is among the most challenging areas of automotive service and encompasses many different theories and service aspects. A thorough understanding of the theory and application of engine mechanics, chemistry, mathematics, electricity, physics, computer operation, safety, and more must be practiced regularly. Part of your responsibility as a professional is to remain current in advancing service technology, tool and equipment usage, as well as any regulatory practices that you need to adhere to including any local, state, or federal licensing or certification mandates. Performing safe, accurate, effective, and efficient service on an automobile—specifically in the area of engine performance—will require a wide array of tools and equipment. Tools are designed to be work savers. When used as designed, they will reduce the amount of time or effort used in performing a service operation. In addition, many tool manufacturers are producing hand tools that are ergonomically designed and comfortable to use to help reduce stress and fatigue on muscles and joints. You also need various types of specialty tools in order to perform different diagnostic tests. Tool and equipment manufacturers are continually upgrading and introducing new products to keep pace with automobile changes. You may invest in some of these tools and diagnostic equipment, but some will belong to the shop due to their cost. The following brief list provides you, the technician, with both common and specialty tools that are typically used in the area of automotive engine performance.

1. Basic hand tools. The basic hand tool set generally includes common tool favorites such as the following:
   - A variety of metric and SAE-size wrenches and sockets
   - Screwdrivers, Torx® drivers, and hex-key wrenches
   - Hammers, chisels, and punches
   - Ratchets and extensions
   - An assortment of various pliers
   - A gear puller set, files, and saws
   - A stethoscope
   - Power tools and torque wrenches

2. Safety and protection equipment:
   - Gloves and hand protection
   - Safety glasses and earplugs
   - Fender and seat covers
   - First-aid kit

3. Measuring devices:
   - Feeler gauges and a spark plug gapping tool
   - Fuel pressure gauges and vacuum gauges
   - Thermometer
   - Various micrometers and dial indicators
   - An engine compression tester

4. Electrical tools:
   - A test light and jumper leads
   - A circuit tester, a logic probe, and test probes
   - A soldering gun and a wire repair kit
   - Crimping and stripping tools
   - A battery tester and battery service tools

Power tools are operated by outside power sources.

5. Diagnostic test equipment:
   - A graphing and/or digital multi-meter (DMM) and a lab scope
   - A scan tool
   - An infrared exhaust gas analyzer
   - A cylinder leakage tester and a vacuum hand pump
   - A fuel injector tester, injector test lights, and a spark tester
   - Test probes and a charging system tester

6. Service information:
   - Online service information for repair, parts and labor time guides
   - Reference manuals and hotline support assistance

EMPLOYER AND EMPLOYEE OBLIGATIONS

When you begin employment, you enter into a business agreement with your employer. A business agreement involves an exchange of goods or services that have value. Although the automotive technician may not have a written agreement with his or her employer, the technician exchanges time, skills, and effort for money paid by the employer. Both the employee and the employer have obligations. The automotive technician's obligations include the following:

1. *Quality.* Each repair job should be a quality job! Work should never be done in a careless manner. Nothing improves customer relations like quality workmanship.
2. *Productivity.* As an automotive technician, you have a responsibility to your employer to make the best possible use of time on the job. Each job should be done in a reasonable length of time. Employees are paid for their skills, effort, and time.
3. *Teamwork.* The shop staff is a team, and everyone, including technicians and management personnel, are team members. You should cooperate with, and care about, other team members. Each member of the team should strive for harmonious relations with fellow workers. Cooperative teamwork helps to improve shop efficiency, productivity, and customer relations. Customers may be turned off by bickering between shop personnel.
4. *Honesty.* Employers and customers expect and deserve honesty from automotive technicians. Honesty creates a feeling of trust between technicians, employers, and customers.
5. *Loyalty.* As an employee, you are obliged to act in the best interests of your employer, both on and off the job.
6. *Attitude.* Employees should maintain a positive attitude at all times. As in other professions, automotive technicians have days when it may be difficult to maintain a positive attitude. For example, there will be days when the technical problems on a certain vehicle are difficult to solve. However, developing a negative attitude certainly will not help the situation. A positive attitude has a positive effect on the job situation as well as on your customers and employer.
7. *Responsibility.* You are responsible for your conduct on the job and your work-related obligations. These obligations include always maintaining good workmanship and customer relations. Attention to details, such as always placing fender and seat covers on customer vehicles prior to driving or working on the vehicle, greatly improves customer relations.
8. *Following directions.* All of us like to do things our way. Such action may not be in the best interests of the shop, and, as an employee, you have an obligation to follow the supervisor's directions.

9. *Punctuality and regular attendance.* Employees have an obligation to be on time for work and to be regular in attendance on the job. It is very difficult for a business to operate successfully if it cannot count on its employees to be on the job at the appointed time.

10. *Regulations.* Automotive technicians should be familiar with all state and federal regulations pertaining to their job situation such as the OSHA and hazardous waste disposal laws. In Canada, employees should be familiar with WHMIS.

Employer-to-employee obligations include the following:

1. *Wages.* The employer has a responsibility to inform the employee regarding the exact amount of financial remuneration he or she will receive and when that wage will be paid.

2. *Fringe benefits.* A detailed description of all fringe benefits should be provided by the employer. These benefits may include holiday pay, sickness and accident insurance, and pension plans.

3. *Working conditions.* A clean, safe workplace must be provided by the employer. The shop must have adequate safety equipment and first-aid supplies. Employers must be certain that all shop personnel maintain the shop area and equipment to provide adequate safety and a healthy workplace atmosphere.

4. *Employee instruction.* Employers must provide employees with clear job descriptions and be sure that each worker is aware of his or her obligations.

5. *Employee supervision.* Employers should inform their workers regarding the responsibilities of their immediate supervisors and other management personnel.

6. *Employee training.* Employers must make sure that employees are familiar with the safe operation of all the equipment that they are required to use in their job situation. Because automotive technology is changing rapidly, employers should provide regular update training for their technicians. Under the right-to-know laws, employers are required to inform all employees about hazardous materials in the shop. Employees should be familiar with the MSDS (or, in Canada, the WHMIS), which detail the labeling and handling of hazardous waste and the health problems if exposed to hazardous waste.

An automotive technician has specific responsibilities regarding each job performed on a customer's vehicle:

1. Do every job to the best of your ability. There is no place in the automotive service industry for careless workmanship! Automotive technicians and students must realize they have a very responsible job. During many repair jobs, you, as a student or technician working on a customer's vehicle, actually have the customer's life and the safety of his or her vehicle in your hands. For example, if you are doing a brake job and leave the wheel nuts loose on one wheel, that wheel may fall off the vehicle at high speed. This could result in serious personal injury or death for the customer and others, plus extensive vehicle damage. If this type of disaster occurs, the individual who worked on the vehicle and the shop may be involved in a very expensive legal action. As a student or technician working on customer vehicles, you are responsible for the safety of every vehicle that you work on. Even when careless work does not create a safety hazard, it leads to dissatisfied customers who often take their business to another shop, and nobody benefits when that happens.

2. Treat customers fairly and honestly on every repair job. Do not install parts that are unnecessary to complete the repair job.

3. Use published specifications; do not guess at adjustments.

4. Follow the service procedures in the service manual provided by the vehicle manufacturer or an independent manual publisher.

5. When the repair job is completed, always be sure the customer's complaint has been corrected.
6. Do not be too concerned with work speed when you begin working as an automotive technician. Speed comes with experience.

NATIONAL INSTITUTE FOR AUTOMOTIVE SERVICE EXCELLENCE (ASE) CERTIFICATION

The Advanced Engine Performance Test is best known as the L1 test.

The **National Institute for Automotive Service Excellence (ASE)** has provided voluntary testing and certification of automotive technicians on a national basis for many years. The image of the automotive service industry has been enhanced by the ASE certification program. More than 350,000 ASE-certified automotive technicians now work in a wide variety of automotive service shops. The ASE provides certification in these nine basic areas of automotive repair:

1. Engine repair
2. Automatic transmissions and transaxles
3. Manual drivetrain and axles
4. Suspension and steering systems
5. Brake systems
6. Electrical systems
7. Heating and air-conditioning systems
8. Engine performance
9. Light vehicle diesel

In addition to these nine basic tests, a technician can take advanced tests (e.g., Advanced Engine Performance L1 or Light Duty Hybrid/Electric Vehicle Specialist Test (L3)) and become certified in that area. The advanced tests require the technician to have particular basic certifications before attempting to be certified at an advanced level.

A technician may take the ASE test and become certified in any or all of the nine basic areas. When a technician passes an ASE test in one of the nine areas, an automotive technician's shoulder patch is issued by the ASE. If a technician passes the A1 to A8 regular tests (light vehicle diesel (A9) is not required for Master Technician Certification), he or she receives a Master Technician's shoulder patch (**Figure 1-27**). Retesting at 5-year intervals is required to remain certified.

The certification test in each of the nine areas contains 40 to 80 multiple-choice questions. The test questions are written by a panel of automotive service experts from

Figure 1-27 ASE certification patches.

TEST SPECIFICATIONS AND TASK LIST ENGINE PERFORMANCE (TEST A8)

Content area	Questions in test	Percentage of test
A. General diagnosis	12	24.0%
B. Ignition system diagnosis and repair	8	16.0%
C. Fuel, air induction, and exhaust system diagnosis and repair	9	18.0%
D. Emissions control system diagnosis and repair (Including OBD II) 1. Positive crankcase ventilation (1) 2. Exhaust gas recirculation (2) 3. Secondary air injection (AIR) and catalytic converter (2) 4. Evaporative emissions controls (3)	8	16.0%
E. Computerized engine controls diagnosis and repair (Including OBD II)	13	26.0%
Total	**50**	**100%**

Figure 1-28 Engine performance test specifications.

various areas of automotive service such as automotive instructors, service managers, automotive manufacturers' representatives, test equipment representatives, and certified technicians. The test questions are pretested and checked for quality by a national sample of technicians. Most questions have the *Technician A* and *Technician B* format similar to the questions at the end of each chapter in this book. ASE regulations demand that each technician must have 2 years of working experience in the automotive service industry prior to taking a certification test or tests. However, relevant formal training may be substituted for one of the years of working experience. Contact the ASE for details regarding this substitution. The contents of the Engine Performance test are listed in **Figure 1-28**.

Shops that employ ASE-certified technicians display an official **ASE blue seal of excellence**. This blue seal increases the customer's awareness of the shop's commitment to quality service and the competency of certified technicians.

ASE-STYLE REVIEW QUESTIONS

1. *Technician A* says it is recommended that you wear shoes with nonslip soles in the shop.
Technician B says shoes worn in the shop should have steel toes.
Who is correct?
   A. A only
   B. B only
   C. Both A and B
   D. Neither A nor B

2. *Technician A* says that some machines can be routinely used beyond their stated capacity.
Technician B says that a power tool can be left running unattended if the technician puts up a power "on" sign.
Who is correct?
   A. A only
   B. B only
   C. Both A and B
   D. Neither A nor B

3. *Technician A* ties his long hair behind his head while working in the shop.
Technician B covers her long hair with a brimless cap.
Who is correct?
   A. A only
   B. B only
   C. Both A and B
   D. Neither A nor B

4. *Technician A* uses compressed air to blow dirt from his clothes and hair.
Technician B says this should only be done outside.
Who is correct?
   A. A only
   B. B only
   C. Both A and B
   D. Neither A nor B

5. While discussing the proper way to lift heavy objects,

   Technician A says you should bend your back to pick up a heavy object.

   Technician B says you should bend your knees to pick up a heavy object.

   Who is correct?

   A. A only
   B. B only
   C. Both A and B
   D. Neither A nor B

6. While discussing shop cleaning equipment safety,

   Technician A says some hot tanks contain caustic solutions.

   Technician B says some metals such as aluminum may dissolve in hot tanks.

   Who is correct?

   A. A only
   B. B only
   C. Both A and B
   D. Neither A nor B

7. While discussing shop rules,

   Technician A says breathing carbon monoxide may cause arthritis.

   Technician B says breathing carbon monoxide may cause headaches.

   Who is correct?

   A. A only
   B. B only
   C. Both A and B
   D. Neither A nor B

8. While discussing air quality,

   Technician A says a restricted chimney on a shop furnace may cause carbon monoxide gas in the shop.

   Technician B says monitors are available to measure the level of carbon monoxide in the shop air.

   Who is correct?

   A. A only
   B. B only
   C. Both A and B
   D. Neither A nor B

9. While discussing air quality,

   Technician A says a battery gives off hydrogen gas during the charging process.

   Technician B says a battery gives off oxygen gas during the charging process.

   Who is correct?

   A. A only
   B. B only
   C. Both A and B
   D. Neither A nor B

10. While discussing parts cleaning,

    Technician A says thermal cleaning relies on a chemical reaction to break down dirt and deposits.

    Technician B says using abrasive cleaning to clean the surface relies on physical abrasion from a brush, glass bead blasting, or similar methods.

    Who is correct?

    A. A only
    B. B only
    C. Both A and B
    D. Neither A nor B

Name _____   **Date** _____

SHOP SAFETY SURVEY

As a professional technician, safety should be one of your first concerns. This job sheet will increase your awareness of shop safety rules and safety equipment. As you survey your shop area and answer the following questions, you will learn how to evaluate the safeness of your workplace.

ASE Education Foundation Correlation

This job sheet addresses the following **RST** tasks: 1. Shop and Personal Safety

Task #1	Identify general shop safety rules and procedures.
Task #2	Utilize safe procedures for handling of tools and equipment.
Task #6	Identify marked safety areas.
Task #7	Identify the location and the types of fire extinguishers and other fire safety equipment; demonstrate knowledge of the procedures for using fire extinguishers and other fire safety equipment.
Task #8	Identify the location and use of eye wash stations.
Task #9	Identify the location of the posted evacuation routes.
Task #10	Comply with the required use of safety glasses, ear protection, gloves, and shoes during lab/shop activities.
Task #11	Identify and wear appropriate clothing for lab/shop activities.
Task #15	Locate and demonstrate knowledge of material safety data sheets (MSDS).

Procedure

Your instructor will review your progress throughout this worksheet and should approve the sheet when you complete it.

1. Before you begin to evaluate your work area, evaluate yourself. Are you dressed to work safely? ☐ Yes   ☐ No

   If no, what is wrong? _____

2. Are your safety glasses OSHA approved? ☐ Yes   ☐ No

   Do they have side protection shields? ☐ Yes   ☐ No

3. Watch around the shop, and note any area that poses a potential safety hazard or is an area that you should be aware of.

   _____

   _____

   _____

   Any true hazards should be brought to the attention of the instructor immediately.

4. Are there safety areas marked around grinders and other machinery? ☐ Yes   ☐ No

5. What is the line air pressure in the shop? _____ psi

   What should it be? _____ psi

6. Where are the tools stored in the shop?

_____

7. If you could, how would you improve the tool storage area?

_____

_____

8. What types of hoists are used in the shop?

_____

9. Ask your instructor to demonstrate the proper use of the hoist.

   A. Describe the placement of the hoist pads.

   _____

   B. Describe how your instructor verified that the vehicle would be lifted safely.

   _____

10. Where is the first-aid kit(s) kept in the work area?

_____

_____

11. What is the shop's procedure for dealing with an accident?

_____

_____

12. Have your instructor supply you with a vehicle make, model, and year. Using the appropriate service manual, find the location of the correct lifting points for that vehicle. Below, draw a simple figure showing where these lift points are.

Instructor's Response

_____

_____

_____

Name _____   Date _____

WORKING SAFELY AROUND AIR BAGS

Upon completion of this job sheet, you should be able to work safely around and with air bag systems.

ASE Education Foundation Correlation

This job sheet addresses the following **RST** task: 1. Shop and Personal Safety

Task #13   Demonstrate awareness of the safety aspects of supplemental restraint systems (SRS), electronic brake control systems, and hybrid vehicle high-voltage circuits.

Tools and Materials

- A vehicle with air bags
- Service manual for the above vehicle
- Component locator for the above vehicle
- Safety glasses
- A DMM

Describe the vehicle being worked on:

Year _____   Make _____   Model _____

VIN _____   Engine type and size _____

Procedure

1. Locate the information about the air bag system in the service manual. How are the critical parts of the system identified in the vehicle?

   _____

   _____

2. List the main components of the air bag system and describe their location.

   _____

   _____

   _____

3. There are some very important guidelines to follow when working with and around air bag systems. These are listed below with some key words left out. Read through these, and fill in the blanks with the correct words.

   a. Wear _____ when servicing an air bag system and when handling an air bag module.

   b. Wait at least _____ minutes after disconnecting the battery before beginning any service. The reserve _____ module is capable of storing enough energy to deploy the air bag for up to _____ minutes after battery voltage is lost.

   c. Always handle all _____ and other components with extreme care. Never strike or jar a sensor, especially when the battery is connected. This can cause deployment of the air bag.

   d. Never carry an air bag module by its _____ or _____, and when carrying it, always face the trim and air bag _____ from your body. When placing a module on a bench, always face the trim and air bag _____ .

e. Deployed air bags may have a powdery residue on them. _____ is produced by the deployment reaction and is converted to _____ when it comes in contact with the moisture in the atmosphere. Although it is unlikely that harmful chemicals will still be on the bag, it is wise to wear _____ and _____ when handling a deployed air bag. Immediately wash your hands after handling a deployed air bag.

f. A live air bag must be _____ before it is disposed. A deployed air bag should be disposed of in a manner consistent with the _____ and manufacturer's procedures.

g. Never use a battery- or AC-powered _____, _____, and other type of test equipment in the system unless the manufacturer specifically says to. Never probe with a _____ for voltage.

Instructor's Response

_____

_____

_____

Name _____ Station _____ Date _____

HIGH VOLTAGE HAZARDS IN TODAY'S VEHICLES

Upon completion of this job sheet, you will be able to describe some of the necessary precautions to take in performing work around high-voltage hazards such as hybrid and HID headlamps. Ignition and fuel injection systems cautions are covered later.

ASE Education Foundation Correlation

This job sheet addresses the following **RST** tasks: 1. Shop and Personal Safety

Task #13   Demonstrate awareness of the safety aspects of supplemental restraint systems (SRS), electronic brake control systems, and hybrid vehicle high-voltage circuits.

Task #14   Demonstrate awareness of the safety aspects of high-voltage circuits (such as high intensity discharge (HID) lamps, ignition systems, injection systems, etc.).

Hybrid High-Voltage Hazards

Tools and Materials

Appropriate service information

AUTHOR'S NOTE According to vehicle manufacturers, a technician should not work on hybrid vehicles without taking specific training, which is beyond the scope of this job sheet. This job sheet assumes that the student will be accessing information only, not actually working on live high-voltage vehicles.

Protective Gear

Goggles or safety glasses with side shields
High-voltage gloves with properly inspected liners
Orange traffic cones to warn others in the shop of a high-voltage hazard

Describe the vehicle selected:

Year _____ Make _____ Model _____

Engine type and size _____

Describe general operating condition.

_____

Procedure

1. Special gloves with liners are to be inspected before each use when working on a hybrid vehicle.

   a. How often should the gloves be tested (at a lab) for recertification?

   _____

   b. What precautions must be taken when storing the gloves?

   _____

   _____

   _____

   c. When must the gloves be worn?

   _____

   _____

   _____

2. What color are the high-voltage cables on hybrid vehicles?

_____

_____

_____

3. What must be done _before_ disconnecting the main voltage supply cable?

_____

_____

_____

4. Describe the safety basis of the "one hand rule."

_____

_____

_____

5. Explain the procedure to disable high voltage on the vehicle you selected.

_____

_____

_____

6. Explain the procedure to test for high voltage to ensure that the main voltage is disconnected.

_____

_____

_____

HID Headlamp Precautions

Describe the vehicle selected: Leave blank if the same vehicle as previous.

Year _____ Make _____ Model _____

Engine type and size _____

Describe general operating condition.

_____

7. List three precautions a technician should observe when handling HID headlamp bulbs.

_____

_____

_____

8. A technician must never probe with a test lamp between the ballast and the bulb. Explain why.

_____

_____

_____

Instructor's Response

_____

_____

_____

CHAPTER 2

TYPICAL SHOP PROCEDURES AND EQUIPMENT

Upon completion and review of this chapter, you should be able to:

- Understand basic diagnostic procedures.
- Describe customer service expectations.
- List the basic units of measure for length, volume, and mass in the two measuring systems.
- Name the diagnostic tools and equipment commonly used in vehicle repair work.
- Describe the basic applications and operation of these tools.
- Explain the operation of a basic infrared analyzer.
- Describe what a waveform on an oscilloscope displays.
- Describe the major test capabilities of an engine analyzer.

- Explain the use of a logic probe.
- Describe how you would find specific information in a manufacturer's service information.
- Explain how you would locate specifications and special tool lists in a manufacturer's service information.
- Describe the information contained in an electronic data system, and explain how this information is displayed.
- Describe the purposes of a repair order, and explain how this order is processed by various personnel in an automotive shop.
- Explain the purpose of an engine tune-up.

Terms To Know

After top dead center (ATDC)

Ammeter

Before top dead center (BTDC)

Breakout box

Combustion chamber

Compression gauge

Continuity tester

Coolant hydrometer

Cooling system pressure tester

Cylinder leakage

Data link connector (DLC)

Digital multimeter (DMM)

Displacement

Exhaust gas analyzer

Exhaust gas recirculation (EGR) valve

Feeler gauge

Flat-rate manuals

High input impedance

Hyperlinked

Injector balance tester

International System of Units (SI)

Kilopascals (kPa)

Multi-channel oscilloscope

Multimeter

Noid light

Ohmmeter

Oil pressure gauge

Oscilloscope

Parts per million (ppm)

Pounds per square inch (psi)

Pressure gauge

Pulse

Revolutions per minute (rpm)

Scan tool

Trace

United States customary (USC) system

Vacuum

Volt/ampere tester (VAT)

Voltmeter

Waveform

INTRODUCTION

During the last decade, advances in technology have made today's vehicles quite advanced machines. Each system used to be treated as a separate, independent system isolated from the others. An engine-related problem, for example, was usually confined to just the engine, whereas transmission-related problems were confined to the transmission, and so on. In the case of a vehicle with black smoke spewing from the tailpipe, the focus was on the fuel delivery system. A misfire was usually associated with the ignition system with its respective components. Although still somewhat applicable to today's systems, there is a major difference in how the technician must approach the problem. This chapter will cover typical diagnostic procedures, shop procedures, and technician responsibilities and behaviors. Additionally, it will cover general tools and equipment application procedures specific to engine performance.

Modern vehicles are managed by several computers that integrate the operations of many systems. The trend in automotive computers has changed over the last several years from the powertrain control module (PCM) that controlled both the engine and transmission operation, although there are some new vehicles and many older vehicles on the road that still use a PCM, to more specialized control modules. Generically speaking, the engine control module (ECM) controls the ignition, fuel, and emission systems related to the engine. The automatic transmission is controlled by the transmission control module (TCM), while the antilock brakes are controlled by the electronic brake control module (EBCM) and entertainment, body control functions, and climate controls are handled by the body control module (BCM). Although these systems still operate independently, they are also integrated to each other via the computer network. The technician must determine early in the diagnostic process what system could be affecting another, seemingly unrelated system. Today's electronically controlled vehicle appears to be very complex, but it can be looked at in relatively simple, easy-to-understand segments. It is critical that the technician understands the basic theory and operating principles of the systems in the vehicle—mechanical or electrical. In doing so, his or her insights for accurate diagnosis are formed. Regardless of the symptom, concern, or system, a specific guideline for a uniform diagnostic procedure must be established. Although finding the problem in a reasonable time frame is important, the accuracy of the diagnosis is *more* important. As the skill level and experience of the technician improve over time, so will the speed at which problems are diagnosed. Technicians should practice the "fix-it-right-the-first-time" concept, thus eliminating a frustrated customer returning repeatedly for the same problem.

A Systematic Approach for Effective Diagnosis

The diagnostic procedure has evolved and is now a result of a critical thought process instead of a mechanical process. In other words, rather than assuming a specific component is faulty and then replacing it, the diagnostic thought process proves *why* you are replacing it based on factual test results. Using a symptom to establish a diagnosis is important; however, acting on the symptom alone without performing qualifying tests can lead you down a costly and incorrect path. Assumptions must be proved out. "Magic quick fixes" do not always work. Certain steps need to be followed when addressing any problem with the vehicle. The following systematic approach for an effective diagnosis gives the technician a **strategy** and process to follow to put things in order. It can apply to just about anything related to the vehicle, from fluid leaks to drivability problems. Throughout this book, you may find yourself revisiting this section for reference. A typical technician's workflow is illustrated in **Photo Sequence 1** on page 70.

1. *Questions and answers.* This is the time that the customer will communicate the symptom that has led to the vehicle being brought to you. Usually, a specific event occurred that appeared out of the ordinary to the customer. How the customer tells

Customer worksheet for driveability problems

Please fill out the following worksheet to the best of your ability and attach to the repair order. This form will be used by the technicians to better understand your problem.

Please circle all that apply

1. Is the check engine light:
    • always on
    • on sometimes
    • never on

2. If the check engine light does come on:
    • the engine runs noticeably worse
    • the engine runs fine even when the light is on

3. What is the nature of the problem you are having?
    • Hard starting
    • Odors   Please describe _____
    • Fuel economy
    • Spark knock or pinging on acceleration
    • Rough idle
    • Hesitation (slow from stops)
    • Surging
    • Engine noises
    • No power
    • Engine miss

    When does the problem occur?

    | Always | Hot | Deceleration/coasting |
    | Usually | Cold | Wet weather |
    | Seldom | During acceleration | Under load |

Figure 2-1 A customer worksheet can help a technician duplicate a customer concern.

you of the problem may require you to be a bit of an interrogator and a detective. During your questioning, you may find an important piece of information that the customer may not have known was relevant. Even minor information can be valuable. Also, refrain from letting the customer diagnose the problem for you. They are not always correct, or they could be making an assumption based on partial experience or facts. Use a worksheet to remind you of the specific questions you need to ask (**Figure 2-1**).

2. *Clarify the complaint.* Repeat to the customer what you understand of the complaint. Use specific explanations in the dialog, and avoid using technical jargon that may confuse the customer. Although many terms are commonly used throughout our industry, few customers actually understand their true meaning. For example, customers may request an alignment because they know the car's problem has something to do with the wheels and steering, but what if the customer asked for an alignment because the car vibrated at speeds over 40 mph? Obviously, an alignment would not help this situation. Imagine the multitude of descriptors that could be used to explain engine- or drivability-related problems, or anything else concerning the vehicle.

3. *Climate, speed, time, distance, and action.* These five categories can further aid your diagnosis by helping you understand exactly what was happening at the time the problem occurred. If necessary, ask specific questions about the climate. Was it a hot or cold day? Was it raining? Ask about the vehicle speed. Was the vehicle idle, or at city or cruise speed? Next, find out when it happened. Did the problem occur at the first start of the day, or after a period of driving? How often did it happen? How far did the customer drive the vehicle? Did the problem begin just out of the driveway or after a long distance? What action was the vehicle performing during the problem? Was the vehicle stopped, accelerating, turning, going uphill, pulling a load, or something else? Also, if applicable, find out if any aftermarket equipment or accessories are installed or if other service has been done recently.

4. *Verify the complaint.* This step is nothing more than confirming that the condition is occurring at the present time. This verifies that the problem exists and matches the description. It may require a test drive, perhaps with the customer, to duplicate the symptom. It may be necessary to attempt to duplicate the conditions as experienced by the customer when the problem occurred. Of course, depending on the situation, the vehicle may have to be left for a period of time in order to recreate the condition and problem, perhaps even overnight.

5. *Use the resources that are available to you.* Whenever possible, you should use all available information and other resources to effectively diagnose the problem, the least of which is your own experience. Consult the service information to know how to apply specific tests to the vehicle. In most cases it can be very beneficial to research any technical service bulletins (TSBs) that may apply to the vehicle. Also, as simple as it may sound, do not overlook the owner's manual that came with the vehicle. Many problems are related to an owner's misunderstanding of how a device or feature operates. Other resources will include proper tools and specific test equipment needed to perform tests and repairs. As you work into a more complex test procedure, it is best to follow the recommended factory test procedure for that particular vehicle, which on occasion may require specialty tools of some type. Do not hesitate to ask for assistance.

6. *Begin with an area test.* The area test gives you the broadest and quickest overview and assessment of the systems in question. You may choose a test that was based on a symptom or a suspected fault. This gives you the most general of starting points. The purpose is to isolate one system from another. For example, in the case of a "no start" complaint, you need to determine if the problem is mechanical, electrical, ignition related, or fuel related. By performing an area test, you will determine the systems that are working and those that are not. Therefore, through an elimination process, you can begin to focus on the specific system that is most likely causing the complaint or symptom. The objectives are to exclude all other components that have nothing to do with the problem and determine the root cause.

7. *Next, perform the system test.* Now that you know what is and is not working, your attention will focus on the system that is initially determined to be at fault. This is because you tested the other systems and found no problems with them during the area test. The system you will be testing is probably made up of several components or subsystems. In the case of an ignition system problem, you will need to perform a few tests to determine which part of the ignition system is not functioning. For example, is the problem in the primary or secondary ignition? As another example, suppose our discussion was related to the mechanical condition of the engine. If an area test revealed low engine vacuum, the system test may include a compression test to determine if an internal or external problem exists.

8. *Pinpoint or component test.* After determining the system fault, this diagnostic step will lead you to the end result, or very close to it. With the problem isolated to a

given system, the pinpoint test determines the component that is most likely the reason for the failure. It is nothing more than continuing the elimination process. Depending on the situation, some tests are done with the component installed in the vehicle, while others need to be tested independently, with the component removed from the vehicle. These tests may be as simple as a visual inspection of a component after disassembly, a precise measurement of a part, or a dynamic test with a test instrument. This testing process helps to validate the performance potential of the unit or part in question. As you can see, the situations will vary, but the concept and process are the same, regardless of the problem.

9. *Approval and repair.* After determining the problem, advise the customer of the cost and get approval for the completion of the repair. Explain what is needed and why. As complex as the systems are, it is possible to uncover other items that need to be addressed before, during, or after the repair of the original problem. Be sure you communicate this to the customer ahead of time.

10. *Confirm and retest.* Make it a mandatory practice to confirm your repairs. Prove the repair you made solved the problem. The repairs you make should restore the vehicle to designed performance levels. If other maintenance is required to prevent a repeat of the failure, it is best to do it before returning the vehicle to the customer.

The above procedure helps establish consistency in the way you begin the diagnosis and how you communicate with the customer. Even if you are not in direct contact with the customer, you still need to be involved with the service writer to impart clear information to the customer.

Customer Service and Vehicle Care

Customers are the purpose of your business. They ultimately pay the bills and your wages. Because of basic human nature, you will encounter a variety of people and behaviors. They are human beings, just like you. The type of customer service you deliver can make the difference between a good shop and a great shop. Obviously, customers want to be treated with courtesy and respect. They expect proper and thorough service to be performed on their vehicle at a fair price, within a reasonable time frame. No one wants to pay for something that is not needed. When discussing a problem with customers, be understanding of their needs. Taking a few extra moments or a few extra steps will win over customers. Be accurate in your diagnosis, the estimated charges for the repair, and when the vehicle will be ready. If you run into an unforeseen problem, be upfront and honest with customers immediately. They would rather know about it early than be unpleasantly surprised later.

You should follow some commonsense procedures when handling a customer's vehicle. The following are some simple yet important things to remember when servicing a customer's vehicle:

- Perform a quick safety check before test driving the vehicle (brakes, steering, etc.).
- Use seat covers, steering wheel covers, and floor mats while test driving.
- Observe all speed limits.
- Do not operate the vehicle beyond its designed capacity or ability.
- Upon entering the shop, roll down the driver's side window to avoid accidentally locking the keys in the car.
- Use clean fender covers.
- Remove any tools that may be in your pockets before getting into the vehicle.
- Do not smoke in or around the vehicle.
- Do not eat or drink in the vehicle.
- Do not change any radio settings (unless, of course, you are working on it).

- Return the seat to its original position if you moved it.
- Remove any evidence of work being performed (remove tools, papers, parts boxes, etc.).
- Reset the radio, clock, and other memory-related devices if the battery was disconnected, or use auxiliary power to keep memories alive.
- If it is a practice in your service facility, have the vehicle cleaned when repairs are completed, or at least make sure that greasy fingerprints are not left behind.
- Secure and lock the vehicle when work is completed—bring the keys inside unless otherwise instructed by the customer.
- Promptly call the customer when repairs are completed.

Being a technician does not begin with opening the hood, nor does it end with closing the hood. Your demonstration of quality customer service and vehicle care throughout the entire process, in addition to your technical aptitude, appearance, and work ethic, defines true professionalism.

Repairing the modern automobile requires the use of various tools. Many of these tools are common hand and power tools used every day by a technician. Other tools are very specialized and are only for specific repairs on specific systems and/or vehicles. Because units of measurement play such an important part in tool selection and in diagnosing automotive problems, this chapter begins with a presentation of measuring systems. Descriptions of the measuring systems and some important measuring tools follow.

UNITS OF MEASURE

The **United States customary (USC) system** is a measuring system that is commonly referred to as the "English system."

Two systems of weights and measures are commonly used in the United States. One system of weights and measures is the **United States customary (USC) system**. Well-known measurements for length in the USC system are the inch, foot, yard, and mile. In this system, the quart and gallon are common measurements for volume, and the ounce, pound, and ton are measurements for weight. A second system of weights and measures is the **International System of Units (SI)**, which is also referred to as the "metric system."

The International System of measurement is more commonly known as the metric system.

In the USC system, the basic linear measurement is the yard, whereas the corresponding linear measurement in the metric system is the meter (**Figure 2-2**). Each unit of measurement in the metric system is related to the other metric units by a factor of 10. Thus, every metric unit can be multiplied or divided by 10 to obtain larger units (multiples) or smaller units (submultiples). For example, the meter may be divided by 10 to obtain centimeters (1/100 meter) or millimeters (1/1,000 meter).

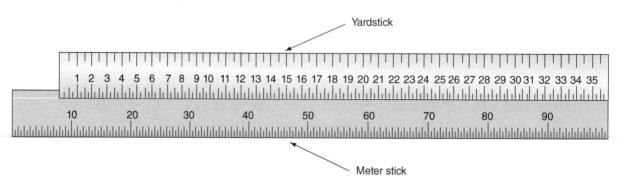

Figure 2-2 A meter is slightly longer than a yard.

Service technicians must be able to work with both the USC and the metric system. One meter (m) in the metric system is equal to 39.37 inches (in.) in the USC system. Some common equivalents between the metric and USC systems are these:

One meter is a little more than three inches longer than a yard.

1 meter (m) = 39.37 inches
1 centimeter (cm) = 0.3937 inch
1 millimeter (mm) = 0.03937 inch
1 inch = 2.54 cm
1 inch = 25.4 mm

In the USC system, phrases such as "⅛ inch" are used for measurements. The metric system uses a set of prefixes. For example, in the word *kilometer*, the prefix *kilo* indicates 1,000, and this prefix indicates there are 1,000 meters in a kilometer. Common prefixes in the metric system follow:

Name	Symbol	Meaning
mega	M	one million
kilo	k	one thousand
hecto	h	one hundred
deca	da	ten
deci	d	one tenth of
centi	c	one hundredth of
milli	m	one thousandth of
micro	μ	one millionth of

SERVICE TIP A metric conversion calculator provides fast, accurate metric-to-USC conversions or vice versa.

Measurement of Mass

In the metric system, mass is measured in grams, kilograms, or tonnes. One thousand grams (g) = one kilogram (kg). In the USC system, mass is measured in ounces, pounds, or tons. When converting pounds to kilograms, 1 pound = 0.453 kilogram.

Measurement of Length

In the metric system, length is measured in millimeters, centimeters, meters, or kilometers. Ten millimeters (mm) = one centimeter (cm). In the USC system, length is measured in inches, feet, yards, or miles. When distance conversions are made between the two systems, some of the conversion factors are these:

1 inch = 25.4 mm
1 foot = 30.48 cm
1 yard = 0.91 m
1 mile = 1.60 km

Measurement of Volume

In the metric system, volume is measured in milliliters, cubic centimeters, and liters. One cubic centimeter = one milliliter. If a cube has a length, depth, and height of 10 centimeters (cm), the volume of the cube is $10\,\text{cm} \times 10\,\text{cm} \times 10\,\text{cm} = 1,000\,\text{cm}^3 = 1$ liter.

Volume conversions can be made between the two systems: 1 cubic inch = 16.38 cubic centimeters. If an engine has a **displacement** of 350 cubic inches, $350 \times 16.38 = 5{,}733$ cubic centimeters, and $5{,}733/1{,}000 = 5.7$ liters.

ENGINE DIAGNOSTIC TOOLS

As the trend toward the integration of ignition, fuel, and emission systems progresses, diagnostic test equipment must also keep up with these changes. New tools and techniques are constantly being developed to diagnose electronic engine control systems.

Today's technician must not only keep up with changes in automotive technology, but also keep up with new testing procedures and specialized diagnostic equipment. To be successful, a shop must continuously invest in this equipment and the training necessary to troubleshoot today's electronic engine systems.

Regardless of how automated new test equipment can be or how new tests are developed for new systems, one thing will probably never change—the need to understand and perform fundamental tests. Because of all the advances in the automobile, sometimes technicians make problems more difficult than they have to be. Basic engine, electrical, fuel, and ignition tests still need to be performed. New test equipment and analyzers can make the job go faster and easier and even help the technician's diagnostic accuracy, but when beginning any diagnostic procedure, *do not forget the basics!*

Compression Testers

A compression gauge is used to check the sealing ability of the combustion chamber by measuring the pressure developed.

Internal combustion engines depend on compression of the air-fuel mixture to maximize the power produced by the engine. The upward movement of the piston on the compression stroke compresses the air and fuel mixture within the **combustion chamber**. The air-fuel mixture becomes hotter as it is compressed. The hot mixture is easier to ignite, and when ignited it will generate much more power than the same mixture at a lower temperature.

If the combustion chamber leaks, some of the air-fuel mixture will escape when it is compressed, resulting in a loss of power and a waste of fuel. Cylinder heat is also lost. The leaks can be caused by burned valves, a blown head gasket, worn rings, slipped timing belt or chain, worn valve seats, a cracked head, or other reasons.

An engine with poor compression (lower compression pressure due to leaks in the cylinder) will not run correctly and cannot be tuned to factory specifications. To see if a drivability problem is caused by poor compression, a compression test is performed.

A **compression gauge** is used to check cylinder compression. The dial face on the typical compression gauge indicates pressure in both **pounds per square inch (psi)** and **kilopascals (kPa)**. The range is usually 0 to 300 psi and 0 to 2,100 kPa.

There are two basic types of compression gauges: the push-in gauge (**Figure 2-3**) and a screw-in gauge.

The push-in type has a short stem that is either straight or bent at a 45-degree angle. The stem ends in a tapered rubber tip that fits any size of spark plug hole. The rubber tip is placed in the spark plug hole after the spark plugs have been removed and is held there while the engine is cranked through several compression cycles. Although simple to use, the push-in gauge may give inaccurate readings if it is not held tightly in the hole.

The screw-in gauge has a long, flexible hose that ends in a threaded adapter (**Figure 2-4**). This type of compression tester is often used because its flexible hose can reach into areas that are difficult to reach with a push-in tester. The threaded adapters are changeable and come in several thread sizes to fit 10 mm, 12 mm, 14 mm, and 18 mm diameter holes. The adapters screw into the spark plug holes in place of the spark plugs.

A compression gauge is used to check the sealing ability of the combustion chamber by measuring the pressure developed.

Pounds per square inch is the measurement of pressure in the English system.

A kilopascal is the measurement of pressure in the metric system, often abbreviated kPa.

A compression test checks the cylinder's sealing ability.

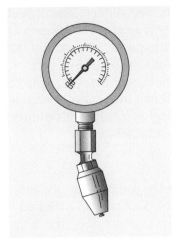

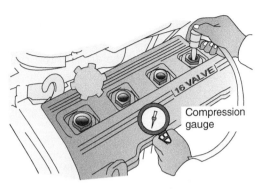

Figure 2-4 A compression tester in place.

Figure 2-3 Push-in compression gauge.

Most compression gauges have a vent valve that holds the highest pressure reading on its meter. Opening the valve releases the pressure when the test is complete.

Cylinder Leakage Tester

If a compression test shows that any of the cylinders are leaking, a **cylinder leakage** test can be performed to measure the percentage of compression lost and help locate the source of leakage.

A cylinder leakage tester (**Figure 2-5**) applies compressed air to a cylinder through the spark plug hole. Before the air is applied to the cylinder, the piston of that cylinder must be at top dead center (TDC) on its compression stroke. A threaded adapter on the end of the air pressure hose screws into the spark plug hole. The source of the compressed air is normally the shop's compressed-air system. A pressure regulator in the tester controls the pressure applied to the cylinder. An analog gauge registers the percentage of air pressure lost from the cylinder when the compressed air is applied. The scale on the dial face reads 0–100 percent. A reading of 10 means there is 10 percent cylinder leakage, and so on.

Cylinder leakage is the amount of air or volume lost from a sealed cylinder; it is measured in percentage.

A cylinder leakage test determines the point of cylinder leakage.

Figure 2-5 A cylinder leakage tester is connected to the cylinder.

The piston needs to be precisely at TDC to keep the applied air pressure from turning the crankshaft.

A 0 (zero) reading means that there is no leakage from the cylinder. Readings of 100 percent would indicate that the cylinder will not hold any pressure. The location of the compression leak can be found by listening and feeling around various parts of the engine (more on these tests can be found in Chapter 3). Most vehicles, even new cars, experience some leakage around the rings. Up to 20 percent is considered acceptable (although it is generally much less) during the leakage test. When the engine is actually running, the rings will seal much better and the actual percentage of leakage will be lower. However, there should be no leakage around the valves or the head gasket.

Vacuum Gauge

Vacuum is best defined as any pressure lower than atmospheric pressure. Atmospheric pressure is the pressure of the air around us.

Measuring intake manifold vacuum is another way to diagnose the condition of an engine. Manifold vacuum is tested with a vacuum gauge (**Figure 2-6**). **Vacuum** is formed on a piston's intake stroke. As the piston moves down, it lowers the pressure of the air in the cylinder if the cylinder is sealed. This lower cylinder pressure is called *engine vacuum*. If there is a leak, atmospheric pressure will force air into the cylinder and the resultant pressure will not be as low. The reason atmospheric pressure will enter is simply that whenever there is a low and high pressure, the high pressure will always move toward the low pressure.

The vacuum gauge measures the difference in pressure between intake manifold vacuum and atmospheric pressure. If the manifold pressure is lower than the atmospheric pressure, a vacuum exists. Vacuum is measured in inches of mercury (in. Hg), kilopascals (kPa), or millimeters of mercury (mm Hg).

To measure vacuum, a flexible hose on the vacuum gauge is connected to a source of manifold vacuum, either on the manifold or at a point below the throttle plates. Sometimes this requires removing a plug from the manifold and installing a special fitting.

The test is made with the engine cranking and/or running. A good vacuum reading is typically at least 16 in. Hg. However, a reading of 15 to 20 in. Hg (50 to 65 kPa) is normally acceptable. Since the intake stroke of each cylinder occurs at a different time, the production of vacuum occurs in pulses. If the amount of vacuum produced by each cylinder is the same, the vacuum gauge will show a steady reading. If one or more cylinders are producing different amounts of vacuum, the gauge will show a fluctuating reading.

Low or fluctuating readings can indicate many different problems. For example, a low, steady reading might be caused by incorrect valve timing. A sharp vacuum drop at regular intervals will be caused by a cylinder with low compression.

Figure 2-6 Vacuum gauge with line adapters.

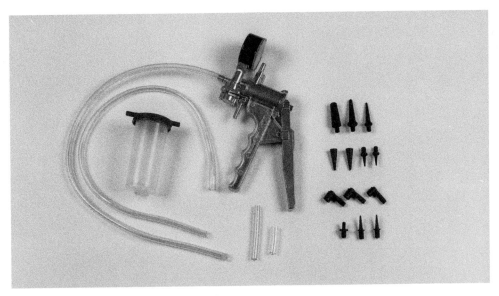

Figure 2-7 Typical vacuum pump with accessories.

Vacuum Pumps

There are many vacuum-operated devices and vacuum switches in cars. These devices use engine vacuum to cause a mechanical action or to switch something on or off. The tool used to test vacuum-actuated components is the vacuum pump (**Figure 2-7**). There are two types of vacuum pumps: an electrical operated pump and a handheld pump. The handheld pump is most often used for diagnostics. A handheld vacuum pump consists of a hand pump, a vacuum gauge, and a length of rubber hose used to attach the pump to the component being tested. Tests with the vacuum pump can usually be performed without removing the component from the car.

When the handles of the pump are squeezed together, a piston inside the pump body draws air out of the component being tested. The partial vacuum created by the pump is registered on the pump's vacuum gauge. While forming a vacuum in a component, watch the action of the component. The vacuum level needed to actuate a given component should be compared to the specifications given in the factory service manual.

The vacuum pump is also commonly used to locate vacuum leaks. This is done by connecting the vacuum pump to a suspect vacuum hose or component and applying vacuum. If the needle on the vacuum gauge begins to drop after the vacuum is applied, a leak exists somewhere in the system.

Vacuum Leak Detector

Low compression pressures might be revealed by a compression check, a cylinder leak down test, or a manifold vacuum test. However, finding the location of a vacuum leak can often be very difficult.

A simple but time-consuming way to find leaks in a vacuum system is to check each component and vacuum hose with a vacuum pump. Simply apply vacuum to the suspected area and watch the gauge for any loss of vacuum. A good vacuum component will hold the vacuum that is applied to it.

Another method of leak detection is done by using an ultrasonic leak detector (**Figure 2-8**). Air rushing through a vacuum leak creates a high-frequency sound, higher than the range of human hearing. An ultrasonic leak detector is designed to hear the frequencies of the leak. When the tool is passed over a leak, the detector responds to the high-frequency sound by emitting a warning beep. Some detectors also have a series of light-emitting diodes (LEDs) that light up as the frequencies are received. The closer the

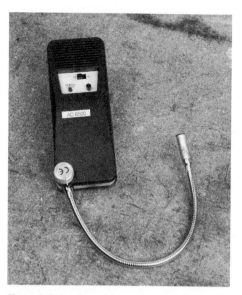

Figure 2-8 An ultrasonic vacuum leak detector.

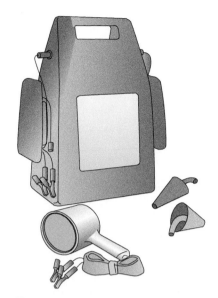

Figure 2-9 Finding small leaks can be made easier with a smoke machine.

detector is moved to the leak, the more LEDs light up or the faster the beeping occurs. This allows the technician to zero in on the leak. An ultrasonic leak detector can sense leaks as small as 1/500 inch and accurately locate the leak to within 1/16 inch.

Smoke Tester

A popular way of detecting vacuum is the smoke leak tester (**Figure 2-9**). A hose is connected from the machine to a hose on the intake manifold. The machine manufactures a nontoxic chemical smoke that can be introduced into the manifold. A high-intensity lamp helps to spot small leaks. The smoke machines are also popular for leak detection tasks in the evaporative emission system, which can set a "check engine" lamp with a leak as small as 0.020 inch in diameter. These testers can also find leaks in several other systems, such as the exhaust system.

Cooling System Pressure Tester

A **cooling system pressure tester** contains a hand pump and a pressure gauge. A hose is connected from the hand pump to a special adapter that fits on the radiator filler neck (**Figure 2-10**). This tester is used to pressurize the cooling system and check for coolant

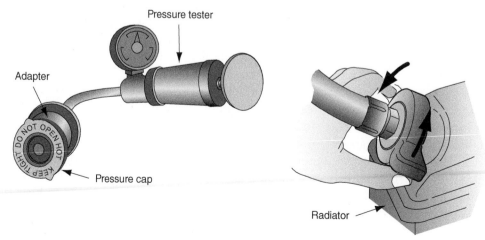

Figure 2-10 Cooling system pressure tester.

leaks. Additional adapters are available to connect the tester to the radiator cap. With the tester connected to the radiator cap, the pressure relief action of the cap may be checked.

Coolant Hydrometer

A **coolant hydrometer** is used to check the effectiveness of antifreeze in the coolant. This tester contains a pick-up hose, coolant reservoir, and squeeze bulb. The pick-up hose is placed in the radiator coolant. When the squeeze bulb is squeezed and released, coolant is drawn into the reservoir. As coolant enters the reservoir, a pivoted float moves upward with the coolant level. A pointer on the float indicates the freezing point of the coolant on a scale located on the reservoir housing (**Figure 2-11**).

Refractometer

The refractometer can be used to determine the freezing-point protection of antifreeze as well as the specific gravity of battery electrolyte. A small sample of the fluid is placed beneath the viewing window of the refractometer, and the meter is held up to the light while the technician looks through the eyepiece and reads the protection on a scale inside (**Figure 2-12**).

Oil Pressure Gauge

The **oil pressure gauge** may be connected to the engine to check the oil pressure (**Figure 2-13**). Various fittings are usually supplied with the oil pressure gauge to fit different openings in the lubrication system.

Stethoscope

A stethoscope is used to locate the source of engine and other noises. The stethoscope pick-up is placed on the suspected component, and the stethoscope receptacles are placed in the technician's ears (**Figure 2-14**). Amplified stethoscopes are also available.

Feeler Gauge

A **feeler gauge** is a thin strip of metal with a precision thickness. Feeler gauges with different thicknesses are usually sold in sets. These gauges are pivoted on one end and are mounted inside a metal holder. When a specific feeler gauge is required, that feeler gauge is pivoted out of the set (**Figure 2-15**).

A **feeler gauge** is a device of a specific thickness used to measure the distance between two components.

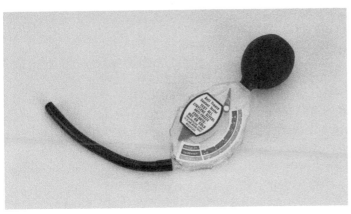

Figure 2-11 Coolant hydrometer.

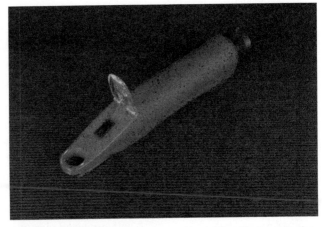

Figure 2-12 A refractometer can be used to check antifreeze protection as well as battery-specific gravity.

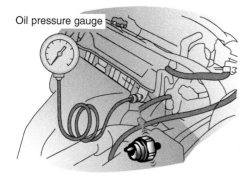

Oil pressure gauge

Figure 2-13 Oil pressure gauge threaded into the oil pressure sending unit's bore.

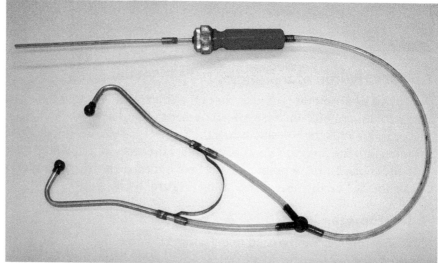

Figure 2-14 Stethoscope.

Modern iridium plugs should not be gapped. The electrode is very fragile and make break during this service.

A circuit tester is commonly called a *test light*.

Figure 2-15 Flat feeler gauge set.

Figure 2-16 Round wire-type feeler gauge set for spark plugs.

⚠ **CAUTION**

Do not use a test lamp or self-powered test lamp to diagnose air bag systems or computer circuits. Use only the vehicle manufacturer's recommended equipment on these systems. Many different tools are used to diagnose electrical problems.

Those discussed here are the most common. Always use the correct tool and procedure for testing an electrical or electronic circuit.

A **self-powered test light** is called a **continuity tester**.

Feeler gauges are used to make precision measurements of small gaps. Individual feeler gauges are used for certain service operations such as measuring piston clearance in a cylinder. Feeler gauge thickness in the English system is measured in thousandths of an inch, whereas metric feeler gauge thickness is measured in millimeters (mm). Round wire feeler gauges are recommended for measuring spark plug gaps (**Figure 2-16**).

ELECTRICAL DIAGNOSTIC TOOLS

Circuit Testers

Circuit testers (**Figure 2-17**) are used to identify shorted and open circuits in an electrical circuit. Low-voltage testers are used to troubleshoot 6- to 12-volt circuits. High-voltage circuit testers diagnose primary and secondary ignition circuits.

A circuit tester looks like a stubby ice pick. Its handle is transparent and contains a light bulb. A probe extends from one end of the handle, and a ground clip and wire from the other end. When the ground clip is attached to a good ground and the probe is touched to a live connector, the bulb in the handle will light up. If the bulb does not light, voltage is not available at the connector.

A self-powered circuit tester also known as a **continuity tester** (**Figure 2-18**) is used on open circuits. This tester also has a clip and a probe end, and it has a small internal battery. When the ground clip is attached to the negative side of a component and the

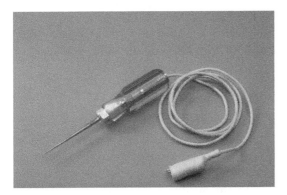

Figure 2-17 Typical circuit tester, commonly called a test light.

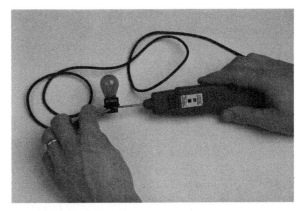

Figure 2-18 Typical self-powered circuit tester, commonly called a continuity tester or a self-powered test lamp.

probe is touched to the positive side, the lamp will light if there is continuity in the circuit. If an open circuit exists, the light will not be illuminated.

ELECTRICAL TEST EQUIPMENT

Several meters are used to test and diagnose electrical systems. These are the logic probe, voltmeter, ohmmeter, and ammeter.

Logic Probe

The logic probe (**Figure 2-19**) can be used on sensitive electronic circuits with very low amperage. The logic probe requires a connection to both the positive and negative cables of the battery. If a voltage detected is over 10 volts, the red LED comes on; if under 4 volts, the green LED comes on. If the voltage changes or pulses, the yellow LED flashes along with the red or green LED.

> **AUTHOR'S NOTE** The logic probe consumes very little current, which makes it ideal for checking sensitive circuits. One downside to this is that it takes very little current to make the LED come on. A poor ground or power circuit (not completely open) may still make the LED light.

Voltmeter

The **voltmeter** measures the voltage available at any point in an electrical system. For example, it can be used to measure the voltage available at the battery. It can also be used to test the voltage available at the terminals of any component or connector. A voltmeter can also be used to test voltage drop across an electrical circuit, component, switch, or connector.

Typically, the negative side of a component is called the *ground side*, while the positive side is called the *power side* or *feed side*.

A **voltmeter** is used to measure available voltage in a circuit.

⚠ **CAUTION**

Do not use a conventional 12-volt test light or self-powered test light to diagnose components and wires in computer systems. The current draw of these test lights may damage computers and computer system components. High-impedance test lights are available for diagnosing computer systems. Always be sure the test light you are using is recommended by its manufacturer for testing computer systems.

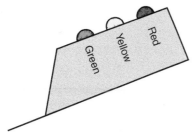

Figure 2-19 The logic probe will turn the red LED on with a positive voltage over 10; the green LED with a voltage under 4.

A voltmeter has two leads: a red positive lead and a black negative lead. The red lead should be connected to the positive side of the circuit or component. The black should be connected to ground or to the negative side of the component. Voltmeters should always be connected across the circuit being tested.

Ohmmeter

An **ohmmeter** is used to measure electrical resistance in a circuit.

An **ohmmeter** measures resistance to current flow in a circuit. In contrast to the voltmeter, which uses the voltage available in the circuit, the ohmmeter is battery powered. The circuit being tested must be open. If the power is "on" in the circuit, the ohmmeter will be damaged.

The two leads of the ohmmeter are placed across or in parallel with the circuit or component being tested. The red lead is placed on the positive side of the circuit, and the black lead is placed on the negative side of the circuit. The meter sends current through the component and determines the amount of resistance based on the voltage dropped across the load. The scale of an ohmmeter reads from 0 to infinity. A 0 reading means there is no resistance in the circuit and may indicate a short in a component that should show a specific resistance. An infinity reading indicates a number higher than that the meter can measure. This is usually an indication of an open circuit or faulty component.

> **AUTHOR'S NOTE** An ohmmeter is an excellent tool to detect continuity, but it can be limited in checking components such as an ignition coil, relay coils or fuel injector, and especially grounds. Voltage drops are better because the component is running at normal amperage. The ohmmeter only tests the circuit with a very low amount of current, and may not reflect "real-world" performance.

Ammeter

An **ammeter** is used to measure electrical current flow in a circuit.

An **ammeter** measures current flow in a circuit. Current is measured in amperes. Unlike the voltmeter and ohmmeter, the ammeter must be placed into the circuit or in series with the circuit being tested. This normally requires disconnecting a wire or connector from a component and connecting the ammeter between the wire or connector and the component. The red lead of the ammeter should always be connected to the side of the connector closest to the positive side of the battery, and the black lead should be connected to the other side. Do not connect an ammeter into a circuit that may have more amps than the meter is rated. Most meters are protected with a fuse; however, it is better to avoid the risk of meter damage if you are not sure about the circuit you are testing.

A **volt/ampere tester (VAT)** is commonly used to measure high currents in battery, starter, and charging circuits.

It is much easier to test current using an ammeter with an inductive pickup. The pickup clamps around the wire or cable being tested. The ammeter determines amperage based on the magnetic field created by the current flowing through the wire. This type of pickup eliminates the need to separate the circuit to insert the meter.

Volt/Ampere Tester

 CAUTION

A high-impedance digital multimeter must be used to test the voltage of some components and systems such as an oxygen (O_2) sensor circuit. If a low-impedance analog meter is used in this type of circuit, the current flow through the meter is high enough to damage the sensor. Always use the type of meter specified by the vehicle manufacturer.

A **volt/ampere tester (VAT)** is used to test batteries, starting systems, and charging systems. The tester contains a voltmeter, ammeter, and carbon pile. The carbon pile is a variable resistor. A knob on the tester allows the technician to vary the resistance of the pile. When the tester is attached to the battery, the carbon pile will draw current out of the battery. The ammeter will read the amount of current draw. When testing a battery, the resistance of the carbon pile must be adjusted to match the ratings of the battery. Newer VATs have more automated capabilities for quicker test results (**Figure 2-20**). These menu-driven units feature reduced testing time on batteries in various stages of charge, and automatic loading of charging systems. They work on all 12- to 30-volt systems. Results can be printed for comparison testing or customer review.

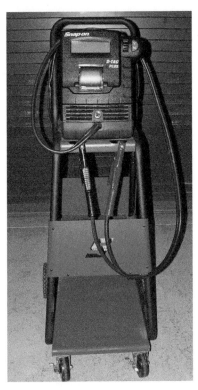

Figure 2-20 Battery, starting, and charging system tester.

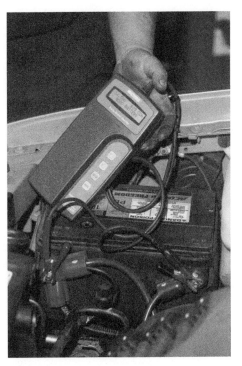

Figure 2-21 Typical battery tester.

Figure 2-21 shows a small battery tester. This battery tester does not have a carbon pile. However, it can be used to test circuit voltage and to perform battery load tests on 6- and 12-volt batteries. It can also be used to check the condition of battery cables and connectors and to check running voltage.

Multimeter

It is not necessary for a technician to own separate voltmeters, ohmmeters, and ammeters. These meters are combined in a single tool called a multimeter. A **multimeter** is one of the most versatile tools used in diagnosing engine performance and electrical systems. The most commonly used multimeter is the **digital multimeter (DMM)**, or digital volt ohmmeter. This meter does not use a sweeping needle and scales to display the measurement. Rather, it displays the measurement digitally on the meter (**Figure 2-22**).

Multimeters provide these readings on several different scales:

1. DC volts
2. AC volts
3. Ohms
4. Amperes
5. Milliamperes
6. Diode
7. Frequency (some models)

Because multimeters do not have heavy leads, the highest ammeter scale on this type of meter is normally 10 amperes. A control knob on the front of the multimeter must be rotated to the desired reading and scale (**Figure 2-23**). Some multimeters are auto-ranging, which means the meter automatically switches to a higher scale if the reading goes above the value of the scale being used. For example, if the meter is reading on the 10-volt scale and the leads are connected to a 12-volt battery, the meter

A **multimeter** is a volt, ohm, and ammeter combined into a single unit.

A **digital multimeter (DMM)** is a digital version used on vehicle circuits.

⚠ **CAUTION**

If a multimeter is going to be used on hybrid vehicles make sure it is rated as a Category (CAT) III meter. This is essential to preventing personal injury when working on high-voltage circuits.

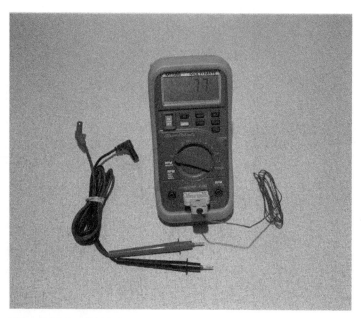

Figure 2-22 The digital voltmeter or ohmmeter can be used to test voltage and resistance.

Figure 2-23 This multimeter has 18 test ranges that have to be selected by the technician as it is not auto-ranging.

⚠ **CAUTION**

Always be sure that the proper scale is selected on the multimeter and that the correct lead connections are completed for the component or system being tested. Improper multimeter lead connections or scale selections may blow the internal fuse in the meter or cause meter damage.

automatically changes to the next highest scale. If the multimeter is not auto-ranging, the technician must select the proper scale for the component or circuit being tested. When using a meter that is not auto-ranging, being on the wrong scale can create some confusion. An easy way to get to the correct scale is a method called *scaling down*. Simply make your digital volt ohm meter (DVOM) connection and begin with the highest scale. Begin to scale down by turning the scale selector knob one range at a time. When the meter displays an "over-limit" indicator, scale up to the next highest range.

Top-of-the-line multimeters are multifunctional. Most test DC and AC volts, ohms, and amperes. Several test ranges are usually provided for each of these functions. Many meters have a diode-testing function. This typically uses a similar setup as an ohmmeter test. During the diode test, the meter applies a small voltage to the diode to act on the boundary layer in the diode. That is why there is usually a separate meter selection for this

System/Component	Measurement Type			
	Voltage Presence and Level	Voltage Drop	Current	Resistance
Charging System				
Alternators	•		•	
Regulators	•			
Diodes		•		•
Connectors	•	•		•
Starting System				
Batteries	•	•	•	
Starters	•	•	•	
Solenoids	•	•		•
Connectors	•	•		•
Cables	•	•	•	•
Ignition System				
Coils	•			•
Connectors	•	•		•
Distributor Caps				•
Plug Wires				•
Rotors				•
Magnetic Pickup	•			•

Figure 2-24 Electrical testing with a multimeter.

test. In addition to these basic electrical tests, some multimeters also test engine **revolutions per minute (rpm)**, diode condition, frequency, and even temperature. **Figure 2-24** shows some of the many tests that can be performed with a multimeter.

Although they are not as popular as they once were, analog multimeters are still available. The problem with analog meters is their low internal resistance (input impedance). The low input impedance allows too much current to flow through circuits and should not be used on delicate electronic devices.

Digital multimeters have **high input impedance**, usually at least 10 megohms. Metered voltage for resistance tests is well below 5 volts, reducing the risk of damage to sensitive components and delicate computer circuits.

Graphing Digital Multimeter

Other DMMs offer enhanced diagnostic features that help the technician capture and read data more than one way. The graphing multimeter can create a waveform or display a signal's history over a given time span. **Figure 2-25** is a unit that displays measurements graphically, digitally, or through a history graph. It has a diagnostic database including various pattern samples for reference. Its testing capabilities include AC and DC volts, amps and ohms, vacuum and pressure, primary and secondary ignition, frequency, and pulse width.

IGNITION SYSTEM TOOLS

Some of the tools used to troubleshoot ignition systems have been used for years. Others are relatively new, having been developed in response to electronic ignition and computer-controlled ignition systems.

If the multimeter has a diode testing feature, the reading you see is the voltage drop across the diode when the diode is forward biased.

For our purposes, impedance is best defined as operating resistance.

Distributorless ignitions and many distributor ignitions do not have adjustable timing.

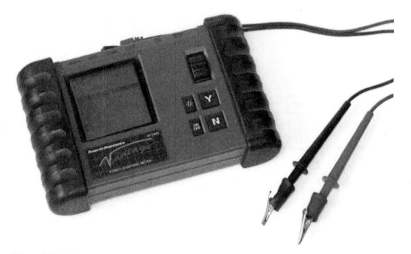

Figure 2-25 Graphing digital multimeter.

Timing Light

Top dead center TDC is the highest level reached by the piston in its bore.
Before top dead center is before the piston reaches this point. After top dead center is after TDC.

A timing light is used on older distributor ignitions for checking the ignition timing in relation to the crankshaft position. Two leads on the timing light must be connected to the battery terminals with the correct polarity. Timing lights have an inductive clamp that fits over the number-one spark plug wire (**Figure 2-26**). A trigger on the timing light acts as an off-on switch. When the trigger is pulled with the engine running, the timing light emits a beam of light each time the spark plug fires.

The timing marks are usually located on the crankshaft pulley or on the flywheel. A stationary pointer, line, or notch is positioned above the rotating timing marks. The timing marks are lines on the crankshaft pulley or flywheel that represent various degrees of crankshaft rotation when the number one piston is **before top dead center (BTDC)** on the compression stroke. The TDC crankshaft position and the degrees are usually identified in the group of timing marks. Some timing marks include degree lines representing the **after top dead center (ATDC)** crankshaft position (**Figure 2-27**).

Figure 2-26 A timing light with an inductive pickup.

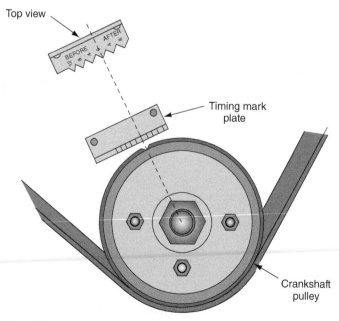

Figure 2-27 Timing reference marks.

Before checking the ignition timing, complete all the vehicle manufacturer's recommended procedures.

Spark Testers

The spark tester shown in **Figure 2-28** is used to stress the ignition system for testing in the shop. A well-performing ignition system should be able to jump the gap of the tester. This eliminates having to use a grounded screwdriver or other makeshift device to check for sufficient spark. Remember, there are several types of spark testers from small engine to electronic ignition, so make sure you have the right tester for the vehicle.

FUEL SYSTEM TOOLS

The proper amount of fuel is essential to the proper operation of an engine. Therefore, misadjusted or faulty fuel system components will adversely affect engine performance. The following tools are used to test the parts of the fuel system.

Pressure Gauge

A **pressure gauge** (**Figure 2-29**) is used to measure the pressure in the fuel system. This tester is very important for diagnosing fuel injection systems. These systems rely on very high fuel pressures, from 35 to well over 2,000 psi on some direct fuel-injected vehicles. A drop in fuel pressure will reduce the amount of fuel delivered to the injectors and result in a lean air-fuel mixture.

A fuel pressure gauge is used to check the discharge pressure of fuel pumps, the regulated pressure of fuel injection systems, and injector pressure drop. This test can identify faulty pumps, regulators, or injectors and can identify restrictions present in the fuel delivery system. Restrictions are typically caused by dirty fuel filters, collapsed hoses, or damaged fuel lines.

Some fuel pressure gauges also have a valve and outlet hose for testing fuel pump discharge volume. The manufacturer's specification for discharge volume will be given as a number of pints or liters of fuel that should be delivered in a certain number of seconds.

> A **pressure gauge** measures pressure in pounds per square inch or kilopascals.

> Late model vehicles may have fuel pressure readings that can be taken with a scan tool.

⚠️ **WARNING** **While testing fuel pressure, be careful not to spill gasoline. Gasoline spills can cause explosions and fires, resulting in serious personal injury and property damage.**

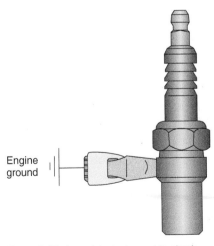

Figure 2-28 A spark tester is used to check coil output.

Engine ground

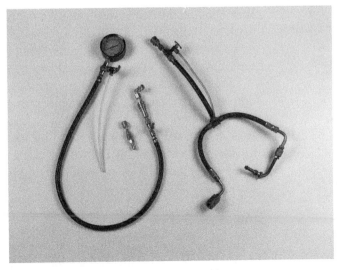

Figure 2-29 Fuel pressure gauge and various adapters.

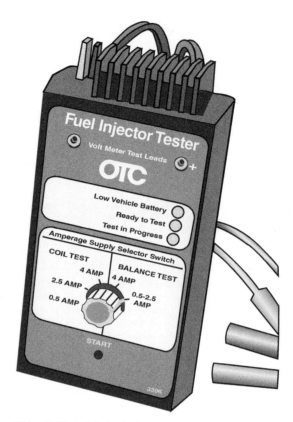

Figure 2-30 An injector tester.

The balance test energizes the injectors 100 times each to check the pintle operation.

 CAUTION

Electronic fuel injection systems are pressurized, and these systems require depressurizing prior to fuel pressure testing and other service procedures.

Injector Balance Tester

The **injector balance tester** is used to test the injectors in a port fuel-injected engine for proper operation. A fuel pressure gauge is also used during the injector balance test. The injector balance tester contains a timing circuit, and some injector balance testers have an off-on switch. A pair of leads on the tester must be connected to the battery with the correct polarity (**Figure 2-30**). The injector terminals are disconnected, and a second double lead on the tester is attached to the injector terminals.

Before the injector balance test, the fuel pressure gauge is connected to the Schrader valve on the fuel rail, and the ignition switch should be cycled two or three times until the specified fuel pressure is indicated on the pressure gauge. When the tester push button is depressed, the tester energizes the injector winding for a specific length of time and the technician records the pressure decrease on the fuel pressure gauge. This procedure is repeated on each injector.

Some vehicle manufacturers provide a specification of 3 psi (20 kPa) maximum difference between the pressure readings after each injector is energized. If the injector orifice is restricted, there is not much pressure decrease when the injector is energized. Acceleration stumbling, engine stalling, and erratic idle operation are caused by restricted injector orifices. The injector plunger is sticking open if excessive pressure drop occurs when the injector is energized. Sticking injector plungers may result in a rich air-fuel mixture.

Injector Tester

If the injector tester has the capability of testing the voltage drop on each injector's coil, move the selector on the tool to "coil test" and select the correct amperage level from the manufacturer's specification. Connect the multimeter leads to the appropriate sockets on

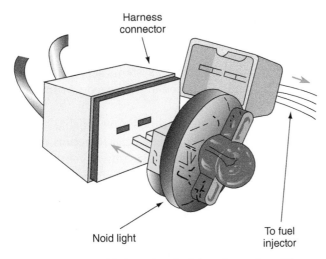

Figure 2-31 A noid light is used to check for voltage pulses at the fuel rail harness.

the tester. Hit the button on the tester, and the tester will turn on the injector and leave it on instead of pulsing as it does for the balance test. If the coil is shorting or opening intermittently, this test should find the fault. Record the voltage drop on the injector coil, and compare it to specifications.

Injector Circuit Test Light

A special test light called a **noid light** can be used to determine if a fuel injector is receiving its proper voltage **pulse** from the computer. The wiring harness connector is disconnected from the injector, and the noid light is plugged into the connector (**Figure 2-31**). After disabling the ignition to prevent starting, the engine is turned over by the starter motor. The noid light will flash rapidly if the voltage signal is present. No flash usually indicates an open in the power feed or ground circuit to the injector. It should be noted that there are several types of noid lights for specific applications. The connectors are different from one injector to the next. The noid lights may also have different resistance values to match the fuel injector's resistance. Check the manufacturer's application instructions.

A **noid light** is used to determine if a fuel injector is receiving the proper voltage pulse.

Fuel Injector Cleaner

Fuel injectors spray a certain amount of fuel into the intake system. If the fuel pressure is low, not enough fuel will be sprayed. This is also true if the fuel injector is dirty. Normally, clogged injectors are the result of inconsistencies in gasoline detergent levels or impurities in the gasoline. When these sensitive fuel injectors become partially clogged, fuel flow is restricted. Spray patterns are altered, causing poor performance and reduced fuel economy.

The solution to a dirty and/or plugged fuel injector is to clean it, not replace it. There are several kinds of fuel injector cleaners. One is a pressure tank. A mixture of solvent and unleaded gasoline is placed in the tank, following the manufacturer's instructions for mixing, quantity, and safe handling. The vehicle's fuel pump must be disabled, and on some vehicles the fuel line must be blocked between the pressure regulator and the return line. Then, the hose on the pressure tank is connected to the service port in the fuel system. The in-line valve is then partially opened, and the engine is started. It should run at approximately 2,000 rpm for about 10 minutes to thoroughly clean the injectors.

Another type is a pressurized canister (**Figure 2-32**) in which the solvent solution is premixed. Use of the canister-type cleaner is similar to this procedure but does not require mixing or pumping.

The canister is connected to the injection system's servicing fitting, and the valve on the canister is opened. The engine is started and allowed to run until it dies. Then, the canister is discarded.

Several other fuel injector cleaning methods are available. Check with the vehicle manufacturer for their recommended process. Fuel injection cleaning methods have improved over the past few years due to new cleaning agents and service equipment. Carbon buildup can be as much of a problem as dirty injectors. Carbon can absorb fuel and affect fuel mixture management by the computer. The unit in **Figure 2-33** cleans the fuel rail, injectors, intake manifold, plenum, intake runners, and throttle body. In addition, it cleans the carbon on the back of the intake valve, inside the combustion chamber, and on the oxygen sensor. This method is an alternative to the physical disassembly of an engine to remove carbon deposits.

SCAN TOOLS

A **scan tool** may be called a scanner or a scan tester.

Technicians need tools capable of troubleshooting electronic engine, transmission, antilock brake, and body control systems. A variety of computer scan tools are available today to do just that. A **scan tool** (**Figure 2-34**) is a microprocessor designed to

Figure 2-32 Fuel injector cleaner using the shop's compressed air.

Figure 2-33 A fuel injector and carbon cleaning system.

Figure 2-34 A typical scan tool and accessories.

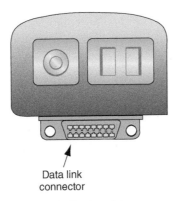

Data link
connector

Figure 2-35 The data link connector is used to connect a scan tool to the vehicle's computer network.

communicate with the vehicle's computer. Scan tools come in two varieties, factory or aftermarket. The factory scan tool is designed to work on specific vehicles from a particular manufacturer, while most aftermarket scan tools can be tailored to fit almost any vehicle with software packages available. Connected to the computer through the **data link connector (DLC)** (**Figure 2-35**), which is standard for all vehicles manufactured after 1996, a scan tool can access trouble codes, run tests to check system operations, perform module reprogramming, and monitor the system's activity. Trouble codes and test results are displayed on a screen or can be printed out on the scanner printer. Scan tools have become smaller and more capable as computer technology has advanced. A handheld scanner (**Figure 2-36**) can be used and stored easily in a technician's toolbox. Some scan tools can use a Bluetooth wireless device

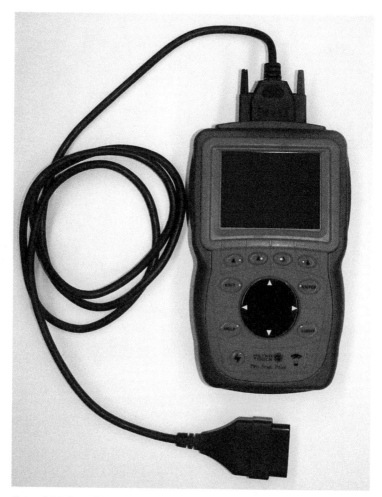

Figure 2-36 A small handheld scan tool.

(**Figure 2-37**) to plug into the vehicle's DLC. This allows the technician to use the scan tool without the burden of cables.

Most scan tools can be updated over the Internet or by a manufacturer's representative. Factory scan tools have all the necessary programming to service a particular manufacturer's vehicles. Some aftermarket scan tools can be upgraded with programming to cover engine diagnostics, and other systems such as stability control may be available for an additional cost.

Scan testers have the capability to store, or "freeze," data into the tester during a road test and play back this data when the vehicle is returned to the shop. The technician can set up the scan tool to take a "snapshot" whenever a button is pressed on the scan tool, or when a code sets. The ECM will automatically take a "freeze frame" when a code sets that will turn on the malfunction indicator lamp (MIL). The freeze-frame data can be read by a scan tool to aid in diagnosis. The ECM stores relative data to help determine the conditions present when the MIL came on. The data helps the technician to verify the repair because he or she can drive the vehicle under similar conditions to see if a code resets.

Programming (Reflashing)

Whenever a module is replaced, many must be programed before it will function, customizing it to the vehicle's network. This includes engine control modules (ECMs), powertrain control modules (PCMs), transmission control modules (TCMs), instrument panel

Figure 2-37 A special connector can be purchased to connect a scan tool via Bluetooth wireless.

Figure 2-38 A pass-through programming tool that connects a PC to the vehicle's computer.

clusters (IPCs), body control modules (BCMs), door modules, remote door lock modules, and more. To perform this programming or reprogramming it is necessary to have either a scan tool with remote programming capabilities, or a laptop and "pass-through" J2534 module (**Figure 2-38**) capable of connecting the laptop to the vehicle's computer network. Updates to module software are performed by reflashing (reprogramming) of the module software. These updates are often addressed in technical bulletins.

Breakout Box

A **breakout box** allows the technician to check voltage and resistance readings between specific points within the computer's wiring harness (**Figure 2-39**).

A **breakout box** simplifies testing at a major component such as a computer or module.

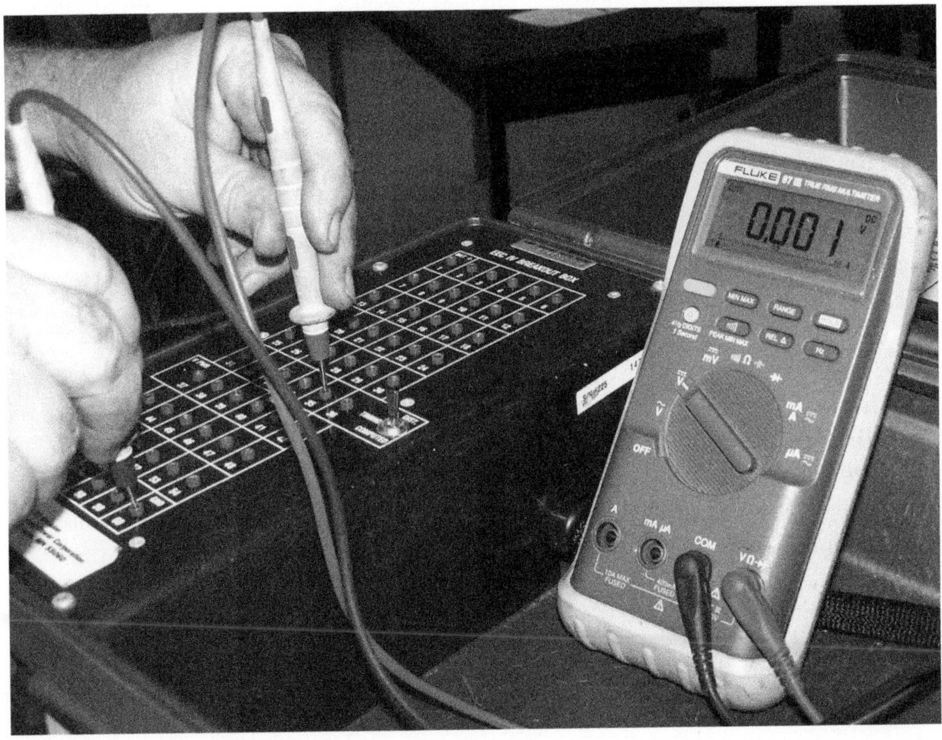

Figure 2-39 The wiring harness breakout box allows for detailed testing of individual circuits.

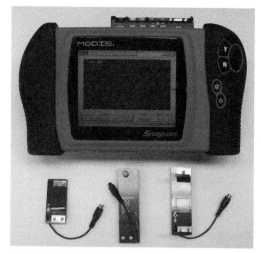

Figure 2-40 A handheld ignition or lab scope.

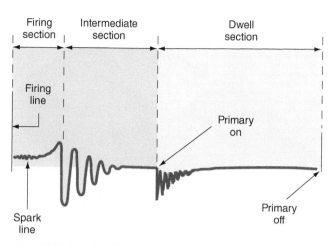

Figure 2-41 Typical oscilloscope waveform pattern.

OSCILLOSCOPES

An **oscilloscope** is a type of voltmeter that displays voltage over time graphically with a trace.

A **waveform** or **trace** is the graphical representation of a voltage signal produced by an oscilloscope.

The term *oscilloscope* is usually replaced with the word *scope*.

A **multi-channel oscilloscope** can display two or more patterns at the same time.

A low-voltage scope is normally called a *lab scope*.

The **oscilloscope** (**Figure 2-40**) is one of the most important diagnostic tools and performs some very valuable tests on today's electronic systems. An oscilloscope converts electrical signals into a visual image representing voltage changes over a specific period of time. This information is displayed on a cathode ray tube (CRT) in the form of a continuous voltage line called a **waveform** pattern or **trace** (**Figure 2-41**).

Although there are many scope designs, all automotive scopes can be classified into one of two categories: low voltage or high voltage. High-voltage scopes are used to monitor the activity of the ignition system and are quite valuable diagnostic tools. Low-voltage scopes are also extremely valuable diagnostic tools. They are used to monitor the activity of the inputs and outputs of a computerized system.

Multi-channel oscilloscopes (**Figure 2-42**) can display two or more different waveform patterns at the same time. For example, multi-channel oscilloscopes can show how fuel injector pulse width affects the oxygen sensor voltage signal and other highly useful cause-and-effect and relational electrical tests.

An oscilloscope may be regarded as a very fast-reacting voltmeter that reads and displays voltages in a system. These voltage readings appear as a voltage trace on the oscilloscope screen.

An upward movement of the voltage trace on an oscilloscope screen indicates an increase in voltage, and a downward movement of this trace represents a decrease in voltage. As the voltage trace moves across an oscilloscope screen, it represents a specific length of time (**Figure 2-43**).

The size and clarity of the displayed waveform are dependent on the voltage scale and the time reference selected by the operator. Most scopes are equipped with controls that allow voltage and time interval selection. It is important when choosing the scales to remember that a scope displays voltage over time.

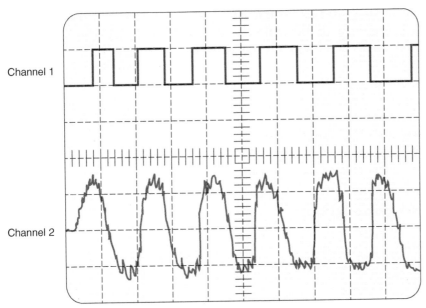

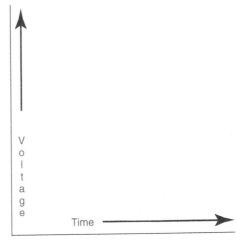

Figure 2-43 Voltage and time axis on a scope's screen.

Figure 2-42 Two wave forms from a dual-trace lab scope.

Figure 2-44 Exhaust emission analyzer (5 gas).

EXHAUST ANALYZERS

Exhaust analyzers (**Figure 2-44**) are very valuable diagnostic tools. By looking at the quality of an engine's exhaust, a technician is able to look at the effects of the combustion process. Any defect can cause a change in exhaust quality. The amount and type of change serve as the basis of diagnostic work.

Five gas analyzers measure hydrocarbons (HC), carbon monoxide (CO), carbon dioxide (CO_2), oxygen (O_2), and oxides of nitrogen (NO_x). All of these gases are affected by the combustion process, so the levels of these gases can reveal combustion problems. Hydrocarbons in the exhaust are raw, unburned fuel. HC emissions indicate that fuel is leaving the engine unburned. Emissions analyzers measure HC in **parts per million (ppm)** or grams per mile (g/mi). Carbon monoxide is an odorless, toxic gas that

Exhaust analyzers are often called *infrared testers*. This is because many analyzers use infrared light to analyze the exhaust gases.

is the product of incomplete combustion and is typically caused by a rich fuel mixture. CO is measured as a percentage of the total exhaust. Carbon monoxide can be referred to as partially burned fuel as carbon monoxide can still be burned to form CO_2 in the converter.

CO_2 is actually an indication that the engine efficiency is good. CO_2 is measured on most analyzers in percentages. Oxygen in the exhaust, while not a pollutant, can indicate a lean condition. NO_x can be caused by overheating, lean conditions, or a defective **exhaust gas recirculation (EGR) valve**.

You can probably begin to see that the **exhaust gas analyzer** can be used to diagnose many drivability problems, such as the following:

- Rich or lean mixtures
- Catalytic converter malfunction
- Faulty injectors
- Leaking EGR valves
- Leaks or restrictions in the exhaust system
- Blown head gaskets
- Intake manifold leaks
- Excessive misfire

PHOTO SEQUENCE 1
Typical Technician Workflow

P1-1 The service advisor gathers as much information as possible from the customer. The service advisor is finding out when and under which conditions the concern occurs.

P1-2 The service advisor translates the customer complaint into a work order on the shop management computer.

P1-3 The technician carefully reads the complaint, and then drives the vehicle to duplicate the customer concern. If the concern cannot be duplicated, the technician can go to the service advisor for clarification.

P1-4 The technician performs a quick visual inspection since the problem has been confirmed. If the problem is not found right away, then the technician will need to do more research.

P1-5 Technicians have many resources at their disposal, including the shop management system, so that previous repairs can be checked for relevance. Service information systems can also be consulted for diagnosis, and technician service bulletins can be searched.

P1-6 The technician will use a scan tool and the information gathered to help diagnose the vehicle's problem.

PHOTO SEQUENCE 1 (CONTINUED)

P1-7 Once the technician has made a diagnosis the next job is to determine the cost and availability of parts.

P1-8 The service advisor and technician consult with the flat-rate manual to determine the amount of time to charge for a particular repair. The flat-rate manual is a standard guide that lists automotive repair jobs and the amount of time to complete them.

P1-9 The service advisor prepares an estimate for the repairs needed, based on input from the technician.

P1-10 The technician makes the needed repairs after authorization from the customer has been given.

P1-11 The repair is verified by the technician by performing a thorough test drive.

P1-12 The vehicle is delivered back to the customer. The advisor is there to answer any questions as necessary and the customer is thanked for his or her business.

MISCELLANEOUS ENGINE PERFORMANCE TOOLS

Oxygen Sensor Service Tool

Most oxygen sensors are difficult to access with an open-end wrench, and the wiring prevents the use of a normal deep socket. Some special sockets have been developed especially for the oxygen sensor, as shown in **Figure 2-45**.

Spark Plug Thread Repair

Spark plug thread repair tools are meant to repair damaged threads in a cylinder head. They can also help repair damaged O_2 sensor threads. No one wants to use these tools, but sometimes spark plugs seize in the cylinder head if the technician before you was not too careful about how the spark plugs were installed. A set of spark plug tools is shown in **Figure 2-46**.

Static Protection Straps

Sensitive electronic components can be destroyed by static electricity. Many technicians have unknowingly damaged PCM's digital dashes, radios, and control modules. Sometimes the damage is not immediately apparent. In order to avoid static electricity damage, it is wise to wear a static strap. The static strap drains any buildup of electrical charge on your body. A picture of the wrist-strap style is shown in **Figure 2-47**. The grounding clip is connected to a good ground on the vehicle.

Figure 2-45 Oxygen sensor socket.

Figure 2-46 Spark plug thread tools.

Figure 2-47 A grounding bracelet is attached to a vehicle ground and the technician's wrist to bleed off possible static charges while working on sensitive electronics.

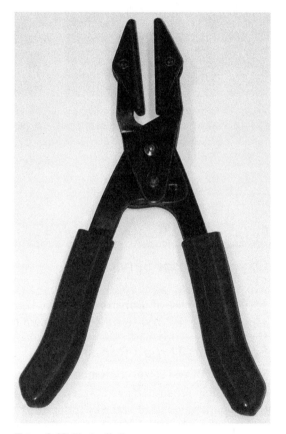

Figure 2-48 Pinch-off pliers.

Pinch-Off Pliers

Pinch-off pliers can be used to close off vacuum lines and rubber fuel lines. Never use these pliers on rubber brake lines or nylon fuel lines because they will be damaged. **Figure 2-48** shows a pair of pinch-off pliers.

The most important tool of the trade is information. There are many different sources of information that should be available to automotive technicians.

Vehicle Identification

One of the first steps you need to take when servicing any vehicle is to find out exactly what you are working on. The year, make, and model are a good start, but due to the various engines, transmissions, accessories, and mid-year production changes, you need to find out about the *exact* vehicle. The vehicle identification number (VIN) contains all the data about that specific vehicle (**Figure 2-49**). It indicates the manufacturer, engine size and type, vehicle model information, and serial number. Most any service information system will usually reference at least a portion of the VIN. Other vehicle information can be found in several other places. The owner's manual has a great deal of useful information regarding accessory operation, how to reset memory devices, recommended service schedules, and even part numbers for service parts. The under hood decal (**Figure 2-50**) also has information about the emission control system and perhaps some minor specifications. The driver's door placard contains information about the build date, vehicle weight, tire size, and inflation information. Elsewhere in the vehicle may be an accessory and equipment build code placard. These are not always out in the open where they are readily visible. It contains a summary, by code number, of all the standard equipment and options the vehicle was built with. For example, it will include the actual type of entertainment system, the exact type of springs used, and so on.

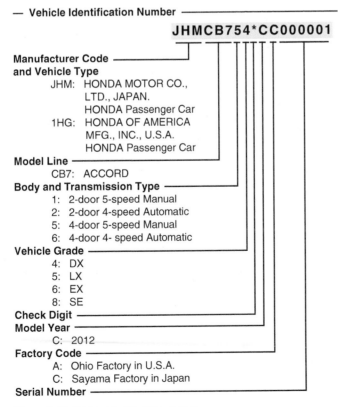

Figure 2-49 Vehicle identification number.

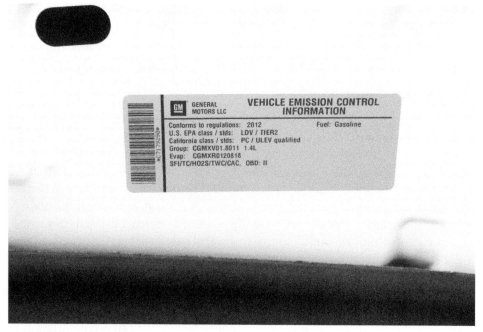

Figure 2-50 Underhood decal.

Automotive Service Information

Most service information that manufacturers release is located on the Internet and accessed by the technician at the dealership. The format is generally similar to that which the manufacturer previously used in its service manuals. By releasing new information on the Internet, there is no lag between finding new service techniques and implementing them in the service information.

The service information includes specifications and procedures for all service requirements on the vehicle. Torque specifications, measurements, and service techniques are included in the service information, but theory is generally excluded from the information because it is commonly assumed that technicians using the information are experienced.

There are as many types of service information as there are manufacturers, so we will not attempt to describe them in detail. Usually the electronic service information has the technician choose a vehicle from a drop-down menu, then proceed to the specific vehicle system in question. The pages of the service manual are usually **hyperlinked**, meaning that relevant information located elsewhere can be quickly brought onto the screen instead of turning through pages in a book (**Figure 2-51**).

A **hyperlink** is a shortcut to additional relevant information in an electronic document or to a website address.

⚠ WARNING **Always follow the service procedures in the vehicle manufacturer's service manual for the vehicle on which you are working. The use of improper service procedures may result in personal injury and/or vehicle damage.**

Automotive manufacturers may also publish a series of technician reference books. The publications provide general instructions on the service and repair of their vehicles with their recommended techniques.

General repair manuals may be referred to as *specialty manuals* or by the publisher of the manual: Mitchell, Motor, Chilton, and so on.

Aftermarket Suppliers' Guides and Catalogs

Many of the larger parts manufacturers have excellent guides on the various parts they manufacture or supply. They also provide updated service bulletins on their products.

PCM Quick Test

1. Complete preliminary checks

Visual inspection for obvious problems
Electrical connections, vacuum lines, air intake system for
leaks and restrictions, quality of fuel and cooling system
for proper operation

2. Search the TSB database for any related concerns

3. If the scan tool cannot access the PCM, go to ET1

Were any DTCs present?

Yes	No
If engine runs rough at idle and DTCs are present: GO to Section 4 (Powertrain DTC charts) for direction to service DTCs after noting the following: Any DTCs Retrieve any freeze frame information	GO TO SYMPTOM PRESENT NO DTC SET CHARTS

Figure 2-51 Typical electronic information page showing hyperlinks in blue.

General and Specialty Repair Manuals

These are published by independent or aftermarket companies rather than the manufacturers. However, they pay for and receive most of their information from the car makers. They may contain component information, diagnostic steps, repair procedures, and specifications for several car makes in one book. Information is usually condensed and is more general in nature than the manufacturer's manuals. The condensed-format allows for more coverage in less space and therefore is not always specific. They also may contain several years of models as well as several car makes in one book.

Flat-Rate Manuals

Flat-rate manuals contain figures relating to the length of time a specific repair is supposed to require. Normally they also contain a parts list with approximate or exact parts prices. They are excellent for making cost estimates and are published by the manufacturers and independents.

> Flat-rate manuals give the service advisor or service manager a labor charge based on the time it takes to complete a repair.

Electronic Data Systems

Electronic data systems replace publications such as service manuals, flat-rate manuals, and service bulletins. In an electronic data system, the information is accessed by Internet subscription. A typical electronic data system contains flat-rate information, parts information, service bulletins, and service information. Can you imagine the number of pages required to publish this information for 20 model years on domestic vehicles and most imported vehicles?

The service bulletins in an electronic data system cover all automotive systems for many model years, including the current model year. Domestic and imported vehicle information is included in these systems. Much of the service information is in the form of diagnostic charts. Most electronic information systems are available on a subscription basis to ensure timely and regular updates.

Information may be printed out by the printer connected to the computer. The technician simply enters the make of vehicle, model year, type of engine, and any other requested information to obtain data on a particular vehicle. Some electronic data systems allow the shop to order parts electronically from a supplier with a compatible interconnected computer.

One great advantage of this system is that finding information is easier and quicker. The disks are normally updated monthly and contain not only the most recent service bulletins but also engineering and field service fixes. Other sources for up-to-date technical information are trade magazines and trade associations.

Hotline Services

Many companies provide online help to technicians. As the complexity of the automobile grows, so does the popularity of these services. Two of the most commonly used hotlines are the Delphi TechSource Assistance Center and Identifix Telediagnosis, which have ASE-certified specialists available for inquiries. These experts are technicians who are familiar with the different systems of certain manufacturers. Armed with various factory service information, general service manuals, service bulletins, and electronic data sources, these brand experts give information to technicians to help them through diagnostic and repair procedures.

ENGINE TUNE-UP

The term *engine tune-up* is often misunderstood. Even to automotive technicians, the words *engine tune up* do not always mean the same thing. The tune-up procedure may vary depending on the make of vehicle, type of test equipment in the shop, and shop policy. Many customers may not fully understand what is done to their vehicle during an engine tune-up.

In general terms, the purpose of an engine tune-up is to restore or maintain the original performance and economy of the engine. When an engine tune-up is performed, the customer and the technician are basically concerned about obtaining satisfactory performance and economy from the engine. The technician and the customer should understand the limitations of an engine tune-up. After an engine tune-up, they can only expect to obtain the original performance and economy designed into the vehicle by the manufacturer. Even the original performance and economy may be difficult to obtain because of normal wear on engine components. In some cases, the customer has a performance or an economy problem, and the technician performs an engine tune-up to restore the original performance and economy of the engine. Some customers may have an engine tune-up performed as preventive maintenance. In this situation, their vehicle operation is satisfactory, but they have a tune-up performed to maintain the engine in good running condition.

As you are aware, an electronics revolution has occurred in the automotive industry, and the pace of this revolution continues to accelerate. Automotive manufacturers can now design and build a new model in less than 3 years. A few years ago, the manufacturers required about 7 years to complete the same job. This accelerated production speed is possible because of computer-assisted engineering and the team concept in which all the engineers, including those from suppliers, work as a project team. Vehicle manufacturers are now able to introduce more new and revised models in a shorter time period, and each of these new models is equipped with some innovative electronic components and systems.

This revolution in automotive engineering and manufacturing has brought about a simultaneous revolution in the automotive service industry, especially in the tune-up area. In the 1960s, the average recommended spark plug replacement interval was 10,000 miles (16,000 km). The use of unleaded fuel and the more accurate air-fuel mixture supplied by electronic fuel injection systems have greatly reduced spark plug carbon deposits. In the 1980s, average spark plug replacement interval was 30,000 miles (48,000 km). More recently, new ignition system designs and the use of platinum-tipped spark plugs have again increased the service intervals anywhere from 60,000 miles to 100,000 miles. Of course, problems with components still can occur, and when they do, service is needed immediately.

There is no standard, universal tune-up procedure. As mentioned previously, tune-up procedures vary for a number of reasons. Many shops advertise a special price for a specific tune-up procedure, and the technicians in the shop follow the advertised procedure. If the technicians perform more tests or services than advertised, they take longer to complete the job, and profit margins are reduced. Therefore, the technicians are restricted to the advertised procedure, and any additional service must be approved by the customer and considered a separate operation. The average tune-up includes all or some of these tests and services:

CASE STUDY

The word *diagnosis* is commonly used in automotive textbooks and service manuals. However, it is a term that is seldom explained. It is more than following a series of interrelated steps in order to find the solution to a specific condition. Diagnosis is a way of looking at systems that are not functioning the way they should and finding out why. It is knowing how the system should work and deciding whether or not it is working correctly. Through an understanding of the purpose and operation of the car's system, a technician can accurately diagnose problems.

Most good diagnosticians use the same basic diagnostic procedure. Because of its logical approach, this procedure can quickly lead a technician to the cause of a problem. Accurate diagnostics include the following steps:

1. Gather information about the condition or problem.
2. Verify that the condition exists.
3. Thoroughly define what the problem is and when it occurs.
4. Determine the possible causes of the problem.
5. Isolate the problem by general testing.
6. Test to pinpoint the cause of the problem.
7. Repair the problem, and verify the repair.

Most service information contains diagnostic charts that help technicians pinpoint the cause of the problem. The steps of diagnostic charts or trees are designed to be followed in order. These charts contain the most probable causes of a particular problem and simple tests that lead to the exact cause.

1. Road test the vehicle if necessary to verify certain problems.
2. Inspect hoses and belts, and replace as necessary.
3. Check all fluid levels.
4. Check cooling fan operation.
5. Inspect the general condition of vacuum hoses and electrical wires.
6. Clean battery terminals, and check cable condition.
7. Replace spark plugs.

8. Check spark plug wires, and replace as required.
9. Test and inspect the distributor cap and rotor, and replace as necessary (in a distributor-type ignition system).
10. Check the air filter.
11. Check air inlet ducts for damage.
12. Inspect the fuel system for leaks.
13. Check the engine computer system for fault codes.
14. Road test the vehicle to be sure there are no performance problems.

ASE-STYLE REVIEW QUESTIONS

1. While discussing infrared analyzers:

   Technician A says hydrocarbon (HC) emissions are the result of unburned fuel.

   Technician B says carbon monoxide (CO) emissions are caused by partially burned fuel.

   Who is correct?

   A. A only   C. Both A and B

   B. B only   D. Neither A nor B

2. While discussing spark plug replacement intervals,

   Technician A says vehicles in the 1960s required spark plug replacement every 100,000 miles.

   Technician B says late-model engines require a tune-up every 30,000 miles.

   Who is correct?

   A. A only   C. Both A and B

   B. B only   D. Neither A nor B

3. While discussing injector balance testers,

   Technician A says the injector balance tester contains its own battery to eliminate battery connection.

   Technician B says the injector balance tester contains a timing circuit.

   Who is correct?

   A. A only   C. Both A and B

   B. B only   D. Neither A nor B

4. While discussing injector cleaning,

   Technician A says the fuel return line must be blocked during the injector-cleaning process on some vehicles.

   Technician B says the electric fuel pump must be disabled during the injector-cleaning process.

   Who is correct?

   A. A only   C. Both A and B

   B. B only   D. Neither A nor B

5. While discussing circuit testers,

   Technician A says a conventional 12-volt test light may be used to diagnose automotive computer circuits.

   Technician B says a self-powered test light may be used to diagnose an air bag circuit.

   Who is correct?

   A. A only   C. Both A and B

   B. B only   D. Neither A nor B

6. While discussing oscilloscopes,

   Technician A says the upward voltage traces on an oscilloscope screen indicate a specific length of time.

   Technician B says the oscilloscope is like a very fast-reacting voltmeter.

   Who is correct?

   A. A only   C. Both A and B

   B. B only   D. Neither A nor B

7. While discussing the purpose of engine tune-ups,

   Technician A says the purpose of an engine tune-up is to improve the vehicle manufacturer's original performance and economy.

   Technician B says in some cases the purpose of an engine tune-up is to maintain the engine in satisfactory running condition and prevent engine performance and economy problems in future driving miles.

   Who is correct?

   A. A only   C. Both A and B

   B. B only   D. Neither A nor B

8. While discussing service information,

   Technician A says that most service information is accessed by Internet subscription.

   Technician B says manufacturers always include theory of operation in their service information?

   Who is correct?

   A. A only

   B. B only

   C. Both A and B

   D. Neither A nor B

9. While discussing electronic data systems,

   Technician A says an electronic data system contains information only on vehicles in the current model year.

   Technician B says an electronic data system may contain flat-rate information, parts information, service bulletins, and service information from an Internet connection.

   Who is correct?

   A. A only

   B. B only

   C. Both A and B

   D. Neither A nor B

10. *Technician A* says that a push-in compression gauge must be held tightly in the spark plug hole to achieve an accurate reading.

    Technician B says the gauge will read a vacuum when the cylinder is on the compression stroke.

    Who is correct?

    A. A only

    B. B only

    C. Both A and B

    D. Neither A nor B

Name _____   Date _____

USE OF A VOLTMETER

Upon completion of this job sheet, you should be able to measure available voltage and voltage drop.

ASE Education Foundation Correlation

This job sheet addresses the following **AST/MAST** task: VI. Electrical/Electronics Systems Diagnosis; A. General Electrical Diagnosis,

Task #3   Demonstrate proper use of a digital multimeter (DMM) when measuring source voltage, voltage drop (including grounds), current flow, and resistance. **(P-1)**

Tools and Materials

• A vehicle
• A DMM
• Wiring diagram for vehicle
• Basic hand tools

Procedure

Task Completed

1. Set the DMM to the appropriate scale to read 12 volts DC.   ☐

2. Connect the meter across the battery (positive to positive and negative to negative). What is your reading on the meter? _____ volts

3. With the meter still connected across the battery, turn the vehicle's headlights on. What is your reading on the meter? _____ volts

4. Keep the headlights on. Connect the positive lead of the meter to the point on the vehicle where the battery's ground cable attaches to the frame. Keep the negative lead where it is.

   What is your reading on the meter? _____ volts

   What is being measured? _____

5. Disconnect the meter from the battery, and turn off the headlights.   ☐

6. Refer to the correct wiring diagram, and determine what wire at the right headlight delivers current to the lamp when the headlights are on and low beams selected. What is the color of the wire? _____

7. From the wiring diagram, identify where the headlight is grounded. Place of ground: _____

8. Connect the negative lead of the meter to the point where the headlight is grounded.   ☐

9. Connect the positive lead of the meter to the power input of the headlight.   ☐

10. Turn the headlights on.

    What is your reading on the meter? _____ volts

    What is being measured? _____

11. What is the difference between the reading here and the battery's voltage? _____ volts

12. Explain why there is a difference.

    _____

    _____

    _____

    _____

Instructor's Response

_____

_____

_____

Name _____   Date _____

USE OF AN OHMMETER

Upon completion of this job sheet, you should be able to check continuity of a circuit and measure resistance on a variety of components.

ASE Education Foundation Correlation

This job sheet addresses the following **AST/MAST** tasks: VI. Electrical/Electronics Systems Diagnosis; A. General Electrical Diagnosis,

Task #3	Demonstrate proper use of a digital multimeter (DMM) when measuring source voltage, voltage drop (including grounds), current flow, and resistance. **(P-1)**
Task #4	Demonstrate knowledge of the causes and effects from shorts, grounds, opens, and resistance problems in electrical/electronic circuits. **(P-1)**
Task #9	Inspect and test fusible links, circuit breakers, and fuses; determine necessary action. **(P-1)**

Tools and Materials

• A vehicle
• A DMM
• Wiring diagram for the vehicle

Note: An ohmmeter works by sending a small amount of current through the path to be measured. Because of this, all circuits and components being tested must be disconnected from power. An ohmmeter must never be connected to an energized circuit; doing so may damage the meter. The safest way to measure ohms is to disconnect the negative battery cable before taking resistance readings.

Procedure   **Task Completed**

1. Locate the fuse panel or power distribution box.   ☐

2. With no power to the fuses, check the resistance of each fuse. Summarize your findings:

   _____

   _____

   _____

   _____

3. Connect the leads of the digital meter across the negative battery cable. Your reading is _____ ohms.

4. Disconnect the wires leading to the ignition coil. Connect the leads of the digital meter across the terminals of the coil. Your reading is _____ ohms.

5. Reconnect the wires to the coil. Carefully remove one spark plug wire from the spark plug and the ignition coil or distributor cap. Connect the leads of the digital meter across the wire. Your reading is _____ ohms.

6. Carefully reinstall the spark plug wire. Locate the cigarette lighter inside the vehicle. Connect the leads of the digital meter from the heating coil to its case. Your reading is _____ ohms.

7. Refer to the service manual, and find out how to remove the bulb in the dome light. Remove it. Connect the leads of the digital meter across the bulb. Your reading is _____ ohms.

8. Reinstall the bulb. Remove the rear-brake light bulb. Connect the leads of the digital meter across the bulb. Your reading is _____ ohms.

9. Reinstall the bulb. Disconnect the wire connector to one of the headlights. From the wiring diagram, identify which terminals are for low-beam operation. Connect the leads of the digital meter across the low-beam terminals. Your reading is _____ ohms.

10. From the wiring diagram, identify which terminals are for high-beam operation. Connect the leads of the digital meter across the high-beam terminals. Your reading is _____ ohms.

11. You measured the resistance across several different light bulbs. On each you should have read a different amount of resistance. Based on your findings, which light bulb would be the brightest and which would be the dimmest? Explain why.

    _____

    _____

    _____

Instructor's Response

_____

_____

_____

Name _____   Date _____

GATHERING VEHICLE INFORMATION

Upon completion of this job sheet, you will be able to gather service information about a vehicle and its engine and related systems.

ASE Education Foundation Correlation

This job sheet addresses the following **MLR** task: VIII. Engine Performance; A. General,

Task #1   Research vehicle service information, including fluid type, vehicle service history, service precautions, and technical service bulletins. **(P-1)**

This job sheet addresses the following **AST/MAST** task:

Task #2   Research vehicle service information, including fluid type, vehicle service history, service precautions, and technical service bulletins. **(P-1)**

Tools and Materials

- Appropriate service information
- Computer
- Protective clothing
- Goggles or safety glasses with side shields

Describe the vehicle being worked on:

Year _____ Make _____ Model _____

VIN _____ Engine type and size _____

Procedure

1. Using the service information or other source, describe what each letter and number in the VIN for this vehicle represents.

   _____

   _____

   _____

   _____

2. While looking in the engine compartment, did you find a label regarding the specifications for the engine? Summarize the information contained on the label.

   _____

   _____

   _____

   _____

3. Find the fluid type specifications for the:

   Engine oil:

   Transmission:

   Engine coolant:

   Differential:

   Brake:

4. Using the service information, locate the information about the vehicle's engine. List the major components of the system and describe the primary characteristics of the engine (number of valves, shape/configuration, firing order, etc.).

_____

_____

_____

_____

5. Describe the engine's support systems (fuel, air, ignition, exhaust, and emissions) and their control system(s).

_____

_____

_____

6. Using the service information, locate and record all service precautions regarding working on the engine and its systems as noted by the manufacturer.

_____

_____

_____

_____

7. Using the information that is available, locate and record the vehicle's service history.

_____

_____

_____

_____

8. Using the information sources that are available, summarize all Technical Service Bulletins for this vehicle that relate to the engine and its systems.

_____

_____

_____

_____

Instructor's Response

_____

_____

_____

_____

Name _____ Date _____

INTERPRETING CODES FROM AN ENGINE CONTROL SYSTEM

Upon completion of this job sheet, you will be able to diagnose the causes of emissions or drivability concerns resulting from the failure of the computerized engine controls with stored diagnostic trouble codes.

ASE Education Foundation Correlation:

This job sheet addresses the following **MAST** task:

B.5   Diagnose the causes of emissions or drivability concerns with stored or active diagnostic trouble codes (DTC); obtain, graph, and interpret scan tool data. **(P-1)**

Tools and Materials
- Scan tool
- Digital multimeter
- Service information
- Protective clothing
- Goggles or safety glasses with side shields

Describe the vehicle being worked on:

Year _____ Make _____ Model _____

VIN _____ Engine type and size _____

Procedure

1. Refer to the listing of codes given in the service information. Briefly describe how the codes are grouped by letter and number.

   _____

   _____

   _____

2. Describe what is indicated by the following codes:

   P0037: _____

   P0143: _____

   P0010: _____

   P0100: _____

   P2195: _____

3. Conduct all preliminary checks of the engine, electrical system, and vacuum lines. Did you find any problems? If so, what were they?

   _____

   _____

   _____

4. Connect the scan tool to the DLC. Enter the vehicle information into the scan tool. Retrieve the Diagnostic Trouble Code (DTC) with the scan tool. What are they?

   _____

   _____

   _____

5. Refer to the service information and locate the description of the DTCs retrieved from the computer. What do they signify?

_____

_____

_____

_____

_____

6. For each of the codes, summarize what steps the manufacturer recommends for locating the exact cause of the problem.

_____

_____

_____

_____

_____

7. Follow those steps and summarize what you found.

_____

_____

_____

_____

8. After correcting the problem, the vehicle should be test driven under the conditions necessary to complete the enable criteria specified for the DTC you are trying to repair. Recheck the DTCs to make sure you properly corrected the problem. What did you find?

_____

_____

_____

_____

Instructor's Response

_____

_____

_____

_____

CHAPTER 3

GENERAL ENGINE CONDITION DIAGNOSIS

Upon completion and review of this chapter, you should be able to:

- List the steps in a general diagnostic procedure that can be used in any diagnostic situation.
- Diagnose fuel leaks and determine needed repairs.
- Diagnose engine oil leaks and determine needed repairs.
- Diagnose engine coolant leaks and determine necessary repairs.
- Diagnose engine exhaust odor, color, and noise, and determine needed repairs.
- Diagnose engine noises and vibration problems, and determine the necessary repairs.
- Test engine oil pressure and determine the cause of low oil pressure.

- Diagnose engine overheating problems.
- Pressure test the cooling system.
- Diagnose engine defects from intake manifold vacuum readings.
- Perform an engine power balance test and determine the needed repairs.
- Diagnose engine problems with an exhaust gas analyzer.
- Perform an engine compression test and determine the necessary repairs.
- Perform a cylinder leakage test and determine the needed repairs.
- Perform valve adjustments on mechanical and hydraulic lifters.
- Check valve timing.

Basic Tools
Basic tool set
Appropriate service manuals
Fender covers
Safety glasses

Terms To Know

Blowby	Hydraulic valve lifters	Radiator shroud
Blowgun	Hydrometer	Recovery tank
Bore	Lobes	Refractometer
Carbon dioxide (CO_2)	Mechanical valve lifters	Ring ridge
Compression ratio	Oil pump	Tachometer
Coolant hydrometer	Overhead camshafts	Valve overlap
Cooling system pressure tester	Pinging	Valve-stem installed height
Detonation	Positive crankcase ventilation (PCV)	Wet compression test
Engine analyzer	Power balance test	

GENERAL DIAGNOSTIC PROCEDURE

As you will probably continue to witness, automotive technology will become more complex with every model year. An automotive service technician's skill and knowledge must grow with these changes. The fundamentals of effective diagnostic skills become increasingly important.

The purpose of this chapter is to give you the skills to test the engine. If we begin diagnosing a drivability problem with the assumption that the engine is fine when it is not fine, we will waste a great deal of time and will probably become frustrated. The key to diagnostics is to know what tests to conduct, and when to conduct them. To know this, you must understand the system and test.

ENGINE LEAK DIAGNOSIS

Engine leaks not only cause unsightly messes on the engine, engine compartment, and driveway, but also can be an indication of a bigger problem. This big problem can cause safety, drivability, and durability concerns. It is extremely important that you are able to identify the type of fluid that is leaving the mess and the possible causes for the leakage. Leaks can and do cause drivability problems by contaminating sensors, wiring and spark plug wires, or boots.

⚠️ WARNING **Never attempt to repair a metal fuel tank until it is drained, removed, and steamed out for at least 30 minutes. Gasoline fumes left in gasoline tanks are extremely dangerous. When ignited, these fumes will cause a very serious explosion and fire, resulting in personal injury and property damage.**

⚠️ WARNING **Always store gasoline drained from a fuel tank in approved gasoline safety cans.**

⚠️ WARNING **While removing and replacing a gasoline tank and handling gasoline, be sure there are no sources of ignition in the area. Do not smoke while performing these operations, and be sure nobody else smokes in the area. Do not drag the gasoline tank on the floor.**

⚠️ WARNING **If gasoline is spilled on the shop floor, place approved absorbent material on it immediately. Dispose of the absorbent material in approved waste containers.**

Fuel Leaks

Engine fuel leaks are expensive and dangerous, and they should be corrected immediately when they are detected. Fuel leaks are expensive because they waste fuel. Because gasoline is very explosive, fuel leaks are extremely dangerous—any spark near a fuel leak may start a fire or cause an explosion. If gasoline odor occurs inside or near a vehicle, it should be inspected for fuel leaks immediately. To locate a fuel leak, inspect these fuel system components:

1. Fuel tank. If a fuel tank is leaking, it must be removed and repaired. Most fuel tanks do not have a drain plug, so the gasoline must be pumped from the tank with a hand-operated pump.
2. Fuel tank filler cap.
3. Fuel lines and filter.
4. Vapor recovery system lines.
5. Fuel rail, and injectors

Engine Oil Leak Diagnosis

Engine oil leaks may cause an engine to run out of oil, resulting in serious engine damage. Therefore, oil leaks should not be ignored. If an engine has an oil leak, the oil level should be checked often and oil should be added as required until the necessary repairs

An exhaust gas analyzer can be used to find the area of a fuel leak. Slowly move the probe along the fuel lines. A high hydrocarbon (HC) reading indicates the area of the leak.

are completed. When it is difficult to locate the exact cause of an oil leak, the engine should be cleaned and then operated while the technician watches for the source of the oil leak. Oil leaks may be difficult to locate because the oil runs down the engine surfaces. For example, the oil from a leak at a rocker arm cover may run down the side of the engine and appear as an oil pan leak. However, closer examination of the engine will indicate that the oil leak is coming from the rocker arm cover. Another method used to detect oil leaks is to add a special dye to the crankcase. The dye may be easier to see because it is colored. Another type of dye that is available for oil leak detection can be easily seen with a black light. It may be necessary to run the vehicle for a period of time, perhaps having the customer return in a few days, to allow the leak to show up adequately. Whenever any additive is used, be sure to follow the manufacturer's instructions and verify that its use will not interfere with any vehicle warranties. The possible causes of oil leaks are as follows:

Classroom Manual
Chapter 3, page 70

1. Rear main crankshaft bearing seal.
2. Expansion plug in rear camshaft bearing.
3. Rear oil gallery plug.
4. Oil pan.
5. Oil filter.
6. Rocker arm covers.
7. Intake manifold front and rear gaskets (V-type engines).
8. Timing gear cover or seal.
9. Front main bearing.
10. Oil pressure sending unit.
11. Engine casting porosity.
12. Oil cooler or lines (where applicable).

Engine Coolant Leak Diagnosis

When a coolant leak causes low coolant level, the engine quickly overheats and severe engine damage may occur. Therefore, coolant leaks must not be ignored. Coolant leaks may occur at these locations:

A blown head gasket allows combustion gases into the coolant, seriously reducing the effectiveness of the coolant.

1. Upper radiator hose.
2. Lower radiator hose.
3. Heater hoses.
4. Bypass hose.
5. Water pump.
6. Engine core plugs or block heater.
7. Radiator.
8. Thermostat housing.
9. Heater core.
10. Internal engine leaks caused by leaking intake manifold gasket, head gasket, or cracked engine components.

Classroom Manual
Chapter 3, page 71

⚡ **WARNING** **Never loosen a radiator cap on a warm or hot engine. When the cap is loosened, the pressure on the coolant is released and the coolant suddenly boils. This action may cause a technician or anyone standing nearby to be seriously burned.**

A pressure tester can be used to determine if a coolant leak is present and to locate the source of the leak. Remove the radiator cap and install the pressure tester on the radiator filler neck (**Figure 3-1**). Operate the pump on the tester until the rated pressure on the radiator cap appears on the tester gauge. Leave this pressure applied to the cooling

Special Tools

Cooling system pressure tester

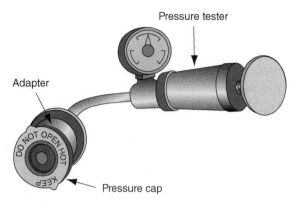

Figure 3-1 Cooling system pressure tester.

system for several minutes, then check the gauge pressure. If there is no drop in gauge pressure during this time, the cooling system does not have a leak. If the pressure drops, the cooling system has a leak. Normally the source of the leak is evident by coolant spraying out of the leak. Inspect the entire cooling system, and repair any leaks as required. Check the floor of the vehicle for any indication of coolant leaking out of the heater core and dripping onto the floor. Heater core leaks usually cause an odor inside the passenger compartment and fogging of the windshield.

A leaking head gasket will cause the engine to overheat not only because of the loss of coolant but because the gas can flow into the cooling system, thereby reducing its effectiveness.

The radiator pressure cap can be connected to the cooling system pressure tester with a special adapter. When the tester pump is operated, the gauge on the tester indicates the pressure required to open the cap pressure relief valve. This pressure should be the same as the pressure rating stamped on the cap.

If the pressure drops but no external leaks are evident, the spark plugs should be removed to check for coolant in the cylinders. After the spark plugs are removed, disable the injection and ignition systems. Crank the engine while checking for any sign of coolant discharging from the spark plug openings. If coolant is discharged from a spark plug opening, the head gasket, cylinder head, or cylinder wall is leaking or cracked.

When there are coolant leaks in the combustion chamber, bubbles may appear in the coolant at the radiator filler neck with the engine running. An infrared exhaust gas analyzer can be used to check for coolant leaks in the combustion chamber. With the radiator cap removed, place the sample probe of the exhaust analyzer at the top of the radiator filler neck (**Figure 3-2**) or at the top of the coolant **recovery tank** opening if the radiator has no cap. Use extreme caution when opening the radiator cap, especially if the engine is already hot. If there is pressure in the system, allow it to cool down before doing this procedure. Be certain the sample probe does not come in contact with any liquid by inadvertently immersing it in the coolant. The analyzer can be damaged if coolant is sucked into the unit. Once the probe is in place, run the engine at 1,500 to 2,000 rpm and observe the readings. Allow a few seconds for the sample to be read and displayed on the unit. Look for HCs to show a reading. You will notice an increase in HC if there is a combustion leak. Check also for an increase in **carbon dioxide (CO_2)**. CO_2 is produced as a result of combustion, and if it is present in the cooling system, it is there because of an internal combustion chamber leak.

If the coolant level continues to go down with no indication of external leaks, the coolant may be leaking through a cracked block into the oil pan. If this type of leak is suspected, check the dipstick first. Drain the engine oil and check for signs of coolant in the oil if further inspection is needed.

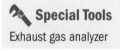

Special Tools

Exhaust gas analyzer

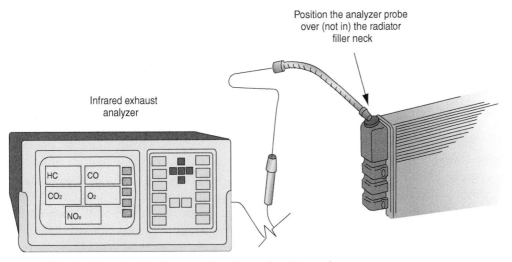

Position the analyzer probe over (not in) the radiator filler neck

Infrared exhaust analyzer

HC | CO
CO₂ | O₂
NOₓ

Figure 3-2 Testing for combustion chamber leaks with an exhaust gas analyzer.

ENGINE NOISE DIAGNOSIS

There are usually warning noises from the engine long before a serious engine failure occurs. When engine problems are detected early, the engine repairs may be much less expensive. Engine defects such as damaged pistons, worn rings, loose piston pins, worn crankshaft bearings, worn camshaft **lobes**, and loose and worn valve-train components usually produce their own peculiar noises. Certain engine defects also cause a noise under specific engine operating conditions. A technician must be able to diagnose engine noises accurately to avoid unnecessary engine repairs.

As it is sometimes difficult to determine the exact location of an engine noise, a stethoscope can be useful in diagnosing noise locations. The stethoscope probe is placed on or near the suspected component, and the ends of the stethoscope are installed in your ears. The stethoscope amplifies sound to assist in noise location. When the stethoscope probe is moved closer to the source of the noise, the sound is louder in your ears. If a stethoscope is not available, a length of hose placed from the suspected component to your ear amplifies sound and helps to locate the cause of the noise.

⚠ WARNING **When placing a stethoscope probe in various locations on a running engine, be careful not to catch the probe or your hands in moving components such as cooling fan blades and belts.**

Lack of lubrication is a common cause of engine noise, so always check the engine oil level and condition prior to noise diagnosis. Carefully observe the oil for contamination by coolant or gasoline. During the diagnosis of engine noises, always operate the engine under the conditions when the noise occurs. Remember that aluminum engine components such as pistons expand more when heated than cast-iron alloy components. Therefore, a noise caused by a piston defect may occur when the engine is cold but disappear when the engine reaches normal operating temperature.

Main Bearing Noise

Loose crankshaft main bearings cause a heavy thumping knock for a brief time when the engine is first started after it has been shut off for several hours. This noise may also be noticeable on hard acceleration depending on the amount of bearing wear. Worn main bearings or crankshaft journals or a lack of lubrication causes this problem. When the thrust surfaces of a crankshaft bearing are worn, the crankshaft has excessive endplay.

The exhaust gas analyzer probe can be placed above the radiator filler neck with the engine running to check for a combustion chamber leak.

 **Special Tools**
Stethoscope

Under this condition, a heavy thumping knock may occur at irregular intervals on acceleration. The main bearings must be replaced to correct this problem, and the crankshaft main bearing journals may require machining. If the journals are severely scored or burned, the crankshaft must be replaced.

Connecting Rod Bearing Noise

Loose connecting rod bearings cause a lighter, rapping noise at speeds above 35 mph (21 kph). The noise from a loose connecting rod bearing may vary from a light to a heavier rapping sound depending on the amount of bearing looseness. Connecting rod bearing noise may occur at idle if the bearing is very loose. When the spark plug is shorted out in the cylinder with the loose connecting rod bearing, the noise usually decreases. Worn connecting rod bearings or crankshaft journals or a lack of lubrication may cause this problem. To correct this noise, the connecting rod bearings require replacement, and the crankshaft journals should be machined. Severely worn or burned crankshaft journals require crankshaft replacement. The connecting rod alignment should also be checked.

Piston Slap

Piston slap is noticeable only when the engine is cold because aluminum expands at twice the rate of steel.

A piston slap causes a hollow, rapping noise that is most noticeable on acceleration when the engine is cold. Depending on the piston condition, the noise may disappear when the engine reaches normal operating temperature. A collapsed piston skirt, a cracked piston, excessive piston-to-cylinder wall clearance, lack of lubrication, or a misaligned connecting rod are causes of piston slap. Correcting this problem requires piston or connecting rod replacement and possible re-boring of the cylinder.

Piston Pin Noise

A loose, worn piston pin causes a sharp, metallic rapping noise that occurs with the engine idling. This noise may be caused by a worn piston pin, worn pin bores, a cracked piston in the pin **bore** area, a worn rod bushing, or a lack of lubrication. An oversized pin can be installed to correct the problem. Severely worn or cracked pistons must be replaced.

Piston Ring Noise

If the piston rings are excessively loose in the piston ring grooves, a high-pitched clicking noise is noticeable in the upper cylinder area during acceleration. To correct this noise, the rings, and possibly the pistons, must be replaced. If the cylinders are severely worn, cylinder re-boring is necessary.

Ring Ridge Noise

When the top piston rings are striking the **ring ridge** on the cylinder wall (**Figure 3-3**), a high-pitched clicking noise is heard. This noise intensifies when the engine is accelerated. Ring ridge noise can be caused by the installation of new rings without removing the ring ridge at the top of the cylinder. This noise may also be caused by a loose connecting rod bearing or piston pin, which allows the piston to move above its normal travel in the cylinder. To correct this problem, remove the ring ridge and check the connecting rod bearing and piston pin.

Valve-Train Noise

Valve-train noise, such as valve lifters or rocker arms, appears as a light clicking noise when the engine is idling. Valve-train noise is slower than piston or connecting rod noise because the camshaft is turning at one-half the crankshaft speed. This noise is less noticeable when the engine is accelerated. Sticking or worn valve lifters may provide an irregular clicking noise that is more likely to occur when the engine is first started. Valve-train noise may be caused by a lack of lubrication or by contaminated oil. To correct valve-train noise, check the valve adjustment where applicable, and check the valve lifters, pushrods, and rocker arms.

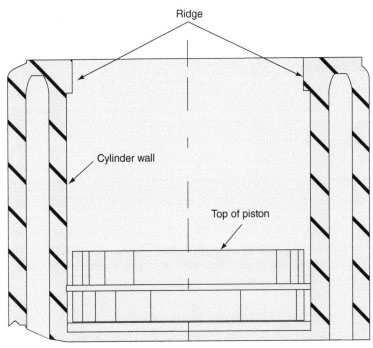

Figure 3-3 When the piston strikes the ridge at the top of the cylinder, a high-pitched rapping or clicking sound is heard.

Timing Chains and Tensioners

A worn, loose timing chain, sprockets, or chain tensioners cause a whirring and light rattling noise when the engine accelerates and decelerates. Severely worn timing chains, sprockets, and tensioners may cause these noises at idle speed. The timing chain, sprockets, and tensioners must be checked and replaced as necessary to correct this noise.

Combustion Noises

The most common abnormal combustion noise is caused by detonation in the cylinders. **Detonation** occurs in a cylinder when the air-fuel mixture suddenly explodes rather than burning smoothly. This action drives the piston against the cylinder wall and suddenly forces the connecting rod insert against the crankshaft journal. Under this condition, a **pinging** noise occurs that is similar to marbles rattling inside a metal can. The pinging noise usually occurs when the engine is accelerated. Detonation may be caused by excessive ignition spark advance, low-octane fuel, higher-than-normal engine temperature, and hot carbon spots on top of the piston. A lean air-fuel mixture or an inoperative exhaust gas recirculation (EGR) valve contributes to detonation; detonation at a steady cruising speed may be caused by a defective EGR valve or related controls. To correct the detonation problem, check the air-fuel mixture, engine compression, engine temperature, and EGR valve.

Pinging is the sound used to describe the noise of spark knock.

Flywheel, Flexplate, and Vibration Damper Noise

A loose flywheel or cracked flexplate causes a thumping noise at the back of the engine. This noise varies depending on whether the engine is equipped with a flywheel or a flexplate. The noise also varies depending on the looseness of the flywheel. To correct this problem requires transmission removal and tightening of the flywheel-to-crankshaft bolts. If the bolt holes in the flywheel are damaged, flywheel replacement is necessary. A knocking sound can also be produced by loose flexplate to torque converter bolts. These can generally be accessed by removing the torque converter cover. The noise sounds similar to a rod bearing. Always check any noise thoroughly before you tear down the engine.

A loose vibration damper on the front of the crankshaft causes a rumbling or thumping noise at the front of the engine, and this noise may be accompanied by engine vibrations. When the engine is accelerated under load, the noise is more noticeable. The vibration damper must be replaced to correct this problem.

Drive Belt or Accessory Noise

Many noises thought to be from the engine actually originate from the drive belt or a driven accessory, such as the alternator, water pump, and power steering pump. Make sure to remove the drive belt to see if the noise goes away. Just do so for a short period of time if the water pump is driven by the belt.

ENGINE EXHAUST DIAGNOSIS

⚡ **WARNING** **Vehicle exhaust contains carbon monoxide (CO), which is a poisonous gas. Breathing CO results in health problems, and a concentrated mixture of CO and air is fatal! Do not operate a vehicle in the shop unless a shop exhaust hose is connected to the tailpipe.**

Bluish-white exhaust indicates excessive amounts of engine oil entering the combustion chambers.

Some engine problems can be diagnosed by the color, smell, or sound of the exhaust. If the engine is operating normally, the exhaust should be colorless. In severely cold weather, it is normal to see a swirl of white vapor coming from the tailpipe, especially when the engine and exhaust system are cold. This vapor is moisture in the exhaust, a normal by-product of the combustion process.

If the exhaust is bluish white, excessive amounts of oil are entering the combustion chamber and are being burned with the fuel. When the bluish-white smoke in the exhaust is more noticeable on deceleration, the oil is likely getting past the rings into the cylinder. Vacuum in the cylinder is high on deceleration, and this high vacuum is pulling oil past the rings into the combustion chamber. If the bluish-white smoke appears in the exhaust immediately after a hot engine is restarted, the oil is likely leaking down the valve guides. In this case, the seals, guides, or valve stems may be worn.

Black exhaust is caused by a rich air-fuel mixture.

If black smoke appears in the exhaust, the air-fuel mixture is too rich. This condition can be caused by a defect in the fuel injection system. In a fuel injection system, a defective pressure regulator, injectors, or input sensors may cause a rich air-fuel mixture. A restriction in the air intake, such as a plugged air filter, may be responsible for a rich air-fuel mixture. The evaporative emission controls may also be at fault.

If gray-white smoke is present in the exhaust immediately after the engine is started, coolant is probably entering the combustion chamber.

Gray-white smoke in the exhaust can be caused by a coolant leak into the combustion chambers. If a coolant leak is the cause of gray-white smoke in the exhaust, this smoke is usually most noticeable when the engine is first started after it has been shut off for over half an hour.

Exhaust Noise

⚡ **WARNING** **Exhaust components may be extremely hot. If you must touch them, wear recommended protective gloves to avoid burns.**

When the engine is idling, the exhaust at the tailpipe should have a smooth, even sound. If the exhaust in the tailpipe has a puff sound at regular intervals, a cylinder may be misfiring. When this sound is present, check the engine's ignition and fuel systems and the engine's compression.

If the vehicle has excessive exhaust noise, check the exhaust system for leaks. The noise from exhaust leaks is most noticeable when the engine is accelerated. A small exhaust leak may cause a whistling noise when the engine is accelerated.

If the exhaust system produces a rattling noise when the engine is accelerated, check the muffler and catalytic converter for loose internal components. When this problem is suspected, rap on the muffler and converter with the engine off. If a rattling noise is heard, replace the appropriate component. This rattling noise in the exhaust system may also be caused by loose exhaust system hangers or an exhaust system component hitting the chassis.

When the engine has a wheezing noise at idle or with the engine running at higher rpm, check for a restricted exhaust system. Connect a vacuum gauge to the intake manifold and observe the vacuum with the engine idling. Accelerate the engine to 2,500 rpm and hold this speed for 2 minutes. If the exhaust is restricted, the vacuum will slowly drop below the recorded vacuum at idle. When the car is driven under normal loads, this vacuum drop is even more noticeable.

DIAGNOSIS OF OIL CONSUMPTION

With advances in piston ring technology, modern engines consume very little oil. These engines can be driven for 1,000 to 3,000 miles (1,600 to 4,800 km) without adding oil to the engine. If an engine uses excessive oil, it may be leaking from the engine or burning in the combustion chambers. A plugged **positive crankcase ventilation (PCV)** system causes excessive pressure in the engine, and this pressure may force oil from some of the engine gaskets. If excessive amounts of oil are burned in the combustion chambers, the exhaust contains bluish-white smoke and the spark plugs may be fouled with oil. Excessive oil burning in the combustion chambers can be caused by worn rings and cylinders or worn valve guides and valve seals.

CUSTOMER CARE Always inform the customer if you notice problems, such as engine noises, with his or her car. Most engine failures are evidenced by unusual noises for a considerable length of time prior to the actual failure. If you inform the customer about a noise that may indicate a serious engine problem, the customer will likely consent to further diagnosis of the problem. When the diagnosis is complete, the customer will likely approve the necessary repairs, and these repairs will likely be less expensive than the repairs of a complete engine failure. The alternative procedure is to let the customer drive the car until the engine noise develops into a serious engine failure. This problem may occur when the customer is miles from home. The customer will likely appreciate your diagnosis of the impending engine failure before it develops into a major problem at an inconvenient time.

Special Tools
Oil pressure gauge and adapters

ENGINE OIL PRESSURE TESTS

If the low oil pressure warning light or pressure gauge indicates low oil pressure, the engine oil level should be checked immediately. The oil should also be checked for contamination with gasoline or coolant. If the oil is contaminated with gasoline, the oil is thinner and oil pressure is reduced. When the oil level and quality are satisfactory but the oil pressure warning light or gauge indicates low pressure, the oil pressure should be checked. If the engine runs for a few minutes without proper lubrication, the crankshaft bearings and cylinder walls may be severely scored. A vehicle should not be driven if the oil pressure warning light or gauge indicates low oil pressure. Low engine oil pressure can be caused by a worn **oil pump**, a plugged pick-up screen, a leaking pick-up pipe, loose main bearings, loose connecting rod bearings, worn/loose camshaft bearings, and worn valve-lifter bores.

Classroom Manual
Chapter 3, page 70

A pressure gauge must be connected to the main oil gallery to check engine oil pressure. The oil pressure sending unit can be removed from the main oil gallery to install the pressure gauge (**Figure 3-4**). **Photo Sequence 2** shows a typical procedure for testing engine oil pressure.

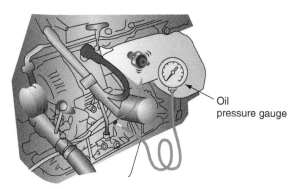

Oil
pressure gauge

Figure 3-4 Oil pressure test gauge connected to the opening of the oil pressure gauge sending unit.

PHOTO SEQUENCE 2
Typical Procedure for Testing Engine Oil Pressure

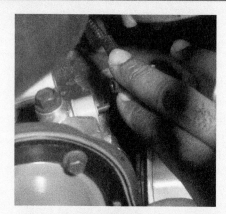

P2-1 Disconnect the wire from the oil sending unit, and remove the oil sending unit.

P2-2 Thread the fitting on the oil pressure gauge hose into the oil sending unit opening, and tighten the fitting.

P2-3 Place the oil pressure gauge where it will not contact rotating or hot components.

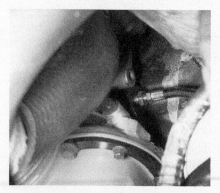

P2-4 Start the engine and check for leaks at the oil pressure gauge fittings.

P2-5 Operate the engine until it reaches normal operating temperature, and observe the oil pressure on the gauge at idle speed and 2,000 rpm.

P2-6 Compare the oil pressure gauge reading to the vehicle manufacturer's specifications.

PHOTO SEQUENCE 2 (CONTINUED)

P2-7 Remove the oil pressure gauge from the oil sending unit opening.

P2-8 Install the oil sending unit, tighten it to the specified torque, and install the sending unit wire.

P2-9 Start the engine and check for oil leaks at the oil sending unit.

The engine should be at normal operating temperature before checking the oil pressure. Engine oil pressure should be checked with the engine idling and with the engine operating at 2,000 rpm. If the oil pressure is equal to the vehicle manufacturer's specifications but the oil pressure warning light or gauge indicated low oil pressure, the sending unit is probably defective. When the oil pressure is lower than specified, one of these problems may exist:

1. Worn oil pump.
2. Plugged oil pump pick-up screen.
3. Leaking oil pump pick-up tube.
4. Sticking pressure regulator valve in the pump.
5. Broken oil pump drive.
6. Loose main and connecting rod bearings.
7. Worn camshaft bearings.
8. Worn valve-lifter bores.
9. Missing or leaking oil gallery plugs.

On many engines, when the oil pressure is lower than specified, the oil pan must be removed to check the cause of the low pressure.

ENGINE TEMPERATURE TESTS

Lower-than-normal coolant temperature reduces engine efficiency and fuel economy. If the engine temperature is too low, combustion temperatures and therefore emissions will be affected. The engine has to be hot enough to keep fuel from condensing out on the cylinder walls during the intake stroke and increasing hydrocarbon emissions. Lower-than-normal engine temperature also prevents the boiling off of excessive water and unburned fuel from the crankcase. This water and fuel mixture tends to form an acidic sludge that can dramatically affect engine life. Cooling system temperature is more critical today due to computerized engine controls. Many functions are determined in part by the engine's operating temperature. The correct temperature must be maintained precisely throughout the entire operating cycle, from start-up to shut down. For this reason, the

Classroom Manual
Chapter 3, page 72

A trouble code PO 128 will set in the engine control module if the thermostat is too slow to open.

There are many causes of engine overheating. During the diagnosis of an overheating situation, all these causes should be checked until the source of the problem is found. Begin the diagnosis with the quickest tests or checks.

engine control module (ECM) monitors the engine thermostat function, and sets trouble code P0128 if proper engine temperature is not maintained. If the coolant temperature is higher than the rated temperature of the thermostat, the engine may overheat.

The thermostat's operating temperature can be checked while installed in the engine or removed from it. You will need to monitor the temperature as part of the test, regardless. Engine coolant temperature can be monitored directly by a scan tool. Another choice is to install a radiator thermometer directly in the coolant at the radiator fill neck. Some digital multimeters (DMMs) have a thermometer accessory that connects to the unit that indicates the temperature on the DMM's readout display. There are also non-contact, infrared thermometers available that display an object's temperature by aiming the unit at the object and reading the unit's display. Start the engine and run it at a fast idle for approximately 15 minutes or until the coolant starts to move in the radiator. Observe the temperature reading when the coolant begins to flow in the radiator. This is the temperature at which the thermostat started to open. The temperature observed should be within ±5° of the thermostat's rated temperature.

The thermostat can also be checked when removed from the vehicle. Monitor the temperature and thermostat as described earlier. After removing the thermostat from the engine, place it in a container of water suspended so that it does not come in contact with the container walls or bottom (**Figure 3-5**).

Heat the water, and note the temperature of the water when the thermostat begins to open. Depending on the rating, most thermostats will open from 185°F to 196°F and should be fully open at 212°F, the boiling point of water at sea level. If the thermostat opens much earlier or later than its rating, it should be replaced. Obviously, if the thermostat is stuck open or closed, regardless of the temperature, it needs to be replaced. When replacing or installing the thermostat, be certain to thoroughly scrape and clean the mating surfaces of the thermostat housing and block, and use a new gasket or sealant as recommended by the manufacturer. Also, look for a recess in the block where the thermostat must be seated (**Figure 3-6**).

If the thermostat is cocked or not seated during installation it will not seal, and you may break the thermostat housing while torque tightening the thermostat housing bolts.

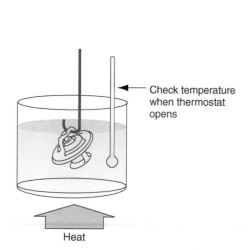

Check temperature when thermostat opens

Heat

Figure 3-5 Checking thermostat operation.

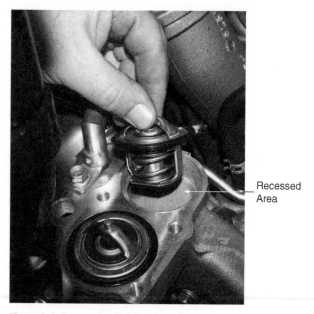

Recessed Area

Figure 3-6 Be sure that the thermostat fits into the recess in the housing or manifold.

When the engine operating temperature is higher than the rated temperature of the thermostat, check these causes of overheating:

1. Loose or slipping belt.
2. Electric-drive cooling fan not operating at the proper temperature.
3. Defective viscous-drive fan clutch.
4. Restricted air passages in radiator core or other restricted airflow to the radiator.
5. Partially plugged coolant passages in radiator core.
6. Restricted or collapsed radiator hoses.
7. Thermostat not opening at the proper temperature.
8. Thermostat improperly installed.
9. Leaking cylinder head gasket.
10. External cooling system leaks causing low coolant level.
11. Leaking radiator pressure cap.
12. Defective water pump, loose impeller.
13. Improper mixture of antifreeze and water.
14. Excessive engine load.
15. Lean air-fuel mixture, engine vacuum leaks.
16. Automatic transmission overheating.
17. Dragging brakes.
18. Broken, improperly positioned radiator shroud.

COOLING SYSTEM INSPECTION AND DIAGNOSIS

When a cooling system is suspected as being faulty, it should be inspected and tested. Keep in mind that an engine and its systems will be efficient only when the engine is at normal operating temperatures (**Photo Sequence 3**).

Classroom Manual
Chapter 3, page 51

PHOTO SEQUENCE 3
Typical Procedure for Testing a Cooling System

 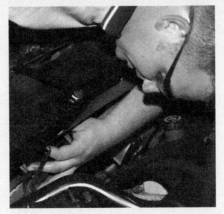 

P3-1 After allowing the engine to cool, the technician starts his or her coolant service by doing a visual inspection of the belts, hoses, and any obvious signs of leakage. Also check for any blockages in front of the radiator.

P3-2 If the vehicle has electric fans or a viscous fan clutch, check its condition.

P3-3 The technician opens the radiator cap and inspects the level and condition of the coolant. Is it rusty or contaminated in appearance? Check the coolant freezing and boiling points with a refractometer.

PHOTO SEQUENCE 3 (CONTINUED)

P3-4 Pressure-test the cooling system using a tester. Make sure the system is able to hold pressure rated on the radiator cap.

P3-5 The radiator pressure cap should also be checked. It must hold the pressure that it is rated.

P3-6 If during inspection any work was required, make sure you double-check your repairs. Remember that many vehicles have special fill procedures to prevent air locks.

VISUAL INSPECTION

⚡ **WARNING** Use caution and wear protective gloves when touching cooling system components. If the engine has been running, these components may be very hot, and touching them could result in severe burns.

⚡ **WARNING** Keep hands, tools, and clothing away from electric-drive cooling fans. On many vehicles, an electric-drive fan may start turning even with the ignition switch off.

All cooling system hoses should be inspected for soft spots, swelling, hardening, chafing, leaks, and collapsing. If any of these conditions are present, hose replacement is necessary. Hose clamps should be inspected to make sure they are tight. Some radiator hoses contain a wire coil inside them to prevent hose collapse due to water pump suction. Remember to include heater hoses and the bypass hose in the hose inspection.

⚙ **SERVICE TIP** If the upper radiator hose collapses when the coolant is cold, the vacuum valve may be sticking in the radiator pressure cap or the hose to the coolant recovery container may be restricted.

The radiator should be inspected for restrictions such as bugs and debris in the air passages through the core. An air hose and air gun can be used to blow bugs and debris from the core. The radiator should also be inspected for leaks and loose brackets. Radiator repairs are usually done in a radiator repair shop. When the radiator cap is removed, the openings in some of the radiator tubes are visible through the filler neck. These tube openings should be free from rust and corrosion. If the tubes are restricted, cooling system flushing may remove the restriction. When flushing does not clean the radiator tubes, the

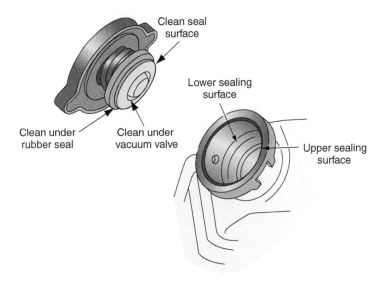

Figure 3-7 Inspecting the radiator pressure cap and filler neck.

radiator must be cleaned at a radiator shop. Any restrictions in the radiator air passages or coolant tubes result in higher engine operating temperature and possible overheating.

Many radiators have a shroud behind the radiator to concentrate the flow of air through the radiator core. Inspect the **radiator shroud** for looseness and cracks or missing pieces. If the shroud is loose, improperly positioned, or broken, airflow through the radiator is reduced and engine overheating may result.

The radiator cap should be inspected for a damaged sealing gasket or vacuum valve (**Figure 3-7**). Check the pressure rating stamped on top of the cap, and be sure this is the same as the radiator cap specification in the vehicle manufacturer's service information. Replace the cap if the pressure rating does not meet the vehicle manufacturer's specified pressure. If the sealing gasket or vacuum valve is damaged, replace the cap. Check the seat in the filler neck for damage or burrs, and remove any rough spots with fine emery paper. Visually inspect the filler neck for rust accumulation, which may indicate a contaminated cooling system. If the pressure cap sealing gasket or seat is damaged, the engine will overheat and coolant will be lost to the coolant recovery system. Under this condition, the coolant recovery container becomes overfilled with coolant.

If the cap vacuum valve is sticking, a vacuum may occur in the cooling system after the engine is shut off and the coolant temperature decreases. This vacuum may cause collapsed cooling system hoses.

The **cooling system pressure tester** is used to pressurize the entire cooling system to check the sealing integrity of the system, including the radiator pressure cap. Pressurizing the system will make it easier to detect any coolant leaks. After installing the pressure tester to the radiator filler neck, the technician manually pumps the tester to a specific pressure equal to the operating pressure of the vehicle's cooling system (see Figure 3-1). If the coolant is low in the radiator, top it off with water, and then perform the test. The pressure will hold if no leaks are present. If the pressure drops steadily over several minutes, there is enough pressure loss to indicate a cooling system leak. Inspect the entire cooling system for any evidence of coolant leaks. Be sure to include the radiator, hoses, water pump, heater core, and water jacket expansion plugs in the engine block and cylinder heads. Also, do not overlook the less obvious possibilities such as coolant in the engine's crankcase or in a cylinder. Do not forget to check the rear seat heating system components, if so equipped. If you added water to the radiator, be sure to add the proper amount of coolant to bring the mixture to the proper protection level after the repair of any leak.

A **radiator shroud** directs air through the radiator.

Make sure that the shrouds around the radiator and the air dams under the vehicle are intact. These are critical to proper air flow through the radiator.

Classroom Manual
Chapter 3, page 51

Defective radiator pressure caps may cause excessive coolant loss to the coolant recovery system, engine overheating, and collapsed hoses.

Figure 3-8 Radiator pressure tester and adaptors.

The pressure tester is also used to test the radiator cap (**Figure 3-8**). An adapter is used to connect the cap to the pressure tester. The pressure tester should be pumped up to build enough pressure to overcome the pressure rating of the radiator cap. If the cap is good, it should hold pressure equal to its rating. Any pressure beyond the specified rating should force open the spring-loaded seal in the cap, allowing the excess pressure to escape. This is how it would function on the radiator in the event that the pressure is too high in the cooling system. The excess coolant under higher pressure is captured in the recovery tank. This is a cyclic occurrence and is normal, especially during hot operating conditions.

A **refractometer** (**Figure 3-9**) or a **hydrometer** (**Figure 3-10**) can be used to measure antifreeze protection. A refractometer is usually considered to be more accurate and does not require temperature correction. Most refractometers can also show the specific gravity of battery cells. A drop of fluid is placed on the viewing screen, and the refractometer is held up to the light. The light rays passing through the refractometer are bent according to the density, and therefore the concentration of coolant can be measured and can be read directly from the screen.

A **coolant hydrometer** is used to check the specific gravity of the coolant mixture in the cooling system. A sample of the coolant is drawn into the hydrometer (following the

A **refractometer** is used to measure the coolant's freezing level.

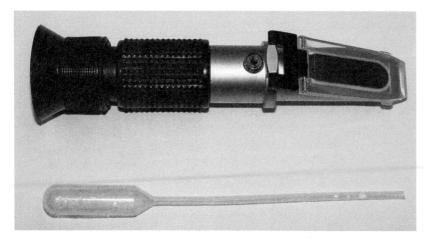

Figure 3-9 A coolant refractometer.

Figure 3-10 A coolant hydrometer.

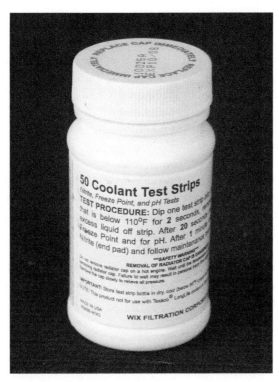

Figure 3-11 Coolant test strips available to measure the freeze protection and acidity of coolant.

tool manufacturer's directions) from the radiator. The coolant will work against a calibrated float device inside the hydrometer once the proper level is drawn in. The float device will indicate the temperature protection value of the coolant in degrees. When adding or replacing coolant, follow the manufacturer's instructions for the proper mixture of coolant and water. Actually, pure coolant offers no protection; it needs to be mixed with water to be effective. The mix is usually 50 percent coolant and 50 percent water for most applications. Otherwise, check the label on the coolant container for specific protection mixtures. Usually the 50-50 mix of coolant and water gives freezing protection to approximately −34°F (−37°C) and boiling protection up to 265°F (129°C). Testing strips are available that can measure the acidity content levels and freeze protection levels (**Figure 3-11**). Follow the vehicle manufacturer's specific recommendations for cooling system servicing, which may include a system flush at certain mileage or time intervals. It should be noted that not all coolant and antifreeze compounds are compatible with each other.

For example, General Motors warns against mixing DEX-COOL™ (silicate-free) coolant with other types of coolant. The special properties of DEX-COOL™ are unique for its application and therefore not compatible with other types of coolant. The manufacturer states that this coolant is designed to remain in the vehicle for 5 years or 150,000 miles. Furthermore, it warns that if DEX-COOL™ is mixed with other types of coolant, cooling system or engine damage may occur, jeopardizing the new vehicle warranty. As mentioned above, all coolants should be mixed with water to be effective. Coolant that is not mixed with water actually has less protection against boiling and freezing than an appropriate mix. It is not beneficial to increase the proportion of coolant to water to more than 70 percent. It is best to use distilled or purified water to mix with any coolant to prevent mineral buildup in the cooling system.

Silicates cause damage to seals in the cooling system.

Classroom Manual Chapter 3, page 72

⚡ WARNING **Used engine coolant is a hazardous waste, and pollution laws prohibit the dumping of coolant in sewers. Coolant recycling equipment is available to clean contaminants from the coolant so it may be reused.**

Clutch Fan and Water Pump

On vehicles with a clutch fan attached to the water pump, with the engine off, grasp the fan blades or the water pump hub and try to move the blades from side to side. This action checks for looseness in the water pump bearing. If there is any side-to-side movement in the bearing, water pump replacement is required. Check the fan blade for cracks or a bent condition. A cracked fan blade causes a clicking noise at idle speed. If the fan blades are cracked or bent, replace the fan blade assembly. Bent fan blades cause imbalance and vibration, which damage the water pump bearing.

⚠ **WARNING If a cracked or bent fan blade is discovered, replace the fan blade immediately. Do not start the engine. If a fan blade breaks while the engine is running, the blade becomes a very dangerous projectile that may cause serious personal injury or vehicle damage.**

⚠ **WARNING Do not attempt to straighten bent fan blades. This action weakens the metal and may cause the blade to break suddenly, resulting in serious personal injury or vehicle damage.**

Serpentine Belts

The majority of vehicles produced are equipped with one or possibly two serpentine belts (**Figure 3-12**). Serpentine belts are usually tensioned automatically; some are tensioned manually. If the belt slips or squeaks, make sure the tensioner is operating correctly. The serpentine belt should be inspected regularly, basically anytime a technician is servicing a customer's vehicle. With earlier neoprene serpentine belt look for fraying, excessive cracking, or missing sections of the ribs (**Figure 3-13a**). Later model belts are made of ethylene propylene diene monomer (EPDM). The EPDM belts do not tend to crack and split like the earlier neoprene belts, but do wear down over time, making the grooves on the belt wider and deeper. (**Figure 3-13b**). There are special depth gauges used to judge the amount of wear on the belt.

It is also important to make sure the belt is aligned properly on the pulleys. In most cases, the tensioner can be manipulated with a socket and drive tool. Some tensioners are designed to accept a ⅜ or ½ inch drive tool such as a ratchet in order to remove the tension from the belt. There are also tools specifically made for serpentine belt service (**Figure 3-14**).

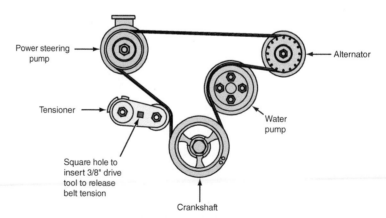

Figure 3-12 A serpentine belt routing diagram, showing tensioner.

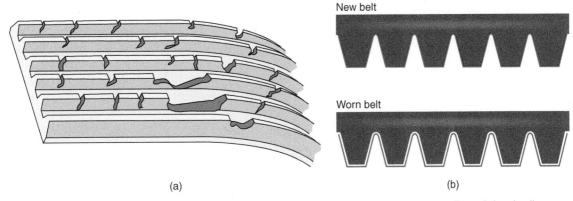

New belt

Worn belt

(a)   (b)

Figure 3-13 (a) Earlier neoprene serpentine belt in need of replacement. (b) EPDM belts do not normally crack, but the ribs can wear causing slipping.

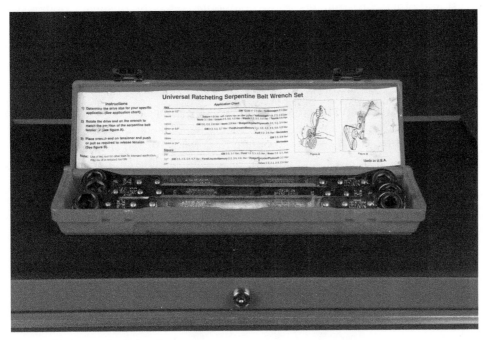

Figure 3-14 A serpentine belt tool.

Cooling System Checks

Check for coolant leaks at the water pump drain hole in the bottom of the pump and at the inlet hose connected to the pump. When coolant is dripping from the pump drain hole, replace the pump. If coolant is leaking around the inlet hose connection on the pump, check the hose and clamp.

Check the coolant recovery container and hose for leaks. Observe the level of coolant in the recovery container. In cooling systems with a coolant recovery system, coolant is added to the recovery container. Most of these containers have cold and hot coolant level marks. The coolant level should be at the appropriate mark on the recovery container, depending on the engine temperature.

The thermostat must be removed to inspect it. Some of the coolant must be drained from the cooling system before the thermostat is removed. Loosen the radiator drain plug and drain the coolant into a clean coolant drain pan. Remove the thermostat housing bolts and remove the housing. Check the thermostat valve and replace the thermostat if the valve is stuck open. Visually inspect the thermostat housing area for rust and contaminants. If excessive rust and contaminants are present, flush the cooling system and replace or recycle the coolant.

If the opening temperature of the thermostat is higher than the rated temperature, the engine will overheat. When the thermostat opening temperature is lower than the rated temperature, the engine runs cooler than normal, providing a reduction in engine efficiency and performance. Lower-than-normal coolant temperature changes some input sensor signals and reduces fuel economy.

Many thermostats are marked for proper installation. The pellet end of the thermostat faces toward the engine. Be sure the thermostat fits properly in the recess in the upper or lower part of the housing. Clean the mating surfaces carefully with a scraper to remove the old gasket. Install a new housing gasket, and tighten the housing bolts to the specified torque.

If the radiator tubes and coolant passages in the block and cylinder head are restricted with rust and other contaminants, these components may be flushed. Cooling system flushing equipment is available for this purpose. Always operate the flushing equipment according to the equipment manufacturer's directions, and be sure that your service procedure conforms to pollution laws in your state. Coolant-reconditioning machines are available to remove harmful particles and restore corrosion additives so the coolant can be returned to the cooling system (**Figure 3-15**).

The **viscous-drive fan clutch** should be visually inspected for leaks. If oily streaks are radiating outward from the hub shaft, the fluid has leaked out of the clutch.

Classroom Manual
Chapter 3, page 79

With the engine off, rotate the cooling fan by hand. When the engine is cold, the viscous clutch should offer a slight amount of resistance. The clutch should offer more resistance to turning when the engine is hot. If the viscous clutch allows the fan blades to rotate easily both hot and cold, the clutch should be replaced. A slipping viscous clutch results in engine overheating. Push one of the fan blades in and out, and check for looseness in the clutch shaft bearing. If any looseness is present, replace the viscous clutch.

Figure 3-15 An engine coolant (antifreeze)–recycling machine.

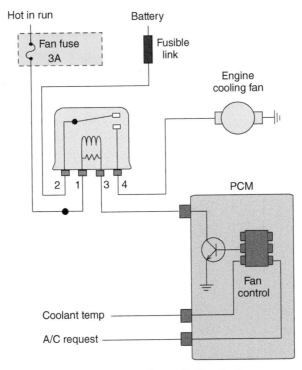

Figure 3-16 A computer-controlled cooling fan circuit.

Electric-Drive Cooling Fan Circuit Diagnosis

Electric-drive cooling fan circuits vary depending on the vehicle and the model year. Always follow the diagnostic procedure in the vehicle manufacturer's service manual. If the electric-drive cooling fan does not operate at the coolant temperature specified by the vehicle manufacturer, engine overheating will result, especially at idle and lower speeds when airflow through the radiator is reduced (**Figure 3-16**). Following is a typical electric cooling fan diagnostic procedure:

1. Check both the fusible link and the fuse.
2. Do a visual check of the relay and connections.
3. Make sure conditions are right for the fan to operate (or use a scan tool to command the coolant fan on). Check for power at the relay terminals 1, 2, and 4, and check terminal 3 for continuity to ground with a DMM. (Do not use a test lamp or self-powered test light.)

   ■ If all terminals show correct voltage, check the fan motor wiring and fan motor.
   ■ If there is no power on terminal 1, check the power feed from the fuse block to the fan relay.
   ■ If there is no power on terminal 2, check the circuit from the fusible link to the fan relay.
   ■ If there is no ground on terminal 3, use the scan tool to see if the fan is showing the correct coolant temperature. If not, run the diagnostics for the coolant temperature sensor. If the coolant temp sensor is OK, then check the wiring between the PCM and relay. If the wiring is OK, replace the PCM.
   ■ If there is no power at terminal 4, remove the relay and, using a fused jumper wire, jumper terminals 2 and 4. If the fan runs replace the relay or if the fan is still inoperative, repeat steps 1 and 2.

Special Tools
12-volt test light
Jumper wire
DMM

PSIA = pounds/in.² absolute. *Absolute* means the gauge reads atmospheric pressure when disconnected from a pressure source. At sea level, a PSIA gauge would read 14.7 pounds/in.²

Special Tools

Vacuum gauge
Various lengths of vacuum hose
Tee-fittings

Vacuum is any pressure lower than atmospheric pressure.

Classroom Manual
Chapter 3, page 43

VACUUM TESTS

A vacuum test on an engine is one of those quick and easy tests that will tell you quite a bit about an engine. By monitoring the vacuum of an engine, a technician is able to see how well each of the engine's cylinders seals and how well the engine's systems support the intake stroke of each cylinder.

Each cylinder of an engine is a vacuum pump. A vacuum is produced during each intake stroke. Ideally, all cylinders of the engine operate under the same conditions and produce the same amount of power. When this does not exist, the extra power produced by the stronger cylinders will be partially consumed by the weaker cylinders. The cylinders' function as a vacuum pump should be equal. Each cylinder should produce the same amount of vacuum during its intake stroke. In order to produce the same amount of power, each cylinder must first produce equal amounts of vacuum.

Vacuum is formed during each piston's intake stroke. The better the piston is sealed in the cylinder, the lower the pressure will be in the cylinder at the end of the intake stroke. A poor-sealing cylinder will allow air to leak into the cylinder, keeping the pressure high. Remember, intake air and fuel enter the cylinder because of the pressure differential between the pressure inside the cylinder and the pressure of the air outside the engine. When a cylinder leaks, vacuum is low and the incorrect amount of air and fuel will enter the cylinder.

The amount and quality of the air present in the cylinder after the exhaust stroke also determine how much vacuum will be formed on the intake stroke. If high pressure remains after the exhaust stroke, vacuum in the cylinder will be low.

To measure engine vacuum, connect a vacuum gauge (**Figure 3-17**) to the intake manifold. An ideal reading on the gauge, with the engine idling, is a steady reading of 16 or more inches of mercury (in. Hg). Any movement of the gauge's needle indicates the cylinders are not producing the same amount of vacuum. The more the needle fluctuates, the greater the problem. If all of the cylinders are producing the same amount of vacuum but the amount is too low, the engine could be worn, or the valve timing could be off due to a damaged timing belt or chain. Refer to **Figure 3-18**.

EXHAUST GAS ANALYZER

An exhaust gas analyzer is sometimes called an *infrared emissions analyzer*.

An **exhaust gas analyzer** is another diagnostic tool that does not take too much time and is easy to use. The analyzer looks at the results of the combustion process. Since anything that affects the combustion process will affect the exhaust, the entire engine can be analyzed with the exhaust. The key to using an exhaust gas analyzer is the proper interpretation of the measured gases. It is important to note, that a working catalytic converter will burn most any emissions from the engine, so your readings may be lower than actual.

Since the most common, and sometimes most misunderstood, pollutants are carbon monoxide (CO) and hydrocarbons (HC), we will start with these two. Hydrocarbons are

Figure 3-17 Typical vacuum gauge.

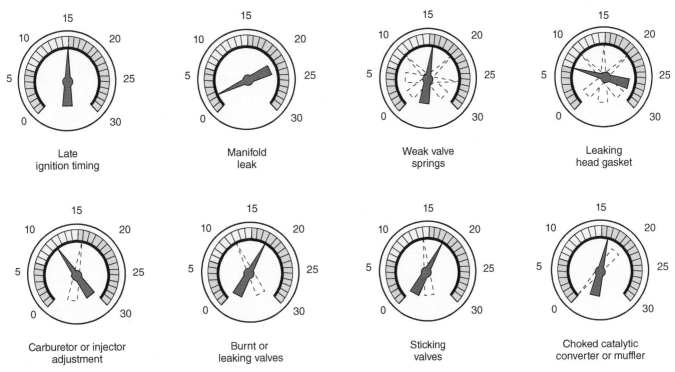

Figure 3-18 Various vacuum gauge readings and what the readings indicate.

unburned gasoline, so anything that keeps fuel from burning causes an excess of HC. Carbon monoxide is partially burned fuel. Partial burning is usually brought about by a lack of oxygen (or too much fuel). There are two ways to think about a rich and lean condition. A lean condition is a lack of fuel, or you could say too much oxygen. Of course, a rich condition is just the opposite.

As stated above, anything that keeps fuel from burning causes high HC emission. Some of these causes are ignition misfires, improper timing, excessively rich or lean mixtures, and low compression including any engine parts that could cause low compression. If the mixture is too lean, the tiny fuel droplets are too far apart for the spark to burn more than a few of them. This is called *lean misfire*. The unburned fuel leaves the vehicle as HC. If the mixture is too rich, there is not enough oxygen to burn the fuel. The combustion is basically smothered out. Once again, the HC goes out the tailpipe.

Carbon monoxide is associated with too much fuel and not enough air. This condition is caused by such things as too much fuel, a dirty air filter, leaking injectors, too much fuel pressure, defective sensors, or oil contaminated with gasoline. One way to think is that CO is formed by a lack of oxygen. If there was plenty of oxygen, CO would not form, and CO_2 would be produced instead. CO means that there was not enough oxygen to go around; the carbon took one oxygen atom instead of two.

Each condition listed in the following chart can affect one or more gases measured by the infrared exhaust gas analyzer. This chart is designed to be used as a quick reference guide to help understand the relationship between the five gases that are typically read by the analyzer. Depending on the problem or condition, the particular exhaust gas may increase, decrease, or stay about the same when one compares normal and abnormal readings. The information here is not intended to replace actual vehicle specifications. All tests are performed with the engine at normal operating temperature. Some computer controls can override the root fault conditions, therefore causing no apparent difference in the gas readings. Since different vehicles use various computerized engine management strategies, it is possible for the conditions listed below to cause varying or subtle test results.

Five-Gas Emissions Interrelationship Chart	Exhaust Gas Components As Sampled from the Tailpipe				
Condition	**HC ppm**	**CO %**	**O_2 %**	**CO_2 %**	**NO_x ppm**
Mechanical (Compression related)	▲	▼	▲	▼	▼
Electrical (Ignition Misfire)	▲	▼	▲	▼	▼
Rich Air/Fuel Mixture Condition	▲	▲	▼	▼	▼
Lean Air/Fuel Mixture Condition	▲	▼	▲	▼	▲
Engine Ping; Detonation; High Combustion Temperatures	▲	▼	▲	▼	▲
Cooler than Normal Operating Temperature	▲	▲	▼	▼	▼
EGR Malfunction	=	=	=	=	▲
Catalytic Converter Inactive	▲	▲	▲	▼	▲
Typical average preferred readings at idle speed with a normally functioning catalytic converter*	<50 ppm	<0.50%	<1.0%	>12%	See NO_x note below

Key: ▲ = increase in number; ▼ = decrease in number; = = no significant change in number

NO_x *Note:* Readings should be <400 ppm at idle unloaded and <900 ppm at 2,000 rpm loaded. The readings should be baselined with several known good vehicles with your own shop equipment. Refer to equipment manufacturer's procedures. NO_x readings may vary with equipment used and method of test performed. Check with the specific vehicle manufacturer and any state or federal regulations regarding specifications and testing.

Lower-than-normal CO_2 levels normally accompany higher-than-normal CO levels. Therefore, all of the possible causes for high CO should be suspect for lower-than-normal CO_2 levels. This problem may also be caused by exhaust gas sample dilution because of a leaking exhaust system and/or an excessively rich air-fuel ratio.

When the O_2 readings are lower than normal and CO readings are higher than normal, check for a rich air-fuel ratio, defective injectors (pintles not seating properly and dripping fuel), higher-than-specified fuel pressure, defective input sensors, a restricted PCV system, and/or a carbon canister purging at idle and low speeds.

When the O_2 readings are higher than normal and CO readings are lower than normal, suspect a lean air-fuel ratio, a vacuum leak, lower-than-specified fuel pressure, defective injectors, and defective input sensors.

Higher-than-normal NO_x levels are caused by excessive combustion chamber temperatures. This can be caused by anything from an overheated engine to an excessively lean air-fuel mixture. NO_x is formed by combustion; therefore, NO_x levels will tend to be lower with increased amounts of HC in the exhaust.

To pinpoint the exact cause of the improper exhaust gas analyzer readings, perform detailed tests of the individual systems and components that could cause the abnormal readings. The cause for readings that could be caused by the engine—lack of sealing, improper operating temperatures, and so on—should be checked prior to testing other systems.

Exhaust gas analyzers have a warm-up and calibration period that is usually about 15 minutes. Modern analyzers perform this calibration automatically. Always be sure the analyzer is calibrated properly so it provides accurate readings.

On most exhaust gas analyzers, a warning light is illuminated if the exhaust flow through the analyzer is restricted because of a plugged filter, probe, or hose. This warning light must be off before proceeding with the infrared exhaust analysis.

Always follow the recommended procedure in the vehicle manufacturer's service manual or the equipment operator's manual.

Quick Tests Using the Infrared Exhaust Gas Analyzer

The infrared exhaust gas analyzer can be used as a diagnostic tool as well as an emission-testing tool. Following is a list of quick tests using the five-gas infrared. Generally, these tests should be done with the engine at normal operating temperature with the transmission in park (neutral for manual), the parking brake set, and the wheels blocked. Please note that there are no hard-and-fast rules to these tests. They are intended as quick tests only and are not to be used in place of specific manufacturer's procedures where applicable.

1. *Engine manifold vacuum leaks.* Place the sample probe of the analyzer in the tailpipe, and observe the readings. In the case of a vacuum leak, you may notice an increased amount of O_2 and/or HC and a decrease of CO_2. To pinpoint the source of the vacuum leak, run the engine, and carefully release a small metered amount of propane along the intake manifold and vacuum-operated accessories. When the propane is drawn into the area of the leak, the infrared will detect the change. Give the process several seconds to occur. At the time the propane reacts on the vacuum leak, it momentarily enriches the mixture. You will notice a decrease of O_2 and HC, and perhaps a momentary increase in CO_2 and CO.

2. *Leaking injectors.* If you suspect the fuel injectors are leaking fuel after the engine is shut down, use the infrared exhaust analyzer to confirm this. With the engine off, place the sample probe in the intake manifold past the throttle plates (in fuel-injected engines). Observe the HC readings. You will notice the HC levels will rise as the infrared draws any excess fuel from the manifold at the start of the test. Gradually, the HC levels will decrease and should stay low. If the numbers start to increase again or otherwise remain high, fuel is continuing to enter the manifold. This could be due to a leaking injector.

3. *Fuel combustion efficiency test.* Use the infrared exhaust gas analyzer to monitor exhaust gases while running the engine at various speeds. You should expect CO and HC levels to go lower as rpm increases. Start at idle speed and note the readings of CO, HC, O_2, and CO_2. Allow the readings to stabilize. Next, increase to and maintain an engine speed of 1,500 rpm, allow the analyzer to stabilize, and note the readings. Now slowly increase the engine speed from 1,500 rpm to 2,500 rpm and maintain that speed, allowing readings to stabilize, then note the readings again. Compare the three test speed results. CO and HC should decrease with each increase of engine speed. CO_2 levels may increase, indicating an efficient combustion. If the CO and HC levels increase with higher engine rpm, there may be a fuel delivery or fuel control problem. This vehicle may have poor fuel economy and sluggish performance due to the mixture being too rich at city and highway speeds.

4. *Contaminated motor oil test.* If the crankcase becomes contaminated with gasoline (HC), premature engine wear and performance problems will result. Gasoline in a large volume can be the result of a rich mixture, leaking injectors, an engine misfire, mechanical problems, or the engine becoming flooded. Suspect a contaminated crankcase anytime these conditions exist. Of course, you will need to address the original problem first, then change the engine oil and oil filter afterward. To check for crankcase contamination, bring the engine to operating temperature and shut the engine off. Remove the oil filler cap and place the sample probe into the engine. Be sure to keep the probe from any oil in the engine. Allow the reading to stabilize, and observe the HCs. An allowable amount will be anywhere from 0 to 300 parts per million (ppm) HC. If the reading is 400 ppm or higher, the engine oil and oil filter need to be changed due to contamination. Please note that this is not intended to determine oil change intervals. It is merely a quick-test gauge to indicate crankcase contamination as a result of another problem.

5. *PCV test.* To test the PCV system, place the sample probe in the tailpipe and run the engine at an idle. Remove the PCV valve from the engine, and place your thumb over the end of the valve. The PCV valve is actually a calibrated vacuum leak. If you block off the valve, you should see a slight increase in CO. This quick test is checking airflow through the valve and hose. It does not confirm that the valve is the correct one.

6. *General emissions test.* Many states have an emission-testing program of one type or another. This quick test will usually not take the place of the state test; however,

it can help to identify a problem at a glance. Place the sample probe into the tail-pipe, and observe the readings while at idle and at 2,000 rpm. Ideally, the reading with a catalytic converter should be as follows: CO = <0.5 percent, HC = <50 ppm, O_2 = 0.5 to 3.0 percent, and CO_2 = >12 percent.

7. *Fuel enrichment test*. This test will verify proper fuel enrichment during a snap acceleration test. Place the sample probe in the tailpipe, and snap the throttle to wide-open throttle (WOT). Watch for at least a rise in CO without the O_2 rising prior to the CO increasing. The increasing CO indicates the extra fuel from the snap was burned. If you notice a weak CO increase or a rise in O_2 before the CO increased, the engine could be hesitating on acceleration. The hesitation could be due to dirty injectors or a lazy oxygen sensor.

8. *Combustion chamber leaks*. As mentioned earlier, the exhaust analyzer can be used to check for coolant leaks in the combustion chamber. Use extreme caution when working on a hot engine. Because you will be opening the radiator cap, it may be a good idea to wait for the engine to cool down first. Remove the radiator cap, and place the sample probe above the radiator filler neck. Do not allow the sample probe to come in contact with the coolant. Run the engine at a fast idle, and observe the CO_2 reading. If CO_2 is present in the cooling system, it means combustion gases are escaping into the cooling system, probably due to a leaking head gasket or similar failure. A cooling system pressure test can assist in locating the source of the leak.

9. *Locating a fuel leak*. Sometimes fuel leaks can be difficult to locate, especially when the fuel lines are routed in hard-to-see areas. The infrared exhaust gas analyzer can help locate the area of the fuel leak. With the engine off, guide the sample probe tip along the fuel line(s) from the fuel tank to the engine. Observe the HC as you move the sample probe along the route. A fuel leak will cause the HC reading to increase. This test can be particularly helpful if gasoline odor is detected but there are no visible signs of a leak. Be sure to check as far on top of the fuel tank as possible, and check the evaporative emission canister.

10. *Excessive valve-guide wear*. Worn internal engine components can cause compression-related problems and other symptoms such as excessive **blowby** or oil consumption. Although there are no firm numbers, CO_2 detected in the valve cover area while the engine is running is a good indication there is something worn internally. Test several known good engines of similar types to establish what may be considered a normal reading. The lower, the better. CO_2 is a by-product of combustion, and if it is present in the valve cover area it is because of worn valve guides or blowby from worn piston rings. These conditions usually have other symptoms associated with them; therefore, use other available testing methods to determine the source of the actual problem.

ENGINE POWER BALANCE TEST

The engine **power balance test** checks the efficiency of individual cylinders (see **Figure 3-19**). The power balance test can determine which cylinder is weak, and whether the problem is mechanical, electrical, or fuel related.

There are basically two ways to perform a cylinder balance test: manually, and with a scan tool using the capability of the onboard diagnostics.

Traditionally, the power balance test was performed manually using a **tachometer** while disconnecting plug wires or injectors one at a time. When a cylinder that was not contributing to cylinder power disconnected, the tachometer reading did not change. Once the problem cylinder was identified, the technician could begin to further isolate the problem by testing for fuel injection problems and ignition problems. Since 1996, OBD II vehicle have the capability to detect cylinder misfires as well as which cylinder is

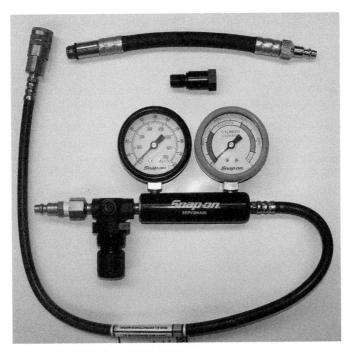

Figure 3-19 A cylinder leakage tester.

misfiring, so the easiest way to determine which cylinder is missing is to use the scan tool. The ECM or PCM can determine which cylinder is missing because it has a crank sensor that watches crankshaft speed and position. When a cylinder misfires, the crankshaft slows down slightly. The ECM or PCM can determine this slowing of the crankshaft, and determine which cylinder caused the crankshaft to slow down. A trouble code is set in the computer's memory that identifies the problem cylinder. Additionally, the scan tool can also be used in most cases to shut injectors down one at a time while comparing the rpm readings.

COMPRESSION TESTS

⚡ WARNING **If the injection system is not disabled during a compression test on a gasoline fuel-injected engine, the injectors continue injecting fuel into the intake ports during the compression test. These fuel vapors are discharged from the spark plug openings and, if ignited, may cause a serious explosion resulting in personal injury and/or property damage.**

Special Tools
Compression gauge
Remote starter switch
Squirt can of engine oil

The compression test checks the sealing qualities of the rings, valves, and combustion chamber. Operate the engine until it reaches normal operating temperature prior to conducting a compression test. **Photo Sequence 4** shows the typical procedure for conducting a compression test. The following are some guidelines to follow when conducting this test:

1. Disable the ignition system by disconnecting the positive primary wire from the ignition coil on a distributor ignition. Insulate this wire with electrician's tape so it does not contact the vehicle ground. On distributorless ignition systems, disconnect the complete coil connector on the coil or coil pack.
2. Disable the fuel injection system by shutting off the fuel pump. On many vehicles, disconnect the wires from the fuel pump relay to shut off the fuel pump. On General Motors products, disconnect the fuel pump wire at the fuel tank.

Classroom Manual
Chapter 3, page 43

Caution

Because a diesel engine has much higher compression than a gasoline engine, a special compression tester is required for a diesel engine.

Caution

If the injection system is not disabled during a compression test on a diesel engine, the injectors continue injecting fuel into the combustion chambers during the test. On a diesel engine, the compression is high enough to self-ignite the fuel in the cylinder in which the compression is tested. This self-ignition destroys the hose on the compression tester and possibly the tester.

3. Loosen the spark plugs, and blow any dirt from the plug recesses with an air **blowgun**. Remove all plugs.
4. Ensure that the throttle is propped open during the test.
5. Install the compression tester in a spark plug hole. Some compression testers are threaded into the spark plug openings, whereas other testers lock into position when the compression pressure is applied to them. Other testers must be held in the spark plug openings by hand.
6. Crank each cylinder through at least four compression strokes, noting the compression on the first puff and the final puff. Each of the four compression strokes can be observed as a definite increase on the gauge pointer reading.
7. Release the pressure from the compression tester, and follow the same procedure to obtain the compression reading on each cylinder. Record the reading obtained on each cylinder.

Compression Test Interpretation

Compare the compression readings to the vehicle manufacturer's compression specifications given in the service manual. If the compression readings on all the cylinders are equal to the specified compression, the readings are satisfactory. Higher-than-expected compression readings are typically caused by carbon buildup in the cylinders, which raises the **compression ratio**. If all of the cylinders have low readings, the engine is worn or the timing belt or chain is misaligned. When interpreting the results of this test, keep in mind that the test shows the cylinders' ability to compress air. The better a cylinder is sealed, the more air it can compress. Likewise, the higher the compression ratio is, the more the air in the cylinder will be compressed.

If the compression on one or more cylinders is lower than the specified compression, the valves or rings are likely worn. When the compression readings on a cylinder are low on the first stroke, increase to some extent on the next three strokes, but remain below the specifications, the rings are probably worn. If the reading is low on the first compression stroke and there is very little increase on the following three strokes, the valves are probably leaking. When all the compression readings are even but considerably lower than the specifications, the rings and cylinders are probably worn or the camshaft timing is wrong.

When the compression readings on two adjacent cylinders are lower than specified, the cylinder head gasket is probably leaking between the cylinders. Higher-than-specified compression readings usually indicate excessive carbon in the combustion chamber, such as on top of the piston.

If the compression reading on a cylinder is zero, there may be a hole in the piston, or an exhaust valve may be severely burned. When the zero compression reading is caused by a hole in a piston, there is excessive blowby from the crankcase. This blowby is visible at the PCV valve opening in the rocker cover if this valve is removed. Zero compression can also be caused by an intake valve not opening.

Wet Compression Test

If a cylinder compression reading is below specifications, a wet test can be performed to determine if the valves or rings are the cause of the problem. Squirt about 2 or 3 teaspoons of engine oil through the spark plug opening into the cylinder with the low compression reading. Crank the engine for several revolutions to distribute the oil around the cylinder wall. Repeat the compression test on the cylinder with the oil added in the cylinder. If the compression reading improves considerably, the rings or cylinders are worn. When there is very little change in the compression reading, one of the valves is leaking.

PHOTO SEQUENCE 4
Conducting a Cylinder Compression Test

P4-1 Before conducting a compression test, disable the ignition and the fuel injection system (if the engine is so equipped).

P4-2 Prop the throttle plate into a wide-open position. This will allow an unrestricted amount of air to enter the cylinders during the test.

P4-3 Remove all of the engine's spark plugs.

P4-4 Connect a remote starter button to the starter system.

P4-5 Many types of compression gauges are available. The screw-in type tends to be the most accurate and easiest to use.

P4-6 Carefully install the gauge into the spark plug hole of the first cylinder.

P4-7 Connect a battery charger to the car. This will allow the engine to crank at consistent and normal speeds that are needed for accurate test results.

P4-8 Depress the remote starter button and observe the gauge's reading after the first engine revolution.

P4-9 Allow the engine to turn through five revolutions, and observe the reading after the fifth. The reading should increase with each revolution.

PHOTO SEQUENCE 4 (CONTINUED)

P4-10 Readings observed should be recorded. After all cylinders have been tested, a comparison of cylinders can be made.

P4-11 Before removing the gauge from the cylinder, release the pressure from it using the release valve on the gauge.

P4-12 Each cylinder should be tested in the same way.

P4-13 After completing the test on all cylinders, compare them. If one or more cylinders is much lower than the others, continue testing those cylinders with the wet test.

P4-14 Squirt a few drops of oil into the weak cylinder(s).

P4-15 Reinstall the compression gauge into that cylinder and conduct the test.

P4-16 If the reading increases with the presence of oil in the cylinder, the most likely cause of the original low readings was poor piston ring sealing. Using oil during a compression test is normally referred to as a wet test.

Running Compression Test

The running compression test is a complement to the cranking compression test. The running compression test is performed on one cylinder at a time, with the compression gauge installed in the same manner as for the cranking compression test. The difference, of course, is that the engine is running at an idle. During a running compression test, the piston is moving approximately three times faster than cranking speed. This causes the compression test pressures to be lower compared to the cranking test. There are no firm specifications for running compression test results, and it is a relative test. The results of one cylinder reading need to be compared to the readings of the other cylinders. During a typical running compression test performed at idle, you could expect to see readings of 50 to 70 psi. If you slowly increase the rpm, the pressure will go down proportionately. The next step is to snap-accelerate the throttle. The rapidly increased throttle opening increases the amount of air into the cylinder, creating a sudden rise in compression pressures. This helps to reinforce the understanding of the principles of volumetric efficiency. The sudden increase in air intake results in higher compression pressures. The increased pressure with added fuel creates more speed. As mentioned, these tests are relative tests compared to the other cylinders. By reviewing test results, you can best determine a potential problem due to volumetric efficiency. For example, typical readings at idle may be at 60 psi and then drop to about 40 psi while holding the engine speed at 1,500 rpm. The snap acceleration reading should be approximately 80 percent of the cranking compression test results. If the snap acceleration readings are low, it could indicate an air intake restriction due to a valve-train problem or carbon buildup on the back of the intake valve. If the snap acceleration reading is high, it may indicate an exhaust-related restriction or exhaust valve problem. The purpose of the cranking compression test is to check for cylinder sealing integrity. The running compression test is done to check volumetric efficiency.

⚙ SERVICE TIP If you are attempting to squirt engine oil through a spark plug opening into the cylinder to perform a **wet compression test** and this opening is difficult to access, place a length of vacuum hose on the end of the squirt can nozzle. Install the other end of this hose in the spark plug opening.

A **wet compression test** is a compression test with a very small amount of oil added to the cylinder. Too much oil can cause the engine to become hydraulically locked and cause engine damage.

Engines have a tendency to rotate unless they are at top dead center (TDC) when the air is applied.

CYLINDER LEAKAGE TEST

The leakage test can be used to determine if a low compression reading is caused by worn rings, burned valves, or other combustion chamber leaks. During this test, a regulated amount of air from the shop air supply is forced into the cylinder with both exhaust and intake valves closed. The gauge on the leakage tester indicates the percentage of leakage in the cylinder. A gauge reading of 0 percent indicates no cylinder leakage; if the reading is 100 percent, the cylinder is not holding any air.

The engine should be at normal operating temperature for conducting the test. Remove the spark plugs from the engine. The cylinder to be tested has to be at top dead center of the compression stroke. If TDC is not reached, the air pressure from the tester will cause the engine to rotate. The rotation of the engine can cause a valve to open and give false readings. The following cylinder leakage test procedure can be used on most engines regardless of the type of ignition system or fuel system:

1. With the pressure regulator valve in the tester in the off position, connect the shop air supply to the tester. Adjust the pressure regulator valve to obtain a zero gauge reading if necessary.
2. Using the same methods described previously in the compression test, disable the ignition system and the fuel injection system.

3. Loosen the spark plug in the cylinder to be tested, and use an air gun to blow any dirt from the plug recess. Remove the spark plug from the cylinder to be tested.
4. Place your thumb over the spark plug hole in the cylinder to be tested. If the spark plug hole is not accessible, install a compression gauge in the spark plug hole of the cylinder to be tested.
5. Connect a remote control switch to the starter solenoid, and crank the engine a very small amount at a time. When compression is available at the spark plug hole, stop cranking the engine.
6. Connect the spark plug removed from the cylinder to the end of the disconnected spark plug wire. Attach a jumper wire from the spark plug case to ground on the engine.

⚠ WARNING When a remote control switch is used to crank the engine slowly with the ignition system and the fuel system in operation, the engine may start. Therefore, keep tools and hands away from the cooling fan and belts to avoid personal injury.

⚠ WARNING Never attempt to rotate the engine by installing a wrench on the crankshaft pulley bolt with the ignition system and fuel system in operation. Under this condition, the engine may start and rotate the wrench with the crankshaft. This action could result in serious personal injury and expensive vehicle damage.

7. Crank the engine a very small amount at a time with the remote control switch until the piston is at or near TDC on the compression stroke with both valves closed. As an alternative, you can place a TDC whistle in place of the spark plug. "Bump" the starter, causing the crankshaft to move small increments at a time. As the piston in the test cylinder begins to move to TDC on the compression stroke, the whistle will sound, indicating the piston is close to the TDC position.
8. Connect the leakage tester air hose to the cylinder (**Figure 3-20**). Various adapters are available to fit different spark plug threads.
9. If the reading exceeds 20 percent, check for air escaping from the tailpipe, the PCV valve opening, and the throttle body (**Figure 3-21**). Air escaping from the tailpipe indicates an exhaust valve leak. When the air is coming out of the PCV valve opening, the piston rings are leaking. An intake valve is leaking if air is escaping from the throttle body. In addition, check for air escaping into the cooling system or into an adjacent cylinder. This can indicate a head gasket or cylinder head leak.
10. Follow the same procedure in steps 2 to 10 to perform a leakage test on the other cylinders as needed.

Classroom Manual
Chapter 3, page 46

VALVE-TIMING CHECKS

If the timing belt or chain has slipped on the camshaft sprocket, the engine may fail to start because the valves are not properly timed in relation to the crankshaft. When the timing belt or chain has slipped only a few cogs on the camshaft sprocket, the engine has a lack of power and fuel consumption is excessive. Follow these steps to check the valve timing:

1. Remove the spark plug from number one cylinder and place your thumb on top of the spark plug hole. If this hole is not accessible, place a compression gauge in the opening. Disable the ignition system.
2. Crank the engine until compression is felt at the spark plug hole.
3. Connect a remote control switch to the starter solenoid terminal and the battery terminal on the solenoid. Slowly crank the engine until the timing mark lines up with the zero-degree position on the timing indicator. The number one piston is now at TDC on the compression stroke. On many engines, the timing mark is on the crankshaft pulley and the timing indicator is mounted above the pulley.

It may be easier to remove the timing belt cover to check valve timing on some overhead cam engines.

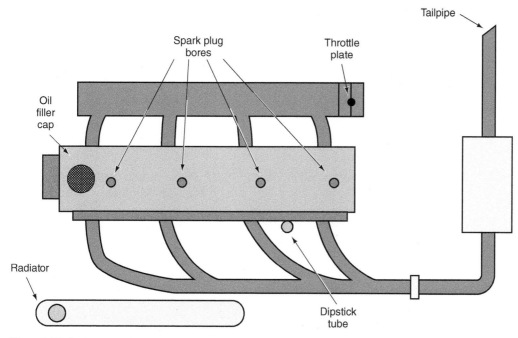

Figure 3-20 During a cylinder leakage test, air may be felt or heard leaking from these areas.

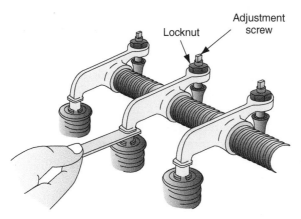

Figure 3-21 Adjusting valves with an adjusting screw and locknut on the pushrod end of the rocker arm.

4. Slowly crank the engine until the number one piston is now at TDC on the exhaust stroke.
5. Remove the rocker arm cover, and install a breaker bar and socket on the crankshaft pulley nut. Observe the valve action while rotating the crankshaft about 30 degrees before and after TDC on the exhaust stroke. In this crankshaft position, the exhaust valve should close a few degrees after TDC on the exhaust stroke, and the intake valve should open a few degrees before TDC on the exhaust stroke. This valve position with the piston at TDC on the exhaust stroke is called **valve overlap**. If the valves do not open properly in relation to the crankshaft position, the valve timing is incorrect. Under this condition, the timing chain or belt cover should be removed to check the position of the camshaft sprocket in relation to the crankshaft sprocket position. When the valve timing is incorrect, the timing chain or belt and/or sprockets must be replaced.

VARIABLE VALVE TIMING CONCERNS

Variable valve timing has the potential to cause drivability problems. The variable valve apply solenoids can fail to work effectively if the apply solenoids do not have clean oil supply. Low oil supply can obviously cause problems as well, with low oil the solenoids cannot function. Electrical wiring from the PCM to the apply solenoids can also cause problems. Of course, the VVT system also has diagnostic trouble codes associated with its operation.

Some generic trouble codes related to the VVT intake camshaft system (exhaust codes are similar) are:

P0010- VVT apply solenoid- Bank 1

P0011- Intake camshaft over-advanced - Bank one

P0012 Intake camshaft over-retarded-Bank one

P0020 VVT apply solenoid- Bank 2

P0021- Intake camshaft over-advanced bank two

P0022- Intake camshaft over-retarded-Bank two

A few possible causes, (not an exhaustive list):

- Apply solenoid wiring
- Faulty Solenoid
- Oil supply problems
- Faulty PCM
- Basic cam/crank timing problem, slipped chain
- Cam timing solenoid stuck mechanically or electrically
- Faulty PCM

VALVE ADJUSTMENT

Mechanical valve lifters are commonly called solid lifters or solid tappets.

The majority of engines do not have a periodic valve adjustment, although many older engines did. Some engines have **mechanical valve lifters** in place of **hydraulic valve lifters**, and these engines require a valve adjustment. Various car manufacturers recommend different valve-adjustment procedures depending on the engine. For example, on some engines the manufacturer recommends adjusting the valves with the engine cold, whereas other manufacturers recommend performing this adjustment with the engine at normal operating temperature. Always follow the vehicle manufacturer's recommended valve-adjustment procedure in the service manual.

The valves in each cylinder must be adjusted with the piston at TDC on the compression stroke and both valves closed.

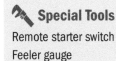

Special Tools

Remote starter switch
Feeler gauge

Some rocker arms have an adjustment screw and a lock screw on the pushrod end of the rocker arm. With the piston at TDC on the compression stroke, place a feeler gauge of the specified thickness between the rocker arm and the valve stem (**Figure 3-22**). If the valve adjustment is correct, the feeler gauge slides between the rocker arm and the valve stem with a light push fit. If necessary, loosen the locknut and turn the adjustment screw to obtain the specified valve clearance. Hold the adjustment screw with a slotted screwdriver and tighten the locknut. On most engines, the specified valve clearance is different on the exhaust valves compared to the intake valves. The reason for this is the fact that the exhaust valves run hotter, therefore they need more clearance to make up for expansion.

In some four-valve per cylinder (4V) engines, the rocker arms are mounted under the **overhead camshafts** and the camshaft lobes contact a friction pad on the top of the rocker arm. One end of the rocker arm contacts the valve stem, and the other end of the rocker arm is mounted in a pivot in the cylinder head. The manufacturer recommends measuring the valve clearance on these engines with the cylinder head temperature less than 100°F (38°C). Follow these steps for valve adjustment:

Adjusting pads, or shims, are located in some mechanical valve lifters to provide a valve adjustment.

1. Remove the rocker arm cover.
2. Disable the ignition system.
3. Crank the engine slowly until the number one piston is at TDC with the marks on the camshaft sprockets pointing upward (Figure 3-22).

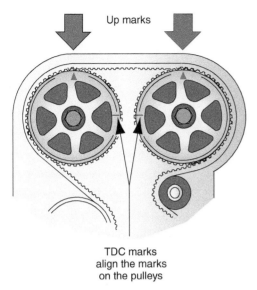

Figure 3-22 Number one piston at TDC and the camshaft sprocket marks facing upward.

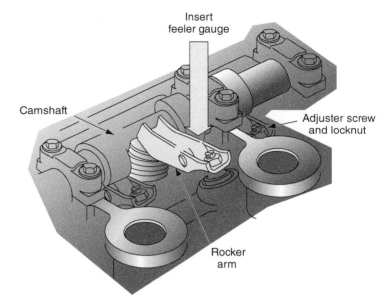

Figure 3-23 Measuring the valve clearance between the camshaft and the rocker arm.

4. Adjust the valves on the number one cylinder. Place the specified feeler gauge between the camshaft lobe and the upper side of the rocker arm (**Figure 3-23**). The feeler gauge should be a light push fit between these components. If the valve clearance requires adjusting, loosen the locknut and turn the adjusting screw in the rocker arm until the specified clearance is obtained. Tighten the locknut.

5. Rotate the crankshaft 180 degrees counterclockwise so the camshaft sprockets move 90 degrees. Under this condition, the marks on the camshaft sprockets face the exhaust side of the cylinder head (**Figure 3-24**). Adjust the valve clearance on the number three cylinder.

6. Rotate the crankshaft 180 degrees counterclockwise so the marks on the camshaft sprockets are facing downward (**Figure 3-25**). Adjust the valves on the number four cylinder.

7. Rotate the crankshaft 180 degrees counterclockwise so the marks on the camshaft gears are facing the intake side of the cylinder head (**Figure 3-26**). Adjust the valves on the number two cylinder.

8. Install a new rocker arm cover gasket and install the rocker arm cover. Tighten all bolts to the specified torque.

Some engines with hydraulic valve lifters do not have a valve adjustment. In these engines, if all the valve-train components are in satisfactory condition, the hydraulic valve lifters maintain zero valve clearance. However, worn or improperly serviced valve-train

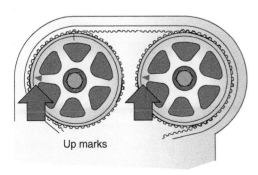

Figure 3-24 Camshaft sprockets positioned to adjust the valves in cylinder number three.

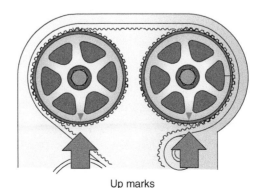

Figure 3-25 Camshaft sprockets positioned to adjust the valves in cylinder number four.

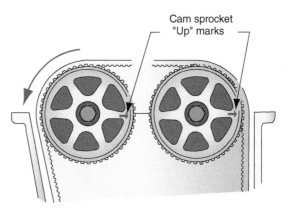

Cam sprocket
"Up" marks

Figure 3-26 Camshaft sprockets positioned to adjust the valves in cylinder number two.

components may affect the valve-lifter operation and result in valve-train noise or improper engine performance.

A valve-clearance measurement with the valve lifter bottomed is recommended on many valve trains that do not have a valve adjustment. With the piston at TDC on the compression stroke, push downward on the pushrod end of the rocker arm until the valve lifter bottoms. Measure the clearance between the end of the rocker arm and the valve stem. If this clearance is more than specified, valve-train components such as the rocker arm, pivot, or pushrod are worn, and this problem may cause a clicking noise at idle and low speed.

When this clearance is less than specified, the installed height of the valve in the cylinder head is too high. Excessive valve stem height is caused by grinding too much material from the valve seat or valve face without removing material from the end of the valve stem when the valves are reconditioned. If the valve-stem height is excessively high, the valve-lifter plunger can be bottomed in the lifter and the valve may not be allowed to close. Under this condition, cylinder misfiring occurs.

Some older but very popular engines have valve trains with hydraulic valve lifters and individual rocker arm pivots retained with self-locking nuts. These valve trains require an initial adjustment of the rocker arm nut to position the lifter plunger. With the valve closed, loosen the rocker arm nut until there is clearance between the end of the rocker arm and the valve stem. Slowly turn the rocker arm nut clockwise while rotating the pushrod (**Figure 3-27**). Continue rotating the rocker arm nut until the end of the rocker arm contacts the end of the valve stem and the pushrod becomes more difficult to turn. Continue turning the rocker arm nut clockwise the number of turns specified in the service manual. In some engines, this specification is one turn plus or minus one-quarter turn.

Valve-stem installed height is the distance from the valve-spring seat to the top of the valve stem when the valve is closed.

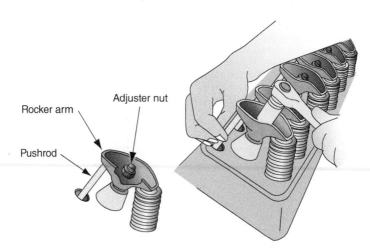

Adjuster nut

Rocker arm

Pushrod

Figure 3-27 Rocker arm adjustment with hydraulic lifters and individual rocker arms.

CASE STUDY

A Nissan was brought into our shop at school with the complaint that the engine had no power over about 3,000 rpm. The vehicle was test-driven; as the driver said, the engine ran fine to 3,000 rpm, but for all practical purposes would run no faster. The vehicle was brought into the shop; air filters, fuel pressure and volume, ignition testing, and a vacuum test were performed. The vacuum was low, only about 14 in. Hg at idle. Some of the students thought the exhaust was restricted, but after checking it was found to be OK. Next a compression test was run, with all cylinders at specification. After some discussion, the students performed a running compression test. The results were very low on all cylinders, especially at higher rpm. It was found that the cam timing was off about one tooth. Some of the students wondered why the regular compression test was fine, actually at specification. The reason: cranking compression is much slower, so even though the valves were not open at exactly the right time, the cylinders had time to fill. The running compression test was a better reflection of actual conditions. Of course, the regular compression test will usually catch valve-timing problems; in this case, the timing was just not that far off, except when pushed to a higher RPM, and cylinder fill became a bigger factor.

ASE-STYLE REVIEW QUESTIONS

1. Two technicians were discussing oil leak diagnosis,

   Technician A says that special dye is available to pinpoint the location of oil leaks.

   Technician B says that the customer may have to drive the vehicle for a few days for the dye to show up.

   Who is correct?

   A. A only

   B. B only

   C. Both A and B

   D. Neither A nor B

2. While discussing engine coolant leaks,

   Technician A says a leaking head gasket lowers engine operating temperature.

   Technician B says a leaking heater core may cause windshield fogging.

   Who is correct?

   A. A only

   B. B only

   C. Both A and B

   D. Neither A nor B

3. While discussing the diagnosis of engine exhaust color,

   Technician A says bluish-white smoke in the exhaust during deceleration may be caused by a rich air-fuel ratio.

   Technician B says bluish-white smoke in the exhaust during deceleration may be caused by loose camshaft bearings.

   Who is correct?

   A. A only

   B. B only

   C. Both A and B

   D. Neither A nor B

4. While discussing exhaust color and odor diagnosis,

   Technician A says black smoke in the exhaust during acceleration indicates a lean air-fuel mixture.

   Technician B says gray-white smoke from the exhaust may be caused by a coolant leak into the combustion chambers.

   Who is correct?

   A. A only

   B. B only

   C. Both A and B

   D. Neither A nor B

5. A technician is performing a compression test. Which statement below is least likely to be true?

   A. All cylinders with higher-than-normal readings could be caused by excessive carbon buildup.

   B. All cylinders reading even but lower than normal may be caused by a slipped timing chain or belt.

   C. Low readings on two adjacent cylinders may be caused by a blown head gasket.

   D. A low reading on one cylinder may be caused by a vacuum leak in that cylinder.

6. While discussing an engine power balance test,
Technician A says the engine may have an ignition problem on the weak cylinder.
Technician B says the engine may have a mechanical condition on the weak cylinder.
Who is correct?
A. A only   C. Both A and B
B. B only   D. Neither A nor B

7. While discussing an engine compression test,
Technician A says the engine should be cranked until five compression strokes occur during the compression test on each cylinder.
Technician B says the gasoline fuel injection system should be left in operation during a compression test.
Who is correct?
A. A only   C. Both A and B
B. B only   D. Neither A nor B

8. While discussing the cylinder leakage test,
Technician A says during a cylinder leakage test, if the air escapes from the tailpipe, an intake valve is leaking.

Technician B says during a cylinder leakage test the piston must be at TDC on the exhaust stroke.
Who is correct?
A. A only   C. Both A and B
B. B only   D. Neither A nor B

9. While discussing engine noise diagnosis,
Technician A says loose main bearings cause a light rapping noise during acceleration.
Technician B says piston slap causes a hollow rapping noise that is most noticeable on acceleration with a cold engine.
Who is correct?
A. A only   C. Both A and B
B. B only   D. Neither A nor B

10. While discussing engine oil pressure,
Technician A says low oil pressure can be caused by a leaking oil pump pickup pipe.
Technician B says low oil pressure can be caused by loose camshaft bearings.
Who is correct?
A. A only   C. Both A and B
B. B only   D. Neither A nor B

ASE CHALLENGE QUESTIONS

1. *Technician A* says an exhaust gas analyzer can be used to find the area of a fuel leak.
Technician B says a high carbon monoxide (CO) reading indicates the area of the leak.
Who is correct?
A. A only   C. Both A and B
B. B only   D. Neither A nor B

2. The customer states the coolant level must be topped off every few days. There are no visible leaks.
Technician A says this may be caused by a bad radiator cap.
Technician B says a stuck-open thermostat will cause this problem.
Who is correct?
A. A only   C. Both A and B
B. B only   D. Neither A nor B

3. Main bearing wear is being discussed.
Technician A says a heavy thumping knock at irregular intervals during acceleration may be caused by worn main thrust bearings.

Technician B says this noise indicates loose or worn main bearings.
Who is correct?
A. A only   C. Both A and B
B. B only   D. Neither A nor B

4. *Technician A* says the vacuum test is one of those quick and easy tests that can tell you a lot about the engine.
Technician B says the vacuum test measures the sealing ability of the engine.
Who is correct?
A. A only   C. Both A and B
B. B only   D. Neither A nor B

5. A vacuum leak is suspected for a rough idle concern. Using a five-gas infrared exhaust analyzer,
Technician A says O_2 will be higher than normal.
Technician B says CO will be higher than normal.
Who is correct?
A. A only   C. Both A and B
B. B only   D. Neither A nor B

Name _____   Date _____

COOLING SYSTEM INSPECTION AND DIAGNOSIS

Upon completion of this job sheet, you should be able to inspect the entire cooling system and determine the cause of abnormal conditions.

ASE Education Foundation Correlation

This job sheet addresses the following **MLR** task: I. Engine repair; C. Lubrication and Cooling Systems,

Task #1   Perform cooling system pressure and dye tests to identify leaks; check coolant condition and level; inspect and test radiator, pressure cap, coolant recovery tank, heater core, and galley plugs; determine needed action. **(P-1)**

This job sheet addresses the following **AST/MAST** task: I. Engine repair; D. Lubrication and Cooling Systems,

Task #1   Perform cooling system pressure and dye tests to identify leaks; check coolant condition and level; inspect and test radiator, pressure cap, coolant recovery tank, heater core, and galley plugs; determine needed action. **(P-1)**

Tools and Materials

- A vehicle
- Clean cloth rag
- Cooling system pressure tester
- Service manual for the above vehicle

Describe the vehicle being worked on:

Year _____   Make _____

Model _____   VIN _____

Procedure   **Task Completed**

1. Inspect the following components and describe their condition:

   Radiator _____

   Radiator hoses _____

   Hose clamps _____

   Heater hoses _____

   Overflow tank _____

   Cooling fan _____

   Drive belts _____

   Galley plugs _____

   Any obvious defects should be corrected before continuing with testing the system.

2. Remove the radiator cap. Make sure the engine is cooled down. *Do not* do this if the   ☐
   engine is warm.

3. With a clean rag, wipe off the radiator filler neck, and then inspect it. Is the sealing area free of accumulated dirt, nicks, or anything that would prevent the radiator cap from sealing good? ☐ Yes ☐ No

   If yes, describe the condition: _____

   _____

   _____

4. Check the locking tabs and/or threads on the filler neck for damage. Describe their condition.

_____

5. Inspect the overflow tank, hose, and connections. Look for cracks or other problems that may not allow the system to seal. Describe the condition of the overflow tank and connecting hose.

_____

_____

6. Run a wire through the overflow tube or hose to clear any obstructions. ☐

7. What is the pressure rating of the radiator cap? _____ psi

8. Using the proper adapter, attach the cooling system pressure tester to the radiator cap. Pump the tester to bring the pressure to the rating of the radiator cap.

   Are you able to bring the pressure to that amount?

   ☐ Yes ☐ No

   Once you are at that amount, does the pressure hold?

   ☐ Yes ☐ No

   If the cap does not hold the specified pressure, it should be replaced. Slowly release the pressure from the tester and cap.

9. Attach the cooling system pressure tester to the radiator filler neck. Pump the tester to bring the pressure to the rating of the radiator cap.

   Are you able to bring the pressure to that amount?

   ☐ Yes ☐ No

   Once you are at that amount, does the pressure hold?

   ☐ Yes ☐ No

   If the system does not hold the specified pressure, the system leaks. Look over the cooling system to find the source of the leak. If the pressure drops quickly to zero, it may be necessary to keep pumping the tester to apply pressure to the system and to find the leak. If the system held pressure, there are no leaks. Then carefully relieve the pressure from the system and tester by following the procedure given by the tester's manufacturer.

10. Top off the coolant to bring it to the proper level. ☐

11. Install the radiator cap. ☐

Instructor's Response

_____

_____

_____

Name _____   Date _____

CONDUCT A CYLINDER POWER BALANCE TEST

Upon completion of this job sheet, you should be able to conduct a cylinder power balance test and accurately interpret the results.

ASE Education Foundation Correlation

This job sheet addresses the following **MLR** task: VIII. Engine Performance; A. General:

Task #3   Perform cylinder power balance test; document results. **(P-2)**

This job sheet addresses the following **AST/MAST** task: VIII. Engine Performance: B. Computerized Engine Controls Diagnosis and Repair,

Task #3   Perform active tests of actuators using a scan tool; determine necessary action. **(P-2)**

This job sheet addresses the following **AST/MAST** task: VIII. Engine Performance: A. General: Engine Diagnosis

Task #6   Perform cylinder power balance test; determine necessary action. **(P-2)**

Tools and Materials

- Scan tool

Describe the vehicle being worked on:

Year _____ Make _____

Model _____ VIN _____

Engine type and size _____ Firing order _____

Procedure

Task Completed

1. Describe the general running condition of the engine.

   _____

   _____

2. Check for any current or history trouble codes. Are any codes present?

   ☐ Yes ☐ No

   If any codes are present, are they misfire codes? ☐ Yes ☐ No

3. Select the scan tool injector balance test. The injector balance test will allow you to isolate individual cylinders' contribution.   ☐

4. Using the scan tool, look at the RPM drop for each cylinder below when the injector for each cylinder is canceled. Record the results below:

Cylinder No.	1	2	3	4	5	6	7	8
RPM loss	___	___	___	___	___	___	___	___

5. Describe what is indicated by the results of this test:

   _____

   _____

6. Briefly explain what is actually being measured by the balance test:

_____

_____

7. Based on the test results, describe the condition of the engine. Does it agree with your original description of the engine?

_____

_____

8. How and why would the readings be different if the camshaft intake lobes for number two cylinder were severely worn?

_____

_____

Instructor's Response

_____

_____

_____

Name _____   Date _____

CONDUCT A CYLINDER CRANKING COMPRESSION, RUNNING COMPRESSION, AND LEAKAGE TEST

Upon completion of this job sheet, you should be able to conduct a cylinder cranking compression and cylinder leakage test on an engine and interpret the results.

ASE Education Foundation Correlation

This job sheet addresses the following **MLR** tasks: VIII. Engine Performance; A. General,

Task #4   Perform cylinder cranking and running compression tests; document results. **(P-1)**

VIII.A.5   Perform cylinder leakage test; document results. **(P-1)**

This job sheet addresses the following **AST/MAST** tasks: VIII. Engine Performance, A. General: Engine Diagnosis

Task #7   Perform cylinder cranking and running compression tests; determine necessary action. **(P-1)**

Task #8   Perform cylinder leakage test; determine necessary action. **(P-1)**

Tools and Materials

• Compression gauge
• Remote starter
• Cylinder leakage tester
• Shop air supply
• Safety glasses

Describe the vehicle being worked on:

Year _____ Make _____

Model _____ VIN _____

Engine size _____ No. of cyl. _____ Compression ratio

_____ : _____

Expected compression readings: _____

Source of information: _____

Describe the general running condition of the engine.

_____

What does a compression test actually measure?

_____

Why would it be wise to connect a battery charger to the vehicle while conducting a compression test?

_____

_____

Procedure

Perform the following tasks exactly as they are written.

1. Using the proper procedures, conduct a dry compression test on the engine and record the results below:

   _____ _____ _____ _____ _____ _____ _____ _____
     1   2   3   4   5   6   7   8

   Summarize and explain the test results.

   _____

   _____

2. Using the proper procedures, conduct a wet compression test on the engine and record the results below:

   _____ _____ _____ _____ _____ _____ _____ _____
     1   2   3   4   5   6   7   8

   Summarize and explain the test results and compare them to the dry results. Why are they the same or different?

   _____

   _____

3. Reinstall all of the spark plugs, except one. Repeat the compression test on the cylinder without a spark plug.

   Your readings now: _____

   Summarize the results, and state why they are different or the same as the previous test.

   _____

4. Running Compression Test

   For this test, you will have to remove the valve stem that holds the compression pressure. Or you can simply relieve the pressure with the pressure relief valve every five or six puffs. With all plugs installed except one, install the gauge and ground the disconnected plug wire and disconnect the injector for the cylinder you are testing.

   Using the proper procedures, conduct a running compression test on the engine at idle and at 1,500 rpm, and record the results below:

   Idle running compression:

   _____ _____ _____ _____ _____ _____ _____ _____
     1   2   3   4   5   6   7   8

   At 1,500 rpm:

   _____ _____ _____ _____ _____ _____ _____ _____
     1   2   3   4   5   6   7   8

5. How can you explain the difference between the normal compression test and the running compression test?

   _____

6. How could a running compression test detect problems not evident in a cranking compression test?

   _____

7. What does a cylinder leakage test actually measure?

    _____

    _____

    Using the proper procedures, conduct a cylinder leakage test on the engine, and then record the results below:

Cylinder No.	Reading	Point of leakage
1	_____	_____
2	_____	_____
3	_____	_____
4	_____	_____
5	_____	_____
6	_____	_____
7	_____	_____
8	_____	_____

    Summarize and explain the results of the test below.

    _____

    _____

8. Based on the results from the compression and leakage tests, describe the general condition of the engine (be sure to explain why).

    _____

    _____

9. Does the above summary agree with your original description of the running condition of the engine? Why or why not?

    _____

    _____

10. How and why would the readings on the compression and leakage be different if the camshaft intake lobes for number two cylinder were severely worn?

    _____

    _____

Instructor's Response

_____

_____

Name _____   Date _____

VERIFYING THE CONDITION OF AND INSPECTING AN ENGINE

Upon completion of this job sheet, you will be able to identify and interpret engine concerns; diagnose engine noises and vibrations; and diagnose the cause of excessive oil consumption, unusual engine exhaust color, odor, and sounds.

ASE Education Foundation Correlation

This job sheet addresses the following **AST/MAST** tasks: VIII. Engine Performance; A. General: Engine Diagnosis,

Task #1	Identify and interpret engine performance concerns; determine necessary action. **(P-1)**
Task #3	Diagnose abnormal engine noises or vibration concerns; determine necessary action. **(P-3)**
Task #4	Diagnose the cause of excessive oil consumption, coolant consumption, unusual exhaust color, odor, and sound; determine needed action. **(P-2)**

Tools and Materials

- A roadworthy vehicle
- Stethoscope
- Protective clothing
- Goggles or safety glasses with side shields

Describe the vehicle being worked on:

Year _____ Make _____

Model _____ VIN _____ Engine type and size _____

Procedure

Verify Engine Condition

1. Verifying the customer's concern is typically the first step you should take when diagnosing a problem. If the owner of the vehicle stated a concern, describe it. If there are no customer complaints, describe the general running condition to the best of your knowledge. The concern may be one of performance, smoke, leaks, or noises. In your answer, be sure to completely describe the condition and state when, where, and how the condition occurs.

   _____

   _____

   _____

   _____

2. Verify the complaint. Describe what you will do in an attempt to duplicate the concern. If necessary, road test the vehicle under the same conditions that are present when the problem normally occurs. Include in this description conditions that will explain when, where, and how the concern occurs.

   _____

   _____

   _____

   _____

3. Often, a customer may notice poor performance only during one condition, even though the problem may exist under other conditions as well. Also, observing the performance of an engine during a variety of modes of operation may let you know what is working fine and what does not need to be tested further. Describe how the engine performed during the following conditions.

A.   Starting:

_____

B.   Idling:

_____

C.   Slow acceleration:

_____

D.   Slow cruise:

_____

E.   Slow deceleration:

_____

F.   Heavy acceleration:

_____

G.   Highway cruise:

_____

H.   Fast deceleration:

_____

I.   Shut down:

_____

4. Based on the results of the preceding checks, what engine systems do you think should be checked to find the cause of the customer's complaint?

_____

_____

5. If an engine uses excessive oil and there is no evidence of leaks, the oil may be burning in the combustion chambers. If excessive amounts of oil are burned in the combustion chambers, the exhaust contains blue smoke, and the spark plugs may be fouled with oil. Excessive oil burning in the combustion chambers may be caused by worn rings and cylinders or worn valve guides and valve seals. Remove the spark plugs and describe the condition of each.

No. 1 _____

No. 2 _____

No. 3 _____

No. 4 _____

No. 5 _____

No. 6 _____

No. 7 _____

No. 8 _____

6. A loss of coolant with no visible leaks can indicate an internal coolant leak. A whitish exhaust may be indicative of this sort of coolant leak if it is caused by a leaking head gasket or cracked engine components. Take a look at the exhaust while the engine is idling and describe it.

_____

_____

Engine Exhaust Diagnosis

1. Some engine problems may be diagnosed by the color, smell, or sound of the exhaust. If the engine is operating normally, the exhaust should be colorless. In severely cold weather, it is normal to see a swirl of white vapor coming from the tailpipe, especially when the engine and exhaust system are cold. This vapor is moisture in the exhaust, which is a normal by-product of the combustion process. Carefully observe the exhaust from your test vehicle and describe what you see and/or smell during the following operating conditions.

    A.   Cold start-up:

    _____

    B.   Cold idle:

    _____

    C.   Warm start-up:

    _____

    D.   Warm idle:

    _____

    E.   Snap-throttle open:

    _____

    F.   Snap-throttle closed:

    _____

2. Based on the preceding observations, what are your conclusions? (Use the explanations that follow to guide your thoughts.)

_____

_____

If the exhaust is blue, excessive amounts of oil are entering the combustion chamber, and this oil is being burned with the fuel. When the blue smoke in the exhaust is more noticeable on deceleration, the oil is likely getting past the rings into the cylinder. If the blue smoke appears in the exhaust immediately after a hot engine is restarted, the oil is likely to be leaking down the valve guides.

• If black smoke appears in the exhaust, the air-fuel mixture is too rich. A restriction in the air intake, such as a plugged air filter, may be responsible for a rich air-fuel mixture.

• Gray smoke in the exhaust may be caused by coolant leaking into the combustion chambers. This may be most noticeable when the engine is first started after it has been shut off for over half an hour.

• A strong sulfur smell in the exhaust indicates a rich air-fuel mixture.

3. You may have noticed a change in sound during the preceding test. If you did, describe the sound change and operating mode in which the sound changed.

_____

_____

_____

_____

4. Use these guidelines to determine the possible cause of the sound. Then, state your best guess for the cause of the noise.

- When the engine is idling, the exhaust from the tailpipe should have a smooth, even sound.
- If, during idle, the exhaust emits a "puff" sound at regular intervals, a cylinder may be misfiring.
- When this sound is present, check the engine's ignition and fuel systems, and the engine's compression.
- If the vehicle makes excessive exhaust noise while the engine is accelerated, check the exhaust system for leaks.
- A small exhaust leak may cause a whistling noise when the engine is accelerated.
- If the exhaust system produces a rattling noise when the engine is accelerated, check the muffler and catalytic converter for loose internal components.
- When the engine makes a wheezing noise at idle or while the engine is running at a higher rpm, check for a restricted exhaust system.

_____

_____

_____

Engine Noise Diagnosis

1. Sounds from the engine itself can help you locate engine problems or help you identify a weakness in the engine before it becomes a big problem. Long before a serious engine failure occurs, the engine usually makes warning noises. Engine defects such as damaged pistons, worn rings, loose piston pins, worn crankshaft bearings, worn camshaft lobes, and loose and worn valve train components usually produce their own peculiar noises. Certain engine defects also cause a noise under specific engine operating conditions. Because it is sometimes difficult to determine the exact location of an engine noise, a stethoscope may be useful. The stethoscope probe is placed on, or near, the suspected component, and the ends of the stethoscope are installed in your ears. The stethoscope amplifies sound to assist in noise location. When the stethoscope probe is moved closer to the source of the noise, the sound is louder in your ears. If a stethoscope is not available, what can be safely used to amplify the sound and help locate the source of the noise?

_____

 Caution

When placing a stethoscope probe in various locations on a running engine, be careful not to catch the probe or your hands in moving components such as cooling fan blades and belts.

2. Because a lack of lubrication is a common cause of engine noise, always check the engine oil level and condition prior to noise diagnosis. Carefully observe the oil for contamination by coolant or gasoline. Check the oil level and condition on your test vehicle, and then record your findings.

_____

_____

3. During the diagnosis of engine noises, always operate the engine under the same conditions as those that are present when the noise ordinarily occurs. Remember that aluminum engine components such as pistons expand more when heated than cast-iron alloy components do. Therefore, a noise caused by a piston defect may occur when the engine is cold but might disappear when the engine reaches normal operating temperature. If the customer has an engine noise concern, describe the noise and state when it occurs.

_____

_____

_____

4. Duplicate the condition at which the noise typically occurs and describe all that you hear as you listen to the engine.

_____

_____

5. If you verified the customer's concern, use a stethoscope to find the spot where the noise is the loudest. Describe where this is.

_____

_____

6. What could be causing the noise to be loud at that spot?

_____

_____

7. To help you understand and use noise as a diagnostic tool, you will be given a description of an engine noise. Using your knowledge and any resources you have handy (such as your textbook), identify the conditions or problems that would cause each of the following noises:

A. A hollow, rapping noise that is most noticeable on acceleration with the engine cold. The noise may disappear when the engine reaches normal operating temperature.

_____

B. A heavy thumping knock for a brief time when the engine is first started after it has been shut off for several hours. This noise may also be noticeable on hard acceleration.

_____

C. A sharp, metallic, rapping noise that occurs with the engine idling.

_____

D. A thumping noise at the back of the engine.

_____

E. A rumbling or thumping noise at the front of the engine, possibly accompanied by engine vibrations. When the engine is accelerated under load, the noise is more noticeable.

_____

F. A light rapping noise at speeds above 35 mph (21 kph). The noise may vary from a light to a heavier rapping sound, depending on the severity of the condition. If the condition is very bad, the noise may be evident when the engine is idling.

_____

G. A high-pitched clicking noise is noticeable in the upper cylinder area during acceleration.

_____

H. A heavy clicking noise is heard with the engine running at 2,000 to 3,000 rpm. When the condition is severe, a continuous, heavy clicking noise is evident at idle speed.

_____

I. A whirring and light rattling noise when the engine is accelerated and decelerated. Severe cases may cause these noises at idle speed.

_____

J. A light clicking noise with the engine idling. This noise is slower than the piston or connecting rod noise and is less noticeable when the engine is accelerated.

_____

K. A high-pitched clicking noise that intensifies when the engine is accelerated.

_____

L. A noise that is similar to marbles rattling inside a metal can. This noise usually occurs when the engine is accelerated.

_____

Instructor's Response

_____

_____

_____

_____

Name _____   Date _____

VERIFYING ENGINE TEMPERATURE

Upon completion of this job sheet, you will be able to verify the operating temperature of an engine and diagnose the cause of abnormal temperatures.

ASE Education Foundation Correlation

This job sheet addresses the following **MLR** task: VIII. Engine Performance; A. General

Task #6   Verify engine operating temperature. **(P-1)**

This job sheet addresses the following **AST/MAST** task: VIII. Engine Performance; A. General Engine Diagnosis,

Task #10   Verify engine operating temperature; determine needed action. **(P-1)**

Tools and Materials

- Scan tool
- Service information
- Pyrometer
- Protective clothing
- Goggles or safety glasses with side shields

Describe the vehicle being worked on:

Year _____ Make _____

Model _____ VIN _____

Engine type and size _____

Procedure

1. Engines are designed to run within a narrow range of temperatures. When they operate within that range, they are running efficiently. Typically, abnormal temperatures are only noticed by the owner when there is insufficient heat inside the vehicle during cold weather or when steam rolls out from under the hood. Why would an engine that is running at lower than normal temperatures be less efficient?

   _____

   _____

   _____

2. Name the part of the cooling system that has the responsibility for maintaining and regulating the temperature of the coolant.

   _____

   _____

3. Most vehicles built after 2002 have a monitor for the engine thermostat. What are the conditions for causing the monitor to set a trouble code? List your findings here.

   _____

   _____

4. Higher than normal operating temperatures will also affect the engine's efficiency. Name five problems that may cause the engine to run hot.

_____

_____

_____

_____

_____

5. There are times when the engine seems to be running colder or hotter than normal, but the temperature gauge reads normally. In these cases, the engine's coolant temperature should be measured and compared to the reading on the vehicle's temperature gauge. The temperature of an engine should also be verified when the activity of the powertrain computer seems to indicate that it is responding to a temperature that is different than what the temperature gauge and/or engine coolant temperature sensor indicates. Engine temperature is commonly measured with a pyrometer. Before measuring, look up the temperature specification for the engine's thermostat. The specification is:

_____

The source of the specification was:

_____

6. Operate the engine for 15 minutes at idle speed. Then, read the operating temperature reading on the scan tool. The reading is:

_____

7. Read the temperature indicated on the thermometer. The reading measured on the pyrometer is:

_____

8. Compare the reading with the specification and state the difference. Explain why it may be different.

_____

_____

_____

Instructor's Response

_____

_____

_____

_____

Name _____   Date _____

VERIFYING CAMSHAFT TIMING

Upon completion of this job sheet, you will be able to verify camshaft timing according to the manufacturer's specifications.

ASE Education Foundation Correlation

This job sheet addresses the following **AST/MAST** task: VIII. Engine Performance; A. General: Engine Diagnosis,

Task #11   Verify correct camshaft timing including engines equipped with variable valve timing systems (VVT). **(P-1)**

Tools and Materials

- Spark plug socket
- Miscellaneous hand tools
- Compression gauge
- Large breaker bar
- Remote starter switch
- New rocker arm or camshaft cover gasket
- Flashlight
- Protective clothing
- Goggles or safety glasses with side shields

Describe the vehicle being worked on:

Year _____ Make _____

Model _____ VIN _____

Engine type and size _____

For vehicles with variable valve timing (VVT) start here. If not equipped, start at the next section.

1. Are there any Technical service bulletins listed for the VVT on this vehicle? If so list them below

    _____

2. What are the symptoms of a failing VVT system?

    _____

    _____

3. How can a lack of maintenance cause failure of the VVT system?

    _____

    _____

4. Is the VVT system covered by DTCs? If so list them here?

    _____

Procedure

Task Completed

1. If the timing belt or chain has slipped on the camshaft sprocket, the engine may fail to start because the valves are not properly timed in relation to the crankshaft. (Sometimes the engine turns over faster than normal due to a lack of compression.) When the timing belt or chain has only slipped a few cogs on the camshaft sprocket, the engine has a lack of power, and fuel consumption is excessive.   ☐

2. What results of another engine performance test could suggest you need to physically verify camshaft timing?

    _____

_____

_____

_____

3. To check the valve timing, begin by removing the spark plug from the number one cylinder. ☐

4. Disable the ignition and fuel injection system. (See the service information.) Block the drive wheels and apply the emergency brake; manual transmissions should be in neutral. ☐

5. Connect a remote control switch to the starter solenoid terminal and the battery terminal on the solenoid. ☐

6. Place your thumb on top of the spark plug hole at cylinder number one. If this hole is not accessible, place a compression gauge in the opening. ☐

7. Crank the engine until compression is felt at the spark plug hole. Then, slowly crank the engine until the timing mark lines up with the zero-degree position on the timing indicator. (If the engine does not use a timing indicator, you may have to remove the timing belt cover to verify camshaft-to-crankshaft timing.) The number one piston is now at TDC on the compression stroke. On many engines, the timing mark is on the crankshaft pulley, and the timing indicator is mounted above the pulley. ☐

8. Slowly crank the engine for one revolution until the timing mark lines up with the zero-degree position on the timing indicator. The number one piston is now at TDC on the exhaust stroke. ☐

9. Remove the rocker arm or camshaft cover and install a breaker bar and socket on the crankshaft pulley nut. Observe the valve action while rotating the crankshaft about 30 degrees before and after TDC on the exhaust stroke. In this crankshaft position, the exhaust valve should close a few degrees after TDC on the exhaust stroke, and the intake valve should open a few degrees before TDC on the exhaust stroke. Is this what you observed?

_____

10. If the valves did not open properly in relation to the crankshaft position, the valve timing is not correct. What should you do to correct it?

_____

_____

_____

11. If the timing was correct, reinstall the rocker arm or camshaft cover with a new gasket. Tighten the attaching bolts to the proper specification. The recommended torque is:

_____

12. Reinstall the spark plug and tighten it to the proper specification. The recommended torque is:

_____

Instructor's Response

_____

_____

_____

_____

CHAPTER 4

BASIC ELECTRICAL TESTS AND SERVICE

Upon completion and review of this chapter, you should be able to:

- Diagnose electrical problems by logic and symptom description.
- Perform troubleshooting procedures using meters, test lights, and jumper wires.
- Inspect and repair wiring harness and connectors.
- Explain the proper use of an oscilloscope.
- Explain the proper use of a logic probe.
- Perform general battery service such as cleaning battery terminals and a battery case as well as removing and replacing a battery.
- Perform battery hydrometer test and determine battery condition.

- Perform battery capacity test and determine battery condition.
- Perform battery-charging procedures.
- Connect booster battery cables properly.
- Measure battery drain accurately.
- Perform starter current draw test.
- Perform voltage drop tests on starter and starter control circuits.
- Inspect AC generator belt condition, and check and adjust belt tension.
- Perform an AC generator output test and determine AC generator condition.
- Check AC generator regulator voltage.

> ### Basic Tools
> Hand tools
> Service manual
> Jumper wires
> DMM

Terms To Know

American Wire Gauge (AWG)	Electromotive force (EMF)	Pulse width
Average responding	Frequency	Radiofrequency interference (RFI)
Chassis ground	Glitches	
Current ramp	Logic probe	Root mean square (RMS)
Digital storage oscilloscope (DSO)	Low-amp probes	Sinusoidal
	Nonsinusoidal	Specific gravity
Duty cycle	Parasitic drain	

BASIC ELECTRICAL DIAGNOSIS

When electrons are able to flow along a path (wire) between two points, an electrical circuit is formed. An electrical circuit is considered complete when there is a path that connects the positive and negative terminals of the electrical power source. Somewhere in the circuit there must be a *load* or resistance to control the amount of current in the circuit. When the resistance decreases, the current increases. And when the resistance decreases, the current flow increases. Most automotive electrical circuits use the chassis as the path to the negative side of the battery. Electrical components have a lead that connects them to the chassis. These are called the **chassis ground** connections. In a complete circuit, the flow of electricity can be controlled and applied to do useful work, such as light

> The **chassis ground** is commonly referred to as the ground of the circuit.

Battery cables need to be large-gauge wires capable of carrying high current for the starter motor.

a headlamp or turn over a starter motor. Components that use electrical power put a load on the circuit and consume electrical energy.

Two basic types of wire are used in automobiles: solid and stranded. Solid wires are single-strand conductors. Stranded wires are made up of a number of small solid wires twisted together to form a single conductor. Stranded wires are the most commonly used type of wire in an automobile. Electronic units, such as computers, use specially shielded twisted cable for protection from unwanted induced voltages that can interfere with computer functions. In addition, some solid-state components use printed circuits.

The current-carrying capacity and the amount of voltage drop in an electrical wire are determined by its length and gauge (size). The wire sizes are established by the Society of Automotive Engineers (SAE), which is the **American Wire Gauge (AWG)** system. Sizes are identified by a numbering system ranging from number 0000 to 50, with number 0000 having the largest cross-sectional area and number 50 the smallest. Most automotive wiring ranges from number 10 to 18 with battery cables that are at least number 4 gauge.

The **American Wire Gauge (AWG)** system designates wire sizes established by the SAE.

> ⚙ SERVICE TIP Keep in mind that electrical symbols are not standardized throughout the automotive industry. Different manufacturers may have different methods of representing certain components, particularly the less common ones. Always refer to the symbol reference charts, wire color code charts, and abbreviation tables listed in the vehicle's service manual to avoid confusion when reading wiring diagrams.

In the metric system, wiring size is identified by the cross-sectional area of the wire. Metric wire size is expressed in square millimeters, so the larger the number, the larger the wire. A chart that compares metric and standard wire sizes and the amperage they are capable of carrying is shown in **Figure 4-1**.

Electrical Wiring Diagrams

Wiring diagrams are used to show how circuits are wired and how the components are connected. A typical service manual contains dozens of wiring diagrams vital to the diagnosis and repair of the vehicle.

Metric Size (mm²)	AWG (Gauge) Size	Ampere Capacity
0.5	20	4
0.8	18	6
1.0	16	8
2.0	14	15
3.0	12	20
5.0	10	30
8.0	8	40
13.0	6	50
19.0	4	60

Figure 4-1 Metric-to-AWG wire sizes.

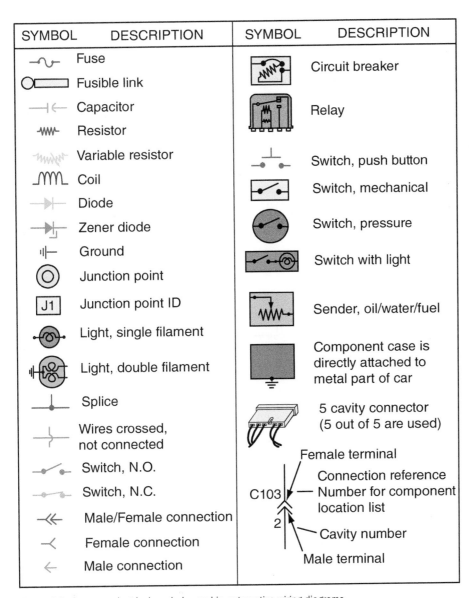

SYMBOL	DESCRIPTION	SYMBOL	DESCRIPTION
	Fuse		Circuit breaker
	Fusible link		Relay
	Capacitor		
	Resistor		Switch, push button
	Variable resistor		Switch, mechanical
	Coil		Switch, pressure
	Diode		
	Zener diode		Switch with light
	Ground		
	Junction point		Sender, oil/water/fuel
J1	Junction point ID		
	Light, single filament		Component case is directly attached to metal part of car
	Light, double filament		
	Splice		5 cavity connector (5 out of 5 are used)
	Wires crossed, not connected		Female terminal
	Switch, N.O.	C103	Connection reference Number for component location list
	Switch, N.C.		
	Male/Female connection	2	Cavity number
	Female connection		Male terminal
	Male connection		

Figure 4-2 Common electrical symbols used in automotive wiring diagrams.

A wiring diagram does not show the actual position of the parts on the vehicle or their appearance, nor does it indicate the length of the wire that runs between components. It usually indicates the color of the wire's insulation and sometimes the wire gauge size. The first letter of the color coding is a combination of letters usually indicating the base color. The second letter usually refers to the stripe color (if any). Tracing a circuit through a vehicle is basically a matter of following the colored wires.

Many different symbols are also used to represent components such as motors, batteries, switches, transistors, and diodes. Common symbols are shown in **Figure 4-2**. Part of a typical wiring diagram is shown in **Figure 4-3**. Notice that the components are also labeled.

Wiring diagrams can become quite complex. To avoid this, the vehicle's electrical system may be divided into many diagrams, each illustrating only one system, such as the backup light circuit, oil pressure indicator light circuit, or wiper motor circuit. In more complex ignition, electronic fuel injection, and computer control systems, one diagram may be used to illustrate only part of the entire circuit.

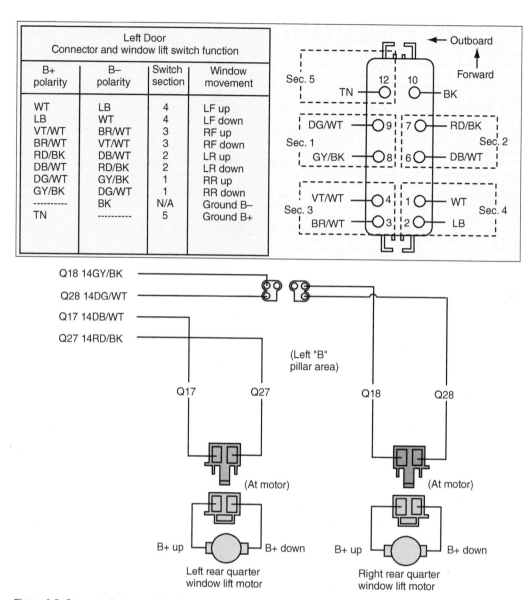

Figure 4-3 One part of a large wiring diagram.

Electrical Problems

All electrical problems can be classified into one of three categories: opens, shorts, and high-resistance problems. Identifying the type of problem allows you to identify the correct tests to perform when diagnosing an electrical problem.

An *open circuit* occurs when a circuit has a break in the wire. Without a completed path, current cannot flow and the load or component cannot work. A disconnected wire, a broken wire, or a switch in the off position can cause an open circuit. Although voltage will be present up to the open point, there is no current flow. Without current flow, there are no voltage drops across the various loads.

A *short circuit* results from an unwanted path for current. Shorts cause an increase in current flow. This increased current flow can burn wires or components. Sometimes two circuits become shorted together. When this happens, one circuit powers another. This may result in strange happenings, such as the horn blasting every time the brake pedal is depressed. In this case, the brake light circuit is shorted to the horn circuit (**Figure 4-4**). Improper wiring and damaged insulation are the two major causes of short circuits.

An open circuit is often called a *break in the circuit's continuity*.

A short to ground will cause fuses to fail.

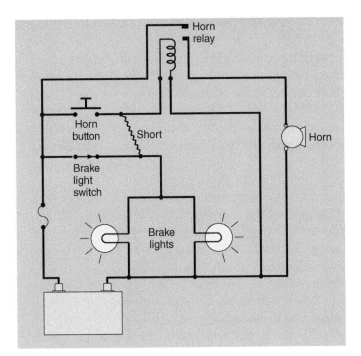

Figure 4-4 A wire-to-wire short.

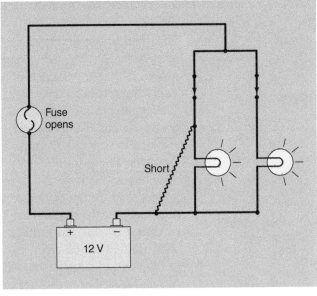

Figure 4-5 A short to ground.

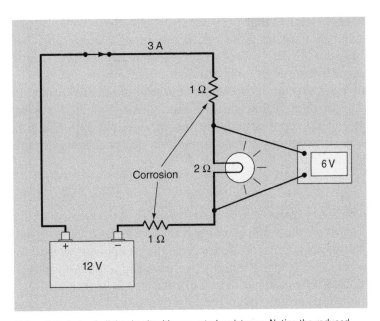

Figure 4-6 A simple light circuit with unwanted resistance. Notice the reduced voltage drop across the lamp and the reduced circuit current.

A short can also be an unwanted short to ground (**Figure 4-5**). This problem provides a low-resistance path to ground. This problem causes extremely high current flow that will damage wires and electrical components.

High-resistance problems occur when there is unwanted resistance in the circuit. The higher-than-normal resistance causes the current flow to be lower than normal and the components in the circuit are unable to operate properly (**Figure 4-6**). A common cause of this type of problem is corrosion at a connector. The corrosion becomes an additional load in the circuit. This load not only decreases the circuit's current, but also uses some of the circuit's voltage, which prevents full voltage to the normal loads in the circuit. In this case, the bulb would be dim.

A circuit with a short to ground is sometimes called a *grounded circuit*.

ELECTRICAL TEST EQUIPMENT

To troubleshoot a problem, always begin by verifying the customer's complaint. Then operate the systems for a complete understanding of the problem. Often there are other problems that are not as evident or bothersome to the customer that will provide helpful information for diagnostics. Refer to the correct wiring diagram and study the circuit that is affected. From the diagram you should be able to identify testing points and probable problem areas. Then test and use logic to identify the cause of the problem. Several meters are used to test and diagnose electrical systems. These are the voltmeter, ohmmeter, and ammeter. These should be used along with test lights and jumper wires.

Electrical current is a term used to describe the movement or flow of electricity. The greater the number of electrons flowing past a given point in a given amount of time, the more current the circuit has. This current, like the flow of water or any other substance, can be measured. The unit for measuring electrical current is the ampere. The instrument used to measure electrical current flow in a circuit is called an *ammeter*.

When any substance flows, it meets resistance. The resistance to electrical flow can be measured. The resistance to current flow produces heat. A unit of measured resistance is called an *ohm*. Resistance can be measured by an instrument called an *ohmmeter*.

Voltage is electrical pressure. Voltage is the force developed by the attraction of electrons to protons. The more positive one side of the circuit is, the more voltage is present in the circuit. Voltage does not, flow; rather, it is the pressure that causes current flow.

To have electricity, some force is needed to move the electrons between atoms. This **electromotive force (EMF)** is the pressure that exists between a positive and negative point within an electrical circuit. This force is measured in units called *volts*. One volt is the amount of pressure (force) required to move 1 ampere of current through a resistance of 1 ohm. Voltage is measured by an instrument called a *voltmeter*.

The resistance in that circuit determines the amount of current that flows. As resistance goes up, the current goes down. The energy used by a load is measured in volts. Amperage stays constant in a circuit, but the voltage is dropped as it powers a load. Measuring voltage drop determines the amount of electrical energy changed to another form of energy by the load.

Voltmeter

A voltmeter can be used to measure the voltage available at the battery. It can also be used to test the voltage available at the terminals of any component or connector. A voltmeter can also be used to test voltage drop across an electrical circuit, component, switch, or connector.

A voltmeter has two leads: a red positive lead and a black negative lead. The red lead should be connected to the positive side of the circuit or component. The black lead should be connected to ground or to the negative side of the component. Voltmeters should always be connected across the circuit being tested.

The loss of voltage due to resistance in wires, connectors, and loads is called *voltage drop* (**Figure 4-7**). Voltage drop is the amount of voltage expended when current is pushed through a resistance. Voltage drop that exceeds specifications for a given circuit indicates an undesired load in the circuit. Too much unwanted resistance or measurable voltage drop may reduce the amount of voltage, causing the component (load) to work improperly. The procedure for measuring a voltage drop is shown in Photo Sequence 5.

A voltmeter can also be used to check for proper circuit grounding. For example, if a voltmeter reading indicates full voltage at the lights but no lighting is seen, the bulbs or sockets could be bad or the ground connection is faulty.

An easy way to check for a defective bulb is to replace it with one known to be good. You can also use an ohmmeter to check for electrical continuity through the bulb.

If the bulbs are good, the problem lies in either the light sockets or ground wires. Connect the voltmeter to the ground wire and a good ground. If the light socket is

An ampere is usually called an *amp*.

Electromotive force (EMF) is also known as voltage.

Classroom Manual
Chapter 4, page 92

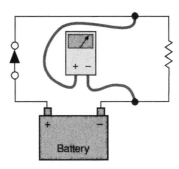

Figure 4-7 A voltmeter is connected in parallel or across the component or part of the circuit being tested for voltage drop.

defective, the voltmeter would read 0 volts. If the socket was not defective but the ground wire was broken or disconnected, the voltmeter would read very close to battery voltage. In fact, any voltage reading would indicate a bad or poor ground circuit. The higher the voltage, the greater the problem.

Generally speaking, voltage drops of 0.10 volts across a connection, 0.20 volts across a wire or cable, 0.30 volts across a switch, and 0.10 volts at a ground are considered maximum values.

PHOTO SEQUENCE 5
Performing a Voltage Drop Test

P5-1 The tools required to perform a voltage drop test are a voltmeter and fender covers.

P5-2 Set the voltmeter to its highest DC voltage scale, and then move down in scale until you get the best reading.

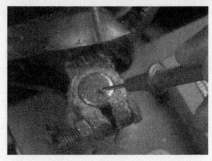

P5-3 To test the voltage drop of an entire power feed circuit, connect the red (positive) lead to the positive terminal of the battery.

P5-4 Connect the black (negative) lead to the power wire at the connector for the load. A low-beam headlight circuit is used in this photo. Turn the circuit on.

P5-5 Read the voltmeter. If there are no unwanted resistances in the circuit, the voltmeter will read less than 0.5 volt. If the reading is greater than 0.5 volt, wire or connector problems are indicated. To find the source of the problem, move the black test lead to another circuit connector closer to the battery.

⚠ Caution

Never place the leads of an ammeter across the battery or a load. This puts the meter in parallel with the circuit and will blow the fuse in the ammeter or possibly destroy the meter.

⚠ Caution

When testing for a short, always use a fuse. Never bypass the fuse with a wire. The fuse should be rated at no more than 50 percent higher capacity than specifications. This offers circuit protection and provides enough amperage for testing. After the problem is found and corrected, be sure to install the specified rating of fuse for circuit protection.

A low-amp probe allows an inductive measurement of small currents. A current ramp is a waveform of amperage flow.

Ammeters

An ammeter measures current flow in a circuit. An ammeter must be placed into or in series with the circuit being tested (**Figure 4-8**). Normally, this requires disconnecting a wire or connector from a component and connecting the ammeter between the wire or connector and the component. The red lead of the ammeter should always be connected to the side of the connector closest to the positive side of the battery, and the black lead should be connected to the other side.

It is much easier to test current using an ammeter with an inductive pickup. The pickup clamps around the wire or cable being tested. These ammeters measure amperage based on the magnetic field created by the current flowing through the wire. This type of pickup eliminates the need to separate the circuit to insert the meter.

Because ammeters are built with very low internal resistance, connecting them in series does not add any appreciable resistance to the circuit. Therefore, an accurate measurement of the current flow can be taken.

For example, assume that a circuit normally draws 5 amps and is protected by a 6-amp fuse. If the circuit constantly blows the 6-amp fuse, a short exists somewhere in the circuit. Mathematically, each light should draw 1.25 amps (5 ÷ 4 = 1.25). To find the short, disconnect all lights by removing them from their sockets. Then, close the switch and read the ammeter. With the load disconnected, the meter should read 0 amps. If there is any reading, the wire between the fuse block and the socket is shorted to ground.

If 0 amps were measured, reconnect each light in sequence. The reading should increase 1.25 amps with each bulb. If when making any connection the reading is higher than expected, the problem is in that part of the light circuit.

Current Probes

Current probes have been used for many years, especially for larger amperages such as battery and starting. In recent years, the trend has been toward **low-amp probes** (**Figure 4-9**) that are small and are designed to measure small currents in the milliampere range. As stated earlier, the amp probe actually measures the magnetic field around the wire. This eliminates the need for back probing and possibly damaging the wiring and its connectors. A typical use for the low-amp probe is shown in **Figure 4-10**. The diagram is a rough drawing of a pair of transmission shift solenoids. Battery positive is supplied to the solenoids which are wired in parallel. In order to confirm that the solenoids were being activated by the PCM at the appropriate time (the solenoids are difficult to access), a low-amp probe was installed on the B+ wire into the transmission. It was possible to determine when both, one, or none of the solenoids were active when commanded on by measuring the current flow as shown in the diagram. Of course, it would also be possible to determine if the solenoids or the wiring were shorted or open by the amount of current flow. Also, the circuit is running at normal amperage which is a better check of component health than using an ohmmeter to check winding resistance. Current probes can also be used to check electric motors, coils, and solenoids for

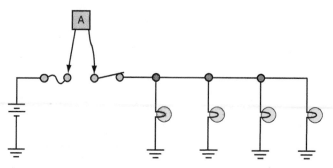

Figure 4-8 Measuring current with an ammeter. Notice this meter must be placed in series with the circuit.

Figure 4-9 A low-amp probe.

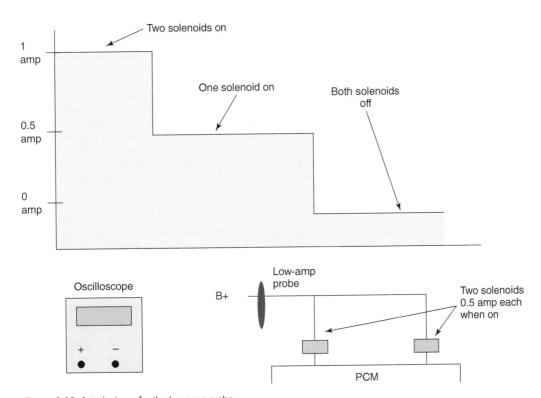

Figure 4-10 A typical use for the low-amp probe.

a correct **current ramp**, which is a characteristic waveform that uses amperage instead of voltage to examine the waveform of the component.

Ohmmeters

An ohmmeter measures resistance to current flow in a circuit. In contrast to the voltmeter, which uses the voltage available in the circuit, an ohmmeter is battery powered. The circuit being tested must be open. If the power is on in the circuit, the ohmmeter will be damaged.

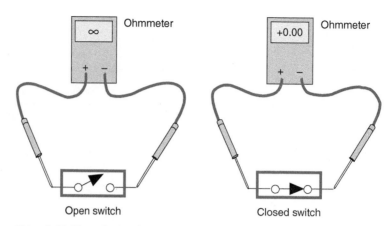

Figure 4-11 Measuring continuity with an ohmmeter.

The two leads of the ohmmeter are placed across or in parallel with the circuit or component being tested after the power to the circuit has been removed (**Figure 4-11**). The red lead is placed on the positive side of the circuit, and the black lead is placed on the negative side of the circuit. The meter sends current through the component and determines the amount of resistance based on the voltage drop across the load. The scale of an ohmmeter reads from 0 to infinity (∞). A 0 reading means there is no resistance in the circuit and may indicate a short in a component that should show a specific resistance. An infinity reading indicates a number higher than the meter can measure. This usually is an indication of an open circuit.

Ohmmeters are also used to trace and check wires or cables. Assume that one wire of a four-wire cable is to be found. Connect one probe of the ohmmeter to the known wire at one end of the cable, and touch the other probe to each wire at the other end of the cable. Any evidence of resistance indicates the correct wire. You can use this same method to check a suspected defective wire. If resistance is shown on the meter, the wire is said to have continuity. If no resistance is measured, the wire is defective (open). If the wire is OK, continue checking by connecting the probe to other leads. Any indication of resistance indicates that the wire is shorted to one of the other wires and that the harness is defective.

The power has to be removed from the circuit or component being tested before the ohmmeter can be used. Erroneous readings and possible meter damage can result.

AUTHOR'S NOTE Many technicians have asked the question "If I have an ohmmeter to measure resistance, why would I have to perform a voltage drop to find unwanted resistance in a circuit?" A DVOM calculates resistance based on Ohm's law. A current is applied through the DVOM's internal circuitry from the meter battery to the circuit or component being tested. This current is usually very, very small compared to the amount of current normally running in the same circuit. This small current does not really load the circuit at enough amperage to determine if there is a problem at operating current. Since we cannot use the ohmmeter to measure current with the power applied, think of voltage drop as a dynamic test of circuit resistance. Consider this: A battery cable that has most of its strands cut through would still read "good" on a resistance test with an ohmmeter. The same cable under the considerable load of a starter motor would quickly heat up and show a significant "voltage drop." While the ohmmeter is an indispensable tool, the service technician must remember that it does not "dynamically" test a circuit or component. Current flow or voltage drop can often expose a component or circuit that "tested OK but still does not work."

DIGITAL MULTIMETER USAGE

A multimeter is one of the most versatile tools used in diagnosing engine performance and electrical systems. **Analog meters** have low input impedance, but most digital meters have high input impedance, usually at least 10 megohms. Metered voltage for resistance tests is well below 5 volts, reducing the risk of damaging sensitive components and delicate computer circuits.

Multimeters either have an auto range feature, in which the appropriate scale is automatically selected by the meter, or they must be set to a particular range. In either case, you should be familiar with the ranges and the different settings available on the meter you are using. To designate particular ranges and readings, meters display a prefix before the reading or range. If the meter has a setting for mAmps, this means the readings will be given in milliamps, or 1/1,000th of an amp. Ohmmeter scales are expressed as a multiple of tens or use the prefix K or M. K stands for *kilo*, or one thousand (1,000). A reading of 10K ohms equals 10,000 ohms. An M stands for *mega*, or one million (1,000,000). A reading of 10M ohms equals 10,000,000 ohms. When using a meter with an auto range, make sure you note the range being used by the meter. There is a big difference between 10 ohms and 10,000,000 ohms. The common abbreviations and symbols used on multimeters are shown in **Figure 4-12**.

There are two ways multimeters display AC voltage: **root mean square (RMS)** and **average responding**. When the AC voltage signal is a true sine wave, both methods will display the same reading. Since most automotive sensors do not produce pure sine wave signals, it is important to know how the meter will display the AC voltage reading when comparing measured voltage to specifications. RMS meters convert the AC signal to a comparable DC voltage signal. Average responding meters display the average voltage peak. Always check the voltage specification to see if the specification is for RMS voltage. If this is the case, use an RMS meter.

When using the ohmmeter function, the DMM will show 0 or close to 0 when there is good continuity. If the continuity is very poor, the meter will display an infinite reading. This reading is usually shown as a blinking "1.000," a blinking "1," or an "OL."

Some multimeters also feature a min/max function. This function displays the maximum, minimum, and average voltage the meter recorded during the time of the test. This feature is valuable when checking sensors or when looking for electrical noise. Noise is primarily caused by **radio frequency interference (RFI)**, which may come from the ignition system. RFI is an unwanted voltage signal that rides on a signal. This noise can cause intermittent problems with unpredictable results. The noise causes slight increases and decreases in the voltage. When a computer receives a voltage signal with noise, it will

10 megohms equals 10,000,000 ohms.

⚠ **Caution**

Many digital multimeters (DMMS) with auto range display the measurement with a decimal point. Make sure you observe the decimal and the range being used by the meter **(Figure 4-13)**. A reading of 0.972 K ohms equals 972 ohms. If you ignore the decimal point, you will read 972,000 ohms.

Root mean square (RMS) meters convert the AC signal to DC voltage signal.

Average responding meters show the average voltage peak.

Radio frequency interference (RFI) is unwanted electrical noise that can affect the operation of computer controlled devices.

PREFIX	SYMBOL	RELATION TO BASIC UNIT
Mega	M	1,000,000
Kilo	K	1000
Milli	m	0.001 or $\frac{1}{1000}$
Micro	μ	0.000001 or $\frac{1}{1000000}$
Nano	n	0.000000001
Pico	p	0.000000000001

Figure 4-12 Common prefixes used on meters.

$$0.345 \text{ K}\Omega = 345\,\Omega$$

$$1025 \text{ mAmps} = 1.025 \text{ Amps}$$

Figure 4-13 Placement of decimal and scale should be noticed when measuring with a meter with auto range.

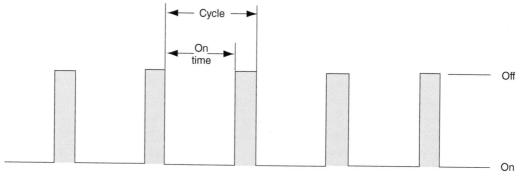

Figure 4-14 Duty cycle.

try to react to the minute changes. As a result, the computer responds to the noise rather than the voltage signal.

Multimeters may also have the ability to measure duty cycle, pulse width, and frequency. All of these represent voltage pulses caused by the turning on and off of a circuit or the increase and decrease of voltage in a circuit. **Duty cycle** is a measurement of the amount of time something is on compared to the time of one cycle (**Figure 4-14**). Duty cycle is measured in percentage. A 60-percent duty cycle means that a device is on 60 percent of the time and off 40 percent of one cycle. Duty cycle is similar to dwell, which is normally expressed in degrees. For example, a 30-degree dwell on a 60-degree scale equals a 50-percent duty cycle.

Pulse width is similar to duty cycle except that it is the exact time something is turned on (**Figure 4-15**). Pulse width is normally measured in milliseconds. When measuring duty cycle, you are looking at the amount of time something is on during one cycle. When measuring pulse width, you are looking at the amount of time something is on.

The number of cycles that occur in one second is called the **frequency** (**Figure 4-16**). The higher the frequency, the more cycles occur in a second. Frequencies are measured

Duty cycle is a measurement of electrical on time.

Pulse width is the amount of time an electrical device is turned on.

Frequency is the number of cycles that occur in one second.

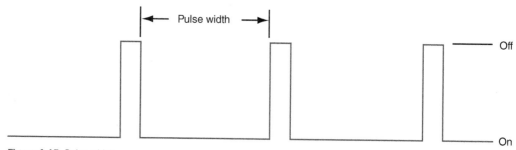

Figure 4-15 Pulse width.

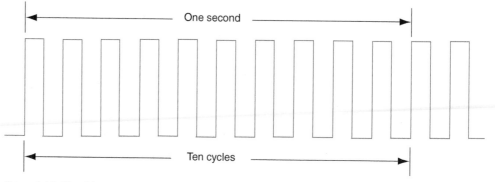

Figure 4-16 Signal frequency.

in hertz. One hertz is equal to one cycle per second. The example in **Figure 4-16** equals 10 hertz (10 Hz).

To accurately measure duty cycle, pulse width, and frequency, the meter's trigger level must be set. The trigger level tells the meter when to start displaying a signal. Trigger levels can be set at certain voltage levels or at a rise or fall in the voltage. Normally, meters have a built-in trigger level that corresponds with the voltage range setting. If the voltage does not reach the trigger level, the meter will not begin to recognize a cycle. On some meters you can select between a rise or fall in voltage to trigger the cycle count. A rise in voltage is a positive increase in voltage. This setting is used to monitor the activity of devices whose power feed is controlled by a computer. A fall in voltage is negative voltage. This setting is used to monitor ground-controlled devices.

Graphing Multimeter

A power-graphing multimeter has a history feature that can capture signals during an operator-selectable or preset timeframe. This can be particularly helpful in locating intermittent problems or performing dynamic tests on various components. The graphing multimeter is connected to the device or circuit you want to test and is set up to capture the signals. Various configurations can be set, including specific triggering voltage, minimum and maximum limits, and time scale. When you are done with the dynamic portion of the test, simply review the signal information recorded or stored in the graphing multimeter. The playback feature allows you to easily observe the results and spot glitches. Some units will have a library of signals, patterns, or other databases that can help you determine the results.

LAB SCOPE USAGE

An oscilloscope is a visual voltmeter. The oscilloscope has become the diagnostic tool of choice for many good technicians.

An oscilloscope allows a technician to see voltage changes over time (**Figure 4-17**). Voltage is displayed across the screen of the scope as a waveform. By displaying the waveform, the oscilloscope shows the slightest changes in voltage. This is a valuable feature for a diagnostic tool. Measuring voltage is a common test for diagnostics. Precise measurement is possible with a scope. When measuring voltage with a voltmeter, the meter only displays the average values at the point being probed. Digital voltmeters simply sample the voltage several times each second and update the meter's reading at a particular rate. If the voltage is constant, good measurements can be made with both types of voltmeters. A scope will display any change in voltage as it occurs. This is especially important for diagnosing intermittent problems.

The screen of a lab scope is divided into small divisions of time and voltage (**Figure 4-18**). These divisions set up a grid pattern on the screen. The horizontal movement of the waveform represents time. Voltage is measured with the vertical position of the waveform. As the scope displays voltage over time, the waveform moves from the left (the beginning of measured time) to the right (the end of measured time). The value of the divisions can be adjusted to improve the view of the voltage waveform. For example, the vertical scale can be adjusted so that each division represents 0.5 volts, and the horizontal scale can be adjusted so that each division equals 0.005 seconds (5 milliseconds). This allows the technician to view small changes in voltage that occur in a very short period of time. The grid serves as a reference for measurements.

The divisions on a scope are called *graticules.*

Since a scope displays actual voltage, it will display any electrical noise or disturbances that accompany the voltage signal (**Figure 4-19**). Noise is generally caused by RFI. Noise on a voltage signal causes the signal to change. These slight changes cause the computer to react, making changes to system operations when they are not necessary.

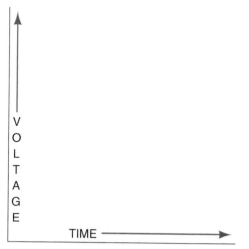

Figure 4-17 Voltage and time axis on a scope's screen.

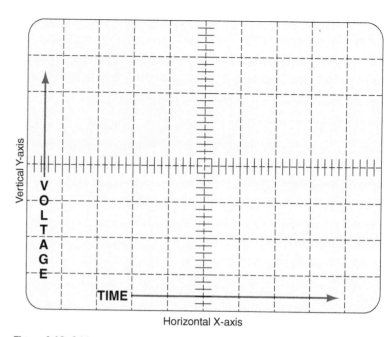

Figure 4-18 Grids on a scope screen that serve as a time and voltage reference.

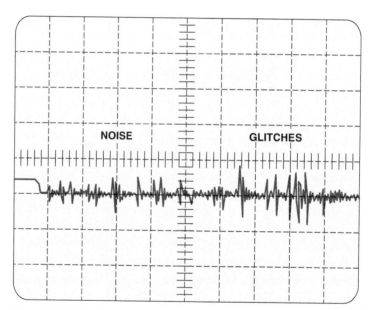

Figure 4-19 RFI noise and glitches may appear on a voltage signal.

Electrical disturbances, or **glitches**, are momentary changes in the signal. These can be caused by intermittent shorts to ground, shorts to power, or opens in the circuit. These problems can occur for only a moment or may last for some time. A lab scope is handy for finding these and other causes of intermittent problems. By observing a voltage signal and wiggling or pulling a wiring harness, any looseness can be detected by a change in the voltage signal.

A glitch is an electrical disturbance.

A DSO converts electrical voltages into waveforms digitally.

Analog versus Digital Scopes

A digital scope, commonly called a **digital storage oscilloscope (DSO)**, converts the voltage signal into digital information and stores it into its memory. Some DSOs use a laptop or desktop computer loaded with the software to store and display the information.

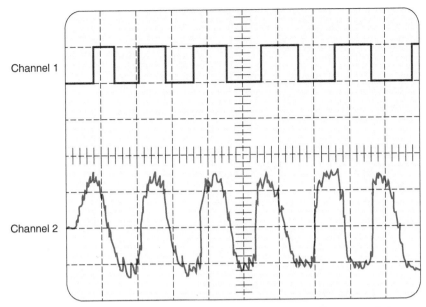

Figure 4-20 Some scopes can display two or more waveforms at once. These are called multichannel scopes.

To help in diagnostics, a technician can freeze the captured signal for close analysis. DSOs used to operate slower than real-time analog scopes, but as computers are getting faster, DSOs operate in virtual real time.

This slight delay is actually very slight. Most DSOs have a sampling rate of 1 million samples per second. This is quick enough to serve as an excellent diagnostic tool. This fast sampling rate allows slight changes in voltage to be observed. Slight and quick voltage changes cannot be observed on an analog scope.

A DSO uses an analog-to-digital (A/D) converter to digitize the input signal. Since digital signals are based on binary numbers, the trace appears slightly choppy when compared to an analog trace. However, the voltage signal is sampled more often, resulting in a more accurate waveform. The waveform is constantly being refreshed as the signal is pulled from the scope's memory. Remember, the sampling rate of a DSO can be over 1 million times per second.

Both an analog and a digital scope can be multiple-trace scopes (**Figure 4-20**). This means they both have the capability of displaying two or more traces at one time. By watching several traces simultaneously, you can watch the cause and effect of a sensor as well as compare a good or normal waveform to the one being displayed.

Waveforms

A waveform represents voltage over time. Any change in the **amplitude** of the trace indicates a change in the voltage. When the trace is a straight horizontal line, the voltage is constant (**Figure 4-21**). A diagonal line up or down represents a gradual increase or decrease in voltage. A sudden rise or fall in the trace indicates a sudden change in voltage. A **sinusoidal** wave represents an equal rise and fall from positive to negative. However, if the rise and fall are different, the wave is called **nonsinusoidal**.

Scopes can display AC and DC voltage, either one at a time or both together, as in the case of noise caused by RFI. Noise results from AC voltage riding on a DC voltage signal. The consistent change of polarity and amplitude of the AC signal causes slight changes in the DC voltage signal. A normal AC signal changes its polarity and amplitude over a period of time. The waveform created by AC voltage is typically

A sinusoidal wave has an equal rise and fall from positive to negative, a nonsinusoidal wave does not.

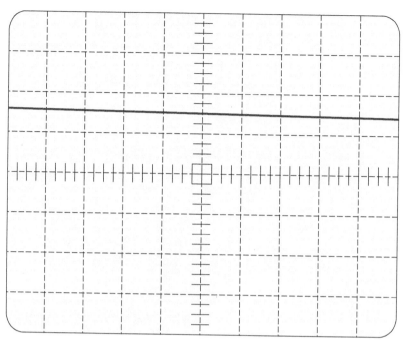

Figure 4-21 A constant voltage waveform.

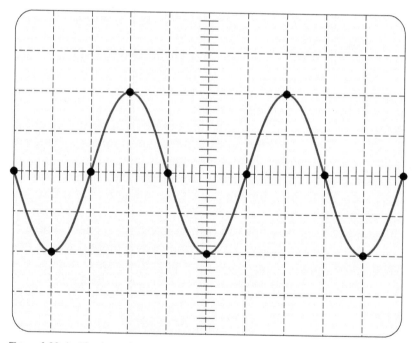

Figure 4-22 An AC voltage sine wave.

called a *sine wave* (**Figure 4-22**). One complete sine wave shows the voltage moving from 0 to its positive peak, then moving down through 0 to its negative peak and returning to 0.

One complete sine wave is a **cycle**. The number of cycles that occur per second is the frequency of the signal. Checking frequency or cycle time is one way of checking the operation of some electrical components. Input sensors are the most common components that produce AC voltage. Permanent magnet voltage generators produce an

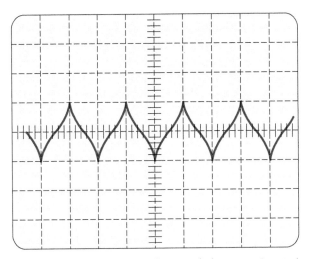

Figure 4-23 An AC voltage trace from a typical permanent magnet generator-type pickup or sensor.

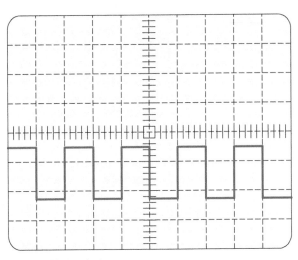

Figure 4-24 A typical square wave.

AC voltage that can be checked on a scope (**Figure 4-23**). AC voltage waveforms should also be checked for noise and glitches. These may send false information to the computer.

DC voltage waveforms may appear as a straight line or a line showing a change in voltage. Sometimes a DC voltage waveform will appear as a square wave, which shows voltage making an immediate change (**Figure 4-24**). Square waves are identified by having straight vertical sides and a flat top and bottom. This type of wave represents voltage being applied (circuit being turned on), voltage being maintained (circuit remaining on), and no voltage applied (circuit is turned off). A DC voltage waveform may also show gradual voltage changes.

Scope Controls

Depending on the manufacturer and model of the scope, the type and number of its controls will vary.

Depending on the manufacturer, there are many types of controls and features on an oscilloscope. Make sure to read your owner's manual for specific details. There are some examples of screens here that should help you get started. The very best way to get familiar with your scope is to use it whenever you get an opportunity.

The vertical scale is voltage (**Figure 4-25**). There are two common methods of setting the voltage scale depending on the manufacturer. Some oscilloscopes use a voltage per division method. For example, if you wanted to measure a signal in the vicinity of 5 volts, and the scope is set at 0.5 volts (500 millivolts) per division, you would need ten divisions to measure the voltage. The same 5 volts with the scope set at 1 volt per division would need five divisions to cover 5 volts. The difference is the size of the pattern on the screen. A pattern at 0.50 volts per division would be twice the size of a pattern at 1 volt per division. Some scopes allow the user to move the 0 voltage point up or down on the screen. This is useful for viewing A/C signals. Some oscilloscopes just ask for the voltage to be input, and the divisions are automatically divided and displayed. If the pattern goes off the top of the screen, then the voltage per division (or the total voltage) was set too low in an injector pattern shown in **Figure 4-26**. **Figure 4-27** shows the same injector pattern at 200 volts total, while **Figure 4-28** shows the pattern at 100 volts total. As you can see, the 100-volt pattern is double the size of the 200-volt one because the voltage per division is double.

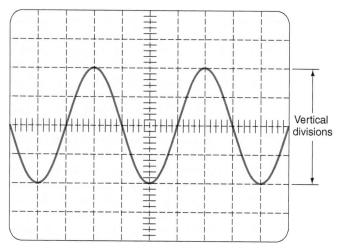

Figure 4-25 Vertical divisions represent voltage.

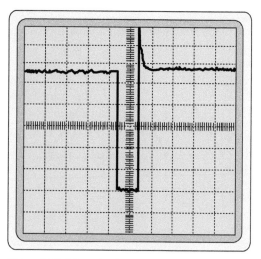

Figure 4-26 Voltage level set too low.

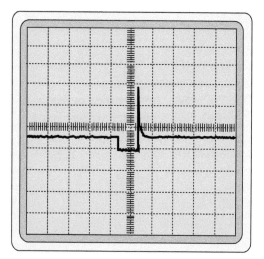

Figure 4-27 Injector pattern at 200 volts total, or 20 volts per division.

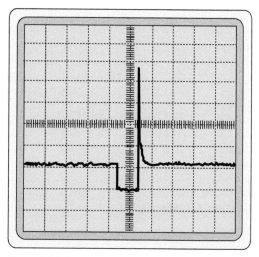

Figure 4-28 Same injector pattern at 100 volts total, or 10 volts per division.

The horizontal scale is time (**Figure 4-29**). As with the voltage setting, the time setting also needs to be correct for the pattern viewed. If the pattern goes off the scale (**Figure 4-30**) or the pattern quickly moves across the screen in a flash, then the time scale is set too low. If there are several small patterns (**Figure 4-31**) or the pattern is too small or updates too slow, then the time scale is set too high.

Trigger controls tell the scope when to begin a trace across the screen. Setting the trigger is important when trying to observe the timing of something. Proper triggering will allow the trace to repeatedly begin and end at the same points on the screen. There are typically numerous trigger controls on a scope. The trigger mode selector has a norm and auto position. In the norm setting, no trace will appear on the screen until a voltage signal occurs within the set time base. The auto setting will display a trace regardless of the time base.

Slope and level controls are used to define the actual trigger voltage. The slope determines whether the trace will begin on a rising or falling edge of the voltage signal

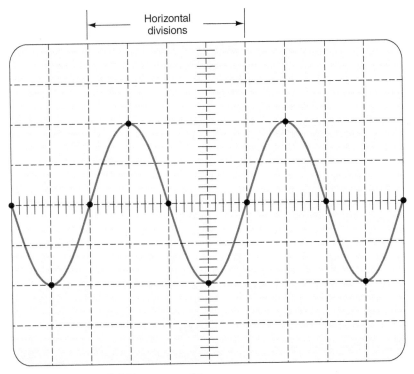

Figure 4-29 Horizontal divisions represent time.

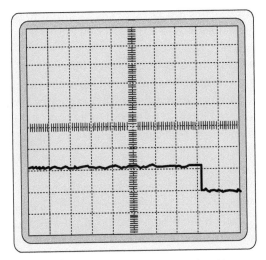

Figure 4-30 Injector pattern with time scale set too short.

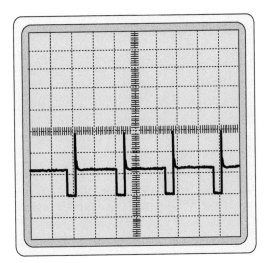

Figure 4-31 Injector pattern with time scale set too long.

(**Figure 4-32**). The level control determines where the time base will be triggered according to a certain point on the slope.

AUTHOR'S NOTE The biggest problem most students have with the oscilloscope is being intimidated by the seeming complexity of the tool. The best way to get past the hesitation is to practice using the scope. Study the manual, and practice using the scope every time you get a chance. Look for good as well as bad patterns. Many technicians like to keep a database of patterns to refer to later.

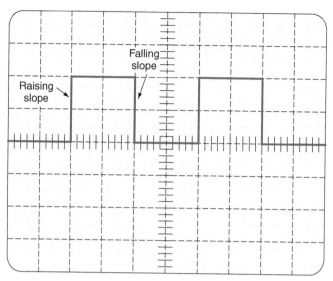

Figure 4-32 The trigger can be set to start the trace with a rise or fall of the voltage.

LOGIC PROBES

In some circuits, pulsed or digital signals pass through the wires. These on-off digital signals either carry information or provide power to drive a component. Many sensors used in a computer-control circuit send digital information back to the computer. To check the continuity of the wires that carry digital signals, a logic probe can be used (**Figure 4-33**).

A **logic probe** is similar in appearance to a test light. It contains three different colored light-emitting diodes (LEDs). A red LED lights when there is high voltage at the point being probed. A green LED lights to indicate low voltage. A yellow LED indicates the presence of a voltage pulse. The logic probe is powered by the circuit and reflects only the activity at the point being probed. When the probe's test leads are attached to a circuit, the LEDs display the activity.

If a digital signal is present, the yellow LED will turn on. When there is no signal, the LED is off. If voltage is present, the red or green LEDs will light depending on the amount of voltage. When there is a digital signal and the voltage cycles from low to high, the yellow LED will be lit and the red and green LEDs will cycle indicating a change in the voltage. The logic probe has a bicolor LED inside the indicator. When current flows through the probe, it can find a path for current flow no matter which way the probe is connected. The

The logic probe is used to indicate the presence of voltage and the direction of current flow.

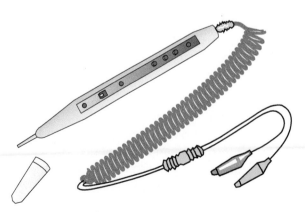

Figure 4-33 A typical logic probe.

LED lights green in one direction, and red in the opposite direction. With an A/C voltage, both LEDs light and produce a different color.

BASIC ELECTRICAL TROUBLESHOOTING

Troubleshooting electrical problems involves the same tools and methods regardless of what circuit has the problem. All electrical circuits must have voltage, current, and resistance. Therefore, testing and measuring these in addition to comparing your measurements to specifications are the keys to effective diagnosis.

A shorted circuit decreases the resistance of the circuit. This happens by shorting across to another circuit or by shorting to a ground. When there is a circuit-to-circuit short, one of the circuits is not controlled by its switch. The shorted-to circuit becomes a new parallel leg to the shorted circuit. With this type of problem, many strange things can happen. When a circuit is shorted to ground, a new parallel leg is present. This new leg has very low resistance and causes the current in the circuit to increase drastically. High-resistance problems can occur anywhere in the circuit. However, the effect of high resistance is the same regardless of where it is. Additional or unwanted resistance in series with a circuit will always reduce the current in the circuit and will reduce the amount of voltage drop by the component in the circuit.

To troubleshoot a problem, always begin by verifying the customer's complaint. Then operate the affected system and others. There are often other problems that are not as evident or bothersome to the customer but that will provide helpful information for diagnostics. With the correct wiring diagram for the car, study the circuit that is affected. From the diagram, you should be able to identify testing points and probable problem areas. Then test and use logic to identify the cause of the problem.

An ammeter and a voltmeter connected to a circuit at different locations shown in **Figure 4-34** should give readings as indicated when there are no problems in the circuit.

If there is an open anywhere in the circuit, the ammeter will read 0 current. If the open is in the 1-ohm resistor, a voltmeter connected from C to ground will read 0. However, if the resistor is open and the voltmeter is connected to points B and C, the reading will be 12 volts. The reason is that the battery, ammeter, voltmeter, 2-ohm resistor, and 3-ohm resistor are all connected together to form a series circuit. Because of the open in the circuit, there is only current flow in the circuit through the meter, not the rest of the circuit. This current flow is very low because the meter has such high resistance. Therefore, the voltmeter will show a reading of 12 volts, indicating little, if any, voltage drop across the resistors.

Classroom Manual
Chapter 4, page 91

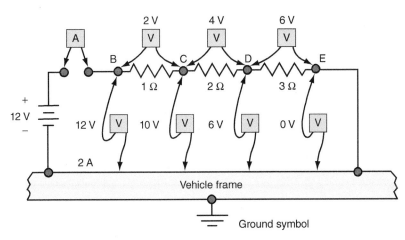

Figure 4-34 A basic circuit being tested with an ammeter and a voltmeter.

Remember Ohm's law when trying to understand circuit defects: high resistance = low current flow; and low resistance = high current flow.

To understand this concept, look at what happens if the 2-ohm resistor was open instead of the 1-ohm resistor. A voltmeter connected from point C to ground would indicate 12 volts. The 1-ohm resistor in series in the high resistance of the voltmeter would have little effect on the circuit. If an open should occur between point E and ground, a voltmeter connected from points B, C, D, or E to ground would read 12 volts. A voltmeter connected across any one of the resistors, from B to C, C to D, or D to E, would also read 0 volts because there will be no voltage drops if there is no current flow.

Excessive current and/or abnormal voltage drops would indicate a short. These examples illustrate how a voltmeter and ammeter may be used to check for problems in a circuit. An ohmmeter also may be used to measure the values of each component and compared to specifications. If there is no continuity across a part, it is open. If there is more resistance than called for, there is high internal resistance. If there is less resistance than specified, the part is shorted.

According to many manuals, the maximum allowable voltage drop for an entire circuit, except for the drop across the load, is 10 percent of the source voltage. Although 1.2 volts is the maximum acceptable amount, it is still too much. Many good technicians use 0.5 volts as the maximum allowable drop. However, there should be no more than 0.1 volts dropped across any one wire or connector. This is the most important specification to consider and remember. A very low current circuit such as a computer circuit requires a very good ground.

BASIC ELECTRICAL REPAIRS

⚠ **Caution**

Always follow the vehicle manufacturer's recommended procedure for disabling the ignition system. If these instructions are not followed, electronic components may be damaged.

Many automotive electrical problems can be traced to faulty wiring. Loose or corroded terminals; frayed, broken, or oil-soaked wires; and faulty insulation are the most common causes. Wires, fuses, and connections should be checked carefully during troubleshooting. Keep in mind that the insulation does not always appear to be damaged when the wire inside is broken. Also, a terminal may be tight but still may be corroded.

Wiring Harness and Terminal Diagnosis and Repair

Wires should be checked for a burned or melted condition (**Figure 4-35**). Connector ring terminals should be checked for loose retaining nuts, which cause high resistance or intermittent open circuits (**Figure 4-36**). An open circuit may be caused by a terminal that is backed out of the connector (**Figure 4-37**). Terminals that are bent or

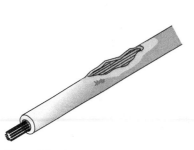

Figure 4-35 Damaged wire insulation.

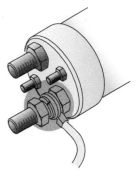

Figure 4-36 Loose retaining nuts on a ring-type terminal can cause high resistance or an intermittent open.

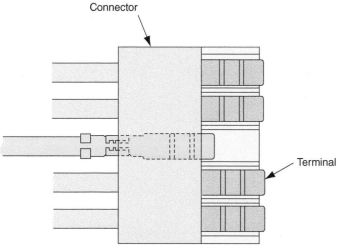

Connector

Terminal

Figure 4-37 Terminals that are backed out of their connector can cause an open circuit.

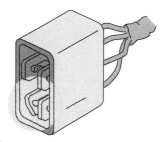

Figure 4-38 Bent or damaged component terminals can result in a shorted or open circuit.

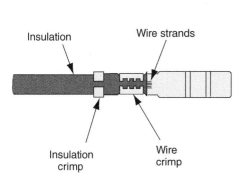

Insulation

Wire strands

Insulation crimp

Wire crimp

Figure 4-39 An open circuit occurs when a terminal is crimped over the insulation of the wire.

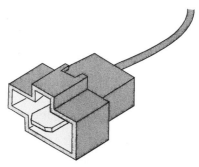

Figure 4-40 A greenish-white film of corrosion will cause high resistance and may cause open in the circuit. It can also cause arcing across the terminals.

damaged may cause shorts or open circuits (**Figure 4-38**). An open circuit occurs when the terminal is crimped over the insulation instead of the wire core (**Figure 4-39**). A greenish-white corrosion on terminals results in high resistance or an open circuit (**Figure 4-40**).

Wire end terminals are connecting devices. They are generally made of tin-plated copper and come in many shapes and sizes. They may be either soldered or crimped in place. When installing a terminal, select the appropriate size and type of terminal. Be sure it fits the unit's connecting post or prongs and it has enough current-carrying capacity for the circuit. Also, make sure it is heavy enough to endure normal wire flexing and vibration.

Some wiring harnesses contain a drain wire. Splice the drain wire separately (**Figure 4-41**), and then place Mylar tape over the spliced area with a winding motion. If more than one splice is required, stagger the wiring splices, and tape the splices with Mylar tape. **Figure 4-42** shows how to crimp a terminal. Be sure to use the proper crimping tool and to follow the tool manufacturer's instructions.

⚠ **Caution**

Do not crimp a terminal with the cutting edge of a pair of pliers. While this method may crimp the terminal, it also weakens it.

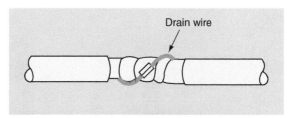

Figure 4-41 Separate splice for a drain wire in a wiring harness.

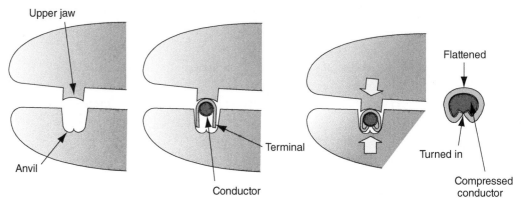

Figure 4-42 Properly crimping a connection with a crimping tool.

General procedures for crimping on a connector are as follows:

1. Use the correct size stripping opening on the crimping tool, and remove enough insulation to allow the wire to completely penetrate the connector.
2. Place the wire into the connector, and crimp the connector. To get a proper crimp, place the open area of the connector facing toward the anvil. Make sure the wire is compressed under the crimp.
3. If connecting two wires, insert the stripped end of the other wire into the connector, and crimp in the same manner.
4. Use electrical tape or heat-shrink tubing to tightly seal the connection. This will provide good protection for the wire and connector.

The preferred way to connect wires or to install a connector is by soldering with rosin-core solder; never use acid-core solder.

 Caution

Never use acid-core solder. It creates corrosion and can damage electronic components. Always use rosin-core solder.

Splicing Copper Wire with Splice Clips

Follow this procedure to splice copper wire with splice clips:

1. Cut the tape off the harness if necessary. Be careful not to cut the wire insulation.
2. Cut the wire to be repaired as necessary. Do not cut more than necessary off the wire.
3. Use the appropriate stripper opening in a pair of wire strippers to strip the proper amount of insulation off the wire ends. Enough insulation must be stripped off the wire ends to allow the wire ends to fit into the splice clip. A small amount of wire should be visible on each side of the clip.
4. Place the wire ends in the splice clip so the clip is centered on the wire ends. Be sure all the wire strands are in the clip.

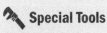 **Special Tools**

Wire strippers
Splice clip
Crimping tool
Soldering gun
Rosin-core solder
Splicing tape

5. Place the proper size opening in the crimping tool over the splice clip so this tool is even with one edge of the clip. Apply steady pressure on the crimping tool to crimp the clip onto the wire.
6. Repeat step 5 on the opposite end of the clip.
7. Heat the splice clip with a soldering gun and melt rosin-core solder into the clip.
8. Use splicing tape to cover the splice, and do not flag the tape.
9. If the wire is installed in a harness, tape the complete harness in the area where the tape was removed. When the wire remains by itself, wrap splicing tape over the spliced area and the wire insulation near the spliced area.

The term *tape flagging* refers to a large piece of excess splicing tape hanging from one side of a wire.

Splicing Copper Wire with a Crimp and Seal Splice Sleeves

Crimp and seal splice sleeves may be used to splice copper wire. These sleeves have a heat-shrink sleeve over the outside of the sleeve. The stripped wire ends are placed in the sleeve until they contact the stop in the center of the sleeve. A wire-crimping tool is used to crimp the sleeve near each end. After the crimping procedure, the sleeve is heated slightly with a heat torch to shrink the insulating heat-shrink sleeve onto the terminal. When the shrinking process is complete, sealant should appear at each end of the sleeve (**Figure 4-43**).

When working with wiring and connectors, never pull on the wires to separate the connectors. This can loosen the connector and cause an intermittent problem that may be very difficult to find later. Always follow the correct procedures and use the tools designed for separating connectors.

Check all connectors for corrosion, dirt, and looseness. Nearly all connectors have pushdown release-type locks (**Figure 4-44**). Make sure these are not damaged when disconnecting the connectors. Many connectors have covers over them to protect them from dirt and moisture. Make sure these are properly installed to provide for that protection.

Never reroute wires when making repairs. Rerouting wires can result in induced voltages from nearby components. These stray voltages can interfere with the function of electronic circuits.

Dielectric grease should be used to moisture-proof and to protect connections from corrosion. If the manufacturer specifies that a connector be filled with grease, make sure it is. If the old grease is contaminated, replace it. Some car manufacturers suggest using petroleum jelly to protect connection points.

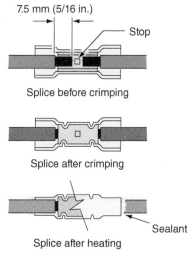

7.5 mm (5/16 in.)

Stop

Splice before crimping

Splice after crimping

Sealant

Splice after heating

Figure 4-43 Crimp and seal splice sleeve after crimping and heating.

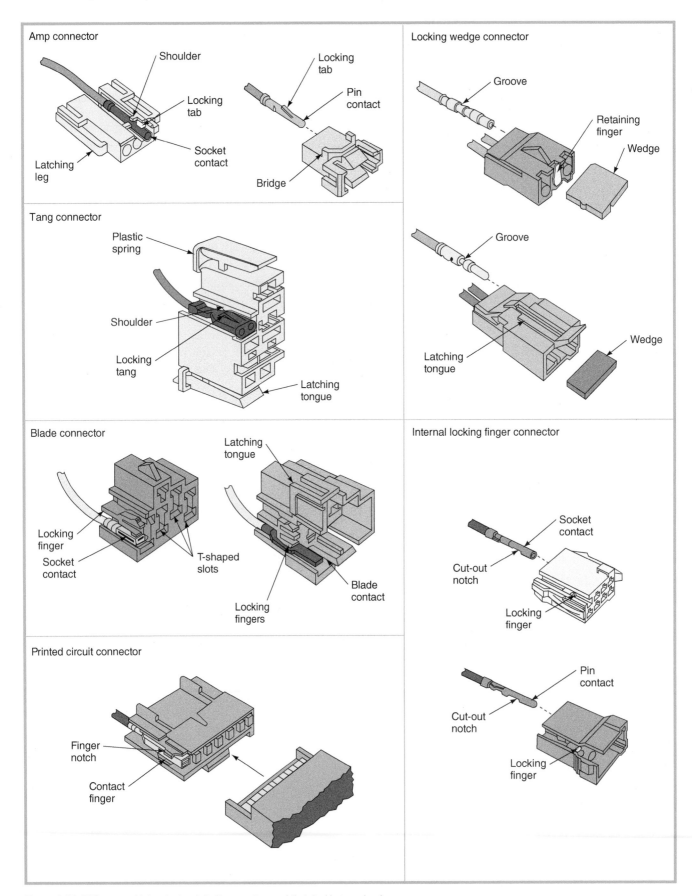

Figure 4-44 Different multiple-wire hard shell connectors and their locking mechanisms.

TESTING BASIC ELECTRICAL COMPONENTS

The testing of individual electrical components requires knowledge of the component's operation and the various diagnostic tools, and a thorough understanding of basic electricity. This section looks at some of the methods to test common electrical components.

Fuses

A protection device is designed to turn the system off whenever excessive current or an overload occurs (**Figure 4-45**). Modern vehicles use blade-style fuses for most applications, and some vehicles use a bolt-down-style fuse for high-amperage applications. The fuse block itself uses mini or regular fuses. To check the fuse, pull it from the fuse panel and look at the fuse element through the transparent plastic housing. Look for internal breaks and discoloration. All types of fuses can be checked with an ohmmeter or test light. If the fuse is good, there will be continuity through it.

Fuses are rated by the current at which they are designed to blow. A three-letter code is used to indicate the type and size of fuses. Blade fuses have codes ATC or ATO.

The current rating for blade fuses is indicated by the color of the plastic case (**Figure 4-46**). In addition, it is usually marked on the top. The insulator on ceramic fuses is color coded to indicate different current ratings.

Fusible Links

Fuse (or fusible) links are used in circuits where limiting the maximum current is not extremely critical. They are often installed in the positive battery lead to the ignition switch and other circuits that have power with the key off. Fusible links are normally found in the engine compartment near the battery. Fusible links are also used when it would be awkward to run wiring from the battery to the fuse panel and back to the load.

Because a fuse link is a lighter gauge of wire than the main conductor, it melts and opens the circuit before damage can occur in the rest of the circuit. Fuse link wire is covered with a special insulation that bubbles when it overheats, indicating that the link has melted. If the insulation appears good, pull lightly on the wire. If the link stretches, the wire has melted. When it is hard to determine if the fuse link is burned out, check for continuity through the link with a test light or ohmmeter.

To replace a fuse link, cut the protected wire where it is connected to the fuse link. Then tightly crimp or solder a new fusible link of the same rating as the original link. Since the insulation on the manufacturer's fuse links is flameproof, never fabricate a fuse link from ordinary wire because the insulation may not be flameproof.

⚡ WARNING **Always disconnect the battery ground cable prior to servicing any fuse link.**

Fuses and other protection devices do not normally wear out. They go bad because something went wrong. Never replace a fuse or fusible link or reset a circuit breaker without finding out why it went bad.

A blade-type fuse is called a *spade fuse.*.

A fuse link is commonly called a *fusible link.*

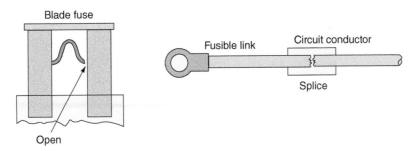

Figure 4-45 A protection device can have a fault that is not seen by a technician. They should always be tested with a volt or ohmmeter.

Blade Fuse Color Coding	
Ampere Rating	**Housing Color**
4	pink
5	tan
10	red
15	light blue
20	yellow
25	natural
30	light green

Fuse Link Color Coding	
Wire Link Size	**Insulation Color**
20 GA	blue
18 GA	brown or red
16 GA	black or orange
14 GA	green
12 GA	gray

Maxi-Fuse Color Coding	
Ampere Rating	**Housing Color**
20	yellow
30	light green
40	amber
50	red
60	blue

Figure 4-46 Typical color coding of protection devices.

Maxi-Fuses

Most late-model vehicles use *maxi-fuses* instead of fusible links. Maxi-fuses look and operate like two-prong, blade, or spade fuses except they are much larger and can handle more current. Typically, a maxi-fuse is four to five times larger. Maxi-fuses are located in their own underhood fuse block.

Maxi-fuses are easier to inspect and replace than fuse links. To check a maxi-fuse, look at the fuse element through the transparent plastic housing. If there is a break in the element, the maxi-fuse has blown. To replace it, pull it from its fuse box or panel. Always replace a blown maxi-fuse with a new one having the same ampere rating.

Maxi-fuses allow the vehicle's electrical system to be broken down into smaller circuits that are easy to diagnose and repair. For example, in older vehicles a single fusible link controls one-half or more of all circuitry. If it burns out, many electrical systems are lost. By replacing this single fusible link with several maxi-fuses, the number of systems lost due to a problem in one circuit is drastically reduced. This makes it easy to pinpoint the source of trouble. Using maxi-fuses also eliminates the need to perform a wiring repair when a fusible link burns out.

Circuit Breakers

Some circuits are protected by circuit breakers (**Figure 4-47**). Like fuses, they are rated in amperes. A circuit breaker can be cycling or must be manually reset. In the cycling type, the bimetal arm will begin to cool once the current to it is stopped. Once it returns to its

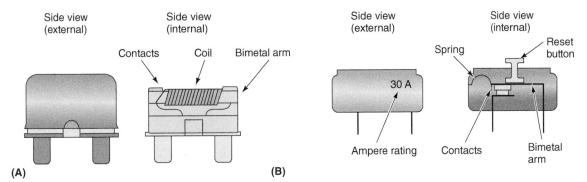

Figure 4-47 Two basic types of circuit breakers.

original shape, the contacts are closed and power is restored. If the current is still too high, the cycle of breaking the circuit will be repeated.

Two types of noncycling or resettable circuit breakers are used. The positive temperature coefficient (PTC) circuit breaker is made of a solid-state material that increases resistance as current flow increases. When current flow gets to the preset point, the resistance increases to effectively open the circuit. The PTC is reset by removing the power from the circuit. PTCs are often used on power window and door lock circuitry. The other type of resettable circuit breaker has a button on it to reset the circuit breaker.

A circuit breaker is typically abbreviated *c.b.* in the fuse chart of a service manual.

> **⚙ SERVICE TIP** Before using a test light, it is good practice to check the tester's lamp. To do this, simply connect the test light across the battery. The light should come on.

A visual inspection of a fuse or fusible link will not always determine if it is blown. To accurately test a circuit protection device, use an ohmmeter, voltmeter, or test light.

With the fuse or circuit breaker removed from the vehicle, connect the ohmmeter's test leads across the protection device's terminals (**Figure 4-48**). On its lowest scale, the ohmmeter should read 0 ohms. If it reads infinity, the protection device is open. Test a fusible link in the same way. Before connecting the ohmmeter across the fusible link, make sure there is no current flow through the circuit. To be safe, disconnect the negative cable of the battery.

To test a circuit protection device with a voltmeter, check for available voltage at both terminals of the unit (**Figure 4-49**). If the device is good, voltage will be present on both sides. A test light can be used in place of a voltmeter.

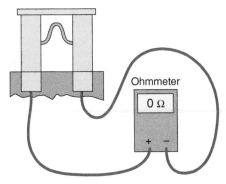

Figure 4-48 A good fuse will have zero resistance across it.

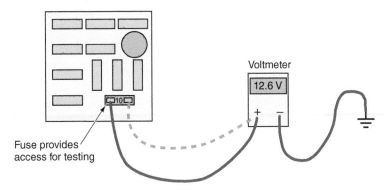

Figure 4-49 Circuit protection devices can be tested with a voltmeter. Make sure there is voltage present on both sides of the device.

Measuring voltage drop across a fuse or other circuit protection device will tell you more about its condition than just if it is open. If a fuse, a fuse link, or circuit breaker is in good condition, a voltage drop of 0 will be measured. If 12 volts is read, the fuse is open. Any reading between 0 and 12 volts indicates some voltage drop. If there is voltage drop across the fuse, it has resistance and should be replaced. Make sure you check the fuse holder for resistance as well.

Switches

Classroom Manual
Chapter 4, page 96

 Caution

Before replacing a switch or any component connected to the vehicle's control computer, disconnect the negative cable from the battery.

Classroom Manual
Chapter 4, page 97

To check a switch, disconnect the connector to the switch. With an ohmmeter, check for continuity between the terminals of the switch (**Figure 4-50**) with the switch moved to the on position and to the off position. With the switch in the off position, there should be no continuity between the terminals. With the switch on, there should be good continuity between the terminals. If the switch is activated by something mechanical and does not complete the circuit when it should, check the adjustment of the switch. (Some switches are not adjustable.) If the adjustment is correct, replace the switch. Another way to check the clutch switch is to simply bypass it with a fused jumper wire. If the component works when the switch is jumped, a bad switch is indicated.

Voltage drop across switches should also be checked. Ideally, when the switch is closed there should be no voltage drop. Any voltage drop indicates resistance and the switch should be replaced.

Relays

A relay can be checked with a jumper wire, voltmeter, ohmmeter, or test light. If the terminals are accessible, a jumper wire and test light will be the quickest method.

Check the wiring diagram for the relay being tested to determine if the control is through an insulated or ground switch. If the relay is controlled on the ground side, follow this procedure to test the relay.

1. Use the test light to check for voltage at the battery side of the relay. If voltage is not present, the fault is in the battery feed to the relay. If there is voltage, continue testing.
2. Probe for voltage at the control terminal. If voltage is not present, the relay coil is faulty. If voltage is present, continue testing.
3. Use a jumper wire to connect the control terminal to a good ground. If the relay works, the fault is in the control circuit. If the relay does not work, continue testing.
4. Connect the jumper wire from the battery to the output terminal of the relay. If the device operated by the relay works, the relay is bad. If the device does not work, the circuit between the relay and the device's ground is faulty.

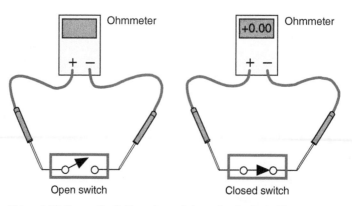

Figure 4-50 The continuity through a switch can be checked with an ohmmeter. With the switch closed, there should be zero resistance. With the switch open, there should be infinite resistance.

If a computer controls the relay, do not use a test light. Rather, use a high-impedance voltmeter set to the 20-volt DC scale, then:

1. Connect the negative lead of the voltmeter to a good ground.
2. Connect the positive lead to the output wire. If no voltage is present, continue testing. If there is voltage, disconnect the ground circuit of the relay. The voltmeter should now read 0 volts. If it does, the relay is good. If voltage is still present, the relay is faulty and should be replaced.
3. Connect the positive voltmeter lead to the power input terminal. Voltage should be very close to battery voltage. If not, the relay is faulty. If the correct voltage is present, continue testing.
4. Connect the positive meter lead to the control terminal. Voltage should be very close to the battery voltage. If not, check the circuit from the battery to the relay. If the correct voltage is there, continue testing.
5. Connect the positive meter lead to the relay ground terminal. If more than 1 volt is present, the circuit has a poor ground.

If the relay terminals are not accessible, remove the relay from its mounting and test it on a bench. Use an ohmmeter to test for continuity between the relay coil terminals. If the meter indicates an infinity reading, replace the coil. If there is continuity, use a pair of fused jumper wires to energize the coil. Check for continuity through the relay contacts. If there is an infinity reading, the relay is faulty. If there is continuity, the relay is good and the circuits need to be tested carefully.

Be sure to check your service information for resistance specifications and compare the relay to them. It is easy to check for an open coil; however, a shorted coil will also prevent the relay from working. Low resistance across a coil would indicate that is shorted. Too low of a resistance may also damage the transistors and/or driver circuits because of the excessive current that would result.

Stepped Resistors

The best method for testing a stepped resistor is to use an ohmmeter. To do this, remove the resistor from its mounting. Connect the ohmmeter leads to the two ends of the resistor. Compare the results against specifications. Make sure the ohmmeter is set to the correct scale for the anticipated amount of resistance.

A stepped resistor can also be checked with a voltmeter. By measuring the voltage after each part of the resistor block and comparing the readings to specifications, you can tell if the resistor is good or not.

Variable Resistors

One way to test a variable resistor is with an ohmmeter. However, one can also be effectively checked dynamically by observing its output voltage.

To test a rheostat, identify the input and output terminals and connect the ohmmeter across them. Rotate the control while observing the meter. The resistance value should remain within the limits specified for the switch. If the resistance values do not match the specified amounts or if there is a sudden change in resistance as the control is moved, the unit is faulty. When using a voltmeter, the readings should be smooth and consistent as the control is moved.

To test a potentiometer, connect an ohmmeter across the resistor. The readings should be within the range listed in the specifications. If they are not, the resistor needs to be replaced. Then move the leads to the input and output of the resistor. The readings should sweep evenly and consistently within the specified resistance values. The condition of a potentiometer can also be checked with a voltmeter or a lab scope. The lab scope can be used to check signal integrity as well as voltage values. The DMM can show dynamic voltage changes and minimum and maximum voltages. As it is moved through its dynamic range, the lab scope can show any erratic voltage changes or "noise."

Classroom Manual
Chapter 4, page 95

Potentiometers are used as computer inputs. The most common example is a **throttle position sensor (TP sensor)**.

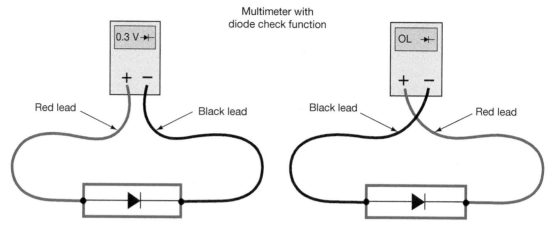

Figure 4-51 Use a multimeter to test a diode for an open and/or short.

Diodes

Regardless of the bias of the diode, it should allow current flow in one direction only. To test a diode, use the diode check feature of the multimeter. Disconnect the diode from the circuit. Connect the meter's leads across the diode (**Figure 4-51**). Observe the reading on the meter, and then reverse the meter's leads and observe the reading on the meter. The resistance in one direction should be very high or infinity, many meters read "OL" for out of limits and in the other direction a reading of 0.5 to 0.8 indicates a good silicon diode and 0.2 to 0.3 indicates a good germanium diode. The diode check function actually checks for the voltage drop across the diode. If any other readings were observed, the diode is bad. A diode that has low resistance in both directions is shorted. A diode that has high resistance or an infinity reading in both directions is open.

Problems may be encountered when checking a diode with a high-impedance digital ohmmeter. Since many diodes will not allow current flow through them unless the voltage is at least 0.6 volts, a digital meter may not be able to forward bias the diode. This will result in readings that indicate the diode is open, when in fact it may not be. Because of this problem, many multimeters are equipped with a diode testing feature. This test allows for increased voltage at the test leads. Some meters will display the voltage required to forward bias the diode. If the diode is open, the meter will display "OL" or another reading to indicate *infinity* or *out of range*. Some meters during a diode check will make a beep noise when there is continuity.

Diodes may also be tested with a voltmeter. Using the same logic as when testing with an ohmmeter, test the voltage drop across the diode. The meter should read low voltage in one direction and near source voltage in the other direction.

Classroom Manual
Chapter 4, page 101

⚠ **Caution**

Battery electrolyte contains sulfuric acid, a strong, corrosive acid. Electrolyte is very damaging to vehicle paint, upholstery, and clothing. Do not allow the electrolyte to contact these items.

BATTERY DIAGNOSIS AND SERVICE

⚡ **WARNING** Battery electrolyte is very harmful to human skin and eyes. Always wear face and eye protection, protective gloves, and protective clothing when handling batteries or electrolyte. If electrolyte contacts your skin or eyes, flush with clean water immediately and obtain medical help.

⚡ **WARNING** Never smoke or allow other sources of ignition near a battery. Hydrogen gas discharged while charging the battery is explosive (**Figure 4-52**)! A battery explosion may cause personal injury or property damage. Even when the battery has not been charged for several hours, hydrogen gas still may be present under the battery cover.

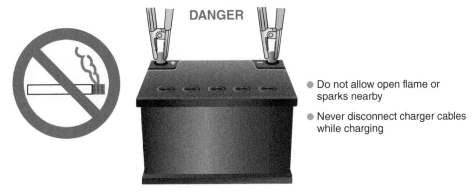

Figure 4-52 Never smoke or allow a spark or excessive heat near a battery.

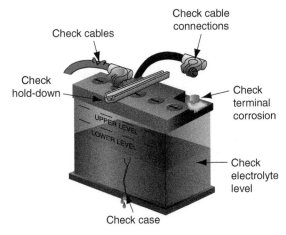

Figure 4-53 Guidelines for the inspection of a battery.

A battery should be checked for terminal corrosion, a loose or broken hold-down, frayed or corroded cables, a cracked case, and electrolyte level (**Figure 4-53**). If the battery has removable vent caps, the electrolyte level may be checked through the vent cap openings. When the vent caps are not removable, the electrolyte level may be checked by observing the level through the translucent plastic case. The battery may be removed from the vehicle, washed with a baking soda and water solution, and rinsed with clean water. (Be careful not to get the baking soda solution into the battery.)

Open-Circuit Voltage Test

The battery test can begin with an open-circuit voltage (OCV) test by connecting a voltmeter to the battery terminals. This can be done in place of a battery hydrometer test. A static voltage reading of the battery can give an indication of the battery's condition and state of charge. The battery temperature should be between 60°F and 100°F (15.5°C and 37.7°C). Before performing the test, the battery voltage must stabilize for 10 minutes. If the battery is coming off a charge, apply a load for 15 seconds at half of the cold cranking amperage (CCA), allow the battery to stabilize, and then test. An acceptable reading is 12.4 volts (see **Figure 4-54**). If the OCV indicates a charge below 75 percent of a full charge, the battery needs to be charged and load tested.

Battery Conductance Testing

Many new car and battery manufacturers require the use of a battery conductance tester (**Figure 4-55**) to test batteries. The tester can measure the remaining CCA of the battery by measuring the usable area of the battery's plates. These testers can determine if a

Classroom Manual
Chapter 4, page 110

Do not allow dirt and corrosion to accumulate on the top and sides of the battery; they can conduct current between the positive and negative terminals, causing a current leakage. This drain can be measured by placing a voltmeter black lead on the negative terminal of the battery and touching the battery case at various points with the positive lead.

Open Circuit Voltage (volts)	State of Charge
12.6 or greater	100%
12.4 to 12.6	75–100%
12.2 to 12.4	50–75%
12.0 to 12.2	25–50%
11.7 to 12.0	0–25%
11.7 or less	0%

Figure 4-54 Battery open-circuit voltage as an indicator of state of charge.

Figure 4-55 A battery conductance tester.

battery needs to be charged and retested, has failed, is weak, or is usable even with a very low state of charge. This can save time in the shop charging and testing a failed battery.

> ⚙ **SERVICE TIP** Always disconnect the negative battery cable before the positive cable. If the positive cable is disconnected first, the wrench may slip and ground this cable to the chassis. This action may result in severe burns or a battery explosion.

Some memory savers have enough power to deploy the air bags. Some manufacturers do not recommend their use.

 Caution

Do not hammer or twist battery cable ends to loosen them. This action will loosen the terminal in the battery cover and cause electrolyte leaks.

The **specific gravity** of a liquid is the weight of a liquid in relation to the weight of an equal volume of water. Water has a specific gravity of 1.000, sulfuric acid has a specific gravity of 1.835, and the electrolyte in a fully charged battery has a specific gravity of 1.265.

⚡ **WARNING** **On air bag–equipped vehicles, the negative battery cable should be disconnected and the technician should wait for 1 minute before working on the vehicle electrical system to avoid accidental air bag deployment. The wait time may vary depending on the vehicle manufacturer. Always follow the instructions in the vehicle manufacturer's service information.**

Disconnecting the battery in a vehicle may cause these problems, unless the vehicle has nonvolatile RAM for the following memories:

1. Erases adaptive memories in various computers.
2. Erases memory in memory seats and memory mirror systems.
3. Erases station memory programmed in the radio.

To avoid these problems, some tool manufacturers supply a dry cell voltage source that may be plugged into the DLC prior to disconnecting the battery.

After the battery terminal nuts have been loosened, the battery cables should be removed from the battery with a terminal puller. Never pry the cable off—the battery may be damaged this way. Battery terminals and cable ends may be cleaned with a battery terminal cleaner (**Figure 4-56**).

Hydrometer Testing

Although most batteries made today are maintenance-free, some may still have cell caps that are removable, allowing access for a hydrometer test. The hydrometer test measures the **specific gravity** of the electrolyte. Remove the vent caps, and insert the hydrometer pickup tube into the electrolyte. Squeeze the hydrometer bulb, and release it to draw electrolyte into the hydrometer until the float is floating freely (**Figure 4-57**).

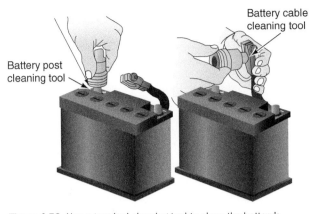

Figure 4-56 Use a terminal cleaning tool to clean the battery's terminals and the cable ends.

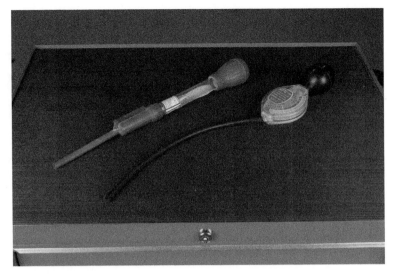

Figure 4-57 Two types of battery hydrometers.

Observe the electrolyte level on the hydrometer float. Read the electrolyte specific gravity on the float at the electrolyte level. A fully charged battery has a specific gravity of 1.265. If the battery specific gravity is less than 1.270, the battery requires charging. A battery with a 1.155 specific gravity is 50 percent charged (**Figure 4-58**). A voltmeter may be connected across the battery terminals to check the open-circuit voltage.

Since the specific gravity is affected by temperature, some correction of the specific gravity reading in relation to temperature is required. A thermometer in the lower part of some hydrometers measures electrolyte temperature. For every 10°F below 80°F, subtract 4 points from the hydrometer reading, and add 4 points to the reading for each 10°F above 80°F.

Battery Capacity Test

A battery must have a minimum specific gravity of 1.190 prior to undergoing a capacity test. Prior to running this test, disconnect and remove the battery cables. The battery capacity test procedure may vary depending on the type of test equipment and whether the tester has digital or analog meters. Always follow the test equipment manufacturer's instructions.

Follow these steps for the capacity test with an analog-type capacity tester:

1. Rotate the capacity tester load control to the off position, and set the voltmeter control to the 18-volt position.
2. Check the mechanical zero on each meter, and adjust if necessary.
3. Connect the tester leads to the battery terminals with the proper polarity. The red lead is always connected to the positive terminal, and the black lead is always connected to the negative terminal (**Figure 4-59**).
4. Observe the battery open-circuit voltage, which should be above 12.4 volts.
5. Set the test selector to the number two charging position, and adjust the ammeter pointer to the zero position with the electrical zero adjust control.
6. Connect the ammeter inductive clamp over the negative tester cable.
7. Set the test selector to the number one starting position.
8. Rotate the load control until the ammeter reads one-half the cold cranking rating or three times the ampere-hour rating.
9. Maintain this load position for 15 seconds. If necessary, adjust the load control slightly to maintain the proper ampere reading.

Special Tools
Battery cable puller
Battery terminal cleaner
Hydrometer

⚠ Caution
Always tighten the battery hold-down bolts to the specified torque. Overtightening these bolts may damage the battery cover and case.

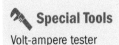

Special Tools
Volt-ampere tester

STATE OF CHARGE	SPECIFIC GRAVITY
100%	1.265
75%	1.225
50%	1.190
25%	1.155
DEAD	1.120

Figure 4-58 The state of charge of a battery in relation to its specific gravity.

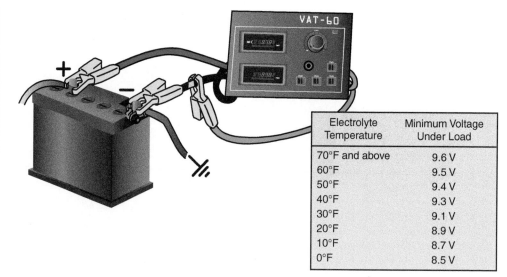

Electrolyte Temperature	Minimum Voltage Under Load
70°F and above	9.6 V
60°F	9.5 V
50°F	9.4 V
40°F	9.3 V
30°F	9.1 V
20°F	8.9 V
10°F	8.7 V
0°F	8.5 V

Figure 4-59 Volt-ampere tester connections for testing the battery.

10. After 15 seconds are completed, observe the voltmeter reading and immediately rotate the load control to the off position.

11. If the battery temperature is 70°F (21°C) and the voltage is above 9.6 volts after 15 seconds on the capacity test, the battery is satisfactory and may be returned to service. Since temperature affects battery capacity, the minimum voltage at the end of the capacity test varies depending on the temperature.

If the battery voltage is less than the minimum load voltage at the end of the load test, the battery should be charged and the load test repeated. After charging and repeating a load test, if the battery voltage is less than the minimum load voltage, the battery should be replaced. As discussed in the steps above, the minimum voltage the battery must maintain during the load test is 9.6 volts. However, most healthy batteries will be able to maintain a higher voltage. Although the battery you are testing may be in borderline condition (age and performance) and may barely be able to maintain minimum voltage, consider discontinuing its service. As mileage and vehicle age continue to increase over time, the electrical demands of the vehicle do not decrease. Therefore, a marginal battery nearing the end of its life cycle should be a candidate for replacement. Extreme hot or cold temperatures can suddenly cause a weak battery to fail. Photo Sequence 6 shows a typical procedure for testing battery capacity.

Battery open-circuit voltage is the voltage of the battery with no load on it.

⚙️ **SERVICE TIP** On automotive electrical test equipment such as battery testers and chargers, the positive cable clamp is red and the negative cable clamp is black.

CHARGING A BATTERY

There are two basic ways to recharge a battery: slow or fast. Knowing when to fast-charge or slow-charge a battery is important to your own safety and to the life of the battery.

Fast Charging

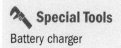

Special Tools

Battery charger

Before charging the battery, the battery case and terminals should be cleaned and the vent caps should be removed. Check the electrolyte level, and add distilled water as required. If the battery is in the vehicle, disconnect both battery cables. Determine the

PHOTO SEQUENCE 6
Typical Procedure for Testing Battery Capacity

P6-1 Check the battery open-circuit voltage before testing. If the voltage is not 12.6 volts, then charge the battery before testing.

P6-2 Be sure the load control is in the off position, and set the voltmeter control to 18 volts.

P6-3 Connect the volt-ampere tester cables to the battery terminals with the correct polarity.

P6-4 Connect the ammeter clamp over the negative tester cable.

P6-5 Set the tester control to the battery test position.

P6-6 Rotate the load control until the ammeter reads one-half the cold cranking ampere rating of the battery.

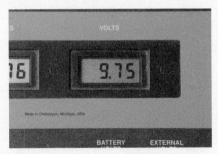

P6-7 Hold the load control in this position for 15 seconds, and read the battery voltage with the load still applied. If the battery temperature is 70°F, the voltage should be above 9.6 volts.

P6-8 Immediately rotate the load control to the off position, and disconnect the tester cables.

charging rate and time (**Figure 4-60**), and then follow these steps to properly and safely charge the battery:

⚡ **WARNING** Never connect or disconnect battery charger cables from the battery terminals with the charger in operation. This procedure causes sparks at the battery terminals, which may result in a battery explosion, personal injury, and/or property damage.

Reserve Capacity Rating	20-Hour Rating	5 Amperes	10 Amperes	20 Amperes	30 Amperes	40 Amperes
75 minutes or less	50 ampere-hours or less	10 hours	5 hours	2 1/2 hours	2 hours	
Above 75 to 115 minutes	Above 50 to 75 ampere-hours	15 hours	7 1/2 hours	3 1/4 hours	2 1/2 hours	2 hours
Above 115 to 160 minutes	Above 75 to 100 ampere-hours	20 hours	10 hours	5 hours	3 hours	2 1/2 hours
Above 160 to 245 minutes	Above 100 to 150 ampere-hours	30 hours	15 hours	7 1/2 hours	5 hours	3 1/2 hours

Figure 4-60 Battery charging time in relation to the battery's rating.

Special Tools

Jumper wire

Caution

When jump starting a vehicle, always turn off all electrical accessories in the vehicle being boosted and the boost vehicle. Electrical accessories left on during the boost procedure may be damaged.

Caution

While recharging a battery, watch for excessive gassing and observe the battery case temperature. Immediately slow the charge rate if the battery is boiling or the temperature goes over 125°F. If the open-circuit voltage (OCV) is less than 11 volts, the battery will take several hours to charge. It may be preferable to begin with a slow rate of charge.

1. Be sure the charger timer switch and main switch are off.
2. Connect the negative charger cable to the negative battery terminal and the positive charger cable to the positive battery terminal.
3. Connect the charger power cable to an electrical outlet.
4. Set the charger voltage switch to the proper battery voltage then turn on the main switch.
5. Set the timer to the required battery charging time.
6. Adjust the current control to obtain the proper charging rate. Some battery chargers have a reverse polarity protector that prevents charger operation if the charger cables are connected to a battery with reversed polarity. If the charger cables are connected properly and the battery voltage is very low, the reverse polarity protector may not allow the charger to begin charging the battery. Many chargers have a bypass button that must be pressed to bypass the reverse polarity protector when this condition occurs.
7. Shut the charger off while testing the battery. Continue charging the battery until the specific gravity is above 1.250 and the open-circuit voltage is above 12.6 volts.
8. When the battery is above 1.225 specific gravity, a high charging rate causes excessive gassing. Reduce the charging rate as required to prevent excessive gassing.
9. Shut the timer, current control, and main switch off, and then disconnect the charger cables from the battery.

Slow Charging

A battery may be slow-charged at one-tenth of the ampere-hour rating. For example, an 80 ampere-hour battery should be charged at 8 amps. Some slow chargers are capable of charging several batteries connected in series. A current control on the charger is rotated until the desired charging rate is obtained on the charger ammeter. The battery should be charged until the specific gravity is above 1.250 or until there is no further increase in specific gravity or battery voltage in 1 hour.

Charging Maintenance-Free Batteries

Maintenance-free batteries require different charging rates compared to low-maintenance batteries. Some vehicle manufacturers recommend using the reserve capacity rating of the battery to determine the ampere-hours charge required by the battery. For example, a battery with a reserve capacity of 75 minutes requires 25 amps × 3 hours charge = 75 ampere-hours. Continue charging the battery until the green dot appears in the hydrometer and the battery passes a capacity test. Always use the vehicle or battery manufacturer's recommended charging rate.

Parasitic Drain Testing

Parasitic drain is the current flow from the battery with the key off. Some parasitic drain is normal due to the on-board computers' memories staying active when the key is off. But if parasitic drain is excessive, the vehicle's battery may run down after sitting for a period of time. The normal amount of drain is in the range of 25 to 40 milliamps. Many computers have a higher drain for a short period of time after the ignition switch is turned off and then the drain is reduced. It is very important for technicians to understand how to perform an accurate battery drain test and determine if the drain is normal or excessive. To do this:

1. Be sure the ignition switch is off. Make sure the glove compartment and trunk lights are off and the doors are closed.
2. Disconnect the negative battery cable and connect a 12-gauge jumper wire between the cable end and the negative battery terminal. Do not attempt to start the engine once this jumper is connected.
3. Connect the ammeter on a multimeter in series between the negative battery cable end and the negative battery terminal (**Figure 4-61**). Set the ammeter scale on 20 amperes.
4. Turn the ignition switch on, and then turn on computer-controlled accessories, such as the A/C and radio or stereo.
5. Turn off all electrical accessories and the ignition switch.
6. Disconnect the 12-gauge jumper wire.
7. If the ammeter reading is very low, switch the ammeter scale to the lowest milliampere scale and observe the reading. It may take several minutes for all the computers to enter a resting mode and the current flow to fall significantly.
8. Check the vehicle manufacturer's specifications for the required time for the computers to reduce the amount of drain (**Figure 4-62**).
9. Compare the drain on the ammeter to the vehicle manufacturer's specifications. If the drain is excessive, disconnect the fuses or circuit breakers in various circuits to locate the source of the high drain.
10. Disconnect the multimeter, and then tighten the negative battery cable.

Parasitic drain refers to the current flow from the battery when the ignition switch is off.

Classroom Manual
Chapter 4, page 91

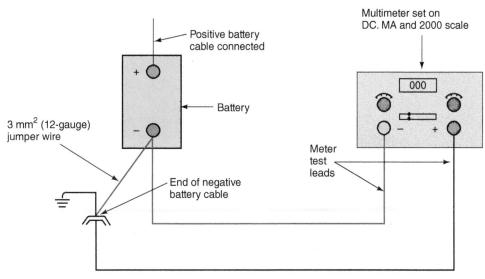

Figure 4-61 Jumper wire and ammeter connected to test battery drain.

Component Module	Current Draw (mA)	
	Typical	Max
ELC (after 7 min. time out)		1.0
Generator Heated windshield control module	2.0 0.3	2.0 0.4
Heated seat control modules		0.5
Instrument panel digital cluster	4.0	6.0
Light control module	0.5	1.0
Multi-function chime module (key removed from ignition)	1.0	1.0
Pass key decoder module	.75	1.0
PCM	5.0	7.0
Radio	7.0 1.8	8.5 3.5
RAC (theft deterrent) (illuminated entry) (auto door locks)		3.8

Figure 4-62 Computer drain specifications for different common accessories (typical).

STARTER DIAGNOSIS AND SERVICE

Check the condition and ensure that the battery is fully charged before the starter current draw test is performed. Always follow the starter current draw test procedure in the vehicle manufacturer's service manual. The following is a typical starter current draw test procedure:

1. Connect the ammeter inductive clamp on the negative battery cable or cables (**Figure 4-63**), including any chassis or body ground coming off the negative cable.
2. Be sure all the electrical accessories on the vehicle are off and the doors are closed.

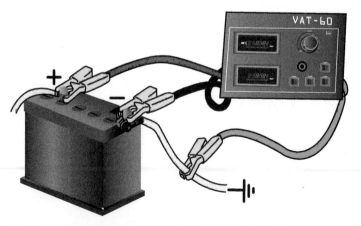

Figure 4-63 Volt-ampere tester connections for testing the starter and charging system.

3. Set the test selector to the starting position, if applicable.
4. Disable the ignition system. Always follow the vehicle manufacturer's recommended procedure for disabling the ignition system.
5. Crank the engine and observe the ammeter and voltmeter readings. The starter current draw should equal the specifications, and the voltage should be above the minimum cranking voltage.
6. Disconnect the tester cables and reconnect the positive primary coil wire.

Starter Current Draw Test Results

High current draw and low cranking speed usually indicate a defective starter. High current draw may also be caused by internal engine problems, such as partial bearing seizure. A low cranking speed and low current draw with high cranking voltage usually indicate excessive resistance in the starter circuit, such as cables and connections. Always remember that the battery and battery cable connections must be in satisfactory condition to obtain accurate starter current draw test results.

Voltage Drop Test, Starting Motor Circuit Insulated Side

The resistance in an electrical wire may be checked by measuring the voltage drop across the wire with normal current flow in the wire. To measure the voltage drop across the positive battery cable, connect the positive voltmeter lead to the positive battery cable at the battery, and connect the negative voltmeter lead to the other end of the positive battery cable at the starter solenoid (**Figure 4-64**). Set the voltmeter selector switch to the lowest scale and disable the ignition system. Crank the engine. The voltage drop indicated

Special Tools
Volt-ampere tester

A starter that draws too much current and cranks the engine slowly may be called a *dragging starter*.

⚠ **Caution**
Always follow the vehicle manufacturer's wiring and terminal repair procedure given in the service manual. On some components and circuits, manufacturers recommend component replacement rather than wiring repairs. For example, some manufacturers recommend replacing air bag system components such as sensors if the wiring or terminals are damaged on these components.

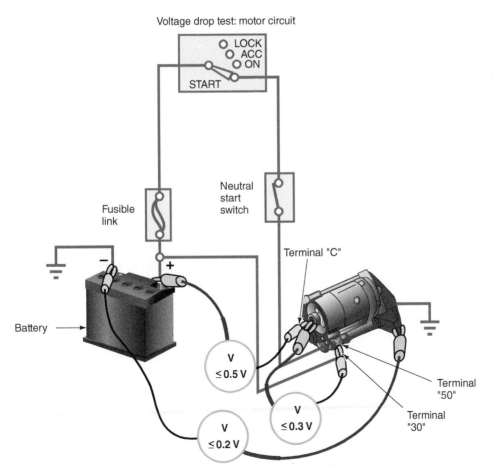

Figure 4-64 Voltmeter connections for measuring voltage drops in the starter and starter control circuits.

on the meter should not exceed 0.5 volts. If the voltage reading is above this figure, the cable has excessive resistance. If the cable ends are clean and tight, replace the cable.

Connect the positive voltmeter lead to the positive battery cable on the starter solenoid, and connect the negative voltmeter lead to the starting motor terminal on the other side of the solenoid. Leave the voltmeter on the lowest scale and crank the engine. If the voltage drop exceeds 0.3 volts, the solenoid disc and terminals have excessive resistance.

Voltage Drop Test, Starting Motor Ground Side

Connect the positive voltmeter lead to the starter case, and connect the negative voltmeter lead to the negative battery cable on the battery. Leave the ignition system disabled, and place the voltmeter selector on the lowest scale. If the voltage drop reading exceeds 0.2 volts while cranking the engine, the negative battery cable or ground return circuit has excessive resistance.

Voltage Drop Tests, Control Circuit

Connect the positive voltmeter lead to the positive battery cable at the battery, and connect the negative voltmeter lead to the solenoid winding terminal on the solenoid. Leave the ignition system disabled, and place the voltmeter selector on the lowest scale. If the voltage drop across the control circuit exceeds 1.5 volts while cranking the engine, individual voltage drop tests on control circuit components are necessary to locate the high-resistance problem.

 SERVICE TIP A lower reading is desirable in *any* voltage drop test.

Connect the voltmeter leads across individual control circuit components such as the ignition switch, neutral safety switch, and starter relay contacts to measure the voltage drop across these components while cranking the engine. In most starting motor control components, if the voltage drop exceeds 0.2 volts, the component is defective. The voltmeter leads may also be connected across individual wires in the starter control circuit to test voltage drop in the wires. Before any starting motor circuit component is replaced, always remove the negative battery cable. High resistance in starting motor control components and wires may cause the starting circuit to be inoperative. High resistance in the solenoid disc and terminals may cause a clicking action when the ignition switch is turned to the start position.

Classroom Manual
Chapter 4, page 117

AC GENERATOR DIAGNOSIS AND SERVICE

AC generator belt condition and tension are extremely important for satisfactory AC generator operation. A loose belt causes low AC generator output and a discharged battery. A loose, dry, or worn belt may cause squealing and chirping noises, especially during engine acceleration and cornering.

The AC generator belt should be checked for oil soaking, worn or glazed edges, or signs of abrasion. If any of these conditions are present, belt replacement is necessary.

Most AC generators are driven by a serpentine belt. The serpentine belt should be checked to make sure it is installed properly on each pulley in the belt drive.

Serpentine belts have a spring-loaded tensioner pulley that automatically maintains **belt tension** (**Figure 4-65**). As the belt wears or stretches, the spring moves the tensioner pulley to maintain the belt tension. Some of these tensioners have a belt length scale that indicates new belt range and used belt range. If the indicator on the tensioner is out of the used belt length range, belt replacement is required. Many belt tensioners have a

A ribbed V-belt may be called a *serpentine belt*.

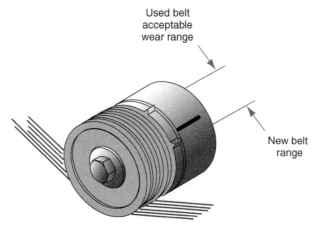

Figure 4-65 Serpentine belt tensioner.

one-half-inch drive opening in which a ratchet or flex handle may be installed to move the tensioner pulley off the belt during belt replacement.

Belt pulleys must be properly aligned to minimize belt wear. The edges of the pulleys must be in line when a straightedge is placed on the pulleys.

AC Generator Output Test

The AC generator belt must be in satisfactory condition before the output test is performed. Always follow the vehicle recommended procedure for testing AC generator output. Photo Sequence 7 covers a typical AC generator output test procedure.

Because the ammeter inductive pickup is installed on the negative battery cable, the ammeter only indicates the AC generator current flow through the battery during the output test. The current is supplied directly from the AC generator to the ignition system and AC generator field, so the current draw for these two systems must be added to the AC generator output to obtain an accurate output reading. If the AC generator output is not within 10 percent of the specified output, further diagnosis is required.

AC Generator Regulator Voltage Test

To check the AC generator for proper regulator voltage, leave the volt-ampere tester connected as in the output test. Operate the engine at 2,000 rpm until the battery is sufficiently charged to provide a charging rate of 10 amperes or less. Observe the voltmeter reading. The voltage should be at the specified regulator voltage.

If the AC generator output is zero, the AC generator field circuit may have an open circuit. The most likely place for an open circuit in the AC generator field is at the slip rings and brushes. When the AC generator voltage is normal but the amps output is zero, the fuse link between the AC generator battery terminal and the positive battery cable may be open. With the ignition switch off, battery voltage should be available on each end of the fuse link. If the voltage is normal on the battery side of the fuse link and zero on the AC generator side, the fuse link is open and must be replaced.

If the AC generator output is less than specified, always be sure that the belt and belt tension are satisfactory. When the belt and belt tension are satisfactory and the AC generator output is less than specified, the AC generator is defective.

Battery overcharging causes excessive gassing of the battery. Overcharging may be caused by a defective voltage regulator, which allows a higher-than-specified charging voltage.

Battery undercharging eventually results in a discharged battery and a no-start problem. Undercharging may be caused by a loose or worn belt, a defective AC generator, low regulator voltage, high resistance in the battery wire between the AC generator and the battery, or an open fuse link.

 Caution

Never allow the AC generator voltage to go above 15.5 volts. High voltage may damage computers and other electrical and electronic equipment on the vehicle.

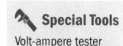 **Special Tools**

Volt-ampere tester

 Caution

Never disconnect the circuit between the AC generator and the battery with the engine running. This action may cause extremely high voltage momentarily, which causes electronic component damage.

P7-1 Testing the charging system begins with a visual inspection of the battery and its cables. Make sure they are clean and free of corrosion. The AC generator drive belt should also be inspected and its tension checked.

P7-2 To test the charging system, the tester's ampere pickup probe must be positioned around the battery's negative cable and its cables connected to the battery.

P7-3 To observe the tester's meters during this test, the tester should be placed where it can easily be seen from the driver's seat.

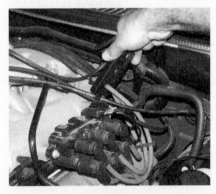

P7-4 It is recommended that engine speed be monitored. Therefore, connect a tachometer to the engine. Some testers, such as the one shown, are equipped with a tachometer that monitors ignition pulses at the plug wires.

P7-5 At about 2,000 rpm, the charging system will deliver a maximum amount of charge to the battery. This charging system is providing 14.6 volts at 52.5 amps. Readings should be compared to specifications. Engine rpm does not show in the photo because the photo was taken after the test, not during it.

P7-6 If the readings are low, this may indicate that there is a fault in the charging system or that the battery is fully charged and not allowing the AC generator to work at its peak. To determine the cause, the voltage regulator can sometimes be bypassed or commanded to full output according to the procedure outlined in service information.

P7-7 With the regulator bypassed, this AC generator increased its voltage output. This indicates that the regulator had been regulating voltage output. The amperage reading is also well within specifications. Therefore, the charging system is functioning properly.

AC Generator Output Requirement Test

AC generators are rated according to their output capacity in amperes. On occasion, it may be necessary to choose the proper AC generator based on the vehicle's electrical load requirements. If an undersized AC generator is installed on a vehicle containing many electrical components and accessories, the battery will likely fail under certain conditions because it is undercharged. In addition, the AC generator may be stressed to keep up with the electrical demands and may also experience problems. To determine the correct ampere output AC generator, perform the following:

1. Connect the amp probe from the tester to the negative battery cable and zero the ammeter if necessary.
2. Switch the ignition to the "on" position. Turn on all the accessories in the vehicle. Be sure to include lights, rear window defroster, heater blower motor, brake lamps, and so on.
3. If applicable, add an approximate estimated amperage value for the electric cooling fan(s) and electric fuel pump.
4. Total all live and calculated ampere values.
5. As a rule of thumb, the AC generator should be rated approximately 5 amperes over the maximum calculated ampere requirement.

J1930 naming uses the term *AC generator* instead of the commonly used term *alternator*.

> ✿ **SERVICE TIP** It is good practice to check the output waveform of an AC generator any time an electronic component fails. Because the electronics of the vehicle cannot accept AC current, the damage to the faulty component may have been caused by a bad diode, allowing AC noise to be present in the charging system's output.

Diode Pattern Testing

One of the final tests that should be performed on the AC generator is the diode or diode–stator test. This test checks the AC component of the DC voltage after rectification. This is measured with an AC voltmeter. Some charging system testers show this as a separate step in the test sequence, while other testers automatically perform it. There are several good reasons to check this part of the AC generator. If one of the six diodes in the AC generator circuit fails, the AC generator can lose at least 17 percent of its output. Often, this can take place without any instrument-warning lamp alert to the driver. Depending on the type of diode problem, results can include a loss of output, AC generator noise being heard through the entertainment system, or the generator-warning lamp being illuminated dimly while running *or* with the engine off. In addition, an excessive amount of AC voltage output from the AC generator due to a failed diode can cause false or phantom signals to be generated and may influence computer input and output signals or the ignition system. To perform the diode or diode–stator test, simply have the tester connected with the selector set and run the engine at 2,000 rpm. Load the system to about 12.5 volts while maintaining 2,000 rpm, and observe the tester reading. It usually shows a go or no-go test result. Some computerized engine analyzers also perform this test during what is referred to as a high cruise test. During this test, there is usually an electrical load placed on the system by turning on the headlamps and high-speed blower. This works the AC generator at a high cruise speed, causing it to put out a higher charge. The diode ripple voltage is displayed either digitally or as a waveform. The analyzer will usually flag an

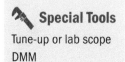

Special Tools
Tune-up or lab scope
DMM

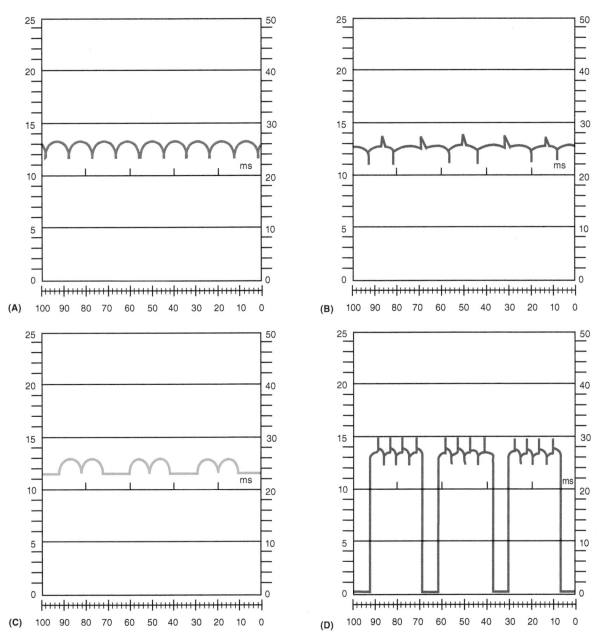

Figure 4-66 AC generator oscilloscope patterns: (A) good AC generator under full load, (B) good AC generator under no load, (C) shorted diode and/or stator winding under full load, and (D) open diode in diode trio.

AC generator ripple greater than 0.12 volts, just as you would if you were using any DMM. Most oscilloscopes can be set up to display an AC generator diode pattern. Connect the appropriate leads either on the positive battery terminal or preferably on the generator's output terminal. Set the scope controls for this test. Start the engine, and hold it at about 2,000 rpm. Next, load the AC generator and observe the pattern. A representation of both good and problem patterns is illustrated in **Figure 4-66**. To perform this test using a DMM, set the meter to low AC voltage, place the leads across the battery terminals, and run the engine as stated earlier. Again, the reading should not exceed 0.12 AC volts. If the readings or diode patterns are unsatisfactory, a teardown or replacement of the AC generator is necessary.

CASE STUDY

A customer complained about a no-start problem on a Chevrolet Malibu. When the service writer questioned the customer about the problem, the customer indicated the battery seemed to be discharged and would not crank the engine. The customer also indicated this problem occurred about once a month.

When the technician lifted the hood, she noticed the battery, AC generator, and starting motor had been replaced, which indicated that someone else had likely been trying to correct this problem. The technician checked the belt and all wiring connections in the starting and charging systems without finding any problems. Next, the technician performed starter current draw and AC generator output tests, and these systems performed satisfactorily. The technician checked the regulator voltage and found it was within specifications. After the battery was charged, the technician performed a capacity test and the battery performance was satisfactory. Since the battery, starting system, and charging system tests were normal, the technician concluded there must be a battery drain problem. The technician connected an ammeter and jumper wire in series between the negative battery cable and terminal, and proceeded with the drain test. When the jumper wire was connected, the ammeter indicated a continual 1 amp drain. The technician began removing the fuses one at a time and checking the ammeter reading. When the courtesy lamp fuse was disconnected, the drain disappeared.

The technician checked the wiring diagram for the vehicle and found the vehicle had a remote keyless entry module. When the courtesy lamp fuse was installed and the remote keyless entry module disconnected, the drain disappeared on the ammeter. This module was draining current through it to ground, but the current flow was not high enough to illuminate the courtesy lights.

A new remote keyless entry module was installed, and the drain was rechecked. The drain was now within the vehicle manufacturer's specifications.

ASE-STYLE REVIEW QUESTIONS

1. While discussing AWG wiring numbering systems:

   Technician A says that most wiring used in vehicles has an AWG number from 10 to 18 gauge.

   Technician B says that on the AWG scale, 0 is the smallest and 20 is the largest.

   Who is correct?

   A. A only
   B. B only
   C. Both A and B
   D. Neither A nor B

2. While discussing resistance,

   Technician A says current will increase with a decrease in resistance.

   Technician B says current will decrease with an increase in resistance.

   Who is correct?

   A. A only
   B. B only
   C. Both A and B
   D. Neither A nor B

3. While discussing oscilloscope patterns,

   Technician A says frequency is measured by counting the number of cycles in 1 second.

   Technician B says a frequency of 10 Hz means there are ten cycles per second.

   Who is correct?

   A. A only
   B. B only
   C. Both A and B
   D. Neither A nor B

4. While discussing electricity,

   Technician A says a short results from an unwanted path for current.

   Technician B says high-resistance problems cause increased current flow.

   Who is correct?

   A. A only
   B. B only
   C. Both A and B
   D. Neither A nor B

5. While discussing battery service,

   Technician A says if a battery is disconnected, computer adaptive memories are erased.

   Technician B says if a battery is disconnected, the memory seat and memory mirror positions are erased.

   Who is correct?

   A. A only   C. Both A and B

   B. B only   D. Neither A nor B

6. While discussing the battery capacity test,

   Technician A says the battery should be discharged at one-third of the cold cranking ampere rating.

   Technician B says a satisfactory battery at 70°F has at least 9.6 volts at the end of the 15-second capacity test.

   Who is correct?

   A. A only   C. Both A and B

   B. B only   D. Neither A nor B

7. While discussing AC generator output,

   Technician A says zero AC generator output may be caused by an open field circuit.

   Technician B says zero AC generator amperage output may be caused by an open AC generator fuse link.

   Who is correct?

   A. A only   C. Both A and B

   B. B only   D. Neither A nor B

8. While discussing battery drain testing,

   Technician A says computers have a low-milliampere drain with the ignition off.

   Technician B says the fuses may be disconnected one at a time to locate the cause of a battery drain.

   Who is correct?

   A. A only   C. Both A and B

   B. B only   D. Neither A nor B

9. While discussing the starter current draw test,

   Technician A says the battery should be at least 50 percent charged before the starter draw test is performed.

   Technician B says the electrical accessories on the vehicle should be on during the starter draw test.

   Who is correct?

   A. A only   C. Both A and B

   B. B only   D. Neither A nor B

10. While discussing wiring harness and terminal repairs,

    Technician A says acid-core solder should be used for soldering electrical terminals.

    Technician B says sealant should appear at both ends of a crimp and seal sleeve after it is heated.

    Who is correct?

    A. A only   C. Both A and B

    B. B only   D. Neither A nor B

ASE CHALLENGE QUESTIONS

1. A parasitic battery drain test is being performed.

   Technician A says the ammeter must be connected in series between the battery's negative post and the negative cable.

   Technician B says all circuits must be off (open) for this test to be valid.

   Who is correct?

   A. A only   C. Both A and B

   B. B only   D. Neither A nor B

2. *Technician A* says a rheostat cannot be tested with an ohmmeter.

   Technician B says a potentiometer cannot be tested with an ohmmeter.

   Who is correct?

   A. A only   C. Both A and B

   B. B only   D. Neither A nor B

3. A jumper wire has been used to energize the coil of a relay, and an ohmmeter has been connected across the contacts of the relay.

   Technician A says an infinity resistance reading across the contacts indicates a good relay.

   Technician B says a zero resistance reading across the contacts indicates the problem is elsewhere in the circuit.

   Who is correct?

   A. A only   C. Both A and B

   B. B only   D. Neither A nor B

4. A diode is being tested with a digital multimeter equipped with a diode check function. The meter reads very high resistance in one direction and 0.20 in the other direction. This indicates a(n):

   A. shorted diode.

   B. open diode.

   C. normal germanium diode.

   D. additional testing is needed.

5. *Technician* A says that the before checking a charging system, the battery should be charged to provide a charging rate of 20 amps or less.

   Technician B says a charging rate of zero may indicate an open in the AC generator field.

   Who is correct?

   A. Tech A   C. Both Techs

   B. Tech B   D. Neither Tech

Name _____   **Date** _____

USING A DSO ON SENSORS AND SWITCHES

Upon completion of this job sheet, you should be able to connect a DSO and observe the activity of various sensors and switches.

ASE Education Foundation Correlation

This job sheet addresses the following **AST/MAST** task: VIII. Engine Performance; A. General: Engine Diagnosis

Task #9   Diagnose engine mechanical, electrical, electronic, fuel, and ignition concerns; determine necessary action. **(P-2)**

This job sheet addresses the following **MAST** task: Engine Performance; **B. Computerized Engine Controls Diagnosis and Repair**

Task #7   Inspect and test computerized engine control system sensors, powertrain/ engine control module (PCM/ECM), actuators, and circuits using a graphing multimeter (GMM)/digital storage oscilloscope (DSO); perform necessary action. **(P-2)**

Tools and Materials

- A vehicle with accessible sensors and switches
- Service manual for the above vehicle
- Component locator manual for the above vehicle
- A DSO
- A DMM

Describe the vehicle being worked on:

Year _____ Make _____

Model _____ VIN _____

Procedure

1. Connect the DSO across the battery. Make sure the scope is properly set. Observe the trace on the scope. Is there evidence of noise? Explain.

   _____

   _____

   _____

   _____

2. Locate the A/C compressor clutch control wires. Start and run the engine. Connect the DMM to read available voltage. Observe the meter, and then turn the compressor on. What happened on the meter?

   _____

   _____

   Now connect the DSO to the same point with the compressor turned off. Observe the waveform, and then turn the compressor on. What happened to the trace?

   _____

   _____

3. Turn the engine off but keep the ignition on. Locate the TP sensor, and identify the purpose of each wire to it. List each wire, and describe the purpose of each.

_____

_____

4. Connect the DMM to read reference voltage at the TP sensor. What do you read?

_____

_____

Now move the leads to read the output of the sensor. Starting with the throttle closed, slowly open the throttle until it is wide open. Watch the voltmeter while doing this. Describe your readings.

_____

_____

5. Now connect the DSO to read the reference voltage at the TP sensor. What do you see on the trace?

_____

_____

Now move the leads to read the output of the sensor. Starting with the throttle closed, slowly open the throttle until it is wide open. Watch the trace while doing this. Describe your readings.

_____

_____

6. Now run the engine. Locate the oxygen sensor, and identify the purpose of each wire to it. Connect the DMM to read voltage generated by the sensor. (To do this, you may use an electrical connector for the O_2 sensor that is positioned away from the hot exhaust manifold.) Watch the meter, and describe below what happened.

_____

_____

_____

_____

7. Now connect the DSO to read voltage output from the sensor. Watch the trace, and describe what happened.

_____

_____

_____

_____

8. Explain what you observed as the differences between testing with a DMM and a DSO.

_____

_____

_____

_____

Instructor's Response

_____

_____

_____

Name _____   Date _____

INSPECTING A BATTERY

Upon completion of this job sheet, you should be able to visually inspect a battery.

ASE Education Foundation Correlation

This job sheet addresses the following **AST/MAST** tasks: VI. Electrical/Electronic Systems; A. General Electrical Diagnosis

Task #8   Diagnose the cause(s) of excessive key-off battery drain (parasitic draw); determine necessary action. **(P-1)**

B. Battery System Service,

Task #4   Inspect and clean battery; fill battery cells; check battery cables, connectors, clamps, and hold-downs. **(P-1)**

Tools and Materials

- A vehicle with a 12-volt battery
- A DMM

Procedure   Task Completed

1. Describe the general appearance of the battery.

   _____

   _____

   _____

2. Describe the general appearance of the cables and terminals.

   _____

   _____

   _____

3. Check the tightness of the cables at both ends. Describe their condition.

   _____

   _____

   _____

4. Connect the positive lead of the meter (set on DC volts) to the positive terminal of the   ☐
   battery.

5. Put the negative lead on the battery case, and move it all around the top and sides of
   the case. What readings do you get on the voltmeter?

   _____

   _____

   _____

6. What is indicated by the readings?

   _____

   _____

   _____

7. Measure the voltage of the battery. Your reading was _____ volts.

8. What do you know about the condition of the battery based on the visual inspection and above tests?

_____

_____

_____

Instructor's Response

_____

_____

_____

Name _____   Date _____

TESTING THE BATTERY'S CAPACITY

Upon completion of this job sheet, you should be able to test a battery's capacity.

ASE Education Foundation Correlation

This job sheet addresses the following **AST/MAST** task: Electrical/Electronic Systems; Battery Diagnosis and Service

Task #2   Confirm proper battery capacity for vehicle application; perform battery capacity and load test; determine needed action. **(P-1)**

Tools and Materials

• A vehicle with a 12-volt battery
• Service manual for the above vehicle
• Starting or charging system tester

Describe the vehicle being worked on:

Year _____ Make _____

Model _____ VIN _____

Procedure   **Task Completed**

1. Perform a battery state-of-charge test:

   Record the specific gravity readings for each cell (if applicable):

   (1) _____   (2) _____   (3) _____   (4) _____   (5) _____   (6) _____

   If the battery is a maintenance-free type of battery, what is the open-circuit voltage? _____ volts

2. Summarize the battery's state of charge from the above, and indicate the percentage of charge.

   _____

   _____

   _____

3. Connect the starting or charging system tester to the battery.   ☐

4. Locate the CCA rating of the battery. What is the rating? _____

5. Based on the CCA, how much load should be put on the battery during the capacity test? _____ amps

6. Conduct the battery load test.

   Battery voltage decreased to _____ volts after _____ seconds.

7. Describe the results of the battery load (capacity) test. Include in the results your service recommendations and the reasons for them.

_____

_____

_____

_____

_____

_____

Instructor's Response

_____

_____

_____

Name _____   Date _____

TESTING CHARGING SYSTEM OUTPUT

Upon completion of this job sheet, you should be able to measure the output of the charging system.

ASE Education Foundation Correlation

This job sheet addresses the following **AST/MAST** task: Electrical/Electronic Systems; D. Charging System Diagnosis and Repair

Task #2   Diagnose (troubleshoot) charging system for causes of undercharge, no-charge, or overcharge conditions. **(P-1)**

Tools and Materials

- A vehicle
- Service information for the above vehicle
- Starting or charging system tester

Describe the vehicle being worked on:

Year _____ Make _____

Model _____ VIN _____

Procedure

Task Completed

1. Identify the type and model of AC generator. What type is it? _____

   What are the output specifications for this AC generator? _____ amps and

   _____ volts at _____ rpm

2. Connect the starting or charging system tester to the vehicle.   ☐

3. Start the engine, and run it at the specified engine speed.   ☐

4. Apply the load to the electrical system in order to obtain the highest amp reading.   ☐

5. Observe the output to the battery. The meter readings are:

   _____ amps and _____ volts (after load is removed and voltage stabilizes)

6. Compare the readings to specifications, and give recommendations.

   _____

   _____

   _____

7. If readings are outside of the specifications, refer to the service manual to find the proper way to full-field the AC generator. Describe the method.

   _____

   _____

   _____

8. Full-field the generator, and observe the output to the battery. The meter readings are:
   _____ amps and _____ volts

9. Compare readings to specifications, and give recommendations.

_____

_____

_____

Instructor's Response

_____

_____

_____

CHAPTER 5

INTAKE AND EXHAUST SYSTEM DIAGNOSIS AND SERVICE

Upon completion and review of this chapter, you should be able to:

- Service and replace air filters.
- Inspect and troubleshoot vacuum and air induction systems.
- Visually inspect the exhaust system and determine needed repairs.
- Diagnose catalytic converters.
- Remove, replace, and service exhaust manifolds.
- Remove, replace, and service mufflers, pipes, and catalytic converters.
- Perform a turbocharger inspection.

- Diagnose intake and exhaust leaks that affect turbocharger operation.
- Test turbocharger boost pressure.
- Measure axial shaft movement in a turbocharger.
- Measure waste-gate stroke.
- Inspect a turbocharger and describe some common turbocharger problems.
- Explain supercharger operation and identify common supercharger problems.

Terms To Know

Boost pressure	Oxygen (O_2) sensor	Pyrometer
Downstream	Polyurethane air cleaner	Slitting tool

INTRODUCTION

An internal combustion engine requires air to operate. This air supply is drawn into the engine by the vacuum created during the intake stroke of the pistons. The air is mixed with fuel and is delivered to the combustion chambers. Controlling the flow of air and the air-fuel mixture is the job of the induction system.

Prior to the introduction of emission control devices, the induction system was quite simple. It consisted of an air cleaner housing mounted on top of the engine with a filter inside the housing. Its function was to filter dust and grit from the air being drawn into the engine.

AUTHOR'S NOTE If the mass airflow (MAF) is mounted on a remote air cleaner box, any leak in the air intake hose from the MAF to the intake results in unmetered air entering the engine, resulting in a very poorly running engine. This is usually most noticeable on acceleration when a cracked hose tends to open with engine movement.

Modern air induction systems do much more than simply filter the air (**Figure 5-1**). Ducts channel cool air from outside the engine compartment to the throttle plate assembly. The air filter has been moved to a position away from the top of the engine to allow for

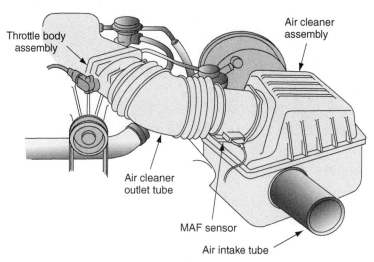

Figure 5-1 Typical air intake system for a late-model engine.

aerodynamic body designs. Electronic meters measure airflow, temperature, and density. These components allow the air induction system to perform the following functions:

- Provide the air that the engine needs to operate.
- Filter the air to protect the engine from wear.
- Monitor airflow temperature and density for more efficient combustion and a reduction of hydrocarbon (HC) and carbon monoxide (CO) emissions.
- Operate with the positive crankcase ventilation (PCV) system to burn the crankcase fumes in the engine.
- Provide air for some air injection systems.

AIR FILTERS

Classroom Manual
Chapter 5, page 128

If the air filter becomes very dirty, the dirt can block the flow of air into the fuel-charging assembly. Without enough air, the engine will constantly receive a rich air-fuel mixture. The use of extra fuel means poor fuel economy and a lack of power.

The manufacturer's recommended air filter replacement interval is usually specified in the vehicle owner's manual. However, replacement of the air filter might be required on a more frequent schedule if the vehicle is subjected to continuous operation in an extremely dusty or severe off-the-road environment. A damaged air filter may cause increased wear on cylinder walls, pistons, and piston rings.

Follow these steps for air filter service or replacement:

Some popular reusable air filters will pass water, so use with care.

1. Remove the wing nut or retaining bolts or clips on the air cleaner cover and remove the cover to access the air filter element (**Figure 5-2**).
2. Remove the air filter element from the air cleaner and be sure that no foreign material, such as small stones, drops into the throttle body while removing the element. If the air filter assembly is remote, remove all dirt, rocks, and so on from the air filter housing and ductwork.
3. Visually inspect the air filter for pin holes in the paper element and damage to the paper element, sealing surfaces, or metal screens on both sides of the element. If the element is damaged or contains pin holes, replace the element.
4. Place a trouble light on the inside surface of the air filter and look through the filter element to the light. The light should be visible through the paper element but no holes should appear in the element. If the paper element is plugged with dirt or oil,

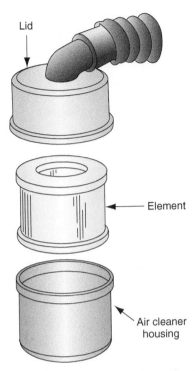

Figure 5-2 To remove the air filter, separate the cover (lid) from the housing.

Lid

Element

Air cleaner housing

Figure 5-3 Inspecting an air cleaner element. Make sure the housing is wiped clean when installing a new filter.

the light is not visible through the element. When the element is plugged with dirt or oil, replace the element. If the air filter element is contaminated with oil, excessive blowby or a defective PCV system is causing a pressure buildup in the engine.

5. If there is any dirt in the air cleaner body around the air filter, remove the air cleaner body and use a clean shop towel to remove the foreign material from the air cleaner body (**Figure 5-3**).

6. Install the air filter element and be sure the seal on the lower side of the element fits evenly against the matching surface on the air cleaner body.

⚠ **Caution**

If air is directed from an air gun against the outside of an air filter element, the blast of air may blow dirt particles through the element and create pin holes in the element.

Figure 5-4 An air filter condition indicator.

7. Install the air cleaner cover and be sure the cover sealing area fits against the element.
8. Install and tighten the cover retainers.
9. Be sure the PCV hose and any other hoses or sensors are properly connected to the air cleaner.

An air gun on a shop air hose may be used to blow the dirt and foreign material from the air filter element. Always keep the air gun 6 inches away from the air filter and direct the air against the inside of the air filter. While blowing dirt out of an air filter, do not allow the air pressure to exceed 30 psi (207 kPa). After blowing the dirt from the air filter, re-inspect the air filter element for pin holes and remaining dirt with a shop light.

Some air filters have an indicator built into the air cleaner assembly. When the filter is in need of replacement, the indicator turns red. The indicator works by measuring the difference in the pressure on both sides of the air cleaner with the engine running (**Figure 5-4**).

 WARNING Do not direct air from an air gun against any part of your body. If air penetrates the skin and enters the bloodstream, it will cause serious personal injury or death.

Heavy-Duty Air Cleaner Service

Some heavy-duty air filter elements have a polyurethane cover over the top of the paper air filter element. This polyurethane cover may be removed and washed in an approved solvent. After washing the cover, the excess solvent should be squeezed from the element. A light coating of ordinary crankcase oil should be placed on the polyurethane cover, and the cover may be squeezed to distribute the oil evenly.

Some heavy-duty air filter elements have a heavy paper element encased between two metal end caps. This type of air filter element may be cleaned and reused. Before cleaning the element, it should be inspected for tears, punctures, pin holes, and bent end caps. If any of these conditions are present, replace the element.

Other heavy-duty air filter elements may be washed in an approved filter-cleaning solution for 15 minutes or more. The filter may be rinsed with low-pressure water from a water hose and then allowed to dry. Always follow the vehicle manufacturer's air filter element-cleaning instructions in the service information.

Always take care when blowing off an air filter, as you may make small holes that can allow dirt to enter the engine!

Caution

Do not wring or twist a polyurethane air cleaner element cover. This action stretches the cover and makes it useless.

A few heavy-duty air cleaners have a pre-cleaner containing a series of tubes and fins. These pre-cleaners should be inspected for dirt accumulation. A brush with stiff nylon or fiber bristles may be used to clean the tubes and fins in the pre-cleaner.

ENGINE VACUUM

Vacuum is measured in relation to atmospheric pressure. Atmospheric pressure is caused by the weight of the surrounding air. At sea level, the pressure exerted by the atmosphere is 14.7 psi. Most pressure gauges ignore atmospheric pressure and read zero under normal conditions. The normal measure of vacuum is in inches of mercury (in. Hg) instead of psi. Other units of measurement for vacuum are kilopascals and bars. Normal atmospheric pressure at sea level is about 1 bar (one bar is actually 14.5 psi) or 100 kilopascals.

The downward movement of the piston creates vacuum in any four-stroke engine during the intake stroke (**Figure 5-5**). With the intake valve open and the piston moving downward, a partial vacuum is created within the cylinder and intake manifold. The air passing the intake valve does not move fast enough to fill the cylinder, thereby causing the lower pressure. This partial vacuum is continuous in a multi-cylinder engine since at least one cylinder is always at some stage of its intake stroke.

The amount of vacuum created is partially related to the positioning of the throttle plate. The throttle plate not only admits air or air-fuel into the intake manifold, but also helps control the amount of vacuum available during engine operation. At closed throttle idle, the vacuum available is usually between 17 and 22 inches. At wide-open throttle acceleration, the vacuum can drop to zero. A vacuum gauge will show a low but steady vacuum at idle if there is a vacuum leak. A good way to find a vacuum leak is with a smoke test, such as that shown in Photo Sequence 8 in this chapter.

An intake vacuum leak causes a low steady reading on a vacuum gauge connected to the intake manifold.

Intake Manifold Tuning Valve

The intake manifold tuning valve (IMTV) does not have any codes, because it does not have to be monitored according to the U.S. Environmental Protection Agency (EPA). If the vehicle seems to lack power at low speeds, check to ensure that the valve is not stuck open. If power seems to be lower than normal, then the valve may not be opening at high RPM.

Classroom Manual
Chapter 5, page 131

Intake Manifold Runner Control

The powertrain control module (PCM) uses the signals from the throttle position (TP) sensor and the revolutions per minute (RPM) signal to open the butterfly valves. The PCM

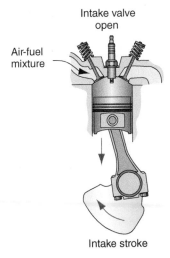

Intake valve
open

Air-fuel
mixture

Intake stroke

Figure 5-5 Vacuum is formed on the engine's intake stroke.

uses a sensor to determine if the intake manifold runner control (IMRC) is operating as designed. Diagnostic trouble codes (DTCs) are available for any valve that is stuck open, closed, or intermittent.

VACUUM PROBLEM DIAGNOSIS

Vacuum Routing Schematic

The vacuum in the intake manifold may operate the brake booster.

An engine emissions vacuum schematic is required to be displayed on all domestic and import vehicles. It is located on an under-hood decal. This schematic shows the vacuum hose routings and vacuum source for all emissions-related equipment. The vacuum schematic shown in **Figure 5-6** shows the relationship and position of components as they are mounted on the engine. It is important to remember that these schematics only show the vacuum-controlled parts of the emission system. Hose routing and positions for vacuum devices not related to the emissions control system can be found in the service information.

Diagnosis and Troubleshooting

Vacuum system problems such as leaks can produce or contribute to the following drivability symptoms:

- Stalls
- No start (cold)
- Hard start (hot soak)
- Backfire (deceleration)
- Rough idle
- Poor acceleration
- Rich or lean stumble
- Overheating
- Detonation, or knock or pinging
- Poor fuel economy

Figure 5-6 Vehicle emissions label information.

As a routine part of problem diagnosis, a technician suspecting a vacuum problem should first visually inspect all vacuum hoses and vacuum-operated components. The following are some guidelines for inspecting the vacuum systems:

1. Check all vacuum hoses to be sure they are properly routed and connected. The under-hood vacuum decal indicates proper vacuum hose routing.
2. Inspect all vacuum hoses for proper tight fit between the hoses and nipples.
3. Inspect all vacuum hoses for kinks, breaks, and cuts.
4. Be sure vacuum hoses are not burned because they are positioned near hot components such as exhaust manifolds or exhaust gas recirculation (EGR) tubes.
5. Inspect all vacuum-operated components for damage.
6. Check for evidence of oil on vacuum hose connections. Oil in a vacuum hose may contaminate vacuum-operated components or plug the vacuum hose.

Broken or disconnected hoses allow vacuum leaks that admit more air into the intake manifold than the mass air flow sensor has measured. The most common result is a rough running engine due to the leaner air-fuel mixture created by the excess air.

Kinked hoses can cut off vacuum to a component, thereby disabling it. For example, if the vacuum hose to the EGR valve is kinked, vacuum cannot be used to move the diaphragm and the valve will not open.

To check vacuum controls, refer to the service information for the correct location and identification of the components. Typical locations of vacuum-controlled components are shown in **Figure 5-7**.

Tears and kinks in any vacuum line can affect engine operation. Any defective hoses should be replaced one at a time to avoid misrouting. Original equipment manufacturer (OEM) vacuum lines are installed in a harness consisting of ⅛-inch or larger outer diameter and 1/16-inch inner diameter nylon hose with bonded nylon or rubber connectors. Occasionally, a rubber hose might be connected to the harness. The nylon connectors have rubber inserts to provide a seal between the nylon connector and the component connection (nipple). Although there are still many vacuum-controlled EGR systems on the road, the advent of electronically controlled EGR valves has eliminated many of the old vacuum controls used previously.

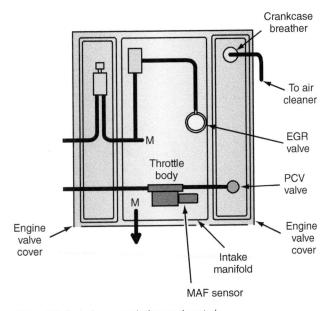

Figure 5-7 Typical vacuum devices and controls.

Vacuum Test Equipment

The vacuum gauge is one of the most important engine diagnostic tools used by technicians. With the gauge connected to the intake manifold and the engine warm and idling, watch the action of the gauge's needle. A healthy engine will give a steady, constant vacuum reading between 17 and 22 in. Hg cylinder. **Figure 5-8** shows some of the common

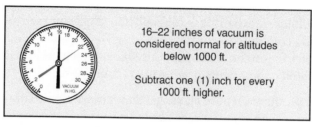

(A) Normal engine vacuum

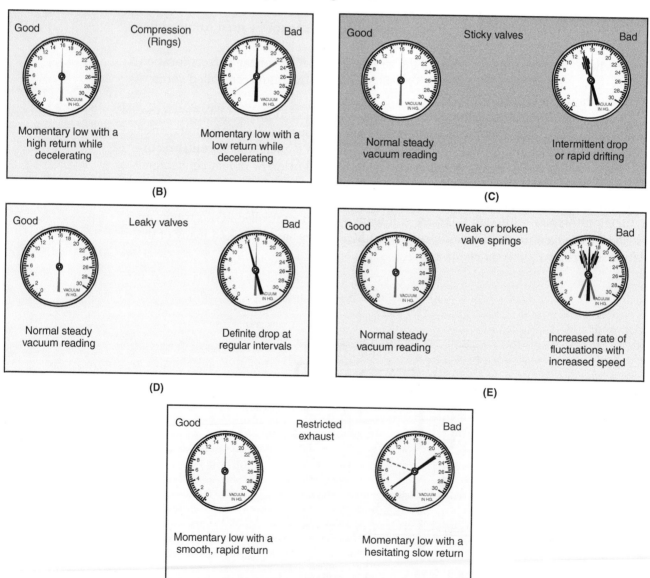

Figure 5-8 Common vacuum gauge readings and what they indicate.

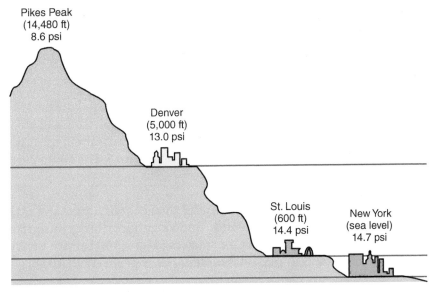

Figure 5-9 Atmospheric pressure changes with changes in elevation.

readings and what engine malfunctions they indicate. With high-performance engines, a slight flicker of the needle can also be expected. Keep in mind that a normal vacuum gauge reading will be 1 inch lower for each 1,000 feet above sea level. This is caused by the decrease in atmospheric pressure at higher altitudes (**Figure 5-9**).

Vacuum Leak Diagnosis

Broken or disconnected vacuum hoses may cause an air leak into the intake manifold, resulting in a lean air-fuel mixture. When the air-fuel mixture is leaner than normal, engine idle operation is erratic. On fuel-injected engines, an intake manifold vacuum leak may cause the engine to idle faster than normal. An intake manifold vacuum leak may cause a hesitation on low-speed acceleration. If the vacuum leak is positioned in the intake manifold so it leans the air-fuel mixture on one cylinder more than the other cylinders, the cylinder may misfire at idle and lower engine speeds. Once the engine speed increases, the reduced intake manifold vacuum does not pull as much air through the vacuum leak and the cylinder stops misfiring.

A vacuum gauge connected to the intake manifold indicates a low steady reading if there is a vacuum leak into the intake manifold. Intake manifold mounting bolts should be checked periodically for proper torque. Loose intake manifold bolts cause leaks between the intake manifold and the cylinder head.

On some V-type engines, the intake manifold covers the valve lifter chamber. If the intake manifold gaskets are leaking on the underside of the intake manifold on these applications, oil splash from the valve lifter chamber is moved past the intake manifold gasket into the cylinders. This action results in oil consumption and bluish-white in the exhaust.

Kinked vacuum hoses shut off the vacuum to a component, making it inoperative. For example, if the vacuum hose to the EGR valve is kinked, vacuum is not supplied to this valve and the valve remains closed. The EGR valve recirculates some exhaust into the intake manifold. Since this exhaust gas contains very little oxygen, it does not burn in the combustion chambers and combustion temperature is reduced, which lowers oxides of nitrogen (NO_x) emissions. If the EGR valve does not open, combustion chamber temperatures are higher than normal and the engine may detonate. When the EGR valve is inoperative, NO_x emissions are high.

Vacuum will also vary according to altitude.

⚡ **WARNING** **When using a propane cylinder to check for intake system leaks, do not smoke and do not place the end of the hose near any source of ignition. Failure to observe this precaution may result in an explosion, causing personal injury and/or property damage.**

A propane cylinder with a metering valve and hose may be used to check the intake system for leaks. With the engine idling, open the metering valve a small amount and position the end of the hose near the locations of any suspected leaks in the intake manifold. When a leak is present, the engine speed changes (**Figure 5-10**). Close the metering valve when the test is completed.

⚡ **WARNING** **When using a propane cylinder to check for intake system leaks, maintain the cylinder in an upright position and keep the cylinder away from rotating components, hot components, or sources of ignition. Failure to observe these precautions may result in personal injury and/or property damage.**

Special Tools

Propane bottle
Metering valve

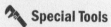

Special Tools

Compressed air
Air nozzle
Soap

When smoke testing, always use nitrogen gas instead of shop air for pressurization when testing fuel system components like fuel tanks for leaks.

Another way to locate vacuum leaks is with an electronic leak detector (**Figure 5-11**). With the engine running, simply move the tester along suspected areas. The tester will emit an audible sound near the location of the leak. The point where the sound is the greatest is the point of the leak.

Smoke Testing

Vacuum leaks can cause a variety of drivability problems. Fuel injection systems are consistently adjusting the air-fuel mixture according to the oxygen content in the exhaust. The PCM relies on the O_2 sensor for mixture information.

A vacuum leak will keep the O_2 sensor's signal voltage low. The PCM would then compensate by adding fuel. To correct the problem, the vacuum leak must be found. A propane bottle or an electronic leak detector can be used to find a vacuum leak; however, there are some new methods of vacuum leak testing that are actually easier to perform.

One of the best methods to find leaks of all sorts is by smoke testing as in **Photo Sequence 8**. The smoke tester uses a regulated pressure and special smoke solution. Some of the smoke can be used along with an ultraviolet lamp to make small leaks more visible. If leaks in the fuel system are being tested, nitrogen gas should be used instead of shop air because of the possibility of producing an explosive mixture.

Figure 5-10 Using propane to locate a vacuum leak.

Figure 5-11 Electronic vacuum leak detector.

P8-1 A truck is brought into the shop with a suspected vacuum leak.

P8-2 One way to pinpoint vacuum leaks is with a "smoke machine." Inert gas such as nitrogen is mixed with a nonflammable smoke and lightly pressurized.

P8-3 The diagnostic smoke output hose is attached to a vacuum line.

P8-4 Several adaptors are included with the smoke testing machine to block off normally open ports, such as the throttle body and the hoses from the intake manifold air duct. There is also a large adaptor for use to detect leaks in the exhaust system.

P8-5 Here the throttle body and PCV intake hose from the air cleaner are blocked off to prevent smoke from exiting the air cleaner.

P8-6 The smoke is turned on and after a few moments, the smoke is seen escaping from the vicinity of the idle air control.

Another way to locate the vacuum leak is using compressed air, an air nozzle, and a squirt bottle full of bubble soap. Begin by finding a large vacuum hose, such as the PCV hose or power brake booster hose. Hook up the air nozzle to the source of compressed air and insert the nozzle into the hose. Allow air to enter the hose and fill the intake manifold. The positive pressure of the air will try to exit the manifold. Where the air escapes is where the vacuum leak is. Spray the manifold and vacuum lines with the soap. Look for bubbles. The location of the bubbles is the source of the vacuum leak.

EXHAUST SYSTEM SERVICE

Exhaust system components are subject to both physical and chemical damage. Any physical damage to an exhaust system part that causes a partially restricted or blocked exhaust system usually results in loss of power or backfire up through the throttle plate(s).

Classroom Manual
Chapter 5, page 134

In addition to improper engine operation, a blocked or restricted exhaust system causes increased noise, exhaust emissions, and engine overheating.

⚠ WARNING **Exhaust gas contains poisonous carbon monoxide (CO) gas. This gas can cause illness and death by asphyxiation. Exhaust system leaks can be dangerous for customers and for technicians.**

⚠ WARNING **Exhaust system components may be extremely hot if the engine has been running. Allow the engine and exhaust system to cool down prior to exhaust system service, and wear protective gloves.**

Exhaust System Problems

Exhaust system components are subject to rust from inside and outside the components. Since exhaust components are exposed to road splash, they tend to rust on the outside. Stainless-steel exhaust system components have excellent rust-resistant qualities. Exhaust system components are subject to large variations in temperature, which form condensation inside these components. Condensation inside the exhaust system components rusts them on the inside. A vehicle that is driven short distances and then shut off tends to have more condensation in the exhaust system than a vehicle that is driven continually for longer time periods. If a vehicle is driven for a short distance, the exhaust system heat does not have sufficient time to vaporize the condensation in the system. The rust erodes exhaust system components, causing exhaust leaks and excessive noise.

Exhaust components, particularly mufflers, catalytic converters, and pipes, may become restricted. Mufflers and catalytic converters may become restricted by loose internal components. Pipes may become restricted by physical damage or collapsed inner walls of double-walled pipes. This exhaust restriction causes a loss of engine power and increased fuel consumption.

Exhaust system components cause a rattling noise if they are touching, or almost touching, chassis components. Loose internal structure in mufflers and catalytic converters may cause a rattling noise, especially when the engine is accelerated.

Exhaust System Inspection

Most parts of the exhaust system, particularly the exhaust pipe, muffler, and tailpipe, are subject to rust, corrosion, and cracking. Broken or loose clamps and hangers can allow parts to separate or hit the road as the car moves.

⚠ WARNING **During all exhaust inspection and repair work, wear safety glasses or equivalent eye protection.**

Any inspection of the exhaust system should include listening for hissing or rumbling sound that would result from a leak in the system. An on-lift inspection should pinpoint any damage. All of the exhaust system components should be inspected for physical damage, kinks, dents, holes, rust, loose internal components, loose clamps and hangers, proper clearance, and improperly positioned shields. Physical damage on exhaust system components includes flattened areas, abnormal bends, and scrapes. Flattened exhaust system components cause restrictions. Components with this type of damage should be replaced.

Loose or broken hangers (**Figure 5-12**) and clamps may allow exhaust system components to contact chassis components, resulting in a rattling noise. Broken hangers cause excessive movement of exhaust system components, and this action may result in premature breaking of these components.

Push upward on the muffler and catalytic converter to check for weak, rusted inlet and outlet pipes. Check the entire exhaust system for leaks. With the engine running, hold

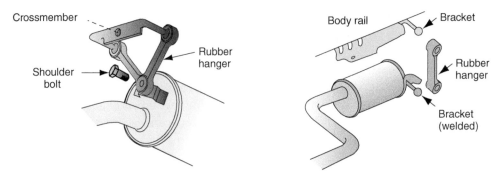

Figure 5-12 Various exhaust hangers used on today's vehicles.

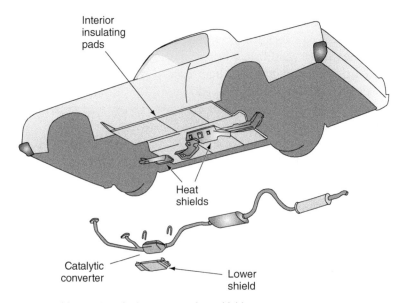

Figure 5-13 Location of exhaust system heat shields.

a length of hose near your ear and run the other end of the hose around the exhaust system components. The hose amplifies the sound of an exhaust leak and helps to locate the source of a leak.

Inspect all exhaust system shields for proper position, and check for the correct clearance between the exhaust system components and the shields (**Figure 5-13**). Check the catalytic converter for overheating, which is indicated by bluish or brownish discoloration.

The system should also be tested for blockage and restrictions using a vacuum and/ or pressure gauge and a tachometer. Before beginning work on the system, be sure it is cool to the touch. Some technicians disconnect the battery ground to avoid short-circuiting the electrical system. Soak all rusted nuts, bolts, and so on in a good-quality penetrating oil. Finally, check the system for critical clearance points so they can be maintained when new components are installed.

CATALYTIC CONVERTER DIAGNOSIS

The catalytic converter can fail in several ways: in efficiency, the measurement of the ability of the converter to reduce emissions; restriction of the exhaust due to plugging of the converter; or rattling due to the breaking up of the converter substrate.

 Caution

Avoid prolonged engine compression measurement or spark jump testing. They can allow fuel to collect in the converter. When you need to conduct these tests, do them as rapidly as possible.

Classroom Manual
Chapter 5, page 136

A catalytic converter's efficiency is checked by the PCM by using an **oxygen (O_2)** **sensor** upstream and downstream of the converter. If the converter is operating properly, the O_2 sensor **downstream** will show very little oxygen compared to the pre-converter sensor. The converter should have used most of the oxygen to oxidize the HC and CO.

A plugged converter or any exhaust restriction can cause loss of power at high speeds, stalling after starting (if totally blocked), a drop in engine vacuum as engine rpm increases, or sometimes popping or backfiring. Catalytic converters can also be destroyed by overheating, usually caused by a very rich fuel mixture due to a fuel delivery problem.

There are many ways to test a catalytic converter. One of these is to simply strike the converter with a rubber mallet. If the converter rattles, it needs to be replaced and there is no need for further testing. A rattle indicates loose catalyst substrate, which will soon rattle into small pieces. This is one test and is not used to determine if the catalyst is good. Keep in mind a converter may also be bad if it is restricted or not working properly.

A vacuum gauge can be used to watch engine vacuum while the engine is accelerated. A better method is to check for a restricted exhaust or catalyst is to insert a pressure gauge in the exhaust manifold's bore for the O_2 sensor (**Figure 5-14**). It may be necessary to fabricate an adapter for the pressure gauge from an old O_2 sensor. Once the gauge is in place, hold the engine's speed at 2,000 rpm and watch the gauge. The desired pressure reading will be less than 1.25 psi. A very bad restriction will give a reading of over 2.75 psi.

The converter should be checked for its ability to convert CO and HC into CO_2 and water. There are three separate tests for doing this. The first method is the delta temperature test. As with all converter tests, make sure the catalyst is warmed up before conducting the test.

Delta Temperature Test

To conduct this test, use a handheld digital **pyrometer** (**Figure 5-15**). By touching the pyrometer probe to the exhaust pipe just ahead of and just behind the converter, there should be an increase of at least 100°F, or 8 percent above the inlet temperature reading, as the exhaust gases pass through the converter. If the outlet temperature is the same or lower, nothing is happening inside the converter. Before running this test make sure the converter is at operating temperature, and generally works better off idle, around 2,000 RPM or so.

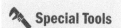
Special Tools

Vacuum gauge
Pressure gauge
Pyrometer
Exhaust gas analyzer

The catalytic converter uses the catalysts to change CO, HC, and NO_x into water vapor, CO_2, N_2, and O_2.

A restricted exhaust causes a loss of engine power.

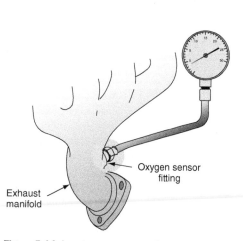

Figure 5-14 Insert pressure gauge in oxygen sensor fitting.

Oxygen sensor fitting

Exhaust manifold

Figure 5-15 A digital pyrometer used to measure inlet and outlet temperatures of a catalytic converter.

Oxygen Storage Test

The next test is called the O_2 storage test and is based on the fact that a good converter stores oxygen. Turn on the gas analyzer and allow it to warm up. Start the engine and allow the converter to warm up. Once the analyzer and converter are ready, hold the engine at 2,000 rpm. Watch the readings on the exhaust analyzer. If the converter was cold, the readings will continue to drop until the converter reaches light-off temperature. Once the numbers stop dropping, check the oxygen level on the gas analyzers. The O_2 readings should be about 0.5 to 1 percent. This shows that the converter is using most of the available oxygen. There is one exception to this, if there is no CO left, there may be a higher amount of oxygen in the exhaust. However, it still should be less than 2.5 percent.

It is important to observe the O_2 reading as soon as the CO begins to drop. Otherwise, good converters will fail this test. The O_2 will go way over 1.25 percent after CO starts to drop. If the O_2 reading is too high and there is no CO in the exhaust, stop the test and make sure the system has control of the air-fuel mixture. If the system is in control, use a propane enrichment tool to bring the CO level up to about 0.5 percent. Now the O_2 level should drop to zero. Once you have a solid oxygen reading, snap the throttle open, and then let it drop back to idle. Check the rise in oxygen. It should not rise above 1.2 percent. If the converter passes these tests, it is working properly. If the converter fails the tests, chances are that it is working poorly or not at all.

This final converter test uses a principle that checks the converter's efficiency. Remember the catalytic converter oxidizes CO and HC into CO_2 and water, and it also chemically reduces NO_x into N_2 and O_2. Before beginning this test, make sure the converter is warmed up. Prepare a five-gas analyzer and insert its probe into the tailpipe. If the vehicle has dual exhaust with a crossover, plug the side that the probe is not in. If the vehicle has a true dual exhaust system, check both sides separately.

Disable the ignition, and then crank the engine for 9 seconds while pumping the throttle. Watch the readings on the analyzer; the CO_2 should be over 11 percent. As soon as you have your readings, reconnect the ignition and start the engine. Do this as quickly as possible to cool off the catalytic converter. If, while the engine is cranking, the HC goes above 1,500 ppm, stop cranking; the converter is not working. Also, stop cranking once the CO_2 readings reach 10 or 11 percent; the converter is good. If the catalytic converter is bad, there will be high HC and low CO_2 at the tailpipe. Do not repeat this test more than one time without running the engine in between.

Catalytic converters are not serviced or repaired; they are generally replaced with another converter.

If a catalytic converter is found to be bad, it is replaced. There are two types of replacement. Installation kits (**Figure 5-16**) include all necessary components for the installation. There are also direct-fit catalytic converters (**Figure 5-17**) (**Photo Sequence 9**).

A **pyrometer** is an electronic device that measures heat.

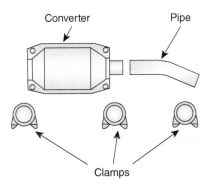

Figure 5-16 Catalytic converter installation kit.

Converter   Pipe

Clamps

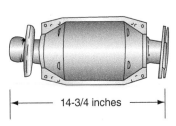

Figure 5-17 A direct-fit replacement catalytic converter.

14-3/4 inches

P9-1 The first method that is commonly used to check the operation of a converter is the temperature difference test, also called a delta temperature test.

P9-2 The converter outlet should be at least 100°F hotter than the inlet with the vehicle at normal operating temperature.

P9-3 The technician holds the engine rpm at 2,000 rpm compared to idle readings, so HC and CO will usually drop. As soon as CO readings start to drop, read the level of O_2. O_2 should be ½ to 1 percent (with 0 percent CO, the O_2 level may be as high as 2½ percent).

P9-4 If the level of O_2 is too high and there is no CO in the exhaust, make sure the vehicle is in control of the air-fuel mixture. If it is, add enough propane to bring the O_2 level to zero.

P9-5 Once the O_2 level is stable, snap the throttle open and allow it to return to idle. The O_2 level should not rise above 1.2 percent. This gives an indication of how well the converter stores oxygen. If the converter fails the test, it should be replaced.

Catalytic Converter Removal and Replacement

Since there are many different catalytic converter mountings, the converter removal-and-replacement procedure varies depending on the vehicle. The following is a typical converter removal-and-replacement procedure for a converter mounted directly to the exhaust manifold:

1. Allow the engine and exhaust system time to cool down before working on the vehicle. Disconnect the negative battery terminal, and wait for 1 minute before working on the vehicle. This gives the air bag module time to power down and prevents accidental air bag deployment.
2. Lift the vehicle on a hoist and disconnect the two bolts and front exhaust pipe bracket. Remove the two bolts holding the front exhaust pipe to the center exhaust pipe.
3. Disconnect the three nuts holding the front exhaust pipe to the front catalytic converter (**Figure 5-18**).
4. Be sure the front catalytic converter is cool, and disconnect the oxygen sensor connector.

5. Remove the bolt, nut, and number one manifold bracket.
6. Remove the bolt, nut, and number two manifold bracket.
7. Remove the two nuts and three bolts holding the catalytic converter to the manifold, and remove the converter. Remove the gasket, retainer, and cushion from the converter. Remove the eight bolts holding the outer heat insulators on the converter, and remove the heat shields (**Figure 5-19**).
8. Install the heat shields on the new catalytic converter and tighten the retaining bolts to the specified torque.
9. Place a new cushion, retainer, and gasket on the catalytic converter (**Figure 5-20**).
10. Install the catalytic converter on the exhaust manifold, and install the three new bolts and two new nuts (**Figure 5-21**). Tighten the bolts and nuts to the specified torque.
11. Install the number one and number two exhaust manifold support brackets, and tighten the fasteners to the specified torque.
12. Place new gaskets on the front and rear of the front exhaust pipe, and install the front exhaust pipe in its proper position. Install the two bolts and nuts holding the front exhaust pipe to the center exhaust pipe.
13. Install the three nuts holding the front exhaust pipe to the converter, and tighten these nuts to the specified torque.
14. Tighten the two bolts and nuts holding the front exhaust pipe to the center exhaust pipe. Install the front exhaust pipe bracket, and tighten the fastener to the specified torque.

Classroom Manual
Chapter 5, page 136

Figure 5-18 Removing the nuts holding the exhaust pipe to the exhaust manifold.

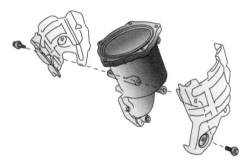

Figure 5-19 Separating the outer heat shields around the catalytic converter.

Gasket
Retainer
Cushion

Figure 5-20 Installing a new cushion, retainer, and gasket on the catalytic converter.

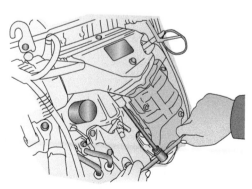

Figure 5-21 Installing the catalytic converter onto the exhaust manifold.

Diesel Exhaust After-Treatment Service

As you recall from the classroom manual, diesel exhaust has several special systems and components to deal with emissions of CO and HC, and in particular for the large amounts of NO_x that diesels produce. Particulate matter is also a concern that is addressed by the diesel particulate filter section of the after-treatment system. There are three basic sections to the diesel after treatment system: the diesel oxidation converter (DOC), the selective catalyst reduction (SCR), and the diesel particulate filter (DPF) (**Figure 5-22**).

Diesel Oxidation Catalyst

The diesel oxidation catalyst is used to oxidize HC and CO into H_2O and CO_2. As with a gasoline converter, in order for the converter to operate effectively, the converter must be hot. For the ECM to determine if the DOC is operating, the system uses exhaust gas temperature sensors in front of and behind the DOC. The DOC also provides the heat necessary for diesel particulate filter to regenerate.

The DOC can be poisoned by using diesel fuel that has more than 15 ppm of sulfur in the fuel, which renders the catalyst ineffective. This is the reason that diesel-fuel sulfur standards were recently enacted (**Figure 5-23**).

Selective Catalyst Reduction

The SCR is the section of the after-treatment deals specifically with NO_x. NO_x is reduced by the injection of diesel exhaust fluid (DEF) into the SCR. The amount of DEF injected is calculated based on engine operating conditions. The necessary amount of DEF fluid is injected into the SCR based on the amount of NO_x detected by the NO_x sensor in the exhaust stream ahead of the SCR, and the result of the DEF treatment can

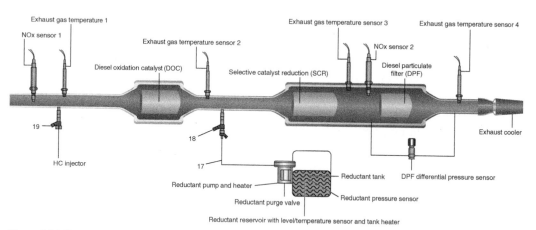

Figure 5-22 Typical diesel exhaust after-treatment system.

Figure 5-23 The diesel oxidation catalyst treats CO and HC.

be evaluated by the NO_x sensor that is mounted downstream of the SCR. The NO_x sensors have to be warm to operate, so they incorporate heaters much like oxygen sensors on a gasoline-fueled vehicle. The NO_x sensors are a feedback system of check and adjust, just as the oxygen sensors do on a gasoline-fueled vehicle. The SCR works best at temperatures about 480°F (250°C). This temperature is monitored by the exhaust gas temperature sensors 2 and 3. If the temperature at the SCR falls below 480°F (250°C), the operation of the SCR can be suspended to prevent the production of sulfates that could poison the catalyst. The ECM has strategies to try maintaining exhaust temperatures at the optimum levels for SCR operation. Do not use silicone lubricants around the exhaust and intake systems that might come into contact with the NO_x sensors or they may be poisoned (**Figure 5-24**).

Diesel Exhaust Fluid

Diesel exhaust fluid (DEF) is a 32.5 percent mixture of urea and deionized water. The heat of the exhaust converts the urea into ammonia (NH_3) and water. DEF allows the conversion of NO_x to N_2 and H_2O, through a chemical reaction; $NO + NO_2 + NH_3 = 2N_2 + 3H_2O$. For the vehicle to pass exhaust emissions, the DEF system has to remain filled with the proper quality and quantity of DEF fluid. The DEF system has to be maintained, or the system will go into a limp home mode (after several warnings) if the DEF fluid is allowed to run out, or the system is disabled or inoperative. One important note on the characteristics of DEF fluid is that it can freeze at low temperatures, so the DEF fluid reservoirs have heaters in the reservoir and lines to allow the fluid to be injected even at low temperatures.

Diesel Particulate Filter

The diesel particulate filter (DPF) (**Figure 5-25**) stores particulate matter until it can be burned off, either through natural or forced regeneration. The regeneration event is

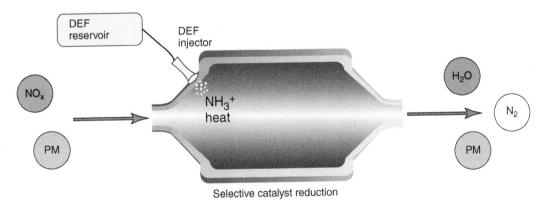

Figure 5-24 The SCR reduces NO_x to N_2 and H_2O.

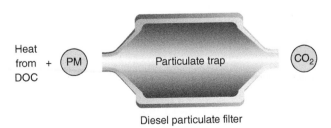

Figure 5-25 Diesel particulate filter.

initiated by the ECM when the exhaust back pressure sensors indicate a need for a regeneration. In order for the regeneration to occur, the vehicle usually has to be driven at highway speeds for at least 30 minutes. This may allow a regeneration because of the heat generated by this driving activity. Ford terms this a *passive regeneration*. If this does not occur and a regeneration is necessary, the vehicle will do a regeneration by raising the exhaust temperature high enough to clean the DPF. Ford calls this an *active regeneration*, but it must occur with the vehicle moving to prevent damage from the hot exhaust pipe. If the vehicle is driven in a manner in which an active or a passive regeneration cannot be accomplished, the driver is alerted of the need for regeneration. If the driver does not drive the vehicle in a manner that will allow the vehicle to do a regeneration, then the vehicle will go into a limp-in mode, which may become more and more speed-limiting until the vehicle will no longer move without the regeneration being completed. If this happens, then the regeneration will have to be done using a scan tool to do a service or manual regeneration outdoors at the shop, away from all flammable materials, and not using any exhaust hoses from the shop, as the heat of regeneration may melt them with a temperature at the tailpipe that may exceed 1,022°F (550°C)! Regeneration temperatures must stay within the guidelines established by the manufacturer to prevent damage to the DPF and DOC, such as melting or fracturing internal components. The heat of the regeneration is measured by a temperature sensor located behind the DPF.

Diesels with the DPF must use a special ash-free oil (CJ-4 API). The ash will plug up the DPF because it cannot be burned off by the DPF.

Exhaust Cooler

Most diesels use a special tailpipe tip to help lower the temperature at the end. This cooler should not be removed for any reason.

REPLACING EXHAUST SYSTEM COMPONENTS

Most exhaust system servicing involves the replacement of parts. When replacing exhaust system components, it is important that original equipment parts (or their equivalent) are used to ensure proper alignment with other parts in the system and to provide acceptable exhaust noise levels. When replacing only one component in an exhaust system, it is not always necessary to take the parts behind it off.

Exhaust Manifold

⚠ **Caution**

Manifolds warp more easily if an attempt is made to remove them while still hot. Remember that heat expands metal, making assembly bolts more difficult to remove and easier to break.

The exhaust manifold gasket seals the joint between the head and exhaust manifold. Many new engines are assembled without exhaust manifold gaskets. This is possible because new manifolds are flat and fit tightly against the head without leaks. Exhaust manifolds go through many heating and cooling cycles. This causes stress and some corrosion in the exhaust manifold. Removing the manifold will usually distort the manifold slightly so it is no longer flat enough to seal without a gasket. Exhaust manifold gaskets are normally used to eliminate leaks when exhaust manifolds are reinstalled (**Figure 5-26**).

Exhaust manifolds can warp because of the heating and cooling cycles they go through. They can also crack because of the high temperatures generated by the engine. This usually occurs after the car passes through a large puddle and cold water splashes on the manifold's hot surface. If the manifold is warped beyond manufacturer's specifications or is cracked, it must be replaced.

To replace an exhaust manifold, remove the exhaust pipe bolts at the manifold flange and disconnect any other components in the manifold, such as an O$_2$ sensor. In some applications, it is easier to disconnect the sensor wires and leave the sensor in the manifold. However, if the exhaust manifold is being replaced, the sensor must be removed and installed in the new manifold. Remove the bolts retaining the manifold to the cylinder head and lift the manifold from the engine compartment. Remove the manifold heat shield.

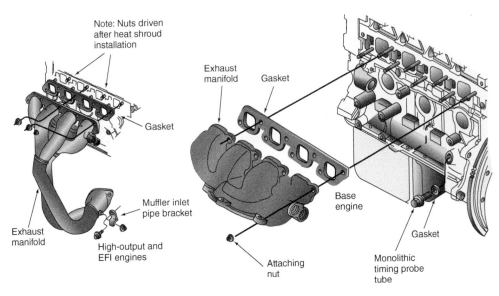

Figure 5-26 Examples of exhaust manifolds for in-line engines.

> **SERVICE TIP** Because exhaust system fasteners are subject to extreme heat, they may be difficult to loosen. A generous application of penetrating solvent on these fasteners makes removal easier.

Use a scraper to clean the matching surfaces on the exhaust manifold and cylinder head. Measure the exhaust manifold surface for warping with a straightedge and feeler gauge. Perform this measurement at three locations on the manifold surface. If the manifold is warped more than specified by the vehicle manufacturer, replace the manifold. Remove any gasket material from the manifold flange. If the flange has a ball connection, be sure the ball is not damaged.

If the manifold has an O_2 sensor, place some anti-seize compound on the threads and install the sensor in the manifold. When this sensor is easy to access, it may be installed after the manifold is installed on the cylinder head. Install a new gasket between the cylinder head and the manifold, and install the manifold against the cylinder head. Install exhaust manifold-to-cylinder head mounting bolts, and tighten these bolts to the specified torque. Many exhaust manifold bolts, nuts, and studs are special heat-resistant fasteners. If exhaust manifold bolts or nuts must be replaced, do not substitute ordinary fasteners for heat-resistant fasteners.

Be sure the exhaust pipe mounting surface that fits against the exhaust manifold flange is in satisfactory condition. If an exhaust manifold flange gasket is required, install a new gasket and tighten the flange bolts to the specified torque (**Figure 5-27**).

Special Tools
Pipe cutter
Slitting tool
Pipe expander

Exhaust Pipes and Mufflers

To replace a damaged exhaust pipe, begin by supporting the converter to keep it from falling. Carefully remove the oxygen sensor if necessary. Remove any hangers or clamps holding the exhaust pipe to the frame. Unbolt the flange holding the exhaust pipe to the exhaust manifold. When removing the exhaust pipe, check to see if there is a gasket. If so, discard it and replace it with a new one. Once the joint has been taken apart, the gasket loses its effectiveness. Disconnect the pipe from the converter, pull the front exhaust pipe loose, and remove it.

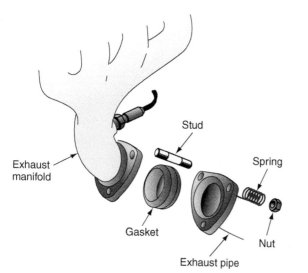

Figure 5-27 Exhaust pipe-to-manifold flange connections.

Figure 5-28 An exhaust pipe-cutting tool.

> **⚙️ SERVICE TIP** An easy way to break off rusted nuts is to tighten them instead of loosening them. Sometimes a badly rusted clamp or hanger strap will snap off with ease. Sometimes the old exhaust system will not drop free of the body because a large part is in the way, such as the differential or the transmission support. Use a large, cold chisel, pipe cutter, hacksaw, muffler cutter, or chain cutter to cut the old system at convenient points to make the exhaust assembly smaller.

Many original mufflers and catalytic converters are integral with the interconnecting pipes. When these components are replaced, they must be cut from the exhaust system with a cutting tool (**Figure 5-28**). The inlet and outlet pipes on the replacement muffler or converter must have a 1.5-inch (3.8-cm) overlap on the connecting pipes. Before cutting the pipes to remove the muffler or converter, measure the length of the new component, and always cut these pipes to provide the required overlap.

If the muffler or converter inlet and outlet pipes are clamped to the connecting pipes, remove the clamp. When the muffler or converter is tight on the connecting pipe, a **slitting tool** and hammer may be used to slit and loosen the muffler pipe. A slitting tool on an air chisel may be used for this job.

The old exhaust pipe might be rusted into the muffler or converter opening. Attempt to collapse the old pipe by using a cold chisel or slitting tool and a hammer (**Figure 5-29**). While freeing the pipe, try not to damage the muffler inlet. It must be perfectly round to accept the new pipe.

Slide the new pipe into the muffler (some lubricant might be helpful). Attach the front end to the manifold. The pipe must fit at least 1½ inches into the converter or muffler. Before tightening the connectors, check the system for alignment. When it is properly aligned, tighten the clamps.

When a new muffler or converter is installed, the connecting pipes may not fit perfectly. A hydraulically operated expanding tool may be used to expand the pipe and provide the necessary fit (**Figure 5-30**). In some cases, adaptors may be used to provide the necessary pipe fit. A sleeve may be used to join two pieces of pipe (**Figure 5-31**). When the new exhaust system components are installed, there must be adequate clearance between all the exhaust components and the chassis. Install and tighten all clamps and hangers securely.

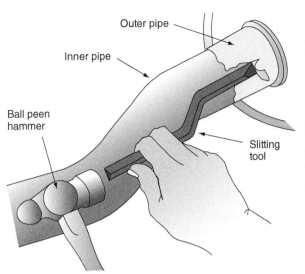

Figure 5-29 Using a slitting tool to separate a muffler from an exhaust pipe.

Figure 5-30 Hydraulically operated pipe-expanding tool.

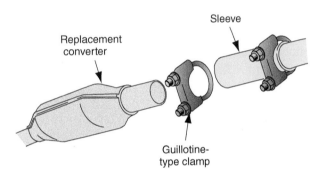

Figure 5-31 Exhaust pipe sleeve installation.

TURBOCHARGER DIAGNOSIS

⚡ WARNING **If the engine has been running, turbochargers and related components are extremely hot. Use caution, and wear protective gloves to avoid burns when servicing these components.**

Classroom Manual
Chapter 5, page 144

The first step in turbocharger diagnosis is to check all linkages and hoses connected to the turbocharger (**Figure 5-32**). Inspect the waste-gate diaphragm linkage for looseness and binding, and check the hose from the waste-gate diaphragm to the intake manifold for cracks, kinks, and restrictions. Check the coolant hoses and oil line connected to the turbocharger for leaks.

Excessive bluish-white smoke in the exhaust may indicate worn turbocharger shaft seals. The technician must remember that worn valve guide seals or piston rings also cause oil consumption and bluish-white smoke in the exhaust. When oil leakage is noted at the turbine end of the turbocharger, always check the turbocharger oil drain tube and the engine PCV system for restrictions.

Check all turbocharger mounting bolts for looseness. Loose turbocharger mounting bolts may cause a rattling noise. Some whirring noise is normal when the turbocharger shaft is spinning at high speed. Excessive internal turbocharger noise may be caused by too much endplay on the shaft, which allows the blades to strike the housings.

Exhaust leaks between the cylinders and the turbocharger decrease turbocharger efficiency.

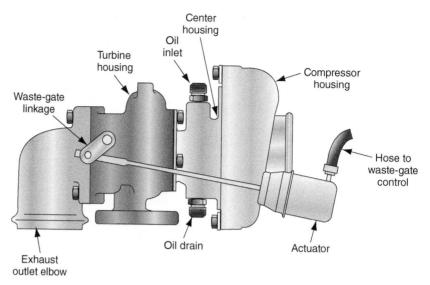

Figure 5-32 A turbocharger assembly.

Check for exhaust leaks in the turbine housing and related pipe connections. If exhaust gas is escaping before it reaches the turbine wheel, turbocharger effectiveness is reduced. Check for intake system leaks. If there is a leak in the intake system before the compressor housing, dirt may enter the turbocharger and damage the compressor or turbine wheel blades. When a leak is present in the intake system between the compressor wheel housing and the cylinders, turbocharger pressure is reduced.

When turbochargers have boost pressure control problems, a DTC is stored in the PCM memory if a fault is present in the boost control solenoid or solenoid-to-PCM wiring.

Common Turbocharger Problems

The turbocharger, with proper care and servicing, will provide years of reliable service. Most turbocharger failures are caused by lack of lubricant, ingestion of foreign objects, or contamination of lubricant. One common cause of turbocharger failure is not allowing the engine a few minutes to cool down before shutting the engine off after a hard run, such as expressway driving. The cooldown period allows the engine oil and coolant to disperse some of the high engine heat. Once the engine is shut down, the oil and coolant stop flowing and the engine temperature actually increases. Refer to **Figure 5-33** and the service information for the symptoms of turbocharger failures and a summary of causes and recommended remedies.

Turbocharger Inspection

To inspect a turbocharger, start the engine and listen to the sound the turbo system makes. As a technician becomes more familiar with this characteristic sound, it will be easier to identify an air leak between the compressor outlet and engine or an exhaust leak between engine and turbo by the presence of a higher pitched sound. If the turbo sound cycles or changes in intensity, the likely causes are a plugged air cleaner or loose material in the compressor inlet ducts or dirt buildup on the compressor wheel and housing.

After listening, check the air cleaner, remove the ducting from the air cleaner to turbo, and look for dirt buildup or damage from foreign objects. Check for loose clamps on the compressor outlet connections and check the engine intake system for loose bolts or leaking gaskets. Then, disconnect the exhaust pipe and look for restrictions or loose material. Examine the exhaust system for cracks, loose nuts, or blown gaskets. Rotate the turbo shaft assembly. Does it rotate freely? Are there signs of rubbing or wheel impact damage?

Special Tools

Stethoscope
Flashlight

TURBOCHARGER TROUBLESHOOTING GUIDE

Condition	Possible Causes Code Numbers	Remedy Description by Code Numbers
Engine lacks power	1, 4, 5, 6, 7, 8, 9, 10, 11, 18, 20, 21, 22, 25, 26, 27, 28, 29, 30, 38, 39, 40, 41, 42, 43	1. Dirty air cleaner element 2. Plugged crankcase breathers 3. Air cleaner element missing, leaking, not sealing correctly; loose connections to turbocharger 4. Collapsed or restricted air tube before turbocharger
Black smoke	1, 4, 5, 6, 7, 8, 9, 10, 11, 18, 20, 21, 22, 25, 26, 27, 28, 29, 30, 38, 39, 40, 41, 43	5. Restricted-damaged crossover pipe, turbocharger to inlet manifold 6. Foreign object between air cleaner and turbocharger 7. Foreign object in exhaust system (from engine, check engine)
Blue smoke	1, 2, 4, 6, 8, 9, 17, 19, 20, 21, 22, 32, 33, 34, 45	8. Turbocharger flanges, clamps, or bolts loose. 9. Inlet manifold cracked; gaskets loose or missing; connections loose
Excessive oil consumption	2, 8, 15, 17, 19, 20, 29, 30, 31, 33, 34, 45	10. Exhaust manifold cracked, burned; gaskets loose, blown, or missing 11. Restricted exhaust system 12. Oil lag (oil delay to turbocharger at startup)
Excessive oil turbine end	2, 7, 8, 17, 19, 20, 22, 29, 30, 32, 33, 34, 45	13. Insufficient lubrication 14. Lubricating oil contaminated with dirt or other material 15. Improper type of lubricating oil used
Excessive oil compressor end	1, 2, 4, 5, 6, 8, 19, 20, 21, 29, 30, 33, 34, 45	16. Restricted oil feed line 17. Restricted oil drain line 18. Turbine housing damaged or restricted 19. Turbocharger seal leakage
Insufficient lubrication	8, 12, 14, 15, 16, 23, 24, 31, 34, 35, 36, 44, 46	20. Worn journal bearings 21. Excessive carbon buildup in compressor housing 22. Excessive carbon buildup behind turbine wheel 23. Too fast acceleration at initial start (oil lag)
Oil in exhaust manifold	2, 17, 18, 19, 20, 22, 29, 30, 33, 34, 45	24. Too little warm-up time 25. Fuel pump malfunction 26. Worn or damaged injectors 27. Valve timing
Damaged compressor wheel	3, 4, 6, 8, 12, 15, 16, 20, 21, 23, 24, 31, 34, 35, 36, 44, 46	28. Burned valves 29. Worn piston rings 30. Burned pistons 31. Leaking oil feed lines
Damaged turbine wheel	7, 8, 12, 13, 14, 15, 16, 18, 20, 22, 23, 24, 25, 28, 30, 31, 34, 35, 36, 44, 46	32. Excessive engine pre-oil 33. Excessive engine idle 34. Coked or sludged center housing 35. Oil pump malfunction 36. Oil filter plugged
Drag or bind in rotating assembly	3, 6, 7, 8, 12, 13, 14, 15, 16, 18, 20, 21, 22, 23, 24, 31, 34, 35, 36, 44, 46	38. Actuator damaged or defective 39. Waste-gate binding 40. Electronic control module or connector(s) defective 41. Waste-gate actuator solenoid or connector defective 42. EGR valve defective
Worn bearings, journals, and bearing bores	6, 7, 8, 12, 13, 14, 15, 16, 23, 24, 31, 35, 36, 44, 46	43. Alternator voltage incorrect 44. Engine shut off without adequate cool-down time 45. Leaking valve guide seals 46. Low oil level
Noisy	1, 3, 4, 5, 6, 7, 8, 9, 10, 11, 12, 13, 14, 15, 16, 18, 20, 21, 22, 23, 24, 31, 34, 35, 36, 44, 46	
Sludged or coked center housing	2, 11, 13, 14, 15, 17, 18, 24, 31, 35, 36, 44, 46	

Figure 5-33 Turbocharger troubleshooting guide.

Visually inspect all hoses, gaskets, and tubing for proper fit, damage, and wear. Check the low-pressure, or air cleaner, side of the intake system for vacuum leaks.

On the pressure side of the system, you can check for leaks by using soapy water. After applying the soap mixture, look for bubbles to pinpoint the source of the leak.

Leakage in the exhaust system upstream from the turbine housing will also affect turbo operation. If exhaust gases are allowed to escape prior to entering the turbine housing, the reduced temperature and pressure will cause a proportionate reduction in boost

Leaks in the air intake system may allow dirt particles to enter the turbocharger and damage the blades.

Special Tools

Pressure gauge
Dial indicator

and an accompanying loss of power. If the waste-gate does not appear to be operating properly (too much or too little boost), check to make sure the connecting linkage is operating smoothly and is not binding. Also, check to make sure the pressure-sensing hose is clear and properly connected. **Photo Sequence 10** shows a typical procedure for inspecting turbochargers and testing boost pressure.

Testing Boost Pressure

Connect a pressure gauge to the intake manifold to check the **boost pressure**. The pressure gauge hose should be long enough so the gauge may be positioned in the passenger compartment. One of the front windows may be left down enough to allow the gauge hose to extend into the passenger compartment. Road test the vehicle at the speed specified by the vehicle manufacturer, and observe the boost pressure. Some vehicle manufacturers recommend accelerating from a stop to 60 mph (96 kph) at wide-open throttle while observing the boost pressure.

 SERVICE TIP If the engine has low cylinder compression, there is reduced airflow through the cylinders, which results in lower turbocharger shaft speed and boost pressure.

⚠ **Caution**

When test driving the vehicle with a scan tool, take an assistant along to drive while you read the scan tool! Do *not* drive and try to operate the scan tool at the same time.

In most turbocharged engines today, the boost pressure can be read from a scan tool. There are sensors that measure intake pressure from the turbocharger and/or the manifold pressure sensor. These systems can also set DTCs to help diagnose the turbocharger and the control system.

Higher-than-specified boost pressure may be caused by a defective waste-gate system that is sticking closed. Low boost pressure may be caused by the waste-gate system or turbocharger defects, such as damaged wheel blades or worn bearings. An engine with low cylinder compression will usually have low boost pressure.

Waste-Gate Service

Waste-gate malfunctions can usually be traced to carbon buildup that keeps the unit from closing or causes it to bind. A defective diaphragm or leaking vacuum hose can result in an inoperative waste-gate (**Figure 5-34**). But before condemning the waste-gate, check the, waste-gate linkage, vacuum hoses, knock sensor, oxygen sensor, and computer to be sure each is operating properly. Check for trouble codes that set for under- or over-boost conditions.

If the waste-gate stroke is reduced, the waste-gate valve opening is decreased and the boost pressure is increased. Connect a hand pressure pump and a pressure gauge to the waste-gate diaphragm. Position a dial indicator against the outer end of the waste-gate diaphragm rod (**Figure 5-35**). Supply the specified pressure to the waste-gate diaphragm and observe the dial indicator movement. If the waste-gate rod movement is less than specified, disconnect the rod from the waste-gate valve linkage and check the linkage for binding. If this linkage moves freely, replace the waste-gate diaphragm. When the waste-gate valve linkage is binding, turbocharger repair or replacement is required. Excessive boost pressure can cause burned pistons due to the extreme heat produced.

⚠ **Caution**

When removing carbon deposits from turbine and waste-gate parts, never use a hard metal tool or sandpaper. Remember that any gouges or scratches on these metal parts can cause severe vibration or damage to the turbocharger. To clean these parts, use a soft brush and a solvent.

Turbocharger Removal

The turbocharger removal procedure varies depending on the engine. On some cars, the manufacturer recommends the engine be removed to gain access to the turbocharger. On other applications, the turbocharger may be removed with the engine in the vehicle.

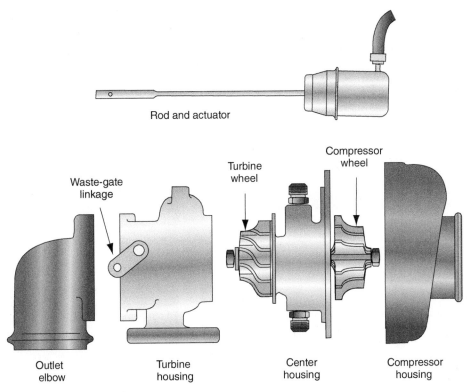

Figure 5-34 Typical turbocharger.

Classroom Manual
Chapter 5, page 144

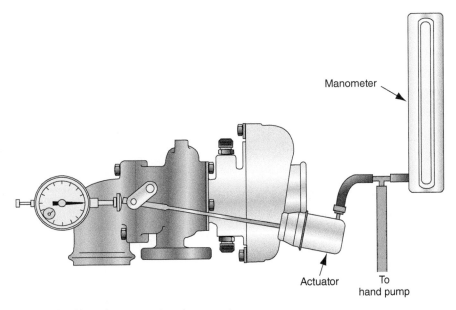

Figure 5-35 Measuring waste-gate rod movement.

Always follow the turbocharger removal procedure in the vehicle manufacturer's service information. The following is a typical turbocharger removal procedure:

> ⚙️ **SERVICE TIP** Excessive boost pressure causes engine detonation and possible engine damage.

The waste-gate opens to reduce boost pressure. A sticking waste-gate could cause engine damage.

1. Disconnect the negative battery cable and drain the cooling system.
2. Disconnect the exhaust pipe from the turbocharger.
3. Remove the support bracket between the turbocharger and the engine block.
4. Remove the bolts from the oil drain back housing on the turbocharger.

PHOTO SEQUENCE 10
Typical Procedure for Inspecting Turbochargers and Testing Boost

P10-1 Check all turbocharger linkages for looseness, and check all turbocharger hoses for leaks, cracks, kinks, and restrictions.

P10-2 Check the level and condition of the engine oil on the dipstick.

P10-3 Check the exhaust for evidence of bluish-white smoke when the engine is accelerated.

P10-4 Check all turbocharger mounting bolts for looseness.

P10-5 Check for exhaust leaks between the engine and the turbocharger, and use a stethoscope to listen for excessive turbocharger noise.

P10-6 Use a propane cylinder, metering valve, and hose to check for intake manifold vacuum leaks.

P10-7 The technician connects a pressure gauge to measure the boost pressure directly. (The gauge can be taped to the windshield to enable reading while driving).

P10-8 With the help of an assistant to drive the vehicle, road-test the vehicle from 0–60 mph while observing the reading on the gauge (Some vehicles have boost pressure that can be read on a scan tool).

P10-9 Back in the shop, the technician can review the service information to help diagnose the turbocharger complaint.

5. Disconnect the turbocharger coolant inlet tube nut at the block outlet and remove the tube-support bracket.
6. Remove the air cleaner element, air cleaner box, bracket, and related components.
7. Disconnect the accelerator linkage, throttle body electrical connector, and vacuum hoses.
8. Loosen the throttle body-to-turbocharger inlet hose clamps and remove the three throttle body-to-intake manifold attaching screws. Remove the throttle body.
9. Loosen the lower turbocharger discharge hose clamp on the compressor wheel housing.

10. Remove the fuel rail-to-intake manifold screws and the fuel line bracket screw. Remove the two fuel rail bracket-to-heat shield retaining clips, and pull the fuel rail and injectors upward out of the way. Tie the fuel rail in this position with a piece of wire.
11. Disconnect the oil supply line from the turbocharger housing.
12. Remove the intake manifold heat shield.
13. Disconnect the coolant return line from the turbocharger and the water box. Remove the line support bracket from the cylinder head and remove the line.
14. Remove the four nuts retaining the turbocharger to the exhaust manifold and remove the turbocharger from the manifold studs. Move the turbocharger downward toward the passenger side of the vehicle, and then lift the unit up and out of the engine compartment.

Measuring Turbocharger Shaft Axial Movement

After the turbocharger is removed from the engine, remove the turbine outlet elbow from the turbine housing. Position a dial indicator against the shaft, and move the shaft inward and outward while observing the dial indicator reading (**Figure 5-36**). The axial shaft movement should not exceed the vehicle manufacturer's specifications. On some turbochargers, the maximum axial shaft movement is 0.003 in. (0.91 mm). If the axial shaft movement exceeds the manufacturer's specifications, the turbocharger must be repaired or replaced. Some manufacturers recommend complete turbocharger replacement, whereas other manufacturers recommend replacing the center housing assembly as a unit. Some turbocharger manufacturers recommend replacement of individual components. Always follow the service procedures in the vehicle manufacturer's service information.

Turbocharger Component Inspection

If the vehicle manufacturer recommends turbocharger disassembly, inspect the wheels and shaft after the end housings are removed. Lack of lubricant or lubrication with contaminated oil results in bearing failure, which leads to wheel rub on the end housings. A contaminated cooling system may provide reduced turbocharger bearing cooling and premature bearing failure. Bearing failure will likely lead to seal damage. Inspect the shaft and bearings for a burned condition (**Figure 5-37**). If the shaft and bearings are burned, replace the complete center housing assembly or individual parts as recommended by the manufacturer.

If the shaft and bearings are in satisfactory condition but the blades are damaged, check the air intake system for leaks or a faulty air cleaner element. When the blades or shaft and bearings must be replaced, always check the end housings for damage (**Figure 5-38**).

Figure 5-37 Inspecting turbocharger shaft and bearings.

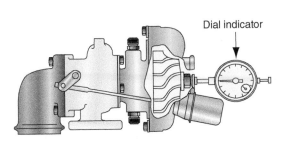

Figure 5-36 Measuring axial movement.

Figure 5-38 Inspecting turbocharger end housings.

When these housings are marked or scored, replacement is necessary. Since turbocharger components are subjected to extreme heat, use a straightedge to check all mating surfaces for a warped condition. Replace warped components as necessary.

Turbocharger Installation

Prior to reinstalling the turbocharger, be sure the engine oil and filter are in satisfactory condition. Change the oil and filter as required, and be sure the proper oil level is indicated on the dipstick. Check the coolant for contamination. Flush the cooling system if contamination is present. Reverse the turbocharger removal procedure explained previously in this chapter to install the turbocharger. Replace all gaskets and be sure all fasteners are tightened to the specified torque. Follow the vehicle manufacturer's recommended procedure for filling the cooling system. On some Chrysler engines, this involves removing a plug on top of the water box and pouring coolant into the radiator filler neck until coolant runs out the hole in the top of the water box (**Figure 5-39**). Install the plug in the top of the water box, and tighten the plug to the specified torque. Continue filling the cooling system to the maximum level mark on the reserve tank (**Figure 5-40**).

⚠ **Caution**

Failure to prelubricate turbocharger bearings may result in premature bearing failure.

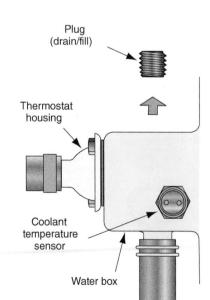

Figure 5-39 Removing the plug in the top of the water box while filling the cooling system.

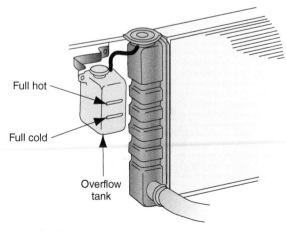

Figure 5-40 Coolant level marks on the reserve tank.

Figure 5-41 Pouring engine oil into the turbocharger oil inlet to pre-lubricate the turbocharger bearings.

The turbocharger bearings must be pre-lubricated before starting the engine to prevent bearing damage. Some vehicle manufacturers recommend removing the turbocharger oil supply pipe and pouring a half pint of the specified engine oil into the turbocharger to pre-lubricate the bearings (**Figure 5-41**).

Other vehicle manufacturers recommend disabling the ignition system and cranking the engine for 10 to 15 seconds to allow the engine lubrication system to lubricate the turbocharger bearings. Always follow the turbocharger pre-lubrication instructions in the vehicle manufacturer's service information.

Turbo Startup and Shutdown

After replacement of a turbocharger or after an engine has been unused or stored, there can be a considerable lag after engine startup before the oil pressure is sufficient to deliver oil to the turbocharger bearings. To prevent this problem, follow these simple steps:

1. When installing a new or remanufactured turbocharger, make certain that the oil inlet and drain lines are clean before connecting them.
2. Be sure the engine oil is clean and at the proper level.
3. Fill the oil filter with clean oil.
4. Leave the oil drain line disconnected at the turbo, and crank the engine without starting it until oil flows out of the turbo drain port.
5. Connect the drain line, start the engine, and operate it at low idle for a few minutes before running it at higher speeds.

AUTHOR'S NOTE Remind your customer to not immediately stop a turbocharged engine after pulling a trailer, or driving at a high speed, or driving uphill. Idle the engine for 20 to 120 seconds. The turbocharger will continue to rotate after the engine oil pressure has dropped to zero, which can cause bearing damage. Also remind the owner of the proper starting procedures and the importance of proper maintenance. Also, remind your customers that it is not a good idea to operate the turbo until 3 to 5 minutes after starting the engine to allow oil to start circulating at full pressure in the lubrication system.

Contaminated oil can cause sludge buildups within the turbo. Check the oil drain outlet for sludge buildup with the oil drain line removed. Failure to follow these steps can result in bearing failure on the turbocharger's main shaft. Remember, shaft rotation speed in modern turbochargers can easily exceed 146,000 revolutions per minute.

A turbocharger should never be operated under load if the engine has less than 30 psi oil pressure. A turbocharger is much more sensitive to a limited oil supply than an engine due to the high rotational speed of the shaft and the relatively small area of the bearing surfaces. Low oil pressure and slow oil delivery during engine starting can destroy the bearings in a turbocharger. During normal engine starting, this should not be a problem. There are, of course, abnormal starting conditions. Oil lag conditions will most often occur during the first engine start after engine oil and filter change. Before the engine is put under load and the turbo activated, the engine should be run for 3 to 5 minutes at idle to prevent oil starvation to the turbo. Similar conditions can also exist if an engine has not been operated for a long period of time. Engine lube systems have a tendency to bleed down. Before allowing the engine to start, the engine should be cranked over until a steady oil pressure reading is observed. This is called priming the lubricating system. The same starting procedure should be followed in cold weather. The thick engine oil will take a longer period of time to flow.

SUPERCHARGER DIAGNOSIS AND SERVICE

Classroom Manual
Chapter 5, page 148

A supercharger (**Figure 5-42**) should be trouble-free for many miles. Since it does not operate in the same high heat as a turbocharger, it tends to be more reliable. However, it does operate at high speeds and at close tolerances (**Figure 5-43**). Proper lubrication is essential to sustained reliability.

The fluid in the front supercharger housing lubricates the rotor drive gears. This fluid does not require changing for the life of the vehicle. However, this fluid level should be checked at 30,000-mile (48,000-km) intervals. To check the fluid level in the front supercharger housing, remove the Allen head plug in the top of this housing. The fluid should be level with the bottom of the threads in the plug opening. If the fluid level is low, add the required amount of synthetic fluid that meets the vehicle manufacturer's specifications.

Supercharger Diagnosis

The supercharger on most vehicles is serviced only as an assembly. Only specially equipped shops rebuild them. Therefore, a technician must diagnose supercharger problems and

Figure 5-42 A supercharged Cadillac engine.

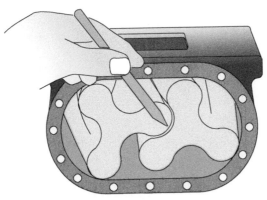

Figure 5-43 Notice the close tolerances between the lobes of the supercharger.

replace the supercharger if necessary. Supercharger problems include low boost, high boost, reduced vehicle response and/or fuel economy, noise, and oil leaks.

A supercharger is often called a blower.

The effectiveness of a supercharger is quickly seen at the dragstrip watching funny cars and top fuel dragsters.

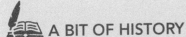 **A BIT OF HISTORY**

The 1906 American Chadwick had a supercharger. Since then, many manufacturers have equipped engines with superchargers. Supercharged Dusenbergs, Hispano-Suizas, and Mercedes-Benzes were giants among luxury-car marques, as well as winners on the race tracks in the 1920s and 1930s. After World War II, supercharging started to fade, although both Ford and American Motors sold supercharged passenger cars into the late 1950s. However, after being displaced first by larger V8 engines, then by turbochargers, the supercharger started to make a comeback with the 1989 models.

Supercharger Removal

Follow these steps for supercharger removal:

1. Disconnect the negative battery cable.
2. Remove the air inlet tube from the throttle body.
3. Remove the cowl vent screens.
4. Drain the coolant from the radiator.
5. Disconnect the spark plug wires from the spark plugs in the right cylinder head, and position them out of the way.
6. Remove the electrical connections from the intake air temperature sensor, throttle position sensor, and idle air control valve.
7. Disconnect the vacuum hoses from the supercharger air inlet plenum.
8. Remove the EGR transducer from the bracket and disconnect the vacuum hose to this component (**Figure 5-44**).
9. Disconnect the PCV tube from the supercharger air inlet plenum.
10. Disconnect the throttle linkage and remove the throttle linkage bracket. Position this linkage and bracket out of the way. Remove the cruise control linkage if equipped.
11. Remove the two EGR valve attaching bolts, and place this valve out of the way.
12. Remove the coolant hoses from the throttle body.
13. Remove the supercharger drive belt.
14. Remove the inlet and outlet tubes from the intercooler (**Figure 5-45**).

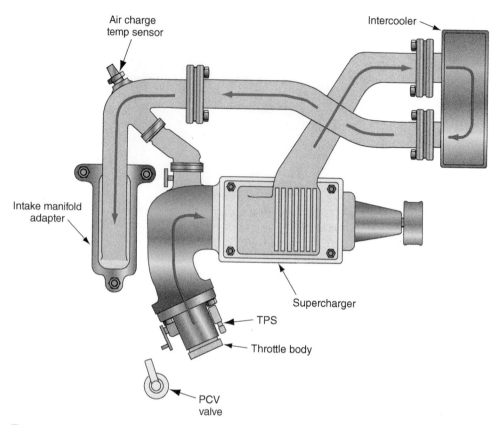

Figure 5-44 Related components of a supercharger.

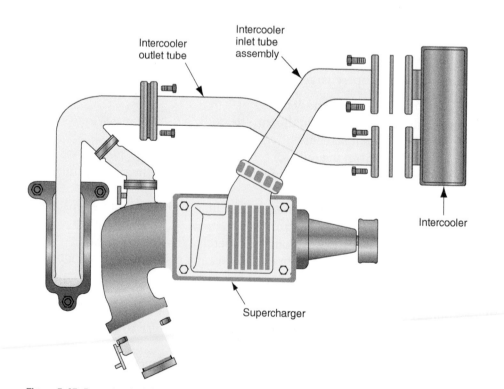

Figure 5-45 Removing the inlet and outlet tubes.

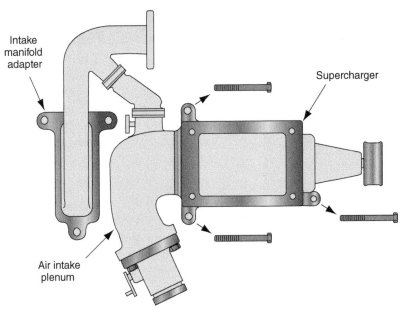

Figure 5-46 Mounting bolts for a supercharger.

15. Remove three intercooler adapter attaching bolts from the intake manifold.
16. Remove three supercharger mounting bolts (**Figure 5-46**).
17. Remove the supercharger and air intake plenum as an assembly.

This procedure should be followed for supercharger installation:

1. Clean all gasket surfaces and inspect these surfaces for scratches and metal burrs. Remove metal burrs as required.
2. Place a new gasket on the intake manifold surface that mates with the supercharger intercooler adapter.
3. Install the supercharger, throttle body, and air intake plenum as an assembly.
4. Install three supercharger mounting bolts and tighten these bolts to the specified torque.
5. Install three bolts that retain the intercooler adapter to the intake manifold and tighten these bolts to the specified torque.
6. Install the intercooler inlet and outlet tubes and tighten all mounting bolts to the specified torque.
7. Install the coolant hoses on the throttle body and tighten these hose clamps.
8. Install a new EGR valve gasket and install the EGR valve. Tighten the EGR valve mounting bolts to the specified torque.
9. Install the throttle linkage and bracket and tighten the bracket bolts to the specified torque.
10. Connect all the vacuum hoses to the original locations on the air intake plenum, and EGR valve.
11. Install the spark plug wires on the spark plugs in the right cylinder head and connect the electrical connectors to the throttle position sensor, intake air temperature sensor, and idle air control valve.
12. Install the cowl covers.
13. Install and tighten the air inlet tube on the throttle body.
14. Install the supercharger drive belt.
15. Refill the cooling system to the proper level.
16. Connect the battery ground cable.

CASE STUDY

A customer complained about lack of power on a turbocharged engine. The customer said that the engine had a much slower acceleration compared to what it had previously.

The technician performed a basic turbocharger inspection and found that all linkages, hoses, and lines were in normal condition. A check for intake and exhaust leaks did not reveal any problems, and the turbocharger did not have an excessive noise level. The technician checked the turbocharger boost pressure during a road test and discovered it was 3 psi below the specified boost pressure. There was some evidence of bluish-white smoke in the exhaust during the road test.

The technician decided to check the engine compression before any further turbocharger service. The compression test revealed that all four cylinders had much lower-than-specified compression. The technician placed a small amount of oil in each cylinder and repeated the compression test. With the oil in the cylinders, the compression pressure improved considerably on each cylinder, indicating worn piston rings and cylinders. The customer was advised that the low engine compression was the cause of the reduced power and that the engine required rebuilding.

ASE-STYLE REVIEW QUESTIONS

1. While discussing intake manifold vacuum leaks,

   Technician A says that an intake manifold leak can be found using a smoke tester.

   Technician B says that an intake manifold vacuum leak causes a low steady reading on a vacuum gauge connected to the intake manifold.

   Who is correct?

   A. A only   C. Both A and B

   B. B only   D. Neither A nor B

2. While discussing exhaust restrictions,

   Technician A says that an exhaust restriction can be detected by measuring backpressure.

   Technician B says that the pressure should be more than 5 psi on a good system.

   Who is correct?

   A. A only   C. Both A and B

   B. B only   D. Neither A nor B

3. While discussing exhaust system diagnosis,

   Technician A says that a rattling noise when the engine is accelerated can be caused by loose internal components in the muffler.

   Technician B says that a lack of engine power can be caused by a restricted exhaust system.

   Who is correct?

   A. A only   C. Both A and B

   B. B only   D. Neither A nor B

4. While discussing catalytic converter diagnosis,

   Technician A says that the temperature should be the same at the catalytic converter inlet and outlet if the converter is working normally.

   Technician B says that an oxygen storage test can be run with an exhaust gas analyzer.

   Who is correct?

   A. A only   C. Both A and B

   B. B only   D. Neither A nor B

5. *Technician A* says that a vacuum leak results in less air entering the engine, which causes a richer air-fuel mixture.

   Technician B says that a vacuum leak anywhere in the system can cause the engine to run poorly.

   Who is correct?

   A. A only   C. Both A and B

   B. B only   D. Neither A nor B

6. *Technician A* says that a catalytic converter oxidizes HC and CO to H_2O and CO_2.

   Technician B says that allowing the converter to overheat because of an overly rich mixture can destroy its usefulness.

   Who is correct?

   A. A only   C. Both A and B

   B. B only   D. Neither A nor B

7. While discussing turbocharger service, *Technician A* says that if the end housings are scored, they must be replaced.

Technician B says that a straightedge should be used to check all turbocharger mating surfaces for a warped condition.

Who is correct?

A. A only

B. B only

C. Both A and B

D. Neither A nor B

8. While discussing turbocharger inspection and diagnosis,

Technician A says that a turbocharger should not be operated until the engine has idled for 3 to 5 minutes.

Technician B says that an intake system air leak upstream from the compressor wheel may allow dirt particles to enter the turbocharger.

Who is correct?

A. A only

B. B only

C. Both A and B

D. Neither A nor B

9. While discussing turbocharger boost pressure, *Technician A* says that low cylinder compression does not affect turbocharger operation.

Technician B says that a waste gate sticking in the closed position decreases boost pressure.

Who is correct?

A. A only

B. B only

C. Both A and B

D. Neither A nor B

10. While discussing turbocharger service, *Technician A* says that turbocharger bearing failure may be caused by a malfunctioning cooling system.

Technician B says that turbocharger bearing failure may be caused by contaminated engine oil.

Who is correct?

A. A only

B. B only

C. Both A and B

D. Neither A nor B

ASE CHALLENGE QUESTIONS

1. A restricted exhaust may cause any of the following *except*:

A. loss of power.

B. increased noise.

C. faster than normal idle speed.

D. overheating.

2. *Technician A* says that catalytic converter operation can be confirmed by the delta temperature test.

Technician B says that the converter operation can be confirmed by an exhaust gas analyzer.

Who is correct?

A. A only

B. B only

C. Both A and B

D. Neither A nor B

3. Turbochargers are being discussed.

Technician A says that worn shaft seals can be indicated by bluish-white smoke in the exhaust.

Technician B says that a clogged PCV valve can cause leakage at the shaft seals.

Who is correct?

A. A only

B. B only

C. Both A and B

D. Neither A nor B

4. A turbocharger with too little boost may be a result of:

A. binding linkage.

B. leaking engine exhaust.

C. worn engine piston rings.

D. all of the above.

5. A turbocharged engine has a burned piston from extreme heat. This may be caused by:

A. excessive turbocharger boost pressure.

B. a leaking engine exhaust system.

C. damaged turbocharger shaft bearings.

D. either A or C.

Name _____   Date _____

CHECK ENGINE MANIFOLD VACUUM

Upon completion of this job sheet, you should be able to measure engine manifold vacuum and determine the condition of the engine.

ASE Education Foundation Correlation

This job sheet addresses the following **AST/MAST** task: VIII. Engine Performance; A. General Engine Diagnosis

VIII.A.5   Perform engine absolute manifold pressure tests (vacuum/boost); determine needed action. **(P-1)**

Tools and Materials

- Vacuum gauge
- Clean rag

Describe the vehicle being worked on:

Year _____   Make _____

Model _____   VIN _____

Engine size _____ # of cyls. _____ compression ratio _____:

_____

Expected manifold vacuum readings at idle: _____

Source of information: _____

Describe the general running condition of the engine:

_____

_____

Procedure

1. What does manifold vacuum represent?

    _____

    _____

2. Connect a vacuum gauge to a manifold vacuum source. Where is the source you used?

    _____

    _____

3. What is the vacuum reading with the engine at idle?

    _____

4. What is the vacuum reading with the engine at 1,500 rpm?

    _____

5. In what unit of measurement is vacuum measured?

    _____

6. Conduct the following tests using the guidelines included in this job sheet, record the readings, and explain what you just found out.

    Cranking vacuum test
    Results and conclusions:

    _____

PCV valve test
Results and conclusions:

_____

Vacuum leak test
Results and conclusions:

_____

Valve action test
Results and conclusions:

_____

Exhaust restriction test
Results and conclusions:

_____

Piston ring test
Results and conclusions:

_____

7. Based on these tests, how would you summarize the condition of this engine?

_____

_____

_____

_____

_____

_____

_____

_____

_____

_____

GUIDELINES FOR VACUUM TESTS

Cranking Vacuum

- Completely close the throttle plate, disable the ignition, connect the vacuum gauge to the manifold, and crank the engine.
- 5 inches or more = a good engine.
- Less than 5 inches = cylinders leak or there is a leak in the intake system.
- Less than 1 inch = possible timing belt or chain problem.

PCV System Test

- With the engine idling and the gauge connected, either pinch the PCV hose shut or hold your finger over the end of the valve. The reading should increase with the valve plugged off.

Vacuum Leak Test

- Gauge connected, engine idling. Look for a steady reading of 16 to 21 inches. Rhythmic drop of 1 to 2 inches = valves not sealing.
- A few inches below normal with needle flutter may indicate worn intake valve guides or a mixture problem.
- A few inches below normal but with a steady needle would indicate a common small leak. Three to nine inches below normal indicates an intake system leak—use oil or carburetor cleaner to identify the location. This can also be caused by low compression.

Valve Action Test

- Run engine.
- If the reading fluctuates between 5 and 7 inches = possible late ignition timing. If the reading drops irregularly from a normal reading = a sticking valve.
- Raise engine speed to 2,000 rpm.
- If the readings fluctuate wildly between 12 and 24 inches = weak valve springs.
- If the readings fluctuate wildly but irregularly between 12 and 24 inches = broken valve springs.

Ignition Timing Test

- Engine idling.
- Readings below normal = late timing.
- Readings above normal = early timing.

Exhaust Restriction Test

- Increase engine speed slowly to 2,000 rpm. Needle should increase over normal. Close throttle quickly; vacuum should return to normal as quickly as it rose.
- Vacuum does not increase with speed = plugged exhaust. Vacuum returns slowly to normal = plugged exhaust.

Piston Ring Test

- All other test results must be satisfactory and engine oil level OK.
- Increase speed to 2,000 rpm and quickly release the throttle.
- Increase of 5 or more inches of vacuum = rings in good shape
- Increase of less than 5 inches of vacuum = suspect worn rings

Instructor's Response

_____

_____

_____

Name _____   Date _____

CHECKING BOOST SYSTEMS

Upon completion of this job sheet, you should be able to test the operation of a turbocharger/supercharger system.

ASE Education Foundation Correlation

This job sheet addresses the following **MAST** task: VIII. Engine Performance; D. Fuel, Air Induction, and Exhaust Systems Diagnosis and Repair

Task #13   Test the operation of turbocharger/supercharger systems; determine needed action. **(P-2)**

Tools and Materials

- Soap and water mixture
- Pressure gauge
- Service information
- Protective clothing
- Goggles or safety glasses with side shields

Describe the vehicle being worked on:

Year _____ Make _____ Model _____

VIN _____ Engine type and size _____

Describe general operating condition:

_____

_____

Procedure

1. Check all linkages and hoses connected to the turbocharger. Record your findings.

   _____

   _____

2. Inspect the waste-gate diaphragm linkage for looseness and binding, and check the hose from the waste-gate diaphragm to the intake manifold for cracks, kinks, and restrictions. Record your findings.

   _____

   _____

3. Check the coolant hoses and oil line connected to the turbocharger for leaks. Record your findings.

   _____

   _____

4. When the engine is running, does the exhaust have an odor or color? Describe the exhaust.

   _____

   _____

5. Check the condition of the engine's oil. Record your findings.

_____

_____

6. Check all turbocharger mounting bolts for looseness. Record your findings.

_____

_____

7. Pay close attention to the sound of the turbocharger when it is operating. Does it make unusual noises? Record your findings.

_____

_____

8. Check for exhaust leaks in the turbine housing and related pipe connections. Record your findings.

_____

_____

9. Check the computer with a scan tool for DTCs. Record your findings.

_____

_____

10. Start the engine and listen to the sound the turbo system makes while you change engine speed. Record your findings.

_____

_____

11. Check the air cleaner and remove the ducting from the air cleaner to turbo and look for dirt buildup or damage from foreign objects. Check for loose clamps on the compressor outlet connections. Record your findings.

_____

_____

12. Check the engine intake system for loose bolts or leaking gaskets. Record your findings.

_____

_____

13. Disconnect the exhaust pipe and look for restrictions or loose material. Examine the exhaust system for cracks, loose nuts, or blown gaskets. Record your findings.

_____

_____

14. Rotate the turbo shaft assembly. Does it rotate freely? Are there signs of rubbing or wheel impact damage? Record your findings.

_____

_____

15. Visually inspect all hoses, gaskets, and tubing for proper fit, damage, and wear. Record your findings.

_____

_____

16. Check the low-pressure, or air cleaner, side of the intake system for vacuum leaks. Record your findings.

_____

_____

17. On the pressure side of the system, you can check for leaks by using soapy water. After applying the soap mixture, look for bubbles to pinpoint the source of the leak. Record your findings.

_____

_____

18. Connect a pressure gauge to the intake manifold to check the boost pressure. Set the gauge so you can see it during a road test. Road test the vehicle at the speed specified by the vehicle's manufacturer and observe the boost pressure. Record your findings.

_____

_____

19. To check the waste-gate, connect a hand pressure pump and a pressure gauge to the waste-gate diaphragm. Position a dial indicator against the outer end of the waste-gate diaphragm rod and supply the specified pressure to the waste-gate diaphragm and observe the dial indicator movement. Record your findings.

_____

_____

20. If the waste-gate rod movement is less than specified, disconnect the rod from the waste-gate valve linkage, and check the linkage for binding. Record your findings.

_____

_____

21. What are your service recommendations?

_____

_____

Instructor's Response

_____

_____

_____

Name _____   Date _____

SERVICING A THROTTLE BODY / REPLACING AN AIR FILTER

Upon completion of this job sheet, you should be able to inspect the throttle body mounting plates, air induction and filtration systems, intake manifolds, and gaskets. You should also be able to remove, service, and install a throttle body.

ASE Education Foundation Correlation

This job sheet addresses the following **MLR** task: VIII. Engine Performance; C. Fuel, Air Induction, and Exhaust Systems

Task #2   Inspect, service, or replace air filters, filter housings, and intake duct work. **(P-1)**

This job sheet addresses the following **AST** tasks: VIII. Engine Performance; D. Fuel, Air Induction, and Exhaust Systems Diagnosis and Repair

Task #4   Inspect, service, or replace air filters, filter housings, and intake duct work. **(P-1)**

Task #5   Inspect throttle body, air induction system, intake manifold and gaskets for vacuum leaks and/or unmetered air. **(P-2)**

This job sheet addresses the following **MAST** tasks: VIII. Engine Performance; D. Fuel, Air Induction, and Exhaust Systems Diagnosis and Repair

Task #5   Inspect, service, or replace air filters, filter housings, and intake duct work. **(P-1)**

Task #6   Inspect throttle body, air induction system, intake manifold and gaskets for vacuum leaks and/or unmetered air. **(P-2)**

Tools and Materials

- 12-volt power supply or memory keeper
- OSHA-approved air nozzle
- Throttle body cleaner
- Hand tools
- Compressed air
- Service information
- Protective clothing
- Goggles or safety glasses with side shields

Describe the vehicle being worked on:

Year _____ Make _____ Model _____

VIN _____ Engine type and size _____

Describe general operating condition:

_____

_____

Procedures			Task Completed
Visually Inspect an EFI System	Yes	No	
1. Is the battery in good condition, fully charged, with clean terminals and connections?	☐	☐	
2. Do the charging and starting systems operate properly?	☐	☐	

	Yes	No	Task Completed
3. Are all fuses and fusible links intact?	☐	☐	
4. Are all wiring harnesses properly routed, with connections free of corrosion and tightly attached?	☐	☐	
5. Are all vacuum lines in sound condition, properly routed, and tightly attached?	☐	☐	
6. Is the PCV system working properly and maintaining a sealed crankcase?	☐	☐	
7. Are all emission control systems in place, hooked up, and operating properly?	☐	☐	
8. Is the level and condition of the coolant/antifreeze good and is the thermostat opening at the proper temperature?	☐	☐	
9. Are the secondary spark delivery components in good shape, with no signs of cross-firing, carbon tracking, corrosion, or wear?	☐	☐	
10. Is the idle speed set to specifications?	☐	☐	
11. Does the air intake ductwork have cracks or tears?	☐	☐	
12. Are all of the induction hose clamps tight and properly sealed?	☐	☐	
13. Are there any other possible air leaks in the crankcase?	☐	☐	
14. Are the electrical connections at the mass airflow sensor or manifold pressure sensor good?	☐	☐	
15. Is there carbon buildup inside the throttle bore and on the throttle plate?	☐	☐	
16. Is there a clicking noise at each injector while the engine is running?	☐	☐	
17. Check the condition of the air filter. Does it show signs of needing replacement? How did you make your determination? _____	☐	☐	
18. Before replacing the air filter (if necessary), make sure the air box is free of debris.			☐

Servicing a Throttle Body

Note: Whenever it is necessary to remove the throttle body assembly for replacement or cleaning, make sure you follow the procedures outlined by the manufacturer. Some throttle bodies have a special coating that should not be cleaned with solvents. Check the service information before starting any work.

	Task Completed
1. Disconnect the negative battery cable.	☐
2. Remove the air ducting from the throttle body.	☐
3. Unbolt and remove the throttle body.	☐
4. Once the assembly has been removed, remove all nonmetallic parts such as the TP sensor, IAC valve, throttle opener, and the throttle body gasket from the throttle body. Never soak any electrical components in solvent.	☐
5. It is now safe to clean the throttle body assembly in the recommended cleaner and blow dry it with compressed air. Make sure you blow out all passages in the throttle body assembly.	☐

Task Completed

6. Before reinstalling the throttle body assembly, check to make sure all metal mating surfaces are clean and free from metal burrs and scratches. ☐

7. With new gaskets and seals, install the assembly. Tighten all fasteners to the recommended torque. ☐

8. After everything that was disconnected is reconnected, reconnect the negative battery cable. ☐

9. After all the parts are reconnected and installed, adjust the throttle linkage according to the manufacturer's recommendations. ☐

10. Relearn idle speeds as necessary. ☐

Instructor's Response

_____

_____

_____

Name _____   Date _____

INSPECT EXHAUST SYSTEM

Upon completion of this job sheet, you should be able to properly inspect exhaust manifolds, exhaust pipes, mufflers, catalytic converters, resonators, tailpipes, and heat shields.

ASE Education Foundation Correlation

This job sheet addresses the following **MLR** tasks: VIII. Engine Performance; C. Fuel, Air Induction, and Exhaust Systems

Task #3	Inspect integrity of the exhaust manifold, exhaust pipes, muffler(s), catalytic converter(s), resonator(s), tailpipe(s), and heat shields; determine necessary action. **(P-1)**
Task #4	Inspect condition of exhaust system hangers, brackets, clamps, and heat shields; determine needed action. **(P-1)**

This job sheet addresses the following **AST** tasks: VIII. Engine Performance; D. Fuel, Air Induction, and Exhaust Systems Diagnosis and Repair

Task #8	Inspect integrity of the exhaust manifold, exhaust pipes, muffler(s), catalytic converter(s), resonator(s), tailpipe(s), and heat shields; determine necessary action. **(P-1)**
Task #9	Inspect condition of exhaust system hangers, brackets, clamps, and heat shields; determine needed action. **(P-1)**

This job sheet addresses the following **MAST** tasks: VIII. Engine Performance; D. Fuel, Air Induction, and Exhaust Systems Diagnosis and Repair

Task #9	Inspect integrity of the exhaust manifold, exhaust pipes, muffler(s), catalytic converter(s), resonator(s), tailpipe(s), and heat shields; determine necessary action. **(P-1)**
Task #10	Inspect condition of exhaust system hangers, brackets, clamps, and heat shields; determine needed action. **(P-1)**

Tools and Materials
- Flashlight or trouble light
- Service information
- Hammer or mallet
- Tachometer
- Lift
- Vacuum gauge
- Protective clothing
- Goggles or safety glasses with side shields

Describe the vehicle being worked on:

Year _____ Make _____ VIN _____

Model _____ Engine type and size _____

Describe general operating condition:

_____

_____

Procedure   **Task Completed**

1. Before doing a visual inspection, listen closely for hissing or rumbling sound that   ☐
    may indicate the beginning of exhaust system failure. With the engine idling, slowly
    move along the entire system and listen for leaks.

    Warning: Be very careful. Remember that the exhaust system gets very hot. Do not
    get your face too close when listening for leaks.

    a. Did you hear any indications of a leak?   ☐ Yes   ☐ No

    b. If so, where?

    _____

    _____

2. Safely raise the vehicle.   ☐

3. With a flashlight or trouble light, check for the following:

    • Holes and road damage   ☐

    • Discoloration and rust   ☐

    • Carbon smudges   ☐

    • Bulging muffler seams   ☐

    • Interfering rattle points   ☐

    • Torn or broken hangers and clamps   ☐

    • Missing or damaged heat shields   ☐

    a. Did you detect any of these problems?   ☐ Yes   ☐ No

    b. If so, where?

    _____

    _____

4. Sound out the system by gently tapping the pipes and muffler with a hammer or   ☐
    mallet. A good part will have a solid metallic sound. A weak or worn-out part
    will have a dull sound. Listen for falling rust particles on the inside of the muffler.
    Mufflers usually corrode from the inside out, so the damage may not be visible from
    the outside. Remember that some rust spots might be only surface rust.

    a. Did you find any weak or worn-out parts?   ☐ Yes   ☐ No

    b. If so, which ones? _____

5. Grab the tailpipe (when it is cool) and try to move it up and down and from side to   ☐
    side. There should be only slight movement in any direction. If the system feels wobbly
    or loose, check the clamps and hangers that fasten the tailpipe to the vehicle.

    a. Did you detect any problems with the clamps or hangers?   ☐ Yes   ☐ No

    b. If so, where? _____

6. Check all of the pipes for kinks and dents that might restrict the flow of exhaust   ☐
    gases.

    a. Did you find any kinks or dents?   ☐ Yes   ☐ No

    b. If so, where? _____

Task Completed

☐

7. Take a close look at each connection, including the one between the exhaust manifold and the exhaust pipe.

   a. Did you find any white powdery deposits?   ☐ Yes   ☐ No

   b. If so, try tightening the bolts or replacing the gasket at that connection.   Not applicable   ☐

   c. Check for loose connections at the muffler by pushing up on the -muffler slightly.

   d. If loose, try tightening them.   Not applicable   ☐

8. If a visual inspection does not identify a partially restricted or blocked exhaust system, perform the following test.

   a. Attach a vacuum gauge to the intake manifold. Connect a tachometer. Start the engine and observe the vacuum gauge. It should indicate a vacuum of 16 to 20 inches of mercury. Does it?   ☐ Yes   ☐ No

   b. Increase the engine's speed to 2,000 rpm and observe the vacuum gauge. Vacuum will decrease when the speed is increased rapidly, but it should stabilize at 16 to 21 inches of mercury and remain constant. If the vacuum does not build up to at least the idle reading, the exhaust system is restricted or blocked. Is the system restricted or blocked?   ☐ Yes   ☐ No

☐

9. Catalytic converters can overheat. Look for bluish or brownish discoloration of the outer stainless-steel shell. Also, look for blistered or burned paint or undercoating above and near the converter.

   Are there any signs of overheating?   ☐ Yes   ☐ No

10. Look up and record the part numbers of any parts discovered to be defective in previous steps.

_____

_____

_____

Instructor's Response

_____

_____

_____

Name _____   Date _____

CHECKING AND REFILLING DIESEL EXHAUST FLUID

Upon completion of this job sheet, you should be able to check and refill the diesel DEF fluid.

ASE Education Foundation Correlation

This job sheet addresses the following **MLR** task: VIII. Engine Performance; C. Fuel, Air Induction, and Exhaust Systems

Task #5   Check and refill diesel exhaust fluid (DEF). **(P-3)**

This job sheet addresses the following **AST** task: VIII. Engine Performance; D. Fuel, Air Induction, and Exhaust Systems Diagnosis and Repair

Task #11   Check and refill diesel exhaust fluid (DEF). **(P-2)**

This job sheet addresses the following **MAST** task: VIII. Engine Performance; D. Fuel, Air Induction, and Exhaust Systems Diagnosis and Repair

Task #12   Check and refill diesel exhaust fluid (DEF). **(P-2)**

Tools and Materials

- Service information
- Diesel vehicle
- DEF (diesel exhaust fluid)
- Protective clothing
- Goggles or safety glasses with side shields

Describe the vehicle being worked on:

Year _____   Make _____

Model _____   VIN _____

Engine type and size _____

Procedure

1. Describe the purpose of the diesel exhaust fluid (DEF) system. Which emission does it reduce?

   _____

   _____

2. What is the capacity of the DEF tank?

   _____

   _____

3. Does the vehicle have a DEF-level sensor in the DEF tank?

   _____

   _____

4. Which component in the diesel exhaust system uses the DEF?

   _____

   _____

5. How does the ECM determine that DEF fluid is needed?

_____

_____

6. At what temperature will DEF freeze?

_____

_____

7. How does the vehicle prevent DEF from freezing?

_____

_____

8. List the warnings that the vehicle gives the driver about the level of DEF in the vehicle.

_____

_____

9. When the DEF tank is empty, what steps does the ECM take to ensure that the driver will not ignore the warning lamp?

_____

_____

10. What happens if the system detects DEF of inferior quality in the DEF tank?

_____

_____

11. How does the ECM monitor the quality of the DEF fluid?

_____

_____

12. When should DEF levels be checked?

_____

_____

13. Describe the procedure to refill the DEF tank.

_____

_____

14. Describe the precautions that must be followed when working around DEF.

_____

_____

_____

Instructor's Response

_____

_____

_____

CHAPTER 6
ENGINE CONTROL SYSTEM DIAGNOSIS AND SERVICE

Upon completion and review of this chapter, you should be able to:

- Retrieve trouble codes from a vehicle.
- Erase fault codes.
- Perform a scan tester diagnosis on various vehicles.
- Perform a cylinder output test.
- Diagnose computer voltage supply and ground wires.
- Test and diagnose switch-type input sensors.
- Test and diagnose engine coolant temperature sensors.
- Test and diagnose intake air temperature sensors.
- Test and diagnose exhaust gas recirculation valve position sensors.
- Test and diagnose oxygen sensors.
- Test and diagnose AF sensors.
- Test and diagnose knock sensors.
- Test and diagnose vehicle speed sensors.
- Test and diagnose manifold absolute pressure (MAP) sensors.
- Test and diagnose mass airflow (MAF) sensors.

 Basic Tools

Basic tool set
DMM
Lab scope
Scanner
Appropriate service manuals

Terms To Know

Air-fuel (AF) ratio
Air-fuel (AF) sensor
Clutch pedal position (CPP)
CPP sensor
Engine coolant temperature (ECT) sensor
Exhaust gas recirculation valve position (EVP) sensors
Fuel tank pressure (FTP) sensor

History code
Hot-wire intake air temperature (IAT) sensors
Key on, engine off (KOEO) test
Key on, engine running (KOER) test
Linear
MAF sensor
Manifold absolute pressure (MAP) sensor

Mass airflow (MAF) sensor
Open loop
Output state tests
Reference voltage
Scan tester
Switch tests
Temperature-responding switches
Vehicle speed sensor (VSS)

AUTHOR'S NOTE The automotive computer is commonly referred to as the engine control module (ECM) or powertrain control module (PCM), depending on the vehicles application. During the history of automotive computer control, the engine emissions and performance were the first to have computer control, called the ECM. Next vehicle automatic transmissions received computer control, so either a transmission control module (TCM) was used in addition to the ECM, or the transmission or engine function was integrated into the PCM. In the past few years, computer networking has tended to the use of many separate modules, so the ECM is beginning to come back to common usage.

INTRODUCTION

A computer is an electronic device that processes and stores data. It is also capable of operating other devices.

This chapter will look at basic computer operation, some manufacturer computer systems, and finally sensor diagnosis. This chapter looks at how an onboard computer processes information and how it uses this information to monitor and control.

For more than three decades, a computer has played an important role in the way an engine runs. The role of the computer has evolved from the control of a single system to the control of nearly all of the engine's systems. Understanding what a computer does and how it works is extremely important to effective diagnosis of drivability problems.

The operation of a computer can be divided into four basic functions:

A program is a set of instructions the computer must follow to achieve the desired results.

1. *Input:* A voltage signal that is sent from an input device. This device can be a sensor or a switch activated by the driver, the technician, or another device.
2. *Processing:* The computer uses the input information and compares it to programmed instructions. The logic circuits process the input signals into output commands.
3. *Storage:* The program instructions are stored into an electronic memory. Some of the input signals are also stored for processing later.
4. *Output:* After the computer has processed the sensor input and checked its programmed instructions, it will put out control commands to various output devices. These output devices may be a system actuator or an indicator. The output of one computer can be used as an input to another computer.

AUTHOR'S NOTE Fail-soft means the computer will substitute a fixed input value if a sensor or its circuit should fail. This provides system operation, but at a limited functioning level. These modes of operation allow the vehicle to continue to operate until the fault can be repaired. An example would be an ECT sensor that failed open, which would be interpreted as a very cold temperature of around 40°F. If the computer did not have a fail-soft strategy, the engine would probably quit running or run very badly, as the fuel delivery would change dramatically. The ECM is programmed to ignore a sensor when it is not rational. If, for instance, the vehicle has been operating for 30 minutes at normal operating temperature, and then suddenly the temperature crashes almost 200°F, the computer would substitute a value for the ECT and continue to run. Of course, the MIL will be illuminated, telling the driver to have the vehicle serviced as soon as possible.

Understanding these four computer functions will help you organize the troubleshooting process. When a system is tested, you are attempting to isolate a problem with one of these functions.

In the process of controlling the various engine systems, the PCM continuously monitors operating conditions for possible system malfunctions. The computer compares system conditions against programmed parameters. If conditions fall outside the limits of these parameters, the computer detects a malfunction. A DTC is set to indicate the portion of the system that is at fault. A technician can access the code as an aid in troubleshooting.

If a malfunction results in improper system operation, the computer may minimize the effects by using fail-soft action. In other words, the computer may substitute a fixed value in place of the real value from a sensor to avoid shutting down the entire system. This fixed value can be programmed into the computer's memory or it can be the last received signal from the sensor prior to failure. This allows the system to operate on a limited basis instead of shutting down completely.

There are several things you need to know prior to learning how to access trouble codes in a computer's memory. You need to become familiar with what you are looking at, and you must follow proper precautions when servicing these systems.

ELECTRONIC SERVICE PRECAUTIONS

A technician must take some precautions before servicing a computer or its circuit. The PCM is designed to withstand normal current draws associated with normal operation. However, overloading the system will destroy the computer. To prevent damage to the PCM and its related components, follow these service precautions:

> Impedance is the combined opposition to current created by the resistance, capacitance, and inductance of a meter.

1. Never ground or apply voltage to any controlled circuit unless the service information instructs you to do so.
2. Use only a high impedance multimeter (10 megohms or higher) to test the circuits. Never use a test light unless instructed to do so in the manufacturer's procedures.
3. Make sure the ignition switch is turned off before disconnecting or connecting electrical terminals at the PCM.
4. Unless instructed otherwise, turn the ignition switch off before disconnecting or connecting any electrical connections to sensors or actuators.
5. Turn the ignition switch off whenever disconnecting or connecting the battery terminals. Also, turn it off when replacing a fuse.
6. Do not connect any electrical accessories to the insulated or ground circuits of computer-controlled systems.
7. Use only manufacturer's specific test and replacement procedures for the year and model of the vehicle being serviced.

Electrostatic Discharge

Some manufacturers mark certain components and circuits with a code or symbol to warn technicians that they are sensitive to electrostatic discharge (**Figure 6-1**). Static electricity can destroy or render a component useless.

> Static electricity can be 25,000 volts or higher.

When handling any electronic part, especially those that are static sensitive, follow the guidelines below to reduce the possibility of electrostatic buildup on your body and the inadvertent discharge to the electronic part. If you are not sure if a part is sensitive to static, treat it as if it is.

1. Always touch a known good ground before handling the part. This should be repeated while handling the part and more frequently after sliding across a seat, sitting down from a standing position, or walking a distance.
2. Avoid touching the electrical terminals of the part unless you are instructed to do so in the written service procedures. It is good practice to keep your fingers off all electrical terminals since the oil from your skin can cause corrosion.
3. When you are using a voltmeter, always connect the negative meter lead first.
4. Do not remove a part from its protective package until it is time to install the part.
5. Before removing the part from its package, ground yourself and the package to a known good ground on the vehicle.

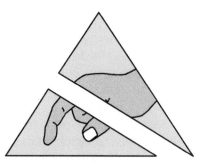

Figure 6-1 GM's electrostatic discharge (ESD) symbol that warns a technician that the component or circuit is sensitive to static.

CUSTOMER CARE It is always an excellent customer service practice to return the vehicle to the customer with no evidence or side effects of being repaired. This applies to seat position, mirror adjustment, cleaning of smudges, and perhaps even cleaning the windshield. Refrain from changing radio stations and volume or tone controls as well as eating, drinking, or smoking in the car. This also includes resetting electronic devices and accessories anytime computer-related work is performed or if an interruption of battery power occurs. Items that will probably need to be reset include the clock, radio (if the stations are known), seat memories, and mirrors. At the very least, offer assistance to the customer at the time of pickup, if possible. He or she may not recall how to reset certain accessories because it is usually not done regularly. Additionally, be sure all DTCs are cleared from the computer's memory.

Many shops now clean all vehicles that are serviced through the day, even for oil changes. Some shops even pick up and deliver vehicles to a customer's home or place of business.

BASIC DIAGNOSIS OF ELECTRONIC ENGINE CONTROL SYSTEMS

Classroom Manual
Chapter 6, page 159

Diagnosing a computer-controlled system is much more than accessing the DTCs in the computer's memory. When diagnosing any system, you need to know what to test, when to test it, and how to test it. Because the capabilities of the vehicle's computers have evolved from simple to complex, it is important to know the capabilities of the systems you are working with before attempting to diagnose a problem. Refer to the service information for help. After you understand the system and its capabilities, begin your diagnosis using your knowledge and logic.

The importance of logical troubleshooting cannot be overemphasized. The ability to diagnose a problem (to find its cause and its solution) is what separates an automotive technician from a parts changer.

There are two logics used to diagnose and service electronic engine controls: computer logic and a technician's logical diagnosis.

Computer Logic Flow

To control an engine system, the computer makes a series of decisions. Decisions are made in a step-by-step fashion until a conclusion is reached. Generally, the first decision is to determine the engine mode. For example, to control air-fuel mixture, the computer first

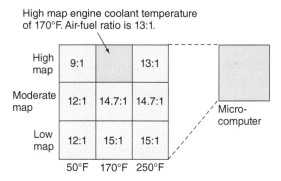

Figure 6-2 A typical computer lookup table.

determines whether the engine is cranking, idling, cruising, or accelerating. Then, the computer can choose the best system strategy for the present engine mode. In a typical example, sensor input indicates the engine is warm, rpm is high, manifold absolute pressure is high, and the throttle plate is wide open. The computer determines the vehicle is under heavy acceleration or wide-open throttle. Next, the computer determines the goal to be reached. For example, with heavy acceleration, the goal is to create a rich air-fuel mixture. When operating in fuel control, the computer uses a lookup table similar to that shown in **Figure 6-2**. At wide-open throttle with high manifold absolute pressure and coolant temperature of 170°F, the table indicates the **air-fuel ratio** should be 13:1 that is, 13 pounds of air for every 1 pound of fuel. An air-fuel ratio of 13:1 creates the rich air-fuel mixture needed for heavy acceleration.

In a final series of decisions, the computer determines how the goal can be achieved. In our example, a rich air-fuel mixture is achieved by increasing fuel injector pulse width. The injector nozzle remains open longer and more fuel is drawn into the cylinder, providing the additional power needed.

Technician's Logical Diagnosis

The best automotive technicians use this logical process to diagnose engine problems. When faced with an abnormal engine condition, they compare clues (such as meter readings, oscilloscope readings, and visible problems) with their knowledge of proper conditions and discover a logical reason for the way the engine is performing. Logical diagnosis means following a simple basic procedure. Start with the most likely cause and work to the most unlikely. In other words, check out the easiest, most obvious solutions first before proceeding to the less likely and more difficult solutions. Do not guess at the problem or jump to a conclusion before considering all of the factors.

The logical approach has a special application to troubleshooting electronic engine controls. Check all traditional non-electronic engine control possibilities before attempting to diagnose the electronic engine control itself. For example, low battery voltage might result in faulty sensor readings. A sensor could also be sending faulty signals to the computer, resulting in improper ignition timing and fuel delivery.

An additional problem that occurs when diagnosing any computer is that the part itself can be difficult to accurately test. Most electronic parts are completely sealed due to complex, interlocking circuitry and cannot be checked at their input and output connections.

This problem can be overcome by using a logical procedure called the process of elimination. In other words, if every related part checks out OK, it must be the electronic part that is bad. This means thoroughly checking every component in a system as well as checking for such basic factors as proper current supply and good grounds.

AUTHOR'S NOTE When beginning a diagnostic procedure, do not skip steps or attempt shortcuts such as assuming all the basic items will check out fine. It is critical to have proper power supply voltage and system ground circuits that are sound. Multiple symptoms and drivability problems that appear to be complex may require minor repairs such as cleaning and tightening a loose ground wire. A recent report from an OEM indicated that less than 2 percent of the PCMs returned for warranty claims were actually verified to be faulty. That means 98 percent of returns were misdiagnosed and had no problems. A detailed systematic approach to effective problem solving and diagnosis was discussed in the beginning of Chapter 2 of this manual.

Isolating Computerized Engine Control Problems

Determining which part or area of a computerized engine control system is defective requires thoroughly understanding how the system works and following the logical troubleshooting process previously explained.

Electronic engine control problems are usually caused by defective sensors, wiring, connections, and output devices. The logical procedure in most cases is to check the input sensors and wiring first, then the output devices and their wiring, and, finally, the computer.

Computerized engine controls have self-diagnosis capabilities. A malfunction in any sensor or output device, or in the computer itself, may be stored in the computer's memory as a trouble code. Stored codes can be retrieved and the indicated problem areas checked further.

Some malfunctions may cause drivability problems without stored codes. These methods can be used to check individual system components:

1. *Verify the concern.* Make certain that you can verify or duplicate the concern before you start checking or replacing parts. You may have to drive the vehicle with the customer in order to really understand the concern.
2. *Visual checks.* This means looking for obvious problems. Any part that is burned, broken, cracked, corroded, or has any other visible problem must be replaced before continuing the diagnosis. Examples of visible problems include disconnected sensor vacuum hoses and broken or disconnected wiring.
3. *Check for Technical Service Bulletins (TSBs).* These can give you a wealth of information about conditions that have been caused by defective parts or tell you about new diagnostic procedures. Always check for TSBs; they will save you much trouble in the long run.
4. *Symptom chart checks.* Many manufacturers have a list of items to check based on specific symptoms. These can give you an idea of which things to check first.
5. *Ohmmeter checks.* Most sensors and output devices can be checked with an ohmmeter. For example, Figure 6-3 shows an ohmmeter used to check a temperature sensor. Output devices such as coils or motors can also be checked with an ohmmeter to determine continuity.
6. *Voltmeter checks.* Many sensors, output devices, and their wiring can be diagnosed by checking the voltage flowing to and from them. Oxygen sensors can be checked in this manner.
7. *Lab scope checks.* The activity of sensors and actuators can be monitored with a lab scope. By watching their activity, you are doing more than testing them. Problems elsewhere in the system will often cause a device to behave abnormally. These situations are identified by the trace on a scope and by the technician's understanding of a scope and the device being monitored.

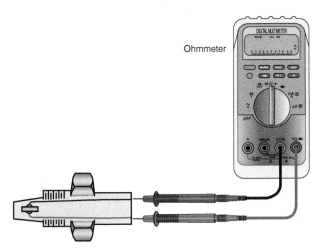

Ohmmeter

Figure 6-3 Checking a temperature sensor with an ohmmeter.

In some cases, a final check on a sensor can be made only by substitution. Substitution is not the most desirable way to diagnose problems. To substitute, replace the suspected part with a known good unit and recheck the system. If the system now operates normally, the original part is defective.

SERVICE BULLETIN INFORMATION

One of the first steps a technician should take during the diagnosis of any problem is checking for TSBs. Technical service bulletins have been developed by the manufacturers after a solution to a difficult problem has been found. Service bulletins address many different aspects of repair. Sometimes a new procedure has been developed, saving time for the technician. New parts may be required due to a production problem. Some bulletins might be strictly informational; running changes might be made, say, to a wiring harness that changes the location of a connector, wire color, splice, and so on. This information can be critical for a proper repair. Fortunately, we have more information available to us than ever before, as service information bulletins are generally included with service information systems on the Internet. At the same time, make sure that the bulletin applies directly to the vehicle you are working with and check identification numbers listed on the bulletin. As an important side note, the public often assumes that a TSB is a "recall." A service bulletin is information designed to help repair the vehicle and may have no effect on the warranty of the vehicle, although this is sometimes done as a goodwill gesture. A "recall" is mandated by the government because it affects safety or emissions and is done regardless of mileage or time.

Reprogramming of the PCM is covered in Chapter 12.

We will discuss an example of how a TSB might help you in a difficult diagnosis. Some 2014 to 2015 Fords with a 1.5-liter engine may set P0036, P0137, P0138, and/or P0141. The service bulletin describes a re-flash programming procedure to fix the problem. While a technician could probably find the cause of the problem through normal testing, the technical bulletin will probably save much time, effort, and probable frustration in testing.

Many bulletins concern "re-flashing," "recalibrating," or reprogramming the PCM. Reprogramming the PCM is revising the software that determines the operation of the PCM. Sometimes drivability problems can be fixed by merely changing a value in the software. The appropriate software is downloaded from the manufacturer via the Internet and then loaded onto a scan tool, and the new calibration is downloaded from the scan tool to the vehicle's PCM.

An example of a reprogramming operation is campaign SA042 concerning a 2010 to 2011 Kia Rio for a transmission shifting concern, fuel efficiency, and cruise control operation. As you can imagine, this would be a very difficult situation to resolve by a technician in the shop. The only way to properly repair the vehicle is by using this campaign. Imagine the frustration of perhaps replacing good parts, not to mention customer dissatisfaction by not checking for TSBs and updated calibrations.

The PCM is programmed to customize itself for the vehicle application. For example, there may be two different vehicles that have the same engine, but one vehicle has an automatic transmission, the other a manual transmission. Both vehicles have the same PCM, but the programming will be custom-tailored to work with the correct transmission. A subscription to the manufacturer's website is required to obtain the correct programming according to the VIN number. PCMs should not be switched between vehicles for this reason.

If any computer is replaced in the vehicle, such as the PCM, BCM, or IPC, it will be necessary to reprogram the module before it will function properly. Many modules, such as the door, antitheft system, sunroof, instrument panel cluster, and so on, are also controlled by modules (computers) that will require reprogramming when they are replaced or as the result of a technical bulletin.

SELF-DIAGNOSTIC SYSTEMS

Today's computer-controlled systems are complicated. It would take endless amounts of time to diagnose these systems without help. For this reason, computerized engine controls have self-diagnostic capabilities. By constantly monitoring the engine, emission, and transmission control systems, the computer is able to evaluate the effectiveness of the electronic control systems, including the computers involved. If problems are found, a DTC is set. There are two types of DTCs: type A and type B. Type A DTCs turn on the malfunction indicator lamp (MIL) after the first failure. Type B DTCs illuminate the MIL after the second failure of a system if the failure occurs under similar conditions. Type B DTCs do set a "pending DTC" that can be read on a scan tool. If the condition that caused the DTC to set passes its monitor three times in a row, then the MIL is turned off, but a code will still be stored in the computer's memory for another 40 warm-up cycles. A pending misfire or fuel trim code can be stored for up to 80 key cycles. Each type of fault or failure is assigned a numerical trouble code that is stored in computer memory (**Figure 6-4**).

A **hard fault** means a problem has been found somewhere in the system by a monitor, and the fault is occurring at the present time the computer is being scanned, or during the last time the monitor for the condition was run. Hard faults are usually easier to diagnose because the problem is current.

A history code is a code that set but the condition is not present at the time the scan tool is connected because the condition is intermittent.

A **history code** can often indicate an intermittent problem. A malfunction occurred (e.g., a poor connection causing an intermittent open or short), but passed the last test of the monitor. Intermittent codes can be difficult to find and repair, because the fault is no longer present. Freeze frame records do store a snapshot of the conditions present at the time of the fault and can be very helpful. The vehicle can be driven during the same conditions that set the code. Nonvolatile RAM allows intermittent faults to be stored for up to a specific number of ignition key on/off cycles. If the trouble does not reappear during that period, it is erased from the computer's memory.

There are various methods of assessing the trouble codes generated by the computer. Manufacturers have diagnostic equipment designed to monitor and test the electronic

EXAMPLE: P0137 LOW VOLTAGE BANK 1 SENSOR 2

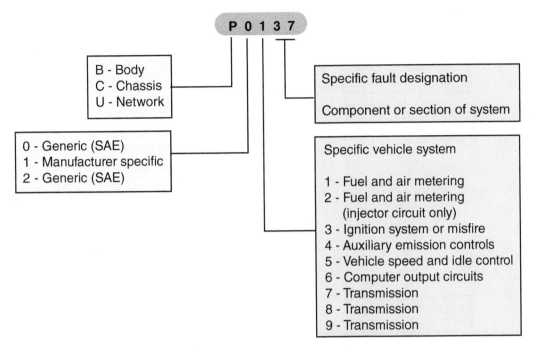

Typical generic code examples:

Mass Airflow	P0102, P0103
Intake Air Temperature	P0112, P0113, P0127
Barometric Pressure	P0106, P0107, P0108, P0109
Engine Coolant Temperature	P0117, P0118
Oxygen Sensor (one of several)	P0131, P0133, P0135, P0136, P0141
Throttle Position	P0121, P0122, P0123
EGR/EVP	P0400, P0401, P0402
Vehicle Speed	P0500, P0501, P0503

Note: Manufacturers will also use specific codes that apply only to their systems.

Figure 6-4 Different systems use different codes to identify problem areas.

components of their vehicles. Aftermarket companies also manufacture scan tools that have the capability to read and record the input and output signals passing to and from the computer.

Visual Inspection

Before reading self-diagnostic or trouble codes, do a visual check of the engine and its systems. This quick inspection can save much time during diagnosis. While visually inspecting the vehicle, make sure to include the following:

1. Inspect the condition of the air filter and related hardware around the filter.
2. Inspect the entire PCV system.
3. Check to make sure the vapor canister is neither saturated nor flooded.
4. Check the battery and its cables, the vehicle's wiring harnesses, connectors, and the charging system for loose or damaged connections. Also, check the connectors for signs of corrosion.

After the 1996 model year, a scan tool is required to read trouble codes; flash codes are no longer available. Since there is so much information to help the technician on the scan tool, a technician would want to use the scan tool anyway.

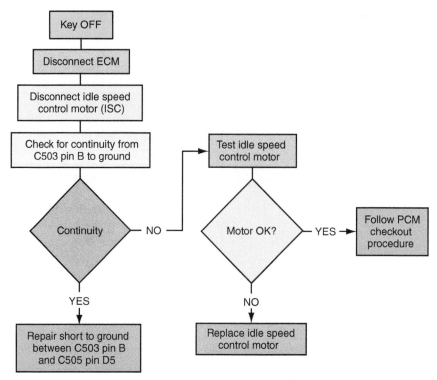

Figure 6-5 A typical diagnostic chart.

5. Check the condition of the battery and its terminals and cables.
6. Make sure all vacuum hoses are connected and are not pinched or cut.
7. Check all sensors and actuators for signs of physical damage.

Accessing Trouble Codes

Although the parts in any computerized system are amazingly reliable, they do occasionally fail. Diagnostic charts in service manuals (**Figure 6-5**) help you through troubleshooting procedures in the proper order. Start at the top and follow the sequence down. There will be branches of the tree—yes or no, on or off, OK or not—to follow after making the check required in each step.

Tools to use when checking circuits, connections, sensors, and signals include the following:

- A digital multimeter (DMM) with an impedance of at least 10 megohms
- Scan tool
- Lab scope (dual trace is preferred)
- Tachometer
- Vacuum gauge and pump
- Multimeter
- Service information with trouble code diagnostic procedures

A spark tester, fuel pressure gauge, a fused jumper wire, and a nonpowered test light can also come in handy.

There are two things that would tell the driver there is a problem with some part of the computer system: the MIL ("check engine," "power loss," or "service engine soon" light) or the car simply does not start.

Trouble codes are referred to as diagnostic trouble codes (DTCs).

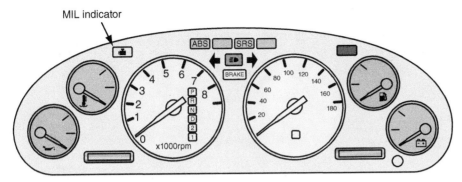

Figure 6-6 A standard MIL.

The MIL (**Figure 6-6**) comes on when the key is turned on and stays on for about 4 seconds, up to 25 seconds on some models. It should go off if there are no codes stored. If it stays on, there is a trouble code stored in memory. If the MIL does not come on, there could be a problem with the computer system or instrument panel cluster (IPC).

USING A SCANNER

Scanners are available to diagnose engine control systems. The exact tester buttons and test procedures vary on these testers, but many of the same basic diagnostic functions are completed regardless of the tester make. When test procedures are performed with a **scan tester**, these precautions must be observed:

1. Always follow the directions in the manual supplied by the scan tester manufacturer.
2. Do not connect or disconnect any connectors or components with the ignition switch on. This includes the scan tester power wires and the connection from the tester to the vehicle diagnostic connector.
3. Never short across or ground any terminals in the electronic system except those recommended by the vehicle manufacturer.
4. If the computer terminals must be removed, disconnect the scan tester diagnostic connector first.

SCAN TESTER FEATURES

Scan testers vary depending on the manufacturer, but many of these testers have the following features:

1. Display window that displays data and messages to the technician. Messages are displayed from left to right. Most scan testers display at least four readings on the display at the same time.
2. Many scan tools are updated via the Internet by a download, in the case of an OEM scan tool, or through the tool dealer for aftermarket scan tools. Even some newer scan tools use a compatibility key to customize the scan tool with the vehicle. Different vehicles use different computer speeds and languages. Scan tools use cartridges or adaptors to match the scan tool to the vehicle.

Some scan tools use compatibility keys because of the different protocols used on earlier OBD II systems.

Figure 6-7 Scan tools usually have simple controls or utilize a touch screen.

⚠ Caution

Never operate a scan tool while driving. Take someone with you to operate the scan tool on the test drive.

3. OBD II vehicles have power for the scanner at the DLC.
4. Keypad that allows the technician to enter data and reply to tester messages. Most scan tools have simple controls, such as a thumb-wheel, select and back buttons. Many also use a touch screen and stylus. Most can store information for review in the shop after the test drive (**Figure 6-7**).

Many manufacturers are using laptop computers that have hardware and software to interface with the computer system for diagnosis.

Scan Tester Initial Entries

Photo Sequence 11 shows a typical procedure for using a scanner in diagnostics. Scan tester operation varies depending on the make of the tester, but a typical example of initial entries follows:

1. With the ignition switch off, connect the scan tool to the DLC. Some vehicles may use a "personality key" to customize the module to the vehicle's serial data system (**Figure 6-8**).
2. Enter the vehicle year. The technician is prompted for the model year and enters it into the scan tool (**Figure 6-9**).
3. Enter the VIN code. This is usually based on the model year and engine type. The technician enters the appropriate information and presses the enter key.

After the scanner has been programmed by performing the initial entries, some entry options appear on the screen. These entry options vary depending on the scan tester and the vehicle being tested. The following is a typical list of initial entry options:

1. Engine
2. Antilock brake system (ABS)
3. Suspension
4. Transmission
5. Data line
6. Test mode

Figure 6-8 Some scan tools use "personality keys" to customize the scan tool to the vehicle communication protocol.

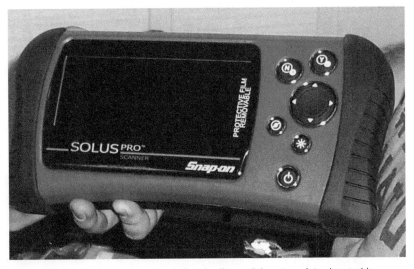

Figure 6-9 Most newer scanners are updated online and do not use interchangeable modules.

The technician enters desired selection to proceed with the test procedure. In the first four selections, the tester is asking the technician to select the computer system to be tested. If "data line" is selected, the scan tester provides a voltage reading from each input sensor in the system.

PHOTO SEQUENCE 11
Trouble Code Diagnosis with a Scan Tool

P11-1 A vehicle is brought into the stop with the service engine soon lamp illuminated.

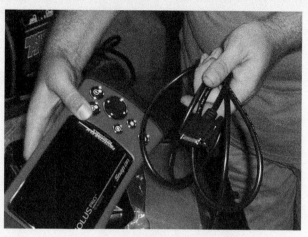

P11-2 A scan tool will be used to help diagnose the problem. The OBD II cable will be attached to the scan tool, along with a personality key on this particular scan tool.

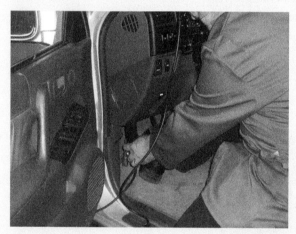

P11-3 The scan tool is plugged into the diagnostic connector, along with the correct personality key if needed.

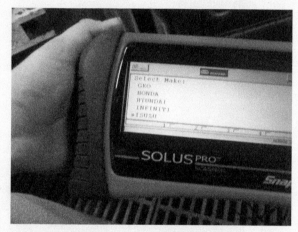

P11-4 The vehicle information is entered into the scan tool.

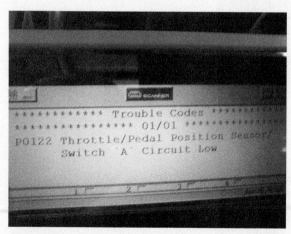

P11-5 The trouble code menu is accessed and a P0122 (throttle position sensor A circuit low) has been set.

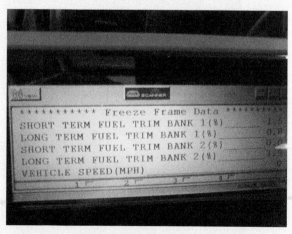

P11-6 The freeze frame is accessed to determine what the conditions were when the P0122 set.

PHOTO SEQUENCE 11 (CONTINUED)

P11-7 The diagnostic procedure for the throttle position is accessed from an electronic information system.

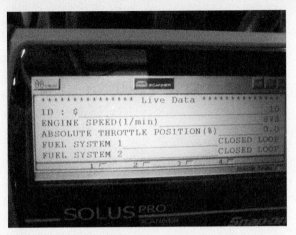

P11-8 The scan tool is used along with the diagnostic chart.

P11-9 The technician performs a thorough inspection and diagnosis and finds a damaged wire.

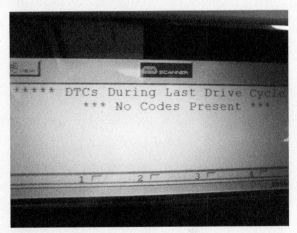

P11-10 The codes were cleared and the vehicle was test driven to verify the repair.

After the test has been selected, the scanner moves on to the actual test selections. These selections vary depending on the scan tester and the vehicle being tested. The following list includes many of the possible test selections and a brief explanation:

1. Fault codes. Displays fault codes on the scan tester display.
2. **Switch tests.** Allows the technician to operate switch inputs such as the brake switch to the PCM. Each switch input should change the reading on the tester.
3. Output tests. The PCM can be commanded to perform certain outputs (depending on the scan tool and vehicle manufacturer) by the scan tool. This capability allows the technician to check the operation of actuators such as EGR solenoids, transmission shift solenoids, fuel pump relay circuits, throttle actuator control motors, and so on.
4. Sensor tests. Provides a voltage reading from each sensor.
5. Solenoid state tests or **output state tests.** Displays the on or off status of each solenoid in the system.
6. **Key on, engine off (KOEO) test.** Allows the technician to perform this test with the scan tester on Ford products. The ECM checks its output devices by commanding them on and off. If any fail this test, then a code is set.

Scan tools with output capabilities are termed "bi-directional."

Key on, engine off
and key on, engine
running tests are
performed with the
scanner on Ford
vehicles.

7. **Key on, engine running (KOER) test.** Allows the technician to perform this test with the scan tester on Ford products.
8. Clear memory or erase codes. Quickly erases fault codes in the PCM memory.
9. Code library. Reviews fault codes.
10. Basic test. Allows the technician to perform a faster test procedure without prompts.

Snapshot and Freeze-Frame Testing

This goes along with a common computer saying: garbage in, garbage out.

Most scan tools have the capability of taking a data "snapshot." The snapshot is a data movie that can be triggered manually or automatically. The technician can test drive the vehicle with the scan tool ready to begin recording. When the vehicle starts to exhibit the symptoms the technician is trying to repair, the snapshot is triggered, and the data is stored into the scan tools memory. Then the scan tool can be brought back to the shop and the technician can replay the event. The advantages of a snapshot are that the technician does not have to watch the scanner and try to drive at the same time. In addition, all available parameters can be recorded, so the technician does not have to be concerned about which screen to watch while the event is happening.

Freeze frame is not a movie but the exact moment a trouble code sets; the PCM stores a screen shot of data for the technician to study later. This data includes information about vehicle speed, load, throttle angle, air-fuel ratio, coolant temperature, and MAP or MAF sensor data. More information is also located in Chapter 10.

> **SERVICE TIP** When diagnosing computer problems, it is usually helpful to ask the customer about service work that has been performed lately on his or her vehicle. If service work has been performed in the engine compartment, it is possible that a computer harness or connector may have been disturbed, causing a problem.

DIAGNOSIS OF COMPUTER VOLTAGE SUPPLY AND GROUND WIRES

A computer cannot operate properly unless it has proper ground connections and a satisfactory voltage supply at the required terminals. A computer wiring diagram for the vehicle being tested must be available for these tests. Back-probe the battery terminal on the computer and connect a pair of digital voltmeter leads from this terminal to ground (**Figure 6-10**). Always ground the black meter lead.

Voltage drop is a better way to check a ground than a resistance check.

The voltage at this terminal should be 12 volts with the ignition switch off. If 12 volts are not available at this terminal, check the computer fuse and related circuit. Turn the ignition switch on, and connect the red voltmeter lead to the other battery terminals at the PCM with the black lead still grounded. The voltage measured at these terminals should be 12 volts with the ignition switch on. When the specified voltage is not available, test the voltage supply wires to these terminals. These terminals may be connected through fuses, fuse links, or relays. Always refer to the vehicle manufacturer's wiring diagram for the vehicle being tested.

Computer ground wires usually extend from the computer to a ground connection on the engine or battery. With the ignition switch on, connect a pair of digital voltmeter leads from the battery ground to the computer ground. The voltage drop across the ground wires should be 30 millivolts or less. If the voltage reading is greater than that or more than that specified by the manufacturer, repair the ground wires or connection.

Not only should the computer ground be checked, but so should the ground (and positive) connection at the battery. Checking the condition of the battery and its cables should always be part of the initial visual inspection before beginning diagnosis of an engine control system.

A voltage drop test is a quick way of checking the condition of any wire. To do this, connect a voltmeter across the wire or device being tested. Then turn the circuit on. Ideally there should be a 0-volt reading across any wire, unless it is a resistance wire that is designed to drop voltage. Even then, check the drop against specifications to see if it is dropping too much.

A good ground is especially critical for all reference voltage sensors. The problem here is not obvious until it is thought about. A bad ground will cause the **reference voltage** (normally 5 volts) to be higher than normal. Normally in a circuit, the added resistance of a bad ground would cause less voltage at a load. Because of the way reference voltage sensors are wired, the opposite is true. If the reference voltage to a sensor is too high, the output signal from the sensor to the computer will also be too high. As a result, the computer will be making decisions based on the wrong information. If the output signal is within the normal range for that sensor, the computer will not notice the wrong information and will not set a DTC.

> The **reference voltage** is a regulated value provided by the PCM that operates vehicle sensors.

To explain why the reference voltage increases with a bad ground, let's look at a voltage divider circuit (**Figure 6-11**). This circuit is designed to provide a 5-volt reference signal off the tap. A vehicle's computer feeds a regulated 12 volts to a similar circuit to ensure the reference voltage to the sensors is very close to 5 volts. The voltage divider circuit consists of two resistors connected in series with a total resistance of 12 ohms. The reference voltage tap is between the two resistors. The first resistor drops 7 volts (**Figure 6-12**) which leaves 5 volts for the second resistor and for the reference voltage tap. This 5-volt reference signal will be always available at the tap as long as 12 volts are available for the circuit.

If the circuit has a poor ground—one that has resistance—the voltage drop across the first resistor will be decreased. This will cause the reference voltage to increase. In **Figure 6-13**, to simulate a bad ground, a 4-ohm resistor was added into the circuit at the ground connection at the battery. This increases the total resistance of the circuit to 16 ohms and decreases the current flowing throughout the circuit. With less current flow through the circuit, the voltage drop across the first resistor decreases to 5.25 volts (**Figure 6-14**). This means the voltage available at the tap will be higher than 5 volts. It will be 6.75 volts.

Poor grounds can also allow electromagnetic interference (EMI) or noise to be present on the reference voltage signal. This noise causes minute changes in the voltage going to the sensor. Therefore, the output signal from the sensor will also have these voltage

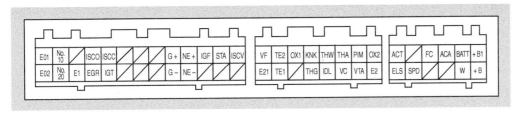

Figure 6-10 Typical computer terminals and their identification.

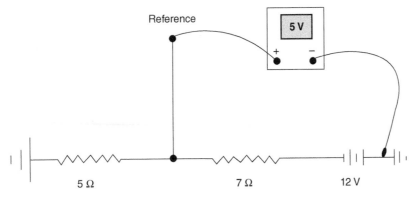

Figure 6-11 Basic voltage divider circuit.

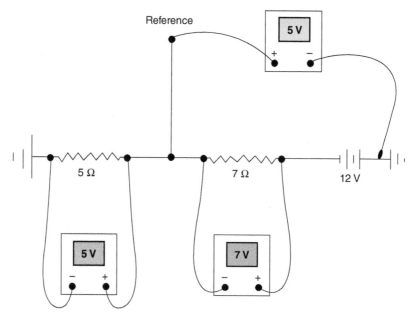

Figure 6-12 Divider circuit with voltage values.

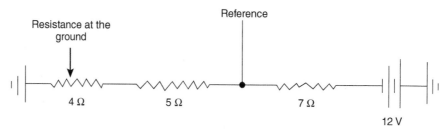

Figure 6-13 Voltage divider circuit with a bad ground.

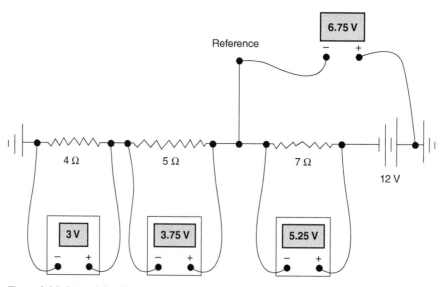

Figure 6-14 Figure 6-2 with voltage readings.

changes. The computer will try to respond to these changes, which can cause a drivability problem. The best way to check for noise is to use a lab scope.

Connect the lab scope between the 5-volt reference signal into the sensor and the ground. The trace on the scope should be flat (**Figure 6-15**). If noise is present, move the

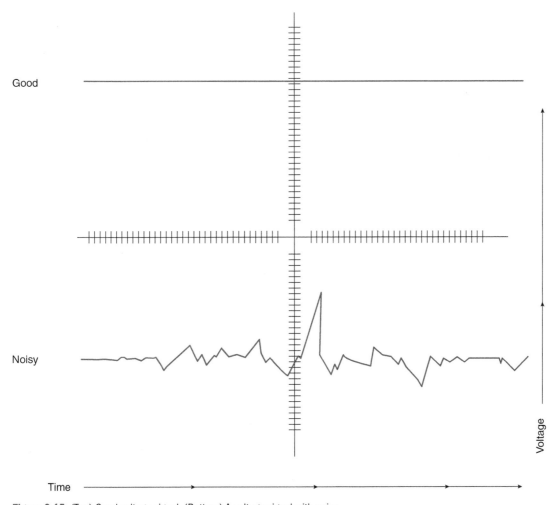

Figure 6-15 (Top) Good voltage signal. (Bottom) A voltage signal with noise.

scope's negative probe to a known good ground. If the noise disappears, the sensor's ground circuit is bad or has resistance. If the noise is still present, the voltage feed circuit is bad or there is EMI in the circuit from another source such as the AC generator. Find and repair the cause of the noise.

Circuit noise may be present in the positive side or the negative side of a circuit. It may also be evident by a flickering MIL, a popping noise on the radio, or an intermittent engine miss. However, noise can cause a variety of problems in any electrical circuit. The most common sources of noise are electric motors, relays and solenoids, AC generators, ignition systems, switches, and A/C compressor clutches. Typically, noise is the result of an electrical device being turned on and off. Sometimes the source of the noise is a defective suppression device. Manufacturers include these devices to minimize or eliminate electrical noise. Some of the commonly used noise suppression devices are resistor-type secondary cables and spark plugs, shielded cables, capacitors, diodes, and resistors. If the source of the noise is not a poor ground or a defective component, check the suppression devices.

Diodes and resistors are the most commonly used noise suppression devices. Resistors do not eliminate the spikes but limit their intensity. If a voltage trace has a large spike and the circuit is fitted with a resistor to limit noise, the resistor may be bad. Clamping diodes are used on devices like A/C compressor clutches to eliminate voltage spikes. If the diode is bad, a negative spike will result (**Figure 6-16**). Capacitors or noise filters are used to control noise from a motor or generator (**Figure 6-17** and **Figure 6-18**).

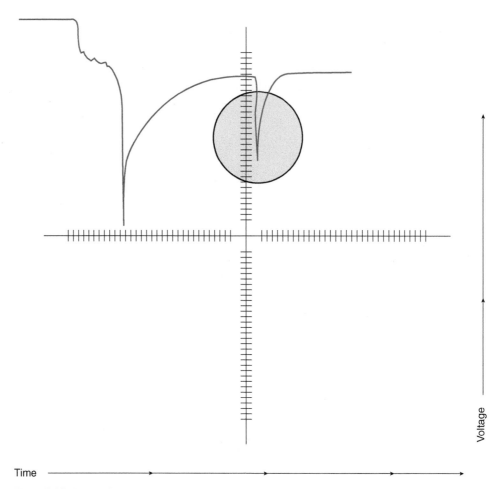

Figure 6-16 A trace of an A/C compressor clutch with a bad clamping diode.

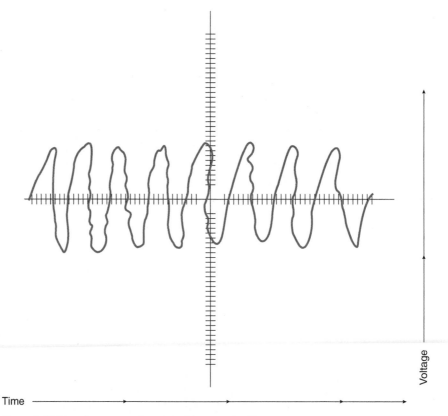

Figure 6-17 The voltage trace of a motor without a noise filter.

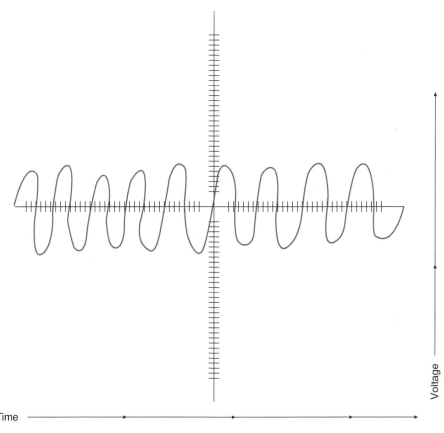

Time

Voltage

Figure 6-18 The voltage trace of a motor with a noise filter.

To avoid added frustration when diagnosing a computer system, especially the inputs, check the integrity of the ground before doing anything else. The test does not take much time and is quite simple. Overlooking the ground circuit will make system diagnosis much more difficult, if not impossible.

TESTING INPUT SENSORS

If a DTC directs you to a faulty sensor or sensor circuit or if you suspect that a sensor is faulty, it should be tested. Testing sensors is included here to orient you to the basic procedures. The recommended procedures given in the service information for testing individual sensors may be different than those described here. Always follow the manufacturer's recommendations. Sensors are tested with a DMM, scanner, lab scope, and/or break-out box. The breakout box can be used along with the DMM and lab scope. The breakout box can be useful in testing circuits, but a specific breakout box is required in most cases. The breakout box has special connectors that allow it to be connected in series with the module being tested. The box has numbered locations that correspond to circuits that can be tested with a DMM or lab scope (**Figure 6-19**). This connection allows the technician to back-probe connections without damaging connectors.

Since the controls are different on the various types of lab scopes that can be used for automotive diagnostic work, the connections and settings for a lab scope are loosely defined in this discussion. Make sure you follow the instructions of the scope's manufacturer when using a lab scope. The lab scope is intimidating to some technicians, but with practice, it becomes easier. You will become more familiar and comfortable with a scope as you use it. If the scope is set wrong, the scope will not break. It just will not show you what you want to be shown. To help with understanding how to set the controls on a

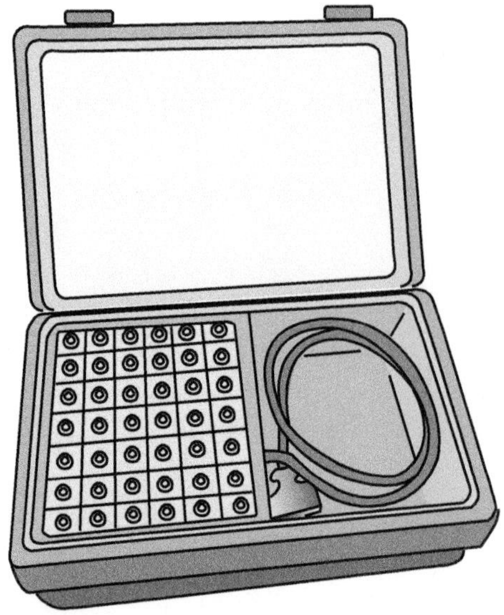

Figure 6-19 A breakout box is connected in series with the PCM to allow meter readings directly from individual circuits.

scope, keep the following things in mind. The vertical voltage scale must be adjusted in relation to the voltage expected in the signal being displayed. The horizontal time base or milliseconds per division must be adjusted so the waveform appears properly on the screen. Many waveforms are clearly displayed when the horizontal time base is adjusted so three waveforms are displayed on the screen.

The trigger is the signal that tells the lab scope to start drawing a waveform. A marker indicates the trigger line on the screen, and minor adjustments of the trigger line may be necessary to position the waveform in the desired vertical position. Trigger slope indicates the direction in which the voltage signal is moving when it crosses the trigger line. A positive trigger slope means that the voltage signal is moving upward as it crosses the trigger line, whereas a negative trigger slope indicates that the voltage signal is moving downward when it crosses the trigger line. Set the trigger to automatic until you get accustomed to the meter.

There are many different types of sensors, and their design depends upon what they are monitoring. Some sensors are simple on-off switches. Others are some form of variable resistor that changes resistance according to temperature changes. Some sensors are voltage or frequency generators, while others send varying signals according to the rotational speed of another device. Knowing what they are measuring and how they respond to changes are the keys to being able to accurately test an input sensor. What follows are typical testing procedures for common input sensors. The sensors are grouped according to the type of sensor they are.

Classroom Manual
Chapter 6, page 159

SWITCHES

Switches are turned on and off through an action of a device or by the actions of the driver. Some of these switches are grounding switches that complete a circuit when they are closed. Others send a 5- or 12-volt signal to the PCM when they are closed. Switches inform the PCM of certain conditions or when another device is being operated.

Grounding switches always send a digital signal to the PCM. The switch's circuit is either on or off. These circuits contain a fixed resistor to limit the circuit's current and prevent voltage spikes in the circuit as the switch opens and closes. When the switch is

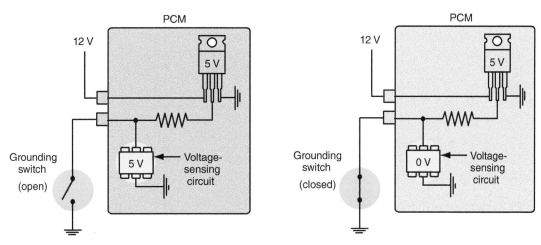

Figure 6-20 Changes in voltage signals as a grounding-type switch opens and closes. The voltage-sensing circuit is somewhat functioning as a voltmeter measuring voltage drop.

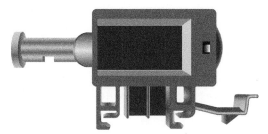

Figure 6-21 A clutch pedal position switch.

closed, the voltage signal to the PCM or ECM is low or zero. When the switch is open, there is a high-voltage signal (**Figure 6-20**).

Some switches are adjustable and must be set so they can close and open at the correct time. An example of this is the **clutch pedal position (CPP)** sensor (**Figure 6-21**). The **CPP sensor** is used to inform the computer when there is no load (clutch engaged) on the engine. The switch is also connected into the starting circuit. The switch prevents the engine from starting unless the clutch pedal is fully depressed. The switch is normally open when the clutch pedal is released. When the clutch pedal is depressed, the switch closes and this acts as an input to the PCM, which in turn controls the starter solenoid. This PCM input also informs the PCM of a low-load condition, which uses this signal to modify the fuel calculations from the MAF and MAP sensors.

Some CPP sensors are varying voltage sensors, which are discussed later in the text.

This switch can be the cause of no-start problems. Because the CPP sensor sends an input to the PCM regarding clutch position, the engine will not crank if the switch is faulty. Some clutch switches are part of the clutch pedal assembly, while others are part of the clutch master cylinder's pushrod.

To check this type of CPP sensor, disconnect the connector to the switch at the clutch pedal assembly. With an ohmmeter, check for continuity between the terminals of the switch with the clutch released and with it engaged. With the clutch released, there should be no continuity between the terminals. With the clutch engaged, there should be good continuity between the terminals. If the switch does not complete the circuit when the clutch is engaged, check the adjustment of the switch (some switches are not adjustable). If the adjustment is correct, replace the switch. Another way to check this type of clutch switch is with a scan tool; look for the clutch engaged-disengaged parameter.

The CPP sensor gives the PCM the location of the clutch pedal for engine load purposes.

If the switch is adjustable, make sure the clutch pedal height and pushrod (free) play are correct. Then check the release point of the clutch. Begin by engaging the parking brake and installing wheel chocks. Start the engine and allow it to idle. Without depressing the clutch pedal, slowly shift the shift lever into reverse until the gears contact. Gradually depress the clutch pedal and measure the stroke distance from the point the gear noise stops (this is the release point) up to the full stroke end position. If the distance is not within specifications, check the pedal height, pushrod play, and pedal free play. If the distance is correct, check the clearance between the switch and the pedal assembly when the clutch is fully depressed. Loosen the switch and adjust its position to provide for the specified clearance.

Pressure Input Sensors

Most grounding switches react to some mechanical action to open or close. However, there are some that respond to changes of condition. These may respond to changes in pressure or temperature. An example of this type of switch is the power steering pressure switch. This switch informs the PCM when power steering pressures reach a particular point. When the power steering pressure exceeds that point, the PCM knows there is an additional load on the engine and will increase idle speed.

To test this type of switch, monitor its activity with a DMM or lab scope. With the engine running at idle speed, turn the steering wheel to its maximum position on one side. The voltage signal should drop as soon as the pressure in the power steering unit has reached a high level. If the voltage does not drop, either the power steering assembly is incapable of producing high pressures or the switch is bad.

Temperature Switches

Most mechanical temperature switches have been replaced by computer-monitored thermistors.

Temperature-responding switches operate in the same way. When a particular temperature is reached, the switch opens. This type of switch is best measured by removing it and submerging it in heated water. Watch the ohmmeter as the temperature increases. A good temperature-responding switch will open (have an infinity reading) when the water temperature reaches the specified amount. If the switch fails this test, it should be replaced.

Voltage Input Signal Switches

Voltage input signal switches send a high-voltage signal to the PCM when they are closed. An example of this type of switch is the A/C compressor switch. This switch lets the PCM know when the extra load on the engine is caused by the air conditioning compressor's clutch engaging. Some A/C switches close when the driver selects A/C. Always check the service manual to determine what action closes this type of switch.

These switches can be tested with a voltmeter. Connect the meter to the output of the switch and to a good ground. When the A/C is turned on, the meter should read 12 volts. If it does not, check the wiring to the switch and the switch itself. Make sure voltage is available to the input side of the switch before condemning the switch. The switch can also be checked with an ohmmeter. Disconnect the wiring to the switch, and connect the meter across the input and output of the switch. The switch should open and close with the cycling of the A/C compressor.

VARIABLE RESISTOR-TYPE SENSORS

A thermistor is a variable resistor made of semiconductor material.

Many sensors send a voltage signal to the PCM in direct response to changes in operating conditions. The voltage signal changes as the resistance in the sensor changes. Most often, these variable resistor-type sensors are thermistors and potentiometers.

Engine Coolant Temperature Sensor

A defective **engine coolant temperature (ECT) sensor** may cause some of the following problems:

1. Hard engine starting
2. Rich or lean air-fuel ratio
3. Improper operation of emission devices
4. Reduced fuel economy
5. Improper converter clutch lockup
6. Hesitation on acceleration
7. Engine stalling
8. Improper activation of emissions control devices

Diagnosing the ECT Sensor

Whenever the ECT is suspected of having a problem, make sure that the coolant level is OK. If coolant does not reach the ECT housing, it may not be reading accurately.

The ECT sensor can be tested with a scan tool, if the vehicle has been sitting overnight. The scan tool reading for the ECT and the **intake air temperature (IAT) sensors** should be very close, if not identical. If the sensors are not the same, you need to do some further testing (**Figure 6-22**).

The ECT sensor may also be removed and placed in a container of water with an ohmmeter connected across the sensor terminals (**Figure 6-23**). A thermometer is also placed in the water. When the water is heated, the sensor should have the specified resistance at any of the given temperatures (**Figure 6-24**). Always use the vehicle manufacturer's specifications. If the sensor does not have the specified resistance, replace the sensor.

With the sensor installed in the engine, the sensor terminals may be back-probed to connect a digital voltmeter to the sensor terminals. The sensor should provide the specified voltage drop at any coolant temperature (**Figure 6-25**).

The computer on some vehicles checks the reading of the ECT ten times per second. If a specified number of samples falls out of a reasonable range, the sensor is flagged as intermittently out of range.

⚠ Caution

Never apply an open flame to an engine coolant temperature (ECT) sensor or intake air temperature (IAT) sensor for test purposes. This action will damage the sensor.

An **engine coolant temperature sensor (ECT)** is a negative temperature coefficient (NTC) device that is affected by temperature.

Special Tools
Container
Heat source
Thermometer

Classroom Manual
Chapter 6, page 174

RPM ___0___	TP Sensor__0_____	APP (%) __0_____

Scan Tool	Value
Engine load	0%
Intake air (F)	39
MAP kPa	99
BARO kPa	100
O$_2$ B1-S1 mV	444

Scan Tool	Value
Coolant (F)	38
MAF g/sec	0.00
MAP (V)	4.95
O$_2$ B1-S2 mV	447

Figure 6-22 Scan tool values from a cold vehicle. Note coolant and intake air temperature values.

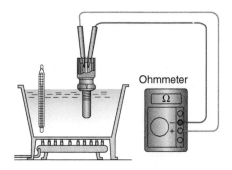

Figure 6-23 Testing an ECT sensor.

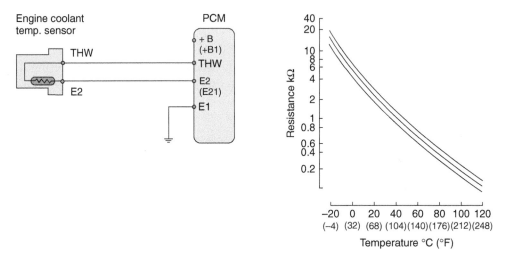

Figure 6-24 Specifications for an ECT sensor.

Cold—10,000-ohm resistor		Hot—909-ohm resistor	
− 20°F	4.7 V	110°F	4.2 V
0°F	4.4 V	130°F	3.7 V
20°F	4.1 V	150°F	3.4 V
40°F	3.6 V	170°F	3.0 V
60°F	3.0 V	180°F	2.8 V
80°F	2.4 V	200°F	2.4 V
100°F	1.8 V	220°F	2.0 V
120°F	1.25 V	240°F	1.62 V

Figure 6-25 Voltage drop specifications for an ECT sensor. A dual-range sensor specification is shown.

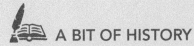

A BIT OF HISTORY

In the earlier days of computer control, technicians would simply unplug the ECT and see if the scan tool indicated −40°F, and then jumper the connections to see if the sensor indicated +260°F. If the ECT circuit indicated these readings, the ECT circuit was operating normally at that time. Today's vehicles use computers with rationality software. This means that the computer is smart enough to realize that if the engine has been running at normal operating temperature, it could not possibly fall to −40°F suddenly. The scan tool might display a fixed value instead of the actual sensor reading. Check on the vehicle manufacturer service information before condemning any parts.

Some computers have internal resistors connected in series with the ECT sensor. The computer switches these resistors at approximately 120°F (49°C). This resistance change inside the computer causes a significant change in voltage drop across the sensor as indicated in the specifications. This is a normal condition on any computer with this feature. This change in voltage drop is always evident in the vehicle manufacturer's specifications.

Intake Air Temperature Sensors

A defective IAT sensor may cause the following problems:

1. Rich or lean air-fuel ratio
2. Hard engine starting
3. Engine stalling or surging
4. Acceleration stumbles
5. Excessive fuel consumption

Classroom Manual
Chapter 6, page 174

As we discussed earlier in this chapter for the ECT, using a scan tool, compare the IAT and ECT sensor readings. If the vehicle has been sitting overnight, then they should be identical or very close to identical. If this is not true, then you need to do further testing. The IAT sensor may be removed from the engine and placed in a container of water with a thermometer. When a pair of ohmmeter leads is connected to the sensor terminals and the water in the container is heated, the sensor should have the specified resistance at any temperature. If the sensor does not have the specified resistance, sensor replacement is required.

With the IAT sensor installed in the engine, the sensor terminals may be back-probed and a voltmeter may be connected across the sensor terminals. The sensor should have the specified voltage drop at any given temperature (**Figure 6-26**). The wires between the air charge temperature sensor and the computer may be tested in the same way as the ECT wires.

Prior to J1930, some IAT sensors were called air charge temperature (ACT) sensors.

| INTAKE AIR TEMPERATURE SENSOR TEMPERATURE VS. VOLTAGE CURVE ||
Temperature	Voltage
−20°F	4.81 V
0°F	4.70 V
20°F	4.47 V
60°F	3.67 V
100°F	2.51 V
140°F	1.52 V
180°F	0.86 V
220°F	0.48 V
260°F	0.28 V

Figure 6-26 Intake air temperature sensor specifications.

Classroom Manual
Chapter 6, page 175

Prior to J1930 naming rules, a TP sensor was called a TPS by almost everyone.

Throttle Position Sensor (Potentiometer Type)

A malfunctioning throttle position (TP) sensor (**Figure 6-27**) may cause acceleration stumbles, engine stalling, and improper idle speed.

Using a scan tool, monitor the TP sensor for dropouts in the signal voltage. If the problem is not found, back-probe the sensor terminals to complete the meter connections. With the ignition switch on, connect a voltmeter from the 5-volt reference wire to ground (**Figure 6-28**). The voltage reading on this wire should be approximately 5 volts. Always refer to the vehicle manufacturer's specifications.

> **SERVICE TIP** When testing the TP sensor voltage signal, use a DSO or graphing DMM because the gradual voltage increase on this wire is easier to monitor. If the sensor voltage increase is erratic, the reading fluctuates.

If the reference wire is not supplying the specified voltage, check the voltage on this wire at the computer terminal. If the voltage is within specifications at the computer but low at the sensor, repair the reference wire. When this voltage is low at the computer, check the voltage supply wires and ground wires on the computer also check the reference voltage with the sensor unplugged. Sometimes a defective sensor can ground the reference wire. If these checks are satisfactory, replace the computer.

With the ignition switch on, connect the voltmeter from the sensor ground wire to the battery ground. If the voltage drop across this circuit exceeds specifications, repair the ground wire from the sensor to the computer.

With the ignition switch on, connect a voltmeter from the sensor signal wire to ground. Slowly open the throttle and observe the voltmeter. The voltmeter reading should increase smoothly and gradually. Typical TP sensor voltage readings would be 0.5 volt to 1 volt with the throttle in the idle position and 4 to 5 volts at wide-open throttle. Always refer to the vehicle manufacturer's specifications. If the TP sensor does not have the specified voltage or if the voltage signal is erratic, replace the sensor.

PCMs are programmed to accept the lowest value seen as 0 percent throttle. Additionally, many vehicles do not start at a 0 percent throttle. Make certain to consult the appropriate service manual before condemning a TP sensor. Check the vehicle

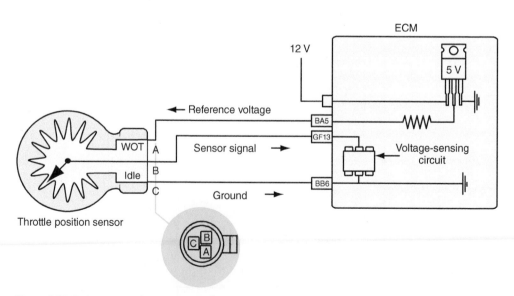

Figure 6-27 Wiring diagram for a TP sensor.

[Circuit Diagram]

[Connection Information]

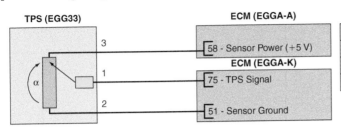

Terminal	Connected to	Function
1	ECM EGGA-K (75)	TPS Signal
2	ECM EGGA-K (51)	Sensor Ground
3	ECM EGGA-A (58)	Sensor Power (+5 V)

[Harness Connector]

EGGA-K

EGGA-A

ECM

EGG33
TPS

Voltage at terminal 2

Closed throttle = 0.2–0.9 volts

Wide-open throttle = minimum 4 volts

Sensor resistance between terminals
1 and 3 = 1.6 K to 2.4 KΩ

Figure 6-28 Typical TP sensor diagnostic chart.

manufacturer's service manual for the TP sensor adjustment procedure. An improper TP sensor adjustment may cause inaccurate idle speed, engine stalling, and acceleration stumbles. Follow these steps for a typical TP sensor adjustment:

> **SERVICE TIP** When the throttle is opened gradually to check the throttle position sensor voltage signal, tap the sensor lightly and watch for fluctuations on the voltmeter or scope, indicating a defective sensor.

1. Back-probe the TP sensor signal wire and connect a voltmeter from this wire to ground.
2. Turn on the ignition switch and observe the voltmeter reading with the throttle in the idle position.
3. If the TP sensor does not register the correct voltage, replace the sensor (**Figure 6-28**).

A TP sensor can be tested with a lab scope. Connect the scope to the sensor's output and a good ground, and watch the trace as the throttle is opened and closed. The resulting trace should look smooth and clean, without any sharp breaks or spikes in the signal (**Figure 6-29**). A bad sensor will typically have a glitch (a downward spike) somewhere in the trace (**Figure 6-30**) or will not have a smooth transition from high to low. These glitches are an indication of an open or short in the sensor.

The EVP sensor voltage signal should change from about 0.8 volt with the EGR valve closed to 4.5 volts with the EGR valve wide open.

Classroom Manual
Chapter 6, page 175

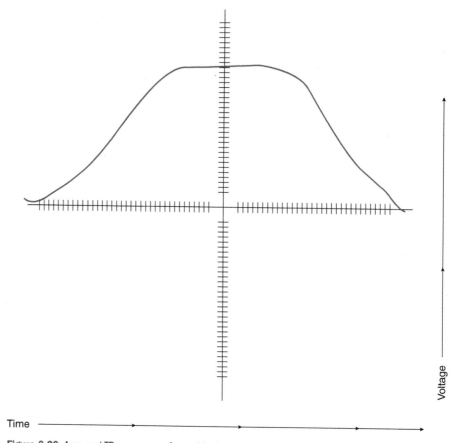

Figure 6-29 A normal TP sensor waveform while it opens and closes.

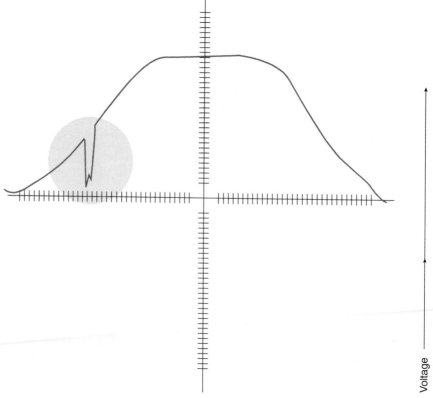

Figure 6-30 The waveform of a malfunctioning TP sensor. Notice the glitch while the throttle opens.

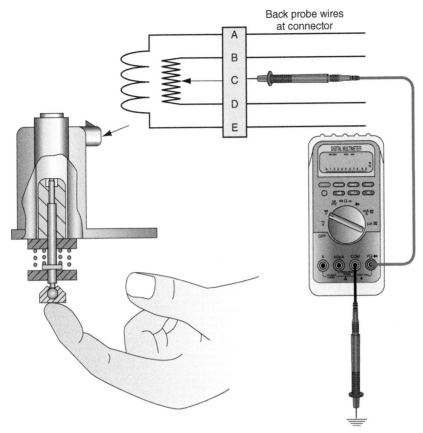

Figure 6-31 Checking a linear exhaust gas recirculation valve position sensor. The ohmmeter should be able to detect the resistance change as the pintle is moved.

Exhaust Gas Recirculation Valve Pintle Position Sensor

Many **exhaust gas recirculation valve position (EVP) sensors** (**Figure 6-31**) have a 5-volt reference wire, a voltage signal wire, and a ground wire. The reference wire and the ground wire may be checked using the same procedure explained previously on TP sensors. Connect a pair of voltmeter leads from the voltage signal wire to ground and turn the ignition switch on. The voltage signal should be approximately 0.8 volt. The EVP sensor voltage signal should gradually increase to 4.5 volts as the pintle is pushed to fully open. Always use the EVP test procedure and specifications supplied by the vehicle manufacturer. If the EVP sensor does not have the specified voltage, replace the sensor. If you have a bi-directional scan tool, you can command the EGR valve open and closed while watching the actual position of the EGR valve. At idle, the valve position should read 0 percent. As you command the valve open, the percentage of EGR valve opening should increase as well. If the valve is opening, the engine should start running very roughly and probably die as the valve opening nears 100 percent. If the engine does not respond, then carbon deposits might be blocking the EGR passages. If this is the case, then mechanically the valve could be operational, but no exhaust would be able to flow through the valve into the intake.

It is good practice to check all variable resistor-type sensors with a lab scope. Any defects in the sensor will show up as glitches in the waveform. These are often unnoticeable on a DMM. The trace from all variable resistors should be clean and smooth.

CPP Sensor (Potentiometer Type)

This type of CPP sensor (**Figure 6-32**) is another version of the potentiometer-type sensor. The ECM sends out a reference voltage of 5 volts. If the clutch pedal is released

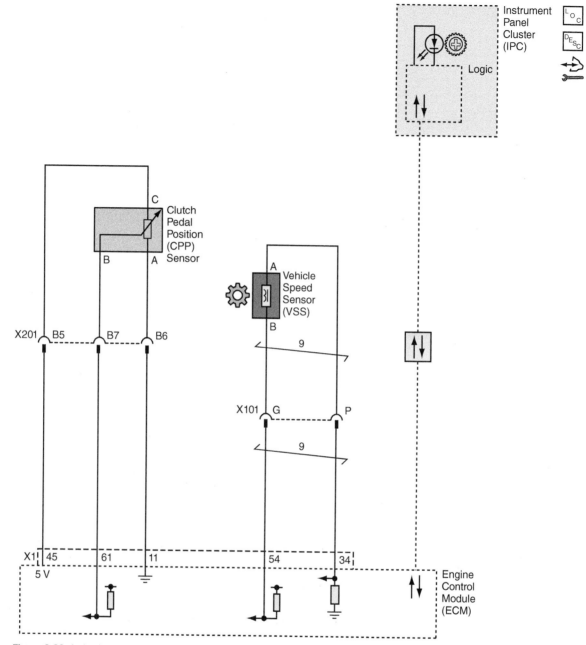

Figure 6-32 A clutch pedal position sensor.

the voltage is near 5 volts, when the clutch pedal is applied, the voltage is around 1.5 volts. If the ECM or the sensor is replaced, the CPP sensor position will have to be "relearned" using a scan tool.

DUAL TP SENSORS USED WITH THROTTLE ACTUATOR CONTROL

Special Tools

DMM

Scan tool

Potentiometer-Type Dual TP Sensors

The dual potentiometer provides increased reliability for systems using throttle actuator control (TAC). The two TP sensors work together, one ranging from low voltage to high voltage and the other from high voltage to low voltage, as shown in **Figure 6-33**. The PCM

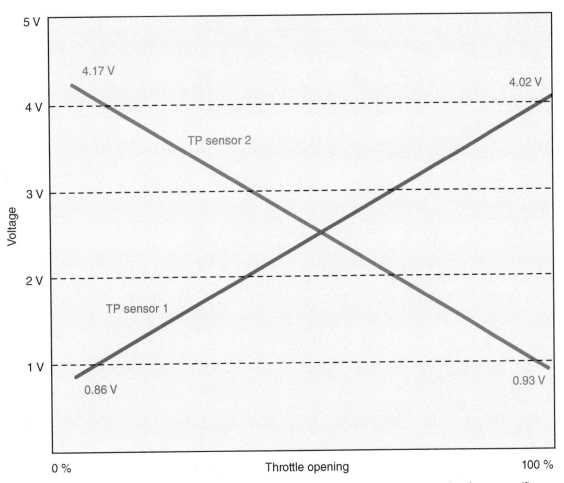

Figure 6-33 Dual TP sensor voltage ranges. (*Note*: There are several types of sensors; always use the chart for your specific vehicle.)

looks at both signals; if one signal is out of specifications, a trouble code is set. The scan tool may display an "absolute" throttle opening and not indicate there are actually two sensors. Most vehicles with TAC will shut down the throttle control if there is any problem in the TAC (which includes the accelerator pedal position circuit). The vehicle would probably have to be towed, but it would allow the vehicle to be moved safely off the road. A "reduced engine power" message will be displayed. Power is reduced because the ECM cannot be certain of the throttle position. In order to prevent a runaway condition, the PCM will allow the vehicle to run only at closed throttle.

> The EGR position sensor is a linear potentiometer.

TIP: If a "reduced power" message is displayed, the vehicle may be able to be driven if the problem is intermittent. Pull the vehicle off the road, then cycle the ignition off, and back on again. If the fault is not currently detected, then the engine will respond normally, until the fault is detected again. Obviously, the vehicle will need to be serviced soon, but the customer might possibly be able to get to a shop.

An example of typical diagnostic routine is found in Figure 6-34. Of course, always refer to the actual diagnostic procedure for the particular vehicle you are servicing.

Noncontact-Type Hall-Effect TP Sensor

A new style of throttle position sensor is being used that does not depend on a potentiometer. This sensor is also utilized on the TAC systems. The Hall-effect TP sensor is a bit different than the shutter style used on most ignition system and crank position sensors. The Hall-effect TP sensor uses two magnets of opposite polarity and two **linear**

> **Linear** refers to a straight line. The linear Hall-effect TP sensor produces a voltage that moves in a straight line from closed to wide-open throttle.

Step	Action	Values	Yes	No
1.	Was the OBD system check performed?		Go to step 2.	Perform the OBD system check.
2.	Has the TAC system check been performed?		Go to step 3.	Go to TAC system check.
3.	Review freeze-frame data, and drive vehicle under the same conditions as those present when the code was set. See if the code has set this ignition cycle.		Go to step 4.	Refer to symptom diagnostics for intermittent code.
4.	Observe the TP sensor reading on the scan tool while slowly opening and closing the throttle. Is the reading consistent and smooth?	TP sensor 1: 8–10% at closed throttle, 90–95% at wide-open throttle (WOT); and TP sensor 2: 90–95% at closed throttle, 8–10% at WOT	Go to step 5.	Go to step 8.
5.	Turn ignition off. Disconnect throttle actuator motor and measure its resistance.	3–10 ohms	Go to step 6.	Go to step 7.
6.	Check the throttle actuator motor wiring and look for wiring concerns.		Repair and confirm fix.	Go to step 8.
7.	Replace the throttle actuator motor.		Repair and confirm fix.	Go to step 6.
8.	With the TP sensor disconnected, look for the TP sensor voltage.	0 volts	Go to step 9.	Go to step 10.
9.	Jumper the reference voltage wires to TP sensor 1 and 2 with a test lamp. Look for the proper voltage.	5 volts	Go to step 12.	Go to step 11.
10.	(a) Look for shorts to voltage, (b) look for high resistance between TP sensor and PCM, and (c) look for poor ground. Repair wiring as necessary.		Verify repair.	Go to step 13.
11.	Check for proper reference voltage or wiring problems.		Verify repair.	Go to step 13.
12.	Replace TP sensor.		Verify repair.	
13	Replace PCM.		Verify repair.	

Figure 6-34 A typical TAC TP sensor diagnostic chart. (*Note:* Always refer to the specific chart for your vehicle.)

Hall-effect sensor integrated circuits: circuit A is the actual throttle position voltage, and circuit B is used to check the integrity of circuit A. The scan tool will show a throttle valve between 10 and 24 percent at idle (depending on the desired idle speed), and a fully open throttle anywhere between 64 and 96 percent. If a failure occurs, the default value is about 16 percent throttle opening. Electrically, the output of the sensor looks much like the action of two potentiometers. This linear Hall-effect TP sensor has the advantage of no contacts to wear or contaminate, and is often referred to as a noncontact TP sensor.

GENERATING SENSORS

Some engine sensors generate a voltage or frequency signal in response to changing conditions. The most common voltage-generating sensors are the oxygen sensor, and permanent magnet-style generators such as the vehicle speed sensor (VSS) and some crankshaft position sensors. Common frequency generator-type sensors are MAF sensors.

Oxygen Sensors

Oxygen (O_2) sensors produce a voltage based on the amount of oxygen in the exhaust. Large amounts of oxygen result from lean mixtures and result in low-voltage output from the O_2 sensor. Rich mixtures have released lower amounts of oxygen in the exhaust—therefore, the O_2 sensor voltage is high. The engine must be at normal operating temperature before the oxygen sensor is tested. Always follow the test procedure in the vehicle manufacturer's service information and use the specifications supplied by the manufacturer.

Testing with a DMM

Connect the voltmeter between the O_2 sensor wire at terminal 14 and a good chassis ground to test this sensor (**Figure 6-35**). Back-probe the connector near the O_2 sensor to connect the voltmeter to the sensor signal wire. If possible, avoid probing through the insulation to connect a meter to the wire. With the engine warm and running around 1,000 rpm, if the O_2 sensor and the computer system are working properly, the sensor voltage should be cycling from low voltage to high voltage. The signal from most O_2 sensors varies between 0 and 1 volt.

If the voltage is continually high, the air-fuel ratio may be rich or the sensor may be contaminated. The O_2 sensor may be contaminated. When the O_2 sensor voltage is continually low, the air-fuel ratio may be lean, the sensor may be defective, or the wire between the sensor and the computer may have a high-resistance problem. If the O_2 sensor voltage signal remains in a midrange position, the computer may be in **open loop** or the sensor may be defective.

The reason for using two sensors in one housing is to provide increased reliability since the throttle is computer controlled.

Classroom Manual
Chapter 6, page 162

Most O_2 sensors are zirconium-type sensors. A zirconium sensor produces between 0 and 1 volt depending on the amount of oxygen it is exposed to.

Most RTV sealant is oxygen-sensor safe, but you should make sure before using it on an engine.

⚠ **Caution**

An oxygen sensor must be tested with a digital voltmeter. If an analog meter is used for this purpose, the sensor may be damaged.

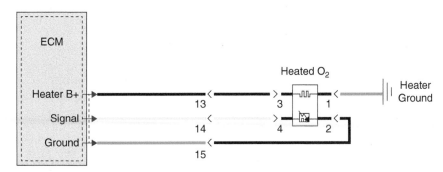

Figure 6-35 Wiring diagram for a single-wire O_2 sensor.

The sensor can also be tested after it is removed from the exhaust manifold. Connect the voltmeter between the sensor wire and the case of the sensor. Using a propane torch, heat the sensor element. The propane flame keeps the oxygen in the air away from the sensor element, causing the sensor to produce voltage. While the sensor element is in the flame, the voltage should be nearly 1 volt. The voltage should drop to 0 immediately when the flame is removed from the sensor. If the sensor does not produce the specified voltage or if the sensor does not quickly respond to the change, it should be replaced.

> ⚙ **SERVICE TIP** A contaminated oxygen sensor may provide a continually high voltage reading because the oxygen in the exhaust stream does not contact the sensor.

Special Tools
Propane bottle
Bottle valve
Hose

If the O_2 sensor voltage signal is higher than specified, the air-fuel ration may be rich or the sensor may be contaminated.

If a defect in the O_2 sensor signal wire is suspected, back-probe the sensor signal wire at the computer and connect a digital voltmeter from the signal wire to ground with the engine idling. The difference between the voltage readings at the sensor and at the computer should not exceed the vehicle manufacturer's specifications. A typical specification for voltage drop across the average sensor wire is 0.2 volt.

Now check the sensor's ground. With the engine idling, connect the voltmeter from the sensor case to the sensor ground wire on the computer. Typically, the maximum allowable voltage drop across the sensor ground circuit is 0.2 volt. Always use the vehicle manufacturer's specifications. If the voltage drop across the sensor ground exceeds specifications, repair the ground wire or the sensor ground in the exhaust manifold.

All OBD II O_2 sensors are heated. If the O_2 sensor heater is not working, the sensor warm-up time is extended and the computer stays in open loop longer. In this mode, the computer supplies a richer air-fuel ratio. As a result, the engine's emissions are high and its fuel economy is reduced. To test the heater circuit, disconnect the O_2 sensor connector and connect a voltmeter between the heater voltage supply wire and ground. With the ignition switch on, 12 volts should be supplied on this wire. If the voltage is less than 12 volts, repair the fuse in this voltage supply wire or the wire itself.

With the O_2 sensor wire disconnected, connect an ohmmeter across the heater terminals in the sensor connector (**Figure 6-36**). If the heater does not have the specified resistance, replace the sensor.

Testing with a Scanner

The output from an O_2 sensor should constantly cycle between high and low voltages as the engine is running in closed loop. This cycling is the result of the computer constantly correcting the air-fuel ratio in response to feedback from the O_2 sensor. When the O_2 sensor reads lean, the computer will richen the mixture. When the O_2 sensor reads rich, the computer will lean the mixture. This enables the computer to control the air-fuel mixture. Many things can occur to take that control away from the computer. One of these is a faulty O_2 sensor.

When the O_2 sensor voltage signal is lower than specified, the air-fuel ratio may be lean or the sensor may be defective.

The activity of the sensor can be monitored on a scanner. The speed at which the oxygen sensor switches from rich to lean is important, as well as the minimum and maximum voltage levels it achieves. By watching the scanner while the engine is running, the O_2 voltage should move to nearly 1 volt then drop back to close to 0 volt. Immediately after it drops, the voltage signal should move back up. This immediate cycling is an important function of an O_2 sensor. If the response is slow, the sensor is lazy and should be replaced. With the engine at about 2,500 rpm, the O_2 sensor should cycle from high to low 10 to 40 times in 10 seconds. The voltage readings shown on the scanner are also an

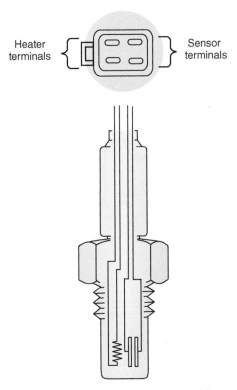

Heater terminals

Sensor terminals

Figure 6-36 Electrical terminals for a heated O_2 sensor.

indicator of how well the sensor works. When testing the O_2 sensor, make sure the sensor is heated and the system is in closed loop.

Testing with a Lab Scope (Including a Graphing Multimeter or Digital Storage Oscilloscope)

A faulty O_2 sensor can cause many different types of problems. It can cause excessively high HC and CO emissions and all sorts of drivability problems. Most computer systems monitor the activity of the O_2 sensor and store a code when the sensor's output is not within the desired range. Again, the normal range is between 0 and 1 volt, and the sensor should constantly toggle from close to 0.2 volt to 0.8 volt then back to 0.2 volt (**Figure 6-37**). If the sensor toggles within the specifications, the computer will think everything is normal and respond accordingly. This does not mean the sensor is working properly.

O_2 signal cross counts (**Figure 6-38**) are the number of times the O_2 voltage signal changes above or below 0.45 volt in a second. If there are not enough cross counts, the sensor is contaminated or lazy. It should be replaced.

If the sensor's voltage toggles between 0 volt and 500 millivolts, it is toggling within its normal range but is not operating normally. It is biased low or lean. As a result, the computer will be constantly adding fuel to try to reach the upper limit of the sensor. Something is causing the sensor to be biased lean. If the toggling only occurs at the higher limits of the voltage range, the sensor is biased rich. In either case, the computer does not have true control of the air-fuel mixture because of the faulty O_2 signals.

The O_2 can be biased rich or lean, not work at all, or work too slowly to ensure good emissions and fuel economy. To test the O_2 sensor for all of these concerns, use a lab scope. Begin by allowing the engine and O_2 sensor to warm up. Insert the hose of a propane enrichment tool into the power brake booster vacuum hose or simply install it

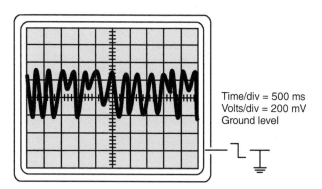

Figure 6-37 A good O₂ sensor trace.

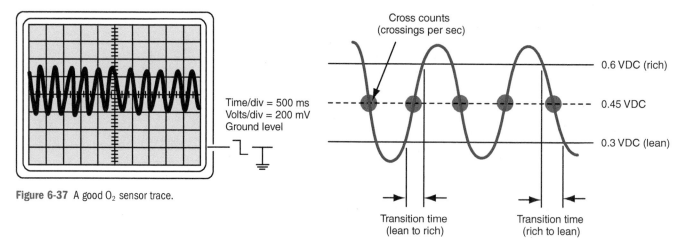

Figure 6-38 O₂ sensor signal cross counts.

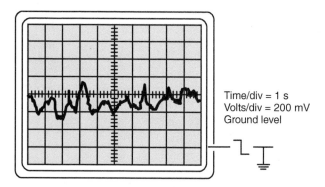

Figure 6-39 O₂ sensor signal caused by a shorted spark plug wire.

Figure 6-40 O₂ sensor signal caused by an open secondary cable.

into the nozzle of the air cleaner assembly. This will drive the mixture rich. Most good O₂ sensors will produce almost 1 volt when driven full rich. The typical specification is at least 800 millivolts.

Connect the lab scope to the sensor and a good ground. Set the scope to display the trace at 200 millivolts per division and 500 milliseconds per division. Inject some propane into the air cleaner assembly. Observe the O₂ signal's trace. The O₂ sensor should show over 800 millivolts. If the voltage does not go high, the O₂ sensor is bad and should be replaced. Now, remove the propane bottle and cause a vacuum leak by pulling off an intake vacuum hose. Watch the scope to see how the O₂ sensor reacts. It should drop to under 175 millivolts. If it does not, replace the sensor. These tests check the O₂ sensor, not the system, therefore they are reliable O₂ sensor checks.

Observing the trace of an O₂ sensor can also help in the diagnosis of other engine performance problems. **Figure 6-39** and **Figure 6-40** show how ignition problems affect the signal from the O₂ sensor. Keep in mind that during complete combustion, nearly all of the oxygen in the combustion chamber is combined with the fuel. This means there will be little O₂ in the exhaust of a very efficient engine. As combustion becomes more incomplete, the levels of oxygen increase. Ignition problems cause incomplete combustion and there is much oxygen in the exhaust. This is also true of lean mixtures or anything else that causes incomplete combustion.

When the mixture is rich, combustion has a better chance of being complete. Therefore, the oxygen levels in the exhaust decrease. The O₂ sensor output will respond to the low oxygen with a high-voltage (**Figure 6-41** and **Figure 6-42**). Remember that the PCM will always try to do the opposite of what it receives from the O₂ sensor. When the O₂ shows

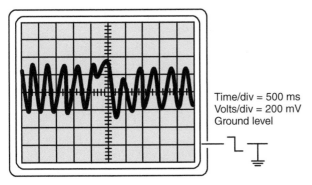

Figure 6-41 Oxygen sensor responding to a lean condition (high oxygen level).

Time/div = 500 ms
Volts/div = 200 mV
Ground level

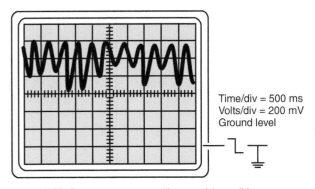

Figure 6-42 Oxygen sensor responding to a rich condition (low oxygen level).

Time/div = 500 ms
Volts/div = 200 mV
Ground level

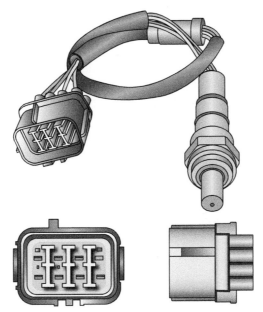

Figure 6-43 An air-fuel ratio sensor. Note the number of connectors.

lean, the PCM goes rich and vice versa. When a lean exhaust signal is not caused by an air-fuel problem, the PCM does not know what the true cause is and will richen the mixture in response to the signal. This may make the engine run worse than it did.

Classroom Manual
Chapter 6, page 173

Air-Fuel Ratio Sensors or Wide-Band Oxygen Sensors

The **air-fuel (AF) ratio sensor** (**Figure 6-43**) (also known as a wide-band oxygen sensor) is diagnosed similarly to the conventional oxygen sensor, although the operation of the two sensors is different. One of the reasons that diagnosis is very similar is that the OBD II legislation had mandated the terms and conditions that a technician would see while doing diagnosis with a scan tool. Although the output of the wide-band sensor had different characteristics and voltage levels, technicians would still see the output in terms of a conventional oxygen sensor.

The air-fuel ratio sensor is more accurate and has a broader operating range than the conventional O_2 sensor.

AUTHOR'S NOTE Many schematic and diagnostic diagrams label the AF ratio sensor as a heated oxygen sensor (**Figure 6-44**). Look at the number of connectors. Most AF ratio sensors may have six or seven connections, although some may actually have four. Most heated oxygen sensors have three or four.

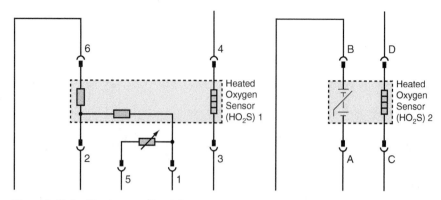

Figure 6-44 An AF ratio sensor (bank 1 sensor 1) and a conventional heated O_2 sensor (bank 1 sensor 2). Note that both are labeled as HO_2S sensors in the schematic.

It is important to note that the AF ratio sensor does not toggle like the conventional oxygen sensor.

Some AF ratio sensors have a trimmer resistor mounted on the connector to the sensor. Be careful not to damage or remove the resistor.

This can be labeled lambda on the scan tool. A lambda reading of one means the air-fuel ratio is at a perfect 14.7:1. A reading of less than one indicates a rich mixture, and a reading higher than one indicates a lean mixture. The recommended way to test an AF ratio sensor is using a scan tool and factory diagnostics. The current and voltage vary in the AF ratio sensor according to the fuel mixture (**Figure 6-45**). The current level is produced inside the ECM and varies according to the current used to keep the Nerst cell inside the AF sensor at a 14.7:1, or a stoichiometric ratio. The actual current used is in the range of about 1.5 mA or even less, so the current is difficult to measure directly, even with a low-amp probe. If the reference voltage is read on the scan tool, it will range around 3.3 or 2.6 volts depending on the manufacturer. When the mixture goes lean the reference voltage on the AF sensor goes up (backward from the narrow-band oxygen sensor), and if the mixture goes rich then the reference voltage goes lower (**Figure 6-46**). The changes in current and voltage happen very rapidly.

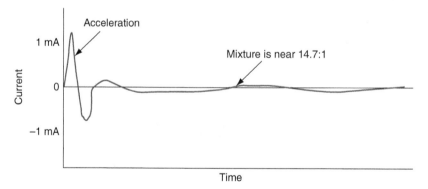

Figure 6-45 Current flow to an AF ratio sensor.

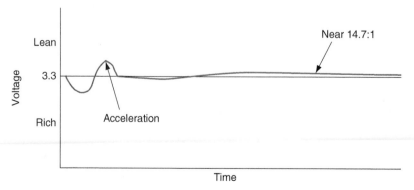

Figure 6-46 Reference voltage changes from an AF ratio sensor based on fuel mixture.

Hall-Effect Switches

To test a shutter-style Hall-effect sensor, disconnect its wiring harness. Connect a voltage source of the correct low-voltage level across the positive and negative terminals of the Hall layer. Then connect a voltmeter across the negative and signal voltage terminals.

Insert a metal feeler gauge between the Hall layer and the magnet. Make sure the feeler gauge is touching the Hall element. If the sensor is operating properly, the meter will read close to battery voltage. When the feeler gauge blade is removed, the voltage should decrease. On some units, the voltage will drop to near zero. Check the service manual to see what voltage you should observe when installing and removing the feeler gauge.

Many crankshaft position sensors are Hall-effect sensors. The PCM uses the signal to determine crankshaft position and engine speed. The computer uses the crankshaft position sensor to build the ignition timing signal it sends to the ignition module. The ignition timing signal is a ground-controlled signal. It appears in a trace as a downward pulse (**Figure 6-47**).

When observing a Hall-effect sensor on a lab scope, pay attention to the downward and upward pulses. These should be straight (**Figure 6-48**). If they appear at an angle (**Figure 6-49**), this indicates the transistor is faulty causing the voltage to rise slowly. This can cause a no-start condition. The entire square wave from a Hall-effect unit should be flat. Any change from a normal trace means the sensor should be replaced (**Figure 6-50**).

Magnetic Pulse Generators

Distributor ignition pickup coils, wheel speed sensors, and vehicle speed sensors are permanent magnet voltage-producing sensors. The procedures for testing each of these are similar.

A pickup coil is a common AC voltage source. The frequency and amplitude of the output signal from an ignition pickup coil will depend on the numbers of cylinders and

According to J1930, the PIP sensor should be called the crank-shaft position (CKP) sensor.

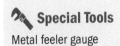

Special Tools

Metal feeler gauge

Many crank sensors require a relearn procedure after replacement of the sensor, PCM, or engine.

Classroom Manual
Chapter 6, page 176

Magnetic pulse generators use the principles of magnetic induction to produce a voltage signal.

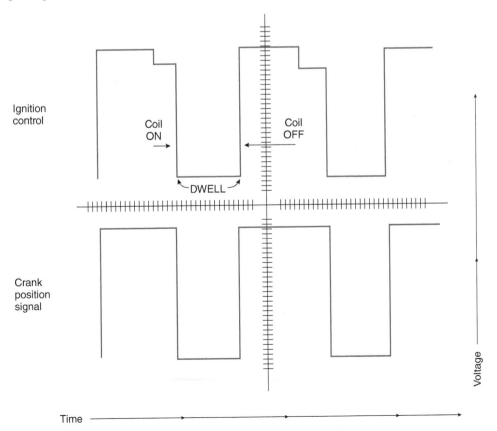

Figure 6-47 Hall-effect crankshaft position and ignition control signals.

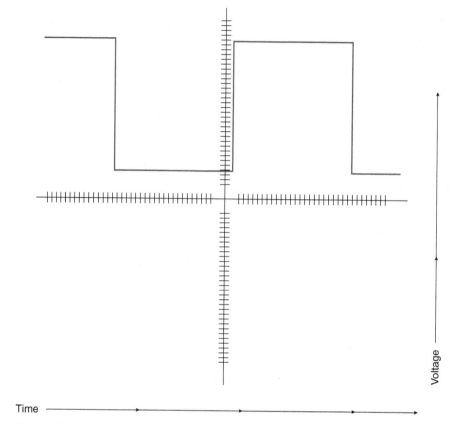

Figure 6-48 A good Hall-effect switch signal.

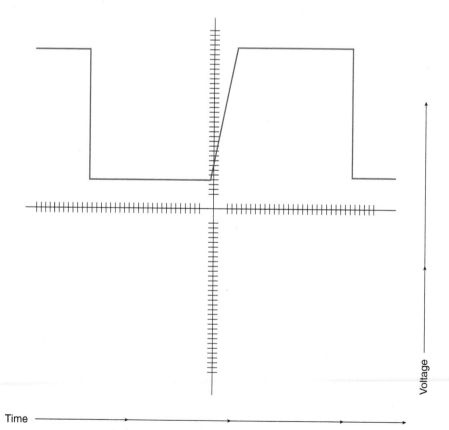

Figure 6-49 A Hall-effect switch with a defective transistor.

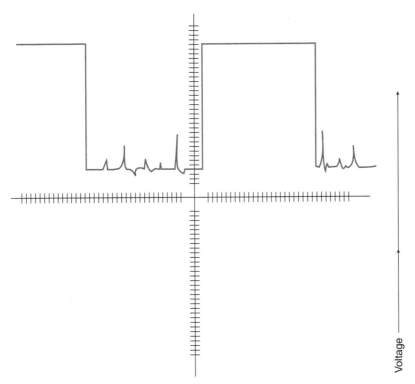

Figure 6-50 Check the square wave for any and all glitches.

the speed of the engine. A good pickup coil will have even peaks that reach at least 300 millivolts when the engine is cranking (**Figure 6-51**).

The waveforms from most crankshaft position sensors will have a number of equally spaced pulses and one double pulse or sync signal, as shown in **Figure 6-52**. The number of evenly spaced pulses equals the number of cylinders that the engine has. Carefully examine the trace. Any glitches indicate a problem with the sensor or sensor circuit.

Classroom Manual
Chapter 6, page 161

Vehicle Speed Sensor

A defective vehicle speed sensor may cause different problems depending on the computer output control functions. A defective **vehicle speed sensor (VSS)** may cause the following problems:

1. Improper converter clutch lockup
2. Inaccurate speedometer operation
3. Inaccurate transmission shift points

The computer controls the torque converter clutch lockup, cruise control, and electronic speedometer using the **vehicle speed sensor (VSS)**.

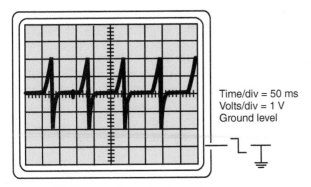

Figure 6-51 The trace of a good PM generator.

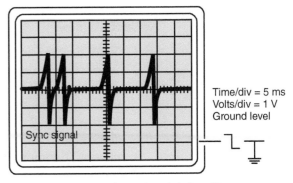

Figure 6-52 The trace of a good crankshaft position sensor.

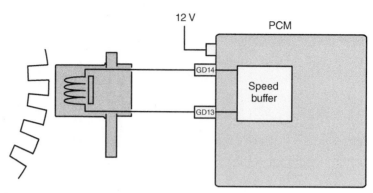

Figure 6-53 Wiring diagram for a VSS.

Prior to VSS diagnosis, the vehicle should be lifted on a hoist so the drive wheels are free to rotate. Back-probe the VSS output wire (**Figure 6-53**), and connect the voltmeter leads from this wire to ground. Select the 20-volt AC scale on the voltmeter, and then start the engine.

The VSS produces an analog A/C voltage.

Place the transaxle in drive and allow the drive wheels to rotate. If the VSS voltage signal is not 0.5 volt or more, replace the sensor. When the VSS provides the specified voltage signal, back-probe the VSS terminal at the PCM and repeat the voltage signal test with the drive wheels rotating. If 0.5 volt is available at this terminal, the trouble may be in the PCM.

When 0.5 volt is not available at this terminal, turn the ignition switch off and disconnect the wire from the VSS to the PCM. Connect the ohmmeter leads across the wire. The meter should read 0 ohms. Repeat the test with the ohmmeter leads connected to the VSS ground terminal and the PCM ground terminal. This wire should also have 0-ohm resistance. If the resistance in these wires is more than specified, repair the wires.

The condition of a speed sensor can be checked by going through the diagnostic routines for the system. If this test indicates that a sensor is faulty, the sensor should be replaced. Speed sensors can also be checked with an ohmmeter. Most manufacturers list a resistance specification. The resistance of the sensor is measured across the sensor's terminals. The typical range for a good sensor is 800 to 1,400 ohms of resistance.

Knock Sensors

A defective knock sensor may cause engine detonation or reduced spark advance and fuel economy. When a knock sensor is removed and replaced, the sensor torque is critical. The procedure for checking a knock sensor varies depending upon the vehicle make and year. Always follow the vehicle manufacturer's recommended test procedure and specifications. Typically, a scan tool is connected to the vehicle and the selection for the knock sensor is brought up. A wrench is used to simulate spark knock (usually on a bracket in the engine compartment). The signal should be evident on the scan tool. If the signal is not found on the scan tool, further diagnosis is necessary. These steps are an example of a typical knock sensor diagnosis:

Classroom Manual Chapter 6, page 181

1. Disconnect the knock sensor wiring connector and turn the ignition switch on.
2. Connect a voltmeter from the disconnected knock sensor wire to ground. The voltage should be 4 to 6 volts. If the specified voltage is not available at this wire, back-probe the knock sensor wire at the computer and read the voltage at this terminal (**Figure 6-54**). If the voltage is satisfactory at this terminal, repair the knock sensor wire. When the voltage is not within specifications at the computer terminal, replace the computer.
3. Connect an ohmmeter from the knock sensor terminal to ground. Some knock sensors should have 3,300 to 4,500 ohms. If the knock sensor does not have the specified resistance, replace the sensor.

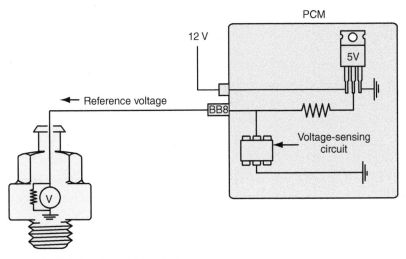

Figure 6-54 Wiring diagram for a knock sensor.

Manifold Absolute Pressure Sensors

A defective **manifold absolute pressure (MAP) sensor** may cause a rich or lean air-fuel ratio, excessive fuel consumption, and engine surging. This diagnosis applies to MAP sensors that produce an analog voltage signal. With the ignition switch on, back-probe the 5-volt reference wire and connect a voltmeter from the reference wire to ground (**Figure 6-55**).

If the reference wire is not supplying the specified voltage, check the voltage on this wire at the computer. If the voltage is within specifications at the computer but low at the sensor, repair the reference wire. When this voltage is low at the computer, check the voltage supply wires and ground wires on the computer. If these wires are satisfactory, replace the computer.

With the ignition switch on, connect the voltmeter from the sensor ground wire to the battery ground. If the voltage drop across this circuit exceeds specifications, repair the ground wire from the sensor to the computer.

Classroom Manual
Chapter 6, page 177

The **manifold absolute pressure (MAP) sensors** monitor engine loads. They are connected to the engine vacuum.

> ⚙️ **SERVICE TIP** Manifold absolute pressure sensors have a much different calibration on turbocharged engines than on non-turbocharged engines. Be sure you are using the proper specifications for the sensor being tested.

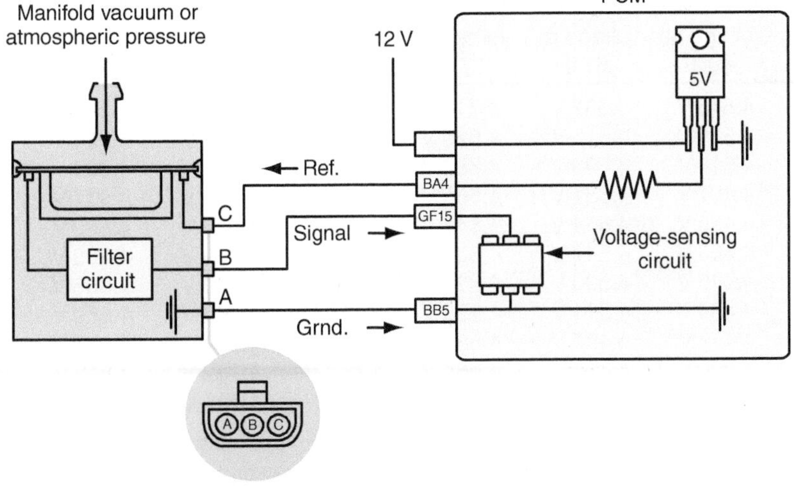

Figure 6-55 Wiring diagram for a MAP sensor.

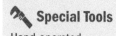
With any 5-volt reference signal sensor, if the 5-volt reference is not present at the sensor, make sure the reference wire has not been grounded by a component or damage to the harness. If the 5-volt reference signal is grounded, all 5-volt reference sensors will fail to produce a signal. The sensors are located on the affected circuit, possibly resulting in a no start and multiple DTCs.

Back-probe the MAP sensor signal wire and connect a voltmeter from this wire to ground with the ignition switch on. The voltage reading indicates the barometric pressure signal from the MAP sensor to the computer. Many MAP sensors send a barometric pressure signal to the computer each time the ignition switch is turned on and each time the throttle is in the wide-open position. If the voltage supplied by the barometric pressure signal in the MAP sensor does not equal the vehicle manufacturer's specifications, replace the MAP sensor (see **Photo Sequence 12**).

The barometric pressure voltage signal varies depending on altitude and atmospheric conditions. Follow this calculation to obtain an accurate barometric pressure reading:

1. Check the weather using the Internet and obtain the present barometric pressure reading for your location; for example, 29.85 inches. The pressure they quote is usually corrected to sea level.
2. Multiply your altitude by 0.001; for example, 600 feet × 0.001 = 0.6.
3. Subtract the altitude correction from the present barometric pressure reading: 29.85 − 0.6 = 29.79 inches.
4. Check the vehicle manufacturer's specifications to obtain the proper barometric pressure voltage signal in relation to the present barometric pressure (**Figure 6-56**).

To check the voltage signal of a MAP sensor, turn the ignition switch on and connect a voltmeter to the MAP sensor signal wire. Connect a vacuum hand pump to the MAP sensor vacuum connection and apply 5 inches of vacuum to the sensor. On some MAP sensors, the sensor voltage signal should change from 0.7 to 1.0 volt for every 5 inches of vacuum change applied to the sensor. Always use the vehicle manufacturer's specifications. If the barometric pressure voltage signal was 4.5 volts with 5 inches of vacuum applied to the MAP sensor, the voltage should be 3.5 to 3.8 volts. When 10 inches of vacuum is applied to the sensor, the voltage signal should be 2.5 to 3.1 volts. Check the MAP sensor voltage at 5-inch intervals from 0 to 25 inches. If the MAP sensor voltage is not within specifications at any vacuum, replace the sensor.

To check a MAP sensor with a lab scope, connect the scope to the MAP output and a good ground. When the engine is accelerated and returned to idle, the output voltage should increase and decrease (**Figure 6-57**). If the engine is accelerated and the MAP sensor voltage does not rise and fall or if the signal is erratic (**Figure 6-58**), the sensor or sensor wires are defective.

Absolute BARO reading	Lowest allowable voltage at −40°F	Lowest allowable voltage at 257°F	Lowest allowable voltage at 77°F	Designed output voltage	Highest allowable voltage at 77°F	Highest allowable voltage at 257°F	Highest allowable voltage at −40°F
31.0"	4.548 V	4.632 V	4.716 V	4.800 V	4.884 V	4.968 V	5.052 V
30.9"	4.531 V	4.615 V	4.699 V	4.783 V	4.867 V	4.951 V	5.035 V
30.8"	4.514 V	4.598 V	4.682 V	4.766 V	4.850 V	4.934 V	5.018 V
30.7"	4.497 V	4.581 V	4.665 V	4.749 V	4.833 V	4.917 V	5.001 V
30.6"	4.480 V	4.564 V	4.648 V	4.732 V	4.816 V	4.900 V	4.984 V
30.5"	4.463 V	4.547 V	4.631 V	4.715 V	4.799 V	4.883 V	4.967 V
30.4"	4.446 V	4.530 V	4.614 V	4.698 V	4.782 V	4.866 V	4.950 V
30.3"	4.430 V	4.514 V	4.598 V	4.682 V	4.766 V	4.850 V	4.934 V
30.2"	4.413 V	4.497 V	4.581 V	4.665 V	4.749 V	4.833 V	4.917 V
30.1"	4.396 V	4.480 V	4.564 V	4.648 V	4.732 V	4.816 V	4.900 V
30.0"	4.379 V	4.463 V	4.547 V	4.631 V	4.715 V	4.799 V	4.883 V

Figure 6-56 Barometric pressure voltage signal specifications at different barometric pressures.

PHOTO SEQUENCE 12
Testing a Map Sensor

P12-1 This particular MAP sensor is attached directly to the intake manifold.

P12-2 Power and ground to the sensor are checked at the connector, using care not to damage the terminals. The technician found 4.97 volts available, which is within specifications.

P12-3 The MAP sensor is removed from the manifold for testing.

P12-4 The MAP sensor is connected to a hand-operated vacuum pump. At 5 in. Hg of vacuum, the voltage reading is 3.93 volts, which is within specification.

P12-5 At 10 in. Hg of vacuum, the voltage reading is 2.93 volts, which is also within specification.

P12-6 Finally, at 20 in. Hg, 1.3 volts is obtained and the sensor passes at all vacuum levels.

P12-7 Another way to check the MAP sensor is using the scan tool and the vacuum pump. At 0 in. vacuum, the reading on the scan tool is 29.8 in. Hg. This is actually the barometric pressure in the area. Because the MAP sensor is an "absolute" pressure sensor, it starts measuring at atmospheric pressure or 0 in. Hg of vacuum.

P12-8 At 10 in. Hg of vacuum, the scan tool reads at 19.8 in. Hg, the pressure left over after the 10 in. Hg is pulled off by the vacuum pump. The reading is correct at this point.

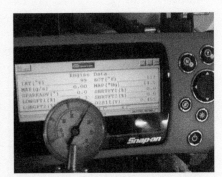

P12-9 The gauge shows 15 in. Hg of vacuum, and the scan tool reading is 14.5 in. Hg. About 15 in. Hg of the original 29.8 in. Hg is pulled off as a vacuum, so the remaining pressure is 14.5 in. Hg. The MAP sensor is working as designed.

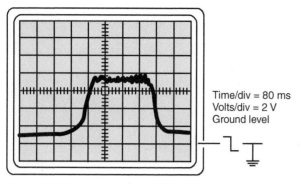

Figure 6-57 The trace of a normal MAP sensor.

Time/div = 80 ms
Volts/div = 2 V
Ground level

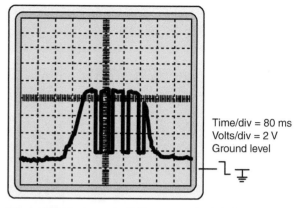

Time/div = 80 ms
Volts/div = 2 V
Ground level

Figure 6-58 The trace of a MAP sensor or circuit failing.

MASS AIRFLOW SENSORS

The mass airflow (MAF) sensor measures the mass of the air being drawn into the engine. The signal from the MAF sensor is used by the PCM to calculate injector pulse width. Two types of MAF sensors are commonly used: grid and hot-wire types. A faulty MAF sensor will cause drivability problems resulting from incorrect ignition timing and improper air-fuel ratios.

The mass airflow (MAF) sensor monitors incoming airflow, including speed, temperature, and pressure.

Classroom Manual
Chapter 6, page 181

The resistance of the heated wire in a hot-wire MAF sensor changes depending on airflow.

The output of some MAF is analog and some output a varying frequency signal. Check with service information to make sure you are measuring the correct signal.

Hot-Wire-Type MAF Sensors

The test procedure for **heated resistor** and **hot-wire-type MAF sensors** (**Figure 6-59**) varies depending on the vehicle make and year. Always follow the test procedure in the vehicle manufacturer's service manual. The following test procedure is based on the use of a scan tool to test the MAF sensor.

1. Connect a scan tool and make sure it can communicate with the PCM; if it cannot, then follow the manufacturer's recommendation for checking the vehicle network and PCM.
2. Read the trouble codes from the scan tool and record them on the repair order. Do not clear the codes, because you will also lose any diagnostic information that could be stored in the scan tool.

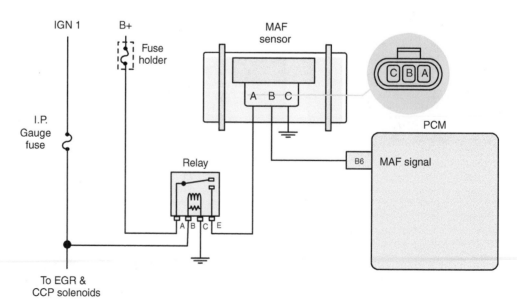

Figure 6-59 A wiring diagram for an MAF sensor.

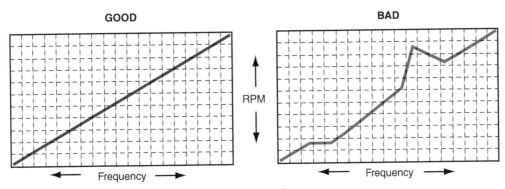

Figure 6-60 Satisfactory and unsatisfactory MAF sensor frequency readings.

3. Look at the freeze frame for the DTC and see when the failure occurred. This will help you operate the vehicle under similar conditions to help you duplicate the concern, and will help you verify the repair later.
4. Once the DTC for a MAF sensor is confirmed, check the air ducts from the MAF sensor (if applicable) for any damage that may allow air to be drawn into the clean air duct downstream of the MAF sensor. This air would not be measured by the MAF and would cause a false reading.
5. Increase the engine speed, and record the meter reading at various speeds.
6. Graph the frequency readings. The MAF sensor frequency should increase smoothly and gradually in relation to the engine speed. If the MAF sensor frequency reading is erratic, replace the sensor (**Figure 6-60**).

While diagnosing a MAF with a scanner, the one-test mode displays grams per second from the MAF sensor. This mode provides an accurate test of the MAF sensor. The grams-per-second reading should be 4 to 7 with the engine idling. This reading should gradually increase as the engine speed increases. When the engine speed is constant, the grams-per-second reading should remain constant. If the grams-per-second reading is erratic at a constant engine speed or if this reading varies when the sensor is tapped lightly, the sensor is defective. A MAF sensor fault code may not be present with an erratic grams-per-second reading, but the erratic reading indicates a defective sensor.

Frequency varying types of MAF sensors can also be tested with a lab scope. The waveform should appear as a series of square waves (**Figure 6-61**). When the engine speed and intake airflow increases, the frequency of the MAF sensor signals should increase smoothly and proportionately to the change in engine speed. If the MAF sensor or connecting wires are defective, the trace will show an erratic change in frequency (**Figure 6-62**).

ACCELERATOR PEDAL POSITION SENSOR

The accelerator pedal position (APP) sensor (OBD II) is composed of three individual sensors in one case. Each of these sensors has a dedicated ground, power, and sensor return wiring. The APP system is only used on those vehicles with "fly-by-wire" throttle actuator control (TAC) that use an electric motor instead of a throttle cable to control the accelerator.

Each sensor has a unique voltage curve. Sensor 1 goes from below 1 volt at closed throttle to about 2 volts at wide-open throttle. Sensor 2 uses a signal that decreases from about 4 volts at closed throttle to 2.9 volts at wide-open throttle. Sensor 3 voltage is from 3.8 volts at closed throttle to about 3.1 volts at wide-open throttle. See **Figure 6-63**. If any of the voltages are incorrect, the TAC is shut down to a default value. Some manufacturers have the dash display "Reduced Engine Power" when a fault in the TAC occurs. The individual sensors could be measured with a voltmeter, but it may be more effective to read the sensors with the scan tool.

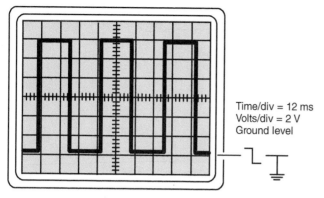

Figure 6-61 A normal trace for a frequency varying MAF sensor.

Time/div = 12 ms
Volts/div = 2 V
Ground level

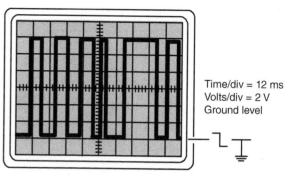

Figure 6-62 The trace of a defective frequency varying MAF sensor.

Time/div = 12 ms
Volts/div = 2 V
Ground level

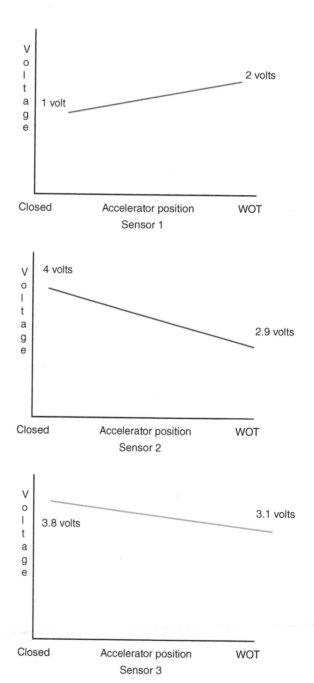

Figure 6-63 The accelerator pedal position is made up of three sensors in one housing.

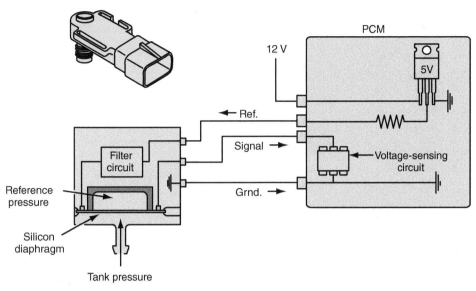

Figure 6-64 The fuel tank pressure sensor circuit.

Fuel Tank Pressure Sensor

The **fuel tank pressure (FTP) sensor** (OBD II) is used by the enhanced emission system to detect fuel tank pressure and vacuum. The system is used by the evaporative emission control system (EVAP) system to monitor for leaks in the sealed system. The fuel tank pressure sensor voltage varies between 0 and 5 volts depending on the amount of pressure in the tank. The voltage increases as fuel tank pressure increases on most vehicles. The MIL is armed the first time a failure is detected and is illuminated on the second consecutive failure. See **Figure 6-64**.

> The fuel tank pressure sensor is used by the EVAP system to check for leaks in the system.

CASE STUDY

A customer brought his car in for a severe hesitation on acceleration. The customer had already had the fuel injectors cleaned at another shop, but the problem still existed. After confirming the problem, the technician checked the fuel pressure (which was also in specification), and a quick visual check did not give any clues to the cause of the problem.

The technician decided to take the car for a test drive with the scan tool. Using the scan tool's snapshot function to avoid trying to read the scan tool and drive, the technician duplicated the condition again and stored the event for reading in the shop.

On returning to the shop, the technician began to analyze the information he had gathered on the test drive. He watched the TP sensor reading rise as the vehicle was accelerated, and the readings looked normal. Next, he watched the TP sensor and oxygen sensors together and noticed that on acceleration, the mixture was running very lean. The oxygen sensor was suspect, but the technician noticed that it seemed to have a good range of activity and switched normally at idle.

Next, the technician checked the MAF sensor against the TP sensor and noticed that there was very little rise in airflow shown under hard acceleration. The technician had already checked the air filter and the inlet hose during his visual inspection but decided to check the MAF sensor more closely. When the MAF sensor was removed from the vehicle, it was noticed that the hot wire was contaminated. When the hot wire was carefully cleaned, the vehicle was test driven and the problem was corrected.

The technician explained to the customer that the contamination had insulated the MAF sensor hot wire and thus the MAF sensor was slow to pick up on a sudden increase in airflow. Without the correct input, the computer did not add enough fuel to support a heavy acceleration. The PCM did not set a trouble code because the MAF readings did not go beyond the specified range.

The technician did an efficient job of repair by using the large amount of information available on a modern vehicle. More importantly, the technician was able to use the information to repair the vehicle.

ASE-STYLE REVIEW QUESTIONS

1. While discussing O_2 sensor diagnostics,
   Technician A says that the speed of switching from rich to lean is important.
   Technician B says that the voltage levels attained (toggling from high to low) are also important.
   Who is correct?
   A. A only
   B. B only
   C. Both A and B
   D. Neither A nor B

2. While discussing ECT sensor diagnosis,
   Technician A says that a defective ECT sensor may cause hard cold engine starting.
   Technician B says that a defective ECT sensor may cause improper operation of emission control devices.
   Who is correct?
   A. A only
   B. B only
   C. Both A and B
   D. Neither A nor B

3. While discussing trouble codes,
   Technician A says that hard failures are those that have occurred in the past, but were not present during the last test of the PCM.
   Technician B says that history or intermittent codes are those that were detected the last time the PCM tested the circuit.
   Who is correct?
   A. A only
   B. B only
   C. Both A and B
   D. Neither A nor B

4. While discussing the testing of sensors,
   Technician A says that sensors can be tested with a scan tool.
   Technician B says that a breakout box is available for some vehicles to test sensor outputs.
   Who is correct?
   A. A only
   B. B only
   C. Both A and B
   D. Neither A nor B

5. While discussing magnetic pulse generator tests,
   Technician A says that an ohmmeter can be used to test the resistance of the coil.
   Technician B says that the voltage generated by the sensor can be measured by connecting a voltmeter across the sensor's terminals.
   Who is correct?
   A. A only
   B. B only
   C. Both A and B
   D. Neither A nor B

6. *Technician A* says that the intake air temperature (IAT) sensor voltage should be about 0.28 volt at 260°F.
   Technician B says that the voltage should be about 2.51 volts at 100°F.
   Who is correct?
   A. A only
   B. B only
   C. Both A and B
   D. Neither A nor B

7. *Technician A* says that some TP sensors used with TAC systems are actually two or three TP sensors in one housing.
   Technician B says that some TP sensors are non-contact type Hall-effect sensors.
   Who is correct?
   A. A only
   B. B only
   C. Both A and B
   D. Neither A nor B

8. *Technician A* says that if the engine rpm is constant, the scan tool should show a constant MAF grams-per-second reading.
   Technician B says that the MAF sensor can be frequency-producing sensor.
   Who is correct?
   A. A only
   B. B only
   C. Both A and B
   D. Neither A nor B

9. *Technician A* says that air-fuel (AF) ratio sensors can be diagnosed with a voltmeter.
   Technician B says that the air-fuel (AF) ratio sensor is a Hall-effect sensor.
   Who is correct?
   A. A only
   B. B only
   C. Both A and B
   D. Neither A nor B

10. While discussing vehicle speed sensors (VSS),
    Technician A says that the resistance of the VSS coil can be checked with an ohmmeter.
    Technician B says that the VSS produces a DC voltage signal that is sent to the PCM.
    Who is correct?
    A. A only
    B. B only
    C. Both A and B
    D. Neither A nor B

ASE CHALLENGE QUESTIONS

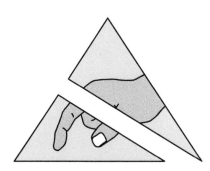

1. The code or symbol shown above depicts:
   A. danger of electrical shock.
   B. nonconductive footwear and gloves should be worn.
   C. antistatic measures and devices should be utilized.
   D. high voltage is present.

2. While discussing the term *fail soft*,
   Technician A says that a fail-soft problem exists when a sensor fails and the ECM is able to substitute a value for the sensor and the vehicle is able to be driven until it can be repaired.
   Technician B says that a fail soft is a failure of the system software.
   Who is correct?
   A. A only
   B. B only
   C. Both A and B
   D. Neither A nor B

3. EMI is being discussed.
   Technician A says that EMI can cause drivability problems.
   Technician B says that EMI is worse with poor grounds.
   Who is correct?
   A. A only
   B. B only
   C. Both A and B
   D. Neither A nor B

4. While discussing a clutch pedal position (CPP) sensor,
   Technician A says that the CPP sensor tells the ECM if the clutch is engaged or disengaged.
   Technician B says that a defective CPP sensor also affects MAF sensor calculations.
   Who is correct?
   A. A only
   B. B only
   C. Both A and B
   D. Neither A nor B

5. A noncontact-style TP sensor is used on some vehicles.
   Technician A says that electrically, the output of the sensor looks much like the action of two potentiometers.
   Technician B says that the linear Hall-effect TP sensor has the advantage of no contacts to wear or contaminate.
   Who is correct?
   A. A only
   B. B only
   C. Both A and B
   D. Neither A nor B

Name _____   Date _____

CHECKING THE OPERATION OF CRANKSHAFT AND CAMSHAFT POSITION SENSORS

Upon completion of this job sheet, you should be able to diagnose a crankshaft and camshaft position sensor.

ASE Education Foundation Correlation

This job sheet addresses the following **AST/MAST** task: VIII. Engine Performance; B. Computerized Engine Controls Diagnosis and Repair

Task #7   Inspect and test computerized engine control system sensors, powertrain/engine control module (PCM/ECM), actuators, and circuits using a graphing multimeter (GMM)/digital storage oscilloscope (DSO); perform needed action. **(P-2)**

Tools and Materials

- Appropriate service manual and test equipment for the vehicle
- Scan tool

Describe the vehicle being worked on:

Year _____ Make _____

Model _____ VIN _____

Procedure

1. Using service information, look up the wiring diagram for the crankshaft position sensor circuit. Describe how the crankshaft sensor is connected to the PCM.

   _____

   _____

   _____

   _____

2. What type of sensor is the *crankshaft* position sensor on your vehicle (Hall-effect, voltage generating, etc.)?

   _____

3. What type of sensor is the *camshaft* position sensor (Hall-effect, voltage generating, etc.)?

   _____

4. Describe the recommended procedure to test the crankshaft position (CKP) sensor.

   _____

   _____

5. Describe the recommended procedure to test the camshaft position (CMP) sensor.

   _____

   _____

6. Following the recommended procedure, check the output of the crankshaft and camshaft position sensors. Describe your results.

Output of crankshaft sensor

_____

_____

Output of camshaft sensor

_____

_____

Instructor's Response

_____

_____

Name _____   Date _____

TEST AN ECT SENSOR

Upon completion of this job sheet, you should be able to check the operation of an engine coolant temperature sensor.

ASE Education Foundation Correlation

This job sheet addresses the following **MAST** task: VIII. Engine Performance;
B. Computerized Engine Controls Diagnosis and Repair

Task #7   Inspect and test computerized engine control system sensors, powertrain/
engine control module (PCM/ECM), actuators, and circuits using a graphing
multimeter (GMM)/digital storage oscilloscope (DSO); perform needed
action. **(P-2)**

Tools and Materials

• DMM

Describe the vehicle being worked on:

Year _____ Make _____

Model _____ VIN _____

Procedure

1. Describe the location of the ECT sensor.

   _____

   _____

   _____

2. What color of wires are connected to the sensor?

   _____

   _____

3. Record the resistance specifications for a normal ECT sensor for this vehicle.

   _____

   _____

   _____

4. Using a scan tool, read and record the engine coolant temperature: _____ °F.

   _____

   _____ Disconnect the electrical connector to the sensor.

5. Measure the resistance of the sensor _____ ohms at approximately
   _____ °F.

6. Connect a DSO or graphing multimeter to the coolant temperature sensor. Start the
   engine and watch the action of the sensor as the vehicle warms up.

   Describe the action of the sensor.

   _____

   _____

   _____

_____

_____

_____

7. Conclusions:

_____

_____

_____

_____

_____

_____

Instructor's Response

_____

_____

_____

Name _____   Date _____

CHECK THE OPERATION OF A TP SENSOR

Upon completion of this job sheet, you should be able to test the operation of a throttle position sensor with a variety of test instruments.

ASE Education Foundation Correlation

This job sheet addresses the following **MAST** task: VIII. Engine Performance;
B. Computerized Engine Controls Diagnosis and Repair

Task #7   Inspect and test computerized engine control system sensors, powertrain/
engine control module (PCM/ECM), actuators, and circuits using a graphing
multimeter (GMM)/digital storage oscilloscope (DSO); perform needed
action. **(P-2)**

Tools and Materials

- DMM

Describe the vehicle being worked on:

Year _____ Make _____

Model _____ VIN _____

Procedure

Task Completed

Perform test with the engine off.

1. Connect the lab scope across the TP sensor.   ☐

2. With the ignition on, move the throttle from closed to fully open and then allow it to   ☐
   close slowly.

3. Observe the trace on the scope while moving the throttle. Describe what the trace
   looked like.

   _____

   _____

   _____

4. Based on the waveform of the TP sensor, what can you tell about the sensor?

   _____

   _____

5. With a voltmeter, measure the reference voltage to the TP sensor. The reading should
   be _____ volts. The reading is _____ volts.

6. What is the output voltage from the sensor when the throttle is closed?
   _____ volts

7. What is the output voltage of the sensor when the throttle is opened?
   _____ volts

8. Move the throttle from closed to fully open and then allow it to close slowly. Describe
   the action of the voltmeter.

   _____

   _____

Task Completed

9. Conclusions:

_____

_____

_____

_____

Instructor's Response

_____

_____

_____

Name _____   **Date** _____

TEST AN O₂ SENSOR

Upon completion of this job sheet, you should be able to test an oxygen sensor with a voltmeter and a lab scope.

ASE Education Foundation Correlation

This job sheet addresses the following **MAST** task: VIII. Engine Performance; B. Computerized Engine Controls Diagnosis and Repair

Task #7   Inspect and test computerized engine control system sensors, powertrain/ engine control module (PCM/ECM), actuators, and circuits using a graphing multimeter (GMM)/digital storage oscilloscope (DSO); perform needed action. **(P-2)**

Tools and Materials

- DMM
- Scan tool
- Propane torch

Describe the vehicle being worked on:

Year _____ Make _____

Model _____ VIN _____

Procedure

Task Completed

1. Connect the voltmeter between the O_2 sensor wire and ground. Back-probe the connector near the O_2 sensor to connect the voltmeter to the sensor signal wire.   ☐

2. With the engine idling, record and describe the voltmeter readings:

   _____

   _____

3. What does this test tell you about the sensor?

   _____

   _____

4. Remove the sensor from the exhaust manifold.   ☐

5. Connect the voltmeter between the sensor signal wire and the case of the sensor.   ☐

6. Using a propane torch, heat the sensor element.   ☐

7. Observe and record the voltmeter reading:

   _____

   _____

8. Your conclusions from this test:

   _____

   _____

9. Back-probe the sensor signal wire at the computer and connect a digital voltmeter from the signal wire to ground with the engine idling.   ☐

10. Record the voltmeter readings:

    _____

    _____

11. Connect the voltmeter from the sensor case to the sensor ground wire on the computer.   ☐

12. Record the voltmeter readings:

_____

_____

13. Your conclusions from these two tests:

_____

_____

14. Connect the scan tool to the DLC.   ☐

15. Observe and record what happens to the voltage reading from the sensor.

_____

_____

16. How many cross counts were there?

_____

17. Your explanation of the above test:

_____

_____

18. Connect a lab scope to the sensor and observe the trace.   ☐

19. Observe and record what happens to the voltage reading from the sensor.

_____

20. How many cross counts were there?

_____

21. Your explanation of the above test:

_____

_____

22. Using all the tests you performed, what is the condition of the oxygen sensor?

_____

_____

Instructor's Response

_____

_____

_____

Name _____   Date _____

TESTING A MAP SENSOR

Upon completion of this job sheet, you should be able to test a manifold absolute pressure sensor in a variety of ways.

ASE Education Foundation Correlation

This job sheet addresses the following **MAST** task: VIII. Engine Performance;
B. Computerized Engine Controls Diagnosis and Repair

Task #7   Inspect and test computerized engine control system sensors, powertrain/ engine control module (PCM/ECM), actuators, and circuits using a graphing multimeter (GMM)/digital storage oscilloscope (DSO); perform needed action. **(P-2)**

Tools and Materials

* Hand-operated vacuum pump
* Lab scope
* DMM

Describe the vehicle being worked on:

Year _____ Make _____

Model _____ VIN _____

Procedure   **Task Completed**

1. If the MAP sensor produces an analog voltage signal, follow this procedure.   ☐

2. With the ignition switch on, back-probe the 5-volt reference wire.   ☐

3. Connect a voltmeter from the reference wire to ground. The reading is _____ volts.

4. If the reference wire is not supplying the specified voltage, what should be checked next?

   _____

   _____

   _____

5. With the ignition switch on, connect the voltmeter from the sensor ground wire to the battery ground.   ☐

6. What is the measured voltage drop? _____ volts

7. What does this indicate?

   _____

   _____

8. Back-probe the MAP sensor signal wire and connect a voltmeter from this wire to ground with the ignition switch on.   ☐

9. What is the measured voltage? _____ volts

10. What does this indicate?

    _____

    _____

    _____

11. How do you determine the barometric pressure based on these voltage readings?

_____

_____

12. Turn the ignition switch on and connect a voltmeter to the MAP sensor signal wire. ☐

13. Connect a vacuum hand pump to the MAP sensor vacuum connection and apply 5 inches of vacuum to the sensor. Record the voltage reading: _____ volts

Note: On some MAP sensors, the sensor voltage signal should change 0.7 to 1.0 volt for every 5 inches of vacuum change applied to the sensor. Always use the vehicle manufacturer's specifications. If the barometric pressure voltage signal was 4.5 volts with 5 inches of vacuum applied to the MAP sensor, the voltage should be 3.5 to 3.8 volts. When 10 inches of vacuum is applied to the sensor, the voltage signal should be 2.5 to 3.1 volts. Check the MAP sensor voltage at 5-inch intervals from 0 to 25 inches.

If the MAP sensor voltage is not within specifications at any vacuum, replace the sensor.

14. Record the results of all vacuum checks:

_____

_____

15. What did these tests indicate?

_____

_____

16. Connect the scope to the MAP output and a good ground. ☐

17. Accelerate the engine and allow it to return to idle. Observe and describe the trace:

_____

_____

18. What did the trace show about the sensor?

_____

_____

Instructor's Response

_____

_____

Name _____   Date _____

NO-CODE DIAGNOSTICS

Upon completion of this job sheet, you should be able to diagnose emissions or drivability concerns when there are no stored diagnostic codes.

ASE Education Foundation Correlation

This job sheet addresses the following **MAST** task: VIII. Engine Performance; B. Computerized Engine Controls Diagnosis and Repair

Task #6   Diagnose emissions or drivability concerns without stored or active diagnostic trouble codes; determine needed action. **(P-1)**

Tools and Materials

- Service information
- Digital multimeter
- Lab scope
- Wiring diagram for the vehicle
- Protective clothing
- Goggles or safety glasses with side shields

Describe the vehicle being worked on:

Year _____ Make _____

Model _____ VIN _____

Engine type and size _____

Procedure

1. Determining which part or area of a computerized engine control system is defective requires having a thorough knowledge of how the system works and following a logical troubleshooting process. Electronic engine control problems are usually caused by defective sensors and, to a lesser extent, output devices. The logical procedure in most cases is, therefore, to check the input sensors and wiring first, then the output devices and their wiring, and, finally, the computer. Most late-model computerized engine controls have self-diagnosis capabilities. A malfunction recognized by the computer is stored in the computer's memory as a trouble code. Stored codes can be retrieved and the indicated problem areas checked further. When there is a problem but no codes are stored, the cause of the problem is something not monitored by the computer or is something working within the acceptable range as seen by the computer. List the components and systems that are monitored by the computer of the vehicle you are working on.

   _____

   _____

   _____

   _____

2. State the major problem you are diagnosing and describe the general condition of the vehicle.

   _____

   _____

3. Diagnostics should begin with a visual inspection. Check the condition of the air filter and related hardware around the filter. Summarize your findings.

_____

_____

_____

_____

4. Check the battery and its cables, and the vehicle's wiring harnesses, connectors, and charging system for loose, corroded, or damaged connections. Summarize your findings.

_____

_____

5. Make sure all vacuum hoses are connected and are not pinched or cut. Summarize your findings.

_____

_____

6. Check all sensors and actuators for signs of physical damage. Summarize your findings.

_____

_____

7. Based on the symptoms, describe the systems and components that you feel could be the cause of the problem if the cause was not discovered in the visual inspection.

_____

_____

8. All PCMs cannot operate properly unless they have good ground connections and the correct voltage at the required terminals. A wiring diagram for the vehicle being tested must be used for these tests. Back-probe the battery terminal at the PCM with the ignition switch off and connect a digital voltmeter from this terminal to ground. How many volts did you measure?

_____

_____

9. The voltage at this terminal should be 12 volts. If 12 volts are not available at this terminal, check the computer fuse and related circuit. Is the fuse good?

_____

_____

10. Turn on the ignition switch and connect the red voltmeter lead to the other battery terminals at the PCM with the black lead still grounded. The voltage measured at these terminals should also be 12 volts with the ignition switch on. How many volts did you measure?

_____

_____

11. If the specified voltage is not available, test the voltage supply wires to these terminals. These terminals may be connected through fuses, fuse links, or relays. If you needed to check these, what were the results?

_____

_____

12. Computer ground wires usually extend from the computer to a ground connection on the engine or battery. With the ignition switch on, connect a digital voltmeter from the battery ground to the computer ground. The voltage drop across the ground wires should be 30 millivolts or less. If the voltage reading is greater than that or more than that specified by the manufacturer, repair the ground wires or connection. Summarize the results of this check.

_____

_____

13. Check the negative and positive terminals and cables of the battery. Conduct a voltage drop test across each. Summarize the results of this check.

_____

_____

14. Conduct a voltage drop test for the grounds at five of the system's sensors and actuators. List the five components whose ground you checked and summarize the results of each check.

_____

_____

15. Check the ground of the same system components with a lab scope. Poor grounds can allow EMI or noise to be present on the reference voltage signal. This noise causes small changes in the voltage going to the sensor. Therefore, the output signal from the sensor will also have these voltage changes. The computer will try to respond to these small rapid changes, which can cause a drivability problem. Summarize the results of these checks.

_____

_____

16. The traces on the scope should have been flat. If noise is present, move the scope's negative probe to a known good ground. If the noise disappears, the sensor's ground circuit is bad or has resistance. If the noise is still present, the voltage feed circuit is bad or there is EMI in the circuit from another source, such as the AC generator. Summarize the results of these checks.

_____

_____

17. If a voltage trace has a large spike and the circuit is fitted with a resistor to limit noise, the resistor may be bad. Clamping diodes are used on devices such as A/C compressor clutches to eliminate voltage spikes. If the diode is bad, a negative spike will result. Capacitors or chokes are used to control noise from a motor or generator. Test the suspected component and summarize the results of that check.

_____

_____

18. Most sensors and output devices can be checked with an ohmmeter. List the ones you suspect as being faulty and that you will test with an ohmmeter.

_____

_____

19. For those components, list the resistance specifications and your measurements. Then, state your conclusions.

_____

_____

_____

_____

20. Many sensors, output devices, and circuit wiring can be diagnosed by checking the voltage to and from them. List the ones you suspect as being faulty and that you will test with a voltmeter.

_____

_____

21. For those components, list the voltage specifications and your measurements. Then, state your conclusions.

_____

_____

22. The activity of sensors and actuators can be monitored with a lab scope. By watching their activity, you are doing more than testing them. Often, problems elsewhere in the system will cause a device to behave abnormally. These situations are identified by the trace on a scope and by the technician's understanding of a scope and the device being monitored. List the ones you suspect as being faulty and that you will test with a lab scope.

_____

_____

23. Summarize the results of your lab scope checks, and then state your conclusions.

_____

_____

Instructor's Response

_____

_____

_____

_____

Name _____   Date _____

USING A SCAN TOOL TO TEST ACTUATORS

Upon completion of this job sheet, you should be able to check the operation of various actuators with a scan tool.

ASE Education Foundation Correlation

This job sheet addresses the following **AST/MAST** task: VIII. Engine Performance; B. Computerized Engine Controls Diagnosis and Repair

Task #3   Perform active tests of actuators using a scan tool; determine needed action. **(P-2)**

Tools and Materials

- Service information
- Scan tool
- Vehicle with OBD II
- Protective clothing
- Goggles or safety glasses with side shields

Describe the vehicle being worked on:

Year _____ Make _____

Model _____ VIN _____

Engine type and size _____

Procedure

Note: The actuators or computer outputs that can be tested directly with a scan tool varies with vehicle model and accessories the vehicle is equipped with. Refer to the service information whenever conducting these tests.

1. What type of scan tool are you using?

   _____

   _____

2. Refer to the operating manual for the scan tool and list the actuators that can be controlled directly by the tool.

   _____

   _____

   _____

   _____

3. Which of these will you check? Also, describe the special procedures for checking that actuator.

   _____

   _____

   _____

4. Connect the scan tool to the DLC. Turn the ignition switch on. Start the engine and allow it to warm up. Select the actuator from the ACTIVE TEST menu on the scan tool. Activate the actuator and listen or look for the specified changes. What did you observe?

_____

_____

_____

5. Refer to the service information and list the actuators that can be monitored by observing the data stream. List them here.

_____

_____

_____

6. Choose one of those actuators and list it here with the specified test conditions and results.

_____

_____

_____

7. Select that data on the scan tool and observe its activity under the specified conditions. Describe your observations.

_____

_____

_____

Instructor's Response

_____

_____

_____

_____

Name _____   Date _____

INTERPRETING CODES FROM AN ENGINE CONTROL SYSTEM

Upon completion of this job sheet, you should be able to diagnose the causes of emissions or drivability concerns resulting from the failure of the computerized engine controls with stored diagnostic trouble codes.

ASE Education Foundation Correlation

This job sheet addresses the following **MAST** task: VIII. Engine Performance;
B. Computerized Engine Controls Diagnosis and Repair

Task #5   Diagnose the causes of emissions or drivability concerns with stored or active diagnostic trouble codes (DTC); obtain, graph, and interpret scan tool data. **(P-1)**

Tools and Materials

- Scan tool
- Digital multimeter
- Service information
- Protective clothing
- Goggles or safety glasses with side shields

Describe the vehicle being worked on:

Year _____ Make _____

Model _____ VIN _____

Engine type and size _____

Procedure

1. Refer to the listing of codes given in the service information. Briefly describe how the codes are grouped by letter and number.

   _____

   _____

   _____

   _____

2. Describe what is indicated by the following codes:

   P0037:

   _____

   P0143:

   _____

   P0010:

   _____

   P0100:

   _____

   P2195:

   _____

3. Conduct all preliminary checks of the engine, electrical system, and vacuum lines. Did you find any problems? What were they?

_____

_____

_____

4. Connect the scan tool to the DLC. Enter the vehicle information into the scan tool. Retrieve the DTCs with the scan tool. What are they?

_____

_____

_____

5. Refer to the service information and locate the description of the DTCs retrieved from the computer. What do they signify?

_____

_____

_____

_____

_____

_____

6. For each of the codes, summarize what steps the manufacturer recommends for locating the exact cause of the problem.

_____

_____

_____

_____

_____

_____

7. Follow those steps and summarize what you found.

_____

_____

_____

_____

_____

8. Enable criteria are the conditions necessary before a code will set or a monitor is ran. Each code may have different enable criteria which should be listed in the vehicle's service information. After correcting the problem, drive the vehicle through the necessary enable criteria to set the code. Recheck the DTCs to make sure you corrected the problem properly. What did you find?

_____

_____

_____

_____

Instructor's Response

_____

_____

_____

_____

Name _____   Date _____

EVAPORATIVE EMISSION CONTROL SYSTEM DIAGNOSIS

Upon completion of this job sheet, you should be able to diagnose emissions and drivability problems resulting from the failure of the evaporative emission control system, and you should also be able to inspect and test the components and the hoses of the evaporative emission control system.

ASE Education Foundation Correlation

This job sheet addresses the following **AST** tasks: Engine Performance; E. Emissions Control Systems Diagnosis and Repair

Task #7   Inspect and test components and hoses of evaporative emission control system; perform needed action. **(P-1)**

Task #8   Interpret diagnostic trouble codes (DTCs) and scan tool data related to the emission control systems; determine needed action. **(P-3)**

This job sheet addresses the following **MAST** tasks: Engine Performance; E. Emissions Control Systems Diagnosis and Repair

Task #5   Diagnose emissions and drivability concerns caused by the evaporative emissions control (EVAP) system; determine needed action. **(P-1)**

Tools and Materials

- Service information
- EVAP leak tester
- Digital multimeter
- Hand-operated vacuum pump
- Scan tool
- Fuel cap tester
- EVAP tester
- Protective clothing
- Goggles or safety glasses with side shields

Describe the vehicle being worked on:

Year _____ Make _____

Model_____ VIN _____

Engine type and size _____

Procedure

Note: EVAP system diagnosis varies depending on the vehicle make and model year. Always follow the service and diagnostic procedure in the vehicle manufacturer's service information.

1. Use a scan tool and check for any EVAP-related DTCs. If an EVAP DTC is set, always correct the cause of this code before further EVAP system diagnosis. Summarize the results of your check.

   _____

   _____

2. Describe the EVAP system found on this vehicle.

   _____

   _____

3. How does this system check for leaks?

_____

_____

4. Check and describe the status of the EVAP monitor.

_____

_____

5. How many trips are necessary to run the monitor?

_____

_____

6. Observe the Short Term Fuel Trim (STFT). How does this relate to vapor purge?

_____

_____

7. When the engine is idling, what is the status of the purge solenoid?

_____

_____

8. Keep the scan tool connected and take the vehicle for a road test that includes all required enable criteria. *Let someone else drive* while you monitor the system from the passenger seat. What conditions does this include?

_____

_____

9. If the purge solenoid does not turn on during the road test, what is indicated?

_____

_____

10. After the road test, check all EVAP hoses for leaks, restrictions, and loose connections. Describe your findings.

_____

_____

11. Check the canister for cracks or damage. Describe what you found.

_____

_____

12. The electrical connections in the EVAP system should be checked for looseness, corroded terminals, and worn insulation. Summarize the results of your check.

_____

_____

13. Check for vacuum leaks in the EVAP system. A vacuum leak in any of the evaporative emission components or hoses can cause starting and performance problems, as can any engine vacuum leak. It can also elicit complaints of fuel odor. Summarize the results of your check.

_____

_____

14. Close the purge and intake ports at the canister and apply low air pressure (about 2.8 psi [19.6 kPa]) to the vent port. Was the canister able to hold that pressure for at least one minute? What does this indicate?

_____

_____

15. Measure the voltage drop across the canister purge solenoid and compare your readings to specifications. If the readings are outside specifications, replace the charcoal canister assembly.

_____

_____

16. The canister purge solenoid winding may be checked with an ohmmeter. Find the specifications in the service information and compare your measurements to them. Summarize the results of your check.

_____

_____

17. Remove the tank pressure control valve. Try to blow air through the valve with your mouth from the tank side of the valve. What happens and what does this indicate?

_____

_____

18. Connect a vacuum pump to the vacuum fitting on the pressure control valve and apply 10 in. Hg to the valve. Now try to blow air through the valve from the tank side. What happens and what does this indicate?

_____

_____

19. Allow the engine to run until it reaches normal operating temperature. Connect the purge flow tester's flow in series with the engine and evaporative canister. Calibrate the gauge of the tester to zero with the engine off, and then start the engine. With the engine at idle, turn on the tester and record the purge flow rate and accumulated purge volume. What did you measure?

_____

_____

20. Gradually increase the engine speed to about 2,500 rpm and record the purge flow. What did you measure?

_____

_____

21. What can you conclude from the two previous tests?

_____

_____

22. How much fuel is in the vehicle?

_____

_____

23. Use the pressure chart that accompanies the EVAP leak tester and determine how much pressure the tester should be set at. Record this amount.

_____

_____

24. Connect the tester to the EVAP service port. Where is this located?

_____

_____

25. Set the scan tool for an EVAP test. Adjust the output pressure from the tester and observe how much pressure the system can hold. What does this indicate?

_____

_____

26. What method will you use to identify the location of a leak? Briefly describe the procedure for doing this.

_____

_____

27. What equipment will you use to check the fuel cap? Briefly describe the procedure for doing this.

_____

_____

28. Remove the fuel cap and inspect the filler neck. Describe its condition.

_____

_____

29. If the fuel tank has a pressure and vacuum valve in the filler cap, check these valves for dirt contamination and damage. The cap may be washed in clean solvent. When the valves are sticking or damaged, replace the cap. Summarize the results of your check.

_____

_____

30. What are your conclusions about this EVAP system?

_____

_____

Instructor's Response

_____

_____

_____

CHAPTER 7

FUEL SYSTEM DIAGNOSIS AND SERVICE

Upon completion and review of this chapter, you should be able to:

- Conduct a visual inspection on a fuel system.
- Test alcohol content in the fuel.
- Relieve fuel system pressure.
- Inspect and service fuel tanks.
- Remove, inspect, service, and replace electric fuel pumps and gauge sending units.
- Inspect and service fuel lines and tubing.

- Remove and replace fuel filters.
- Conduct a pressure and volume output test on an electric fuel pump.
- Service and test electric fuel pumps.
- Understand pulse width modulated fuel systems.
- Service and diagnose fuel injection systems.

Terms To Know

"Banjo" fitting
Displacement
Double flare
Electronic fuel injection (EFI)
EVAP
Fuel pressure test port

High-pressure pump
Hydrocarbons (HC)
Lift pump
Low-pressure pump
Malfunction indicator lamp (MIL)

Mass
Onboard refueling vapor recovery (ORVR)
Port fuel injection (PFI)
Schrader valve

INTRODUCTION

One of the key requirements for an efficient running engine is the correct amount of fuel. Although the fuel injection system is responsible for the fuel delivery into the cylinders, many other components are responsible for the proper delivery of fuel. The fuel must be stored, pumped out of storage, piped to the engine, and filtered. All of this must be accomplished in an efficient and safe manner.

The fuel system should be checked whenever there is evidence of a fuel leak or fuel smell. Leaks are not only costly to the customer, but also very dangerous. The fuel system should also be checked whenever basic tests suggest there is too little or too much fuel being delivered to the cylinders. Lean mixtures are often caused by insufficient amounts of fuel being drawn out of the fuel tank.

No-start conditions can be caused by a lack of fuel. When no fuel is delivered to the engine, the engine will not run. Remember, there must be four factors to have combustion: air, fuel, compression, and spark. To check for fuel, connect a fuel pressure gauge to the fuel line or rail and observe the fuel pressure while cranking the engine. More pressure testing details are included later in this chapter. However, if there is no fuel pressure while cranking, there is no fuel being delivered to the engine.

There are many components in the fuel system. These components can be grouped into two categories: fuel delivery or fuel injection. Diagnosis and basic service to both of these are covered in this chapter. All of the tests in this chapter assume that the fuel is good and not severely contaminated. Obviously, water and contaminants do not burn as well as gasoline. Therefore, water in the fuel tank can cause a drivability problem. If water is mixed with the gasoline, drain the tank and refill it with fresh clean gasoline. Also keep in mind that after gasoline has sat for a while, it becomes less volatile. There are additives available to revitalize the fuel. However, if the fuel is so stale that an engine will not run, drain and refill the tank with fresh gasoline.

ALCOHOL IN FUEL TEST

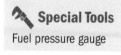

Special Tools
100-milliliter
 graduated cylinder
Approved gasoline
 containers

Classroom Manual
Chapter 7, page 195

Some gasoline may contain a small quantity of alcohol. The percentage of alcohol mixed with the fuel does not usually exceed 10 percent. However, an excessive amount of alcohol mixed with gasoline may result in fuel system corrosion, fuel filter plugging, deterioration of rubber fuel system components, and a lean air-fuel ratio. These fuel system problems caused by excessive alcohol in the fuel may cause drivability complaints such as lack of power, acceleration stumbles, engine stalling, and no-start. If the correct amount of fuel is being delivered to the engine and there is evidence of a lean mixture, check for air leaks in the intake, then check the gasoline's alcohol content.

Some vehicle manufacturers supply test equipment to check the level of alcohol in the gasoline. The following alcohol-in-fuel test procedure requires only the use of a calibrated cylinder:

1. Obtain a 100-milliliter (mL) cylinder graduated in 1-mL divisions.
2. Fill the cylinder to the 90-mL mark with gasoline.
3. Add 10 mL of water to the cylinder so it is filled to the 100-mL mark.
4. Install a stopper in the cylinder and shake it vigorously for 10 to 15 seconds.
5. Carefully loosen the stopper to relieve any pressure.
6. Install the stopper and shake vigorously for another 10 to 15 seconds.
7. Carefully loosen the stopper to relieve any pressure.
8. Place the cylinder on a level surface for 5 minutes to allow liquid separation.
9. Any alcohol in the fuel is absorbed by the water and settles to the bottom. If the water content in the bottom of the cylinder exceeds 10 mL, there is alcohol in the fuel.

For example, if the water content is now 15 mL, there was 5 percent alcohol in the fuel. If the water level goes to 20 mL, then there is 10 percent alcohol in the fuel. Since this procedure does not extract 100 percent of the alcohol from the fuel, the percentage of alcohol in the fuel may be higher than indicated.

Kent-Moore makes an electronic alcohol-in-fuel tester J44175 that is used in conjunction with a multimeter.

FUEL SYSTEM PRESSURE RELIEF

Special Tools
Fuel pressure gauge

Because **electronic fuel injection (EFI)** systems have a residual fuel pressure, this pressure must be relieved before disconnecting any fuel system component. Most **port fuel injection (PFI)** systems have a **fuel pressure test port** on the fuel rail (**Figure 7-1**). Follow this procedure for fuel system pressure relief:

⚠ WARNING **Failure to relieve the fuel pressure on electronic fuel injection (EFI) systems prior to fuel system service may result in gasoline spills, serious personal injury, and expensive property damage.**

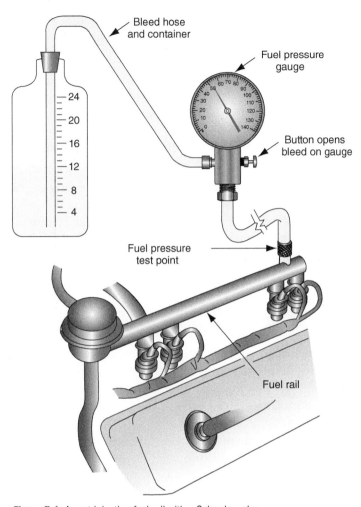

Figure 7-1 A port injection fuel rail with a Schrader valve.

⚡ WARNING **Gasoline direct fuel injection (DFI) systems (and some diesels) have pressures well over 2,000 psi. This pressure has more than enough force to penetrate the skin and cause serious injury or even death! There are special procedures to follow to release this pressure. Do not remove any fuel system component before following the pressure relief methods recommended by the manufacturer.**

1. Disconnect the negative battery cable to avoid fuel discharge if an accidental attempt is made to start the engine.
2. Loosen the fuel tank filler cap to relieve any fuel tank vapor pressure.
3. Wrap a shop towel around the fuel pressure test port on the fuel rail, and remove the dust cap from this valve.
4. Connect the fuel pressure gauge to the fuel pressure test port on the fuel rail.
5. Install the bleed hose on the gauge in an approved gasoline container and open the gauge bleed valve to relieve fuel pressure from the system into the gasoline container. Be sure all the fuel in the bleed hose is drained into the gasoline container.

On EFI systems that do not have a fuel pressure test port, follow these steps for fuel system pressure relief:

1. Loosen the fuel tank filler cap to relieve any tank vapor pressure.
2. Remove the fuel pump fuse.
3. Start and run the engine until the fuel is used up in the fuel system and the engine stops.

4. Engage the starter for 3 seconds to relieve any remaining fuel pressure.
5. Disconnect the negative battery terminal to avoid possible fuel discharge if an accidental attempt is made to start the engine.

Classroom Manual
Chapter 7, page 205

⚡ WARNING **Always wear eye protection and observe all other safety rules to avoid personal injury when servicing fuel system components.**

Fuel Pressure Relief: Direct Fuel Injection–Equipped Vehicles

The recommended procedures for relieving fuel pressure varies somewhat between manufacturers, but it seems that most recommend either disconnecting the fuel pump driver module (FPDM) or removing the fuse from the FPDM, starting the vehicle and waiting for the vehicle to run until the pressure falls so low it is no longer a factor. This will remove the pressure from both the low- and high-pressure fuel systems of the direct fuel injection (DFI) system. As always refer to the latest service information for specifics on the vehicle concerned.

FUEL TANKS

The fuel tank should be inspected for leaks, road damage, corrosion, and rust on metal tanks; loose, damaged, or defective seams; or loose mounting bolts and damaged mounting straps. Leaks in the fuel tank, lines, or filter may cause gasoline odor in and around the vehicle, especially during low-speed driving and idling. In most cases, the fuel tank must be removed for servicing.

Leaks in the metal fuel tank can be caused by a weak seam, rust, or road damage. The best method of permanently solving this problem is to replace the tank.

⚙ SERVICE TIP When a fuel tank must be removed, if possible, ask the customer to bring the vehicle to the shop with a minimal amount of fuel in the tank.

Holes in a plastic tank can sometimes be repaired by using a special tank repair kit. Be sure to follow manufacturer's instructions when doing the repair.

If a tank has contaminated fuel in it or water, the tank must be cleaned and repaired or replaced.

Fuel Tank Draining

The fuel tank must be drained prior to tank removal. If the tank has a drain bolt, this bolt may be removed and the fuel drained into an approved container. If the fuel tank does not have a drain bolt, follow these steps to drain the fuel tank:

1. Remove the negative battery cable.
2. Raise the vehicle on a hoist.
3. Install a ground wire from the tank to the pump. This helps to prevent arching from the tank to the pump.
4. Install the pump hose through the filler pipe into the fuel tank. (On some vehicles, it may be necessary to remove the fuel filler pipe from the tank to install the pump hose.)
5. Install the discharge hose from the hand-operated or air-operated pump into an approved gasoline container, and operate the pump until all the fuel is removed from the tank.

⚠ **Caution**

Always drain gasoline into an approved container, and use a funnel to avoid gasoline spills.

⚡ WARNING **When servicing fuel system components, always place a Class B fire extinguisher near the work area.**

Fuel Tank Service

The fuel tank removal procedure varies depending on the vehicle make and year. Always follow the procedure in the vehicle manufacturer's service manual. The following is a typical procedure:

1. Disconnect the negative terminal from the battery.
2. Relieve the fuel system pressure, and drain the fuel tank.
3. Raise the vehicle on a hoist, or lift the vehicle with a floor jack and lower the chassis onto jack stands.
4. Use compressed air to blow dirt from the fuel line fittings and wiring connectors.
5. Remove the fuel tank wiring harness connector from the body harness connector.
6. Remove the ground wire retaining screw from the chassis if used.
7. Disconnect the fuel lines from the fuel tank. If these lines have quick-disconnect fittings, follow the manufacturer's recommended removal procedure in the service manual. Some quick-disconnect fittings are hand releasable, and others require the use of a special tool (**Figure 7-2**).

⚡ WARNING **Abide by local laws for the disposal of contaminated fuels. Be sure to wear eye protection when working under the vehicle.**

8. Wipe the filler pipe and vent pipe hose connections with a shop towel, and then disconnect the hoses from the filler pipe and vent pipe to the fuel tank.
9. Unfasten the filler from the tank. If it is a rigid one-piece tube, remove the screws around the outside of the filler neck near the filler cap. If it is a three-piece unit, remove the neoprene hoses after the clamp has been loosened.
10. Place a transmission jack under the fuel tank for support. Loosen the bolts holding the fuel tank straps to the vehicle (**Figure 7-3**) until they are about two threads from the end.

⚡ WARNING **Do not heat the bolts on the fuel tank straps in order to loosen them. The heat could ignite the fumes.**

Fuel tanks with onboard refueling vapor recovery (ORVR) systems may be impossible to siphon.

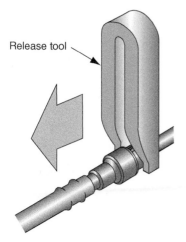

Figure 7-2 Quick-disconnect fuel line tool.

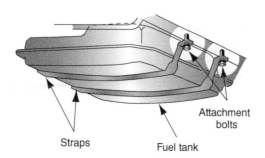

Figure 7-3 Fuel tank strap mounting bolts.

Make certain that the tank insulators are in place to prevent excessive fuel pump noise being transmitted into the vehicle.

 Caution

Use a drain pan to catch any spilled fuel. If fuel does go on the floor, make sure to clean it up immediately.

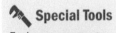 **Special Tools**

Fuel pressure gauge
Fuel pump retaining
   ring remover

Classroom Manual
Chapter 7, page 204

11. Holding the tank securely on the transmission jack, remove the strap bolts and lower the tank to the ground. When lowering the tank, make sure all wires and tubes are unhooked. Keep in mind that small amounts of fuel might still remain in the tank.

12. To reinstall the new or repaired fuel tank, reverse the removal procedure. Be sure that all the rubber or felt tank insulators are in place. Then, with the tank straps in place, position the tank. Loosely fit the tank straps around the tank, but do not tighten them. Make sure that the hoses, wires, and vent tubes are connected properly. Check the filler neck for alignment and for insertion into the tank. Tighten the strap bolts and secure the tank to the car. Install all of the tank accessories (vent line, sending unit wires, ground wire, and filler tube). Fill the tank with fuel and check it for leaks, especially around the filler neck and the pickup assembly. Reconnect the battery and check the fuel gauge for proper operation.

Electric Fuel Pump Removal and Replacement

1. Remove the fuel tank from the vehicle.
2. Follow the vehicle manufacturer's recommended procedure to remove the fuel pump and gauge sending unit from the fuel tank. In many cases, a special tool must be used to remove this assembly (**Figure 7-4**).
3. Check the filter on the fuel pump inlet. If the filter is contaminated or damaged, replace the filter.
4. Inspect the fuel pump inlet for dirt and debris. Replace the fuel pump if these foreign particles are found in the pump inlet.
5. If the pump inlet filter is contaminated, it will be necessary to have the tank cleaned.
6. Check all fuel hoses and tubing on the fuel pump assembly. Replace fuel hoses that are cracked, deteriorated, or kinked. When fuel tubing on the pump assembly is damaged, replace the tubing or the pump.
7. Be sure that the sound insulator sleeve is in place on the electric fuel pump, and check the position of the sound insulator on the bottom of the pump.
8. Clean the pump and sending unit mounting area in the fuel tank with a shop towel, and install a new gasket or O-ring on the pump and sending unit. Install the fuel pump and gauge sending unit assembly in the fuel tank, and secure this assembly in the tank using the vehicle manufacturer's recommended procedure.

The easiest way to remove a sending unit retaining ring is to use a special tool designed for this purpose. This tool fits over the metal tabs on the retaining ring, and after about a quarter turn, the ring comes loose and the sender unit can be removed. If the special tool is not available, a brass drift punch and ball peen hammer usually do the job (**Figure 7-5**). If the lock ring holding the sender in place is badly rusted or damaged during removal, it will have to be replaced.

When removing the sending unit from the tank, be very careful not to damage the float arm, the float, or the fuel gauge sender. Check the unit carefully for any damaged components. Make sure the float arm is not bent. It is necessary to replace the strainer and O-ring before replacing the unit. Check the fuel gauge as described in the service manual.

FUEL LINES

Fuel lines can be made of either metal tubing or flexible nylon or synthetic rubber hose. The latter must be able to resist gasoline. It must also be impermeable, so gas and gas vapors cannot evaporate through the hose. Ordinary rubber hose, such as that used for vacuum lines, deteriorates when exposed to gasoline. Only hoses made for fuel systems

Figure 7-4 Special tool for removing a fuel pump and gauge sending unit assembly from a fuel tank.

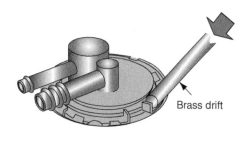

Brass drift

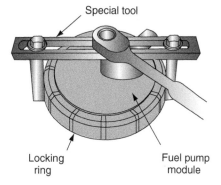

Special tool

Locking ring

Fuel pump module

Figure 7-5 Tightening the fuel pump locking ring with a special tool or brass drift.

should be used for replacement. Similarly, vapor vent lines must be made of material that resists attack by fuel vapors. Replacement vent hoses are usually marked with the designation **EVAP** to indicate their intended use. The inside diameter of a fuel delivery hose is generally larger (5/16 to 3/8 inch) than that of a fuel return hose (1/4 inch).

The fuel lines carry fuel from the fuel tank to the fuel pump, fuel filter, and fuel injection assembly. These lines are usually made of rigid metal, although some sections are constructed of rubber hose to allow for car vibrations. This fuel line, unlike filler neck or vent hoses, must work under pressure or vacuum. Because of this, the flexible synthetic hoses must be stronger. This is especially true for the hoses on fuel injection systems, where pressures reach 60 psi or more. For this reason, flexible fuel line hoses must also have special resistance properties. Many auto manufacturers recommend that flexible hose only be used as a delivery hose to the fuel **metering** unit in fuel injection systems. It should not be used on the pressure side of the injector systems. This application requires a special high-pressure hose. On DFI systems, the high-pressure stainless steel lines must be replaced if they are removed. These lines have to hold almost 2,200 psi.

All fuel lines should occasionally be inspected for holes, cracks, leaks, kinks, or dents. Many fuel system troubles that occur in the lines are blamed on the fuel pump or injectors. For instance, a small hole in the fuel line admits air but does not necessarily show any drip marks under the car. Air can then enter the fuel line, allowing the fuel to gravitate back into the tank. Then, instead of drawing fuel from the tank, the fuel pump sucks only air through the hole in the fuel line. When this condition exists, the fuel pump is frequently tested and, if there is insufficient fuel, it is considered faulty when in fact there is nothing wrong with it. If a hole is suspected, remove the coupling at the tank and the pump, and pressurize the line with air. The leaking air is easily spotted.

As the fuel is under pressure, leaks in the line between the pump and injectors are relatively easy to recognize. When a damaged fuel line is found, replace it with one of similar construction—steel with steel or the flexible with nylon or synthetic rubber. When installing flexible tubing, always use new clamps. The old ones lose some of their tension when they are removed and do not provide an effective seal when used on the new line.

EVAP is the evaporative emissions control system for containment of gasoline vapors.

Classroom Manual
Chapter 7, page 208

Nylon fuel lines have to be replaced, not repaired.

 Caution

Do not substitute aluminum or copper tubing for steel tubing. Never use hose within 4 inches of any hot engine or exhaust system component. A metal line must be installed.

Fuel feed
line

Clip

Fuel return
line

Forward   Frame

Clip

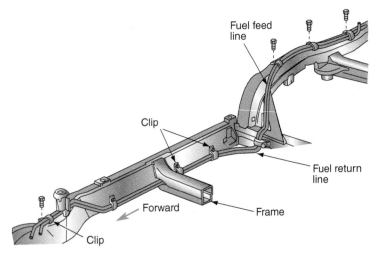

Figure 7-6 Gaps between the frame and tank are joined with flexible hose.

Crack   Leakage

Deformation

Figure 7-7 Steel fuel tubing should be inspected for leaks, kinks, and other damage.

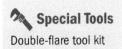
Special Tools

Double-flare tool kit

⚠ **Caution**

O-rings in fuel line fittings are usually made of fuel-esistant Viton. Other types of O-rings must not be substituted for fuel fitting O-rings.

Double flare refers to the type of mating surfaces in fittings.

The angle of a fuel line flare is generally 45 degrees not 60 degrees as in a brake line flare.

Fuel supply lines from the tank to the injectors are routed to follow the frame along the under-chassis of vehicles. Generally, rigid lines are used extending from near the tank to a point near the fuel pump. To absorb engine vibrations, the gaps between the frame and tank or fuel pump are joined by short lengths of flexible hose (**Figure 7-6**).

Steel tubing should be inspected for leaks, kinks, and deformation (**Figure 7-7**). This tubing should also be checked for loose connections and proper clamping to the chassis. If the fuel-tubing threaded connections are loose, they must be tightened to the specified torque. Some threaded fuel line fittings contain an O-ring. If the fitting is removed, the O-ring should be replaced.

Any damaged or leaking fuel line—either a portion or the entire length—must be replaced. To fabricate a new fuel line, select the correct tube and fitting dimension and start with a length that is slightly longer than the old line. With the old line as a reference, use a tubing bender to form the same bends in the new line as those that exist in the old. Although steel tubing can be bent by hand to obtain a gentle curve, any attempt to bend a tight curve by hand usually kinks the tubing. To avoid kinking, always use a bending tool like the ones shown in **Figure 7-8**.

The most-used tubing fitting is the **double flare**. The double flare is made with a special tool that has an anvil and a cone (**Figure 7-9**). The double-flaring process is performed in two steps. First, the anvil begins to fold over the end of the tubing. Then, the cone is used to finish the flare by folding the tubing back on itself, doubling the thickness and creating two sealing surfaces.

The angle and size of the flare are determined by the tool. Careful use of the double flaring helps to produce strong, leak-proof connections. **Figure 7-10** shows other metal fuel line connections that are used by vehicle manufacturers.

The flare tool can also be used to make sure that nylon and synthetic rubber hoses stay in place. That is, to make sure the connection is secure, put a partial double-lip flare on the end of the tubing over which the hose is installed. This can be done quickly with the proper flaring tool by starting out as if it was going to be a double flare but stopping halfway through the procedure (**Figure 7-11**). This provides an excellent sealing ridge that does not cut into the hose. A clamp should be placed directly behind the ridge on the hose caused by the raised section on the metal line.

There are a variety of clamps used on fuel system lines, including the spring and screw types (**Figure 7-12**). The crimp clamps shown in **Figure 7-13** are used most for metal tubing, but they require a special tool to install.

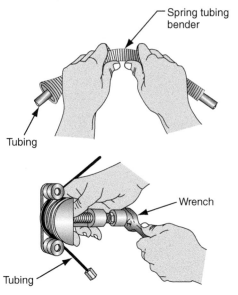

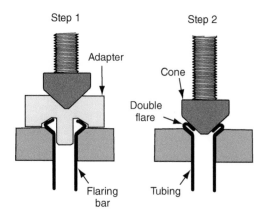

DOUBLE FLARE

Figure 7-9 Forming a double-flare fitting.

Figure 7-8 Two types of steel tubing bending tools.

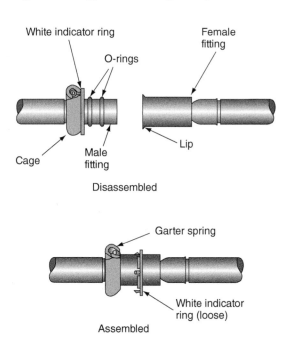

Disassembled

Assembled

Figure 7-10 Common metal fuel line connections.

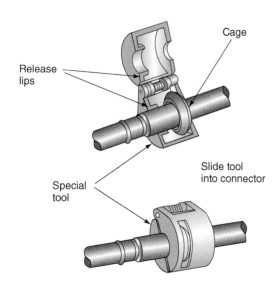

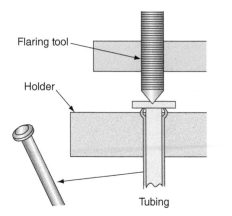

Figure 7-11 Using a flaring tool to secure hose connections.

Figure 7-12 Various clamps used on fuel lines.

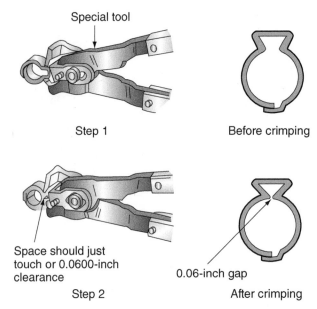

Figure 7-13 Crimp clamps require a special tool for installation.

To control the rate of vapor flow from the fuel tank to the vapor storage tank, a plastic or metal restrictor may be placed either in the end of the vent pipe or in the vapor-vent hose itself. When the latter hose must be replaced, the restrictor must be removed from the old vent hose and installed in the new one.

⚡ WARNING **Always cover a nylon fuel pipe with a wet shop towel before using a torch or other source of heat near the line. Failure to observe this precaution may result in fuel leaks, personal injury, and property damage.**

Nylon fuel pipes should be inspected for leaks, nicks, scratches and cuts, kinks, melting, or loose fittings. If these fuel pipes are kinked or damaged in any way, they must be replaced. Nylon fuel pipes must be secured to the chassis at regular intervals to prevent fuel pipe wear and vibration.

> **AUTHOR'S NOTE** Always verify the manufacturer's recommendation for fuel line servicing to determine if it can be repaired or if replacement is required.

Nylon fuel pipes provide a certain amount of flexibility and can be formed around gradual curves under the vehicle. Do not force a nylon fuel pipe into a sharp bend because this action may kink the pipe and restrict the flow of fuel. When nylon fuel pipes are exposed to gasoline, they may become stiffer, making them more susceptible to kinking. Be careful not to nick or scratch nylon fuel pipes.

Rubber fuel hose should be inspected for leaks, cracks, cuts, kinks, oil soaking, and soft spots or deterioration. If any of these conditions are found, the fuel hose should be replaced. When rubber fuel hose is installed, the hose should be installed to the proper depth on the metal fitting or line (**Figure 7-14**).

The rubber fuel hose clamp must be properly positioned on the hose in relation to the steel fitting or line, as illustrated in Figure 7-14. Fuel hose clamps can be spring type or screw

Some gasoline direct fuel injection lines must be replaced if they are removed because of the extremely high pressures involved.

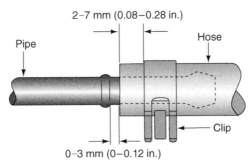

Figure 7-14 Rubber fuel hose installation on a steel fitting or tube.

type. Screw-type fuel hose clamps must be tightened to the specified torque. Of course, these types of clamps are generally suited only for low-pressure applications.

FUEL FILTERS

Automobiles and light trucks usually have an in-tank strainer and a gasoline filter (**Figure 7-15**). The strainer, located in the gasoline tank, is made of a finely woven fabric. The purpose of this strainer is to prevent large contaminant particles from entering the fuel system where they could cause excessive fuel pump wear or plug fuel-metering devices. Servicing of the fuel tank strainer is required when replacing the fuel pump.

   The gasoline filter was usually located under the vehicle, and might require service on a regular basis. The most common types of gasoline filters are in-line filters and out-pump filters. Many returnless systems, especially those after 2005, have "lifetime" fuel filters that can be replaced only by removing the fuel pump from the fuel tank. The returnless system actually pumps less fuel through the filter, due to the fact that no fuel is returned to the tank. Fuel that is returned to the tank via the return line will pass through the filter again on the way to the fuel rail.

Classroom Manual
Chapter 7, page 212

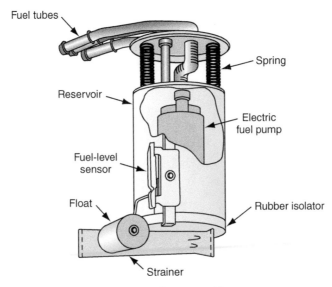

Figure 7-15 A fuel strainer at the fuel pump pickup.

Servicing Filters

Fuel filters and elements are serviced by replacement only. Most fuel filters are part of the fuel sender module assembly and may require replacement of the entire module if dirty or contaminated fuel is placed in the fuel tank. A plugged fuel filter may cause the engine to surge and cut out at high speed or hesitate on acceleration. A restricted fuel filter causes low fuel pump pressure and volume.

The fuel filter replacement procedure varies depending on the make and year of the vehicle and the type of fuel system. Always follow the filter replacement procedure in the appropriate service manual. The following is a typical filter replacement procedure on a vehicle with EFI:

1. Relieve the fuel system pressure as mentioned previously.
2. Disconnect the negative battery cable.
3. Raise the vehicle on a hoist.
4. Flush the quick connectors on the filter with water, and use compressed air to blow debris from the connectors.
5. Disconnect the inlet connector first. Grasp the large connector collar, twist in both directions, and pull the connector off the filter.
6. Disconnect the outlet connector using the same procedure used on the inlet connector (**Figure 7-16**).
7. Loosen and remove the filter mounting bolts, and remove the filter from the vehicle.
8. Use a clean shop towel to wipe the male tube ends of the new filter.
9. Apply a few drops of clean engine oil to the male tube ends on the filter.
10. Check the quick connectors to be sure the large collar on each connector has rotated back to the original position. The springs must be visible on the inside diameter of each quick connector.
11. Install the filter on the vehicle in the proper direction, and leave the mounting bolt slightly loose.
12. Install the outlet connector onto the filter outlet tube, and press the connector firmly in place until the spring snaps into position. Grasp the fuel line and try to pull this line from the filter to be sure the quick connector is locked in place.
13. Do the same with the inlet connector.
14. Tighten the filter retaining bolt to the specified torque.
15. Lower the vehicle, start the engine, and check for fuel leaks at the filter.

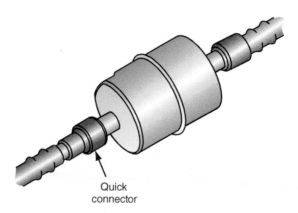

Quick
connector

Figure 7-16 A fuel filter with quick-disconnect fittings.

Typical Procedure for Relieving Fuel Pressure and Servicing the Fuel Tank

P13-1 To begin, relieve any fuel pressure from the system.

P13-2 Disconnect the negative battery cable.

P13-3 If the tank has a significant amount of fuel inside, pump the fuel out of the tank with a gas caddy. (It may be difficult to insert a siphon if the tank has a check valve at the bottom of the fuel filler pipe.)

P13-4 Raise the vehicle on the lift.

P13-5 Disconnect electrical connections from the fuel pump, fuel sender, and vapor pressure sensor.

P13-6 Remove fuel filler tube and vapor and fuel lines from the fuel tank.

P13-7 Position a transmission jack under the fuel tank and remove the tank hold-down straps.

P13-8 Slowly lower the fuel tank from the vehicle; sometimes it is necessary to maneuver the fuel tank around the exhaust system.

P13-9 Once the fuel tank is down, clean the top of the fuel tank with compressed air to prevent introducing dirt into the fuel tank.

P13-10 Remove the fuel sender hold-down ring.

P13-11 Carefully remove the modular fuel pump assembly, taking care not to damage the fuel sender float.

P13-12 Replace the fuel tank opening O-ring and fuel strainer and install the modular fuel pump assembly into the fuel tank.

P13-13 Now install the fuel tank back into the vehicle, installing straps, reconnecting fuel lines, vapor lines, and electrical connections.

P13-14 Reconnect battery; turn key on to pressurize fuel system. Check for leaks under the vehicle.

P13-15 Test-drive vehicle to confirm repair; double-check for any leaks.

The lift pump can be used to deliver fuel to the high-pressure pump. This saves on the amount of fuel line that has to be under an extreme amount of pressure in the DFI system.

Classroom Manual
Chapter 7, page 213

 Caution

The fuel supply lines can remain pressurized for long periods of time after the engine is shut down. This pressure must be relieved before servicing of the fuel system begins. A valve may be provided on the fuel rail for this purpose, or special adaptors may be needed to attach a fuel pressure gauge.

FUEL PUMPS

Electric Fuel Pumps

Electric fuel pumps have been in use for many years. The electric fuel pump is generally located in the tank with the fuel-sending unit. Some vehicles have a pump inside the fuel tank (called a **lift pump** or **low-pressure pump**) and a main pump located outside the fuel tank. Vehicles with (DFI) have a mechanical **high-pressure pump** that is engine driven off the camshaft. A typical wiring diagram for an electric fuel pump is shown in **Figure 7-17**.

CUSTOMER CARE Advise your customers to avoid leaving the level in the fuel tank extremely low. Vehicles with in-tank electric fuel pumps can actually overheat due to the lack of liquid in the tank because it helps dissipate the heat from the pump. Additionally, sediment or debris is more likely to be drawn into the pump as a result of repeated low fuel levels. Recommend that the fuel level not dip below a quarter tank of fuel.

Troubleshooting

Problems in fuel systems using electric fuel pumps are usually indicated by improper fuel system pressure or a dead or inoperative fuel pump.

Low pressure can be due to a clogged fuel filter, restricted fuel line, weak pump, leaky pump check valve, defective fuel pressure regulator, or dirty strainer in the tank. It is possible to rule out filter and line restrictions as a cause of the problem by making a pressure check at the pump outlet. A higher reading at the pump outlet (at least 5 psi) means there is a restriction in the filter or line. If the reading at the pump outlet is unchanged, then the pump either is weak or is having trouble picking up fuel (clogged strainer in the tank). Either way it is necessary to get inside the fuel tank. If the filter strainer is gummed up with dirt or debris, it is necessary to clean out the tank and replace the in-tank filter.

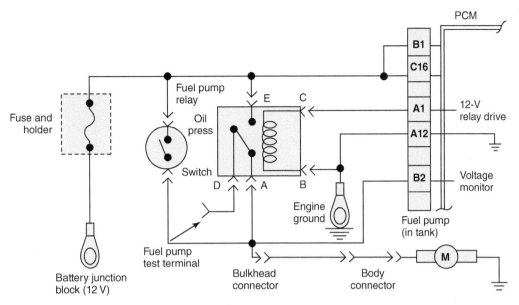

Figure 7-17 Typical wiring diagram for an electric fuel pump.

Another possible source of trouble is the pump check valve. Some pumps have one, and others have two (positive **displacement** roller vane pumps). The purpose of the check valve is to prevent fuel movement through the pump when the pump is off so residual pressure remains at the injectors. This can be checked by watching the fuel pressure gauge after the engine is shut off. If pump pressure bleeds quickly back into the tank when the pump shuts off, the vehicle will be hard to start because the fuel pump will have to re-prime itself on every engine start. Once the engine is running, the vehicle will run fine.

Depending on the type of pump used, the check valve can also prevent the reverse flow of fuel and relieve internal pressure to regulate maximum pump output. Check valves can stick and leak. So, if a pump runs but does not pump fuel, a bad check valve is to blame. Unfortunately, the check valve is usually an integral part of the pump assembly that is sealed at the factory. Therefore, if the check valve is causing trouble, the entire pump must be replaced.

When an engine fails to start because there is no fuel delivery, the first check is the fuel gauge. A defective tank sending unit or miss-calibrated gauge might be giving a false indication. If you have no pressure and the pump appears to be running, try adding some fuel and recheck. If the gauge is faulty, repair or replace it.

Listen for pump noise. When the key is turned on, the pump should run for a couple of seconds to build system pressure. If the fuel pump is not running, then a wiring diagram will be helpful to check the electrical system. On many vehicles, the computer energizes a pump relay when it receives a cranking signal from the crankshaft sensor (**Figure 7-18**). An oil pressure switch might be included in the circuitry for safety purposes and to serve as a backup in case the relay or computer signal fails. Failure of the pump relay or computer driver signal can cause long cranking because the fuel pump does not come on until the engine cranks long enough to build up sufficient oil pressure to close the oil pressure switch.

If the pump does not appear to be running, check for the presence of voltage at the pump electrical connectors. The pump might be good, but if it does not receive voltage and have a good ground, it does not run. To check the ground, connect a test light across the ground and feed wires at the pump to check for voltage, or use a voltmeter to read applied voltage and to check voltage drop to ground. The latter is the better test technique because a poor ground connection or low voltage can reduce pump operating speed and output. If the electrical circuit checks out but the pump does not run, the pump is probably bad and should be replaced.

Many electric fuel pumps in returnless fuel systems have their pressure controlled by the powertrain control module (PCM); therefore, testing these pumps must be done according to the manufacturer's recommendations.

Even though there is a check valve in the pump, it is normal for some of the pressure to bleed down over an hour or so, but most of the fuel should remain in the lines.

Some fuel pumps can be activated with a scan tool.

On vehicles that use an inertia switch, make sure that the switch has not been tripped (opened). These switches can sometimes be activated by hitting large bumps or railroad tracks. Some late-model vehicles with fuel pump driver modules have the ability to shut down the fuel pump when the air bag systems detect an impact, replacing the action of the inertia switch.

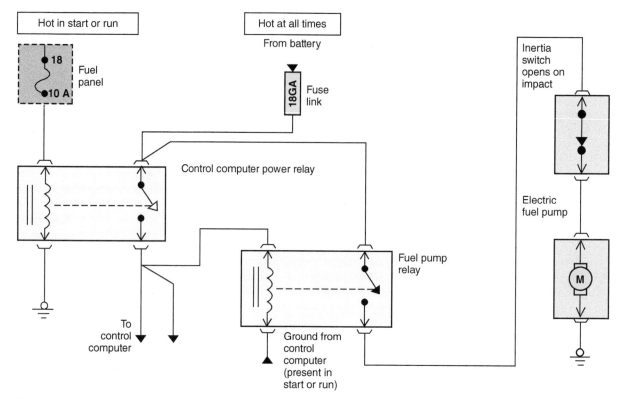

Figure 7-18 Electrical fuel delivery system wiring diagram.

No voltage at the pump terminal when the key is on and the engine is cranking indicates a faulty oil pressure switch, pump relay, relay drive circuit, fuel pump control module, computer, or wiring problem. Check the pump fuse to see if it is blown. Replacing the fuse might restore power to the pump, but until you have found out what caused the fuse to blow, the problem is not solved.

A faulty oil pressure switch can be checked by bypassing it with a jumper wire. If this restores power to the pump and the engine starts, replace the switch. If an oil pressure switch or relay sticks in the closed position, the pump can run continuously whether the key is on or off, depending on how the circuit is wired.

To check a pump relay, use a test light to check across the relays power and ground terminals. This tells if the relay is receiving battery voltage and ground. Next, turn off the ignition, wait for about 10 seconds, and then turn it on. The relay should click and you should see battery voltage at the relay's pump terminal. If nothing happens, repeat the test checking for voltage at the relay terminal that is wired to the computer. The presence of a voltage signal means the computer is doing its job but the relay is failing to close and should be replaced. No voltage signal from the computer indicates an opening in that wiring circuit or a fault in the computer itself.

Fuel pumps can also be checked for amperage draw and current waveform. Check the amperage required for the fuel pump at the fuse block if possible. Make certain that you use a fused multimeter that is capable of measuring at least as much amperage as the vehicle's fuel pump fuse. Measure the amperage draw with the vehicle running, and compare to vehicle specifications. Another method is to use a low-amp probe (**Figure 7-19**) and look at the amperage and/or the waveform the fuel pump produces (**Figure 7-20**). The low-amp probe is popular with technicians because it is not necessary to back-probe or otherwise disturb vehicle wiring; instead, just use a clamp around the wire. This method is described in detail in **Photo Sequence 14**.

⚠ WARNING When testing an electric fuel pump, do not let fuel contact any electrical wiring. The smallest spark (electric arc) could ignite the fuel.

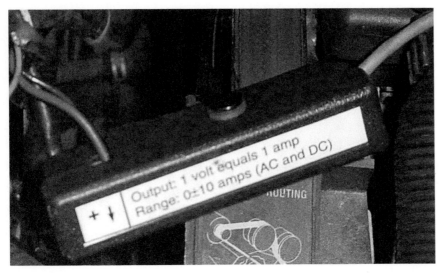

Figure 7-19 A low-amp probe.

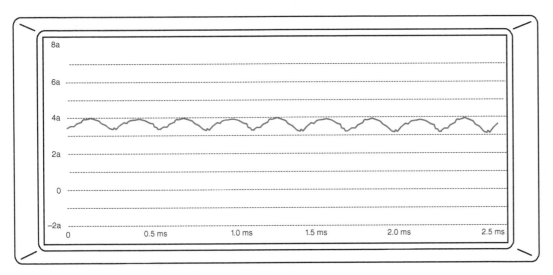

Pattern/Sweep 2.5 ms   Scale 10a

Figure 7-20 Waveform from a low-amp probe of a good fuel pump. Notice even transitions caused by brushes on commutator bars.

To conduct this test on specific fuel injection systems, refer to the service manual for instructions. However, most port fuel injection systems have a *Schrader valve* on the fuel rail that can be used to connect the fuel pressure gauge. If the system does not have a **Schrader valve**, the fuel rail should be relieved of any pressure before loosening a fitting to install the pressure gauge. Special adapters may be needed. Many direct fuel injection systems have a fuel rail pressure sensor that allows a fuel pressure reading with the scan tool.

The following are tips on fuel pressure checking:

- Always verify correct fuel pressure from service information. Do not rely on memory and past experience. Fuel pressures vary greatly between engines, injection types, and makes and model years. Special adaptors are sometimes required to check fuel pressure such as the **"banjo" style** shown in **Figure 7-21**.
- When checking fuel pressure, make sure the specifications are for either the engine running or engine off. If the specifications are for "key on, engine off" and the fuel system has a vacuum-operated pressure regulator, the key-on pressure will be higher than the idle pressure due to vacuum at the pressure regulator. On most returnless fuel systems, the key on, engine off, and engine running pressures will be about the same. Always check service information to be sure.

A Schrader valve is very similar to a tire valve. The center of the valve must be depressed before anything can enter or leave.

A "banjo" fitting is named after the shape of the line attachment.

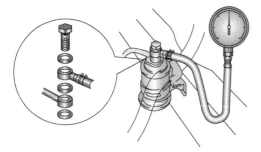

Figure 7-21 A fuel pressure gauge used with a "banjo-style" fuel fitting.

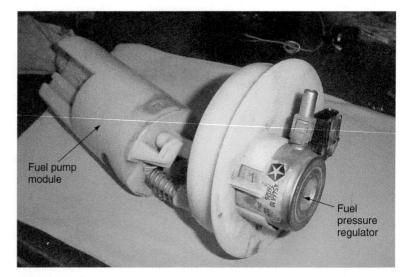

Fuel pump module

Fuel pressure regulator

Figure 7-22 A returnless fuel pump module. Note the fuel pressure regulator location and single outlet.

> ⚠️ **Caution**
>
> The high volatility of fuel and high pressures make careful handling mandatory. Never perform the test near an open flame or spark, and avoid fuel contact with your skin.

> ⚠️ **Caution**
>
> Never attempt to measure fuel volume at the injector outlet of a DFI vehicle.

High fuel pressure can cause the mixture to go richer than normal, while low fuel pressure can cause the mixture to go leaner than normal.

- If pressures are OK at idle, accelerate the engine while observing fuel pressure. The pressure should go up momentarily if the system has a vacuum-operated pressure regulator because of the drop in vacuum. Most returnless systems should remain steady because the fuel goes through the regulator at the fuel pump and is not affected by vacuum. A returnless style fuel pump module is shown in **Figure 7-22**.
- If fuel pressure is too high, either the fuel pressure regulator (on return line style systems) is defective or the return line is restricted (on vehicles that use a return line). If the return line is bypassed with a hose placed into an approved container and the fuel pressure is normal, then the problem is the return line. If the pressure is still too high, then the regulator is defective. Of course, high fuel pressure in a returnless system is the fault of the regulator.
- If fuel volume is low, the pressure is generally low as well, especially when measured under a load. Fuel volume can be measured directly by pumping fuel into a container and comparing the output to specification on a port fuel injection vehicle.

If the gauge goes down on acceleration or is too low at any time, there is not enough volume, the pump is faulty, or the pressure regulator is defective. There is also a chance that a fuel injector could be stuck open. There are special adaptors available that may contain a shut-off valve or are able to be crimped to do this test. If the fuel feed line is restricted after pressure is allowed to build (if possible) and then the pressure on the pressure gauge at the fuel rail holds after key off, the fuel pump check ball or a leak in the system (which could be located in the tank itself) is at fault. If the pressure still falls with the feed line restricted, there is a possibility of an injector sticking on. With a mechanic's stethoscope, listen for fuel escaping through an injector. If you do find an injector bleeding off, unplug the electrical connection at the injector and see if the pressure will hold. If it does, then the ground connection to the injector may be permanently grounded or the PCM is at fault. If the pressure still falls, then the injector itself is at fault. *Note:* The fuel pressure on most fuel-injected vehicles will bleed down after several minutes. It is important to remember that the fuel quantity in the lines is present at start-up, if not full pressure. If the pressure is still too low, inspect the fuel filter and lines for restrictions, and check for proper power and ground feeds at the fuel pump. If none of these seem to be a problem, you will have to remove the tank and inspect the fuel pump, fuel strainer, and condition of the fuel in the tank. Generally speaking, if the pressure regulator or fuel injector is stuck open, the vehicle will have some major drivability issues as well. If the check valve is causing the bleed down, then the vehicle will run fine after starting. High fuel pressure can cause a rich mixture, while lower-than-normal fuel pressure can cause a lean mixture.

PHOTO SEQUENCE 14
Fuel Pump Waveform Diagnosis

P14-1 The electric fuel pump can be diagnosed by using an oscilloscope.

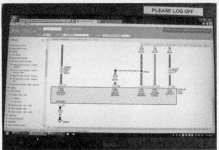

P14-2 First, a schematic of the fuel pump circuit needs to be looked up to find the location of the fuel pump power feed.

P14-3 Find the fuel pump power lead and attach a low-amp probe around the wire. The low-amp probe measures amperage by converting the magnetic field around the wire to voltage. The scope must be set up for the low-amp probe to read milliamps, and then converted to the equivalent amperage reading. Some scopes do this automatically.

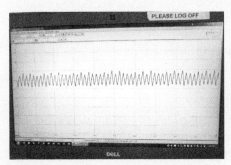

P14-4 The engine is started and the pattern for the fuel pump is shown.

P14-5 The pattern should be consistent, and rises and falls because of the commutator bars in the electric motor. Each commutator bar is connected to a set of windings in the armature. Every time a brush makes contact with a segment of the commutator, current rises and then falls when the brush breaks contact.

P14-6 A general rule of thumb for amperage draw of the fuel pump is generally about 1 amp per 10 psi of fuel pressure. If the amperage is higher, then the fuel flow might be restricted at the fuel filter or crimped line. If the current flow is too low, then look for high resistance in the fuel pump feed circuit. Of course, the source of too high or too low current flow could be a defective pump. Defective bearings could cause the pump to drag, or there could be high internal resistance in the pump.

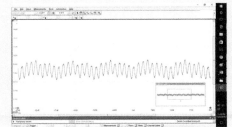

P14-7 With the oscilloscope, a technician can also determine the RPM of the fuel pump by observing the commutator bars, electrical signature. Each commutator has slightly different conductive properties and has its own signature. Most electric fuel pumps (but not all) have eight commutator segments.

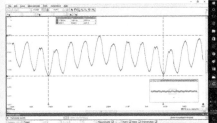

P14-8 Now that we know how long one revolution of the pump takes, we can see how many revolutions the pump makes in one minute, or RPM. Sixty seconds in 1 minute 1,000 milliseconds in one second 60,000 milliseconds in 1 minute 60,000 / 9.6 ms = 6250 rpm.

P14-9 A bad fuel pump will certainly show up on a waveform check, as shown above. Note the very irregular forms.

Some older fuel injection systems had rubber lines that could be carefully crimped with block-off pliers (to perform diagnosis). Many later systems have nylon lines that should never be crimped. Make sure to follow service manual recommendations before crimping lines.

⚠ Caution

Avoid the temptation to test the new pump before reinstalling the fuel tank by energizing it with a couple of jumper wires. Running the pump dry can damage it because the pump relies on fuel for lubrication and cooling.

⚠ Caution

Never turn the ignition switch on or crank the engine with a disconnected fuel line. This action will result in gasoline discharge from the disconnected line, which may result in a fire, causing personal injury and/or property damage.

Direct fuel injection is also called gasoline direct injection or GDI.

Replacement

When replacing an electric pump, be sure that the replacement unit meets the minimum requirements of pressure and volume for that particular vehicle. This information can be found in the service manual. To replace a typical electric fuel pump:

1. Disconnect the ground terminal at the battery.
2. Disconnect the electrical connectors on the electric fuel pump. Label the wires to aid in connecting it to the new pump. Reversing polarity on most pumps destroys the unit.
3. Fuel and vapor lines should be removed from the pump. Label the lines to aid in connecting them to the new pump.
4. The inside tank pump can usually be taken out of the tank by removing the fuel sending unit retaining or lock ring. However, it may be necessary to remove the fuel tank to reach the retaining ring.
5. On a fuel pump that is outside the tank, remove the bolts holding it in place. On in-tank models, loosen the retaining ring. Pull the pump and sending unit out of the tank (if they are combined in one unit), and discard the tank O-ring seal.
6. Most electric fuel pumps are part of a fuel pump module assembly containing the fuel pump, fuel tank pressure (FTP) sensor, and gauge sending unit, and the fuel pump is not serviced separately. In some units, the pump and/or gauge sender can be serviced separately from the fuel module.
7. Compare the replacement pump with the old one. If necessary, transfer any fuel line fittings from the old pump to the new one. Note the position of the filter sock on the pump so you can install a new one in the same relative position. Some pumps require the technician to replace the wiring connector.
8. Install a new filter strainer on the pump inlet and reconnect the pump wires. Be absolutely certain you have correct polarity. Replace the O-ring seal on the fuel tank opening, then put the pump and sender assembly back in the tank and tighten the locking ring by rotating it clockwise. Some pump and sender assembly units are secured by bolts.
9. If the fuel was removed from the vehicle, replace it.
10. Reconnect the electrical connectors.
11. Reconnect the ground terminal at the battery.
12. Start the engine and check all connections for fuel leaks.

DFI High-Pressure Pump System

As we have discussed, the DFI system actually has a conventional electric lift pump (low-pressure fuel system) and a high-pressure pump located on the fuel rail that feeds the injectors (high-pressure fuel system). The high-pressure pump is mechanically driven by the camshaft (**Figure 7-23**). The high-pressure pump controls fuel delivery by varying the input volume of the pump from the low-pressure fuel system through a solenoid on the pump controlled by the PCM (**Figure 7-24**). Never check for injection spray or for fuel volume at the injector outlet of a DFI fuel system. The high pressure can puncture your skin causing severe injury or even death.

The low-pressure fuel system is the same basic system that is found on port fuel injection vehicles, and service is the same on vehicles with mechanical returnless fuel systems. When there is a fault in the fuel pressure system, diagnosis starts by checking the low-pressure system for faults. After the low-pressure system has been confirmed, then the volume-control solenoid and wiring are checked for integrity. Finally, the pump itself is checked for failure.

Figure 7-23 High-pressure DFI pump is driven by the camshaft and has a volume control solenoid to control fuel pressure in the high-pressure system.

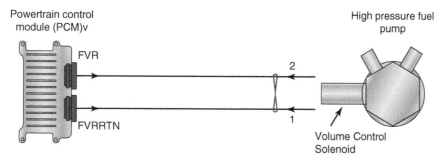

Figure 7-24 The high-pressure pump output pressure is controlled by the volume control solenoid via the PCM

The trouble codes from a Ford fuel volume regulator are as follows:

- *P0001—Fuel volume regulator control circuit/open*. PCM detects too high or low voltage on either the high or low side of the fuel volume regulator (mounted on the high-pressure fuel pump). Basically, this code would be caused by an open circuit on the power or return wires to the solenoid, or the coil itself could be open.
- *P0003—Fuel volume regulator control circuit/low*. PCM detects a grounded power or return circuit from the PCM to the volume control solenoid. This can also damage the solenoid, so it should be checked if this code is set.
- *P0004—Fuel volume control circuit high*. PCM detects either a short to voltage on the control circuit, or the solenoid feed and return signal wires are shorted together.
- *P0087—Fuel rail pressure too low*. The PCM cannot control the fuel rail pressure with the fuel volume solenoid. Of course, if any of the circuit DTCs are set they should be diagnosed first, but this code could be set by a restricted fuel supply line or filter, or even a damaged pump.
- *P0088—Fuel rail pressure too high*. The fuel pressure in the fuel rail is too high and the PCM cannot control the pressure. Once again, this could be caused by a circuit to the solenoid, so any circuit DTCs should be diagnosed first. But this could also be caused by a faulty high-pressure pump.

Fuel Pump Driver Module

Many late-model vehicles use a fuel pump driver module to control fuel pump output speed through the engine control module (ECM). The low side fuel pump driver module and fuel pressure sensor are not exclusive to DFI systems, as they are used on most late-model port fuel systems as well. The fuel pressure sensor reports the fuel pressure to the PCM, which adjusts fuel pressure and pump speed according to anticipated demands of the engine by sending the appropriate commands to the fuel pump driver module.

The low-pressure system had DTCs associated with it, such as:

- *P008A—Low system pressure too low.* This code can indicate several problems. It can be a restricted fuel feed, low fuel, defective electric fuel pump, or a malfunctioning fuel pressure sensor.
- *P008B—Low system pressure too high.* This code could result from a defective pressure sensor, or from a circuit fault that would cause a high fuel pump command.

Electronic Returnless Fuel Pressure Regulation/DFI Systems

The complete electronic returnless system is shown in **Figure 7-25**. Note that two feedback systems are involved: one for the low-pressure system and one for the high-pressure system. These systems may not use a mechanical regulator in the fuel pump module at all. The speed of the fuel pump determines the output pressure of the pump, based on readings from the fuel pressure sensor. The ECM commands are sent to the fuel pump driver module as a duty cycle command, and the driver module adjusts the output speed of the pump based on current flow. The high-pressure system (for DFI) is similar, with the fuel rail pressure sensor sending an output to the PCM, and the PCM, in turn, commands the fuel volume solenoid on the high-pressure pump to affect changes in the fuel rail pressure.

FUEL INJECTION SYSTEMS

Classroom Manual
Chapter 7, page 204

Troubleshooting fuel injection systems requires systematic step-by-step test procedures. With so many interrelated components and sensors controlling fuel injection performance, a hit-or-miss approach to diagnosing problems can quickly become frustrating, time consuming, and costly.

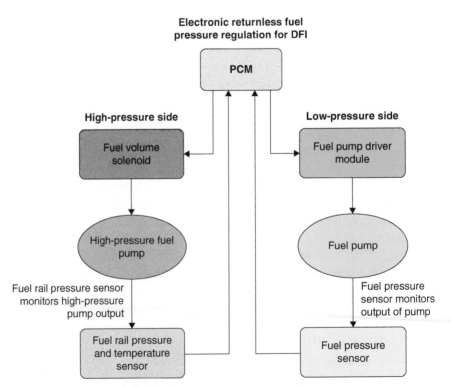

Figure 7-25 Electronic returnless fuel pressure regulation for DFI systems.

Most fuel injection systems are integrated into engine control systems. The self-test modes of these systems are designed to help in engine diagnosis. Unfortunately, when a problem upsets the smooth operation of the engine, many service technicians automatically assume that the computer (PCM) is at fault. But in the vast majority of cases, complaints about drivability, performance, fuel mileage, roughness, or hard starting or no-starting are due to something other than the computer itself, although many problems are caused by sensor malfunctions that can be traced using the self-test mode.

Before condemning sensors as bad, remember that weak or poorly operating engine components can often affect sensor readings and result in poor performance. For example, a sloppy timing chain, bad rings, or valves reduce vacuum and cylinder pressure, resulting in a lower exhaust temperature. This can affect the operation of a perfectly good fuel delivery system.

A problem like an intake manifold leak can cause the manifold absolute pressure (MAP) sensor to adjust engine operation to less than ideal conditions.

One of the basic rules of electronic fuel injection servicing is that EFI cannot be adjusted to match the engine; you have to make the engine match EFI. In other words, make sure the rest of the engine is sound before condemning the fuel injection and engine control components.

Preliminary Checks

The best way to approach a problem on a vehicle with electronic fuel injection is to check out all the basic systems before condemning electronic components. As the previous examples illustrate, any engine is susceptible to problems that are unrelated to the fuel system itself. Unless all engine support systems are operating correctly, the control system does not operate as designed.

Before proceeding with specific fuel injection checks and electronic control testing, be certain of the following:

- The battery is in good condition and fully charged, with clean terminals and connections.
- The charging and starting systems are operating properly.
- All fuses and fusible links are intact.
- All wiring harnesses are properly routed with connections free of corrosion and are tightly attached.
- All vacuum lines are in sound condition, properly routed, and tightly attached.
- The PCV system is working properly and maintaining a sealed crankcase.
- All emission control systems are in place, hooked up, and operating properly.
- The level and condition of the coolant-antifreeze is good, and the thermostat is opening at the proper temperature.
- The secondary spark delivery components are in good shape with no signs of cross-firing, carbon tracking, corrosion, or wear.
- The engine is in good mechanical condition.
- The gasoline in the tank is of good quality and has not been substantially cut with alcohol or contaminated with water.

EFI System Component Checks

In any electronic fuel injection system, three things must occur for the system to operate.

1. An adequate air supply must be supplied for the air-fuel mixture.
2. Fuel at the proper pressure must be delivered correctly to operating injectors.
3. The injectors must receive a pulse from the control computer.

Prior to the J1930 standards, the MIL was most commonly called the *check engine light.*

The **malfunction indicator lamp (MIL)** is controlled by the PCM and is turned on to alert the driver to a problem.

If all of these preliminary checks do not reveal a problem, proceed to test the electronic control system and fuel injection components. Some older control systems require involved test procedures and special test equipment, but most designs have monitors designed to help diagnose the problem. These monitors perform a number of checks on components within the system. Input sensors, output devices, wiring harnesses, and even the electronic control computer itself are among the items tested.

If the **malfunction indicator lamp (MIL)** is on or there are trouble codes stored in the PCM, these can help a technician in diagnosing a problem. Some trouble codes are mandated (also called "generic" codes), and some codes are manufacturer specific. Always check service information and follow the trouble code chart for your particular vehicle.

Always remember that trouble codes only indicate the particular circuit in which a problem has been detected. They do not pinpoint individual components. So if a code indicates a defective oxygen sensor, the problem could be the sensor itself, the wiring to it, or its connector. Trouble codes are not a signal to replace components. They signal that a more thorough diagnosis is needed in the area indicated.

The following sections outline general troubleshooting procedures for the most popular EFI designs in use today.

Air Induction System Checks

In an injection system, particularly designs that rely on mass airflow sensors, all the air entering the engine must be measured. If it is not, the air-fuel ratio becomes overly lean. For this reason, cracks or tears in the plumbing between the airflow sensor and throttle body are potential air leak sources that can affect the air-fuel ratio.

Air that enters the engine unmeasured through a vacuum leak or air duct problem is termed "false air" by many in the field.

Carefully inspect the intake hose on a MAF-equipped vehicle if the vehicle stumbles on acceleration. The movement of the engine will cause the hose to open up on acceleration.

During a visual inspection of the air control system, pay close attention to these areas, looking for cracked or deteriorated ductwork (**Figure 7-26**). Also make sure all induction hose clamps are tight and are properly sealed. Look for possible air leaks in the crankcase, for example, the dipstick tube and oil filter cap. Any extra air entering the intake manifold through the PCV system is not measured either and can upset the delicately balanced air-fuel mixture at idle.

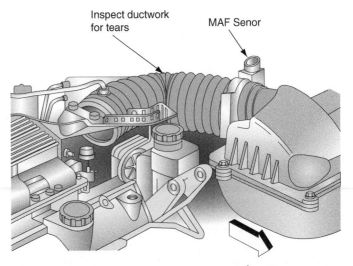

Inspect ductwork for tears

MAF Senor

Figure 7-26 Make sure that the ductwork from the MAF to the throttle body is free from damage and the clamps are tight.

Airflow Sensors

On systems that use a mass airflow meter or manifold pressure sensor to measure airflow, check for good electrical connections. A scan tool can be helpful in this respect. While observing sensor values, manipulate the wiring connector and harness to watch for dropouts in the signal. An oscilloscope can also be helpful in this regard.

Mass airflow (MAF) sensors measure the **mass** of air entering the engine. Cold air has more oxygen content than warm air. Moist air burns differently and is heavier than dry air. The MAF sensor is a very accurate way to measure airflow into the engine. The more accurately the PCM can match the air with the fuel mixture, the more efficient the engine. MAF sensors can be located either on the throttle body or near the air cleaner (**Figure 7-27**). Locating the MAF on the throttle body helps reduce the chance of a split intake hose causing drivability problems, but these defects must still be repaired so that unfiltered air is not entering the engine. Most MAF sensors output a frequency signal that increases with airflow amount. An example of a MAF sensor waveform is shown in **Figure 7-28**. Look for consistency in the waveform. The higher the rpm, the higher the frequency read. The square waves will be smaller and closer together as the rpm increases. A voltmeter will not give you accurate readings on a MAF sensor because the computer is not looking for voltage variation. In fact, the voltage will vary very little from low to high rpm because most digital voltmeters average their voltage readings. The frequency of the MAF is important and changes in relation to airflow entering the engine. If the technician does not have an oscilloscope available, higher-level digital multimeters do have a Hertz (Hz) or frequency measurement. This reading can be watched for smooth transitions during rpm changes.

Mass airflow sensors take into account the volume and weight of the air entering the engine.

> **SERVICE TIP** If the engine does not start or idles poorly, unplug the MAF. If the engine starts or runs better with the sensor unplugged, the sensor is likely malfunctioning and should be replaced.

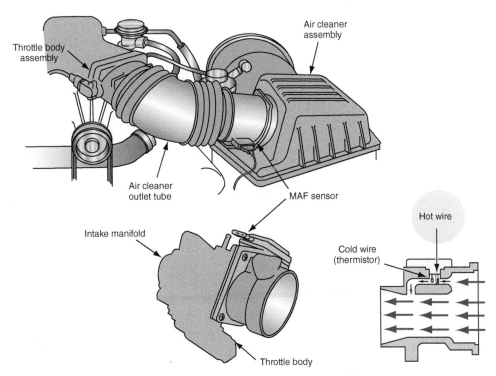

Figure 7-27 MAF sensors can be mounted on the air cleaner or intake manifold.

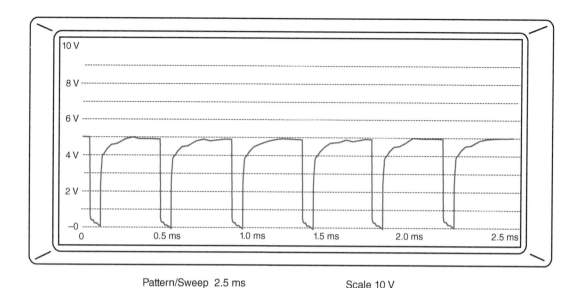

Pattern/Sweep 2.5 ms   Scale 10 V

Figure 7-28 Voltage waveform from a hot-wire MAF sensor.

A hot-wire MAF sensor can be checked for contamination. Just a small amount of contamination will cause the MAF sensor reading to be inaccurate. A contaminated MAF sensor wire will measure less airflow than is actually entering the engine, causing the mixture to go very lean on acceleration and results in a hesitation on acceleration.

The hot wire can be cleaned with contact cleaner.

The MAF sensor can be checked with a scan tool; usually readings are obtained in grams per second of airflow. Look for consistent readings under different operating conditions.

Throttle Body

Remove the air duct from the throttle assembly and check for carbon buildup inside the throttle **bore** and on the throttle plate. Soak a cloth with throttle body cleaning solution and wipe the bore and throttle plate to remove light to moderate amounts of carbon residue. Also, clean the backside of the throttle plate as shown in **Figure 7-29**. Then if equipped, remove the idle air control valve from the throttle body clean any carbon deposits from the pintle tip and the idle air control (IAC) air passage. Some throttle bodies should not be cleaned, so check the service information first.

Fuel System Checks

If the air control system is in working order, move on to the fuel delivery system. It is important to always remember that fuel injection systems operate at high fuel pressure

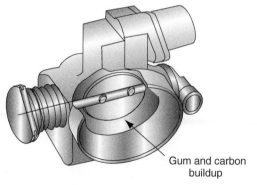

Gum and carbon buildup

Figure 7-29 Gum and carbon can build up on the throttle bore and cause stalling or rough idle.

levels. This pressure must be relieved before any fuel line connections can be broken. Spraying gasoline (under a pressure as high as 2176 PSI on a DFI engine) on a hot engine creates a real hazard when dealing with a liquid that has a flash point of −45°F. Do not attempt to measure fuel volume on a DFI system.

DFI has a high enough fuel pressure inside the fuel rail to penetrate your skin, which can be fatal.

Follow the specific procedures given in the service information when relieving the pressure in the fuel lines. If the system has a return line, one alternative procedure is to apply 20 to 25 inches of vacuum to the externally mounted fuel pressure regulator (with a hand vacuum pump connected to the manifold control line of the regulator), which bleeds fuel pressure back into the tank (**Figure 7-30**). The fuel pump fuse can be pulled and the engine started and let run until it dies. The fuel pump can also be unplugged if necessary.

(Sometimes the fuel pump and ignition are on the same fuse.) In addition, some fuel pressure gauges are equipped with a fuel pressure bleed-off valve (**Figure 7-31**).

Fuel Delivery

It is difficult to visually inspect the spray pattern and volume of port system injectors. However, an accurate indication of their performance can be obtained by performing simple fuel pressure and fuel volume delivery tests. Never attempt to view the spray pattern on a DFI system, the high pressure can cause severe injury or even death.

Low fuel pressure can cause a no-start or poor-run problem. It can be caused by a clogged fuel filter, a faulty pressure regulator, or a restricted fuel line anywhere from the fuel tank to the fuel filter connection.

If a fuel volume test shows low fuel volume, it can indicate a bad fuel pump or blocked or restricted fuel line. The volume specifications for your particular vehicle should be

⚠ Caution

Dispose of the fuel-soaked rag in a fireproof container.

Classroom Manual
Chapter 7, page 214

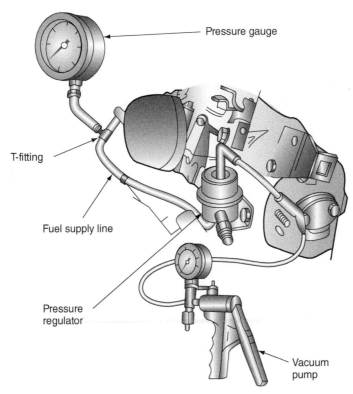

Figure 7-30 With an externally mounted fuel pressure regulator, it is possible to bleed off system pressure into the tank using a hand vacuum pump.

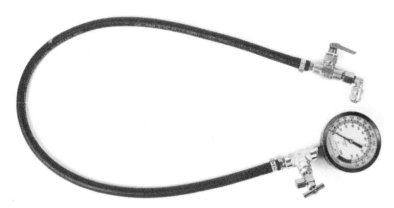

Figure 7-31 A fuel pressure gauge with a bleed-off valve can be used to relieve fuel pressure.

listed in your service information. When performing the test, visually inspect the fuel for signs of dirt or moisture. These indicate the fuel tank needs to be removed and cleaned and the fuel filters replaced.

High fuel pressure readings will result in a rich-running engine. On return-line style systems, a restricted fuel return line to the tank or a bad fuel regulator may be the problem. To isolate the cause of high pressure, relieve system pressure and connect a tap hose to the fuel return line. Direct the hose into a container and energize the fuel pump. If fuel pressure is now within specifications, the fuel return line is blocked. If pressure is still high, the pressure regulator is faulty. On returnless systems, the pressure regulator is the likely culprit for high fuel pressures.

Injector Checks

A fuel injector is nothing more than a solenoid-actuated fuel valve. Its operation is quite basic in that as long as it is held open and the fuel pressure remains steady, it delivers fuel until it is told to stop.

Because all fuel injectors operate in a similar manner, fuel injector problems tend to exhibit the same failure characteristics. An injector that does not open causes hard starts, while an injector that is stuck partially open causes loss of fuel pressure (most noticeably after the engine is stopped and restarted within a short time period) and flooding due to raw fuel dribbling into the engine. In addition to a rich-running engine, a leaking injector also causes the engine to diesel or run on when the ignition is turned off. If an injector leaks badly enough, it can hydraulically lock the engine, resulting in major damage.

Many fuel injection systems will shut down the fuel injector on a cylinder that is missing for any reason. This is part of the strategy of the PCM to save the catalytic converter from damage by raw fuel.

Checking Voltage Signals

When an injector is suspected as the cause of a problem, the first step is to determine if the injector is receiving a signal from the control computer to fire. Fortunately, determining if the injector is receiving a voltage signal is easy and requires simple test equipment. Unfortunately, the location of the injector's electrical connector can make this simple voltage check somewhat difficult.

Once the injector's electrical connector has been removed, check for voltage at the injector using an ordinary test light or a convenient noid light that plugs into the connector (**Figure 7-32**). After making the test connections, crank the engine. A series of rapidly flickering lights indicates the computer is doing its job and supplying voltage or a ground to open the injector.

Figure 7-32 Checking for voltage at the injector using a noid light.

Ohmmeter

Figure 7-33 An ohmmeter can be connected across the injector terminals to test for shorted or open windings.

When performing this test, make sure to keep off the accelerator pedal. On some models, fully depressing the accelerator pedal activates the clear flood mode in which the voltage signal to the injectors is automatically cut off. Technicians who are unaware of this waste time tracing a phantom problem.

If sufficient voltage is present after checking each injector, check the electrical integrity of the injectors themselves. Use an ohmmeter to check each injector winding for shorts, opens, or excessive resistance. Compare resistance readings to the specifications found in the service manual.

An ohmmeter can be used to test the electrical soundness of an injector. Connect the ohmmeter across the injector terminals (**Figure 7-33**) after the wires to the injector have been disconnected. If the meter reading is infinity, the injector winding is open. If the meter shows more resistance than the specifications call for, there is high resistance in the winding. A reading that is lower than the specifications indicates that the winding is shorted. If the injector is even a little bit out of specifications, it must be replaced.

If the injector's electrical leads are difficult to access, an injector power balance test is hard to perform. As an alternative, start the engine and use a technician's stethoscope to listen for correct injector operation (**Figure 7-34**). A good injector makes a rhythmic

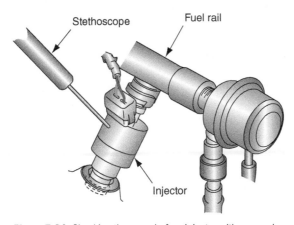

Stethoscope   Fuel rail

Injector

Figure 7-34 Checking the sound of an injector with a sound scope (stethoscope).

clicking sound as the solenoid is energized and de-energized several times each second. If a *clunk-clunk* instead of a steady *click-click* is heard, chances are the problem injector has been found. Cleaning or replacement is in order. If a stethoscope is not handy, use a thin steel rod, wooden dowel, or fingers to feel for a steady on-off pulsing of the injector solenoid.

Another way to isolate an offending cylinder when injector access is limited is to perform the more traditional cylinder power balance test by momentarily grounding each spark plug wire instead of disabling the injectors. Remember to bypass the idle air control if equipped. Following the same warm-up and idle stabilizing procedures, watch the rpm drop and note any change in idle quality as each plug is shorted. Most scan tools can also perform a balance test by turning off injectors and watching the corresponding rpm drop. Another strategy is to look at the misfire data on the scan tool. If a lazy or dead cylinder is located, concentrate efforts on the portion of the fuel or ignition system pertaining to that cylinder, assuming no mechanical problems are present.

> **AUTHOR'S NOTE** With OBD II and even some earlier vehicles, the technician can actually do a balance test with the scan tool. The scan tool can also be checked for cylinder misfire with the weak cylinder identified by the PCM. Sometimes the injectors can be shut down while watching the rpm. Get to know the capability of your scan tool. If this is not possible on your vehicle and the injectors are accessible, the injectors can be unplugged one at a time instead of the spark plug wires. The advantage of these methods is that the unburned fuel from a missing spark does not go through the converter, and the chance of electric shock is also eliminated.

Oscilloscope Checks

An oscilloscope can be used to monitor the injector's pulse width and duty cycle when an injector-related problem is suspected. The pulse width is the time in milliseconds that the injector is energized. The duty cycle is the percentage of on time to total cycle time.

To check the injector's firing voltage on the scope, a typical hookup involves connecting the scope's positive lead to the injector supply wire and the scope's negative lead to an engine ground. Even though these connections are considered typical, it is still a good idea to read the instruction manual provided with the test equipment before making connections.

With the scope set on the low-voltage scale and the pattern adjusted to fill the screen, a square-shaped voltage signal should be present with the engine running or cranking (**Figure 7-35**). If the voltage pattern reads higher than normal, excessive resistance in the injector circuit is indicated. Conversely, a low-voltage trace indicates low circuit resistance. If the pattern forms a continuous straight line, it means the injector is not functioning due to an open circuit somewhere in the injector's electrical circuit.

Injector Replacement

Consult the vehicle's service information for instructions on removing and installing injectors. Before installing the new one, always check to make sure the sealing O-ring is in place (**Figure 7-36**). Also, prior to installation, lightly lubricate the sealing ring with engine oil or automatic transmission fluid (avoid using **silicone grease**, which tends to clog the injectors) to prevent seal distortion or damage.

⚠ **Caution**

Any time cylinders are shorted during a power balance test, make the readings as quickly as possible. Prolonged operation of a shorted cylinder causes excessive amounts of unburned fuel to accumulate inside the catalytic converter and increase the risk of premature converter failure.

The PCM controls fuel mixture by varying injector on time as needed.

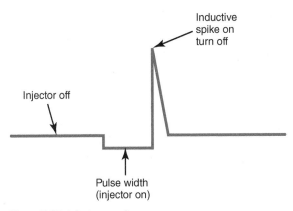

Figure 7-35 Injector waveform.

Inductive spike on turn off

Injector off

Pulse width (injector on)

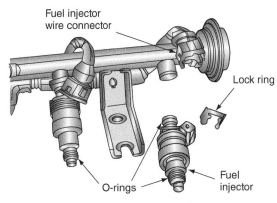

Figure 7-36 Replace the O-rings when replacing an injector.

Fuel injector wire connector

Lock ring

O-rings

Fuel injector

Photo Sequence 15 outlines a typical procedure for removing and installing an injector. Always refer to the service manual for the exact procedure for the engine being serviced.

PHOTO SEQUENCE 15
Removing and Replacing a Fuel Injector on a PFI System

P15-1 Often, an individual injector needs to be replaced. Random disassembly of the components and improper procedures can result in damage to one of the various systems located near the injectors.

P15-2 The injectors are normally attached directly to a fuel rail and inserted into the intake manifold or cylinder head. They must be positively sealed because high-pressure fuel leaks can cause a serious safety hazard.

P15-3 Prior to loosening any fitting in the fuel system, the fuel pump fuse should be removed.

P15-4 As an extra precaution, many technicians disconnect the negative cable of the battery.

PHOTO SEQUENCE 15 (CONTINUED)

P15-5 To remove an injector, the fuel rail must be able to move away from the engine. The rail-holding brackets should be unbolted and the vacuum line to the pressure regulator disconnected.

P15-6 Disconnect the wiring harness to the injectors by depressing the center of the attaching wire clip.

P15-7 The injectors are held to the fuel rail by a clip that fits over the top of the injector. O-rings at the top and at the bottom of the injector seal the injector.

P15-8 Pull up on the fuel rail assembly. The bottoms of the injectors will pull out of the manifold while the tops are secured to the rail by clips.

P15-9 Remove the clip from the top of the injector and remove the injector unit. Install new O-rings onto the new injector. Be careful not to damage the seals while installing them, and make sure they are in their proper locations.

P15-10 Install the injector into the fuel rail and set the rail assembly into place.

P15-11 Tighten the fuel rail hold-down bolts according to manufacturer's specifications.

P15-12 Reconnect all the parts that were disconnected. Install the fuel pump fuse and reconnect the battery. Turn the ignition switch to the run position and check the entire system for leaks. After a visual inspection has been completed, conduct a fuel pressure test on the system.

Idle Adjustment

In a fuel injection system, idle speed is regulated by controlling the amount of air that is allowed to bypass the airflow sensor or throttle plates, or is set by the throttle actuator control system. Idle speeds are not adjustable on the vast majority of fuel-injected vehicles. If a vehicle has a problem with idle speed the cause must be found for the condition.

EVAPORATIVE EMISSIONS

Evaporative emissions, as we have seen previously, are important to reduce hydrocarbon (HC) emissions due to the evaporation of fuel stored in the vehicle. Evaporative emission systems were implemented in 1968. Early systems, even after the advent of computers on vehicles, did not set codes. Since 1996, evaporative emission systems have to be checked for integrity by the PCM (**Figure 7-37**). Several different strategies have been employed. We will try to reproduce a generic system here, not meant to be exactly like any one system but a combination of designs. Make sure to check the latest service information for the specific vehicle before doing any work.

Canister

The canister, often referred to as the charcoal canister, is filled with activated charcoal. Instead of fuel vapors escaping through a tank vent into the air as HC emission, the vapor is stored and later burned in the engine. The canister is purged during driving when conditions are such that the richer mixture produced will not be detrimental to engine operation.

Purge Solenoid

The purge solenoid is used to purge the canister of vapors. The PCM supplies a duty cycle, or pulse width modulated signal, to the purge valve, thereby controlling the amount and timing of purge. Trouble codes PO441 through PO445 (generic) can be set if the purge valve

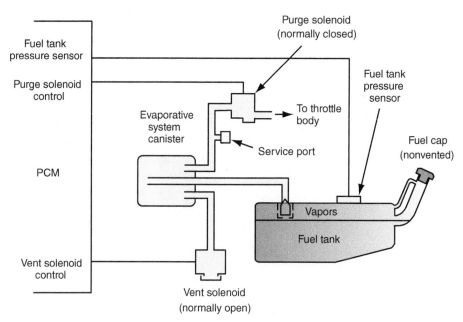

Figure 7-37 Evaporative emission system diagram.

is malfunctioning; PO458 and PO459 cover diagnosis of the purge valve electrical circuit. Manufacturers can also add manufacturer-specific codes for use on their vehicles.

Vent Solenoid

This solenoid allows the PCM to close off the evaporative emission system to check for leaks. Leaks are determined by using engine vacuum that builds with a closed vent solenoid and an open purge solenoid. Generic trouble codes PO446 through PO449 cover the operation of the vent solenoid, its related wiring, and hose connections as applicable.

Fuel Tank Pressure Sensor

The FTP sensor is much like a MAP sensor, but more sensitive. This sensor is capable of measuring the amount of pressure (vacuum) inside the fuel tank. This sensor is a vital part of the PCM's ability to check the system for leaks that are no larger than 0.020 inch in diameter. PO450 to PO454 concern the operation of the fuel tank pressure sensor operation, hoses, and wiring.

Service Port

Some Evaporative emission control systems have connection fittings that can be used by the technician to check the system for leaks in the shop. These connections have a green cap and a Schrader valve to attach special tools to service the system (**Figure 7-38**). The port is convenient when it is available, although many late-model vehicles do not have them installed.

Fuel Cap

The fuel cap must be able to seal the evaporative system against leaks or the MIL will come on for a large leak. Some fuel caps have a bayonet-style attachment that has been designed to make it easier for consumers to tell when the cap is installed properly. PO457 is a code set for a loose or missing fuel cap on late-model vehicles. Many vehicles will have a check gas cap light that will illuminate when the cap is loose, but if they do not tighten the cap soon the warning will generate a code for a large evaporative leak. Problems with customers installing fuel caps has led to many manufacturers eliminating the gas cap and replacing it with a port that the fuel nozzle can be installed directly into.

Figure 7-38 Evaporative service port.

Fuel-Level Sensor

The fuel level of a vehicle is important to the PCM because the evaporative emissions check will not be performed on a tank with more than 85 percent and less than 15 percent of a full tank. The fuel-level sensor can also be used to determine if the fuel tank has been filled recently, and some systems check periodically during driving to see if the fuel tank has been filled with the engine running. This is not recommended for safety reasons, but manufacturers have to take this into account because this will affect the way the PCM runs testing. Note: Some systems will set a trouble code if the tank is filled with the engine running and the PCM is performing its diagnostic test at the same time. P0460 to P0464 are trouble codes set for a fuel-level sensor malfunction.

PCM Action

Depending on the model year and manufacturer, there are several ways to check the system for leaks as mandated by the federal government.

Many systems pull a vacuum on the system during cold engine operation, by opening the purge valve and closing the vent valve to check for large leaks. If the system passes a large leak, then the system will check for smaller leaks by pulling a vacuum and measuring the rate of vacuum leak down. There are also systems that use a small pump to pressurize the system while a pressure sensor checks the system for leaks.

Some systems use the natural volatility of the fuel in the tank, looking for a rise in pressure with the system off.

Make certain that you follow the correct procedures when servicing these systems; including checking for any applicable service bulletins that may apply.

System Service

OBD II systems are required to check system integrity. The main concern for the service technician will be leaks in the system. Evaporative emission systems have to be able to detect a leak as small as 0.020 inch in diameter, so leaks can be difficult to find. Remember also that a kinked hose can also cause a trouble code if vacuum were unable to reach the pressure sensor or if a vent line did not allow the system to vent properly. Faults in the sensors and solenoids required for the PCM to check system integrity also have trouble codes to help the technician determine the point of failure. PO455 is set for a large leak, and code PO456 is set for a small leak.

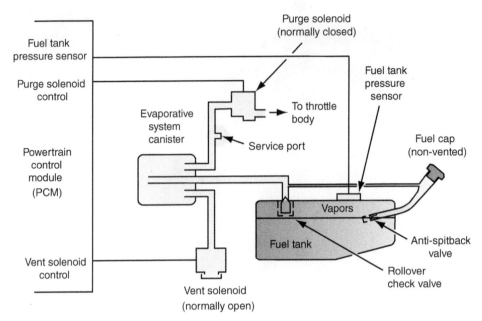

Figure 7-39 Onboard refueling vapor recovery system.

Onboard Refueling Vapor Recovery System

The **onboard refueling vapor recovery (ORVR)** system was implemented in 1998 to control the release of **hydrocarbons (HC)** during vehicle refueling. It was determined that a greater amount of HC was released while refueling than was released during the burning of the tank of gasoline. The vapor is stored in the charcoal canister just like the normal vent vapors, although the canister is larger than systems without ORVR (**Figure 7-39**).

The fill pipe on ORVR-equipped vehicles is much smaller than was previously used, about 1 inch in diameter. The small hose is filled with liquid, and vapor is not blown back out of the tank while refueling. The fill pipe is also equipped with a check valve to prevent fuel from coming back out of the pipe while refueling.

The shutoff assembly actually controls the vent to the fuel tank. The valve floats on the fuel in the tank. When the tank level becomes high enough, the vent is closed. Since there is no vent, no more fuel can be added to the tank and the gas station fill nozzle shuts off. There are no codes for malfunctions in the ORVR system. Obviously, a problem in the vent system will make the fuel tank hard to fill. The shutoff valve also acts as a rollover check valve for the fuel tank vapor line.

> ORVR or onboard refueling vapor recovery systems limit the amount of hydrocarbons escaping in to the air while refueling.

ASE-STYLE REVIEW QUESTIONS

1. *Technician A* says to relieve pressure on an EFI system, connect a pressure gauge to the fuel pressure test port and open the bleed valve on the gauge with the bleed hose installed in an approved container.

   Technician B says that before fuel system components on an EFI system can be removed, the fuel system pressure must be relieved.

   Who is correct?

   A. A only

   B. B only

   C. Both A and B

   D. Neither A nor B

2. While discussing alcohol content in gasoline, *Technician A* says that excessive quantities of alcohol in gasoline may cause fuel filter plugging.

   Technician B says that excessive quantities of alcohol in gasoline may cause lack of engine power.

   Who is correct?

   A. A only

   B. B only

   C. Both A and B

   D. Neither A nor B

3. *Technician A* says that fuel pumps can be checked for amperage draw.

   Technician B says that fuel pumps can also be checked by observing the current waveform.

   Who is correct?

   A. A only   C. Both A and B
   B. B only   D. Neither A nor B

4. *Technician A* says that a trouble code identifies the exact location of a fuel system fault.

   Technician B says that the trouble code only identifies the area of the fault and that the sensor wiring and PCM still have to be checked.

   Who is correct?

   A. A only   C. Both A and B
   B. B only   D. Neither A nor B

5. *Technician A* says a that fuel-level sensor malfunction may cause a trouble code to set.

   Technician B says that the evaporative emissions check test is run only when the fuel tank is empty.

   Who is correct?

   A. A only   C. Both A and B
   B. B only   D. Neither A nor B

6. *Technician A* says that the sending unit O-ring must be replaced when the sending unit is replaced.

   Technician B says that a special tool may be required to remove the pump and sender from the tank.

   Who is correct?

   A. A only   C. Both A and B
   B. B only   D. Neither A nor B

7. *Technician A* says that fuel tanks contain an in-tank strainer to filter large particles of contaminate from the fuel pump.

   Technician B says that some returnless fuel systems have a "lifetime" fuel filter that is not replaced unless it is plugged with debris.

   Who is correct?

   A. A only   C. Both A and B
   B. B only   D. Neither A nor B

8. *Technician A* says that the evaporative emission system controls hydrocarbon emissions.

   Technician B says that the evaporative emission system controls carbon monoxide emissions.

   Who is correct?

   A. A only   C. Both A and B
   B. B only   D. Neither A nor B

9. *Technician A* says that the fuel pump can be checked for fuel volume.

   Technician B says that the fuel pump can be checked for fuel pressure output.

   Who is correct?

   A. A only   C. Both A and B
   B. B only   D. Neither A nor B

10. *Technician A* says that the hot-wire MAF sensor sends a varying voltage signal back to the PCM.

    Technician B says that contamination of the hot-wire MAF sensor can cause drivability problems.

    Who is correct?

    A. A only   C. Both A and B
    B. B only   D. Neither A nor B

ASE CHALLENGE QUESTIONS

1. *Technician A* says that gasoline contains 10 percent alcohol if the water content is 20 mL at the end of an alcohol-in-fuel test.

   Technician B says that during an alcohol-in-fuel test, the alcohol and water remain separated.

   Who is correct?

   A. A only   C. Both A and B
   B. B only   D. Neither A nor B

2. *Technician A* says that a rich air-fuel mixture in EFI systems may be the result of higher-than-normal fuel pressure.

   Technician B says that lower-than-normal fuel pressure in an EFI system causes a lean air-fuel ratio.

   Who is correct?

   A. A only   C. Both A and B
   B. B only   D. Neither A nor B

3. *Technician A* says that the fuel pump should run for a couple of seconds when the key is turned on.

   Technician B says that if the pump is running and there is no fuel pressure, try adding some fuel to the tank and see if it starts.

   Who is correct?

   A. A only   C. Both A and B
   B. B only   D. Neither A nor B

4. *Technician A* says that PO441 through PO445 cover purge valve operation.

   Technician B says that OBD II systems have no codes for the evaporative emissions system.

   Who is correct?

   A. A only

   B. B only

   C. Both A and B

   D. Neither A nor B

5. *Technician A* says that the fuel tank pressure sensor is much like an MAF sensor.

   Technician B says that the FTP sensor measures fuel pressure from the fuel pump.

   Who is correct?

   A. A only

   B. B only

   C. Both A and B

   D. Neither A nor B

Name _____   Date _____

RELIEVING PRESSURE IN AN EFI SYSTEM AND REPLACING A FUEL FILTER

Upon completion of this job sheet, you should be able to relieve pressure from the fuel lines and system on a vehicle equipped with electronic fuel injection.

ASE Education Foundation Correlation

This job sheet addresses the following **MLR** task: Engine Performance; C. Fuel, Air Induction, and Exhaust Systems

Task #1   Replace fuel filter(s) where applicable. **(P-2)**

This job sheet addresses the following **AST** tasks: Engine Performance; C. Fuel, Air Induction, and Exhaust Systems Diagnosis and Repair

Task #1   Check fuel for contaminants; determine needed action. **(P-2)**

Task #3   Replace fuel filter(s) where applicable. **(P-2)**

This job sheet addresses the following **MAST** tasks: Engine Performance; C. Fuel, Air Induction, and Exhaust Systems Diagnosis and Repair

Task #2   Check fuel for contaminants; determine needed action. **(P-2)**

Task #4   Replace fuel filter(s) where applicable. **(P-2)**

Tools and Materials

- Clean shop rags
- Bleed hose
- Service information
- Pressure gauge with adapters
- Approved gasoline container
- Replacement fuel filter

Describe the vehicle being worked on:

Year _____   Make _____

Model _____   VIN _____

Note: Because electronic fuel injection (EFI) systems have a residual fuel pressure, this pressure must be relieved before disconnecting any fuel system component. Failure to relieve the fuel pressure on electronic fuel injection (EFI) systems prior to fuel system service may result in gasoline spills, serious personal injury, and expensive property damage.

Procedure A: For vehicles with a fuel pressure testing port   **Task Completed**

1. Disconnect the negative battery cable to avoid fuel discharge if an accidental attempt is made to start the engine.   ☐

2. Loosen the fuel tank filler cap to relieve any fuel tank vapor pressure.   ☐

3. Wrap a shop towel around the fuel pressure test port on the fuel rail and remove the dust cap from this valve.   ☐

4. Connect the fuel pressure gauge to the fuel pressure test port on the fuel rail.   ☐

5. Install the bleed hose on the gauge in an approved gasoline container and open the gauge bleed valve to relieve fuel pressure from the system into the gasoline container. Be sure all the fuel in the bleed hose is drained into the gasoline container.   ☐

6. Using a shop towel to catch any remaining fuel in the filter, replace the filter according to service information procedure.   ☐

7. Start the vehicle and check for leaks. (It may take a few seconds for the fuel to fill the filter.) ☐

8. Describe any problems encountered while performing this procedure:

_____

Procedure B

On EFI systems that do not have a fuel pressure test port, follow these steps for fuel system pressure relief and fuel filter replacement:

1. Loosen the fuel tank filler cap to relieve any tank vapor pressure. ☐

2. Remove the fuel pump fuse. ☐

3. Start and run the engine until the fuel is used up in the fuel system and the engine stops. ☐

4. Engage the starter for 3 seconds to relieve any remaining fuel pressure. ☐

5. Disconnect the negative battery terminal to avoid possible fuel discharge if an accidental attempt is made to start the engine. ☐

   Check the fuel that has been drained, does the fuel appear to be clean? Allow the container to sit for a few minutes. Is there any indication that there is water in the fuel? Describe what you found below:

   _____

   _____

6. Using a shop towel to catch any remaining fuel, remove and replace the fuel filter. ☐

7. Start the vehicle and check for any fuel leaks. ☐

8. Describe any problems encountered while following this procedure:

_____

Instructor's Response

_____

_____

_____

Name _____   Date _____

TESTING FUEL PRESSURE ON A RETURN-LINE STYLE EFI SYSTEM

Upon completion of this job sheet, you should be able to test fuel pressure on a vehicle equipped with electronic fuel injection and a fuel return line.

ASE Education Foundation Correlation

This job sheet addresses the following **AST** task: VIII. Engine Performance; D. Fuel, Air Induction, and Exhaust Systems Diagnosis and Repair

Task #2   Inspect and test fuel pumps and pump control systems for pressure, regulation, and volume; perform needed action. **(P-1)**

This job sheet addresses the following **MAST** task: VIII. Engine Performance; D. Fuel, Air Induction, and Exhaust Systems Diagnosis and Repair

Task #1   Diagnose (troubleshoot) hot or cold no-starting, hard starting, poor drivability, incorrect idle speed, poor idle, flooding, hesitation, surging, engine misfire, power loss, stalling, poor mileage, dieseling, and emissions problems; determine needed action. **(P-2)**

Task #3   Inspect and test fuel pumps and pump control systems for pressure, regulation, and volume; perform necessary action. **(P-1)**

Tools and Materials

- Clean shop rags
- Approved gasoline container
- Pressure gauge with adapters
- Hand-operated vacuum pump

Describe the vehicle being worked on:

Year _____   Make _____

Model _____   VIN _____

Engine size and type _____

Fuel pump pressure specifications _____ psi

Procedure   Task Completed

1. Carefully inspect the fuel rail and injectors for signs of leaks. Record findings:

   _____

   _____

2. Connect the fuel pressure tester to the Schrader valve on the fuel rail.   ☐

3. Connect a hand-operated vacuum pump to the fuel pressure regulator.   ☐

4. Turn the ignition on, and observe the fuel pressure readings. Your readings are _____ psi.

5. Compare the readings to specifications. What is indicated by the readings?

   _____

   _____

6. Create a vacuum at the pressure regulator with the vacuum pump.   ☐

7. What happened to the fuel pressure?

    _____

    _____

8. Conclusions:

    _____

    _____

Instructor's Response

_____

_____

_____

Name _____   Date _____

TESTING FUEL PRESSURE ON A RETURNLESS STYLE EFI SYSTEM

Upon completion of this job sheet, you should be able to test fuel pressure on a vehicle equipped with electronic fuel injection and a fuel return line.

ASE Education Foundation Correlation

This job sheet addresses the following **AST** task: VIII. Engine Performance; D. Fuel, Air Induction, and Exhaust Systems Diagnosis and Repair

Task #2   Inspect and test fuel pumps and pump control systems for pressure, regulation, and volume; perform needed action. **(P-1)**

This job sheet addresses the following **MAST** task: VIII. Engine Performance; D. Fuel, Air Induction, and Exhaust Systems Diagnosis and Repair

Task #1   Diagnose (troubleshoot) hot or cold no-starting, hard starting, poor drivability, incorrect idle speed, poor idle, flooding, hesitation, surging, engine misfire, power loss, stalling, poor mileage, dieseling, and emissions problems; determine needed action. **(P-2)**

Task #3   Inspect and test fuel pumps and pump control systems for pressure, regulation, and volume; perform needed action. **(P-1)**

Tools and Materials

- Clean shop rags
- Approved gasoline container
- Pressure gauge with adapters

Describe the vehicle being worked on:

Year _____ Make _____

Model _____ VIN _____

Engine size and type _____

Fuel pump pressure specifications _____ psi

Procedures   **Task Completed**

1. Carefully inspect the fuel rail and injectors for signs of leaks. Record your findings:

   _____

   _____

2. Connect the fuel pressure tester to the Schrader valve on the fuel rail.   ☐

3. Turn the key to run and record the fuel pressure. _____ psi

4. Start the vehicle and observe the fuel pressure readings. Your readings are _____ psi.

5. Did the pressure reading change from key on, engine off to engine running? _____ Yes _____ No

6. Compare the readings to specifications. What do these readings indicate?

   _____

   _____

7. While watching the pressure gauge, snap accelerate the engine.   ☐

8. What happened to the fuel pressure?

_____

_____

9. Conclusions:

_____

_____

Instructor's Response

_____

_____

_____

Name _____   Date _____

CHECKING FUEL FOR CONTAMINANTS

Upon completion of this job sheet, you will be able to check the fuel on a vehicle for contaminants and quality.

ASE Education Foundation Correlation

This job sheet addresses the following **AST** task: VIII. Engine Performance; D. Fuel, Air Induction, and Exhaust Systems Diagnosis and Repair

Task #1   Check fuel for contaminants; determine needed action. **(P-2)**

This job sheet addresses the following **MAST** task: VIII. Engine Performance; D. Fuel, Air Induction, and Exhaust Systems Diagnosis and Repair

Task #2   Check fuel for contaminants; determine needed action. **(P-2)**

Tools and Materials
- Calibrated cylinder
- Protective clothing
- Goggles or safety glasses with side shields

Describe the vehicle being worked on:

Year _____   Make _____

Model _____   VIN _____

Engine type and size _____

Procedure   **Task Completed**

1. Pump gasoline may contain a small amount of alcohol, normally up to 10 percent, unless the vehicle is equipped to burn E-85. However, for most vehicles, if the amount is greater than that, problems may result such as fuel system corrosion, fuel filter plugging, deterioration of fuel system components, and a lean air-fuel ratio. The fuel system problems caused by excessive alcohol in the fuel may result in drivability complaints such as lack of power, acceleration stumbles, engine stalling, and no-start. If the correct amount of fuel is being delivered to the engine and there is evidence of a lean mixture, check for vacuum leaks in the intake, and then check the gasoline's alcohol content. Did you find any vacuum leaks in the system? Where?

   _____

   _____

2. Obtain a 100-milliliter (mL) cylinder graduated in 1-mL divisions.   ☐

3. Fill the cylinder to the 90-mL mark with gasoline.   ☐

4. Add 10 mL of water to the cylinder so it is filled to the 100-mL mark.   ☐

5. Install a stopper in the cylinder and shake it vigorously for 10 to 15 seconds.   ☐

6. Carefully loosen the stopper to relieve any pressure.   ☐

7. Install the stopper and shake vigorously for another 10 to 15 seconds.   ☐

8. Carefully loosen the stopper to relieve any pressure.   ☐

9. Place the cylinder on a level surface for 5 minutes to allow liquid separation.   ☐

10. Any alcohol in the fuel is absorbed by the water and settles to the bottom. If the water   ☐
    content in the bottom of the cylinder exceeds 10 mL, there is alcohol in the fuel.

11. How much alcohol was measured?

_____

12. Did any particles settle to the bottom? If so, describe them.

_____

_____

13. What are your service recommendations?

_____

_____

Problems Encountered

_____

_____

_____

Instructor's Response

_____

_____

_____

CHAPTER 8

ELECTRONIC FUEL INJECTION DIAGNOSIS AND SERVICE

Upon completion and review of this chapter, you should be able to:

- Perform a preliminary diagnostic procedure on a fuel injection system.
- Remove, clean, inspect, and install throttle body assemblies.
- Explain the results of incorrect fuel pressure.
- Perform an injector balance test and determine the injector condition.
- Clean injectors.
- Perform an injector sound test.
- Perform an injector ohmmeter test.
- Perform an injector noid light test.
- Perform an injector flow test and determine the injector condition.

- Perform an injector leakage test.
- Check for leakage in the fuel pump check valve and the pressure regulator valve.
- Remove and replace the fuel rail, injectors, and pressure regulator.
- Perform a minimum idle speed adjustment.
- Diagnose causes of improper idle speed on vehicles with an idle air control motor.
- Diagnose idle air control motors and idle air control bypass solenoids.
- Remove, replace, and clean idle air control bypass air motors and related throttle body passages.
- Diagnose idle air control bypass air valves.

Terms To Know

14.7:1 air-fuel ratio
Actuation test mode (ATM)
Air-fuel (A/F) ratio sensor
Bore
Cam
Data link connector (DLC)
Direct fuel injection (DFI)

Electronic fuel injection (EFI)
Feedback
Fuel
Fuel control
Idle air control (IAC)
Idle air control bypass air (IAC BPA)

Long-term fuel trim (LTFT)
Peak and hold injector circuits
Pulse width modulated injector circuit
Stethoscope
Stoichiometry
Variable valve timing

INTRODUCTION

Although fuel injection technology has been around since the 1920s, it was not until the 1980s that manufacturers began to replace carburetors with **electronic fuel injection (EFI)** systems. Many of the early domestic EFI systems were **throttle body injection (TBI)** systems in which the fuel was injected above the throttle plates. Engines equipped with TBI have gradually become equipped with port fuel injection (PFI) systems, which have injectors located in the intake ports of the cylinders. The latest development is **direct fuel injection (DFI)**. Direct fuel injection is predicted to gradually phase out PFI. Since the 1995 model year, all new cars are equipped with an EFI system.

Fuel injection systems are at the heart of engine performance, and good engine performance, fuel economy, and emissions depend on keeping many interrelated systems working together: as shown in **Figure 8-1**. A misfiring engine can have several possible causes; low compression, vacuum leak, open ignition coil, fouled spark plug, injector that has failed electrically or mechanically, or an electrical problem in the ignition of fuel injector circuits. It is the job of the technician to quickly and efficiently find the root cause of

Classroom Manual
Chapter 8, page 221

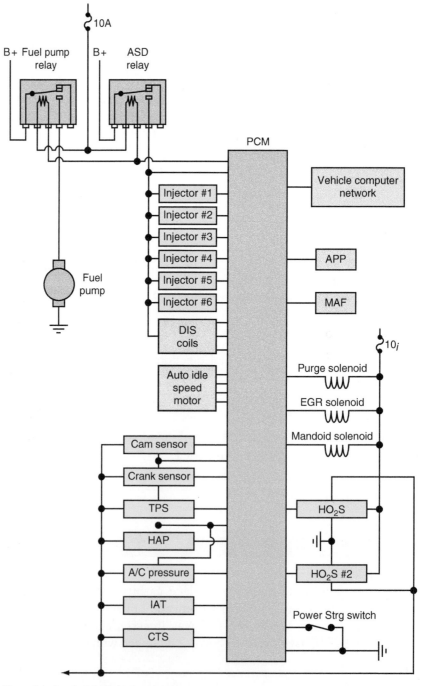

Figure 8-1 A typical fuel injection system.

the problem. Effective diagnosis is worth the extra time and effort to fix the vehicle right the first time. Use your training to check out systems instead of "trying" parts to eliminate problems. A good technician will use all means at their disposal to help them identify the problem. Knowing how to use a vacuum gauge or a multimeter and your experience is just as important as knowing how to use a scan tool. In the case of an engine miss, the malfunction indicator lamp can tell the driver that there is a code stored in the ECM. The code will determine which cylinder is missing, and in the case of an electrical problem in the fuel or ignition circuit may even give a hint as to the location of the problem. The trouble code does not tell the technician the cause of the problem, only the approximate location. The technician might use a multimeter to determine that a fuel injector is open electrically. Of course, the secret to diagnosis is not to instantly know what the problem is, but to understand how the engine systems work, and to put the knowledge to work diagnosing engine performance problems for your customers.

PRELIMINARY CHECKS

The best way to approach a problem on a vehicle with electronic fuel injection is to treat it as though it had no electronic controls at all. Any engine is susceptible to problems that are unrelated to the fuel system itself. Unless all engine support systems are operating correctly, the control system does not operate as designed.

Before proceeding with specific fuel injection checks and electronic control testing, be certain of the following:

- The air cleaner and related components of the air intake system function properly.
- The battery is in good condition and fully charged, with clean terminals and connections.
- The charging and starting systems are operating properly.
- All fuses and fusible links are intact.
- All wiring harnesses are properly routed with connections free of corrosion and are tightly attached.
- All vacuum lines are in sound condition, properly routed, and tightly attached.
- The PCV system is working properly and maintaining a sealed crankcase.
- All emission control systems are in place, hooked up, and operating properly.
- The level and condition of the coolant-antifreeze are good, and the thermostat is opening at the proper temperature.
- The secondary spark delivery components are in good shape with no signs of cross-firing, carbon tracking, corrosion, or wear.
- The base timing and idle speed are set to specifications.
- The engine is in good mechanical condition and has acceptable compression.
- The gasoline in the tank is of good quality and has not been substantially cut with alcohol or contaminated with water.
- The engine is at its normal operating temperature and the cooling system works normally.
- Prior to testing, make sure all accessories are turned off.

Assuring that all basic systems are OK will save many hours of frustration.

Service Precautions

These precautions must be observed when electronic fuel injection systems are diagnosed and serviced:

1. Always relieve the fuel pressure before disconnecting any component in the fuel system (**Figure 8-2**).
2. Never turn the ignition switch on when any fuel system component is disconnected.
3. Use only the test equipment recommended by the vehicle manufacturer.

Direct injection systems have dangerously high fuel pressures.

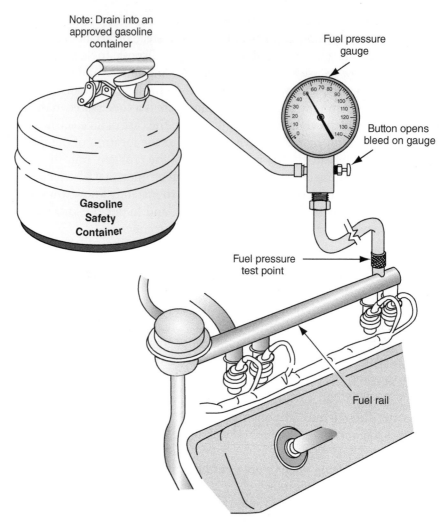

Note: Drain into an
approved gasoline
container

Fuel pressure
gauge

Button opens
bleed on gauge

Gasoline
Safety
Container

Fuel pressure
test point

Fuel rail

Figure 8-2 Relieving fuel pressure from the port fuel injection system.

4. Always turn the ignition switch off before connecting or disconnecting any system component or test equipment.
5. When arc welding is necessary on a computer-equipped vehicle, disconnect both battery cables before welding is started. Always disconnect the negative cable first.
6. Never allow electrical system voltage to exceed 16 volts. This could be done by disconnecting the circuit between the alternator and the battery with the engine running.
7. Avoid static electric discharges when handling computers and modules.

Disconnecting Battery Cables

While servicing electronic fuel injection systems, many procedures indicate the removal of the negative battery cable or both battery cables. If the negative battery cable is disconnected during diagnostic and service procedures, it has these effects:

1. Deprograms the radio.
2. Deprograms other convenience items such as memory seats or mirrors.
3. Erases the trip odometer if the vehicle has digital instrumentation.
4. Erases the adaptive strategy in the computer.

If the adaptive strategy in the computer is erased, engine operation may be rough at low speeds when the engine is restarted simply because the computer must relearn the

If a back up dry cell battery is used to prevent deprogramming, make sure that the air bags are disabled. Some manufacturers are using nonvolatile RAM in many of the areas described, eliminating the need for the memory saver in some cases.

computer system defects. Under this condition, the vehicle should be driven for 5 minutes with the engine at normal operating temperature. Some technicians will attach a "memory saver," which is basically a 9-volt battery. The voltage supplied by the dry cell prevents deprogramming and memory erasing. Some voltage sources designed for this purpose plug into the power circuit socket. There are backup power supplies designed to be attached to the sixteen pin DLC. It is important to note that the air bag modules may be active with these backup systems, and that some manufacturers do not recommend their use.

BASIC EFI SYSTEM CHECKS

As you have already seen in the book, modern fuel injection systems continually check their own operation through a series of self-tests. These self-tests perform a number of checks on components within the system. Input sensors, output devices, wiring harnesses, and even the electronic control computer itself may be among the items tested. Always remember that trouble codes indicate only the particular circuit in which a problem has been detected. In most cases, they do not pinpoint the exact component or failed connection. Therefore, if a code indicates an oxygen sensor fault, the problem could be a fuel control problem, engine problem, the sensor itself, the wiring to it, or its connector. Trouble codes are not a signal to replace components. They signal that a more thorough diagnosis is needed in that area.

In any electronic fuel injection system, five things must occur for the system to operate:

1. An adequate air supply must be supplied for the air-fuel mixture.
2. A pressurized fuel supply must be delivered to properly operating injectors.
3. The injectors must receive a trigger signal from the control computer.
4. There must be adequate spark at the right time to ignite the fuel.
5. There must be adequate compression to be able to burn the fuel.

> **AUTHOR'S NOTE** An old technician's saying is that in order to start and run, the engine must have the "FACS":
> Fuel
> Air
> Compression
> Spark

The following sections cover the testing of common EFI components that are involved with all of these things.

> **CUSTOMER CARE** Electric in-tank fuel pumps can be labor intensive to replace. When replacing a fuel pump, the fuel pump strainer should be replaced as well. The customer may also need to be advised of any debris or contamination within the fuel tank, which warrants cleaning and replacing the fuel filter. Otherwise, there is the risk that the vehicle will return with the same problem.

A common misconception is that the oxygen sensor can detect fuel in the exhaust, but it can only detect oxygen levels that correspond to a rich or lean mixture.

Special Tools

Lab scope
Scan tool

Classroom Manual
Chapter 8, page 230

Oxygen Sensor Diagnosis

Oxygen sensors produce a voltage based on the amount of oxygen (O_2) in the exhaust. Large amounts of oxygen result from lean mixtures and produce a low voltage output from the O_2 sensor. Rich mixtures have released lower amounts of oxygen in the exhaust. Therefore, the O_2 sensor voltage is high. The engine must be at normal operating temperature before the O_2 sensor is tested.

An important thing to remember is that the oxygen sensor is reporting on a rich or lean condition. For instance, high fuel pressure could result in a rich mixture, and low fuel pressure could result in a lean mixture. A coolant temperature sensor that is reading incorrectly might cause the computer to add too much or too little fuel. Check the basics out before condemning the oxygen sensor itself.

An O_2 sensor can be checked with a voltmeter, scan tool, or oscilloscope.

Connect the voltmeter between the O_2 sensor signal wire and ground. The sensor's voltage should be cycling from low voltage to high voltage after warming up for a few minutes. The signal from most O_2 sensors varies between 0 and 1 volt, although some Chrysler vehicles operate between 2 and 3 volts. If the voltage is continually high, the air-fuel ratio (AFR) may be rich or the sensor may be contaminated. When the O_2 sensor voltage is continually low, the air-fuel ratio may be lean, the sensor may be defective, or the wire between the sensor and the computer may have a high-resistance problem. If the O_2 sensor voltage signal remains in a mid-range position, the computer may be in open loop or the sensor may be defective.

The activity of the sensor can be monitored on a scanner. By watching the scanner while the engine is running, the O_2 voltage should move to nearly 1 volt than drop back to close to 0 volt. Immediately after it drops, the voltage signal should move back up. This immediate cycling is an important function of an O_2 sensor. If the response is slow, the sensor is lazy and should be replaced. With the engine at about 2,500 rpm, the O_2 sensor should cycle from high to low 10 to 40 times within a few seconds. When testing the O_2 sensor, make sure the sensor is heated and the system is in closed loop. The activity of an O_2 sensor is best monitored with a lab scope (**Figure 8-3**).

A scan tool is also an excellent way to monitor the fuel control of a fuel-injected engine. There are many factors that determine the pulse width of the injectors, but it should always respond to the O_2 readings. Fuel correction for fuel injection systems is shown on a scan tool.

Fuel Trims

Short-term fuel trim (STFT) is the short-term fuel correction. STFT numbers will vary with engine operating conditions. When there is no correction, the STFT number will be near zero. If fuel is being added, the numbers climb positively. If **fuel** is being subtracted, the numbers will go negative. It is also important to note that the short-term fuel trims will be erased every time the engine is shut off.

Long-term fuel trim (LTFT) is a long-term number for fuel trim. The numbers will rise and fall, but not immediately. Instead, these numbers reflect long-term trends. For instance, if a vacuum line was leaking or there was any condition that would cause a long-term lean condition, the PCM would see lower oxygen sensor readings. The PCM would add fuel to try and bring the oxygen sensor voltage back into **stoichiometry**, or **14.7:1 air-fuel ratio**. This shift would be reflected as positive numbers on the LTFT. On the other hand, if for instance there was a condition that would cause a long-term rich condition, like an injector leaking internally or fuel pressure that is too high, the numbers would go negative, showing that the PCM was subtracting fuel from the engine in an effort to bring the air-fuel mixture into stoichiometry. The LTFT will change in an attempt to keep the STFT near 0. The STFT will move around while driving to

Normal operation

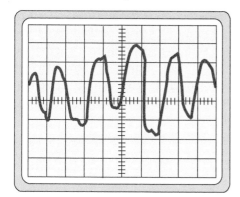

Lack of activity - DTC PO134

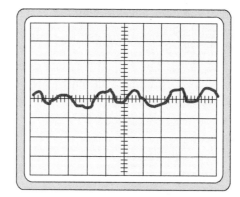

Lean too long - DTC PO171

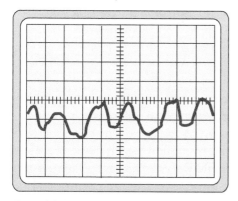

Rich too long - DTC PO172

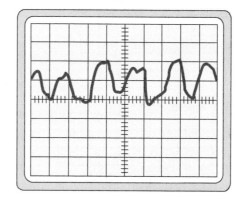

Figure 8-3 Normal and abnormal O_2 sensor waveforms.

correct for temporary changes. But if something happens that affects fuel trim, like a vacuum leak, the STFT will go positive to add fuel; if it stays there for more than a few seconds, the LTFT would kick in and add fuel to make the STFT hover around zero again. This action would show up on the scan tool reading as an LTFT correction. Long-term fuel trim is kept in memory by the PCM and is not lost until the battery was disconnected.

Positive fuel trim numbers mean that the PCM is adding fuel to a mixture that is too lean. Negative fuel trim numbers mean that the PCM is subtracting fuel from a mixture that is too rich.

The PCM will add and subtract fuel from its programmed delivery during its regular operation. This adding and subtracting of fuel are called **fuel control**. The PCM can add or subtract only so much fuel without setting a DTC. If the PCM reaches its programmed limit and is still lean, then a lean mixture code will set. If the PCM has subtracted its programmed limit and the oxygen sensor still shows a rich condition, then a rich code will set. When this condition has occurred, the PCM is said to be out of fuel control.

A technician can benefit from watching the fuel trim numbers. For instance, if the engine hesitates on acceleration, watch the STFT for changes. If it does not change, watch for a change in the throttle position, MAF, or MAP numbers. If a customer complains of poor fuel mileage, look at the LTFT numbers to see if the PCM has been adding or subtracting an excessive amount of fuel.

Modern oxygen sensors use a heater circuit to quickly place the fuel control system in a closed loop. The PCM also monitors this circuit for failures.

Short-term and long-term fuel trims are values the PCM uses for fuel correction.

Short-term fuel trim shows a temporary change in fuel mixtures. Long-term fuel trim is the long-term trend toward correcting a rich or lean condition.

When the PCM is able to control the fuel mixture effectively, it is called **fuel control**.

Air-Fuel Ratio Sensor

One of the later developments in exhaust oxygen detection is the **air-fuel (A/F) ratio sensor** (also called the AF ratio sensor, wide band sensor, AFR sensor or Universal Exhaust Gas Oxygen Sensor (UEGO)). The normal oxygen sensor reads only in a narrow range around 14.7:1. The A/F ratio sensor does a better job of determining just how rich or lean the engine is running. This allows for better fuel control. The PCM can better adjust the mixture over a wider range of conditions not possible with the normal oxygen sensor.

AUTHOR'S NOTE Some technicians call the zirconia oxygen sensor a *narrow band oxygen sensor.*

The oxygen sensor and the A/F sensor both have the same job: telling the computer if the mixture is lean or rich, which allows the computer to take steps to correct the mixture. The AF sensor has different operating characteristics that make it more accurate. The old oxygen sensor could tell the computer *if* the mixture was rich or lean of 14.7:1, whereas the A/F sensor can tell the computer *how* rich or lean the mixture is running: for instance, 17:1 (lean) and 9:1 (rich) is more useful information than simply "rich" or "lean."

The A/F ratio sensor is similar in appearance to the oxygen sensor, with some important differences. The A/F sensor runs at a much hotter 1,200°F. The A/F sensor also alters current output as oxygen content in the exhaust varies. At 14.7:1, there is no current flow from the A/F sensor. A rich mixture produces a negative current flow which measures below 3.3 volts, and a lean mixture produces a positive current flow of over 3.3 volts. Since the current flow is very small, around 1.5 mA or less, the best way to check the A/F sensor is with the aid of a scan tool, rather than direct measurement. The A/F sensor does not swing rich and lean around 0.44 volt as the oxygen sensor does. The scan tool may refer to the A/F sensor or oxygen sensor as a *lambda sensor.* Some manufacturers claim that the A/F sensor cannot be defective without an associated code for the sensor. This implies that the A/F sensor is better able to detect downgraded performance than the older oxygen sensor.

Manufacturers refer to the A/F sensor by many different names; oxygen sensor, HO_2S for heated oxygen sensor, lambda sensor, wide-band oxygen sensor, WRAF for wide-range oxygen sensor, or as a universal oxygen sensor (UGO). J1930 standards for naming refer to the sensor as an *A/F sensor.*

AIR SYSTEM CHECKS

In an injection system, particularly designs that rely on mass airflow (MAF) sensors, all the air entering the engine must be accounted for by the air-measuring device. If it is not, the air-fuel ratio cannot be correct. For this reason, cracks or tears in the plumbing between the airflow sensor and throttle body and other potential air leak sources can affect the air-fuel ratio.

During a visual inspection of the air control system, pay close attention to these areas, looking for cracked or deteriorated ductwork. Also, make sure all induction hose clamps are tight and properly sealed. Look for possible air leaks in the crankcase, gaskets, dipstick tube, and oil filter cap. Any extra air entering the intake manifold through the PCV system can upset the delicately balanced air-fuel mixture and cause poor idle and surging at cruise. What follows are brief discussions on the diagnosis and repair of EFI components related to the air delivery system.

On systems that use a mass airflow meter or manifold pressure sensor to measure airflow, inspect the condition of the air cleaner duct, and check for damaged vacuum lines. Any air that enters the engine without being measured by the MAF can create problems because the computer has problems calculating the amount of fuel needed. If the MAF sensor is suspected of problems, try turning the key off and unplugging the MAF sensor, and see if the condition improves, if it does improve, the MAF sensor could be at fault. Inspect the MAF hot wire for contamination; any substance on the hot wire will cause erroneous readings by insulating the wire. The best diagnostic clues are from a scan tool plugged into the **data link connector (DLC)** (**Figure 8-4**) to check for proper values.

The MAF sensor converts the incoming airflow to grams per second (gps) or a frequency in hertz that the PCM uses for computing fuel and timing requirements. The MAF sensor can be accurately tested by observing the gps values using a scan tool at various engine speeds. Check the readings at idle through 2,500 rpm in 500-rpm increments. The gps value should increase proportionately to the engine rpm without any drastic or sudden jumps in the number. Some manufacturers have MAF gps or frequency values specified in their service information.

Make sure to clean the air filter housing whenever you replace an air filter. Any small particles of grass, leaves, and so on will cause problems on the MAF sensor.

Fuel pressure should be relieved prior to disconnecting fuel system components.

Make sure to determine if the MAF on the vehicle you are testing outputs a frequency or analog signal to the ECM before testing to make sure you are looking for the correct signal.

AUTHOR'S NOTE A vehicle was brought into the shop for a clutch replacement. After the clutch was installed, the technician test drove the vehicle. He came back very concerned that the vehicle had a very violent shutter on hard acceleration. He was sure that the clutch was causing the problem and he would have to remove the transmission again. On further inspection, the MAF ducting was damaged on the bottom and a large gap opened on acceleration when the engine moved. When the duct and the front engine mount were replaced, the vehicle ran fine.

Throttle Body

The throttle body (**Figure 8-5**) controls the amount of air that enters the engine on command from the driver, thereby controlling the speed of the engine. Each type of throttle body assembly is designed to allow a certain amount of air to pass through it at a

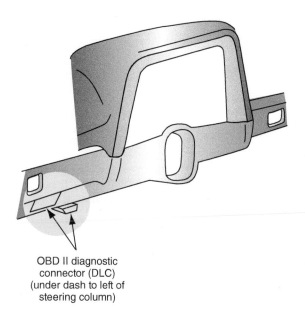

OBD II diagnostic connector (DLC) (under dash to left of steering column)

Figure 8-4 Data link connector (DLC) locations.

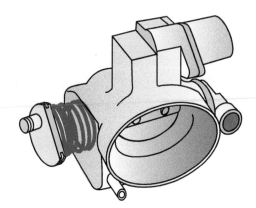

Figure 8-5 A throttle body assembly.

particular amount of throttle opening. If accumulates on the throttle plates or in the throttle **bore**, especially with PFI systems, the amount of air that can pass through is reduced. This will cause stalling on deceleration or idle.

These deposits can be cleaned off the throttle assembly and the airflow through them restored. Begin by removing the air duct from the throttle assembly. This gives access to the plate and bore. The deposits can be cleaned (some manufacturers do not recommend cleaning) with a spray cleaner or wiped off with a cloth. If either of these cleaning methods do not remove the deposits, the throttle body should be removed, disassembled, and placed in an approved cleaning solution.

A pressurized can of throttle body cleaner may be used to spray around the throttle area without removing and disassembling the throttle body. The throttle assembly can also be cleaned by soaking a cloth in carburetor solvent and wiping the bore and throttle plate to remove light to moderate amounts of carbon residue. Also, clean the backside of the throttle plate. Then, on vehicles with idle air control (IAC) remove the idle air control valve from the throttle body (if so equipped) and clean any carbon deposits from the pintle tip and the **idle air control (IAC)** air passage.

Throttle Body Inspection

Throttle body inspection and service procedures vary widely depending on the year and make of the vehicle. However, some components, such as the throttle position (TP) sensor, are found on nearly all throttle bodies. Since throttle bodies have some common components, inspection procedures often involve checking common components. The following throttle body service procedure example is typical:

1. (If equipped with a throttle cable) Check for smooth movement of the throttle linkage from the idle position to the wide-open position. Check the throttle linkage and cable for wear and looseness.
2. Check the vacuum at each vacuum port on the throttle body while the engine is idling and running at a higher speed (**Figure 8-6**).
3. Test the TP sensor output with a voltmeter connected across the appropriate terminals or with a scan tool.
4. The TP sensor is not adjustable on late-model vehicles. The lowest number that is registered in the PCM is taken at zero.
5. Depending on the vehicle, you may need to "relearn" the idle speed; or, on vehicles with IAC, check for minimum air (idle speed) that is set with the IAC at its closed position.

   Most vehicles will not have an adjustable idle or TP sensor.
6. Test drive the vehicle.

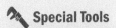
Throttle Body Removal and Cleaning

Follow these steps for throttle body removal:

1. Disconnect the negative battery cable. If the vehicle is equipped with an air bag, wait for 1 minute.
2. Drain the engine coolant from the radiator.
3. Disconnect the air intake temperature sensor connector (if necessary).
4. Loosen the air cleaner hose clamp bolt at the throttle body and disconnect the air cleaner cap clips.
5. Disconnect the air cleaner hose from the throttle body and remove the air cleaner cap, air hose, and resonator (**Figure 8-7**).
6. Disconnect the TP sensor wiring connector.
7. Disconnect the idle air control connector.

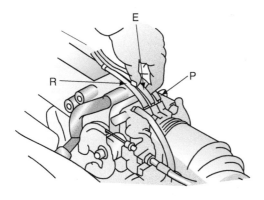

Port name	At idle	Other than idle
P	No vacuum	Vacuum
E	No vacuum	Vacuum
R	No vacuum	No vacuum

Figure 8-6 Throttle body vacuum ports and the specified vacuum at different throttle positions.

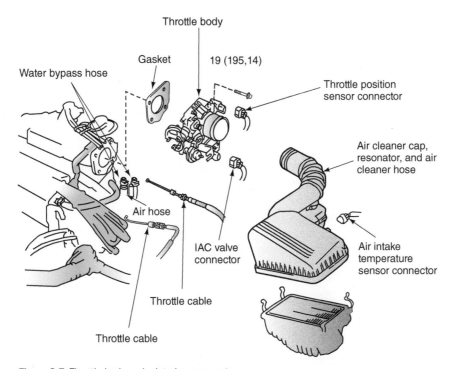

Figure 8-7 Throttle body and related components.

8. Remove the vacuum hoses from the throttle body and note the position of each hose so they may be installed in the same location.
9. Remove the throttle body mounting bolts, and then remove the throttle body and gasket from the intake manifold (**Figure 8-8**).
10. Disconnect the water bypass hoses and air hose from the throttle body. Note the position of each hose so they can be reconnected properly.

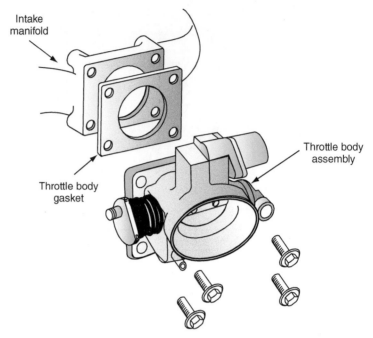

Figure 8-8 Unbolting the throttle body from the manifold.

11. Remove all nonmetallic parts such as the TP sensor, IAC valve, throttle opener, and the throttle body gasket from the throttle body.
12. Clean the throttle body assembly in the recommended throttle body cleaner and blow dry with compressed air. Blow out all passages in the throttle body assembly.

To reinstall the throttle body assembly:

1. Make sure all metal mating surfaces are clean and free from metal burrs and scratches. Install a new IAC valve gasket and install the IAC valve (**Figure 8-9**). Tighten the valve mounting screws to the proper torque (**Figure 8-10**).
2. Install the water bypass hoses and the air hose in their original locations on the throttle body. Be sure the hose clamps are tight.
3. Install a new throttle body gasket.

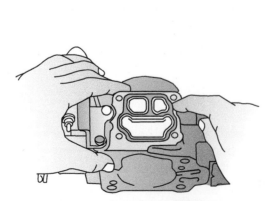

Figure 8-9 Installing a new IAC valve gasket.

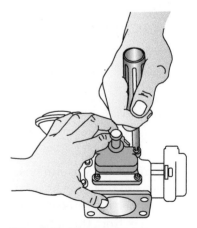

Figure 8-10 IAC valve retaining screws must be tightened evenly.

4. Install the throttle body attaching bolts, and tighten these bolts to the specified torque.
5. Install the vacuum hoses on the throttle body in their original location.
6. Connect the IAC valve and TP sensor wiring connectors.
7. Check the air cleaner element. Replace it if necessary. Inspect the air cleaner box, cap, hose, and resonator for cracks and distortion. Remove any debris from the air cleaner box. Connect the air cleaner hose and tighten the hose clamp. Install the air cleaner cap and the retaining clamps.
8. Connect the air intake temperature sensor connector.
9. Replace the engine coolant.
10. Connect the negative battery cable.

> **⚙ SERVICE TIP** Be sure to consult the service information for the vehicle being serviced. Not all procedures are the same.

Throttle Actuator Control

The throttle actuator control (TAC) system has a few unique features because it is a *throttle by wire system*. The accelerator pedal position (APP) sensor sends the input from the driver to the ECM, the ECM sends the signal to the throttle body module, and the throttle plate is moved according to the driver's input. Of course, since the throttle is a very important input, if the ECM sees any variation from normal in the throttle position or pedal position sensor signal, the vehicle power and speed are limited to around 25 mph or totally eliminated. The driver may see a "Reduced power" message or similar message along with an illuminated MIL. Several trouble codes are associated with the TAC system; a few of them are shown in **Figure 8-11**. **Figure 8-12** shows the correct readings from a scan tool: TP sensor 1 starts out at 0 volt and increases with increasing throttle angle, and TP sensor 2 starts out at 5 volts and goes to 0 volt at full throttle. **Figure 8-13** shows a basic schematic diagram for the system.

P0120	Throttle/Pedal Position Sensor "A" Circuit
P0121	Throttle/Pedal Position Sensor "A" Circuit Range/Performance
P0122	Throttle/Pedal Position Sensor "A" Circuit Low
P0123	Throttle/Pedal Position Sensor "A" Circuit High
P0124	Throttle/Pedal Position Sensor "A" Intermittent
P0220	Throttle/Pedal Position Sensor/Switch "B" Circuit
P0221	Throttle/Pedal Position Sensor/Switch "B" Circuit Range/Performance
P0222	Throttle/Pedal Position Sensor/Switch "B" Circuit Low
P0223	Throttle/Pedal Position Sensor/Switch "B" Circuit High
P0224	Throttle/Pedal Position Sensor/Switch "B" Circuit Intermittent

Figure 8-11 Trouble codes for throttle/pedal position sensor circuits A and B.

Throttle Angle %	TP Sensor 1	TP Sensor 2
0	0.24	4.76
20	.95	4.05
40	1.9	3.1
60	2.86	2.14
80	3.81	1.19
100	4.76	0.24

Figure 8-12 Voltage measurements of the TP sensors 1 and 2 from the scan tool.

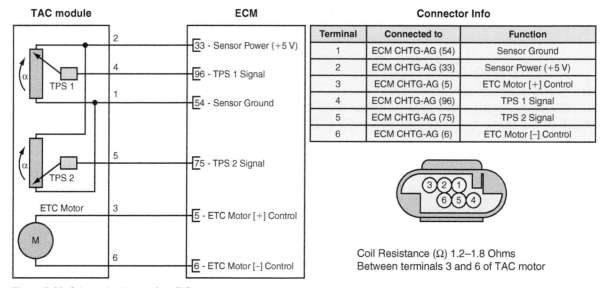

Terminal	Connected to	Function
1	ECM CHTG-AG (54)	Sensor Ground
2	ECM CHTG-AG (33)	Sensor Power (+5 V)
3	ECM CHTG-AG (5)	ETC Motor [+] Control
4	ECM CHTG-AG (96)	TPS 1 Signal
5	ECM CHTG-AG (75)	TPS 2 Signal
6	ECM CHTG-AG (6)	ETC Motor [–] Control

Coil Resistance (Ω) 1.2–1.8 Ohms
Between terminals 3 and 6 of TAC motor

Figure 8-13 Schematic diagram for a TAC system.

DIAGNOSTICS FOR A P0122

Definition of Code

Voltage for circuit "A" is lower than possible for a properly operating circuit. Circuit A is reading below approximately 0.012 volt for 0.005 second.

1. Connect scan tool and observe values for TP sensors 1 and 2. Sensor 1 starts at a low voltage and goes high; sensor 2 starts high and goes low (see Figure 8-12).
    a. If the readings are correct, the code is intermittent. The technician will have to look at the freeze-frame data stored in the ECM and drive the vehicle under the same conditions to duplicate the condition before diagnosis is possible.

2. If the readings are incorrect, then more investigation is in order. The problem could be the TP sensor, the ECM, or the wiring connecting the ECM and the TAC module on the throttle body.

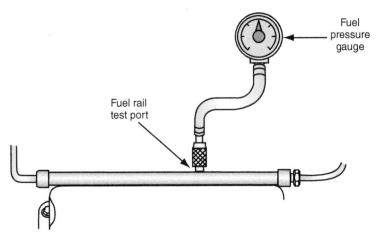

Figure 8-14 Checking the fuel pressure at the fuel rail.

3. Looking at the schematic and connector diagram in **Figure 8-14** can tell us the diagnostic clues needed to repair the vehicle. Notice that Sensors 1 and 2 share ground and power leads from the ECM.

   a. If the ground wire were open, then both sensors would read 5 volts.
   b. If the shared power circuit were open, then the voltage reading for both sensors would be zero.
   c. If the power lead were shorted to battery, the voltage reading would be more than 5 volts.

4. Disconnect the TP sensor 1 connection and jumper harness connections 2 and 4 with the key on; check the reading on the scan tool or pin out terminal 96 of the ECM. If 5 volts is present as a signal for TP sensor 1, replace the TP sensor and retest. If 5 volts is not present, check the wire from terminal 4 of the TP sensor to terminal 96 of the ECM for continuity. If the wiring is showing continuity, then replace the ECM and retest.

   Always use the correct diagnosis for the vehicle you are working on.
5. If the TP sensor, TAC module, or ECM is replaced, then the TP sensor adaptation will have to be relearned.
6. After repairs are complete, test drive the vehicle and make sure no codes or pending codes are present in the PCM.

FUEL DELIVERY

Port Fuel Injection Fuel Pressure Testing

When dealing with an apparent fuel complaint that is preventing the vehicle from starting, one of the first steps is to determine if fuel is reaching the cylinders (assuming there is gasoline in the tank).

Classroom Manual
Chapter 8, page 235

It is difficult to visually inspect the spray pattern and volume of port fuel system injectors. However, an accurate indication of fuel delivery to the fuel rail can be obtained by performing simple fuel pressure and fuel volume delivery tests (Figure 8-14).

Low fuel pressure can cause a no-start or poor-run problem. It can be caused by a clogged fuel filter, a faulty pressure regulator, or a restricted fuel line anywhere from the fuel tank to the fuel filter connection. For a fuel system without a return

If there is no fuel pump pressure, always check the inertia switch, fuse, or fuse link first.

line, the procedure is similar. But there is not normally an inline fuel filter to check for blockage. If the fuel system has a fuel pressure sensor, the pressure can be read on a scan tool.

If a fuel volume test shows low fuel volume, it can indicate a bad fuel pump or blocked or restricted fuel line or pickup screen in the fuel tank. When performing the test, visually inspect the fuel for signs of dirt or moisture. If debris is found in the fuel, it will be necessary to drain and clean the tank and replace all fuel filters.

High fuel pressure readings will result in a rich-running engine and/or rich mixture trouble codes.

INJECTOR CHECKS

A fuel injector is nothing more than a solenoid-actuated fuel valve. Its operation is quite basic in that as long as it is held open and the fuel pressure remains steady, it delivers fuel until it is told to stop.

Because all fuel injectors operate in a similar manner, fuel injector problems tend to exhibit the same failure characteristics.

An injector that does not open causes hard starts on port fuel systems. An injector that is struck partially open causes loss of fuel pressure (most noticeably after the engine is stopped and restarted within a short time period) and flooding due to raw fuel dribbling into the engine. In addition to a rich-running engine, a leaking injector may also cause the engine to diesel or run on when the ignition is turned off. Buildups of gum and other deposits on the tip of an injector can reduce the amount of fuel sprayed by the injector or they can prevent the injector from totally sealing, allowing it to leak. Injectors on port fuel systems are not as subject to heat as the DFI injectors, which require special seals and open with high voltages because of the fact that the tips are actually inside the combustion chambers.

Because an injector adds the fuel part to the air-fuel mixture, any defect in the fuel injection system will cause the mixture to go rich or lean. If the mixture is too rich and the PCM is in control of the air-fuel ratio, a common cause is that one or more injectors are leaking. An easy way to verify this on a fuel-injected engine of any type is with the LTFT on the scan tool. If the LTFT is a positive number, the PCM is adding extra fuel to make up for a lean *condition*. If the LTFT numbers are in the negative numbers, then the PCM is subtracting fuel to make up for a rich *condition*.

It is also important to remember that the fuel trim numbers will vary while driving, and they are seldom at "0" or perfect fuel trim. The system is meant to make up for slight differences between fuel systems, and the effects of wear on the engine.

Diagnosing a Leaking Injector on a Port Fuel Vehicle

With the engine warmed up, turned on, but not running, remove the air duct from the airflow sensor. Insert the gas analyzer's probe into the intake plenum area. Be careful not to damage the airflow sensor or throttle plates while doing this. Look at the hydrocarbon (HC) readings on the analyzer. They should be low and drop as time passes. If an injector is leaking, the HC reading will be high and will not drop. This test does not locate the bad injector but does verify that one or more are leaking.

Diagnosing a Port-Fuel Return Line Style Pressure Regulator

Another cause of a rich mixture is a leaking fuel pressure regulator. If the diaphragm of the regulator is ruptured, fuel will move to the intake manifold through the diaphragm, causing a rich mixture. The regulator can be checked by using two simple tests. After

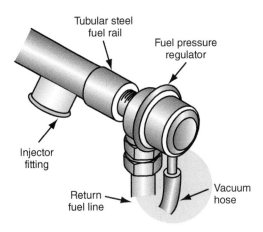

Figure 8-15 Location of a fuel pressure regulator on a fuel rail.

the engine has been run, disconnect the vacuum line to the fuel pressure regulator (**Figure 8-15**). If there are signs of fuel inside the hose or if fuel comes out of the hose, the regulator's diaphragm is leaking. The regulator can also be tested with a hand-operated vacuum pump. Apply 5 in. Hg to the regulator. A good regulator diaphragm will hold that vacuum.

Checking Voltage Signals

When an injector is suspected as the cause of a no-start problem, the first step is to determine if the injector is receiving a signal from the PCM to fire. Fortunately, determining if the injector is receiving a voltage signal is easy and requires simple test equipment. Unfortunately, the location of the injector's electrical connector can make this simple voltage check somewhat difficult.

Once the injector's electrical connector (**Figure 8-16**) has been removed, check for voltage at the injector using an ordinary test light or a convenient noid light that plugs into the connector. After making the test connections, crank the engine. The noid light flashes if the computer is cycling the injector on and off. If the light is not flashing, the

Special Tools
Noid light
Test light

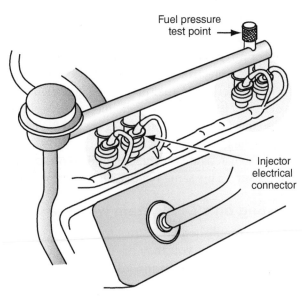

Figure 8-16 Typical location of fuel injector electrical connections.

Ohmmeter

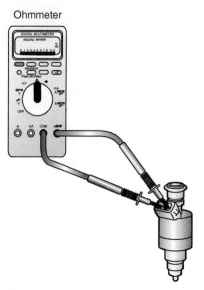

Figure 8-17 An ohmmeter can be connected across the injector's terminals to check injector resistance.

computer or connecting wires are defective. If sufficient voltage is present after checking each injector, check the electrical integrity of the injectors themselves.

An ohmmeter can be used to test the electrical soundness of an injector. Connect the ohmmeter across the injector terminals (**Figure 8-17**) after the wires to the injector have been disconnected. If the meter reading is infinity, the injector winding is open. If the meter shows more resistance than the specifications call for, there is high resistance in the winding. A reading that is lower than the specifications indicates that the winding is shorted. If the injector is even a little bit out of specifications (e.g., as little as 1 ohm), it must be replaced. Refer to the service manual for injector specifications.

The technician should keep in mind that if you check resistance on a hot engine, the resistance will fall as the engine cools down. If it takes a while to do the injector test, the first injectors tested will have more resistance than those tested later.

Port Fuel Injector Balance Test

If the injectors are electrically sound, perform an injector pressure balance test. This test will help isolate a clogged or dirty injector. **Photo Sequence 16** shows a typical procedure for testing injector balance. An electronic injector pulse tester is used for this test. Each injector is energized while observing a fuel pressure gauge to monitor the drop in fuel pressure. The tester is designed to safely pulse each injector for a controlled length of time. The tester is connected to one injector at a time (**Figure 8-18**). The ignition is turned on until a maximum reading is on the pressure gauge. That reading is recorded, and the ignition turned off. With the tester, activate the injector and record the pressure reading after the needle has stopped pulsing. This same relative test is performed on each injector.

Injector Balance Testing on a DFI System with Scan Tool

Injector balance testing is accomplished on a direct fuel system in much the same manner as a port fuel system; the only port you may have to measure fuel might be the low-pressure pump side, so connect a scan tool and take a reading from the fuel rail pressure sensor (FRPS). The scan tool can activate the fuel pump and the appropriate injector. The drop can be read directly off the scan tool. The injectors should be within 20 percent of each other.

Figure 8-18 Electrical setup for performing an injector balance test.

PHOTO SEQUENCE 16
Typical Procedure for Testing Injector Balance

P16-1 Connect the fuel pressure gauge to the Schrader valve on the fuel rail, and then relieve the pressure in the system.

P16-2 Disconnect the number 1 injector and connect the injector pulse tester to the injector's terminals.

P16-3 Connect the injector pulse tester's power supply leads to the battery.

P16-4 Cycle the ignition switch several times until the system pressure is at the specified level.

P16-5 Push the injector pulse tester switch and record the pressure on the pressure gauge. Subtract this reading from the measured system pressure. The answer is the pressure drop across that injector.

P16-6 Move the injector tester to the number 2 injector and cycle the ignition switch several times to restore system fuel pressure.

P16-7 Depress the injector pulse tester's switch and observe the fuel pressure. Again, the difference between the system pressure and the pressure when an injector is activated is the pressure drop across the injector.

P16-8 Move the injector tester's leads to the number 3 injector and cycle the ignition switch to restore system pressure.

P16-9 Depress the switch on the tester to activate that injector and record the pressure drop. Continue the procedure for all injectors, and then compare the results of each to specifications and to each other.

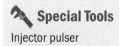 **Special Tools**

Injector pulser

It may be wise to start the engine before testing an injector more than one time to prevent flooding.

An injector balance test may be performed to diagnose restricted injectors on PFI systems. To conduct an injector balance test, follow these steps:

1. Connect the fuel pressure gauge to the Schrader valve on the fuel rail.
2. Connect the injector tester leads to the battery terminals with the correct polarity. Remove one of the injector wiring connectors and install the tester lead to the injector terminals (**Figure 8-19**).
3. Cycle the ignition switch on and off until the specified fuel pressure appears on the fuel gauge. Many fuel pressure gauges have an air bleed button that must be pressed to bleed air from the gauge. Cycle the ignition switch or start the engine to obtain the specified pressure on the fuel gauge and then leave the ignition switch off.
4. Push the timer button on the tester and record the gauge reading. When the timer energizes the injector, fuel is discharged from the injector into the intake port and the fuel pressure drops in the fuel rail.
5. Repeat steps 2, 3, and 4 on each injector, and record the fuel pressure after each injector is energized by the timer.
6. Compare the gauge readings on each injector.

The difference between the maximum and minimum reading is the pressure drop.

Ideally, each injector should drop the same amount when opened. A variation of 1.5 to 2 psi (10 kPa) or more is cause for concern (**Figure 8-20**). If there is no pressure drop or a low pressure drop, suspect a restricted injector orifice or tip. A higher-than-average pressure drop indicates a rich condition. When an injector plunger is sticking in the open position, the fuel pressure drop is excessive. If there are inconsistent readings, the nonconforming injectors have to be either cleaned or replaced.

Injector Power Balance Test

The injector pressure balance test is a good method of checking injectors on vehicles with no-start problems. When the vehicle runs, several other tests are possible. The injector power balance test is an easy method of determining if an injector is causing a miss.

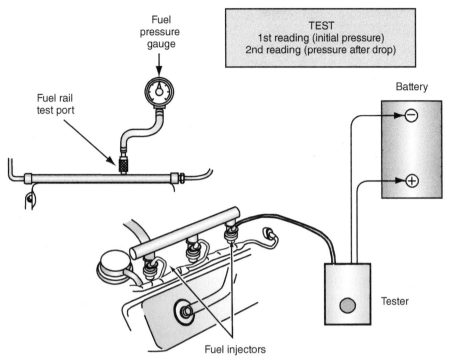

Figure 8-19 Setup for an injector pressure balance test.

CYLINDER	1	2	3	4	5	6
HIGH READING	225	225	225	225	225	225
LOW READING	100	100	100	90	100	115
AMOUNT OF DROP	125	125	125	135	125	110
RESULTS	OK	OK	OK	Faulty, rich (too much fuel drop)	OK	Faulty, lean (too little fuel drop)

Figure 8-20 Pressure readings from an injector pressure balance test.

To perform this test, first hook up a tachometer and fuel pressure gauge to the engine. Once all gauges are in place, start the engine and allow it to reach operating temperature. As soon as the idle stabilizes, unplug each injector one at a time. Note the rpm drop and pressure gauge reading. To ensure accurate test results on many electronic systems, it may be necessary to disconnect some type of idle air control device to prevent the computer from trying to compensate for the unplugged injector.

After recording the rpm and fuel pressure drop, reconnect the injector, wait until the idle stabilizes, and move on to the next one. Note the rpm and pressure drop each time. If an injector does not have much effect on the way the engine runs, chances are it is either clogged or electrically defective. It is a good idea to back up the power balance test with an injector resistance check, pressure balance test, or noid light check.

On many vehicles, an injector balance test can be performed with a scan tool. This is helpful when injectors are difficult to access.

The power balance test can also be run with the scan tool in most instances.

Injector Sound Test

If the injector's electrical leads are difficult to access, an injector power balance test is hard to perform. As an alternative, start the engine and use a technician's **stethoscope** to listen for correct injector operation (**Figure 8-21**). A good injector makes a rhythmic clicking sound as the solenoid is energized and de-energized several times each second. If a *clunk-clunk* instead of a steady *click-click* is heard, chances are the problem injector has been

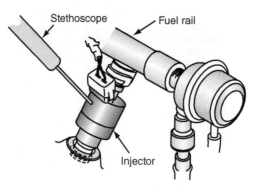

Figure 8-21 Using a stethoscope to check injector action.

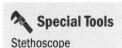
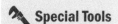
found. Cleaning or replacement is in order. If an injector does not produce any clicking noise, the injector, connecting wires, or PCM may be defective. When the injector clicking noise is erratic, the injector plunger may be sticking. If there is no injector clicking noise, proceed with the injector ohms test and noid light test to locate the cause of the problem. If a stethoscope is not handy, use a thin steel rod, wooden dowel, or your fingers to feel for a steady on-off pulsing of the injector solenoid.

Injector Flow Testing

Some vehicle manufacturers recommend an injector flow test rather than the balance test. Follow these steps to perform an injector flow test:

1. Disconnect the negative battery cable. If the vehicle is equipped with an air bag, wait for 1 minute.
2. Remove the injectors and fuel rail, and place the tip of the injector to be tested in a calibrated container. Leave the injectors in the fuel rail.
3. Connect a jumper wire across the specified terminals in the DLC for fuel pump testing.
4. Turn the ignition switch on.
5. Connect a jumper wire from the terminals of the injector being tested to the battery terminals (**Figure 8-22**).
6. Disconnect the jumper wire from the negative battery cable after 15 seconds.
7. Record the amount of fuel in the calibrated container. Also make note of the injector spray pattern—it should be a nice cone shape, not concentrated into a stream.
8. Repeat the procedure on each injector. If the volume of fuel discharged from any injector varies more than 0.3 cu. in. (5 cc) from the specifications, the injector should be replaced.
9. Reconnect the negative battery cable, and disconnect the 12-volt power supply.

⚙ **SERVICE TIP** Many technicians use a DSO when testing fuel injectors and fuel injector circuits. If the DSO is equipped with a graphing mode, perform a comparative amperage flow test through the fuel injectors. Discrepancies may be easier to observe and may lead to locating or verifying faulty fuel injectors or related circuit problems. The technician can also record and store various problem situations for future reference.

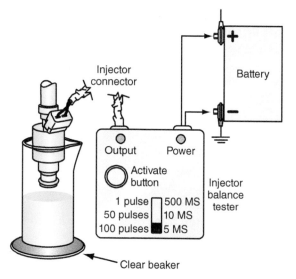

Figure 8-23 Checking for injector leaks with the fuel rail pressurized.

Figure 8-22 Testing an injector's flow rate off the vehicle.

Connect the fuel pressure gauge to the fuel system. While the fuel system is pressurized with the injectors removed from the fuel rail after the flow test, observe each injector for leakage from the injector tip (**Figure 8-23**). Injector leakage must not exceed the manufacturer's specifications. If the injectors leak into the intake ports on a hot engine, the air-fuel ratio may be too rich when a restart is attempted a short time after the engine is shut off. When the injectors leak, they drain all the fuel out of the rail after the engine is shut off for several hours. This may result in slow starting after the engine has been shut off for a longer period of time.

While checking leakage at the injector tips, observe the fuel pressure in the pressure gauge. If the fuel pressure drops off and the injectors are not leaking, the fuel may be leaking back through the check valve in the fuel pump. Repeat the test with the fuel line plugged. If the fuel pressure no longer drops, the fuel pump check valve is leaking. If the fuel pressure drops off and the injectors are not leaking, the fuel pressure may be leaking through the pressure regulator and the return fuel line. Repeat the test with the return line plugged. If the fuel pressure no longer drops off, the pressure regulator valve is leaking.

Oscilloscope Checks/Port Fuel Injection

An oscilloscope can be used to monitor the injector's pulse width and duty cycle when an injector-related problem is suspected. The pulse width is the time in milliseconds that the injector is energized. The duty cycle is the percentage of on-time to total cycle time.

To check the injector's firing voltage on the scope, a typical hookup involves connecting the scope's positive lead to the injector supply wire and the scope's negative lead to an engine ground.

Fuel injection signals vary in frequency and pulse width. The pulse width is controlled by the PCM, which varies it to control the air-fuel ratio. The frequency varies with engine speed. The higher the speed, the more pulses per second there are. Most often, the injector's ground circuit is completed by a driver circuit in the PCM. All of these factors are important to remember when setting a lab scope to look at fuel injector activity. Set the scope to read 12 volts, then set the sweep and trigger to allow you to clearly see the on signal on the left and the off signal on the right. Make sure the entire waveform is clearly seen. Also remember that the setting may need to be changed as engine speed increases or decreases.

Special Tools

Lab scope

On port fuel injection systems, each injector has its own driver circuit and wiring. To check an individual injector, the scope must be connected to that injector. This is great for locating a faulty injector. If the scope has a memory feature, a good injector waveform can be stored and recalled for comparison to the suspected bad fuel injector pattern. To determine if a problem is the injector itself or the PCM and/or wiring, simply swap the injector wires from an injector that had a good waveform to the suspect injector. If the waveform clears up, the wiring harness or the PCM is the cause of the problem. If the waveform is still not normal, the injector is to blame.

There are three different types of fuel injector circuits. In the conventional circuit, the driver constantly applies voltage to the injector. The circuit is turned on when a ground is provided. The waveform for this type of injector circuit is shown in **Figure 8-24**. Notice there is a single voltage spike at the point where the injector is turned off. The total on time of the injector is measured from the point where the trace drops on the left to the point where it rises up next to the voltage spike.

Peak and hold injector circuits use two driver circuits to control injector action. Both driver circuits complete the circuit to open the injector. This allows for high current at the injector, which forces the injector to open quickly. After the injector is open, one of the circuits turns off. The second circuit remains on to hold the injector open. This is the circuit that controls the pulse width of the injector. This circuit also contains a resistor to limit current flow during on time. When this circuit turns off, the injector closes. When looking at the waveform for this type of circuit (**Figure 8-25**), there will be two voltage spikes. One is produced when each circuit opens. To measure the on time of this type of injector, measure from the drop on the left to the point where the second voltage spike is starting to move upward.

A **pulse-modulated injector circuit** uses high current to open the injector. This allows for quick injector firing. Once the injector is open, the circuit ground is pulsed on and off to allow for a long on time without allowing high current flow through the circuit. To measure the pulse width of this type of injector (**Figure 8-26**), measure from the drop on the left to the beginning of the large voltage spike, which should be at the end of the pulses.

For all types of injectors, the waveform should have a clean, sudden drop in voltage when it is turned on. This drop should be close to 0 volt. Typically, the maximum allowable voltage during the injector's on time is 600 millivolts. If the drop is not perfectly vertical, either the injector is shorted or the driver circuit in the PCM is bad. If the voltage does not drop to below 600 millivolts, there is resistance in the ground circuit or the injector is shorted. When comparing one injector's waveform to another, check the height of the voltage spikes. The voltage spike of all injectors in the same engine should have approximately the same height. If there is a variance, the power feed wire to the injector with the variance or the PCM's driver for that injector is faulty.

The side of a V-type engine is commonly referred to as a bank of cylinders.

The conventional driver is also referred to as the saturated driver.

Peak and hold injector circuits incorporate unique circuitry that applies higher amperage to initially open the injector, then lower amperage for the remainder of the pulse width.

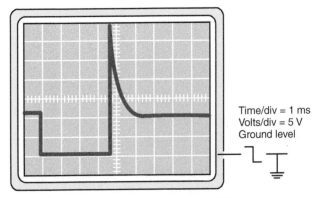

Time/div = 1 ms
Volts/div = 5 V
Ground level

Figure 8-24 The waveform of a conventional fuel injector driver circuit.

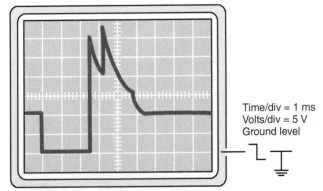

Time/div = 1 ms
Volts/div = 5 V
Ground level

Figure 8-25 The waveform of a peak and hold fuel injector driver circuit.

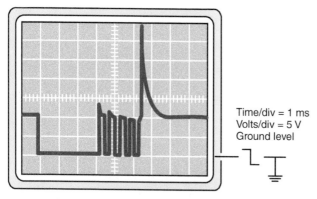

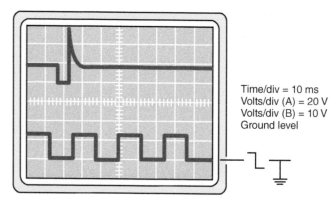

Figure 8-26 The waveform of a pulse-modulated fuel injector driver circuit.

Figure 8-27 Compare the ignition reference signal to the injector's ON signal.

While checking the injectors with a lab scope, make sure the injectors are firing at the correct time. To do this, use a dual-trace scope and monitor the ignition reference signal and a fuel injector signal at the same time. The two signals should have some sort of rhythm between them. For example, there can be one injector firing for every four ignition reference signals (**Figure 8-27**). This rhythm is dependent upon several factors—however, it does not matter what the rhythm is, only that it is constant. If the injector's waveform is fine but the rhythm varies, the ignition reference sensor circuit is faulty and not allowing the injector to fire at the correct time. If the ignition signal is lost because of a faulty sensor, the injection system will also shut down. If the injector circuit and the ignition reference circuit shut down at the same time, the cause of the problem is probably the ignition reference sensor. If the injector circuits shut off before the ignition circuit, the problem is the injector circuit or the PCM. NOTE: On some engine, if a miss is detected on a cylinder by the ECM, the ECM will not attempt to fire the injector until the engine is shut down and then restarted. Remember this when balance testing injectors manually. It might be necessary to restart the engine after reach attempt.

Another use of the oscilloscope for checking injector operation is viewing the amperage waveform. The amperage waveform requires a low-amp probe. Note the coil charge building over time, the turn-off point, and the pintle opening "gull wing" in **Figure 8-28**. The gull wing is produced because of the changing inductance of the pintle moving into the body of the injector. This feature is not always as pronounced as it is in this Honda injector.

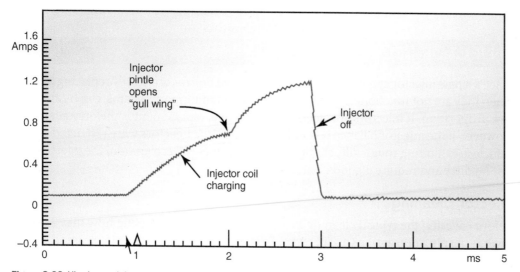

Figure 8-28 Viewing an injector amperage pattern.

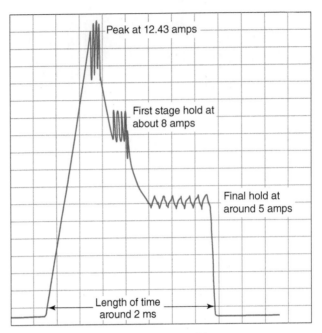

Figure 8-29 Current ramp of a DFI injector firing.

DFI Injector Waveform Testing

The Direct Fuel Injection (DFI) injector operates at a much higher amperage and voltage than the port fuel injector. The injector runs at as much as 80 volts and more than 10 amps. The DFI has to overcome much more pressure; up to 2,200 psi versus about 60 psi in port fuel. The strategy is the same as the peak and hold injectors that are used in port fuel. Notice the waveform in **Figure 8-29**. DFI happens much quicker than port fuel; the duration is from less than about 1 to 5 milliseconds compared to about 5 to 15 milliseconds for a typical port fuel duration. Note that at its peak, current was more than 12 amps, which then fell down quickly. It takes less energy to hold the injector open once it has been opened so the current can be cut down. Because the injector draws so much current, if it were held open at maximum amperage, it would quickly overheat. It seems to be easier to obtain the DFI waveform with the low-amps probe than with the voltage probe. One thing to remember if you are going to try and capture waveforms is that the higher current and voltage of the DFI may be high enough to cause personal injury, so you must exercise caution.

INJECTOR SERVICE

Since a single injector can cost up to several hundred dollars, arbitrarily replacing injectors when they are not functioning properly, especially on multiport systems, can be an expensive proposition. If injectors are electrically defective, replacement is the only alternative. However, if the injector balance test indicated that some injectors were restricted or if the vehicle is exhibiting rough idle, stalling, or slow or uneven acceleration, the injectors may just be dirty and require a good cleaning.

Injector Cleaning

Before covering the typical cleaning systems available and discussing how they are used, several cleaning precautions are in order. First, never soak an injector in cleaning solvent. Not only is this an ineffective way to clean injectors, but also it most likely destroys the injector in the process. Also, never use a wire brush, pipe cleaner, toothpick, or other

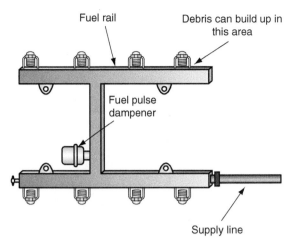

Figure 8-30 Location of a possible returnless fuel system debris problem.

cleaning utensil to unblock a plugged injector. The metering holes in injectors are drilled to precise tolerances. Scraping or reaming the opening results in a clean injector that is no longer an accurate fuel-metering device.

Returnless Fuel Systems Cleaning

Some returnless fuel systems will require a cleaning of the fuel rail if the injector at the opposite of the fuel rail feed line is failing a drop test and conventional cleaning has not been effective. The reason for this is that any debris that accumulates in the fuel rail winds up in the end of the rail, instead of being flushed out with the return line (**Figure 8-30**).

The basic premise of all injection cleaning systems is similar in that some type of cleaning chemical is run through the injector in an attempt to dissolve deposits that have formed on the injector's tip. The methods of applying the cleaner can range from single-shot, premixed, pressurized spray cans to self-mix, self-pressurized chemical tanks resembling bug sprayers. The premixed, pressurized spray can systems are fairly simple and straightforward to use since the technician does not need to mix, measure, or otherwise handle the cleaning agent.

Automotive parts stores usually sell pressurized containers of injector cleaner with a hose for Schrader valve attachment. During the cleaning process, the engine is operated on the pressurized container of unleaded fuel and injector cleaner. Fuel pump operation must be stopped to prevent the pump from forcing fuel up to the fuel rail. The fuel return line is to be plugged to prevent the solution in the cleaning container from flowing through the return line into the fuel tank. Follow these steps for the injector cleaning procedure:

1. Disconnect the wires from the in-tank fuel pump or the fuel pump relay to disable the fuel pump. If you disconnect the fuel pump relay on General Motors products, the oil pressure switch in the fuel pump circuit must also be disconnected to prevent current flow through this switch to the fuel pump.
2. Plug the fuel return line from the fuel rail to the tank.
3. Connect a can of injector cleaner to the Schrader valve on the fuel rail, and run the engine for about 20 minutes on the injector solution.

Other systems require the technician to assume the role of chemist and mix up a desired batch of cleaning solution for each application. The chemical solution then is placed in a holding container and pressurized by hand pump or shop air to a specified operating pressure (**Figure 8-31**). The injector cleaning solution is poured into a canister

If an injector does not respond to cleaning, it will have to be replaced.

Special Tools
Injector cleaning kit
Shop air supply

LTFT can be used to verify clogged injectors. See if the numbers indicate the adding of fuel.

Figure 8-31 Injector cleaner that operates with compressed air.

Figure 8-32 Injector cleaner connected to a fuel rail.

on some injector cleaners and shop air supply is used to pressurize the canister to the specified pressure. The injector cleaning solution contains unleaded fuel mixed with injector cleaner. The container hose is connected to the Schrader valve on the fuel rail (**Figure 8-32**). The procedure for using pressurized injector cleaner follows:

1. Disable the fuel pump according to the car manufacturer's instructions (e.g., pull fuel pump fuse, disconnect lead at pump, and so on). Clamp off the fuel pump return line (if equipped) at the flex connection to prevent the cleaner from seeping into the fuel tank.
2. Before starting the engine, open the cleaner's control valve one-half turn or so to prime the injectors, and then start the engine.
3. If available, set and adjust the cleaner's pressure gauge to approximately 5 psi below the operating pressure of the injection system, and let the engine run at 1,000 rpm for 10 to 15 minutes or until the cleaning mix has run out. If the engine stalls during cleaning, simply restart it.
4. Run the engine until the recommended amount of fluid is exhausted and the engine stalls. Shut the ignition off, remove the cleaning setup, and reconnect the fuel pump.
5. After removing the clamping devices from around the fuel lines, (if used) and start the car. Let it idle for 5 minutes or so to remove any leftover cleaner from the fuel lines.
6. On the more severely clogged cases, the idle improvement should be noticeable almost immediately. With more subtle performance improvements, an injector balance test verifies the cleaning results. Once the injectors are clean, recommend the use of an in-tank cleaning additive or a detergent-laced fuel.

The more advanced units feature electrically operated pumps neatly packaged in roll-around cabinets that are quite similar in design to an A/C charging station (**Figure 8-33**).

After the injectors are cleaned or replaced, rough engine idle may still be present. This problem occurs because the long term fuel tril in the computer has learned previously about the restricted injectors. If the injectors were supplying a lean air-fuel ratio, the computer increased the pulse width to try to bring the air-fuel ratio back to stoichiometric. With the cleaned or replaced injectors, the adaptive computer memory is still supplying

On some vehicles fuel trims can be reset with a scan tool.

Figure 8-33 A fuel injector cleaner cart.

the increased pulse width. This action makes the air-fuel ratio too rich now that the restricted injector problem does not exist. With the engine at normal operating temperature, drive the vehicle for at least 5 minutes to allow the adaptive computer memory to learn about the cleaned or replaced injectors. After this time, the computer should supply the correct injector pulse width and the engine should run smoothly. This same problem may occur when any defective computer system component is replaced.

FUEL RAIL, INJECTOR, AND REGULATOR SERVICE

There are service operations that will require removing the fuel injection fuel rail, pressure regulator, and/or injectors. Most of these are not related to fuel system repair. However, when it is necessary to remove and refit them, it is important that it be done carefully and according to the manufacturer's recommended procedures.

Injector Replacement

Photo Sequence 15 in Chapter 7 outlines a typical procedure for removing and installing a port fuel injector. Consult the vehicle's service manual for instructions on removing and installing injectors. Before installing the new one, always check to make sure the sealing O-ring is in place (**Figure 8-34**). Also, prior to installation, lightly lubricate the sealing ring with engine oil or automatic transmission fluid (avoid using silicone grease, which tends to clog the injectors) to prevent seal distortion or damage.

Fuel Rail, Injector, and Pressure Regulator Removal

The procedure for removing and replacing the fuel rail, injectors, and pressure regulator (**Figure 8-35**) varies depending on the vehicle. On some applications, certain components must be removed to gain access to these components. The following is a typical removal and replacement procedure for the fuel rail, injectors, and pressure regulator on a General Motors V6 engine:

1. Disconnect the battery negative cable. If the vehicle is equipped with an air bag, wait for 1 minute.
2. Bleed the pressure from the fuel system. There are several ways to relieve pressure. One is by removing the fuse or connector and starting the vehicle, letting it run

 Caution

Cap injector openings in the intake manifold to prevent the entry of dirt and other particles. Also, after the injectors and pressure regulator are removed from the fuel rail, cap all fuel rail openings to keep dirt out of the fuel rail.

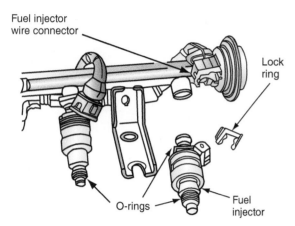

Fuel injector
wire connector

Lock
ring

O-rings

Fuel
injector

Figure 8-34 When replacing or reinstalling an injector, always replace the O-rings.

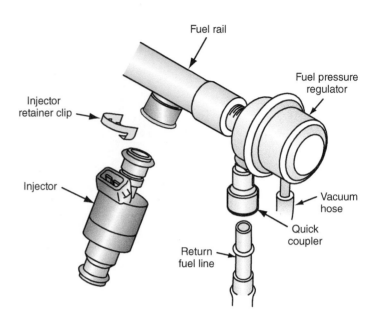

Fuel rail

Fuel pressure
regulator

Injector
retainer clip

Injector

Vacuum
hose

Quick
coupler

Return
fuel line

Figure 8-35 Fuel rail components.

until it dies. Additionally, some scan tools can command the fuel pump relay off, eliminating the need for pulling a fuse; this method is recommended for DFI (vehicles). Make sure to consult service information on DFI vehicles before removing any parts, as DFI has a low- and a high-pressure pump. The high side pressure can be read on the scan tool. The low side pressure can be relieved just like in a PFI vehicle.

3. Wipe excess dirt from the fuel rail with a shop towel.
4. Loosen fuel line clamps on the fuel rail if clamps are present on these lines. If these lines have quick-disconnect fittings, grasp the larger collar on the connector and twist in either direction while pulling on the line to remove the fuel supply and return lines (**Figure 8-36** and **Figure 8-37**).
5. Remove the vacuum line from the pressure regulator on return-line style.
6. Disconnect the electrical connectors from the injectors.
7. Remove the fuel rail hold-down bolts.
8. Pull with equal force on each side of the fuel rail to remove the rail and injectors.

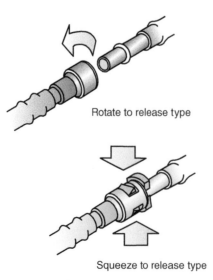

Rotate to release type

Squeeze to release type

Figure 8-36 Quick-disconnect fuel line fittings.

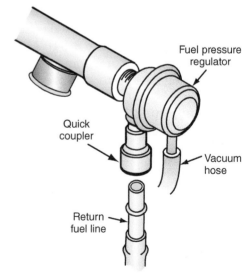

Fuel pressure
regulator

Quick
coupler

Vacuum
hose

Return
fuel line

Figure 8-37 Quick-connector fuel line at fuel rail.

Direct Fuel Injection Injector and Fuel Rail Service

As was mentioned previously, DFI uses injectors with extended tips that are located in the combustion chamber itself. The injectors may be very difficult to remove since carbon may form around the injector tips. A special tool allows the technician to pull the injector straight out of the bore. Do not try to rock the injector back and forth to remove; the extended tip may break off in the cylinder head! Remove the injector with a slide hammer (**Figure 8-38**) if it does not come out easily. Since the injectors have to seal against combustion chamber pressures, the seals are particularly critical. Special tools have been released to replace and resize the seals (**Figure 8-39**). Since the seals are so critical, the injector bore needs to be very clean before an injector is installed into the cylinder head. The injector bore needs to be cleaned with a special cleaning set (**Figure 8-40**) to ensure the seal or injector is not damaged on reinstallation. In addition to replacing the injector seals, the gasket for the high-pressure fuel pump, the high-pressure fuel lines, and the injector hold-downs all have to be replaced when the injectors or fuel rail are serviced.

Do not twist the DFI injector or try to rock it from side to side, as the injector may break off in the bore!

Regulator Cleaning and Inspection: Return-Line Style Fuel Injection

1. Prior to injector and pressure regulator removal, the fuel rail may be cleaned with a spray-type engine cleaner. Approved cleaners are normally listed in the service manual.
2. Pull the injectors from the fuel rail.
3. Use snap ring pliers to remove the snap ring from the pressure regulator cavity. Note the original direction of the vacuum fitting on the pressure regulator, and pull the pressure regulator from the fuel rail (**Figure 8-41**).

On DFI vehicles, the line from the high-pressure pump to the fuel rail must be replaced if it is removed.

⚠ **Caution**

Direct fuel injection vehicles have pressure up to 2,180 psi—enough to puncture skin and cause serious injury or death!

It is very important to relieve fuel pressure before servicing fuel-injected vehicles to prevent fines and personal injury!

⚠ **Caution**

Do not use compressed air to flush or clean the fuel rail. Compressed air contains water, which may contaminate the fuel rail.

Figure 8-38 A removal tool for a DFI injector. The tool is designed to be used with a slide hammer.

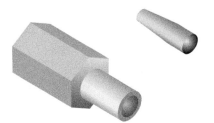

Figure 8-39 Tools used to replace high-pressure seals on DFI injector. The cone is used to slide the seal onto the injector. The other tool is used to resize the seal.

Figure 8-40 Tools used to clean the injector bore of carbon.

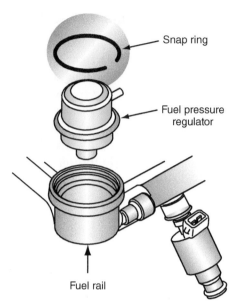

Snap ring

Fuel pressure
regulator

Fuel rail

Figure 8-41 Some fuel pressure regulators are
retained with a snap ring.

4. Clean all components with a clean shop towel. Be careful not to damage fuel rail
   openings and injector tips.
5. Check all injector and pressure regulator openings in the fuel rail for metal burrs
   and damage.

Installation of Fuel Rail, Injectors, and Pressure Regulator

1. If the same injectors and pressure regulator are reinstalled, replace all O-rings and
   lightly coat each O-ring with engine oil.
2. Install the pressure regulator in the fuel rail, and position the vacuum fitting on the
   regulator in the original direction.
3. Install the snap ring above the pressure regulator.
4. Install the injectors in the fuel rail.
5. Install the fuel rail while guiding each injector into the proper intake manifold
   opening. Be sure the injector terminals are positioned so they are accessible to the
   electrical connectors.
6. Tighten the fuel rail hold-down bolts alternately and torque them to specifications.
7. Reconnect the vacuum hose on the pressure regulator.
8. Install the fuel supply and fuel return lines on the fuel rail.
9. Install the injector electrical connectors.
10. Connect the negative battery terminal and disconnect the 12-volt power supply
    from the cigarette lighter.
11. Start the engine, check for fuel leaks at the rail, and be sure the engine operation is
    normal.

IDLE SPEED CHECKS

In PFI systems, idle speed is regulated by controlling the amount of air that is allowed
to bypass the airflow sensor or throttle plates. When presented with a car that tends
to stall, especially when coming to a stop, or idles too fast, look for obvious problems
like binding linkage and vacuum leaks first. If no problems are found, go through the
minimum idle speed check and setting procedure described on the under-hood decal.

The instructions listed on the decal spell out the necessary conditions that must be met prior to attempting an idle adjustment. These adjustment procedures can range from a simple twist of a throttle stop screw to more involved procedures requiring circumvention of idle air control devices, removal of casting plugs, or recalibration of the TP sensor.

Throttle actuator control (TAC), also known as *throttle by wire*, does not need idle speed adjustments.

Minimum Idle Speed Adjustment

The minimum idle speed adjustment may be performed on some MFI or SFI systems with a minimum idle speed screw in the throttle body. This screw is factory adjusted, and the head of the screw is covered with a plug. This adjustment should be required only if throttle body parts are replaced. If the minimum idle speed adjustment is not adjusted properly, engine stalling may result. The procedure for performing a minimum idle speed adjustment varies considerably depending on the vehicle. Always follow the vehicle manufacturer's recommended procedure in the service manual. The following is a typical minimum idle speed adjustment procedure for a General Motors vehicle:

1. Make certain there are no vacuum leaks, throttle is not binding, and PCV system is operating correctly.
2. Be sure the engine is at normal operating temperature, and turn the ignition switch off.
3. Using the outputs function of the scan tool, position the idle speed as low as the IAC will allow, and connect a tachometer from the ignition tach terminal to ground.
4. Disconnect the IAC motor connector.
5. Remove the connection for the scan tool, and start the engine.
6. Place the transmission selector in drive with an automatic transmission or neutral with a manual transmission.
7. Adjust the idle stop screw if necessary to obtain 500 to 600 rpm with an automatic transmission or 550 to 650 rpm with a manual transmission. A plug must be removed to access the idle stop screw (**Figure 8-42**).
8. Turn the ignition switch off and reconnect the IAC motor connector.
9. Turn the ignition switch on and connect a digital voltmeter from the TP sensor signal wire to ground. If the voltmeter does not indicate the specified voltage of 0.55 volt, loosen the TP sensor mounting screws and rotate the sensor until this voltage reading is obtained. Hold the TP sensor in this position and tighten the mounting screws.

Before an IAC motor is installed in a General Motors throttle body on port fuel systems, the distance from the end of the valve to the shoulder on the motor body must not exceed 1.125 inches (28 mm). If the IAC motor is installed with the plunger extended beyond this measurement, the motor may be damaged.

Special Tools
Jumper wires
Tachometer

Classroom Manual
Chapter 8, page 251

Special Tools
Scan tool
Tachometer

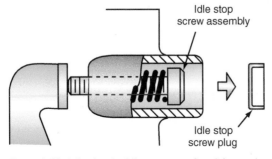

Figure 8-42 Adjusting the idle stop screw for minimum air.

Scan Tester Diagnosis

If the idle speed is not within specifications, the input sensors and switches should be checked carefully with the scan tester. **Photo Sequence 17** shows a typical procedure for performing a scan tester diagnosis of an idle air control motor. If the throttle position sensor voltage is lower than specified at idle speed, the PCM interprets this condition as the throttle being closed too much. Under this condition, the PCM opens the IAC or **idle air control bypass air (IAC BPA)** motor to increase idle speed.

If the engine coolant temperature sensor resistance is higher than normal, it sends a higher-than-normal voltage signal to the PCM. The PCM thinks the coolant is colder than it actually is, and the PCM operates the IAC or IAC BPA motor to increase idle speed. Many input sensor defects cause other problems in engine operation besides improper idle rpm.

Defective input switches result in improper idle rpm. For example, if the A/C switch is always closed, the PCM thinks the A/C is on continually. This action results in the PCM operating the IAC or IAC BPA motor to provide a higher idle rpm. On many vehicles, the scan tester indicates the status of the input switches as closed or open as well as high or low. Most input switches provide a high-voltage signal to the PCM when they are open and a low-voltage signal if they are closed.

On some vehicles, a fault code is set in the PCM memory if the IAC or IAC BPA motor or connecting wires are defective. On other systems, a fault code is set in the PCM memory if the idle rpm is out of range. On Chrysler products, the **actuation test mode (ATM)**, or actuate outputs mode, may be entered with the ignition switch on. The IAC or IAC BPA motor may be selected in the actuate outputs mode, and the PCM is forced to extend and retract the IAC or IAC BPA motor plunger every 2.8 seconds. When this plunger extends and retracts properly, the motor, connecting wires, and PCM are in normal condition. If the plunger does not extend and retract, further diagnosis is necessary to locate the cause of the problem.

Idle air control bypass air (IAC BPA) controls incoming air around the throttle plate to control idle speed.

On some vehicles, the scan tester may be used to force the PCM to operate the IAC or IAC BPA motor while observing the engine rpm.

PHOTO SEQUENCE 17

Performing a Scan Tester Diagnosis of an Idle Air Control Motor

P17-1 Be sure the ignition switch is off.

P17-2 Connect the scan tester leads to the battery terminals with the correct polarity.

P17-3 Turn the key on but do not start the engine.

P17-4 Program the scan tester as required for the vehicle being tested.

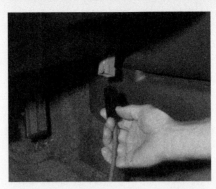

P17-5 Use the proper adapter to connect the scan tester lead to the data link connector (DLC).

P17-6 Select idle air control (IAC) motor test on the scan tester.

P17-7 Press the appropriate scan tester button to increase engine rpm and observe the rpm on the scan tester. The scan tester will automatically perform a nonrunning actuation test of the IAC hardware and software.

P17-8 At the low range, the IAC motor has a target range of 16 steps. The scan tester steps the motor through and monitors those steps.

P17-9 At the high range, the scan tester steps the motor through and monitors 112 steps.

On some systems, the scan tester reads the IAC or IAC BPA motor counts, and the count range is provided in the scan tester instruction manual. Some of the input switches, such as the A/C, may be operated and the scan tester counts should change. If the scan tester counts change when the A/C is turned on and off, the motor, connecting wires, and PCM are operational. When the scan tester counts do not change under this condition, further diagnosis is required.

A scan tool can help detect problems with the IAC valve. If the IAC valve is at or near zero, the idle is too high and the PCM is trying to reduce the idle speed. The IAC valve has totally bottomed out and can travel no farther. If this happens and the idle is still too high, a MIL light and trouble code will result. If the engine is the speed density type (using a MAP sensor instead of a MAF), then the problem could be a manifold vacuum leak or wrong PC valve, or someone has set the minimum air too high. If the IAC counts are higher than they should be, the idle speed is too low. This condition could be caused by a buildup of carbon on the backside of the throttle plate, which cuts down on the amount of air that goes into the engine. If the buildup is bad enough, the engine will die on

The scan tester can also be used in the snapshot mode on most vehicles to record conditions for playback in the shop. Trouble codes set in OBD II will trigger an automatic snapshot (freeze frame) for some code.

The scan tool can also be used to clear any trouble codes set quickly. Remember that the data collected during an OBD II freeze frame will also be cleared.

Some manufacturers do not recommend cleaning of the throttle body. Always check service information.

⚠ **Caution**

IAC BPA motor damage may result if throttle body cleaner is allowed to enter the motor.

deceleration because the IAC valve may not be fast enough to prevent stalling if the throttle is closed too fast.

Another problem that can be detected by the scan tool is intermittent operation of the IAC. Watch the counts at the IAC valve when the engine comes to an idle while warm. Unless the coolant fan or some other load comes on at idle, the IAC counts should be fairly consistent every time you come to a stop, at least within a few counts. If the counts are erratic, then there is a problem with the IAC valve sticking or binding.

Sometimes the complaint is hard starting, but the engine starts easier if the throttle is opened slightly while cranking. Depressing the throttle should not be necessary. If this condition is noted, see if the spring on the IAC valve is overcoming the IAC valve stepper motor when the engine sits for a long period of time. If this happens, the valve will be closed completely, shutting off the bypass air.

IAC Bypass Air Motor Removal and Cleaning

Carbon deposits in the IAC BPA motor air passage in the throttle body or on the IAC BPA motor pintle result in erratic idle operation and engine stalling. Remove the motor from the throttle body and inspect the throttle body air passage for carbon deposits. If heavy carbon deposits are present, remove the complete throttle body for cleaning. Clean the throttle body IAC BPA air passage, motor sealing surface, and pintle seat with throttle body cleaner. Clean the motor pintle with throttle body cleaner.

IAC BPA Motor Diagnosis and Installation

Connect a pair of ohmmeter leads across terminals D and C on the motor connector, and observe the ohmmeter reading (**Figure 8-43**). Repeat this ohmmeter test on terminals A and B on the motor. In each test, the ohmmeter reading should equal the manufacturer's specifications, usually 40 to 80 ohms. If the ohmmeter reading is not within specifications, replace the IAC BPA motor.

If a new IAC BPA motor is installed, be sure the part number, pintle shape, and diameter are the same as those on the original motor. Measure the distance from the end of the pintle to the shoulder of the motor casting (**Figure 8-44**). If this distance exceeds 1.125 inches (28 mm), use hand pressure to push the pintle inward until this specified distance is obtained.

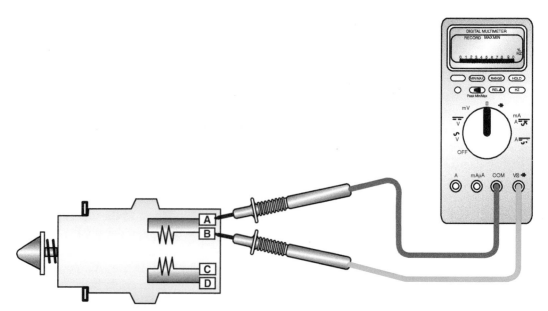

Figure 8-43 Testing the IAC BPC motor with an ohmmeter.

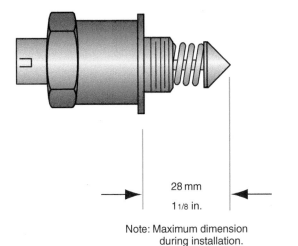

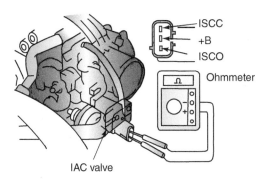

Figure 8-45 Ohmmeter connections to test the IAC BPA valve windings.

28 mm

1 1/8 in.

Note: Maximum dimension during installation.

Figure 8-44 Measuring the distance of extension on an IAC BPA motor.

Install a new gasket or O-ring on the motor. If the motor is sealed with an O-ring, lubricate this ring with transmission fluid and install the motor. If the motor is threaded into the throttle body, tighten the motor to the specified torque. When the motor is bolted to the throttle body, tighten the mounting bolts to the specified torque.

To test the IAC BPA valve, disconnect the IAC BPA valve connector and connect a pair of ohmmeter leads to terminals B+ and ISCC on the valve. Repeat the test between terminals B+ and ISCO on the valve (**Figure 8-45**). In both these tests on the valve windings, the ohmmeter should read 19.3 to 22.3 ohms. If the resistance of the valve windings is not within specifications, replace the valve.

Disconnect all throttle body linkages and vacuum hoses. Remove the throttle body and the IAC BPA valve mounting bolts. Remove the valve and gasket (**Figure 8-46**). The cleaning procedure for the IAC BPA motor may be followed on the IAC BPA valve.

Connect a jumper wire from the battery positive terminal to the valve B+ terminal. Connect another jumper wire from the battery negative terminal to the valve ISCC terminal. Be careful not to short the jumper wires together. With this connection, the valve must be closed (**Figure 8-47**). Leave the jumper wire connected from the battery

⚠ **Caution**

IAC BPA motor damage may occur if the motor is installed with the pintle extended more than the specified distance.

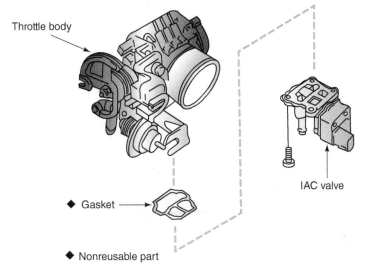

Figure 8-46 IAC BPA valve removed from the throttle body.

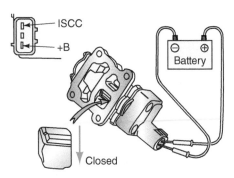

Figure 8-47 Test connections to close the IAC BPA valve.

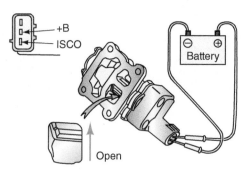

Figure 8-48 Test connections to open the IAC BPA valve.

positive terminal to the valve B+ terminal, and connect the jumper wire from the battery negative terminal to the ISCO terminal. Under this condition, the valve must be open (**Figure 8-48**). If the valve does not open and close properly, replace the valve.

Clean the throttle body and valve mounting surfaces, and install a new gasket between the valve and the throttle body. Install the valve and tighten the mounting bolts to the specified torque. Be sure the throttle body and intake mounting surfaces are clean, and install a new gasket between the throttle body and the intake. Tighten the throttle body mounting bolts to the specified torque. Reconnect the IAC BPA valve wiring connector and all throttle body linkages and hoses.

DIRECT FUEL INJECTION SERVICE

There are many similarities between DFI and PFI. Most of the service techniques are also very similar, so we will concentrate on the major differences here. First and foremost is the amount of pressure that the injectors are supplied. The pressure can be from 400 to 2,100 psi, depending on make, model, and operation. It cannot be stressed enough that service procedures have to be followed closely. Any sealing surfaces, O-rings, and so on must be replaced as necessary. Leaks could pierce the skin with fuel, causing serious injury. The possibility of fire with even small amounts of fuel under such high pressure is also a real possibility.

Direct fuel injection (DFI) system injects fuel directly into the combustion chamber.

High-Pressure Fuel Pump

Obviously, a DFI vehicle needs a high-pressure fuel pump to deliver the required pressure. This is accomplished by using a conventional low-pressure (about 60 electric psi) fuel pump to deliver fuel to a high-pressure fuel pump that is mechanically driven by the camshaft and electrically regulated by the ECM (**Figure 8-49**). Both power and ground are

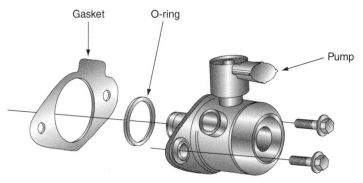

Figure 8-49 High-pressure fuel pump used on a direct fuel injection system.

Code	Problem
P0087	Fuel pressure is 1,500 kPa lower than desired pressure
P0088	Fuel pressure is more than 2,000 kPa above desired pressure
P0089	Fuel pressure regulator performance. ECM is adding or subtracting more than 3,000 kPa
P0090	Fuel pressure regulator voltage is between 1 and 4.5 volts and has been commanded off. The control circuit has high resistance or open connections.
P0091	Fuel pressure regulator voltage is less than 1 volt and has been commanded off. The control circuit may be shorted to ground.
P0092	Fuel pressure regulator voltage is more than 4.5 volts and has been commanded on. The control circuit may be shorted to voltage

Figure 8-50 Some generic trouble codes for the high-pressure fuel regulator circuit.

supplied by the PCM. Correct fuel pressure is verified by a fuel rail pressure (FRP) sensor that gives **feedback** to the PCM. Of course, if the commanded and actual fuel rail pressures are not within range, a DTC is set. There are several codes possible for a pressure malfunction in the high-pressure system (**Figure 8-50**).

Fuel Injectors

The DFI fuel injectors are placed directly into the combustion chamber (**Figure 8-51**). Most port fuel injectors are fired when supplied by a ground through the PCM. The DFI injectors from the Pontiac Solstice receive both power and ground from the PCM (**Figure 8-52**).

Diagnosing an Injector Trouble Code from a DFI Engine

A DFI vehicle has come in with code P0201. The PCM has detected a problem in the high-voltage supply circuit from the ECM to the fuel injector. The PCM can diagnose problems with the injector circuit by looking at the current flow in the circuit. If the wrong amount of current flows in the circuit, the PCM can set a code to help assist in diagnosis.

 CAUTION

Do not back probe a DFI injector due to the high current and voltage present.

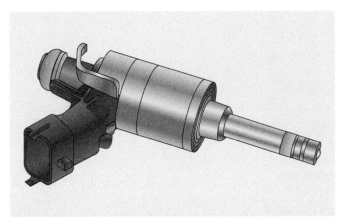

Figure 8-51 A fuel injector used with direct injection. The long tip helps to dissipate combustion chamber heat.

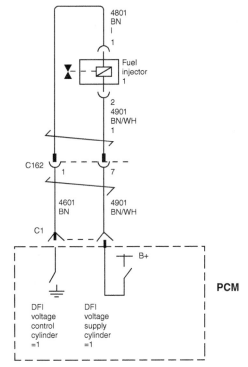

Figure 8-52 This fuel injector used with direct injection receives both power and ground from the PCM.

This situation is based on the generic trouble code; the manufacturer will also provide specific codes that may be more detailed.

1. First, run the diagnostic circuit check according to the manufacturer's recommendation.
2. The diagnostic information says that the ECM looks at the conditions to set this trouble code anytime the ignition voltage is from 8 to 18 volts and the engine is running over 80 rpm. This means that the DTC can set during cranking or running.
3. The code is set when the ECM sees an incorrect voltage in the injector 1 circuit for 4 continuous seconds, or if the fault is intermittent for a cumulative period of 50 seconds in the same drive cycle. This code is a type B code; the code will set as "pending" in the ECM on the first failure, then set as a DTC on the second failure.
4. The diagnosis of the code will have the technician look at misfire charts to determine if the fault is occurring now. If not, test drive the vehicle within the freeze-frame parameters when the code set last time to duplicate the problem.
5. Now that the problem has been verified in the misfire counters for cylinder 1, look at the schematic diagram (**Figure 8-53**) for the fuel injection system. According to the diagnosis, start by disconnecting connector X162 and look for damaged or bad connections, between terminals 1 and 7 (**Figure 8-54**). The connector X162 has been selected for testing because the connector at injector 1 is difficult to access.
6. If the connector is OK, then using a test lamp between terminal 7 and ground, check for a brightly blinking light while the engine is cranked. This will determine if the ECM is sending a voltage signal to the injector.
   a. If the test lamp stays on constantly, check for a short to power from the ECM to the wiring harness. If the harness is OK, then replace the ECM and retest.
   b. If the test light stays off, check for a short to ground or high resistance in the wire from terminal 7 and the ECM. If the circuit is OK, then replace the ECM.

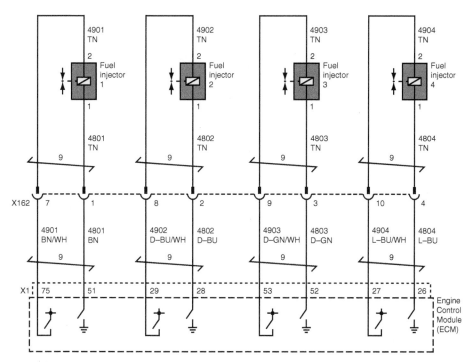

Figure 8-53 Typical wiring diagram for a DFI system.

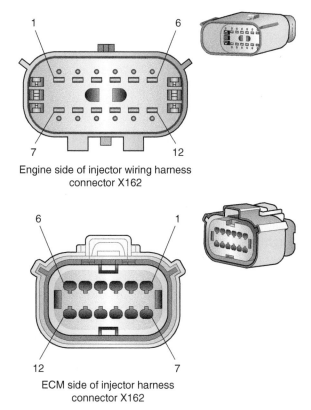

Engine side of injector wiring harness
connector X162

ECM side of injector harness
connector X162

Figure 8-54 Harness connector views for X162 Injector harness
connector.

7. Next, check the ground circuit by placing a test light from power to terminal 1 of connector X162. Crank the engine, the test lamp should flash dimly.

    a. If the test light stays on, check for a short to ground in the circuit from the ECM to connector X162. If normal, replace the PCM.

    b. If the test light stays off, check for a short to power in the circuit from X162 to the PCM. If the wiring is normal, replace the PCM.

    The ground and power control circuits have checked out OK on the ECM side of connector X162; now the remainder of the circuit will be checked on the engine side including the injector.

8. With the ignition off, check for continuity between terminal 7 and each of the following terminals: 8, 9, and 10. All connections should show **Out of Limits** on a multimeter. If any show continuity with 7, then the wiring is shorted together and will need to be repaired.

9. With a test lamp connected to B+, see if the test lamp lights on terminals 7, 8, 9, or 10. If the test lamp lights, then that terminal's wiring is shorted to ground (power side).

10. With a test lamp connected to B+, see if the test lamp lights on terminals 1, 2, 3, or 4. If the test lamp lights, then that terminal's wiring is shorted to ground (ground side).

Variable Valve Timing

Variable valve timing allows the PCM to adjust the valve opening and closing times for optimum fuel economy and power. As mentioned in the classroom manual, variable valve timing fine-tunes the camshaft timing for a best possible idle, power, and fuel economy and reduces emissions over a wide range of engine operation. Additionally, VVT systems can eliminate the EGR and AIR systems by controlling the amount of overlap in the valve timing. The PCM takes into account engine coolant, oil temperature, and airflow to determine the best valve (**cam**) timing for a given engine demand. In most late-model systems, the PCM commands a solenoid through pulse width modulation. The solenoid directs oil through passages in the camshaft itself, or the cam bearing area to an actuator on the camshaft sprocket (**Figure 8-55**). The camshaft is able to move several degrees inside the special sprocket, while the outside of the sprocket is fixed to the timing chain. The outside of the sprocket does not move in relation to the crankshaft, only the camshaft itself. The PCM utilizes the camshaft (CMP) and crankshaft (CKP) sensors to determine the amount of advance or retard. The PCM has several parameter identifications (PIDs) that relate to the operation of the variable valve timing system that can be accessed through the scan tool (**Figure 8-56**). The advance error should be around −5 to +5 degrees on a normally running engine; the error number will increase momentarily when the engine is accelerated until the cam actuator and solenoid can catch up with the computer's commands. After this momentary lapse, the commanded position and actual position should be very close.

Variable valve timing allows the PCM to fine tune the camshaft profiles to meet engine operating conditions.

Variable Valve Timing Engines and Maintenance

The main problems that most manufacturers are having with variable valve timing engines are a lack of oil changes and customers running vehicles low on oil. Dirty oil plays havoc with variable valve timing components. Oil must be free to flow through the solenoids and actuators in order to operate properly. **Figure 8-57** shows a variable valve timing solenoid with contaminated oil in the screen.

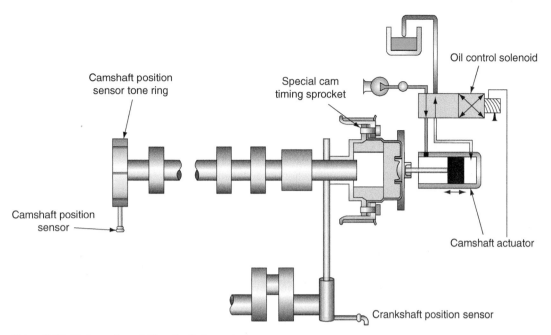

Figure 8-55 Diagram of a variable valve timing system.

Parameter Identification (PID)	Description of PID
VCTADV	Variable cam timing advance in degrees as measured by the cam sensor
VCT_1 F	Shows "fault" or "no fault" to show a VCT concern. Diagnose any cam sensor concerns before VCT concerns
VCTADVERR	Cam advance error in degrees. Normal around –5 to +5 degrees
VCTDC	Cam timing duty cycle from 0–100% (default is 0%)
VCTSYS	Defaults to off in open loop, on in closed loop

Figure 8-56 FORD variable valve timing Scan tool PID chart (Ford refers to this as Variable Valve Timing VCT or EcoBoost).

Figure 8-57 Variable valve timing solenoid with contaminated oil.

CASE STUDY

A customer phoned to say that he was having his PFI Cadillac towed to the shop because the engine had stopped and would not restart. Before the technician started working on the car, he routinely checked the oil and coolant. The engine oil dipstick indicated the crankcase was severely overfilled with oil, and the oil had a strong odor of gasoline. The technician checked the ignition system with a test spark plug and found the system to be firing normally. Of course, the technician thought the no-start problem must be caused by the fuel system and the most likely problems would be the fuel filter or fuel pump.

The technician removed the air cleaner hose from the throttle body and removed the air cleaner element to perform a routine check of the air cleaner element and throttle body. The throttle body showed evidence of gasoline lying at the lower edge of the throttle bore. The technician asked a coworker to crank the engine while he looked in the throttle body. While cranking the engine, gasoline was flowing into the throttle body below the throttle. The technician thought this situation was impossible. A PFI system cannot inject fuel into the throttle body!

The technician began thinking about how fuel could be getting into the throttle body on this SFI system, and he reasoned the fuel had to be coming through one of the vacuum hoses. Next, he thought about which vacuum hose could be a source of this fuel and remembered that the pressure regulator vacuum hose is connected to the intake. The pressure regulator vacuum hose was removed from the throttle body and placed in a container. When the engine was cranked, fuel squirted out of the vacuum hose, indicating the regulator diaphragm had a hole in it.

The technician installed a new pressure regulator and changed the engine oil and filter. After this service, the engine started and ran normally.

ASE-STYLE REVIEW QUESTIONS

1. While discussing fuel injection testing,
   Technician A says that power and ground to the injectors can be tested with a noid light.
   Technician B says that a test light can also be used to test for power and ground at the injector.
   Who is correct?
   A. A only
   B. B only
   C. Both A and B
   D. Neither A nor B

2. *Technician A* says that a buildup of carbon on the backside of a throttle body can cause stalling or low idle speeds.
   Technician B says that some throttle bodies should not be cleaned.
   Who is correct?
   A. A only
   B. B only
   C. Both A and B
   D. Neither A nor B

3. *Technician A* says that DFI injection duration is from less than about 1 to 5 milliseconds.
   Technician B says that Port fuel injector duration is typically about 5 to 15 milliseconds.
   Who is correct?
   A. A only
   B. B only
   C. Both A and B
   D. Neither A nor B

4. *Technician A* says that variable valve timing allows the PCM to adjust the injector opening and closing times for optimum fuel economy and power.
   Technician B says that variable valve timing fine-tunes the camshaft timing for the best possible idle, power, and fuel economy and reduces emissions.
   Who is correct?
   A. A only
   B. B only
   C. Both A and B
   D. Neither A nor B

5. While discussing VVT systems,

    Technician A says that these systems can replace EGR systems.

    Technician B says that VVT systems can eliminate the knock sensor.

    Who is correct?

    A. A only   C. Both A and B

    B. B only   D. Neither A nor B

6. *Technician A* says that DFI injectors can be rocked back and forth as necessary to remove them from the cylinder head.

    Technician B says that DFI injectors require special tools to install new seals on the injectors.

    Who is correct?

    A. A only   C. Both A and B

    B. B only   D. Neither A nor B

7. While discussing direct fuel injection (DFI) systems,

    Technician A says that the DFI injector hold downs can be reused when reinstalling new injectors.

    Technician B says that the DFI high-pressure fuel line from the high-pressure pump to the fuel rail can be reused only twice before replacement.

    Who is correct?

    A. A only   C. Both A and B

    B. B only   D. Neither A nor B

8. While discussing STFT and LTFT numbers,

    Technician A says that a short-term fuel trim (STFT) is a temporary correction to the fuel mixture by the PCM.

    Technician B says that negative numbers on the long-term fuel trim (LTFT) mean the PCM is adding fuel to the mixture to try and reach 14.7:1.

    Who is correct?

    A. A only   C. Both A and B

    B. B only   D. Neither A nor B

9. While discussing the comparison of air-fuel ratio (AFR) sensor to the oxygen sensor,

    Technician A says that the AFR sensor runs at a lower temperature than the oxygen sensor.

    Technician B says that the AFR sensor is more accurate than the oxygen sensor.

    Who is correct?

    A. A only   C. Both A and B

    B. B only   D. Neither A nor B

10. While discussing the causes of a rich air-fuel ratio,

    Technician A says that a rich air-fuel ratio may be caused by low fuel pump pressure.

    Technician B says that a rich or lean air-fuel ratio may be caused by a defective coolant temperature sensor.

    Who is correct?

    A. A only   C. Both A and B

    B. B only   D. Neither A nor B

ASE CHALLENGE QUESTIONS

1. An oxygen sensor detects:

    A. The amount of oxygen in the exhaust.

    B. The amount of fuel in the exhaust.

    C. Only lean mixtures.

    D. Only rich mixtures.

2. *Technician A* says that the DFI injector runs at lower fuel pressures than port fuel injectors.

    Technician B says that the DFI injector may be hard to remove from the cylinder head.

    Who is correct?

    A. A only   C. Both A and B

    B. B only   D. Neither A nor B

3. Which of the following is *false* concerning variable valve timing engines?

    A. They use a crankshaft actuator to change valve timing.

    B. They use a PCM-controlled solenoid that operates an actuator.

    C. They use the camshaft (CMP) and crankshaft (CKP) sensors to determine the amount of advance or retard.

    D. Owners need to pay particular attention to the oil level and condition.

4. *Technician A* says that the current waveform of a port fuel injector shows the injector coil charging over time.

Technician B says that the voltage waveform of a port fuel injector shows a sharp peak after the injector is turned off due to inductance in the coil.

Who is correct?

A. A only

B. B only

C. Both A and B

D. Neither A nor B

5. *Technician A* says that poor idle and surging at cruise may be caused by an air leak in the PCV system.

Technician B says that an MAF sensor failure may be confirmed by unplugging its electrical connection and noting engine performance.

Who is correct?

A. A only

B. B only

C. Both A and B

D. Neither A nor B

Name _____   Date _____

VISUALLY INSPECT AN EFI SYSTEM

Upon completion of this job sheet, you should be able to conduct a preliminary inspection of an electronic fuel injection system.

ASE Education Foundation Correlation

This job sheet addresses the following **AST** tasks: VIII. Engine Performance; D. Fuel, Air Induction, and Exhaust Systems Diagnosis and Repair

Task #4   Inspect, service, or replace air filters, filter housings, and intake duct work. **(P-1)**

Task #5   Inspect throttle body, air induction system, intake manifold, and gaskets for vacuum leaks and/or unmetered air. **(P-2)**

This job sheet addresses the following **MAST** task: VIII. Engine Performance; D. Fuel, Air Induction, and Exhaust Systems Diagnosis and Repair

Task #5   Inspect, service, or replace air filters, filter housings, and intake duct work. **(P-1)**

Describe the vehicle being worked on:

Year _____ Make _____

Model _____ VIN _____

Procedure

1. Is the battery in good condition, fully charged, with clean terminals and connections?   ☐ Yes   ☐ No

2. Do the charging and starting systems operate properly?   ☐ Yes   ☐ No

3. Are all fuses and fusible links intact?   ☐ Yes   ☐ No

4. Are all wiring harnesses properly routed with connections free of corrosion, and are they tightly attached?   ☐ Yes   ☐ No

5. Are all vacuum lines in sound condition, properly routed, and tightly attached?   ☐ Yes   ☐ No

6. Is the PCV system working properly and maintaining a sealed crankcase?   ☐ Yes   ☐ No

7. Are all emission control systems in place, hooked up, and operating properly?   ☐ Yes   ☐ No

8. Is the level and condition of the coolant-antifreeze good, and is the thermostat opening at the proper temperature?   ☐ Yes   ☐ No

9. Are the secondary spark delivery components in good shape with no signs of cross-firing, carbon tracking, corrosion, or wear?   ☐ Yes   ☐ No

10. Are the base timing and idle speed set to specifications (if adjustable)?   ☐ Yes   ☐ No

    Is the engine in good mechanical condition?   ☐ Yes   ☐ No

11. Is the gasoline in the tank of good quality and not been substantially cut with alcohol or contaminated with water?   ☐ Yes   ☐ No

12. Does the air intake duct work have cracks or tears?   ☐ Yes   ☐ No

13. Inspect the air filter, is it clean?   ☐ Yes   ☐ No

14. Are all of the induction hose clamps tight and properly sealed?   ☐ Yes   ☐ No

15. Are there any other possible air leaks in the crankcase?   ☐ Yes   ☐ No

16. Are the electrical connections at the mass airflow meter or
manifold pressure sensor good?   ☐ Yes   ☐ No

17. Is there carbon buildup inside the throttle bore and on the
throttle plate?   ☐ Yes   ☐ No

18. Is there a clicking noise at each injector while the engine
is running?   ☐ Yes   ☐ No

Instructor's Response

_____

_____

_____

Name _____   Date _____

CHECK THE OPERATION OF THE FUEL INJECTORS ON AN ENGINE

Upon completion of this job sheet, you should be able to check the operation of the fuel injectors on an engine using a variety of test instruments and techniques.

ASE Education Foundation Correlation

This job sheet addresses the following **AST** tasks: VIII. Engine Performance; D. Fuel, Air Induction, and Exhaust Systems Diagnosis and Repair

Task #1   Identify and interpret engine performance concerns; determine needed action. **(P-1)**

Task #6   Inspect, test and/or replace fuel injectors. **(P-2)**

This job sheet addresses the following **MAST** task: VIII. Engine Performance; B. Computerized Engine Controls Diagnosis and Repair

Task #6   Diagnose emissions or drivability concerns without stored or active diagnostic trouble codes; determine needed action. **(P-2)**

Tools and Materials

- Noid light
- DMM
- Stethoscope
- Lab scope

Describe the vehicle being worked on:

Year _____   Make _____

Model _____   VIN _____

Procedure   **Task Completed**

1. Remove the electrical connector from each injector.   ☐

2. While cranking the engine, check for voltage at each of the connectors.   ☐

3. What instrument did you use to check for voltage?

    _____

4. Use an ohmmeter to check each injector winding.   ☐

5. Compare resistance readings to the specifications found in the service manual.   ☐

6. What are the specifications for resistance? _____ ohms

7. Share your conclusions from the resistance checks:

    _____

    _____

8. Reconnect the electrical connectors to the fuel injectors and run the engine.   ☐

9. Using a technician's stethoscope, listen for correct injector operation.   ☐

10. Describe what you heard and what is indicated by the noise.

    _____

    _____

11. Connect the positive lead of a lab scope to the injector supply wire and the scope's   ☐
    negative lead to a good ground.

12. Set the scope on a low voltage scale and adjust the time so that the trace fills the screen.   ☐

13. Then observe the waveform of each injector, one at a time.   ☐

14. Record your findings from each injector.

_____

_____

_____

15. Summarize the results of all of these tests.

_____

_____

_____

_____

Instructor's Response

_____

_____

_____

Name _____   Date _____

CONDUCT AN INJECTOR BALANCE TEST

Upon completion of this job sheet, you should be able to perform an injector balance test on a port fuel injected engine.

ASE Education Foundation Correlation

This job sheet addresses the following **AST/MAST** tasks: VIII. Engine Performance; A. General: Engine Diagnosis

Task #1   Identify and interpret engine performance concerns; determine needed action. **(P-1)**

D. Fuel, Air Induction, and Exhaust Systems Diagnosis and Repair

Task #7   Inspect, test and/or replace fuel injectors. **(P-2)**
(AST Task #6)

Tools and Materials

• Fuel pressure gauge
• Injector tester

Describe the vehicle being worked on:

Year _____   Make _____

Model _____   VIN _____

Procedure	**Task Completed**

1. Carefully inspect the fuel rail and injector for fuel leaks. Do not proceed if fuel is leaking.   ☐

2. Connect the fuel pressure gauge to the Schrader valve on the fuel rail.   ☐

3. Disconnect the electrical connector to the number 1 injector.   ☐

4. Connect the injector tester lead to the injector terminals.   ☐

5. Connect the injector tester power supply leads to the battery terminals.   ☐

6. Cycle the ignition switch several times until the specified pressure appears on the fuel pressure gauge. The specified pressure is _____ psi.

   Your reading is _____ psi.

   This indicates:

   _____

   _____

7. Depress the injector tester switch and record the pressure on the pressure gauge. The reading is _____ psi.

8. Move the injector tester to injector 2.   ☐

9. Cycle the ignition switch several times until the specified pressure appears on the fuel pressure gauge.

   Your reading is _____ psi.

   This indicates:

   _____

   _____

10. Depress the injector tester switch and record the pressure on the pressure gauge. The reading is _____ psi.

11. Continue the same sequence until all injectors have been tested.

    Reading on injector 3 is _____ psi.

    Reading on injector 4 is _____ psi.

    Reading on injector 5 is _____ psi.

    Reading on injector 6 is _____ psi.

    Reading on injector 7 is _____ psi.

    Reading on injector 8 is _____ psi.

12. Summarize the results:

    _____

    _____

    _____

    _____

Instructor's Response

_____

_____

_____

Name _____   Date _____

VERIFY IDLE CONTROL OPERATION

Upon completion of this job sheet, you will be able to check the engine idle speed on a late-model vehicle.

ASE Education Foundation Correlation

This job sheet addresses the following AST/MAST tasks: VIII. Engine Performance; D. Fuel, Air Induction, and Exhaust Systems Diagnosis and Repair

Task #7 (AST Task #8)   Verify idle control operation. **(P-1)**

Tools and Materials

- Service information
- Scan tool
- Protective clothing
- Goggles or safety glasses with side shields

Describe the vehicle being worked on:

Year _____ Make _____

Model _____ VIN _____Engine type and size _____

Describe general operating condition:

_____

_____

Procedure

1. Before checking the idle speed, check that the MIL is not lit. If it is, diagnose the cause. What did you find?

   _____

   _____

2. Visually inspect the air cleaner assembly and PCV system for damage. Also, check to make sure there are no apparent ignition problems. What did you find?

   _____

   _____

3. Does the manufacturer call for an idle learn sequence? If so, briefly explain the process.

   _____

   _____

4. Start the engine. Hold the engine at 3,000 rpm without load (in park or neutral) until the radiator cooling fan comes on, then let it idle. Check the idle speed with the headlights, blower fan, radiator fan, and air conditioner off. What was the engine speed?

   _____

   _____

5. Let the engine idle for 1 minute with the heater fan and air conditioner off. Check the idle speed and compare it to specifications. Is the idle speed correct?

_____

_____

6. Let the engine idle for 1 minute with the headlights, blower fan, radiator fan, and air conditioner on. Check the idle speed and compare to specifications. Is the idle speed correct? What should be done to correct it?

_____

_____

7. Turn off the engine. If the idle speed was not correct, check the throttle cable for binding or excessive freeplay. What did you find?

_____

_____

8. If the throttle cable needs adjustment, describe the procedure for adjusting.

_____

_____

9. With the cable properly adjusted, make sure the throttle plate opens fully when you push the accelerator pedal to the floor. Also, make sure it returns to the idle position when the accelerator pedal is released. If the vehicle is equipped with electronic throttle control, watch the throttle while an assistant moves the accelerator pedal. What did you find?

_____

_____

10. If no problems were evident, check the intake manifold and throttle body for vacuum leaks. What did you find?

_____

_____

Instructor's Response

_____

_____

_____

CHAPTER 9

IGNITION SYSTEMS DIAGNOSIS AND SERVICE

Upon completion and review of this chapter, you should be able to:

- Diagnose and determine the cause of a no-start condition on an electronic ignition (EI) system.
- Perform tests on the camshaft and crankshaft sensors as well as EI systems.
- Perform coil tests on an EI system.
- Install and adjust camshaft and crankshaft sensors.
- Perform magnetic sensor tests on EI systems.
- Diagnose problems with compression sense ignition.
- Diagnose problems with ion sense ignition.

Basic Tools

Basic tool set
Tachometer
DMM

Terms To Know

Alternating current (AC)
Automatic shutdown (ASD) relay
Burn time
Compression sense ignition
Current limiting
Display pattern
Distributor ignition (DI)
Dwell section

Electronic distributorless ignition system (EI)
Firing order
Firing voltage
Ignition systems
Intermediate section
Ion-sensing ignition
Kilovolts (kV)
Magnetic sensor

Parade pattern
Raster pattern
Spark duration
Spark line
Test spark plug
Timing meter
Variable rate sensor signal (VRS)

INTRODUCTION

Ignition systems should not overwhelm a knowledgeable technician. Basically, the ignition system itself looks very different, but it still accomplishes the same tasks; deliver a spark of the correct intensity and duration at the proper time. Eliminating the distributor replaces wear-prone parts that can cause inaccuracy. However, the distributor ignition (DI) and electronic ignition (EI) have the same job and actually work in very similar ways. A position sensor gives the position of the engine to the module, which controls the coil firing through the primary ignition system, which controls the secondary, which fires the spark plugs. In this chapter, we will cover DI systems briefly, and then move on to the EI systems.

DISTRIBUTOR IGNITION

The distributor is mechanically indexed with and turns with the camshaft, which is driven by the crankshaft. Therefore it does not need electronic sensors to determine the position of the engine for proper spark distribution. The spark is distributed by a rotor

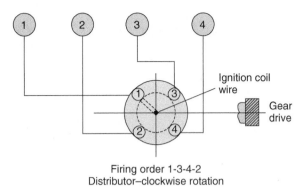

Firing order 1-3-4-2
Distributor–clockwise rotation

Figure 9-1 Drawing of a basic distributor ignition system.

that moves around the distributor cap and is sent to the appropriate cylinder through a spark plug wire (**Figure 9-1**). Of course, if the distributor is not indexed properly, the spark is not delivered to the correct spark plug at the right time, and the vehicle will run poorly, or not at all.

Distributor Cap Inspection

The distributor cap should be properly seated on its base. All clips or screws should be tightened securely. The distributor cap and rotor also should be removed for visual inspection. Physical or electrical damage is easily recognizable. Electrical damage from high voltage can include corroded or burned metal terminals and carbon tracking inside distributor caps (**Figure 9-2**). Carbon tracking indicates that high-voltage electricity has found a low-resistance conductive path over or through the plastic. The result is a cylinder that fires at the wrong time, misfires, or fires at the same time as another cylinder. Check the outer cap towers and metal terminals for defects. Cracked plastic requires replacement of the unit. Damaged or carbon-tracked distributor caps or rotors should be replaced (**Figure 9-3**). Also inspect for moisture.

Classroom Manual
Chapter 9, page 263

If the distributor cap or rotor has a mild buildup of dirt or corrosion, it should be cleaned. If it cannot be cleaned, it should be replaced. Small round brushes are available to clean cap terminals. Wipe the cap and rotor with a clean shop towel, but avoid cleaning these components with solvent or blowing them off with compressed air, which may contain moisture. Cleaning these components with solvent or compressed air may result in high-voltage leaks.

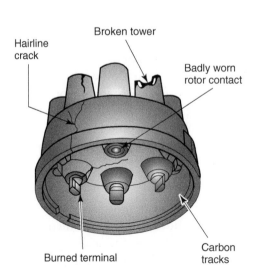

Figure 9-2 Types of distributor cap defects.

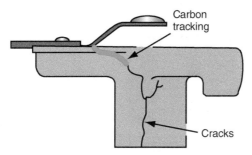

Figure 9-3 Inspect rotor for damage and signs of high-voltage leaks.

Check the distributor cap and housing vents. Make sure they are not blocked or clogged. If they are, the internal ignition module will overheat. It is good practice to check these vents whenever a module is replaced.

Control Modules

Distributor ignitions use transistors as switches. These transistors are contained inside a control module housing that can be mounted to or in the distributor or remotely mounted on the vehicle's fire wall or another engine compartment surface. Control modules should be tightly mounted to clean surfaces. A loose mounting can cause a heat buildup that can damage and destroy transistors and other electronic components contained in the module. Some manufacturers recommend the use of a special heat-conductive silicone grease between the control unit and its mounting (**Figure 9-4**). As noted, this helps conduct heat away from the module, reducing the chance of heat-related failure. During the visual inspection, check all electrical connections to the module. They must be clean and tight.

Sensors

The transistor in the control module is triggered by a voltage pulse from a crankshaft position sensor. In most ignition systems, this sensor is either a magnetic pulse generator or Hall-effect sensor. These sensors are mounted on either the distributor shaft or the crankshaft.

According to SAE J19309 standards that were recently mandated to the automotive industry, the term **distributor ignition** replaces all previous terms for distributor-type ignition systems that are electronically controlled.

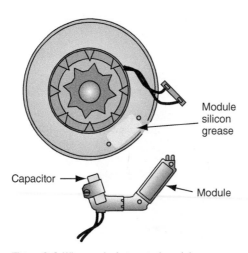

Figure 9-4 When replacing control modules, some manufacturers specify to apply silicon lubricant to the mounting surface.

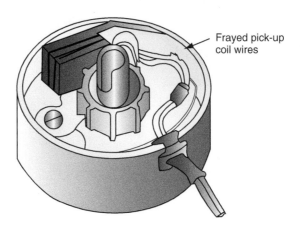

Figure 9-5 Inspect pick-up coil wiring for damage.

Magnetic pulse generators are relatively trouble-free. The reluctor or pole piece is replaced if it is broken or cracked. Magnetic pulse generators can also lose magnetism or the ability to produce the proper voltage needed to trigger the module. Magnets in the magnetic pulse generator are subjected to engine vibration and heat. The magnets can crack or break, affecting voltage output production reliability. In some cases, the unit may check within specifications with an ohmmeter yet is not generating enough AC voltage.

Under unusual circumstances, the nonmagnetic reluctor can become magnetized and upset the pick-up coil's voltage signal to the control module. Use a steel feeler gauge to check for signs of magnetic attraction, and replace the reluctor if the test is positive (**Figure 9-5**).

Hall-effect sensor problems are similar to those of magnetic pulse generators. Although not detected by a visual inspection, the Hall-effect device can become defective and fail to provide a voltage change to the ignition module. This can be checked with a digital multimeter (DMM) or oscilloscope and will be covered later in this chapter.

Photo Sequence 18 details the procedure of indexing the distributor to the engine mechanically.

PHOTO SEQUENCE 18
Timing the Distributor to the Engine

P18-1 Remove the number one spark plug.

P18-2 Place your thumb over the number one spark plug opening and crank the engine until compression is felt.

P18-3 Crank the engine a very small amount at a time until the timing marks indicate that the number one piston is at TDC on the compression stroke.

P18-4 Determine the number one spark plug wire position in the distributor cap.

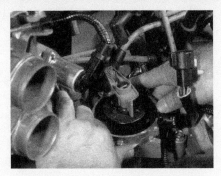

P18-5 Install the distributor with the rotor under the number one spark plug wire terminal in the distributor cap and one of the reluctor high points aligned with the pick-up coil.

P18-6 After the distributor is installed in the block, turn the distributor housing slightly so the pick-up coil is aligned with the reluctor.

P18-7 Install the distributor clamp bolt, but leave it slightly loose.

P18-8 Connect the pickup leads to the wiring harness.

P18-9 Install the spark plug wires in the cylinder firing order and in the direction of distributor shaft rotation. Set ignition timing.

AUTHOR'S NOTE Distributor ignitions keep track of the engine position mechanically. Distributorless ignitions keep track of engine position electronically with sensors.

Distributor Base Timing

Base timing is the timing that an engine has without the computer's input. The computer adds timing as appropriate to the running conditions of the vehicle. Base timing has to be set properly to ensure that computer-controlled timing is accurate. To check the base timing, the manufacturers give specific instructions for readying the engine. Some require disconnecting certain components, such as many Ford and GM systems, which require that a connector located in the engine compartment be disconnected. Toyotas use an under-hood connector that the technician can jump to put the vehicle into base timing. The emission's decal may provide information about the location of this connector. It is important that these instructions be followed so as to disable the computer's control over ignition timing. After these preparation steps are followed, the engine may be timed in the normal manner.

A **timing meter** shows timing degrees and engine rpm.

Making Base Timing Adjustments

Once you have obtained a base timing reading, compare it to the manufacturer's specifications. For example, if the specification reads 10 degrees before top dead center (**Figure 9-6A**) and the reading found is 3 degrees before top dead center (**Figure 9-6B**), the timing is slowed or off by 7 degrees.

This means the timing must be advanced by 7 degrees to make it correct. To do this, rotate the distributor until the timing marks align at 10 degrees (**Figure 9-7**), then retighten the distributor hold-down bolt (**Figure 9-8**).

Computer-Controlled Ignition System Service

> ⚙️ **SERVICE TIP** After adjusting the initial timing, be sure to check for, and clear, any trouble codes that may have been set in the computer's memory. Anytime a connector or component is disconnected for initial timing adjustment, a code will likely be stored.

Begin with a secondary ignition test. Place a spark tester (**Figure 9-9**) in the end of a spark plug wire, and crank the engine. If there is no spark, try another spark plug wire in the event that the first wire or distributor cap is defective. Confirm that there is an adequate

Timing marks aligned at 10 degrees

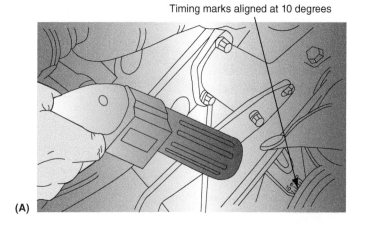

(A)

Timing marks aligned at 3 degrees

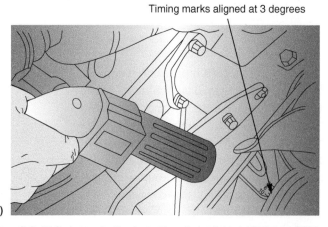

(B)

Figure 9-6 (A) Timing marks illuminated by a timing light at 10 degrees BTDC and (B) timing marks at 3 degrees BTDC.

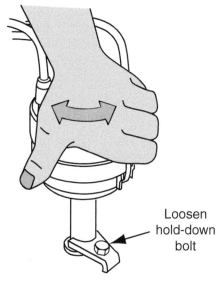

Loosen hold-down bolt

Figure 9-7 Turning the distributor to set timing.

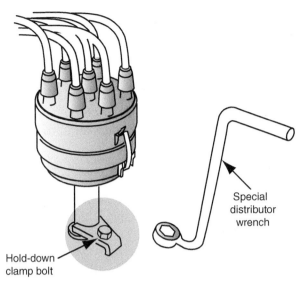

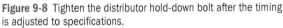

Figure 9-8 Tighten the distributor hold-down bolt after the timing is adjusted to specifications.

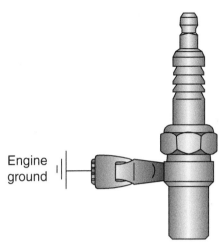

Figure 9-9 Spark plug tester.

spark, which would designate that the ignition is functioning. At times, the ignition coil may be weak but able to deliver adequate spark to occasionally start the engine, depending on the temperature. If there is no spark at the spark plugs, check for coil output at the coil terminal. If spark occurs, the cap and/or rotor could be at fault. If there is no spark at the coil tower, connect a test light to the negative (tach) side of the ignition coil and crank the engine. If the light flashes, the primary is functioning. Therefore, the problem is in the coil. If the test lamp does not flash or there is no dwell reading, check for proper primary feed voltage from the ignition switch to the positive side of the coil.

Always use a spark tester; do not modify a spark plug. The tester is specially calibrated to stress the ignition system.

> ⚙ **SERVICE TIP** Check the fuel injector pulse with a noid light as a quick test. If the injector is pulsing, the primary ignition is probably functioning properly. In most applications, the injector will not pulse if the primary signal is absent.

A Hall-effect switch provides a digital signal to trigger the ignition module. If none is found, check for problems between the coil and ignition switch. If there is power at the positive side of the coil and the test light did not flash or there was no dwell, the problem is narrowed to either the pick-up coil or ignition module. On occasion, a defective ignition coil can prevent the module from triggering. First, check the pick-up coil or Hall-effect switch. The pick-up coil, or magnetic pulse generator, can be checked with an ohmmeter, DMM, or scope. The Hall-effect switch can be checked with a DMM or scope. If the pick-up coil or Hall-effect switch is good, inspect the ignition module. A final check can be made on the module using a special module tester. Be sure to check all wiring connections and grounds. As mentioned earlier in this chapter, an ignition coil can prevent the module from triggering properly, making it appear faulty. Perform a transistor switch test before condemning the module. Simply disconnect the ignition coil, connect the test light clip to the positive battery terminal, and place the probe end of the test light to the coil negative wire harness terminal. Observe the test light while cranking the engine. If the light is flashing, the module is triggering and the coil is at fault. If the light does not flash and all other connections and components are good, then the module or pickup is faulty.

Electronic Ignition

EI systems keep track of engine position using electronic sensors on the crankshaft and camshaft. Instead of using a distributor cap, the EI systems generally have a coil-over-plug (COP) configuration, where there is a coil for each cylinder mounted over the spark plug. In the case of waste spark systems, there is one coil for each pair of spark plugs. The primary windings of the coil still energize the secondary windings of the coil when the primary windings are shut down. The secondary winding of the coils still provides a high-voltage surge that is delivered to the spark plugs.

Waste spark systems use one coil for two spark plugs; coil-near-plug and coil-over-plug systems use a coil for each spark plug. The advantage here is simple: There is more time to saturate the coils, which is important with high-rpm engines where one coil may not have sufficient time to fully charge; turbocharging and supercharging, which effectively raise compression; and spark demand and leaner mixtures, which require a hotter spark.

Diagnosis of the primary and secondary ignition systems is very similar to distributor-based systems with some of the variations explained in this chapter. As stated before, there are many similarities between DI and EI systems, and some differences as well. We will explain some of the differences in diagnosis and scope connections.

CONFIRM THE PROBLEM

Always confirm the problem before you assume that anything is wrong. Many technicians, and some service managers, do not like to return a vehicle back to a customer with a "no problem found" on the repair order. Sometimes technicians feel the pressure to put a part on a car in hopes that it will take care of an intermittent problem. This can often lead to frustration if the vehicle comes back later for a repeat repair. Make sure you can duplicate the problem before trying to fix the problem. Ask the service advisor or manager to contact the customer for more information or take a test drive with him or her if necessary to clarify the complaint.

EFFECTS OF IMPROPER IGNITION OPERATION

The primary circuit controls the secondary circuit. Therefore, the primary circuit controls ignition timing. Most problems within the primary circuit of computer-controlled ignition systems result in starting problems or poor performance due to incomplete combustion. A primary circuit failure can easily cause a no-start condition; any malfunction in the primary triggering, such as a crankshaft position sensor, will result in no fuel or spark delivered to the engine. Secondary ignition system problems usually start out by causing a misfire under a load or high rpm when the system is stressed.

If the engine does not start, it is usually because the engine does not have compression, air, fuel, or spark. To verify if there is spark, observe the spark while cranking the engine. To do this, remove a spark plug wire from a spark plug and insert a spark tester in the spark plug wire terminal. Position the test plug away from the fuel lines, and connect the test plug's grounding clip to a good ground. Crank the engine and observe the spark at the test plug. If the spark is weak, the ignition coil should be checked. If there is no spark, the triggering and switching units of the primary circuit should be checked, and if the problem is not found, then check the secondary circuit. This test will work with either DI or EI systems.

If engine performance is poor, the cause of the problem can be many factors. There can be a problem with the engine such as poor compression, incorrect valve timing, overheating, and so on. It is always a good idea to check for diagnostic trouble codes (DTCs) in the powertrain control module (PCM). See if any DTCs or pending DTCs are available

It is usually a good idea to check for spark at two different cylinders before making a conclusion on the ignition system.

Misfire Monitor		
Cylinder	Current	History
#1	0	2
#2	35	349
#3	0	2
#4	0	6

Figure 9-10 Misfire monitor information.

to help you find the problem. If the engine is running, the technician can often check the misfire monitor and see if a particular cylinder is missing (**Figure 9-10**). If only one cylinder shows an engine miss, then that cylinder can be investigated further. If the scan tool shows a random misfire, then a cause common to more than one cylinder in the engine should be investigated further. Check for any technical service bulletins (TSBs) on the problem to see if there might possibly be a known problem with a fix.

If you suspect a problem with a particular sensor, it is always possible to analyze the system waveforms on an oscilloscope. When anything indicates a problem with the primary ignition circuit, the suspected parts should be tested.

Although an EI system has no timing adjustment, incorrect ignition timing on the EI system could be caused by a defective crank or cam sensor, a slipped timing belt, or a worn timing chain.

VISUAL INSPECTION

All ignition system diagnosis should begin with a visual inspection after verifying the complaint. The entire system should be checked for obvious problems. Although some no-start problems are caused by the primary circuit, the secondary circuit should also be checked anytime the ignition system is suspected of a problem.

1. Spark plug and coil cables should be pushed tightly on the coil (or distributor cap) and spark plugs.
2. Inspect all secondary cables for cracks and worn insulation, which causes high-voltage leaks.
3. Inspect all of the boots on the ends of the secondary wires for cracks and hard, brittle conditions.
4. The ignition coil should be inspected for cracks or any evidence of voltage leaks in the coil tower.
5. Secondary cables must be connected according to the firing order.
6. Spark plug cables from consecutively firing cylinders should cross rather than run parallel to one another.
7. If dealing with a no-start condition, make sure the "Service engine soon" or (MIL) lamp comes on when the ignition is turned to run. If not, check the PCM power and ground.

Primary Circuit

Primary ignition system wiring should be checked for clean, tight connections. Electronic circuits operate on very low voltage. Voltage drops caused by corrosion or dirt can cause running problems. Missing tab locks on wire connectors are often the cause of intermittent ignition problems due to vibration- or thermal-related failure.

Test the integrity of a suspect connection by tapping, tugging, and wiggling the wires while the engine is running. Be gentle. The object is to recreate an ignition interruption, not to cause permanent circuit damage. With the engine off, separate the suspect connectors and check them for dirt and corrosion. Clean the connectors according to the manufacturer's recommendations.

All wiring connections to and from the ignition module should be thoroughly inspected for damage.

Do not overlook the ignition switch as a source of intermittent ignition problems. A loose mounting rivet or poor connection can result in erratic spark output. To check the switch, gently wiggle the ignition key and connecting wires with the engine running. If the ignition cuts out or dies, the problem is located.

Moisture can cause a short to ground or reduce the amount of voltage available to the spark plugs. This can cause poor performance or a no-start condition. Carefully check the ignition system for signs of moisture. **Figure 9-11** shows the common places where moisture may be present in an ignition circuit.

Spark Plug Wire Visual Inspection

Look inside the connections of the plug wires. If the secondary terminal is not clean and bright, but is dark or rusty looking, many times that cable is open electrically. These cables should be replaced, not cleaned and recrimped.

AUTHOR'S NOTE One of my students is employed in a HD truck shop as an apprentice technician. A van was being constantly towed into the shop because it would shut down while driving. Of course, when the van was towed to the shop it would start like there was no problem. The shop foreman advised the technicians in the shop to make an attempt to repair the vehicle, even though the complaint could not be duplicated. The vehicle had the spark plugs, wires, ignition coil, crank sensor, cam sensor, ignition module, and fuel pump replaced previously. The cam sensor was being replaced (the part had to be ordered) just as the owner arrived to pick up his truck. The owner used the truck for his business and was anxious to get it back into service. The technician replaced the sensor the owner left in the truck, and it stalled before he left the parking lot. Needless to say, this is not a good situation! The owner left the truck again; the apprentice found the cause—a poor connection at the coil's positive connection. Imagine how the customer felt about the previous repairs, the shop, and the technicians. Yes, it is easy to be a "Monday morning quarterback." Many times, technicians are pressured into "trying something" in an attempt to repair an intermittent problem. Unless there is a "pattern failure" or TSB on a problem, it is best not to guess on repair parts. And, always perform a test drive before releasing a vehicle to the owner.

Chassis Grounds

Keep in mind that to simplify a vehicle's electrical system, automakers use body panels, frame members, and the engine block as current return paths to the battery.

Unfortunately, ground straps are often neglected or, worse, left disconnected after routine service. With the increased use of plastics in today's vehicles, ground straps may mistakenly be reconnected to a nonmetallic surface. The result of any of these problems is that the current that was to flow through the disconnected or improperly grounded strap is forced to find an alternate path to ground. Sometimes the current attempts to back up through another circuit. This may cause the circuit to operate erratically or fail altogether. The current may also be forced through other components, such as wheel bearings or shift and clutch cables that are not meant to handle current flow, causing them to wear prematurely or become seized in their housing.

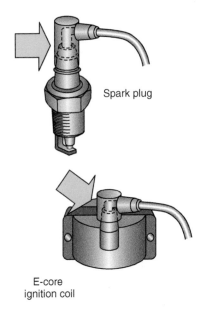

Spark plug

E-core
ignition coil

Water at these points can
cause a short to ground

Figure 9-11 Places to check for moisture
in an ignition system.

Examples of bad ground circuit-induced ignition failures include burnt ignition modules resulting from missing or loose coil ground straps, erratic performance caused by a poor ground, and intermittent ignition operation resulting from a poor ground at the control module.

Control Modules

Electronic control modules for EI systems can be mounted on the engine block, on top of the valve cover, under the coils, or on the firewall. They can also be contained within the PCM itself. Make sure that the mounting of the module is clean and tight, since some modules depend on their mounting surface for a proper ground. During your visual inspection, make sure the wiring connections are also clean and tight. The module for DI system is usually located in the distributor itself, but can be located remotely, or even on the outside of the distributor.

Sensors

The transistor in the control module is activated by a voltage pulse from a crankshaft position sensor. This sensor is either a magnetic pulse generator or Hall-effect sensor. The magnetic reluctor-style crank sensor for an EI system is shown in **Figure 9-12**. The crank sensor is mounted approximately 0.050 inch away from the reluctor cast into the crankshaft. The sensor is not adjustable. The sensor produces a characteristic waveform that can be checked with an oscilloscope and will look similar to that shown in **Figure 9-13**. Every notch produces an A/C voltage that rises and falls as the notch passes, telling the module the position of the crankshaft number one cylinder by referencing the "sync" notch. A magnetic pulse generator for a DI system turns with the distributor. The DI system is shown in **Figure 9-14**, and produces a similar waveform. An A/C signal is produced every time the points of the magnets align. The number of points is equal to the number of cylinders in the engine. The module interprets this signal as a trigger to break the primary circuit and produces a spark. This spark is then "distributed" to the correct spark plug from a single coil by the secondary ignition system. The DI system does not need a sync notch, as the distributor is mechanically indexed to the correct spark plug.

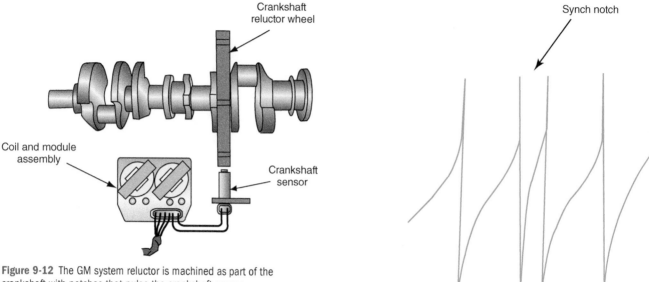

Figure 9-12 The GM system reluctor is machined as part of the crankshaft with notches that pulse the crankshaft sensor.

Figure 9-13 A portion of the magnetic sensor's waveform showing the sync notch.

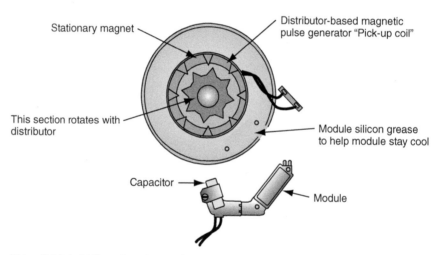

Figure 9-14 A distributor-based magnetic generating sensor, termed a "pick-up coil." Magnetic pulses are converted into an AC signal when the points of the magnets align.

One problem with some EI systems is the difficulty of accessing the primary circuit on waste spark systems.

The oscilloscope can look inside the ignition system by giving the technician a visual representation of voltage changes over time.

Hall-effect sensor problems are similar to those of magnetic sensors. This sensor produces a voltage when it is exposed to a magnetic field. The Hall-effect assembly is made up of a permanent magnet located a short distance away from the sensor. The Hall-effect sensor produces a digital signal in the shape of a square wave. Attached to the harmonic balancer pulley is a shutter wheel. When the shutter is between the sensor and the magnet, the magnetic field is interrupted and voltage immediately drops to zero (**Figure 9-15**). This drop-in voltage is the signal to the ignition module. When the shutter leaves the gap between the magnet and the sensor, the sensor produces voltage again. Hall-effect sensors are used for cam and crankshaft sensors and in both EI and DI systems. The advantage of the Hall-effect switch is that the amplitude, or strength of the signal, is the same regardless of the speed of the crank or cam rotation, making the signal more accurate at low speeds. Also, the pattern of the shutter wheel can be altered to show the ECM the exact position of the camshaft or crankshaft (**Figure 9-16**).

The ignition module switches primary current flow on and off in response to the signal from the pick-up unit. The module does not act alone in determining how long the

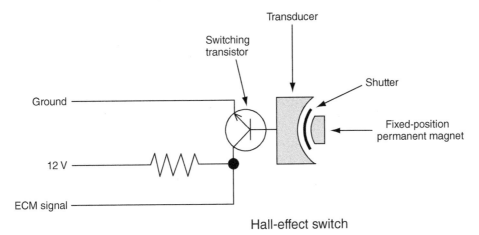

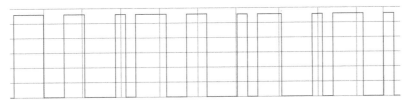

Hall-effect switch

Figure 9-15 The position of the shutter blade in or out of the Hall-effect sensor will produce either an on or an off signal.

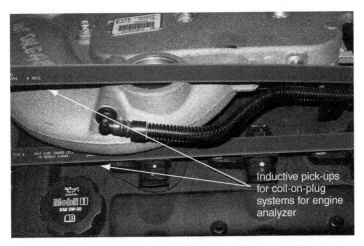

Figure 9-16 A camshaft sensor signal from a Hall-effect switch with varying windows for location identification.

Figure 9-17 COP ignition adapters for diagnostic analyzer connections.

dwell should be or when the firing of a spark plug should occur. The PCM sends a signal to the module to control the actual timing of events.

Scope Testing

The EI system secondary ignition can be tested with an oscilloscope, but with either COP or waste spark there are special adaptors that have to be used to read the secondary pattern. The COP adaptors are shown in **Figure 9-17**, while a typical setup for a waste spark system and an example pattern are shown in **Photo Sequence 19**. The interpretation of the scope patterns is very much like that used for a DI system and is shown in **Figure 9-18**. The primary circuit on waste spark systems is usually hidden by the fact that the coil pack is on

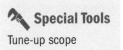

Special Tools

Tune-up scope

Normally we do not recommend testing electronic circuitry with test lamps. In this case, the circuit is accustomed to providing several amps. The test lamp draws only about 0.25 A.

Current ramping with a low-amp probe eliminates the need to "backprobe" connections.

top of the module, and some COP ignitions coils actually contain the ignition coil and feedback circuitry, making it difficult to obtain a primary pattern using conventional methods. Usually a COP coil with three or four wires will be difficult to obtain a good primary signal directly. Those coils with two wires are generally accessible by oscilloscope. On either system, the low-amp probe can be placed on the primary ignition feed wire to the ignition module and a primary current ramp can be watched for irregularities (**Figure 9-19**). A current ramp of the primary ignition on a typical waste spark system is also shown in **Photo Sequence 20**. If the vehicle is a COP ignition, the negative lead on an individual coil can be scoped with a resulting pattern, much like that shown in **Figure 9-20**. The primary can also be checked by using a test lamp watching for a flash as the primary is switched on and off as the engine is cranked, as shown in **Figure 9-21**.

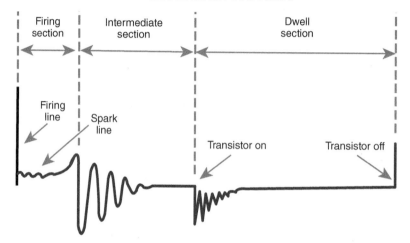

Figure 9-18 Typical secondary pattern.

Figure 9-19 Amperage ramp of a coil primary circuit on an EI system.

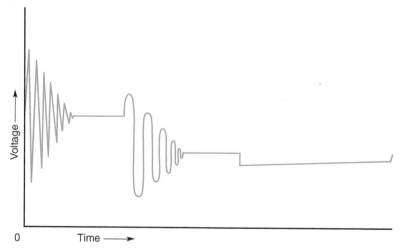

Figure 9-20 The primary voltage signal will mirror the secondary voltage trace, and may be more accessible.

Figure 9-21 Checking coil primary switching with a test lamp.

PHOTO SEQUENCE 19
Coil Current Ramping on a Waste Spark Ignition System

P19-1 For our example, we are connecting a console-style engine analyzer. With this setup, we can check secondary waveforms as well as current ramping the primary. Shown are the necessary levels for the secondary ignition pattern.

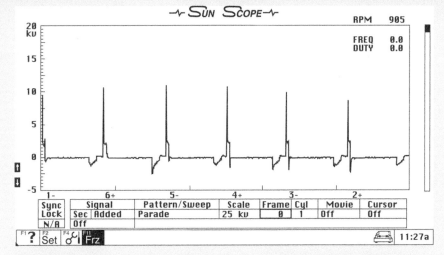

P19-2 There are three leads shown in the previous picture. Half of the spark plugs fire forward (conventionally, as shown by 6+, 4+, and 2+ on the pattern) and half fire backward. The forward leads are connected to a red cable end and the negative (backward, shown as 1-, 3-, and 5-) firing cables are connected to a black cable end. The number one cylinder is also clamped with a "sync" lead. The waste spark is also available for viewing on most testers.

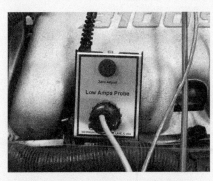

P19-3 Because we have the secondary in place, we can synchronize the current ramp to the correct cylinders. On a COP setup, each coil could be current ramped individually and give the same result.

PHOTO SEQUENCE 19 (CONTINUED)

P19-4 The low-amp probe is connected to the ignition feed at the module. The low-amp probe requires no piercing or backprobing. The amp clamp fits around the wire.

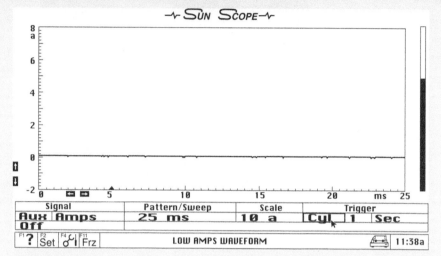

P19-5 Moving to the low-amp probe, we have to set the baseline to zero. Note that because the probe is dedicated to our analyzer, it changes the range to amperage automatically. Most probes require the use to convert from mV to mA.

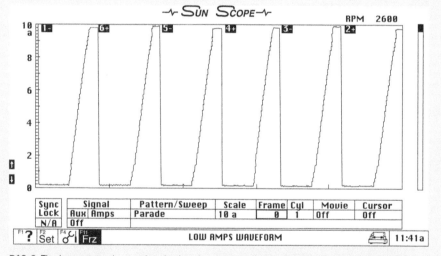

P19-6 The low-amp probe synchronized to the correct cylinders. The slope on the front edge is showing the time it takes for the coil to saturate. The flat tops are due to current limiting.

P19-7 Many low-amp probes are available that can be connected to either an oscilloscope or a DVOM.

GENERAL DIAGNOSTICS

Classroom Manual
Chapter 9, page 276

Standard test procedures using an oscilloscope, ohmmeter, and timing light can be used to diagnose problems in distributorless ignition systems. Keep in mind, however, that some problems involving one cylinder in a waste spark system may also occur in its companion cylinder that fires off the same coil. COP ignition coils have their own system to work from, so coils have no effect on each other unless a problem in the primary switching sensors or modules exists. Some oscilloscopes require their pickups placed on each pair of cylinders to view all patterns. Special adapters are available to make these hookups less troublesome. One of the best ways to test these circuits is to use a low-amp probe to see current flow through the coil. If there is a problem with the coil windings, it should be apparent in the current ramp.

Follow the testing procedures outlined in the vehicle's service manual for the engine control system. Specific computer-generated trouble codes are designed to help troubleshoot ignition problems in these systems. The diagnostic procedure for EI systems varies depending on the vehicle make and model year. Always follow the recommended procedure.

When diagnosing these systems, keep in mind there is a separate primary circuit for each coil. If one coil does not work properly, it may be caused by something common or not common to other coils. Regardless of the system's design, there are common components in all electronic ignitions: an ignition module, crankshaft and/or camshaft sensors, ignition coils, secondary circuit, and spark plugs. Most of these components are common to only the spark plug or spark plugs they are connected to. The components that are common to all cylinders (such as the camshaft sensor) will most often be the cause of no-start problems. The other components such as a single spark plug wire will generally cause misfire problems. There are many other factors that can cause a misfire. Check those before moving on to the ignition system. Test for intake manifold vacuum leaks. Also check the engine's compression and the fuel injection system.

Check Secondary Wiring and Connections

All ignition system diagnosis should begin with a visual inspection. The system should be checked for obvious problems: disconnected, loose, or damaged secondary cables; disconnected, loose, or dirty primary wiring; a worn or damaged primary system switching mechanism; and an improperly mounted ignition module.

On distributorless or direct ignition systems, visually inspect the secondary wiring connections at the individual coil modules. Make sure all of the spark plug wires are securely fastened to the coil and the spark plug. If a plug wire is loose, inspect the terminal for signs of burning. The coils should be inspected for cracks or any evidence of leakage in the coil tower. Check for evidence of terminal resistance. A loose or damaged wire or a bad plug can lead to carbon tracking of the coil, or the spark plug wire boot and spark plug. If a spark plug is carbon tracked as in **Figure 9-22**, always

According to the SAE J1930 standards, the term "electronic ignition" replaces all previous terms for distributorless ignition systems.

Fifteen percent of all drivability complaints are caused by problems in the ignition system.

Seventy percent of all electrical problems are at the connectors.

Figure 9-22 On COP and waste spark systems, carbon tracking of the spark plugs is a common problem.

replace the spark plug wire and spark plug. If this condition exists on the coil itself, the coil must be replaced.

Inspect all secondary cables for cracks and worn insulation, which cause high-voltage leaks. Inspect all of the boots on the ends of the secondary wires for cracks and hard brittle conditions. Replace the wires and boots if they show evidence of these conditions. Some vehicle manufacturers recommend spark plug wire replacement only in complete sets.

If the visual inspection does not reveal obvious problems, test those individual components and circuits that could cause the problem.

Crankshaft and Camshaft Position Sensors

If a crankshaft position (CKP) sensor or in some vehicles the camshaft position (CMP) sensor fails, the engine will not start (**Figure 9-23**). Both of these sensor circuits can be checked. If the sensors are receiving the correct amount of voltage and have good low-resistance circuits, a magnetic reluctance sensor should output an AC signal. The Hall-effect sensor should output a square wave signal on an oscilloscope. Of course, if the crank sensor does not output a signal, the scan tool or the tachometer (if equipped) will not show an RPM signal. If the cam sensor fails, the cam sensor PID on the scan tool will not change while cranking. Some manufacturers have a fail-safe mode where it is possible the engine can start and run with a faulty camshaft sensor. Each manufacturer may use a different strategy and will vary from one engine family to another. Depending on the strategy, a prolonged cranking time may be a symptom, or perhaps it may take two to three times for the engine to start.

No-Start Diagnosis

When an EI or DI system has a no-start problem, begin your diagnosis with a visual inspection of the ignition system. Check for good and tight primary connections. Perform a spark intensity check with a spark tester. A bright, snapping spark indicates that the secondary voltage output is good. If the spark is weak or if there is no spark, check the

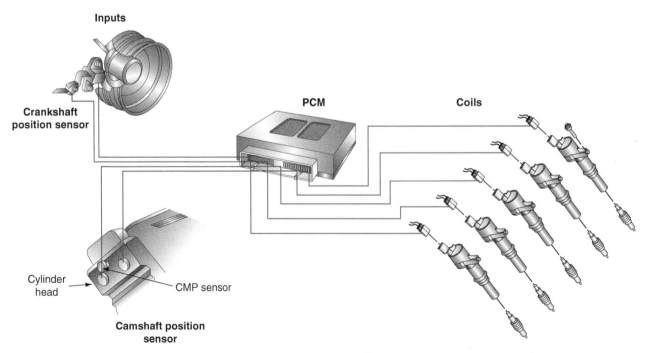

Figure 9-23 Inputs from the camshaft and crankshaft sensors are processed by the PCM to fire the coils.

primary circuit voltage available at the coil. If voltage is not available, check the primary voltage circuits and fuses for problems. If voltage is available, check for a pulsing signal at the coil ground side using a test lamp. If the coil is receiving a signal but not firing, the coil should be checked. If there is no signal to the coil, then the crankshaft position sensor may not be sending a message to the module, or the module may not be sending a signal to the coil.

If a crankshaft or camshaft sensor needs to be replaced, always clean the mounting locations and look at service information for the proper procedure involved.

The base timing cannot be adjusted on an EI system. However, the air gap between the sensor and the trigger wheel can affect the operation of the ignition system. On many EI systems, this gap is not adjustable. This does not mean that the gap is not important and should not be checked. The gap should be checked whenever possible. If there is no provision for adjusting the gap and the gap is incorrect, the sensor should be replaced. If the gap between the blades and the crankshaft sensor is incorrect, the engine may fail to start, stall, misfire, or hesitate on acceleration. It is important to note that the sensor air gap does not affect ignition timing.

The primary can also be checked with a logic probe. There are three lights on a logic probe. The red light illuminates when the probe senses more than 10 volts. When the monitored signal has less than 4 volts, the green light flashes. The yellow light will flash whenever the voltage changes. This light is used to monitor a pulsing signal, such as one produced by a digital sensor such as a Hall-effect switch.

Special Tools
Logic probe

To check the primary circuit with a logic probe, turn the ignition on. Touch the probe to both (positive and negative) primary terminals at the coil. The red light should come on at both terminals (**Figure 9-24**). This indicates that at least 10 volts is available

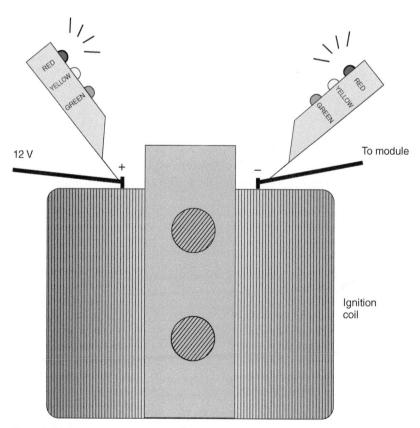

Figure 9-24 The red light on a logic probe should turn on when touched to both sides of the primary winding.

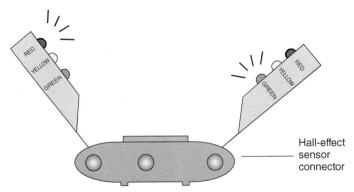

Figure 9-25 One end terminal of a Hall-effect sensor connector should cause the red light to come on; the other end terminal should cause the green light to come on.

Double-check the ground connection for voltage drop when using a logic probe. Since the ground LED comes on with less than 4 volts, the ground could have excessive voltage drop and still light.

to the coil and there is continuity through the coil. If the red light does not come on when the positive side of the coil is probed, check the power feed circuit to the coil. If the light comes on at the positive terminal but not on the negative, the coil has excessive resistance or is open.

Now move the probe to the negative terminal of the coil and crank the engine. The red and green lights should alternately flash. This indicates that over 10 volts is available to the coil while cranking and that the circuit is switching to ground. If the lights do not come on, check the ignition power feed circuit from the starter. If the red comes on but the green light does not, check the crankshaft or camshaft sensor. If these are working properly, the ignition module is probably defective.

A Hall-effect switch is also easily checked with a logic probe. If the switch has three wires, probe the outer two wires with the ignition on (**Figure 9-25**). The red light should come on when one of the wires is probed, and the green light should come on when the other wire is probed. If the red light does not turn on at either wire, check the power feed circuit to the sensor. If the green light does not come on, check the sensor's ground circuit.

The oddly spaced slot is called the synch slot. This function is sometimes achieved with a camshaft sensor.

> ⚙️ **SERVICE TIP** When checking the ignition system with a lab scope, gently tap and wiggle the components while observing the trace. This may indicate the source of an intermittent problem.

Back probe the center wire and crank the engine. All three lights should flash as the engine is cranked. The red light will come on when the sensor's output is above 10 volts. As this signal drops below 4 volts, the green light should come on. The yellow light will flash each time the voltage changes from high to low. If the logic probe's lights do not respond in this way, check the wiring at the sensor. If the wiring is OK, replace the sensor.

Keep in mind that the action of a Hall-effect switch and an inductive sensor can be monitored with a lab scope. The waveform for a normally operating magnetic inductive-type crankshaft sensor is shown in **Figure 9-26**. Carefully examine the waveform for traces of noise and false pulses.

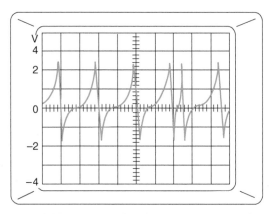

Figure 9-26 Normal waveform for an inductive crankshaft sensor.

⚙⚙ **SERVICE TIP** Another quick test of the primary ignition triggering operation is to check fuel injector pulsing. If a no-start is ignition related, this test will eliminate the camshaft or crankshaft signals in a port fuel injection system as the fault. Connect a fuel injector noid light to the fuel injector connector, and observe the light while cranking the engine. If the injector noid light flashes, the primary triggering circuit is functioning. The injector will not pulse if there is no trigger input to the PCM. This quickly confirms crankshaft sensor operation. If there is no spark and the injector pulses, the fault is probably within the secondary system or ignition module. Remember that the crankshaft position sensor and CMP signals are needed for fuel injection as well as spark. If the crankshaft position sensor does not work, then the PCM does not know the engine is running and therefore will not provide spark or fuel.

The pattern of the trace will vary according to the position and number of slots machined into the trigger wheel. The trigger wheel in **Figure 9-27** has nine slots. Eight of them are evenly spaced, and one slot is placed close to one of the eight. The trace for this type sensor will also have eight evenly spaced pulses. One of the pulses will quickly be followed by another pulse (**Figure 9-28**). Any waveform that does not match the configuration of the trigger wheel indicates a problem with the sensor or its circuit.

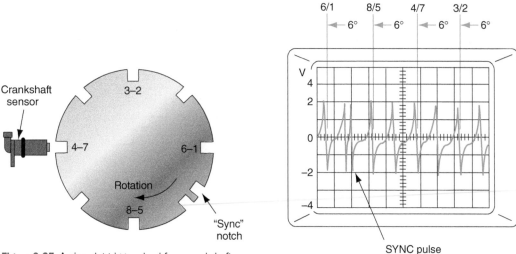

Figure 9-27 A nine-slot trigger wheel for a crankshaft sensor.

Figure 9-28 Waveform for the sensor shown in Figure 9-27.

If the crankshaft sensor is a Hall-effect switch, the scope pattern should reflect a digital signal on the scope (**Figure 9-29**). Each pulse should be identical in spacing, shape, and amplitude. By using a dual trace lab scope, the relationship between the crankshaft sensor and the ignition module can be observed. During starting, the module will provide a fixed amount of timing advance according to its program and the cranking speed of the engine. By observing the crankshaft sensor output and the ignition module, this advance can be observed (**Figure 9-30**). The engine will not start if the ignition module does not provide for a fixed amount of timing advance.

Ground Circuits

Poor grounds can be identified by conducting voltage drop tests and monitoring the circuit with a lab scope (**Figure 9-31**). When conducting a voltage drop test, the circuit must be turned on and have current flowing through it. If the circuit is tested without current flow, the circuit will show zero voltage drop, indicating that it is good regardless of the amount of resistance present. For a computerized ground to be good, the voltage drop should be very low—no more than 0.2–0.3 V.

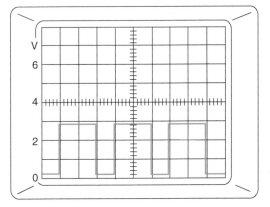

Figure 9-29 Normal waveform for a Hall-effect sensor.

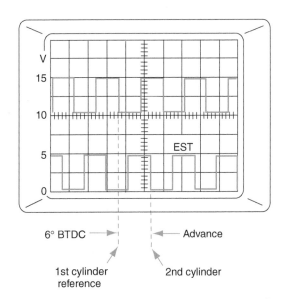

Figure 9-30 EST and crankshaft sensor signals compared on a dual-trace scope.

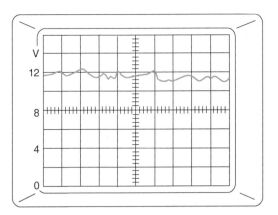

Figure 9-31 A voltage signal caused by a poor ignition module ground.

The same is also true when checking a ground with the lab scope. Make sure the circuit is on. If the ground is good, the trace on the scope should be at zero volt and be flat. If the ground is bad, some voltage will be indicated and the trace will not be flat.

Many service procedures tell the technician to check for grounds by checking the resistance to ground with an ohmmeter. The ohmmeter does not put out enough current to stress the circuit; it is much better to do a dynamic test with the voltmeter using voltage drops.

Often, a bad sensor ground will cause the same symptoms as a faulty sensor. Before condemning a sensor, check its ground with a lab scope. **Figure 9-32** shows the output of a good Hall-effect switch with a bad ground. Depending on the Hall-effect sensor or the circuit design, a potential faulty ground may cause the signal at the top or bottom of the pattern to be abnormal. Check the service information for specific operation.

Electromagnetic Interference

EMI problems can be identified by connecting a lab scope to voltage and ground wires. Common problems, such as poor spark plug wire insulation, will allow EMI. In addition, EMI-induced signals can influence other low-voltage signals, such as inputs to the PCM. These false values can be interpreted by the PCM as an actual signal and cause improper outputs to the engine. **Figure 9-33** shows a voltage trace contaminated with EMI from the ignition system.

To minimize the effects of EMI, check to make sure that sensor wires running to the computer are routed away from potential EMI sources. Rerouting a wire by no more than an inch or two may keep EMI from falsely triggering or interfering with computer operation.

Checking Different Types of Ignition Coils

Some styles of coils in an EI system can be checked in the same way as conventional coils. **Figure 9-34** shows the procedure on a waste spark system. When checking the resistance across the windings, pay particular attention to the meter reading. The procedure to check a distributor-style coil is similar. **Figure 9-35** shows the procedure for checking the primary windings, **Figure 9-36** shows the procedure to check the secondary windings, and **Figure 9-37** shows how to check the primary windings for a short to the case.

To check the primary windings, use an ohmmeter and connect the meter leads to the primary coil terminals to test the winding. An infinity ohmmeter reading indicates an open winding. The winding is shorted if the meter reading is below the specified resistance. Most primary windings have a resistance of 0.5 to 2.0 ohms, but the exact manufacturer's specifications must be compared to the meter readings. It is important

The ohmmeter tests on the ignition coil windings do not check the coil for insulation leakage.

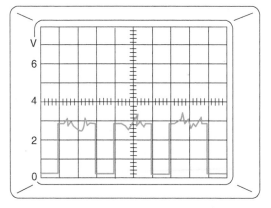

Figure 9-32 A faulty sensor signal caused by a bad ground.

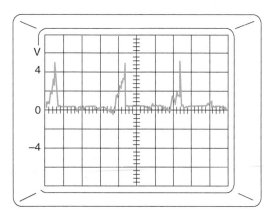

Figure 9-33 A voltage trace with ignition system noise.

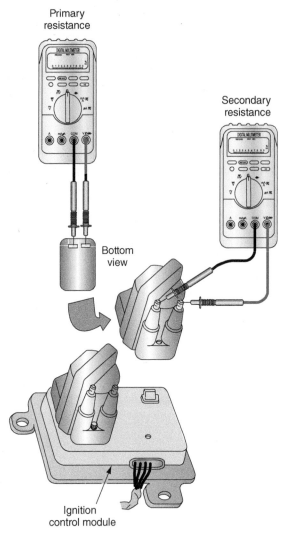

Primary resistance

Secondary resistance

Bottom view

Ignition control module

Figure 9-34 Checking primary and secondary resistance of an a waste spark coil.

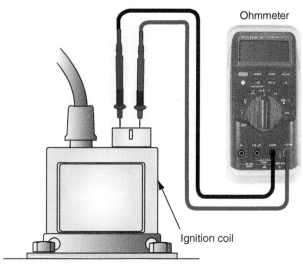

Ohmmeter

Ignition coil

Figure 9-35 Ohmmeter connected to primary coil terminals.

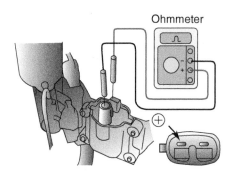

Ohmmeter

Figure 9-36 Ohmmeter connected from one primary terminal to the coil tower to test secondary winding.

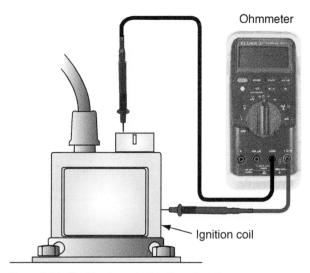

Ohmmeter

Ignition coil

Figure 9-37 Checking for a short to the coil casing.

to note that some COP coils have extra circuitry that may prevent a good ohmmeter test on the coil windings. If you cannot test with the ohmmeter, us a low-amp probe and do a current ramp, or, in some cases, the vehicle may set a code for a coil or coil circuit problem.

To check the secondary winding, place the meter on the correct scale and connect it from the coil's secondary terminal to one of the primary terminals. A meter reading below the specified resistance indicates a shorted secondary winding. An infinity meter reading proves that the winding is open.

In some coils, the secondary winding is connected from the secondary terminal to the coil frame. When the secondary winding is tested in these coils, the ohmmeter must be connected from the secondary coil terminal to the coil frame or to the ground wire extending from the coil frame.

Many secondary windings have 5,000 to 20,000 ohms resistance, but the meter readings must be compared to the manufacturer's specifications. The ohmmeter tests on the primary and secondary windings indicate satisfactory, open, or shorted windings. Ohmmeter tests do not indicate such defects as defective insulation around the coil windings, which cause high-voltage leaks. Therefore, an accurate indication of coil condition is the coil maximum voltage output test with a spark tester connected from the coil secondary wire to ground as explained in the no-start diagnosis.

> **CUSTOMER CARE** When verifying the customer's complaint, be aware of some common terms having more than one meaning that the customer may use. If possible, go for a test drive with the customer in order to witness the symptom together. For instance, the customer may complain of a transmission shifting problem or a shudder at part-throttle. To the customer, this may seem like a transmission-related problem. However, to a trained technician it may be an ignition-related problem. For example, a weak coil that breaks down under load can cause the symptom described. Listen to the customer, but politely decline his or her self-diagnosis unless it is obvious and relevant.

Secondary Circuit

Testing the secondary circuit of an EI system is just like testing the secondary of any other type of ignition system. The spark plug wires and spark plugs should be tested to ensure they have the appropriate amount of resistance. Since the resistance in the secondary dictates the amount of voltage the spark plug will fire with, it is important that secondary resistance be within the desired range.

Excessive resistance is not the only condition that will affect the firing of a spark plug. Spark plug wires and the spark plug itself can allow the high voltage to leak and establish current through another metal object instead of the electrodes of the spark plug. When this happens, the spark plug does not fire and combustion does not take place within the cylinder.

Spark plug wires can be also checked with an ohmmeter. To do this, use an ohmmeter on the K Ohms scale, then connect it across each cable. If the ohmmeter reading is more than that specified by the manufacturer, replace the wire. When replacing spark plug wires, be sure they are routed properly. COP plug systems have a boot that attaches the coil to the spark plug arc-out and carbon track just like spark plug wires.

Ford recommends that an engine be warm to the touch, not hot or cold, before removing spark plugs from the engine.

Plug wire connections must be clean; if they appear dark or rusty, replace the wire.

An ohmmeter reading of infinity is usually displayed as an OL or simply 1. This means that the coil windings are open. A very low reading on the secondary means that the windings are shorted.

Classroom Manual
Chapter 9, page 266

Improperly torqued spark plugs can cause a misfire on EI systems because the circuit is completed through ground.

 Caution

Before removing spark plugs, it is a good idea to clean around the spark plug with compressed air to prevent debris from falling into the combustion chamber.

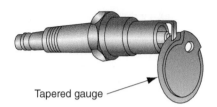

Figure 9-38 Using a taper gauge to check spark plug gap.

Special Tools

Anti-seize compound
Spark plug torque wrench

Most late-model engines are equipped with platinum or iridium tip spark plugs. Extra care must be taken when setting the gap on these plugs. Iridium spark plugs should not be re-gapped because the electrode is very small. A wire gauge is recommended by many manufacturers for checking and adjusting the spark plug gap on platinum-tipped styles. However, the use of more delicate spark plugs has led many manufacturers to recommend the use of a tapered gauge (**Figure 9-38**). The platinum-tipped spark plugs should never be filed to prevent damage to the small platinum pads on the plugs.

Keep in mind while replacing spark plugs that the secondary circuit is completed through the metal of the engine. If the spark plugs are not properly torqued into the cylinder heads, the threads of the spark plug may not make good contact and the circuit may offer resistance. Always tighten spark plugs to their specified torque. Many of today's engines have aluminum cylinder heads. Not only do these heads require different torque specifications than iron heads, they also require extra care when installing the plugs. Make sure the threads of the plugs match the threads of the spark plug bores. Take extra care not to cross-thread them. Repairing damaged threads in a cylinder head can be a costly and time-consuming job.

Never file a platinum-tipped plug.

An easy way to help prevent cross-threading of spark plugs is to insert the porcelain end into a vacuum line and then insert the spark plug. It should be difficult to cross-thread a spark plug with the vacuum hose. This can also be helpful if the spark plug is difficult to reach. No method is fail-safe, but this is safer than trying to start a spark plug with the ratchet and socket.

Most manufacturers recommend the use of an anti-seize compound on spark plug threads. This compound must be applied in the correct amounts and at the correct place (**Figure 9-39**). Too little compound will cause gaps in the contact between the spark plug threads and the spark plug bores. Too much may allow the spark to jump to a buildup rather than the spark plug electrode.

Classroom Manual
Chapter 9, page 267

CHRYSLER EI SYSTEMS

Chrysler's EI systems use the waste spark method for firing the spark plugs. These systems rely on a camshaft and a crankshaft sensor to inform the PCM as to the position of piston number one and time at which the other pistons reach TDC. Basic ignition timing is not adjustable in these systems.

TDC stands for top dead center. TDC is the highest piston position in the block.

If an engine misfire problem is caused by the ignition system, the basic operation of the secondary circuits should be checked. This testing should begin with a complete visual inspection of the circuit. With any waste spark system, the secondary circuit should be looked at as groups of two. Each coil fires two cylinders. Each coil and its associated spark plugs should be checked as a group.

For each group of cylinders, the coil, spark plug wires, and spark plugs should be checked on a scope or with an ohmmeter. Resistance checks of coils and spark plug wires are recommended. Make sure the plug wires are fully seated onto the spark plug and the ignition coil.

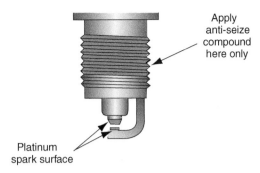

Figure 9-39 Proper placement of anti-seize compound on the threads of a spark plug.

No-Start Diagnosis

Chrysler EI systems use a crankshaft timing and camshaft reference sensor that are modified Hall-effect switches. When the engine fails to start, follow these steps:

1. Check for fault codes, which could be caused by a defective CMP or CKP sensor signal or low primary current in coil number one, two, or three.

> ⚙️ **SERVICE TIP** When using a digital voltmeter to check a crankshaft or camshaft sensor signal, crank the engine a very small amount at a time and observe the voltmeter. The voltmeter reading should cycle from almost 0 volt to a higher voltage of 9 to 12 volts. Because digital voltmeters do not react instantly, it is difficult to see the change in voltmeter reading if the engine is cranked continually.

Prior to checking the voltage signal from a crankshaft or camshaft sensor on an EI system, the sensor voltage supply and ground wires should be checked.

2. With the engine cranking, check voltage from the orange wire to ground on the CKP and the CMP sensors (**Figure 9-40**). Over 7 volts is satisfactory. If the voltage is less than specified, repeat the test with the DMM connected from PCM terminal 7 to ground. If the voltage is satisfactory at terminal 7 but low at the sensor orange wire, repair the open circuit or high resistance in the orange wire. If the voltage is low at terminal 7, replace the PCM. Be sure 12 volts are supplied to PCM terminal 3 with the ignition switch off or on, and 12 volts must be supplied to PCM terminal 9 with the ignition switch on. Check the PCM ground connections on terminals 11 and 12 before replacing the PCM.

3. With the ignition switch on, check the voltage drop across the ground circuit (black or light-blue wire) on the crankshaft timing sensor and the camshaft reference sensor. A reading below 0.2 volt is satisfactory.

4. If the reading in steps 3 and 4 are satisfactory, connect a 12-volt test lamp or a digital voltmeter from the gray or black wire on the CKP sensor and the tan or yellow wire on the camshaft reference sensor to ground. When the engine is cranking, a flashing 12-volt lamp indicates that a sensor signal is present. If the lamp does not flash, sensor replacement is required. Each sensor voltage signal should cycle from low voltage to high voltage as the engine is cranked.

If the sensor tests are satisfactory and the engine will not start, proceed with these coil and PCM tests:

1. Check the spark plug wires with an ohmmeter.

2. With the engine cranking, connect a voltmeter from the dark-green or black wire on the coil to ground. If this reading is below 12 volts, check the **automatic shutdown (ASD) relay** circuit (**Figure 9-41**).

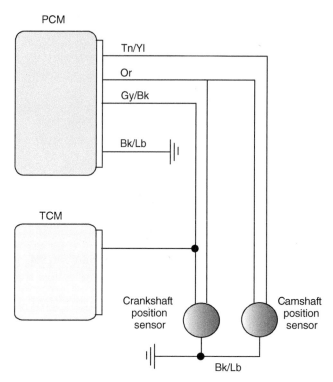

Figure 9-40 CKP and CMP sensor terminals.

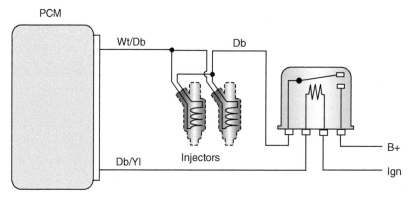

Figure 9-41 ASD relay circuit.

3. If the reading in step 2 is satisfactory, check the primary and secondary resistance in each coil with the ignition switch off. Primary resistance should be from 0.52 to 0.62 ohm, and the secondary resistance from 11,000 to 15,000 ohms (**Figure 9-42**). If these ohm readings are not within specifications, replace the coil assembly.

> **SERVICE TIP** A no-start problem on Chrysler 3.3-liter engines may be a shorted cam or crank sensor. These sensors are three-wire Hall-effect sensors that are fed 8 volts by the PCM. If the power feed is shorted to ground, the PCM shuts off causing the engine to stall and not restart. Because the PCM is not working, a scan tool will not work, nor will the system's self-test. To verify this as the cause of the problem, simply disconnect the sensor and try to retrieve data with a scan tool. If data is now available, the sensor or wire to it is shorted. Voltage to the sensor can also be checked with a voltmeter; again, there should be 8 volts available.

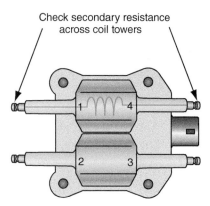

Figure 9-42 Measure across the companion coil towers to measure secondary winding resistance.

SERVICE TIP Some Chrysler vehicles have separate fuel pump and ASD relays. Always use the proper wiring diagram for the vehicle being tested.

Primary resistance on any coil is usually less than 2 ohms, and coil secondary resistance is generally in the range of 5,000 to 20,000 ohms depending on the manufacturer. Always check service information for your particular vehicle.

4. With the ignition switch off, connect an ohmmeter across the three wires from the coil connector to PCM terminals 17, 18, and 19 (**Figure 9-43**). These terminals are connected from the coil primary terminals to the PCM. If an infinity ohmmeter reading is obtained on any of the wires, repair the open circuits.

5. Connect a 12-volt test lamp from the dark-blue or black wire, the dark-blue or gray wire, and the black or gray wire on the coil assembly to ground while cranking the engine. If the test lamp does not flutter on any of the three wires, replace the PCM. Since the crankshaft and camshaft sensors, wires from the coils to the PCM, and voltage supply to the coils have been tested already, the ignition module must be defective. This module is an integral part of the PCM on Chrysler vehicles; thus, the PCM must be replaced.

WARNING Because EI systems have more energy in the secondary circuit, electrical shocks from these systems should be avoided. The electrical shock may not injure the human body, but such a shock may cause you to jump and hit your head on the hood or push your hand into contact with a rotating cooling fan.

6. If the tests in steps 1 to 4 are satisfactory, connect a spark tester to each spark plug wire and ground and crank the engine. If any of the coils do not fire on the two spark plugs connected to the coil, replace the coil assembly.

Sensor Replacement

If the crankshaft timing sensor or the camshaft reference sensor is removed, follow this procedure to replace the sensor:

1. Thoroughly clean the sensor tip and install a new spacer on the sensor tip. New sensors should be supplied with the spacer installed (**Figure 9-44**).
2. Install the sensor until the spacer lightly touches the sensor ring, and tighten the sensor mounting bolt to 105 inch-pounds.

Caution

Do not crank or run an EI-equipped engine with a spark plug wire completely removed from a spark plug. This action may cause leakage defects in the coils or spark plug wires. Always use a spark tester.

Terminal end view

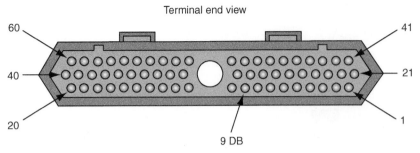

9 DB

Figure 9-43 PCM terminal identification numbers.

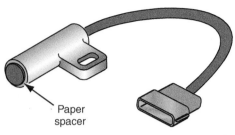

Paper spacer

Figure 9-44 Paper spacer on the tip of a crankshaft sensor.

FORD EI WITH COP IGNITION SYSTEMS

Ford has used several designs of EI through the years. It is important that the design be identified prior to testing the system. Throughout all, certain basic operational features can be applied. The CKP signal is an indication of crankshaft position and engine speed. The CKP signal is sent to both the ignition module and the PCM. The CMP signal is used in conjunction with CKP for the identification of cylinder number one for the PCM.

Many of Ford's recommended diagnostic procedures for their EI systems involve the use of their Integrated Diagnostic System (IDS), and their Vehicle Communications Module (VCM).

1. For this explanation, it is assumed that the misfire monitor has set a code P0301, for a misfire on number one cylinder of a Ford vehicle with a 5.4-liter V8. If any other fuel delivery codes had set, it is recommended that those be diagnosed first, since the other codes may be the cause of the misfire.
2. Do a quick visual inspection of the vehicle, checking connections, oil and coolant levels, and battery cable and condition, besides any obvious wiring or connection problems. The primary circuit for each coil is constantly monitored by the PCM. If a coil circuit code was set, then the technician would be instructed to check for ignition switching and power at the coil connection, then the primary resistance of the coil itself.
3. Before checking the ignition system for spark, disable the fuel supply by turning off the fuel pump inertia switch, which will keep fuel from flooding into the cylinders while testing the ignition system.
4. Remove the coils and check each coil for spark using a spark tester, pay particular attention to number one, as it set the misfire code. Record any coils with weak sparks.
5. On any coils that did not fire or had weak spark, check the resistance of the spark plug from the center terminal to the plug boot connection (**Figure 9-45**). Replace any spark plug that has less than 2,000 or more than 20,000 ohms resistance. Remember that the spark plug does have a resistor assembly built inside that can burn out over time.
   Many manufacturers recommend simply swapping the spark plug out.
6. If the spark plug(s) checks out OK, then check the resistance of the ignition coil secondary from the positive terminal of the coil connector to the spring in the coil boot. The resistance should be about 5,000 to 6,000 ohms (**Figure 9-46**).
7. If the coils and plugs check out OK, then the problem might be an intermittent in the primary triggering circuit. The technician could check the signals from the sensors at the appropriate PCM connections (**Figure 9-47**) while test driving (let someone else drive while the technician looks for problems) to duplicate the problem.

It is important to note that some COP coils cannot be checked for a secondary or primary resistance due to the presence of an internal module. Most two-wire COP coils can be checked, but many three- and four-wire coils cannot be checked in this manner. It may be necessary to check for a DTC concerning the coil, substitute a known good coil, or use an inductive current probe and compare current ramps.

 Caution

Improper sensor installation may cause sensor, rotating drive plate, or timing gear damage.

Classroom Manual
Chapter 9, page 270

 **Special Tools**

Star tester
Breakout box

Figure 9-45 A spark plug from Ford's 5.4-liter V8 should have a resistance from 2,000 to 20,000 ohms from the center electrode to the top electrode.

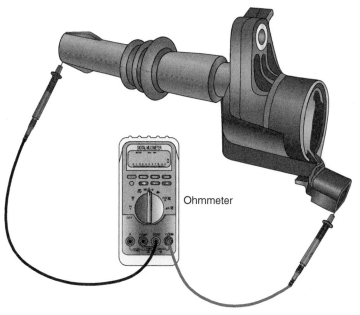

Ohmmeter

Figure 9-46 A coil from Ford's 5.4-liter V8 should have a resistance between 5 and 6 K ohms on the secondary windings.

190 Pin PCM Harness Connector

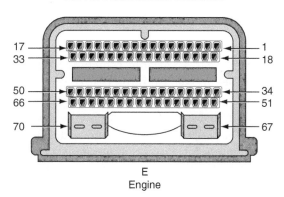

E
Engine

Component	PCM Pin #	Value in Hertz
Crankshaft position sensor (CKP)	E47	350 Hz @ idle 900 Hz @ 55 mph
Camshaft Position sensor #1	E44	25 Hz @ idle 65 Hz @ 55 mph
Camshaft position sensor #2	E45	25 Hz @ idle 65 Hz @ 55 mph

Figure 9-47 Pin outs with expected values for the CKP and both CMP sensors.

FORD WASTE SPARK SYSTEM

Visual Inspection

Ford uses locking tabs on their waste spark ignition coil connections. To remove the secondary cable, squeeze the tabs together and pull straight up (**Figure 9-48**). When reinstalling the wires, make sure the cable is in place by pushing down on the center of the cable end until the cable snaps into place (**Figure 9-49**).

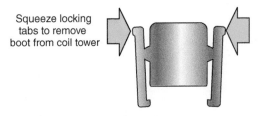

Figure 9-48 Spark plug wire locking tabs.

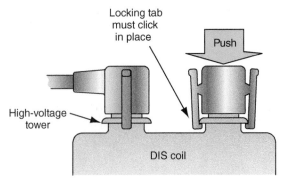

Figure 9-49 When reinstalling the plug wires, firmly push down on the center of the boot to lock it into position.

Make sure the factory-supplied wire separators are used and are in their original position after the spark plug wires are put in place (**Figure 9-50**). The spark plugs and ignition coils can be tested in the same way as other ignition systems.

Camshaft and Crankshaft Sensors

Although the ignition timing on Ford's EI system is not adjustable, the air gaps at the crankshaft and camshaft sensors are critical (**Figure 9-51**). On some engines this gap is adjustable, and on others it is an indication that the sensor should be replaced. When checking the gap, make sure there are no signs of damage to the rotating vane assembly.

Ford's **electronic distributorless ignition system (EDIS)** is a high data rate system. The crankshaft position sensor is a variable reluctance-type sensor triggered by a 36-minus-1 tooth trigger wheel pressed onto the rear of the crankshaft dampener. The signal generated by this sensor is called a **variable rate sensor signal (VRS)**. The VRS signal provides engine position and rpm information to the ignition module. Base timing is 10 degrees BTDC and is not adjustable. The ignition module receives information from the crankshaft position sensor and other sensors to calculate the on and off times for the coils in order to achieve the desired dwell and spark advance. The ignition module also synthesizes a PIP signal for use by the PCM's engine control strategy.

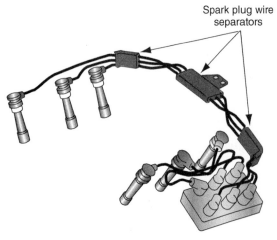

Figure 9-50 Make sure the wire separators are in place when reconnecting the spark plug wires.

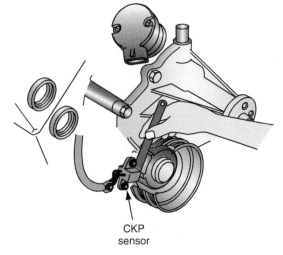

Figure 9-51 CKP sensor gap measurement and alignment.

Figure 9-52 Ford 105 pin breakout box for testing EEC-V systems.

Waste Spark EI Service and Diagnosis

The diagnostic procedure varies according to the system being tested. It is important to follow the correct wiring diagram and procedure for the system being tested. Some systems have an available breakout box (**Figure 9-52**). The breakout box allows a technician to check PCM connections with a DMM or oscilloscope without having to back probe the harness connections.

The technician must follow the test procedures given in the service manual and measure the resistance or voltage at the specified breakout box terminals connected to the ignition system. If a breakout box is not available, the voltage and resistance measurements must be taken at the terminals of the individual components or at the PCM. The following is a general no-start diagnostic procedure for a waste spark system.

1. Connect a spark tester to each of the spark plug wires to ground and crank the engine. If the test plug does not fire on a pair of spark plugs connected to the same coil, test the spark plug wires connected to that coil. If these wires are good, the coil is probably bad. When the spark tester does not fire on any plug, continue testing.
2. Connect a voltmeter from terminal two on each coil pack to ground (**Figure 9-53**). With the ignition switch on, the voltmeter should read 12 volts. If the voltage is less than that, test the wire from the ignition switch to the coils and check the ignition switch.
3. With the ignition off, connect an ohmmeter across the primary terminals of each coil. If the primary winding resistance readings are not within specifications, replace the coil.
4. Connect the ohmmeter across the secondary terminals of each coil pack. Replace the coil pack if the secondary resistance is not within specifications.

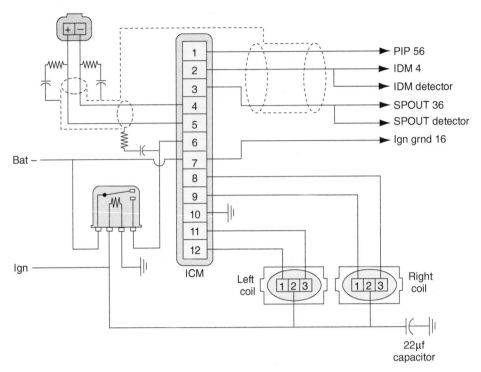

Figure 9-53 Wiring diagram for the EI system on a 4.6-liter engine.

5. Connect the ohmmeter between each primary coil terminal to the terminal on the primary terminal of the ignition module. Each wire should have less than 0.5-ohm resistance. If the resistance is greater than that, replace the wire.

6. Connect the voltmeter from terminal 6 on the ignition module to ground. With the ignition on, the voltmeter should read 12 volts. If the voltage is less than that, test the wire from the power relay to terminal 6 and the power relay itself. Turn the ignition off.

7. Connect the voltmeter from terminal 10 at the ignition module and ground. The voltmeter should read 0.5 volt or less with the ignition on. If the reading is greater than that, repair or place the ground wire.

8. Connect the voltmeter from terminal 7 at the ignition module and ground. With the ignition on, the voltmeter should read less than the specified amount. If the voltage is higher, repair the ground wire.

9. Connect the voltmeter from terminals 8, 9, 11, and 12 at the ignition module to ground, and crank the engine. The voltmeter reading should fluctuate on each wire. If the readings do not fluctuate on one wire, replace the ignition module. If the readings do not fluctuate, proceed testing.

10. With the ignition off and the ignition module disconnected, connect an ohmmeter across terminals 4 and 5 at the harness. If it reads 2,300 to 2,500 ohms, the CKP sensor and connecting wires are OK. If the readings are outside those specifications, repeat the test at the CKP sensor terminals. If the readings are now within specifications, repair the wires. If the readings are still outside the specified range, replace the sensor.

11. Inspect the trigger wheel behind the crankshaft pulley and the CKP sensor for damage.

12. If the voltmeter did not fluctuate during step 9 but the CKP sensor checks out fine, replace the ignition module.

GENERAL MOTORS EI SYSTEMS

GM has used several variations of EI systems. Some are waste spark systems, while others are direct or integrated systems. With each basic design there have been different operational designs as well. Each of these variances has its own diagnostic procedure. Make sure you identify the system being worked on prior to doing any exhaustive diagnostics. Each of the diagnostic procedures is based on the premise that the secondary circuit has been inspected and tested before testing the primary. Obviously, if the engine does not start, the secondary should not be suspect. However, if the engine misfires, the secondary circuit should be looked at.

The spark plug wires should be visually inspected, and their resistance checked with an ohmmeter. Check the spark plug wires across the entire length of the cable, including the cable ends. Some GM engines are equipped with an integrated ignition coil and module assembly. This assembly must be removed (**Figure 9-54**) to access the spark plugs. Once the assembly is removed, all of its components should be inspected and tested with an ohmmeter (**Figure 9-55**). This testing includes the various connectors and wiring harnesses, as well as the ignition coils, mounted to cover for the assembly.

Classroom Manual
Chapter 9, page 277

Ignition Coil Tests

With the coil terminals disconnected, an ohmmeter calibrated on the lowest (if not autoranging meter) scale should be connected to the primary coil terminals to test the primary winding. The primary winding in any EI coil should have 0.35 to 1.50 ohms resistance. An ohmmeter reading below the specified resistance indicates a shorted primary winding. A higher reading indicates excessive resistance, and an infinity meter reading proves that the primary winding is open.

Some COP coils have extra circuitry inside that may prevent a good primary reading with an ohmmeter.

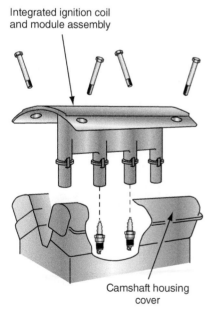

Figure 9-54 Integrated ignition coil and module assembly.

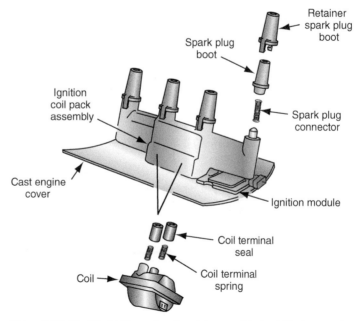

Figure 9-55 All of the ignition coil and module assembly should be inspected.

An ohmmeter should be connected to each pair of secondary coil terminals to test the secondary windings. The coil secondary winding should have 5,000 to 7,000 ohms resistance. If the secondary winding is open, the ohmmeter reading is infinity. A shorted secondary winding provides an ohmmeter reading below the specified resistance.

Hall-Effect Crankshaft Sensors

A basic timing adjustment is impossible on any EI system. However, if the gap between the blades and the crankshaft sensor is incorrect, the engine may fail to start, stall, misfire, or hesitate on acceleration. Follow these steps during the crankshaft sensor adjustment procedure:

1. Install the sensor loosely on the pedestal.
2. Position the sensor and pedestal on the special adjusting tool recommended by GM.
3. Position the adjusting tool on the crankshaft surface (**Figure 9-56**).
4. Tighten the pedestal-to-block mounting bolts to the specified torque.
5. Tighten the pinch bolt to the specified torque.
6. The interrupter rings on the back of the crankshaft pulley should be checked for a bent condition, and the same crankshaft sensor adjusting tool may be used to check these rings. Place the tool on the pulley extension surface and rotate the tool around the pulley (**Figure 9-57**). If any blade touches the tool, replace the pulley.

If a single-slot crankshaft sensor requires adjustment, follow this procedure:

⚠️ WARNING **Always be sure the ignition switch is off before attempting to rotate the crankshaft with a socket and breaker bar. If the ignition switch is on, the engine may start suddenly and rotate the socket and breaker bar with tremendous force. This action may result in personal injury and vehicle damage.**

1. Be sure that the ignition switch is off, and then rotate the crankshaft with a pull handle and socket installed on the crankshaft pulley nut. Continue rotating the crankshaft until one of the interrupter blades is in the sensor and the edge of the interrupter window is at the edge of the defector on the pedestal.

Figure 9-56 Using the special alignment tool to adjust the crankshaft sensor.

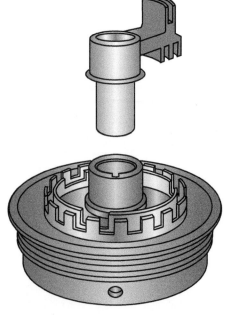

Figure 9-57 Checking the interrupter rings with the crankshaft sensor alignment tool.

2. Insert the special adjustment tool between each side of the blade and the sensor. If the tool does not fit between each side of the blade and the sensor, adjustment is required. The gap measurement should be repeated at all three blades.
3. If a sensor adjustment is necessary, loosen the pinch bolt and insert the adjusting tool between each side of the blade and sensor. Move the sensor as required to insert the gauge.
4. Tighten the sensor pinch bolt to the specified torque.
5. Rotate the crankshaft and recheck the gap at each blade.

No-Start Diagnosis

If the engine fails to start, follow these steps for a no-start ignition diagnosis:

1. Connect a spark tester from each spark plug wire to ground and crank the engine while observing the spark tester.
2. If the spark tester does not fire on any spark plug, check the 12-volt supply wires to the coil module; some coil modules have two fused 12-volt supply wires. Consult the manufacturer's wiring diagrams for the car being tested to identify the proper coil module terminals.
3. If the spark tester does not fire on a pair of spark plugs, the coil connected to that pair of spark plugs is probably defective.
4. If the spark tester did not fire on any of the spark plugs and the 12-volt supply circuits to the coil module are satisfactory, disconnect the crankshaft and camshaft sensor connectors and connect short jumper wires between the sensor connector and the wiring harness connector. Be sure the jumper wire terminals fit securely to maintain electrical contact. Each sensor has a voltage supply wire, a ground wire, and a signal wire on 3.8-liter engines. On the 3.3-liter and 3300 engines, the dual crankshaft sensor has a voltage supply wire, ground wire, crank signal wire, and SYNC signal wire. Identify each of these wires on the wiring diagram for the system being tested.
5. Connect a digital voltmeter to each of the camshaft and crankshaft sensor black ground wires to an engine connection. With the ignition switch on, the voltmeter reading should be 0.2 volt or less. If the reading is above 0.2 volt, the sensor ground wires have excessive resistance.
6. With the ignition switch on, connect a digital voltmeter from the camshaft and crankshaft sensor's white or red voltage supply wires to an engine ground (**Figure 9-58**). The voltmeter readings should be 5 to 11 volts. If the readings are below these values, check the voltage at the coil module terminals that are connected to the camshaft and crankshaft sensor voltage supply wires. When the sensor voltage supply readings are low at the coil module terminals, the coil module should be replaced. If the voltage supply readings are low at either sensor connector but satisfactory at the coil module terminal, the wire from the coil module to the sensor is defective. On the 3.3-liter and 3300 engines, the crankshaft sensor ground wire and voltage supply wire are checked in the same way as explained in steps 5 and 6.
7. If the camshaft and crankshaft sensor ground and voltage supply wires are satisfactory, connect a digital voltmeter to each sensor signal wire and crank the engine. Each sensor should have a 5-volt and 7-volt fluctuating signal. On the 3.3-liter and 3300 engines, test this voltage signal on the crank and SYNC signal wires at the crankshaft sensor. If the signal is less than specified, replace the sensor with the low signal.
8. When the camshaft and crankshaft sensor signals on 3.8-liter engines or crank and SYNC signals on 3.3-liter and 3300 engines are satisfactory and the spark tester did not fire at any spark plug, the coil module is probably defective.

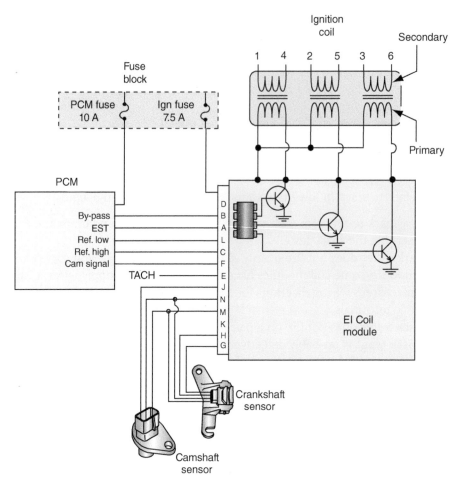

Figure 9-58 Wiring diagram for the crankshaft and camshaft sensors on a GM 3.8-liter engine.

9. On 3.8-liter engines where the coil assembly is easily accessible, the coil assembly screws may be removed and the coil lifted up from the module with the primary coil wires still connected. Connect a 12-volt test lamp across each pair of coil primary wires and crank the engine. If the test lamp does not flicker on any of the coils, the coil module is defective, assuming that the crankshaft and camshaft sensor readings are satisfactory.

Some GM EI systems are referred to as fast-start systems. These systems require additional steps during the diagnosis of a no-start condition. When an EI Type 1 Fast-Start System fails to start, complete steps 1, 2, and 3 in the previous procedure, and then do the following:

1. If the 12-volt supply circuits to the coil module are satisfactory, disconnect the crankshaft sensor connector and connect four short jumper wires between the sensor connector and the wiring harness connector.
2. Connect a digital voltmeter from the sensor ground wire to an engine ground. With the ignition switch on, the voltmeter should read 0.2 volt or less. A reading above this value indicates a defective ground wire.
3. Connect the voltmeter from the sensor voltage supply wire to an engine ground. With the ignition switch on, the voltmeter reading should be 8 to 10 volts. If the reading is lower than specified, check the voltage at coil module terminal N (**Figure 9-59**).

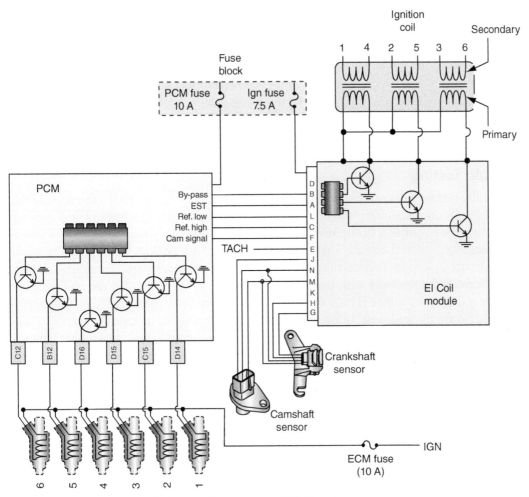

Figure 9-59 Wiring diagram for the fast-start EI system on a GM 3800 engine.

When the voltage at terminal N is satisfactory and the reading at the sensor voltage supply wire is low, the wire from terminal N to the sensor is defective. A low-voltage reading at terminal N indicates a defective coil module.

4. If the readings in steps 2 and 3 are satisfactory, connect a voltmeter from the 3 and 18× signal wires at the sensor connector to an engine ground and crank the engine. The voltmeter reading should fluctuate from 5 to 7 volts. The exact voltage may be difficult to read, especially on the 18× signal, but the reading must fluctuate. If the voltmeter reading is steady on either sensor signal, the sensor is defective.

5. Connect a digital voltmeter from the 18× and 3× signal wires at the coil module to an engine ground, and crank the engine. The voltmeter readings should be the same as in step 4. If these voltage signals are satisfactory at the coil module terminals but low at the sensor, repair the wires between the coil module and the sensor.

6. If the 18× and 3× signals are satisfactory at the coil module terminals, remove the coil assembly-to-module screws and lift the coil assembly up from the module. Connect a 12-volt test lamp across each pair of coil primary terminals, and crank the engine. If the test lamp does not flash on any pair of terminals, the coil module is defective.

> ⚙ SERVICE TIP If the module tests discussed above lead to a replacement of the modules, be sure to thoroughly test the disconnected coil before reinstalling it. It is possible the coil has an internal short that caused the module to fail. By reinstalling the affected coil on the new module, a repeat failure is likely to occur. It is recommended that the coil be tested using an approved spark tester that will adequately stress the coil to simulate real driving conditions.

Module Testing

EI module tests are easy to perform. The EI module can be compared to the regular distributor ignition system module in its operation. Previous testing steps in the diagnostic procedure have probably brought you to this point. It was likely that a cylinder or pair of cylinders matched to the same coil was misfiring or had no spark. Next, through an elimination process, it must be determined whether the module or the coil is at fault. Tests can be performed dynamically with the module installed. Identify the affected coil pack and remove it from the module. This will expose the two connector terminals from the module that slide onto the coil unit. If the system you are working on groups the coils into one unit rather than possessing individual coils, remove the coil pack assembly. This will expose the wires and terminal ends from the module to the coil. Slide the wire connectors off the coil assembly. Be sure to note the orientation of the coil and identify the module wires as you remove them from the coil. Once the connectors are open, install a test light in place of the coil pack across the module terminals. The test light acts as a substitute load to the module. Crank the engine and observe the test light. If the light flashes, the module is switching the primary ignition circuit properly, signifying that the problem is within the coil. If the light does not flash, the module is defective. Remember, it is possible for the module to be defective in such a way that one coil of the set may fire while another may not. It is impossible to repair or service a malfunctioning or inoperative ignition module, therefore the entire unit must be replaced.

EI Systems with Magnetic Sensors

With the wiring harness connector to the **magnetic sensor** disconnected and an ohmmeter calibrated on the (kΩ scale on a non-auto-ranging meter) connected across the sensor terminals, the meter should read 900 to 1,200 ohms on 2.0-, 2.2-, 2.8-, 3.1-, and 3.4-liter engines (**Figure 9-60**).

Meter readings below the specified value indicate a shorted sensor winding, whereas infinity meter readings prove that the sensor winding is open. Since these sensors are mounted in the crankcase, they are continually splashed with engine oil. In some sensor failures, the engine oil enters the sensor and causes a shorted sensor winding. If the magnetic sensor is defective, the engine fails to start.

With the magnetic sensor wiring connector disconnected, an **alternating current (AC)** voltmeter may be connected across the sensor terminals to check the sensor signal while the engine is cranking. On 2.0-, 2.8-, 3.1-, and 3.4-liter engines, the sensor signal should be 100 millivolts (mV) AC. When the sensor is removed from the engine block, a flat steel tool placed near the sensor should be attracted to the sensor if the sensor magnet is satisfactory.

No-Start Diagnosis

When an engine with an EI system and a magnetic sensor fails to start, complete steps 1, 2, and 3 of the previously given no-start diagnostic procedure, and then follow this procedure:

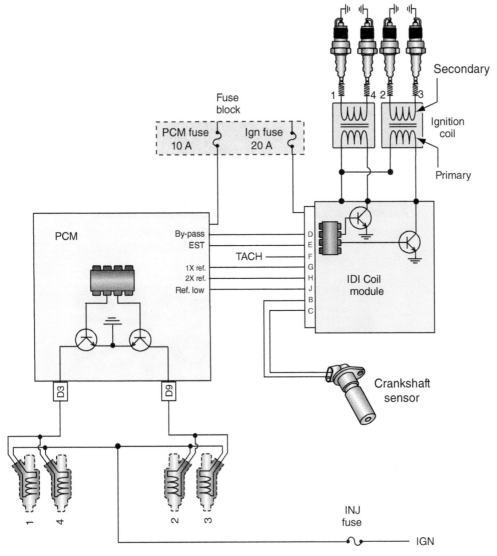

Figure 9-60 Wiring diagram for an integrated EI system.

1. If the spark tester did not fire on any of the spark plugs, check for 12 volts at the coil module voltage input terminals. Consult the wiring diagram for the system being tested for terminal identification.
2. If 12 volts are supplied to the appropriate coil module terminals, test the magnetic sensor.
3. When the magnetic sensor tests are satisfactory and the engine does not start because of an ignition problem, the coil module is probably defective.

GM Coil Near Plug System

GM coil near plug (CNP) systems share most of the same elements as the other systems seen so far. This system is from a 2010 Chevrolet Camaro. The crankshaft and camshaft sensors are both magnetic pulse generating sensors. The crankshaft has a special reluctor ring (**Figure 9-61**) that has 60 minus 2 teeth; the two missing teeth give the PCM a reference pulse to determine the position of the crankshaft. The CKP sensor is located near the rear of the engine block (**Figure 9-62**). The camshaft sensor uses a 4× reluctor made into the camshaft gear (**Figure 9-63**) that tells the PCM the exact location of the camshaft. The camshaft sensor fits into the engine front cover (**Figure 9-64**).

A magnetic pulse generating sensor is another name for a magnetic reluctance sensor.

Figure 9-61 Chevrolet V8 EI system with a CNP ignition system

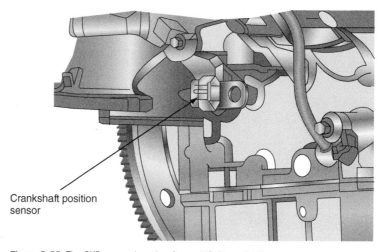

Figure 9-62 The CKP sensor location from a V8 Chevrolet Camaro.

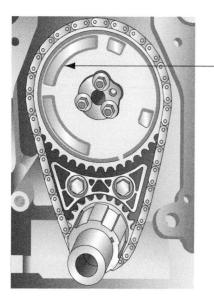

Figure 9-63 The CKP sensor from a Chevrolet V8.

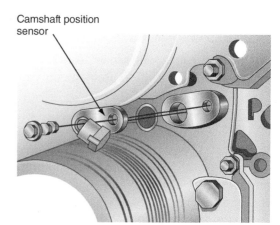

Figure 9-64 The CKP sensor from a Chevrolet V8.

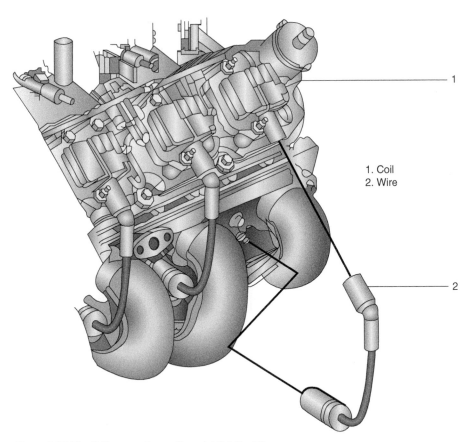

1. Coil
2. Wire

Figure 9-65 The CNP system from a Chevrolet 6.2-liter V8.

Diagnosis of the ignition system is started by doing visual inspection of the short plug wire from the valve cover to the spark plug for obvious problems (**Figure 9-65**), the connections to the primary connection on the coils (**Figure 9-66**). Next, check for any trouble codes that might be current or history. Check for any misfire codes only after diagnosing fuel control or heated oxygen sensor codes if they are present. If a current misfire code is found and the number of misfires is accumulating, then it is recommended to check for spark with a spark plug tester on the affected cylinder. If the spark tester confirms spark, remove the plug and inspect for signs of fouling with fuel, coolant, or gasoline. If the spark plug has no apparent problems, swap the plug with another cylinder. If the problem moves with the spark plug, then replace the spark plug. If the problem stays with the original cylinder, check the resistance of the spark plug wire and replace as necessary. If the problem still has not been found, check the ignition power feed to the injector. Turn the key off, wait about 5 minutes for the computer to shut down, and then measure from the ground terminal (back to the PCM) for resistance. If the resistance is more than 5 ohms, replace and program the PCM. It is important to note that the coils used on this vehicle give the PCM information about spark production and the integrity of the primary circuit, but a code for the affected ignition coil may not set with an intermittent condition. On this vehicle, the manufacturer recommends swapping ignition coils if the primary circuit checks good instead of measuring the coil primary or secondary resistance (the coil primary and secondary resistance might measure accurately with an ohmmeter due to the electronics involved). If the misfire moves with the coil, then the problem has been found, if not, more testing is needed as mentioned previously.

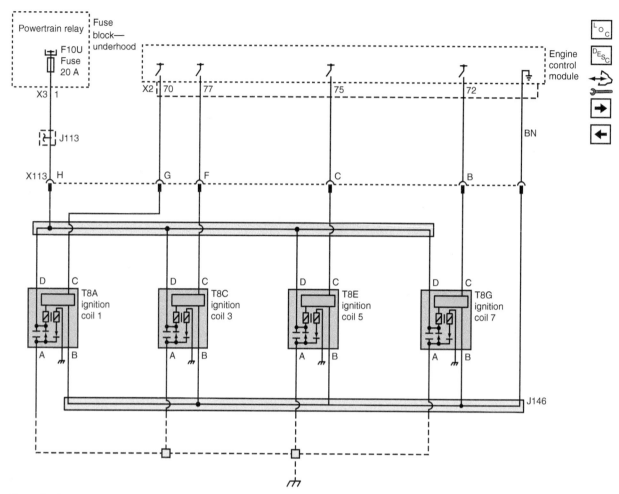

Figure 9-66 The CNP wiring diagram from left bank of Chevrolet 6.2-liter V8.

TOYOTA EI SYSTEMS

Toyota and Lexus engines are equipped with electronic ignition. One of the systems used by Toyota is very similar to the waste spark systems used by everyone else. It uses multiple camshaft sensors, one crankshaft sensor, an ignition module, and one coil for each pair of cylinders. There is one camshaft sensor for each coil. A major difference with this system is that the ignition coil is mounted directly over one spark plug. This eliminates the second spark plug wire from the ignition coil. The other systems have an individual ignition coil mounted over each spark plug.

Toyota uses a special locking feature to secure the spark plug wires to the spark plugs and the ignition coils. To remove the wires, begin by using a pair of needle nose pliers. Disconnect the cable clamp from the engine wire protector (**Figure 9-67**).

Disconnect the cables from the spark plugs by firmly holding the boot and, with a twist-and-pull effort, remove the cable from the plug (**Figure 9-68**).

Using a screwdriver, lift up the locking tab and separate the spark plug wire from the ignition coils (**Figure 9-69**).

On some engines it may be necessary to remove a wire protector assembly to remove the spark plug wires. Once the wires are removed, they can be checked with an ohmmeter (**Figure 9-70**). The maximum allowable resistance across a secondary cable is 26,000 ohms. If the resistance is higher, check the condition of the terminals. If they are fine, replace the cable.

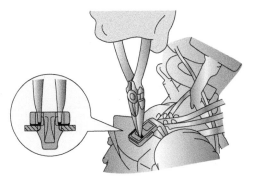

Figure 9-67 Use needle-nose pliers to remove the wires from the spark plug cable protector.

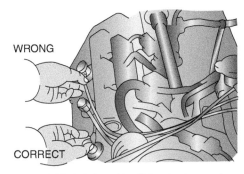

Figure 9-68 Pull the cable off the plug by grasping the boot, not the wire.

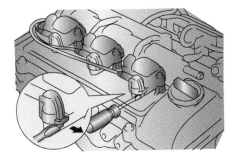

Figure 9-69 Use a screwdriver to lift the lock at the cable's connection to the coil.

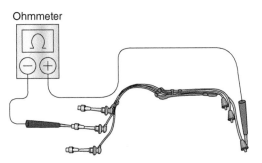

Figure 9-70 Measuring the resistance of the spark plug cable assembly.

To reinstall the spark plug cables, reverse the procedure. Make sure that all locking mechanisms are secure and that the routing of the cables is correct (**Figure 9-71**).

Many Toyota, as well as other import engines, use dual electrode spark plugs. Contrary to common belief, the air gap of this type of plug is very important and is adjustable. If the air gap to one of the ground electrodes is larger than the gap to the other electrode, electrons will always jump the smaller gap. The air gaps to both electrodes must be the same. Check and adjust the gap with a wire gauge (**Figure 9-72**).

Critical to the proper operation of the system is the air gap at the camshaft sensor. Although the gap is not adjustable, it should be checked whenever diagnostics points to a faulty camshaft signal. To measure the gap, use a nonmagnetic feeler gauge. The gap should be between 0.0008 and 0.016 inch. If the gap is not within these specifications, the sensor should be replaced.

The camshaft sensor should also be checked with an ohmmeter (**Figure 9-73**). Connect the negative ohmmeter lead to the ground terminal at the sensor, and measure the resistance between that terminal and the others, one at a time. The resistance should be within the specified range. If not, replace the sensor.

The secondary ignition coil windings may contain a high-voltage diode, which is why Toyota only recommends and gives specifications for measuring the resistance across the primary windings of some of their coils. When the resistance is measurable, the normal secondary winding resistance is 11,000 to 17,500 ohms (11 to 17.5K) (**Figure 9-74**). The specified primary winding resistance is normally less than 1 ohm. Also check the coils for shorts to ground by connecting one ohmmeter lead to a primary or secondary terminal and the other to ground. There should be no continuity, and the meter should read infinity.

Coil over plug coils that have more than two wires at the coil connection usually have an integral module. The presence of this module may make ohmmeter testing of the coil impossible.

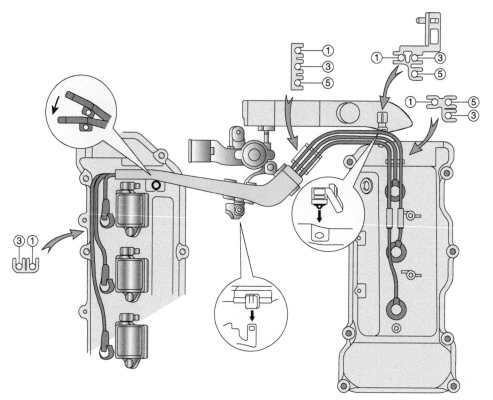

Figure 9-71 Correct routing for the spark plug cables.

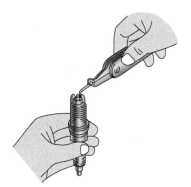

Figure 9-72 Check the air gap at both spark plug electrodes.

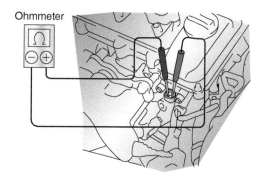

Figure 9-73 Checking a camshaft sensor with an ohmmeter.

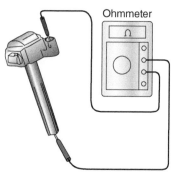

Figure 9-74 Checking the secondary of an ignition coil.

Toyota EI COP Systems

Toyota COP systems are very similar to those already presented in this chapter, and the diagnosis for them also follows closely. The wiring diagram for a 2010 Toyota Prius is shown in **Figure 9-75**. Notice that the coil does have an integral module shown inside each coil. The diagnostic procedure for checking the ignition is to unplug all the injectors, then reconnect one coil at a time and check for spark at the coil with a spark plug tester. If spark is occurring, then check for another cause of the misfire. If there is no spark, substitute a known good plug, then a coil if necessary. If the coil is not receiving power and a switching triggering signal from the ECM, then the wiring from the ECM to the coil should be checked for continuity. If the wiring from the ECM to the coil checks good, then the ECM is replaced and the vehicle is retested.

DIAGNOSIS OF COMPRESSION SENSE IGNITION

Even though the **compression sense ignition (CSI)** system does not use a camshaft position sensor, the code P0340 and P0341 (**Figure 9-76**) can still be displayed on a scan tool if there are problems that keep this ignition system from being able to decipher the

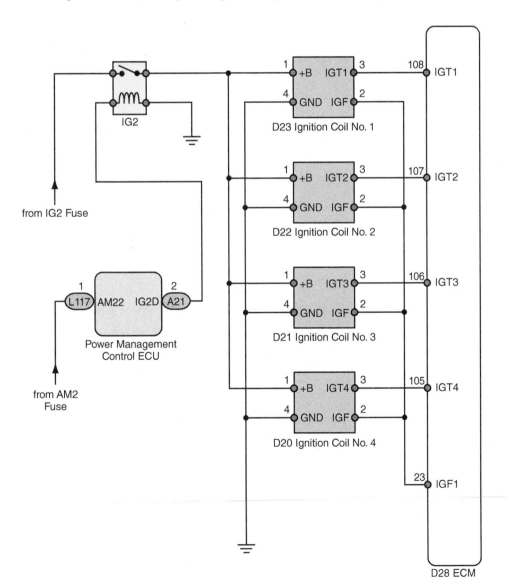

Figure 9-75 Testing a COP ignition diagram from a Toyota Prius.

Compression Sense Trouble Codes

- P0340: PCM does not receive cam signal from the ignition control module for 107 crankshaft revolutions.

- P0341: More than 11 extra camshaft signal transitions.

Figure 9-76 Trouble codes associated with the CMP and CSI.

ignition pulses that the CSI module relies on to produce the cam sensor signal. As you recall from the classroom manual, the CSI module picks up on the higher voltage level of the cylinder on compression of the waste spark. The module uses this information to produce a cam sensor signal, even though there is no actual cam sensor. Anything that affects the plug firing voltages, such as open plug wire connections, defective spark plugs, defective coil, or shorts to ground, or any problem that would affect spark plug firing voltages could result in a camshaft position sensor code. For example, if a spark plug were open electrically, then all the energy from the coil would be expended across the companion spark plug and the voltage would return through the ground connection to the coil. The resulting signal would be a "compression" stroke indication for every coil firing. Of course, if there were problems that resulted in poor compression, burned valve, worn intake cam lobes, poor sealing rings, and so forth, then the CSI may be fooled and send a poor cam signal to the PCM. Of course, there may be additional codes for engine misfire for these problems as well.

Ion-Sensing Ignition

Because an **ion-sensing ignition** depends on the spark plugs for readout of conditions in the combustion chamber, it makes sense that any problem that affects spark plugs, coils, module, or connections would also affect the ability of the ion-sensing ignition system to accurately relay information to the PCM. Remember that the knock sensor's function has been replaced by the ignition module's ability to measure the combustion environment. Additionally, some Saab automobiles have replaced the upstream O_2 sensors with this same combustion chamber information.

PHOTO SEQUENCE 20
Determining the Cause of a High Firing Line

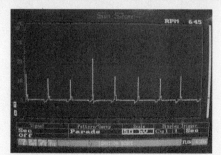

P20-1 Observe the secondary display pattern with the engine at idle.

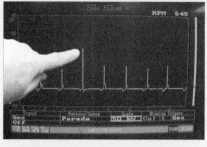

P20-2 Identify the cylinder with the high firing line and record the height of the line on a piece of paper.

P20-3 Turn the engine off and carefully remove the plug wire from the affected cylinder.

P20-4 Connect a jumper wire or grounding probe to the end of the spark plug wire and to a good ground.

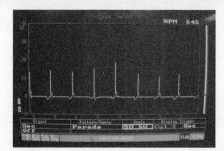

P20-5 Start the engine and observe the height of the firing line now. Do not allow the engine to run very long with the plug bypassed. After you have a reading, turn the engine off.

P20-6 Record the new height and write the difference between the readings off to the side.

P20-7 Move the jumper wire to the correct terminal at the distributor cap.

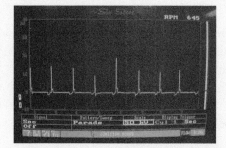

P20-8 Start the engine and observe the height of the firing line. Then turn the engine off.

P20-9 Record the new height and write the difference between the readings off to the side.

P20-10 Reconnect the spark plug wire and start the engine. Allow it to idle while analyzing the results. Compare the drops in required voltage during each phase of this test. Notice when C was bypassed, the voltage dropped considerably more than when other parts were bypassed. This indicates high resistance in the spark plug wire.

OSCILLOSCOPE TESTING OF IGNITION SYSTEMS

No discussion of ignition troubleshooting would be complete without a comprehensive discussion of oscilloscope use. The job of the oscilloscope is to convert the electrical signals of the ignition system into a visual image showing voltage changes over a given period of time. This information is displayed on a screen in the form of a continuous voltage line called a pattern or trace. By studying the pattern, a technician can determine what is happening inside the ignition system.

The information on the design and use of oscilloscopes given in this text is general in nature. Always follow the oscilloscope manufacturer's specific instructions when connecting test leads or conducting test procedures.

> **AUTHOR'S NOTE** In this section on oscilloscope testing, I am describing a generic oscilloscope procedure, because I have no idea which one of the many scopes you might be using in your shop. Hopefully, you have read about and practiced the use of your oscilloscope enough that you are comfortable with these tests. Remember there is no better way to learn than to practice!

Scales

The oscilloscope can look inside the ignition system by giving the technician a visual representation of voltage changes over time.

On the face of the screen is a voltage versus time graph (**Figure 9-77**). The waveform patterns are displayed on the graph. The graph charts the changes in voltage and the time span in which the changes occur.

Normally, vertical scales measure voltage. The vertical scale on the left side of the graph is divided into increments of voltage that are scalable according to the measurement needed. This scale is useful for testing secondary voltage set to a kilovolt scale. It can also be used to measure primary voltage by interpreting the scale in volts rather than **kilovolts (kV)** (0 to 25 volts).

The horizontal portion of the scale is divided for time, and is also adjustable according to the voltage measurement needed.

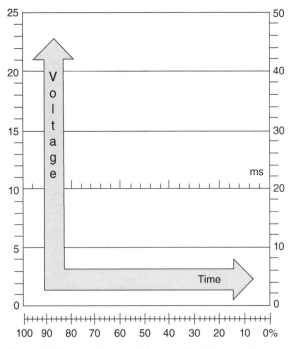

Figure 9-77 An oscilloscope plots a graph of voltage levels versus time.

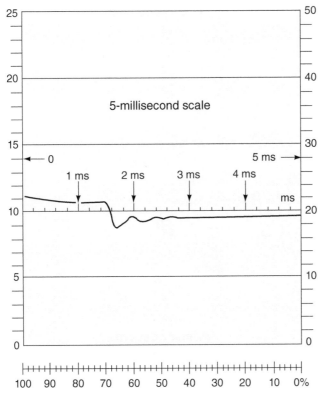

Figure 9-78 A 5-ms pattern showing spark duration.

Depending on the test mode selected by the technician, the millisecond scale shows a range of 0 to 5 milliseconds (ms) or 0 to 25 seconds. The technician has many choices on an oscilloscope, from milliseconds to seconds, depending on the need. Most oscilloscopes read in time/division. The 5-millisecond scale is often used to measure the duration of the spark (**Figure 9-78**). In the 25-millisecond mode, the complete firing pattern can normally be displayed. The 5-millisecond scale selection expands the pattern in order to enhance the detail of the trace. A typical duration should be 1.0 to 3.0 milliseconds.

Many oscilloscopes can have from one to eight channels available for use at the same time. Each channel can be set up with its own voltage scaling.

Pattern Types and Phases

To monitor various phases of ignition system performance, a typical scope pattern is divided into three main sections: firing, intermediate, and dwell. Information received from specific testing in each one of these areas can be used to piece together a complete ignition picture.

Depending on the oscilloscope function selected by the technician, the scope can display the voltage versus time pattern for either the secondary or primary circuit. A typical secondary pattern is shown in **Figure 9-79**. A secondary pattern displays the following types of information:

- Firing voltage
- Spark duration
- Coil and condenser oscillations
- Transistor on-off switching or breaker point close and open
- Secondary circuit accuracy

A typical primary circuit pattern is shown in **Figure 9-80**. Most problems that could occur in the primary pattern will usually also be seen in the secondary pattern. Conversely, some secondary ignition problems can also be seen in primary patterns.

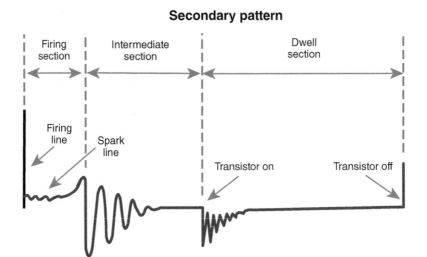

Figure 9-79 Typical secondary pattern.

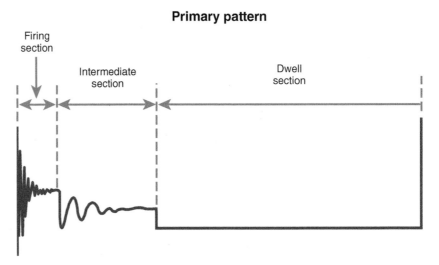

Figure 9-80 Typical primary pattern.

This can be helpful when the secondary connection is limited due to some newer ignition system designs.

Pattern Display Modes

The **display pattern** on an oscilloscope displays cylinder patterns from left to right.

The **parade pattern** starts with the spark line of the number one cylinder and ends with the firing line of the number one cylinder.

Depending on the equipment that you have access to, the oscilloscope has several ways to display the voltage patterns of the primary and secondary circuits. When the **display pattern** is selected, the oscilloscope displays the patterns of all the cylinders in a row from left to right as shown in **Figure 9-81**. This pattern arrangement is also commonly referred to as a **parade pattern**. It refers to the same information using a different name. Each cylinder's ignition cycle is displayed in the engine's **firing order**. In the example of a secondary display pattern shown in figure the firing order is 1, 8, 4, 3, 6, 5, 7, and 2. The pattern begins with the spark line of the number one cylinder and ends with the firing line for the number one cylinder. This display pattern allows the technician to compare the voltage peaks for each cylinder. The **raster pattern**, sometimes referred to as stacked, is nothing more than the repositioning of the parade pattern. This is done by stretching

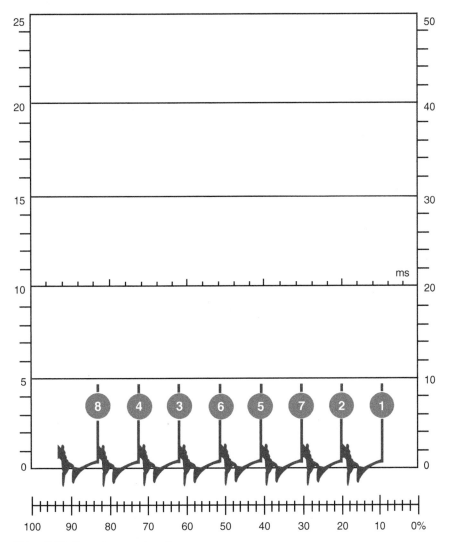

Figure 9-81 Secondary parade pattern.

out each cylinder's pattern from the parade pattern (**Figure 9-82**) to the width of the screen, then stacking them in the firing order beginning at the bottom (**Figure 9-83**). Viewing the pattern in this mode helps to enhance detail and vertically compare the cylinders' patterns.

The **raster pattern** stacks voltage patterns one above the other.

A raster pattern stacks the voltage patterns of the cylinders one above the other. The number one pattern is displayed at the bottom of the screen, and the rest of the cylinder's firing patterns are arranged above it in the engine's firing order.

In a raster pattern, the patterns for each cylinder are displayed across the width of the graph, beginning with the spark line and ending with the firing line. This allows for a much closer inspection of the voltage and time trends than is possible with the display pattern.

Understanding Single-Cylinder Patterns

The following section takes a closer look at the secondary circuit pattern (**Figure 9-84**). As shown, the firing line of the pattern appears at the left side of the screen. An upward line signifies voltage. The firing line indicates the voltage needed to start the spark. The ignition coil's voltage pressure rises up to the point where it overcomes all electrical resistance in the secondary circuit. Typically, it requires around 10,000 volts to overcome this resistance and initiate a spark.

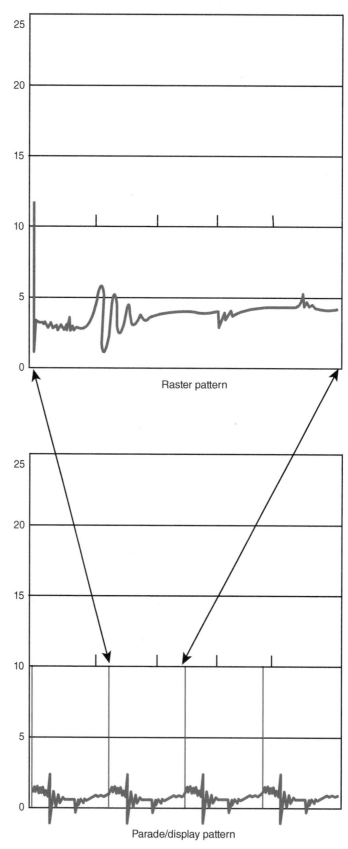

Figure 9-82 Raster pattern from parade pattern.

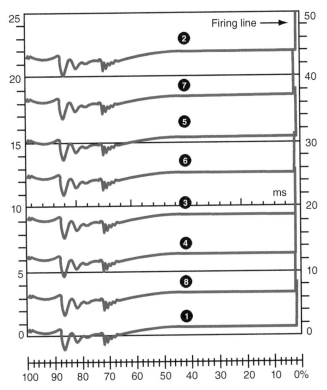

Figure 9-83 Secondary raster pattern.

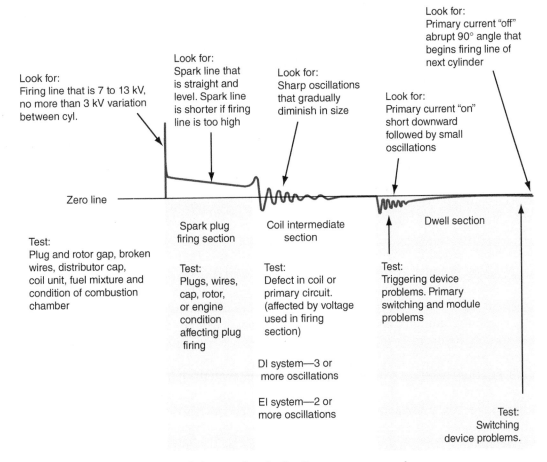

Look for:
Primary current "off" abrupt 90° angle that begins firing line of next cylinder

Look for:
Firing line that is 7 to 13 kV, no more than 3 kV variation between cyl.

Look for:
Spark line that is straight and level. Spark line is shorter if firing line is too high

Look for:
Sharp oscillations that gradually diminish in size

Look for:
Primary current "on" short downward followed by small oscillations

Zero line

Spark plug firing section

Coil intermediate section

Dwell section

Test:
Plug and rotor gap, broken wires, distributor cap, coil unit, fuel mixture and condition of combustion chamber

Test:
Plugs, wires, cap, rotor, or engine condition affecting plug firing

Test:
Defect in coil or primary circuit. (affected by voltage used in firing section)

DI system—3 or more oscillations

EI system—2 or more oscillations

Test:
Triggering device problems. Primary switching and module problems

Test:
Switching device problems.

Figure 9-84 Breakdown of a single-cylinder secondary circuit pattern.

The firing line and **spark line** of the firing section indicate **firing voltage** and spark duration.

Classroom Manual
Chapter 9, page 266

Keep in mind that cylinder conditions have an effect on this resistance. Leaner air-fuel mixtures increase the resistance and increase the required **firing voltage**.

Once secondary resistance is overcome, the spark jumps the plug gap, establishes current flow, and ignites the air-fuel mixture in the cylinder. The length of time the spark actually lasts is represented by the spark line portion of the pattern. The spark line begins at the firing line and continues tracing to the right until the coil's voltage drops below the level needed to keep current flowing across the gap. In other words, the spark is draining the voltage built up in the coil's secondary windings. As long as the coil has sufficient voltage, the spark continues.

The spark line has a slant and length which, combined with the firing line, provide insight into firing performance. The slant of the spark line indicates the amount of voltage required to sustain spark at the spark plug after the coil fires. The steeper the slant, known as "spark kV," the more voltage is required to overcome the resistance. The length of the spark line, which is referred to as "spark duration" and is measured in milliseconds, measures the amount of time the spark remains between the spark plug electrodes. The physics of the dissipation of the coil's energy proportionately brings the three measurements (firing kV, spark kV, and **burn time**) together. Evidence of this can be found in common ignition problems such as a wide-gap/open secondary or close-gap/shorted spark plug or spark plug wire. In the case of the wide-gap condition, the firing line indicates more voltage will be required to initiate the spark to the spark plug, causing a high firing line. Due to the relationship with the spark line, the high resistance added to the circuit will require more voltage to sustain spark across the spark plug. This will result in a steeper slant to the spark line or a higher spark kV. Finally, extra energy from the coil reduces the amount of time the spark can remain between the electrodes of the spark plug, resulting in lower spark duration, when measured as lower millisecond duration. To summarize, if the firing voltage is higher than normal, the burn time will decrease. If the voltage is lower than normal, the burn time will increase.

It is important to remember the elements that affect and determine firing voltage requirements. Internal and external cylinder resistances are two basic areas that influence these requirements. Internal resistance refers to anything inside the cylinder that would cause the firing voltage to increase or decrease.

High firing voltage requirements shorten spark duration. Low firing voltages lengthen spark duration.

A lean mixture (too much air) is harder to ignite than a rich mixture, so firing voltage will be high and duration will be low. A rich mixture (too much fuel) is easy to ignite, so firing voltages are low and spark duration is longer.

This would include compression, mixture, and temperature. For example, high compression adds to cylinder pressure, which increases resistance. Therefore, higher voltage will be required to fire. A lean mixture has more oxygen molecules between the hydrogen chloride (HC) molecules. The additional oxygen does not conduct electricity as easily as HC molecules. Therefore, more voltage will be required to initiate spark and there will be less energy to sustain voltage, so burn times are shorter. Even ignition timing, as well as combustion chamber temperature, will affect the firing voltage requirements.

External resistance refers to anything external to the cylinder and usually includes the secondary ignition components. Examples are spark plugs, wires, distributor caps, rotors, coil wires, and ignition coils. Wide air gaps at the spark plugs or faulty spark plug wires will proportionately increase firing voltage requirements. On vehicles equipped with distributor caps and rotors, an incorrect rotor-to-cap air gap will have the same effect. As discussed earlier in this chapter, it will be necessary to determine if the ignition problem is common to all cylinders or if it affects only one or several cylinders. If all the firing lines indicate high voltage, the problem is common to all cylinders. If only one or a few cylinders have abnormal requirements, then the problem has to do with only those particular cylinders.

Identifying firing-related problems is relatively easy. After determining whether the problem is common to all cylinders or only one, proceed to the next step. For example, if one cylinder has a firing line and burn time different from the other cylinders, the problem is most likely the spark plug, spark plug wire, or distributor cap. It may also be a mixture- or mechanical-related problem within the cylinder that you have identified. To determine how to locate the problem, refer to Photo Sequence 20 on page 486.

Continuing on, once coil voltage drops below the level needed to sustain the spark, the next major section of the scope trace pattern begins. This section is called the **intermediate section** or coil-condenser zone. It shows the remaining coil voltage as it dissipates or drops to zero. Remember, once the spark has ended, there is quite a bit of voltage stored in the ignition coil. The voltage remaining in the coil then oscillates or alternates back and forth within the primary circuit until it drops to zero. Notice that the lines representing the coil-condenser section steadily drop in height until the coil's voltage is zero.

The next section of the trace pattern begins with the primary circuit current on signal. It appears as a slight downward turn followed by several small oscillations. The slight downward curve occurs just as current begins to flow through the coil's primary winding. The oscillations that follow indicate the beginning of the magnetic field buildup in the coil. This curve marks the beginning of a period known as the **dwell section** of zone. The end of the dwell zone occurs when the primary current is turned off by the switching device. The trace turns sharply upward at the end of the dwell zone. Turning the primary current off collapses the magnetic field in the coil and generates another high-voltage spark for the next cylinder in the firing order. Remember, the primary current off signal is the same as the firing line for the next cylinder. The length of the dwell section represents the amount of time that current is flowing through the primary.

Most scope patterns look more or less like the one just described. The patterns produced by some systems have fewer oscillations in the intermediate section. Patterns may also vary slightly in the dwell section. The length of this section depends on when the control module turns the transistor on and off. Several variables may affect this timing. These variables can be categorized as follows:

1. Variable dwell
2. Current limiting
3. Internal circuitry

Most electronic ignition systems have a variable dwell function built into their control modules. In these systems, dwell changes significantly with engine speed. At idle and low-rpm speeds, a short dwell provides enough time for complete ignition coil saturation. The current on and current off signals appear very close to each other, usually less than 20 degrees.

As engine speed increases, the control module lengthens the dwell time (**Figure 9-85**). This, of course, increases the available time for coil saturation.

Although not common, it is possible for the variable dwell function of the control module to fail and still allow the vehicle to run. If testing indicates a lack of variable dwell on a system equipped with it, the control module must be replaced.

Many modern electronic ignitions feature **current limiting**. These systems saturate the ignition coil quickly by passing very high current through the primary winding for a fraction of a second. Once the coil is saturated, the need for high current is eliminated and a small amount of current is used to keep the coil saturated. This type of system extends coil life.

The **intermediate section** shows coil voltage dissipation.

The **dwell section** shows the primary circuit turning on and off. The off signal is the firing line for the next cylinder in the firing order.

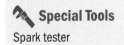
Special Tools
Spark tester

SERVICE TIP Do not attempt to use an ordinary spark plug to make a **test spark plug** by breaking off the side electrode. It will not work the same as an actual test plug. The center electrode is too close to the spark plug shell, and the spark will simply jump there with practically no effective stress test. Several types of spark testers are available, including those adjustable to 40 kV.

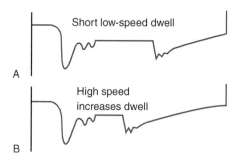

Figure 9-85 Dwell starts earlier at higher rpm.

Open Circuit Precautions

Some technicians check coil output by removing a secondary spark plug wire while keeping it from arcing to ground and reading the coil maximum output. Open circuit testing of the secondary is not recommended.

Features such as variable dwell and current limiting circuitry pull out all primary resistance if a spark does not occur. This causes tremendous amounts of current to flow through the primary circuit, producing an extremely high-voltage spark that must search out a path to ground. A frequent path to ground chosen by the spark is through the side of the coil. This results in an insulation breakdown at the coil and a site where arcing to ground is likely to occur in the future.

To prevent these tremendous voltages from occurring during coil output testing, an appropriate spark tester is used when checking the coil voltage output. The spark tester is connected to the coil wiring running to the distributor or to an ignition cable. A special grounding clip on the spark tester is then connected to a good ground source. (Refer to **Figure 9-9**.)

Some spark testers look like a spark plug, but they have a very large electrode gap. When the engine is cranked to check for voltage output, the ignition coil is forced to produce a higher voltage to overcome the added resistance created by the wide gap. Typically, about 35,000 volts are needed to fire the test plug. This is enough to stress the system without damaging it.

Coil Output and Oscillation Test

To safely perform a coil output test, proceed as follows:

1. Install the test plug in the coil wire or a plug wire if there is not a coil wire.
2. Set the oscilloscope on display and a voltage range of 50 kV.
3. Crank the engine over and note the height of the firing line.

The firing line should exceed 35 kV and be consistent. Lower-than-specified firing line voltages may indicate that the test plug did not fire. This could be the result of lower-than-normal available voltage in the primary circuit. The control module may have developed high internal resistance. The coil or coil cable may also be faulty. Further testing is needed to help pinpoint the problem.

In addition to the coil output test, simply observe the coil oscillations while connected to the scope. The best setting is either raster or 5-millisecond sweep. Coil oscillations begin at the end of the spark line at the first rise. One cycle is counted when the upswing of that oscillation almost reaches the top of the coil pattern as it begins the second oscillation. Simply count the top of each coil oscillation to determine how many there are. Electronic systems will have three or more oscillations. The more coil oscillations, the more coil capacity in reserve. Some EI and other electronic systems may have only two oscillations, because there is only one coil for every one or two cylinders.

Classroom Manual
Chapter 9, page 266

Spark duration, measured in milliseconds, is useful in determining the function of spark plugs.

⚙ **SERVICE TIP** Refer to Figure 9-86. Notice the small, hash-like lines along the spark line. Remember that any vertical line on the scope represents voltage. The hash lines on the spark line are a result of air and fuel movement within the cylinder reacting to a swirl of rapidly changing mixture movement. The molecules are all mixed up and not uniform. This movement, known as combustion turbulence, affects the spark line by literally moving it during the spark duration. This can represent a valuable bit of information in determining an ignition fault. Sometimes a fault can occur in the secondary that will not necessarily greatly affect the firing or spark line. For example, this may be a misfire under load. Look closely at the spark lines while lightly throttling the engine. If all the spark lines except one have combustion turbulence, there is a good possibility that the spark is not jumping across the gap of the spark plug. In this example, the spark may be firing through the spark plug shell to ground, external of the cylinder when the cylinder load increases. Remember, electricity seeks the path of least resistance.

Spark Duration Testing

Spark duration is measured in milliseconds using the millisecond sweep of the oscilloscope. Most vehicles have a spark duration of approximately 1.5 milliseconds (**Figure 9-86**). A spark duration of approximately 0.8 millisecond is too short to provide complete combustion. A short spark also increases pollution and power loss. If the spark duration is too long (over approximately 2.0 milliseconds), the spark plug electrodes might wear prematurely. When the oscilloscope shows a long spark duration, it normally follows a short firing line which may indicate a fouled spark plug, low compression, or a spark plug with a narrow gap.

Spark duration is normally measured two times—during engine cranking and during engine running.

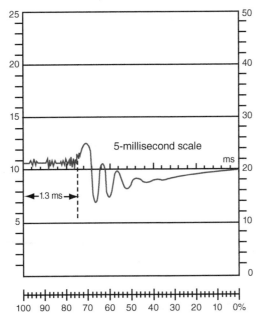

Figure 9-86 A 5-ms pattern showing a spark duration of 1.3 ms.

Spark Plug Firing Voltage

The coil must generate sufficient voltage in the secondary system to overcome the total resistance in the secondary circuit and to establish a spark across the spark plug electrodes. On the oscilloscope, this spark plug firing voltage is seen as the highest line in the pattern. The firing voltage might be affected by the condition of the spark plugs or the secondary circuit, engine temperature, fuel mixture, and compression pressures. To test the spark plug firing voltage on an oscilloscope, observe the firing line of all cylinders for height and uniformity. The normal height of the firing voltages should be between 7 and 13 kV with no more than a 3-kV variation between cylinders.

If during the test one or more of the firing voltages are uneven, low, or high, consult **Figure 9-87** for possible causes and corrections. Refer to Photo Sequence 18 to learn how to identify the cause of an abnormally high firing line. It should be noted that although the photo sequence demonstrates locating a component causing a high firing voltage, the

Condition	Probable Cause	Remedy
Firing voltage lines the same, but abnormally high	1. Retarded ignition timing	1. Rest ignition timing.
	2. Fuel mixture too lean	2. Check for vacuum leak.
	3. High resistance in coil wire	3. Replace coil wire.
	4. Corrosion in coil tower terminal	4. Clean or replace coil.
	5. Corrosion in distributor coil terminal	5. Clean or replace distributor cap.
Firing voltage lines the same, but abnormally low	1. Fuel mixture too rich	1. Check for plugged air filter.
	2. Breaks in coil wire causing arcing	2. Replace coil wire.
	3. Cracked coil tower causing arcing	3. Replace coil.
	4. Low coil output	4. Replace coil.
	5. Low engine compression	5. Determine cause and repair.
One or more, but not all firing voltage lines higher than the others	1. Idle mixture not balanced	1. Readjust idle mixture.
	2. EGR valve stuck open	2. Inspect or replace EGR valve.
	3 High resistance in spark plug wire	3. Replace spark plug wires.
	4. Cracked or broken spark plug insulator	4. Replace spark plugs.
	5. Intake vacuum leak	5. Repair leak.
	6. Defective spark plugs	6. Replace spark plugs.
	7. Corroded spark plug terminals	7. Replace spark plugs.
One or more, but not all firing voltage lines lower	1. Curb idle mixture not balanced	1. Readjust idle mixture.
	2. Breaks in plug wires causing arcing	2. Replace spark plug wires.
	3. Cracked coil tower causing arcing	3. Replace coil.
	4. Low compression	4. Determine cause and repair.
	5. Defective or fouled spark plugs	5. Replace spark plugs.
Cylinders not firing	1. Cracked distributor cap terminal	1. Replace distributor cap.
	2. Shorted spark plug wire	2. Determine cause of short and replace wire.
	3. Mechanical problem in engine	3. Determine problem and correct.
	4. Defective spark plugs	4. Replace spark plugs.
	5. Spark plugs fouled	5. Replace spark plugs.

Figure 9-87 Firing line diagnosis.

same elimination process is used to locate a component causing a low firing voltage. To do this, simply follow the same process but use a spark tester instead of a grounding probe. The spark tester will cause a controlled increase in the firing voltage at the point at which the problem part is eliminated.

Classroom Manual
Chapter 9, page 275

Rotor Air Gap Voltage Test (Distributor Ignitions Only)

When high firing voltages are present in one or more of the cylinders of a distributor ignition, perform a rotor air gap test. The purpose of this test is to determine the amount of secondary voltage required to bridge the rotor air gap.

To perform the rotor air gap voltage test:

1. Set the voltage scale to measure secondary voltage scale to 0 to 25 kV.
2. Start the engine and run it to about 1,000 rpm.
3. Observe the height of the firing lines and make note of any abnormal cylinder voltages.
4. Shut the engine off.
5. Disconnect the spark plug wire of the abnormal cylinder and use a grounding probe to ground the distributor cap terminal.
6. Start the engine and again run the engine at 1,000 rpm.
7. Observe the firing line of the abnormal cylinder previously recorded. There should be a slight drop in the firing line as shown in **Figure 9-88**.

The observed voltage represents voltage needed to overcome the resistance in the coil wire and of the rotor air gap. The rotor air gap measurements should not exceed manufacturer's specifications. If during the test, the firing line remains high or drops to a level that does not meet manufacturer's specifications, the rotor or distributor cap might be defective. Visually inspect both and replace as necessary. A bent distributor shaft or worn shaft bushings will cause excessive rotor air gaps on about half the cylinders.

Secondary Circuit Resistance

Analysis of the spark line of a secondary pattern reveals the condition of the secondary circuit. The amount of secondary circuit resistance is indicated by the slope of the spark line. Excessive resistance in the secondary circuit causes the spark line to have a steep slope with a shorter firing duration. The excessive resistance restricts the current flow necessary to generate a good spark.

Notice the difference in spark duration with the plug wire grounded in Figure 9-88.

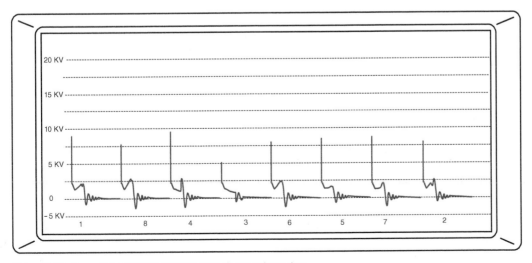

Figure 9-88 Grounded cylinder 3 showing rotor air gap voltage drop.

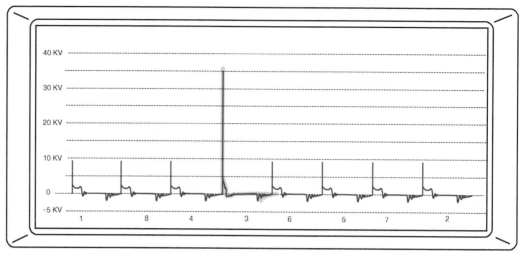

Figure 9-89 High kV showing excessive secondary resistance on cylinder 3. Note firing line height and lack of duration in spark line.

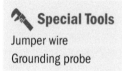 **Special Tools**

Jumper wire
Grounding probe

Refer to **Figure 9-89**. The firing kV is very high and the duration is very short. If you look closely at the pattern for number three compared to the rest of the cylinders, the duration is very short or even non-existent. The reason for this is the fact that the energy from the coil was almost totally spent in jumping the plug gap with very little left for duration. The situation is a plug wire with extremely high resistance and shows the tendency for the spark kV to rise when the spark plug gap is difficult to ionize. The slope on the spark line, as short as it is, is steeply downward.

A good spark line should be relatively even and measure 2 to 4 kV in height. High resistance in the secondary circuit produces a firing line and spark line that are higher in voltage with shorter firing durations. If high resistance is shown during testing, see **Figure 9-90**.

To pinpoint the cause of high resistance, use a grounding probe and jumper wire to bypass each component of the secondary circuit on all abnormal cylinders. Connect one end of the jumper lead to ground and the other end to the large portion of a grounding probe. Start the engine and adjust the speed to 1,000 rpm. Touch each secondary connector with the point of the grounding probe and observe the spark lines.

If after grounding the abnormal spark lines appear normal, the part just bypassed is the cause of the problem. For further diagnosis, see **Figure 9-91**.

While scanning the secondary circuit with the grounding probe, watch the firing lines on the scope. If there is an insulation break in the circuit, the firing lines decrease in height when the high secondary voltage arcs across to the probe.

Any insulated part of the secondary circuit can break down and leak. This includes the coil, coil wire and boots, distributor cap and terminals, and spark plug wires and boots.

Spark Plugs under Load

The voltage required to fire the spark plugs increases when the engine is under load. The voltage increase is moderate and uniform if the spark plugs are in good condition and properly gapped. However, if any unusual characteristics are displayed on the scope patterns when load is applied to the engine, the spark plugs are probably faulty. This condition is most evident in the firing voltages displayed on the scope. To test spark plug operation under load, note the height of the firing lines at idle speed. Then, quickly open and release

Symptom	Probable Cause	Remedy
SECONDARY CIRCUIT RESISTANCE TEST		
High resistance in spark lines	1. Rotor tip burned 2. High resistance in the spark plug wire 3. Distributor cap segments burned 4. Faulty spark plug	1. Visually inspect. 2. Perform ohmmeter test. 3. Visually inspect. 4. Perform secondary resistance test using the grounding probe.
After grounding spark plug, abnormal spark lines appear normal	1. Faulty spark plug	1. Substitute spark plug.
After grounding spark plug, abnormal spark lines still show high resistance	1. Corroded distributor cap towers 2. Distributor cap segments burned 3. High resistance in the spark plug wires	1. Visually inspect. 2. Visually inspect. 3. Perform ohmmeter test.
All the spark lines show high resistance	1. Defective coil wire 2. Rotor tip burned	1. Perform ohmmeter test. 2. Visually inspect.
SPARK PLUGS UNDER LOAD TEST		
One or more of cylinders show a voltage rise over 4 kV	1. Spark plug gap too wide 2. Worn spark plug electrodes 3. Open or high resistance spark plug wire 4. Improper fuel mixture 5. Open spark plug resistor 6. Leakage in vacuum system	1. Check the gap, and then regap the plug. 2. Replace the spark plug. 3. Perform ohmmeter test. 4. Adjust mixture. 5. Substitute spark plug. 6. Check for leaky vacuum hoses and diaphragms or gaskets.
One or more cylinders show a voltage rise less than 3 kV or no voltage rise at all	1. Shorted spark plug wire 2. Broken spark plug insulator 3. Fouled spark plug 4. Low compression	1. Perform secondary insulation test. 2. Replace spark plug. 3. Clean or replace spark plug. 4. Perform compression test.
CYLINDER TIMING ACCURACY TEST		
Transistor turn-off signals exceed specifications for engine being tested	1. Bent distributor shaft 2. Worn distributor bushings 3. Worn gear on distributor 4. Worn camshaft gear 5. Worn timing chain	1. Perform distributor test. 2. Perform distributor test. 3. Visually inspect. 4. Visually inspect. 5. Visually inspect.

Figure 9-90 Troubleshooting spark plugs.

the throttle (snap accelerate) and note the rise in the firing lines while checking the voltages for uniformity. A normal rise would be between 3 and 4 kV upon snap acceleration.

If during testing, one or more of the cylinders show a voltage rise of over 4 kV or if one or more cylinders show a voltage rise less than 3 kV or no voltage rise at all, check Figure 9-90.

Any problem that makes kV higher makes duration shorter and vice versa.

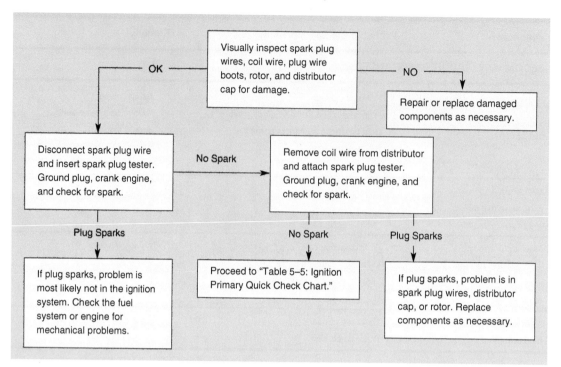

Figure 9-91 Ignition secondary quick-check chart.

Coil Condition

The energy remaining in the coil after the spark plugs fire gradually diminishes in a series of oscillations. These oscillations are observable in the intermediate sections of both the primary and secondary patterns. If the scope pattern shows an absence of normal oscillations in the intermediate section, check for a possible short in the coil by testing the resistance of the primary and secondary windings.

CASE STUDY

A customer complained about a stalling problem on a Pontiac with an EI system. When questioned about this problem, the owner said the engine stalled while driving in the city, but it only happened about once a week. Further questioning of the customer indicated the engine would restart after 5 to 10 minutes, and the owner said the engine seemed to be flooded.

The technician performed voltmeter tests, ohmmeter tests, and an oscilloscope diagnosis on the EI system. There were no defects in the system, and the engine operation was satisfactory. The fuel pump pressure was tested and the filter checked for contamination, but no problems were discovered in these components. The customer was informed that it was difficult to diagnose this problem when the symptoms were not present. The service writer asked the customer to phone the shop immediately the next time the engine stalled without attempting to restart the car. The service writer told the customer that a technician would be sent out to check the problem when she phoned the shop.

Approximately 10 days later, the customer phoned and said her car had stalled. A technician was dispatched immediately, and she connected a spark tester to several of the spark plug wires. The ignition system was not firing any of the spark plugs. The car was towed to the shop without any further attempts to start the engine.

The technician discovered that the voltage supply and the ground wires on the crankshaft sensor were normal, but there was no voltage signal from this sensor while cranking the engine. The crankshaft sensor was replaced. When the customer returned later for other service, she reported the stalling problem had been eliminated.

ASE-STYLE REVIEW QUESTIONS

1. While discussing the primary and secondary ignition circuits,
   Technician A says that the primary controls the secondary.
   Technician B says that the primary circuit also controls ignition timing in the secondary circuit.
   Who is correct?
   A. A only
   B. B only
   C. Both A and B
   D. Neither A nor B

2. While discussing EI systems,
   Technician A says that base timing is not adjustable on EI systems.
   Technician B says that the air gap is adjustable on some crank sensors but timing is not affected.
   Who is correct?
   A. A only
   B. B only
   C. Both A and B
   D. Neither A nor B

3. *Technician A* says that most vehicles will not run without a CKP sensor inputs.
   Technician B says that some cars will run without any CMP sensor inputs.
   Who is correct?
   A. A only
   B. B only
   C. Both A and B
   D. Neither A nor B

4. *Technician A* says that the compression-sensing ignition or CSI uses the lower firing voltage of the plug in the compression stroke to eliminate the MAF sensor.
   Technician B says that ion-sensing ignition vehicles have been able to eliminate the downstream oxygen sensors in Saab vehicles.
   Who is correct?
   A. A only
   B. B only
   C. Both A and B
   D. Neither A nor B

5. *Technician A* says that as engine speed increases, the control module lengthens the dwell time.
   Technician B says that this increases the available time for coil saturation.
   Who is correct?
   A. A only
   B. B only
   C. Both A and B
   D. Neither A nor B

6. While discussing how to test a CKP,
   Technician A says that a logic probe can be used.
   Technician B says that an oscilloscope can be used.
   Who is correct?
   A. A only
   B. B only
   C. Both A and B
   D. Neither A nor B

7. *Technician A* says that poor grounds can be identified by conducting voltage drop tests and monitoring the circuit with a lab scope.
   Technician B says that when conducting a voltage drop test, the circuit must be turned on and have current flowing through it.
   Who is correct?
   A. A only
   B. B only
   C. Both A and B
   D. Neither A nor B

8. While discussing the diagnosis of an EI system,
   Technician A says that the coil primary normally has between 5,000 and 20,000 ohms.
   Technician B says that the coil secondary normally has less than 2 ohms.
   Who is correct?
   A. A only
   B. B only
   C. Both A and B
   D. Neither A nor B

9. *Technician A* says that some spark testers look like a spark plug, but they have a very large electrode gap.
   Technician B says that when the engine is cranked to check for voltage output, the ignition coil is forced to produce a lower voltage to overcome the reduced resistance created by the wide gap.
   Who is correct?
   A. A only
   B. B only
   C. Both A and B
   D. Neither A nor B

10. While testing the coils in an EI system,
    Technician A says that an infinity reading means that the windings have zero resistance and are shorted.
    Technician B says that the primary windings in some (EI) coils cannot be measured with an ohmmeter.
    Who is correct?
    A. A only
    B. B only
    C. Both A and B
    D. Neither A nor B

ASE CHALLENGE QUESTIONS

1. While discussing the comparison between spark duration and firing kV,

   Technician A says that if firing kV is high, duration will be short.

   Technician B says that if firing kV is low, duration will be long.

   Who is correct?

   A. A only
   C. Both A and B
   B. B only
   D. Neither A nor B

2. *Technician A* says that Chrysler EI systems use a crankshaft timing and camshaft reference sensor that are modified Hall-effect switches.

   Technician B says that these sensors should output from 0 volt and 9-12 volts with a DMM when the engine is cranked over in small increments.

   Who is correct?

   A. A only
   C. Both A and B
   B. B only
   D. Neither A nor B

3. *Technician A* says that if the spark plug wire terminal ends are not clean and bright, the terminals should be cleaned and recrimped.

   Technician B says that the spark plug wires need to be replaced.

   Who is correct?

   A. A only
   C. Both A and B
   B. B only
   D. Neither A nor B

4. *Technician A* says that the crankshaft position sensor can be a Hall-effect switch.

   Technician B says that the crankshaft position sensor can be a magnetic pulse generator.

   Who is correct?

   A. A only
   C. Both A and B
   B. B only
   D. Neither A nor B

5. While discussing spark plugs,

   Technician A says that iridium spark plugs should not have the gap checked by the technician.

   Technician B says that platinum spark plugs should not have the electrode filed.

   Who is correct?

   A. A only
   C. Both A and B
   B. B only
   D. Neither A nor B

Name _____   Date _____

SCOPE TESTING AN IGNITION SYSTEM

Upon completion of this job sheet, you should be able to observe the ignition system activity by observing it on a scope.

ASE Education Foundation Correlation

This job sheet addresses the following **AST/MAST** task: VIII. Engine Performance; C. Ignition System Diagnosis and Repair

Task #1   Diagnose (troubleshoot) ignition system–related problems such as no-starting, hard starting, engine misfire, poor drivability, spark knock, power loss, poor mileage, and emissions concerns; determine needed action. **(P-2)**

Tools and Materials

- A vehicle
- Service manual for the above vehicle
- An oscilloscope capable of measuring primary and secondary ignition patterns

Describe the vehicle being worked on:

Year _____ Make _____

Model _____ VIN _____

Engine type _____ Size _____Firing order _____

Procedure   **Task Completed**

1. Describe the general appearance of the engine.

   _____

   _____

   _____

   _____

2. Connect the scope leads to the vehicle.   ☐

3. Set the scope to look at the secondary ignition circuit in the display or parade pattern.   ☐

4. Start the engine and observe the height of the firing lines. Are they within 3 kV of each other? _____ If not, which cylinders are the most different from the rest? _____ Are the heights of the firing lines all below 13 kV? _____ Is the height of the firing lines above 7 kV? _____ Based on these findings, what do you conclude?

   _____

   _____

   _____

   _____

5. Switch the scope to show the patterns in raster. Observe the length, height, and shape of the spark line for each cylinder. Describe them.

   _____

   _____

   _____

   _____

6. Based on these findings, what do you conclude?

_____

_____

_____

_____

7. Describe the general appearance of the intermediate section of the pattern for each cylinder.

_____

_____

_____

_____

8. Based on these findings, what do you conclude?

_____

_____

_____

_____

9. Describe the general appearance of the dwell section for each cylinder.

_____

_____

_____

_____

10. Based on these findings, what do you conclude?

_____

_____

_____

_____

11. Based on the appearance of all sections of the scope pattern for each cylinder, what are your recommendations and conclusions?

_____

_____

_____

_____

Instructor's Response

_____

_____

_____

Name _____   Date _____

TESTING AN IGNITION COIL

Upon completion of this job sheet, you should be able to test an ignition coil with an ohmmeter.

ASE Education Foundation Correlation

This job sheet addresses the following **AST/MAST** task: VIII. Engine Performance; C. Ignition System Diagnosis and Repair

Task #1   Diagnose (troubleshoot) ignition system–related problems such as no-starting, hard starting, engine misfire, poor drivability, spark knock, power loss, poor mileage, and emissions concerns; determine needed action. **(P-1)**

Tools and Materials

• A vehicle or a separate ignition coil

• DMM

• Service information for the above vehicle or ignition coil

Describe the vehicle being worked on:

Year _____ Make _____

Model _____ VIN _____

Procedure   **Task Completed**

1. Describe the general appearance of the coil.

   _____

   _____

2. Locate the resistance specifications for the ignition coil in the service manual.

   The primary winding should have _____ ohms of resistance.

   The secondary winding should have _____ ohms of resistance.

3. If the coil is still in the vehicle, disconnect all leads connected to it.   ☐

4. Connect the ohmmeter from the negative (tach) side of the coil to its container or frame. Observe the reading on the meter.

   The reading was _____ ohms.

   What does this indicate?

   _____

5. Connect the ohmmeter from the center tower of the coil to its container or frame. Observe the reading on the meter.

   The reading was _____ ohms. What does this indicate?

   _____

6. Connect the ohmmeter across the primary winding of the coil.

   The reading was _____ ohms. Compare this to specifications.

   What is indicated by this reading?

   _____

7. Connect the ohmmeter across the secondary winding of the coil. The reading was _____ ohms. Compare this to specifications. What is indicated by this reading?

   _____

8. Connect the scope and set the screen to raster or 5-millisecond sweep. Start the engine and observe the coil oscillations. The number of coil oscillations is _____. Based on your observation, the coil's oscillation performance is _____.

9. Based on the above tests, what is your conclusion about the coil?

_____

Instructor's Response

_____

_____

_____

Name _____   Date _____

INDIVIDUAL COMPONENT TESTING

Upon completion of this job sheet, you should be able to test components of the primary and secondary ignition system.

ASE Education Foundation Correlation

This job sheet addresses the following **AST/MAST** tasks: VIII. Engine Performance; C. Ignition System Diagnosis and Repair

Task #1   Diagnose (troubleshoot) ignition system–related problems such as no-starting, hard starting, engine misfire, poor drivability, spark knock, power loss, poor mileage, and emissions concerns; determine needed action. **(P-1)**

Task #3   Inspect, test, and/or replace ignition control module, powertrain/engine control module; reprogram/initialize as needed. **(P-3)**

Task #4   Remove and replace spark plugs; inspect secondary ignition components for wear and damage. **(P-1)**

Tools and Materials

- 12-volt test light
- DMM
- Appropriate service literature
- Lab scope

Describe the vehicle being worked on:

Year _____ Make _____

Model _____ VIN _____

Type of ignition system _____

Procedure

Task Completed

1. Check the following components:

Ignition Switch

a. Turn the ignition key off and disconnect the wire connector at the module.   ☐

b. Disconnect the S terminal of the starter solenoid to prevent the engine from cranking when the ignition is in the run position.   ☐

c. Turn the key to the run position.   ☐

d. With the test light, probe the red wire connection to check for voltage. Was there voltage?   ☐ Yes ☐ No

e. Check for voltage at the bat terminal of the ignition coil. Was there voltage?   ☐ Yes ☐ No

f. Turn the key to the start position and check for voltage at the start power wire connector at the module. Was there voltage?   ☐ Yes ☐ No

g. Check for voltage at the bat terminal of the ignition coil. Was there voltage?   ☐ Yes ☐ No

Conclusions

_____

_____

h. Turn the ignition switch to the off position.   ☐

i. Install a small straight pin into the appropriate module's input power wire.   ☐

j. Connect the digital voltmeter's positive lead to the straight pin and ground the negative lead to the distributor base.   ☐

k. Turn the ignition to the run position. Your voltage reading is _____.

l. Turn the ignition to the start position. Your voltage reading is _____.

Conclusions

_____

_____

Pick-up Coil

a. Turn the ignition off.   ☐

b. Remove the distributor cap.   ☐

c. Connect the ohmmeter to the pick-up coil terminals.   ☐

d. Record your readings: _____ ohms

e. The specified resistance is _____ ohms.

Conclusions

_____

_____

f. Connect the ohmmeter from one of the pick-up leads to ground.   ☐

g. Record your readings: _____ohms

h. The specified resistance is _____ ohms.

Conclusions

_____

_____

i. Reinstall the distributor cap, and then connect the scope leads to the pick-up coil leads.   ☐

j. Set the scope on its lowest scale.   ☐

k. Spin the distributor shaft by cranking the engine with the ignition disabled.   ☐

l. Describe the trace shown on the scope.

_____

_____

Conclusions

_____

_____

m. Disconnect the scope.   ☐

n. Connect a voltmeter set on its low-voltage scale.   ☐

o. Describe the meter's action.

_____

_____

Conclusions

_____

_____

Hall-Effect Sensors

Task Completed

a. Connect a 12-volt battery across the plus (+) and minus (−) voltage (supply current) terminals of the Hall layer. ☐

b. Connect a voltmeter across the minus (−) and signal voltage terminals. ☐

c. Insert a steel feeler gauge or knife blade between the Hall layer and magnet, then remove the feeler gauge. ☐

d. Describe what happens on the voltmeter.

_____

_____

Conclusions

_____

_____

e. Remove the 12-volt power source and prepare the engine to run. ☐

f. Set a lab scope on its low-scale primary pattern position. ☐

g. Connect the primary positive lead to the Hall-effect signal lead, and the negative lead to ground or the ground terminal at the sensor's connector. ☐

h. Start the engine and observe the scope. ☐

Record the trace on the scope.

_____

_____

Conclusions

_____

_____

Control Module

a. Connect one lead of the ohmmeter to the ground terminal at the module and the other lead to a good engine ground. ☐

Record your readings: _____ ohms

The specified resistance is _____ ohms.

Conclusions

_____

_____

Secondary Ignition Wires

a. Remove the distributor cap with the spark plug wires attached to the cap but disconnected from the spark plugs. ☐

b. Calibrate an ohmmeter on the 1,000 scale. ☐

c. Connect the ohmmeter leads from the end of a spark plug wire to the distributor cap terminal inside the cap to which the plug wire is connected. ☐

d. Record your readings: _____ ohms

e. The specified resistance is _____ ohms.

Conclusions

_____

_____

Spark Plugs

a. Remove the engine's spark plugs. Place them on a bench according to the cylinder number. ☐

b. Carefully examine the electrodes and porcelain of each plug. ☐

c. Describe the appearance of each plug.

_____

_____

_____

Conclusions

_____

_____

_____

d. Measure the gap of each spark plug and record your findings.

_____

_____

_____

e. What is the specified gap? _____inches

Conclusions

_____

_____

_____

Instructor's Response

_____

_____

_____

Name _____   Date _____

SETTING IGNITION TIMING

Upon completion of this job sheet, you should be able to check and set the ignition timing on a distributor-type ignition system.

ASE Education Foundation Correlation

This job sheet addresses the following **AST/MAST** tasks: VIII. Engine Performance; C. Ignition System Diagnosis and Repair

Task #1   Diagnose (troubleshoot) ignition system–related problems such as no-starting, hard starting, engine misfire, poor drivability, spark knock, power loss, poor mileage, and emissions concerns; determine needed action. **(P-1)**

Tools and Materials

- Timing light
- Tachometer
- Appropriate service information

Describe the vehicle being worked on:

Year _____ Make _____

Model _____ VIN _____

Engine type and size _____ Ignition timing specs _____

Source of specifications _____

Conditions that must be met before checking the timing _____

_____

What should the idle speed be? _____ rpm

What is the idle speed? _____ rpm

If not within specifications, correct the idle before proceeding.

Procedure

Task Completed

1. Connect the timing light pickup to the number one cylinder's spark plug wire, and the power supply wires on the light to the battery terminals with the proper polarity.   ☐

2. Start the engine.   ☐

3. The engine must be idling at the manufacturer's recommended rpm, and all other timing procedures must be followed.   ☐

4. Aim the timing light marks at the timing indicator and observe the timing marks. Timing found: _____ degrees.

5. If the timing mark is not at the specified location, rotate the distributor until the mark is at the specified location. Describe any difficulties you had doing this.

_____

_____

6. Tighten the distributor hold-down bolt to the specified torque. What is the specified torque?

_____

7. Reconnect any hoses or components that were disconnected for the timing procedure.   ☐

Instructor's Response

_____

_____

Name _____   Date _____

VISUALLY INSPECT A DISTRIBUTOR IGNITION SYSTEM

Upon completion of this job sheet, you should be able to visually inspect the components of a distributor-type ignition system.

ASE Education Foundation Correlation

This job sheet addresses the following **AST/MAST** tasks: VIII. Engine Performance; C. Ignition System Diagnosis and Repair

Task #1   Diagnose (troubleshoot) ignition system–related problems such as no-starting, hard starting, engine misfire, poor drivability, spark knock, power loss, poor mileage, and emissions concerns; determine needed action. **(P-1)**

Tools and Materials

- Steel feeler gauge set
- A vehicle with a distributor

Describe the vehicle being worked on:

Year _____ Make _____

Model _____ VIN _____

Type of ignition _____

Procedure

1. Are the spark plug and coil cables firmly seated in the distributor cap and coil and onto spark plugs?   ☐ Yes ☐ No

2. Do the secondary cables have cracks or signs of worn insulation?   ☐ Yes ☐ No

3. Are the boots on the ends of the secondary wires cracked or brittle?   ☐ Yes ☐ No

4. Are the secondary cables connected according to the firing order?   ☐ Yes ☐ No

5. Are there any white or grayish powdery deposits on secondary cables?   ☐ Yes ☐ No

6. Are spark plug cables from consecutively firing cylinders running parallel with each other?   ☐ Yes ☐ No

7. Is the coil cracked or showing signs of leakage in the coil tower?   ☐ Yes ☐ No

8. Is there oil around or on the coil?   ☐ Yes ☐ No

9. Is the distributor cap properly seated on its base?   ☐ Yes ☐ No

10. Are the mounting clips or screws securely fastened?   ☐ Yes ☐ No

11. Is there any electrical damage on the rotor or in the distributor cap?   ☐ Yes ☐ No

12. Is there any evidence of carbon tracking in the distributor cap?   ☐ Yes ☐ No

13. Are the distributor cap towers damaged?   ☐ Yes ☐ No

14. Is the rotor discolored or burnt?   ☐ Yes ☐ No

15. Are the distributor cap and housing vents blocked or clogged?   ☐ Yes ☐ No

16. Are the hoses to the vacuum advance unit secured to the unit (if so equipped)?   ☐ Yes ☐ No

17. With the rotor on the distributor shaft, move the rotor on the distributor shaft clockwise and counterclockwise. Does it move in one direction only? ☐ Yes ☐ No

18. Did it spring back into position after it was released? ☐ Yes ☐ No

19. Are the connections of the primary ignition system wiring tight connections? ☐ Yes ☐ No

20. Is the control module tightly mounted to a clean surface? ☐ Yes ☐ No

21. Are the electrical connections to the module corroded? ☐ Yes ☐ No

22. Are the electrical connections to the module loose or damaged? ☐ Yes ☐ No

23. Is the reluctor of the switching unit broken or cracked? ☐ Yes ☐ No

24. Is the insulation of the pick-up coil wire leads worn? ☐ Yes ☐ No

25. Is the nonmagnetic reluctor magnetized? ☐ Yes ☐ No

26. Record your summary of the visual inspection. Include what looked good as well as what looked bad.

_____

_____

_____

27. Based on the visual inspection, what are your recommendations?

_____

_____

_____

Instructor's Response

_____

_____

_____

Name _____   Date _____

VISUALLY INSPECT AND TEST AN ELECTRONIC IGNITION SYSTEM

Upon completion of this job sheet, you should be able to perform a visual inspection of the EI-type system and perform routine tests.

ASE Education Foundation Correlation

This job sheet addresses the following **AST/MAST** tasks: VIII. Engine Performance; C. Ignition System Diagnosis and Repair

Task #1   Diagnose (troubleshoot) ignition system–related problems such as no-starting, hard starting, engine misfire, poor drivability, spark knock, power loss, poor mileage, and emissions concerns; determine needed action. **(P-1)**

Task #2   Inspect and test crankshaft and camshaft position sensor(s); perform needed action. **(P-1)**

Tools and Materials

- Clean rag
- A vehicle equipped with an EI system
- Appropriate service manual
- Scan tool

Describe the vehicle being worked on:

Year _____ Make _____

Model _____ VIN _____

Type of ignition _____

Procedure

1. Are the spark plug cables connected and firmly seated on the spark plugs and coil pack terminals?   ☐ Yes ☐ No

2. Do the spark plug cables show signs of cracked, worn, or oil-soaked insulation?   ☐ Yes ☐ No

3. Are the spark plug cables routed properly and in their support looms?   ☐ Yes ☐ No

4. Are there any white or grayish powdery deposits on the spark plug cables?   ☐ Yes ☐ No

5. Is (are) the primary wire connector(s) firmly inserted into the EI module housing?   ☐ Yes ☐ No

6. Are the connections to the module corroded?   ☐ Yes ☐ No

7. If this system has the coils mounted to the EI module, are the coil(s) tightly bolted and positioned properly?   ☐ Yes ☐ No

8. Is the module or coil assembly properly fastened to the engine mounting brackets or as individual components if so designed?   ☐ Yes ☐ No

9. Are there any cracks or burn marks in the coil towers?   ☐ Yes ☐ No

10. If equipped with a Hall-effect crankshaft sensor, is there evidence of scraping, rubbing, or other contact marks on the interrupter blades?   ☐ Yes ☐ No

11. Are oil, antifreeze, or other contaminants on the crankshaft sensor or wire connector?   ☐ Yes ☐ No

12. Connect a scan tool and check for any ignition-related or misfire-related fault codes. Are any fault codes present?   ☐ Yes ☐ No

13. If the camshaft sensor is based on an old distributor location, explain how you would verify camshaft timing if you were replacing the sensor.

_____

_____

_____

_____

14. Record a summary of the visual inspection and tests performed. Include all findings including items that passed as well as items that failed.

_____

_____

_____

_____

15. Based on the visual inspection, what are your recommendations?

_____

_____

_____

_____

Instructor's Response

_____

_____

_____

Name _____   Date _____

INSPECTING AND TESTING AN IGNITION SYSTEM

Upon completion of this job sheet, you will be able to inspect and test the ignition system pick-up sensor or the triggering devices. You will also be able to inspect and test the ignition primary circuit wiring and its components.

ASE Education Foundation Correlation

This job sheet addresses the following **MLR** task: VIII. Engine Performance; A. General

Task #7   Remove and replace spark plugs; inspect secondary ignition components for wear and damage. **(P-1)**

This job sheet addresses the following **AST/MAST** task: VIII. Engine Performance; C. Ignition System Diagnosis and Repair

Task #4   Remove and replace spark plugs; inspect secondary ignition components for wear and damage. **(P-1)**

Tools and Materials

- Clean rag
- Appropriate service information
- DMM
- Test light
- DSO
- Protective clothing
- Goggles or safety glasses with side shields

Describe the vehicle being worked on:

Year _____ Make _____

Model _____ VIN _____

Engine type and size _____

Ignition type _____

Describe general operating condition.

_____

_____

Describe the type of ignition system found on this vehicle.

_____

_____

Procedure   **Task Completed**

Fill in the following sections, where applicable.

Basic Inspection (Coil-Over-Plug)

1. Inspect the wiring to the coil. What did you find?

_____

_____

2. Remove the coil from the valve cover.   ☐

3. Does the coil boot show signs of damage or being soaked with oil? ☐ Yes ☐ No

4. Is the boot brittle? ☐ Yes ☐ No

5. Are there any white or grayish powdery deposits? ☐ Yes ☐ No

6. Does the coil show signs of leakage in the coil towers? ☐ Yes ☐ No

7. Remove the coil boot. Do the coil towers show any signs of burning? ☐ Yes ☐ No

8. Does the vehicle have an externally mounted module? ☐ Yes ☐ No

9. Is the control module tightly mounted to a clean surface? (If applicable, some modules are part of the ECM.) ☐ Yes ☐ No

10. Are the electrical connections to the module corroded? ☐ Yes ☐ No

11. Are the electrical connections to the module loose or damaged? ☐ Yes ☐ No

Basic Inspection (with Spark Plug Cables)

1. Are the spark plug cables pushed tightly into the coil and onto the spark plugs? ☐ Yes ☐ No

2. Do the secondary cables have cracks or signs of worn insulation? ☐ Yes ☐ No

3. Are the boots on the ends of the secondary wires cracked or brittle? ☐ Yes ☐ No

4. Are there any white or grayish powdery deposits on secondary cables? ☐ Yes ☐ No

5. Do the coils show signs of leakage in the coil towers? ☐ Yes ☐ No

6. Do the coil towers show any signs of burning? ☐ Yes ☐ No

7. Separate the coils and inspect the underside of the coil and the ignition module wires. Are the wires loose or damaged? ☐ Yes ☐ No

8. Are the connections of the primary ignition system wiring tight connections? ☐ Yes ☐ No

9. Is the control module tightly mounted to a clean surface? ☐ Yes ☐ No

10. Are the electrical connections to the module corroded? ☐ Yes ☐ No

11. Are the electrical connections to the module loose or damaged? ☐ Yes ☐ No

For All Ignition Systems

Record your summary of the visual inspection. Include what looked good, as well as what looked bad.

_____

_____

_____

_____

Check the following components:

1. Describe the general appearance of the coil.

_____

_____

2. Locate the resistance specifications for the ignition coil in the service information and record those here.

   Note: Some COP ignition coils have built-in modules and resistance checks are not possible, especially those with three or more wires to each coil. Check with the service information you have for details.

   The primary winding should have _____ ohms of resistance.

   The secondary winding should have _____ ohms of resistance.

3. Disconnect the leads connected to the coil being tested. Connect the ohmmeter from the negative (tach) side of the coil to its container or frame. Observe the reading on the meter. The reading is _____ ohms.

4. What does this indicate?

   _____

   _____

5. Connect the ohmmeter from the center tower of the coil to its container or frame. Observe the reading on the meter. The reading is _____ ohms. What does this indicate?

   _____

   _____

6. Connect the ohmmeter across the primary winding of the coil. The reading is _____ ohms. Compare this to specifications. What does this reading indicate?

   _____

   _____

7. Connect the ohmmeter across the secondary winding of the coil. The reading is _____ ohms. Compare this to specifications. What does this reading indicate?

   _____

   _____

8. Based on the preceding tests, what is your conclusion about the coil?

   _____

   _____

Ignition Switch

1. Turn the ignition key off and disconnect the wire connector at the ignition module. Does the system have an ignition module? If so, where is it located?

   _____

   _____

2. Look at the wiring diagram and identify the colors of wires that represent the battery input, the output to the ignition system, and the output to the starting system. Describe them here.

   _____

   _____

3. Disconnect the S terminal of the starter solenoid that prevents the engine from cranking when the ignition is in the run position. Then, turn the ignition switch to the run position. With the test light, probe the power wire to the switch. Is there voltage? ☐ Yes ☐ No

4. Probe the power output of the switch. Is there voltage? ☐ Yes ☐ No

5. Check for voltage at the battery or positive terminal of an ignition coil. Is there voltage? ☐ Yes ☐ No

6. Turn the key to the start position and check for voltage at the start power wire connector at the module. Is there voltage? ☐ Yes ☐ No

7. Check for voltage at the bat terminal of the ignition coil. Is there voltage? ☐ Yes ☐ No

8. What can you conclude about the ignition switch after conducting these tests?

   _____

   _____

9. Turn the ignition switch to the off position. Identify the power feed to the ignition module(s). Describe the wire and its location.

   _____

   _____

10. Install a small straight pin or wire probe into the appropriate module's power wire. Then, connect the positive lead of the DMM to the straight pin and the negative lead to ground. Turn the ignition switch to the run position. Your voltage reading is _____.

11. Turn the ignition switch to the start position. Your voltage reading is _____.

12. From all of the tests, what are your conclusions about the ignition switch?

   _____

   _____

Secondary Ignition Wires

1. Does the system have separate secondary wires or is the system a COP type where the ignition coil mounts directly over the spark plug?

   _____

   _____

2. If available, remove each secondary wire (one at a time) from the coil and spark plug. Measure the resistance of each wire from end to end. Record your readings here.

   _____

   _____

3. What is the specified resistance? _____ ohms

4. What can you conclude about the condition of the secondary wires?

   _____

   _____

Spark Plugs   **Task Completed**

1. Remove the engine's spark plugs. Place them on a bench arranged according to the cylinder number. Carefully examine the electrodes and porcelain of each plug and describe each here.

   _____

   _____

   _____

   _____

2. What should a normal spark plug look like?

   _____

   _____

3. Measure the gap of each spark plug and record your findings. _Note:_ The gap of iridium spark plugs should not be measured. This can cause damage to the delicate electrode.

   _____

   _____

   _____

   _____

4. What is the specified gap? _____ inches

5. What is the apparent condition of the plugs?

   _____

   _____

6. Measure the resistance between the spark plug terminal and the center electrode. What is your measurement? _____ ohms

7. Why would a spark plug have resistance?

   _____

   _____

8. Summarize your findings.

   _____

   _____

   _____

9. Reinstall the plugs and torque them to specifications. What are the specs?

   _____

10. Reinstall the secondary wires.   ☐

Instructor's Response

_____

_____

_____

CHAPTER 10

EMISSION CONTROL SYSTEM DIAGNOSIS AND SERVICE

Upon completion and review of this chapter, you should be able to:

- Describe the different types of state vehicle emissions (I/M) control programs.
- Diagnose and service evaporative (EVAP) systems.
- Describe the testing of the canister and of the canister purge valve.
- Describe the inspection and replacement of positive crankcase ventilation (PCV) system parts.

- Diagnose engine performance problems caused by improper exhaust gas recirculation (EGR) operation.
- Diagnose and service the various types of EGR valves.
- Diagnose EGR vacuum regulator (EVR) solenoids.
- Check the efficiency of a catalytic converter.
- Diagnose and service secondary air injection systems.

 Basic Tools

Vacuum gauge
Exhaust gas analyzer
Basic tool set
Fender covers
Safety glasses

Terms To Know

Acceleration simulation mode (ASM)
AIR system
Canister purge solenoid
Charcoal canister
Closed loop
Constant volume sampling (CVS)
Diagnostic trouble code (DTC)
Differential pressure feedback (DPFE) EGR system
Digital EGR valve
Dyno
Efficiency

EGR control solenoid
EGR vacuum regulator (EVR)
EGR valve port
EGR vent solenoid (EGRV)
Engine Off Natural Vacuum (EONV)
Enhanced EVAP
EVR valve
Fuel tank level sensor
Fuel tank pressure control valve
Fuel tank pressure sensor
Handheld digital pyrometer

Initial vacuum pull-down
Inspection and maintenance (I/M)
Knock sensor module
Leak detection pump (LDP)
Linear EGR valve
OBD II
Purge valve
Quad driver
Repair grade (RG-240)
Small leak test
Vacuum hold and decay test
Vent valve
Weak vacuum test

INTRODUCTION

The kinds of emissions that are being controlled in gasoline engines today are unburned hydrocarbons, carbon monoxide, and oxides of nitrogen. Additionally, standards have been set for the elimination of greenhouse gases (GHGs), principally carbon dioxide (CO_2) emissions. Also, new fuel mileage standards have been set to cut back on our use of fossil fuel production.

Unburned hydrocarbons (HC) are particles, usually vapors, of gasoline that have not been burned during combustion. They are present in the exhaust and in crankcase vapors. Any raw fuel that evaporates from the fuel system is also classified as HC.

Carbon monoxide (CO) is a poisonous chemical compound of carbon and oxygen. It forms in the engine when there is not enough oxygen to combine with the carbon during combustion. When there is enough oxygen in the mixture, CO_2 is formed. CO_2 is not a pollutant but, as mentioned earlier, had a standard for the first time in 2012.

Oxides of nitrogen (NO_X) are various compounds of nitrogen and oxygen. Both of these gases are present in the air used for combustion. They are formed in the cylinders during combustion and are part of exhaust gas. NO_X is the principle chemical that causes smog.

What comes out of an engine's exhaust depends on two factors. One is the effectiveness of the emission control devices. The other is the effectiveness or **efficiency** of the engine. A totally efficient engine changes all of the energy in the fuel, in this case gasoline, into heat energy. This heat energy is the power produced by the engine. To run, an engine must receive fuel, air, and heat. Our air is primarily nitrogen and oxygen, but the engine only uses the oxygen.

In order for an engine to run efficiently, it must have fuel mixed with the correct amount of air. This mixture must be shocked by the correct amount of heat (spark) at the correct time. All of this must happen in a sealed container or cylinder. If conditions were perfect, a great amount of heat energy is produced and the fuel and air combine to form water and carbon dioxide, and nitrogen leaves the engine unchanged. However, we know that conditions are never perfect, so some of the fuel remains unburned or partially burnt, while some of the nitrogen manages to combine with the oxygen in the air.

When the engine does not receive enough air (or gets too much fuel), it produces less heat energy, water, and carbon dioxide. It also produces carbon monoxide and releases some fuel (HCs) and air that was not burned. As the amount of air delivered to the engine decreases, the production of CO increases and more unburned fuel and air are released in the exhaust.

When the engine receives too much air, the amount of heat energy it produces increases, but complete combustion does not take place and the exhaust has large amounts of fuel and air in it. Because of the increased heat, NO_X is formed. If the cylinder cannot hold the pressures because of a compression leak, the amount of heat it produces decreases and large amounts of fuel and air are released in the exhaust.

Since it is nearly impossible for an engine to receive the correct amounts of everything, a well-running engine will emit some amounts of pollutants. It is the job of the emission control devices to clean them up.

EMISSIONS TESTING

The first Clean Air Act prompted Californians to create the California Air Resources Board (CARB). CARB's purpose was to implement strict air standards. These became the standard for federal mandates. One of the approaches to clean the air by the CARB was to start periodic motor vehicle inspection (PMVI). The purpose for the PMVI, or **inspection and maintenance (I/M)**, is to inspect a vehicle's emission controls once a year, in some locations every other year. This inspection included a tailpipe emissions test and an under-hood inspection. The tailpipe test certifies that the vehicle's exhaust emissions are within the limits set by law. The under-hood and vehicle inspection verifies that the pollution control equipment has not been tampered with or disconnected. Many states now check the status of the OBD2 readiness monitors to determine if the vehicle passes emissions.

Many states now perform PMVIs or I/M inspections, if not in the entire state, then in some of the major urban areas.

Classroom Manual
Chapter 10, page 304

NO_X is commonly pronounced *knocks*.

Some testing of OBDII vehicles is done with a scan tool. If the readiness monitors are set, the vehicle passes. A tailpipe test is not performed.

Inspection and maintenance programs were developed to aid in the reduction of emissions by checking and mandating the maintenance of vehicles on the road.

The 1970 Clean Air Act made it illegal for professional technicians to remove, disconnect, or tamper with factory emission controls. The law did not make it illegal for the owner of the vehicle to make the emission control devices inoperable. The Clean Air Act was revised in 1990, and it made it illegal for anyone to remove, disconnect, or tamper with emission controls on any vehicle that would be used on public roads.

Today, some states have incorporated an emissions test with their annual vehicle registration procedures. Most states have the IM240 or similar program that was first implemented in California. The IM240 tests the emissions of a vehicle while it is operating under a variety of load conditions. This is an improvement over exhaust testing during idle and high speed with no load.

The IM240 test requires the use of a chassis dynamometer, commonly called a **dyno**. While on the dyno, the vehicle is operated for 240 seconds and under different load conditions. The test drive on the dyno simulates both in-traffic and highway driving and stopping. The emissions tester tracks the exhaust quality through these conditions.

The IM240 program also includes a leakage test (**Figure 10-1**) and a functional test (**Figure 10-2**) of the evaporative emission control devices and a visual inspection of the total emission control system. If the vehicle fails the test, it must be repaired and certified before it can be registered.

The exhaust analyzer typically used to certify vehicles is an infrared tester. Specific infrared testers have been approved to certify exhaust levels and must meet specific standards.

> Testing a vehicle while it is driven on a chassis dynamometer is called *transient testing*.

> A **dyno** is used to simulate driving conditions while the vehicle remains stationary for vehicle testing.

Exhaust Analyzers

Testing the quality of the exhaust is both a procedure for testing emission levels and a diagnostic routine. One of the most valuable diagnostic aids is the exhaust analyzer (**Figure 10-3**).

A typical infrared exhaust analyzer has a long sample hose with a probe in the end of the hose. The probe is inserted in the vehicle's tailpipe. When the analyzer is turned on,

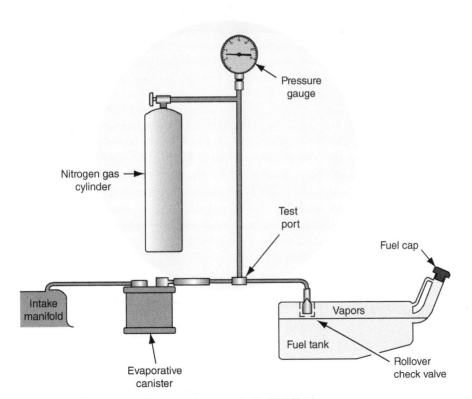

Figure 10-1 Typical evaporative system test setup for the IM 240 test.

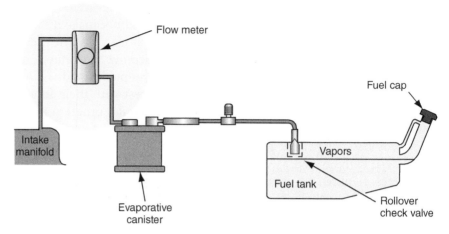

Figure 10-2 Typical evaporative system purge test setup.

Figure 10-3 A typical five-gas analyzer.

an internal pump moves an exhaust sample from the tailpipe through the sample hose and the analyzer. A water trap and filter in the hose removes moisture and carbon particles.

The pump forces the exhaust sample through a sample cell in the analyzer. The exhaust sample is then vented to the atmosphere (**Figure 10-4**). In the sample cell, a beam of infrared light passes through the exhaust sample. The analyzer then determines the quantities of HC, CO, CO_2, O_2, and NO_X. These are called five-gas analyzers.

When using the exhaust analyzer as a diagnostic tool, it is important to realize that the severity of the problem dictates how much higher than normal the readings will be.

Classroom Manual
Chapter 10, page 299

IM240 Test

The IM240 test is a 240-second test of a vehicle's emissions. During this time, the vehicle is loaded to simulate a short drive on city streets and then a longer drive on a highway. The complete test cycle includes acceleration, deceleration, and cruising. During the test, the vehicle's exhaust is collected by a **constant volume sampling (CVS)** system that makes sure a constant volume of ambient air and exhaust pass through the exhaust analyzer. The CVS exhaust hose covers the entire exhaust pipe. Therefore, it collects all of the exhaust, not just a sample as regular gas analyzers do. The hose contains a mixing tee that draws in outside air to maintain a constant volume to the gas analyzer. This is important

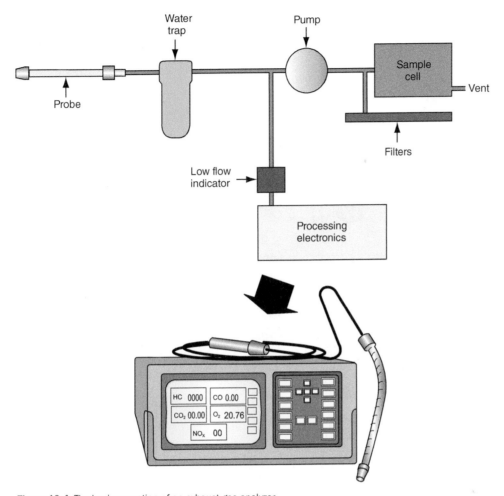

Figure 10-4 The basic operation of an exhaust gas analyzer.

for the calculation of mass exhaust emissions. During an IM240 test, the exhaust gases are measured in grams per mile.

The test is conducted on a chassis dynamometer (**Figure 10-5**), which loads the drive wheels to simulate real-world conditions. The rollers of the dyno are loaded by a computer after an inspector inputs basic information about the vehicle to be tested. The computer's program will adjust the resistance on the rollers according to the vehicle's weight, aerodynamic drag, and road friction.

At most I/M testing stations when a vehicle enters an inspection lane, the driver is greeted by a lane inspector. The inspector receives the needed information from the driver, then directs the driver to the customer waiting area. The inspector moves the vehicle to be visually inspected. Upon completion of the inspection, the inspector enters all pertinent information into the computer. After this information has been entered, the inspector moves the car onto the inertia simulation dynamometer.

Once on the dyno, the hood of the vehicle is raised and a large cooling fan is moved in front of the car. The engine is allowed to idle for at least 10 seconds, then testing begins. The transient driving cycle is 240 seconds and includes periods of idle, cruise, and varying accelerations and decelerations. During the drive cycle, the car's exhaust is collected by the CVS system. The inspector driving the car must follow an electronic, visual depiction of the speed, time, acceleration, and load relationship of the transient driving cycle. For cars equipped with a manual transmission, the inspector is prompted by the computer to shift gears according to a predetermined schedule.

The constant volume sampling method involves collecting everything that comes from the tailpipe and weighing the results.

I/M 240 TESTING

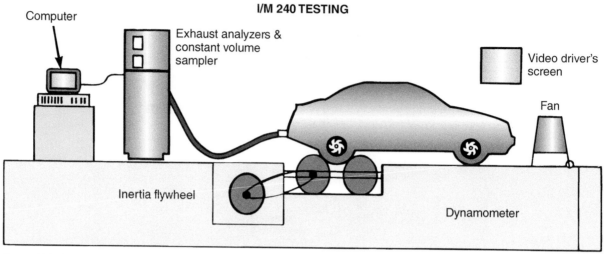

Figure 10-5 Components of a typical IM240 test station.

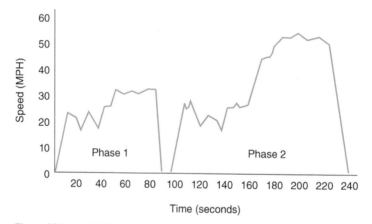

Figure 10-6 An IM240 drive trace.

The IM240 drive cycle is displayed by a trace. The trace is based on road speed versus time. The test is composed of two phases—Phase 1 and Phase 2—combining the results of these two phases results in the Test Composite (**Figure 10-6**). Phase 1 is the first 95 seconds of the drive cycle. The car travels 56 hundredths of a mile. This phase represents driving on a flat highway at a maximum speed of 32.4 miles per hour under light to moderate loads. Phase 2 is from the 96th second of the test to the end, or the 240th second. The vehicle travels 1.397 miles on a flat highway at a maximum speed of 56.7 miles per hour, including a hard acceleration to highway speeds.

The end result of the IM240 test is the measurement of the pollutants emitted by the vehicle during normal, on-the-road driving. After the test, the customer receives an inspection report. If the car failed the emissions test, it must be fixed. This inspection report is a valuable diagnostic tool (**Figure 10-7**). To correct the problems that caused the vehicle to fail, the inspection report must be studied.

The report shows the amount of gases emitted during the different speeds of the test. It also shows the cut-point for the various pollutants. Readings above the cut-point line are above the maximum allowable. Nearly all vehicles will have some speeds and conditions where the emissions levels are above the cut-point. A vehicle that fails the test will have many abnormal speeds or will have an area of very high pollutant output.

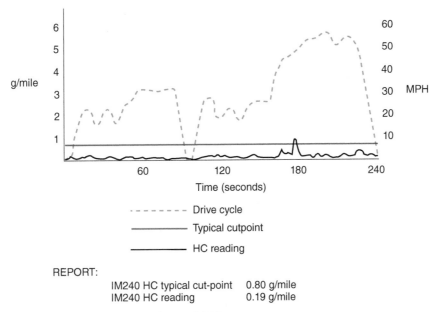

REPORT:

IM240 HC typical cut-point	0.80 g/mile
IM240 HC reading	0.19 g/mile

Figure 10-7 A hydrocarbon trace from an IM240 test.

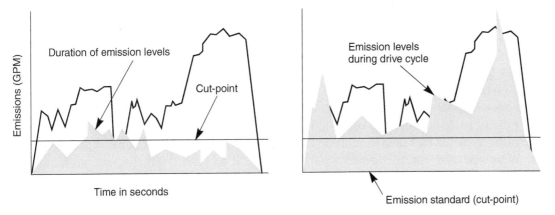

Figure 10-8 *Left*: The emissions level is acceptable. *Right*: The emissions level is unacceptable.

To use the report as a diagnostic tool, pay attention to all of the gases and to the loads and speeds at which the vehicle went over the cut-point (**Figure 10-8**). Think about what system or systems are responding to the load or speed. Would a malfunction of these systems cause this gas to increase? Pay attention not only to the gases that are above the cut-point, but also to those that are below. If the HC readings are above the cut-point at a particular speed and the NO_X readings are slightly below the cut-point at the same speed, fixing the problem causing the high HC will probably cause the NO_X to increase above the cut-point. As combustion is improved, the chances of forming NO_X also improve. Consider all of the correlations between gases when diagnosing a problem.

The IM240 measures HC, CO, NO_X, and CO_2. The job of a technician is to determine what loads and engine speeds the vehicle failed. Then test those systems that could affect engine operation and exhaust output at the failed speeds and loads. These tests could include looking at an O_2 waveform, watching a four-gas analyzer, checking data with a scan tool, reading injector pulse width, or others.

Most of the IM240 drive cycle can be simulated either in the diagnostic bay or on the road. It is hard to duplicate the entire drive cycle on a road test, but you can successfully

The cut-point represents the maximum amount or limit of each gas that the vehicle is allowed to emit.

duplicate part of the test. Hopefully it is the part that failed the vehicle. This may be important for diagnostics and is very important for the verification of the repair. The drive cycle is actually six operating modes. These modes need to be duplicated to identify why the vehicle failed that part of the trace.

- Mode one—idle, no load at 0 miles per hour
- Mode two—acceleration from 0 to 35 miles per hour
- Mode three—acceleration from 35 to 55 miles per hour
- Mode four—a steady cruise at 35 miles per hour
- Mode five—a steady high cruise at 55 miles per hour
- Mode six—decelerations from 35 miles per hour to 0 and from 55 miles per hour to 0

Keep these in mind, along with how many seconds each module lasts, and you will be able to duplicate the actual drive cycle as best as possible with the equipment you have. Once the problem has been diagnosed and repaired, repeat the test you did to find the problem. This will let you know if you fixed the problem.

Recently, many states are implementing other types of testing programs that are in place of the IM240 test.

Other I/M Testing Programs

Testing a vehicle at a constant load on a dyno is called steady-state testing.

Many state I/M programs only measure the emissions output from a vehicle while it is idling. The test is conducted with a certified exhaust gas analyzer. The measurements of the exhaust sample are then compared to standards dictated by the state according to the production year of the vehicle.

Some states also include a preconditioning mode at which the engine is run at a high idle (approximately 2,500 rpm) or the vehicle is run at 30 mph on a dynamometer for 20 to 30 seconds, prior to taking the idle tests. This preconditioning mode heats up the catalytic converter, allowing it to work at its best. Some programs include the measurement of the exhaust gases during a low constant load on the dyno or during a constant high idle. These measurements are taken in addition to the idle tests. A visual inspection and functional test of the emission control devices is part of some I/M programs. If the vehicle has tampered, nonfunctional, or missing emission control devices, the vehicle will fail the inspection.

The California ARB developed a test that incorporates steady-state and transient testing. This test is called the **Acceleration Simulation Mode (ASM)** test. The ASM includes a high-load steady-state phase and a 90-second transient test. This test is an economical alternative to the IM240 test, which requires a dyno with a computer-controlled power absorption unit. The ASM can be conducted with a normal chassis dyno and a five-gas analyzer.

Repair Grade (RG-240) is an economical version of the standard I/M 240 test.

Another alternative to the IM240 test is the **Repair Grade (RG-240)** test. This program uses a chassis dynamometer, constant volume sampling, and a five-gas analyzer (**Figure 10-9**). It is very similar to the IM240 test and is conducted in the same way; however, it is more economical. The primary difference between the IM240 and the RG-240 is the chassis dynamometer. Although they accomplish the same task, the RG-240 dynamometer is less complicated but nearly matches the load simulation of the IM240 dynamometer.

The **OBD II** readiness monitor is performed in conjunction with an exhaust emissions test or as a stand-alone test. The ability of the OBD II system to reveal engine data, performance, and emission control–related history began on all domestic vehicles in 1996, although a few models had some OBD II capability in 1994 and 1995. The **diagnostic trouble codes (DTCs)** and systems monitor data and history are retrieved by connecting a scanner to the under-dash diagnostic connector. When a repeated emissions-related failure occurs within a specific period of time, the powertrain control module (PCM) will turn the malfunction indicator lamp on to warn the driver of a problem. A DTC is then stored in the memory and is used to identify the system at fault.

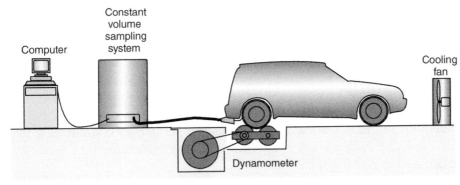

THE RG-240 TESTING SYSTEM

Figure 10-9 Typical components for a RG-240 test station.

EVAPORATIVE EMISSION CONTROL SYSTEM DIAGNOSIS AND SERVICE

⚡ WARNING If gasoline odor is present in or around a vehicle, check the EVAP system for cracked or disconnected hoses and check the fuel system for leaks. Gasoline leaks or escaping vapors may result in an explosion causing personal injury and/or property damage. The cause of fuel leaks or fuel vapor leaks should be repaired immediately. Do not overlook a loose fuel tank cap!

EVAP system diagnosis varies depending on the vehicle make and model year. Always follow the service and diagnostic procedure in the vehicle manufacturer's service manual. If the EVAP system (**Figure 10-10**) is purging vapors from the charcoal canister when the

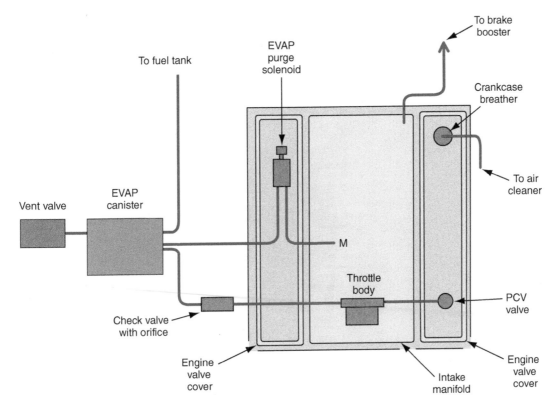

Figure 10-10 A typical EVAP system.

Classroom Manual
Chapter 10, page 304

Evaporative emissions
are hydrocarbon
emissions.

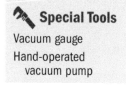
Special Tools

Vacuum gauge
Hand-operated
vacuum pump

engine is idling or operating at very low speed, rough engine operation will occur, especially during warmer weather. Cracked hoses or a canister saturated with gasoline may allow gasoline vapors to escape to the atmosphere, resulting in gasoline odor in and around the vehicle and codes for a leaking canister.

All of the hoses in the EVAP system should be checked for leaks, restrictions, and loose connections. The electrical connections in the EVAP system should be checked for looseness, corroded terminals, and worn insulation. When a defect occurs in the **canister purge solenoid** and related circuit, a DTC is usually set in the PCM memory. If a DTC related to the EVAP system is set in the PCM memory, always correct the cause of this code before further EVAP system diagnosis.

A scan tester may be used to diagnose the EVAP system. In the appropriate tester mode, the tester indicates whether the purge solenoid is on or off. Connect the scan tester to the DLC and start the engine. With the engine idling, the purge solenoid should be off and not allow vacuum to pass through the valve. Leave the scan tester connected and command the purge valve open. Using a hand-operated vacuum pump, confirm that the valve is opening and closing, and not losing its seal. If the purge solenoid is not operating during the test, check the power supply wire to the solenoid, solenoid winding, and the wire from the solenoid to the PCM.

EVAP System Component Preliminary Diagnosis

Check the canister to make sure that it is not cracked or otherwise damaged (**Figure 10-11**). Also make certain that the canister filter is not completely saturated. Remember that a saturated charcoal filter can cause symptoms that can be mistaken for fuel system problems. Rough idle, flooding, and other conditions can indicate a canister problem. A canister filled with liquid or water causes backpressure in the fuel tank. It can also cause richness and flooding symptoms during purge or start-up. (Some hybrid vehicles have intentionally pressurized fuel tank systems. Check the calibration and engine decal before diagnosing.)

To test for saturation, unplug the canister momentarily during a diagnosis procedure and observe the engine's operation. If the canister is saturated, it must be replaced.

A vacuum leak in any of the evaporative emission components or hoses can cause starting and performance problems (as can any engine vacuum leak) or cause a DTC to set. It can also cause complaints of fuel odor. Incorrect connection of the components can cause rich stumble or lack of purging, resulting in fuel odor and DTCs to be set. If the fuel

Figure 10-11 The charcoal canister should be checked for damage.

tank has a pressure and vacuum valve in the filler cap, check these valves for dirt contamination and damage. The cap may be washed in clean solvent. When the valves are sticking or damaged, replace the cap.

Check the service information for other common evaporative emission control system problems.

Classroom Manual
Chapter 10, page 305

Enhanced EVAP

The **vent valve** and tank pressure sensor are used for diagnostic purposes, and the fuel level sensor was added to the list of input sensors to the PCM. With **enhanced EVAP**, the PCM conducts several tests to determine if the system is operational and there are no leaks. Of course, the technician must be able to test these valves, confirm the problem, and verify the repair. The canister purge or vent solenoids (**Figure 10-12**) can usually be commanded on and checked for operation with a scan tool, and the solenoid windings can be checked with an ohmmeter or as directed by service information.

Enhanced EVAP systems have automatic leak detecting systems.

Evaporative emission systems must be able to detect a leak as small as 0.020 inch in diameter and use tests to confirm that there are no leaks present (**Figure 10-13**).

The following information is intended to help you understand the system involved, not to diagnose a particular system, so always use information specific to the vehicle concerned.

The PCM can control the purge and vent valves open and closed. When used in conjunction with the tank pressure sensor, this enables diagnostics on the integrity of the system.

The first test we will describe is the **initial vacuum pull-down**, or the **weak vacuum test**. This will establish if there is a large system leak. Once the enable criteria are set, the PCM will close the vent valve and open the purge valve. This would, of course, cause a vacuum to be indicated on the **fuel tank pressure sensor** if the system were sealed. If the vehicle should fail to pull a vacuum of up to 7 in. Hg, then the MIL would illuminate after the second consecutive failure under similar conditions and set a code PO455, gross evaporative system leak. Some vehicles will set a "Check fuel cap" light and a code PO457 after this test was failed after a refueling event was detected by the PCM using the fuel level sensor. One of the problems associated with the evaporative emissions were customers

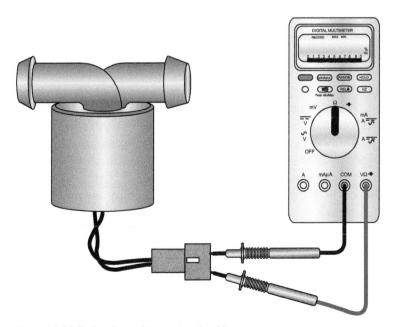

Figure 10-12 Testing the canister purge solenoid.

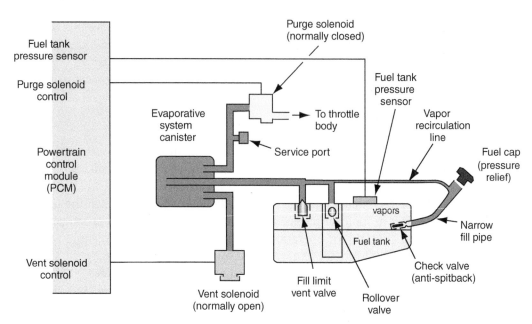

Figure 10-13 The enhanced EVAP system components.

The initial vacuum pull-down test is performed by closing the vent valve and opening the purge valve. If no leaks are present vacuum will be detected by the fuel tank pressure sensor. This is also called the weak vacuum test.

The vacuum hold and decay test is automatically run to check for small leaks in the EVAP system after the large leak test has been passed.

who failed to install their fuel cap tight enough for a good seal, which alerts the PCM of a problem. Some other problems that may be indicated by this failure are as follows:

- A large system leak, such as a vapor line loose or kinked, or a hole in the system
- A stuck-open vent solenoid, or loose vent line (basically the PCM cannot close the valve)

If the initial vacuum pull-down (weak vacuum test) passes, the system assumes there is no large leak, and begins to check for a small leak. This is done through the **vacuum hold and decay test**, or **small leak test**. For this test, the system will look for a leak that is down to 0.020 inch in size. The vent and purge solenoids will close and the PCM will watch for decay in the vacuum being held. If the vacuum drops too quickly, a PO442 is armed and set if the system fails again under similar conditions. With leaks so small, the software in the PCM makes allowance for the buildup of pressure inside the tank due to high-volatility fuel, and/or high ambient temperatures and excessive fuel slosh. For these reasons, the test will abort if fuel level, engine load, or fuel tank pressure should change abruptly. The pressure from high-volatility fuel could reduce the vacuum measured and result in false code triggering.

In order to guard against this false triggering, the vent can be opened, eliminating the vacuum entirely, and then the system vent and purge valves can be closed while the PCM actually measures how much pressure is being added to the tank by the fuel volatility.

Finally, the cold-idle test is used to entirely screen out the effects of fuel volatility by running the test only with the fuel cold (engine running less than 10 minutes) and the vehicle at idle or less than 10 mph. Even this test can be canceled by conditions that may produce excessive fuel slosh conditions.

The sealing of the purge valve is checked by measuring vacuum inside the tank with the vent and purge valve closed. If there is vacuum building under these conditions, it is assumed that the purge valve is leaking, setting code P1443 on Ford vehicles.

The most recent use of the vent valve and tank pressure sensor on EVAP systems is the **Engine Off Natural Vacuum (EONV)** test. The system runs a pressure and vacuum test based on the natural volatility of the fuel. After driving, the fuel will be warm, and if the vent to the tank is closed, the fuel tank pressure will build according to the ambient temperature. After the fuel starts to cool down, again with the vent closed, there

The EONV system runs a pressure and vacuum test based on the natural volatility of the fuel.

will be a natural vacuum build within the tank that the system can measure. The PCM can use this information to determine if there are any small leaks in the system. If the fuel is too volatile, as determined by the rise in pressure after the engine is turned off, the test is canceled. This test is also a small leak test setting PO442 if failed.

Some systems use a **leak detection pump (LDP)** that pressurizes the system and checks for leaks, along with the fuel tank pressure sensor.

The **charcoal canister** is the heart of the EVAP system. The charcoal canister can be located in the engine compartment or near the fuel tank. The charcoal canister has to deal with not only the evaporation of the fuel from the tank but also the generation of vapors while refueling. The EVAP system incorporates the onboard refueling vapor recovery (ORVR) system. The ORVR system has no trouble codes, and is not monitored by the ECM. The canister has to be large enough on these vehicles to store the vapors from refueling. Charcoal has the ability to store the vapors and then release them when the PCM determines that the engine can handle the extra fuel load, usually after the engine is warm. When the PCM decides to purge the canister, fresh air is allowed into the canister and vacuum is applied from the engine through the purge valve or solenoid.

The **purge valve** has undergone many changes since the inception of EVAP. Earlier purge valves were vacuum operated without computer control, but modern purge valves are computer controlled to allow more precise control of the vapors according to the many sensor inputs at the PCM. The purge solenoid can also be remotely mounted from the purge valve. The canister purge valve is normally closed. It opens the inlet to the purge outlet when vacuum is applied to the valve. Purging is done only when conditions warrant. Some of the conditions typically include the following:

- 150 seconds since the PCM entered closed loop
- Coolant temperature above 176°F
- PCM not enabling injector shutdown, such as during a traction control
- Vehicle speed above 20 miles per hour
- Engine speed above 1,100 rpm
- Temperature sensor not indicating overheating
- Low coolant not indicated

The conditions that purge occur will, of course, vary from vehicle to vehicle. If the purge valve opens at the wrong time, or the purge valve is stuck open, it can result in a very rich mixture, with a rough idle and possible stalling and rich mixture codes.

The **vent valve** is normally open to allow fresh air through the charcoal canister. Early canisters did not use a vent valve; the vent was left open to the atmosphere. The vent valve is closed only for OBD II diagnostics. The vent valve can be closed with the purge valve open while the PCM monitors the fuel tank pressure sensor for vacuum. Vacuum should build with the vent closed and the purge open. The vent is checked for blockage during normal purging by checking for excessive vacuum with the vent valve commanded open. If there is vacuum drawn on the fuel tank with the vent and purge valves open, then the vent must be blocked. Finally, the purge valve can be checked for leaks by closing both the purge and vent valves and determining if a vacuum begins to build. If it does, the purge valve is leaking. One common problem with vent valves is that they tend to get blocked by dirt and debris, setting a trouble code.

The fuel tank pressure sensor is mounted on the fuel tank and can measure vacuum and pressure at the tank. The fuel tank pressure sensor is a PCM input.

The **fuel tank level sensor**, which is actually the same sensor used for the fuel gauge, is used on OBD II vehicles to determine whether the fuel tank pressure or vacuum test should be run. Generally, if the tank is not between 15 and 85 percent full the test may not be run. The fuel tank sensor can also be checked at start-up and during driving to detect refueling which could upset the results of the leak test.

The **leak detection pump** is a small vacuum pump used on earlier enhanced EVAP systems to check for leaks in the system.

The **charcoal canister** stores fuel vapors (HC) to prevent them going into the atmosphere.

The **purge valve** opens to allow the charcoal canister to release its stored HC vapors for burning.

The **vent valve** can be opened and closed by the ECM for testing purposes.

The **fuel tank pressure sensor** is used by the ECM to monitor fuel tank pressure. This allows the ECM to detect leaks.

The **fuel tank level sensor** is used by the ECM to detect fuel tank levels. Leak tests are only run by the ECM during certain fuel level conditions, generally between 20 and 80 percent full.

The **fuel tank pressure control valve** can open if fuel pressure inside the tank becomes too great and the pressure is released into the charcoal canister.

The **fuel tank pressure control valve** is designed to control fuel tank pressure while the vehicle is sitting still. Vapors are stored in the tank with the valve closed. If tank pressure builds too high, the valve opens and lets the vapor into the canister. When the engine is running, the valve has vacuum applied, opening the valve and allowing vapors to be stored in the canister until purging.

Some systems use a leak detection pump that pressurizes the system and checks for leaks, along with the fuel tank pressure sensor.

Checking the System for Leaks

The main problem encountered with the EVAP system is leaks. Since a leak of 0.020 inch can set a DTC, and since much of the system may be hard to see, these leaks can be hard to locate without the use of a "smoke tester." The smoke tester (**Figure 10-14**) uses a diagnostic smoke that can be seen by the technician while testing for leaks. A flow meter is incorporated into the design of the tester. If the tester is connected to the EVAP system, the flow meter can detect the flow of nitrogen into the system from the tester (**Figure 10-15**). If the system is leaking, the tester will show flow. If the tester does not show any flow (after the system is initially pressurized), with the vent and purge valves closed via the scan tool, then there is no leak. The technician can also see if the system is operating correctly by commanding the vacuum or vent valve open and closed and watching for flow on the gauge.

PCV System Diagnosis and Service

No adjustments can be made to the PCV system. Therefore, service to this system involves careful inspection, operation, and replacement of faulty parts. Some engines use a fixed orifice tube in place of a valve. These should be cleaned periodically with a pipe cleaner soaked in carburetor cleaner. Although there is no PCV valve, this type of system is diagnosed in the same way as those systems with a valve. When replacing a PCV valve, match the part number on the valve with the vehicle maker's specifications for the proper valve. If the valve cannot be identified, refer to the part number listed in the manufacturer's service manual. Using an exact replacement PCV valve is critical to engine performance and emission control.

⚠ **Caution**

Never use shop air to test the fuel tank or evaporative system for leaks. Air introduced into the fuel system may cause an explosion. Always use nitrogen with the smoke tester, not air.

Figure 10-14 A portable smoke tester used for EVAP system leaks. Note it also has a flow meter to see if a leak is present.

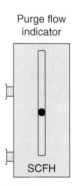

Purge flow indicator

SCFH

Figure 10-15 A flow meter can aid the technician in diagnosing an EVAP system.

PCV System OBD Monitor

One of the mandated OBDII monitors is for the PCV monitor. You may never see any DTCs for the PCV monitor because most manufacturers have made the PCV connection so that it cannot become disconnected, by building it into the intake manifold, making the connection very hard to become accidentally disconnected, or making the vacuum line connections large enough that the engine cannot run or will immediately set a DTC if the valve is disconnected. The PCV valve shown has a quarter-turn twist-lock connection (**Figure 10-16**). One exception to this is on turbo charged vehicles from Ford that have a crankcase pressure (CKCP) sensor in the clean air line. If the clean air line became disconnected under boost, crankcase vapors would be sent into the atmosphere. The CKCP sensor is monitored continuously for disconnection, and for electrical circuit faults. The DTCs are as follows:

P051C—Crankcase Pressure Sensor Circuit Low
P051D—Crankcase Pressure Sensor Circuit High
P051B—Crankcase Pressure Sensor Circuit Range/Performance

If the PCV valve is stuck in the open position, excessive airflow through the valve causes a lean air-fuel ratio and possible rough idle operation or engine stalling. When the PCV valve or hose is restricted, excessive crankcase pressure forces blow-by gases through the clean air hose and filter into the air cleaner. Worn rings or cylinders cause excessive blow-by gases and increased crankcase pressure, which forces blow-by gases through the clean air hose and filter into the air cleaner. A restricted PCV valve or hose may result in the accumulation of moisture and sludge in the engine and engine oil.

Leaks at engine gaskets, such as rocker arm cover or crankcase gaskets, will result in oil leaks and the escape of blow-by gases to the atmosphere. However, the PCV system also draws unfiltered air through these leaks into the engine. This action could result in wear of engine components, especially when the vehicle is operating in dusty conditions. Check all the engine gaskets for signs of oil leaks (**Figure 10-17**). Be sure the oil filler cap fits and seals properly.

The first step of PCV servicing is a visual inspection. The PCV valve can be located in several places. The most common location is in a rubber grommet in the valve or rocker arm cover. It can be installed in the middle of the hose connections as well as directly in the intake manifold.

Classroom Manual
Chapter 10, page 313

Special Tools
Vacuum gauge

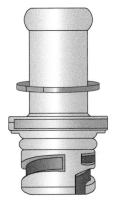

Figure 10-16 A PCV valve with a quarter-turn twist-lock to secure it to the valve cover.

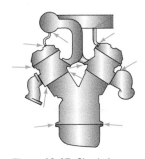

Figure 10-17 Check the engine for signs of oil leaks.

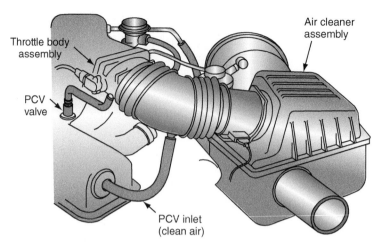

Figure 10-18 Typical PCV system plumbing on a port fuel injection vehicle.

Always use original equipment PCV valves when possible. Some after-market manufacturers take hundreds of original equipment part number PCV values and replace them with just a few part numbers.

Once the PCV valve is located, inspect the system by using the following procedure:

1. Make sure all the PCV system hoses (**Figure 10-18**) are properly connected and that they have no breaks or cracks.
2. Remove the air cleaner and inspect the air filter. Crankcase blow-by can clog these with oil. Oil in the air cleaner assembly indicates that the PCV valve or hoses are plugged or the engine has excessive blow-by. Make sure you check these and replace the valve and clean the hoses and air cleaner assembly. When the PCV valve and hose are in satisfactory condition and there was oil in the air cleaner assembly, perform a cylinder compression test to check for worn cylinders and piston rings.
3. Inspect for deposits that could clog the passages in the manifold. These deposits can prevent the system from functioning properly, even though the PCV valve, valve filter, and hoses might not be clogged.

CUSTOMER CARE You may want to discourage the customer from bringing his or her own parts to you for installation. Even well-meaning customers may not understand what may result. For example, if the customer asks the technician to install something as simple as a PCV valve they had purchased elsewhere, other problems may result. True, replacing the PCV valve is simple, but what if it is the wrong one, even though it may look identical? Or what if it is an inferior component or is otherwise defective? Drivability or idle problems may result if it is the wrong calibration or application. Unless you are confident of the source and quality of the component you are installing, it may be better to avoid the practice of installing customer-purchased parts.

Functional Checks of PCV Valve

A rough idling engine can signal a number of PCV valve problems, such as a clogged valve or a plugged hose. But before beginning the functional checks, double-check the PCV valve part number to make certain the correct valve is installed. If the correct valve is being used, perform the following functional checks:

1. Disconnect the PCV valve from the valve cover, intake manifold, or hose.
2. Start the engine and let it run at idle. If the PCV valve is not clogged, a hissing is heard as air passes through the valve.

3. Place a finger over the end of the valve to check for vacuum. If there is little or no vacuum at the valve, check for a plugged or restricted hose. Replace any plugged or deteriorated hoses.
4. Turn the engine off and remove the PCV valve. Shake the valve and listen for the rattle of the check needle inside the valve. If the valve does not rattle, replace it.

Some vehicle manufacturers recommend removing the PCV valve from the rocker arm cover and the hose. Connect a length of hose to the inlet side of the PCV valve, and blow air through the valve with your mouth while holding your finger near the valve outlet (**Figure 10-19**). Air should pass freely through the valve. If air does not pass freely through the valve, replace the valve. Move the hose to the outlet side of the PCV valve, and try to blow back through the valve (**Figure 10-20**). It should be difficult to blow air through the PCV valve in this direction. When air passes easily through the valve, replace the valve.

⚡ WARNING **Do not attempt to suck through a PCV valve with your mouth. Sludge and other deposits inside the valve are harmful to the human body.**

Proper operation of the PCV system depends on a sealed engine. Remember that the crankcase is sealed by the dipstick, valve cover, gaskets, and sealed filler cap. If oil sludging or dilution is noted and the PCV system is functioning properly, check the engine for oil leaks and correct them to ensure that the PCV system functions as intended. Also, be aware of the fact that a very worn engine may have more blow-by than the PCV system can handle. If there are symptoms that indicate the PCV system is plugged (oil in air cleaner, saturated crankcase filter, etc.) but no restrictions are found, check the wear of the engine.

IGNITION CONTROL SYSTEMS

Through the years, many devices and combination of devices have been used to control ignition timing. Today, most systems rely on the PCM for timing control based on inputs from various sensors. One of these sensors, the knock sensor, has one purpose—control spark knock.

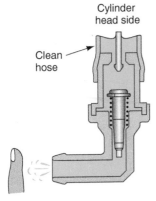

Figure 10-19 Blowing air through the PCV valve from the inlet side.

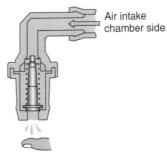

Figure 10-20 Blowing air through the PCV valve from the outlet side.

Classroom Manual
Chapter 10, page 316

Diagnosis of Knock Sensor and Knock Sensor Module

If the knock sensor system does not provide an engine detonation signal to the PCM, the engine detonates, especially on acceleration. When the knock sensor system provides excessive spark retard, fuel economy and engine performance are reduced.

The first step in diagnosing the knock sensor and **knock sensor module** is to check all the wires and connections in the system for loose connections, corroded terminals, and damage. With the ignition switch on, be sure 12 volts are supplied through the fuse to the knock sensor module. Repair or replace the wires, terminals, and fuse as required.

Connect a scan tester to the DLC and check for DTCs related to the knock sensor system. If DTCs are present, diagnose the cause of these codes. When no DTCs related to the knock sensor system are present, the system needs to be checked. **Photo Sequence 21** shows a typical procedure for diagnosing knock sensors and knock sensor modules. The following is a typical diagnostic procedure. However, always follow the recommended procedure of the manufacturer.

1. Connect the scan tester to the DLC and be sure the engine is at normal operating temperature.
2. Operate the engine at 1,500 rpm and observe the knock sensor signal on the scan tester. If a knock sensor signal is present, disconnect the wire from the knock sensor and repeat the test at the same engine speed. If the knock sensor signal is no longer present, the engine has an internal knock or the knock sensor is defective. When the knock sensor signal is still present on the scan tester, check the wire from the knock sensor to the knock sensor module for picking up false signals from an adjacent wire. Reroute the knock sensor wire as necessary.
3. If the knock sensor signal is not indicated on the scan tester in step 2, tap on the engine block near the knock sensor with a small hammer. If the knock sensor signal is now present, the knock sensor system is satisfactory.
4. When a knock sensor signal is not present in step 3, turn the ignition switch off and disconnect the knock sensor module wiring connector. Connect a 12-volt test light from 12 volts to terminal D in this wiring connector (**Figure 10-21**). If the light is off, repair the wire connected from this terminal to ground. When the light is on, proceed to step 5.
5. Reconnect the knock sensor module wiring connector and disconnect the knock sensor wire. Operate the engine at idle speed, and momentarily connect a 12-volt test light from 12 volts to the knock sensor wire. If a knock sensor signal is now generated on the scan tester, there is a faulty connection at the knock sensor or the knock sensor is defective. When a knock sensor signal is not generated, check for faulty wires from the knock sensor to the module or from the module to the PCM. Check the wiring connections at the module. If the wires and connections are satisfactory, the knock sensor module is likely defective.

⚠ **Caution**

Operating an engine with a detonation problem for a sufficient number of miles may result in piston, ring, and cylinder wall damage.

🔧 **Special Tools**

Scanner
12-volt test light

⚙ SERVICE TIP When installing a knock sensor, make sure it is tightened to the proper amount of torque. If the knock sensor torque is more than specified, the sensor may become too sensitive and provide an excessively high-voltage signal, resulting in more spark retard than required. When the knock sensor torque is less than specified, the knock sensor signal is lower than normal, resulting in engine detonation.

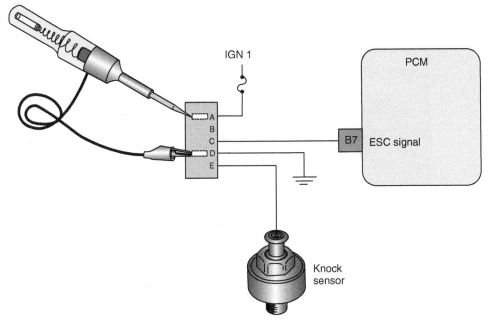

Figure 10-21 Testing the knock sensor circuit.

PHOTO SEQUENCE 21
Typical Procedure for Diagnosing Knock Sensors and Knock Sensor Modules

P21-1 Be sure the engine is at normal operating temperature and the ignition switch is off.

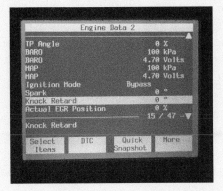

P21-2 Use the proper adaptor to connect the scan tester lead to the data link connector (DLC).

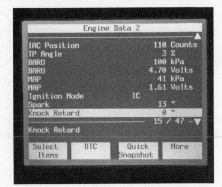

P21-3 Connect the scan tool to the DLC.

P21-4 Program the scan tester for the vehicle being tested.

P21-5 Select knock sensor on the scan tester.

P21-6 Observe the knock sensor signal on the scan tester with the engine running at 1,500 rpm. The knock sensor should indicate no signal.

PHOTO SEQUENCE 21 (CONTINUED)

P21-7 Tap on the right exhaust manifold above the sensor with a small hammer and observe the scan tester reading.

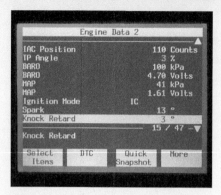

P21-8 The knock sensor should now indicate a signal on the scan tester.

P21-9 Shut the engine off and disconnect the scan tester leads.

EGR SYSTEM DIAGNOSIS AND SERVICE

Manufacturers program the PCM for correct EGR flow. If there is too much or too little, it can cause performance problems by changing the engine breathing characteristics. With too little EGR flow, the engine can detonate, and emit excessive amounts of NO_X. When any of these problems exist and it seems likely that the EGR system is at fault, check the system. Typical problems that show up in EGR systems follow. Check for any applicable DTCs to help you pinpoint the problem.

- *Rough idle.* Possible causes are an EGR valve stuck open, dirt on the valve seat, or loose mounting bolts. Loose mounting causes vacuum leak and a hissing noise.
- *Surge, stall, hesitation on acceleration, or does not start.* Probable cause is the valve stuck open, or opening too soon or at the wrong time.
- *Detonation (spark knock).* Any condition that prevents proper EGR gas flow can cause detonation on acceleration. This includes a valve stuck closed, leaking valve diaphragm, restrictions in flow passages, EGR disconnected, or a problem in the vacuum source.
- *Excessive NO_X emissions.* Any condition that prevents the EGR from allowing the correct amount of exhaust gases into the cylinder can cause this problem. High combustion temperatures allow NO_X to form. Therefore, anything that allows combustion temperatures to rise can cause this problem as well.
- *Poor fuel economy.* This is an EGR condition only if it relates to detonation or other symptoms of restricted or zero EGR flow.

Classroom Manual
Chapter 10, page 317

EGR System Troubleshooting

Before attempting to troubleshoot or repair a suspected EGR system on a vehicle, the following conditions should be checked and be within specifications.

- Engine is mechanically sound.
- Injection system is operating properly.
- Electronic timing advance is operating properly.

If one or more of these conditions is faulty or operating incorrectly, perform the necessary tests and services to correct the problem before servicing the EGR system.

Apart from the electronic control, the system can have all of the problems of any EGR system. Most EGR valves are fully electrically controlled, but some still use a vacuum signal to open and close the valve. The fully electronic EGR system can show the same problems as others, with the exception of vacuum leaks and other vacuum-related problems. Sticking valves, obstructions, and loss of vacuum produces the same symptoms as on Non-electronic controlled systems. If an electronic control component is not functioning, the condition is usually recognized by the computer. Check the service information for instructions on how to use computer service codes. The **EGR vent solenoid (EGRV)** and **EGR control solenoid (EGRC)** (**Figure 10-22**), or the **EGR vacuum regulator (EVR)** should normally cycle on and off very frequently when EGR flow is being controlled (warm engine and cruise rpm). If they do not, it indicates a problem in the electronic control system or the solenoids. Generally, an electronic control failure results in low- or zero-EGR flow and might cause symptoms like detonation and power loss.

> The EGR vent, control solenoid, and vacuum regulator are used on some vehicles by the ECM to control a vacuum-operated EGR valve operation.

Before attempting any testing of the EGR system, visually inspect the condition of all vacuum hoses for kinks, bends, cracks, and flexibility. Replace defective hoses as required. Check vacuum hose routing. (See the under-hood decal or the manufacturer's service information for correct routing.) Correct any misrouted hoses. If the emissions system is fitted with an EVP sensor, the wires routed to it should also be checked.

If the EGR valve remains open at idle and low engine speed, the idle operation is rough and surging occurs at low speed. When this problem is present, the engine may hesitate on low-speed acceleration or stall after deceleration or after a cold start. If the EGR valve does not open, engine detonation occurs. When a defect occurs in the EGR system, a **diagnostic trouble code** is usually set in the PCM memory.

> The PCM uses information from the ECT, TPS, and MAP to operate the EGR valve.

In many EGR systems, the PCM uses inputs from the ECT, TP and MAP sensors to operate the EGR valve. Improper EGR operation may be caused by a defect in any one of these sensors. A scan tester may be connected to the DLC to check for an EGR DTC or a DTC from another sensor which may affect EGR operation. The cause of any DTCs should be corrected before any further EGR diagnosis.

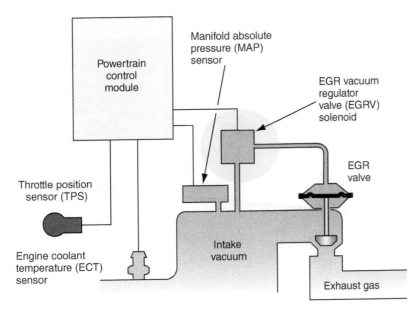

Figure 10-22 An EGR system with an EVR solenoid.

EGR Valves and Systems Testing

1. Connect a scan tool to the vehicle that has bidirectional control available for the EGR valve.
2. Enter the appropriate information into the scan tool, and go to outputs. Select EGR.
3. Pull up the EGR position information on the scan tool or, if that is not available, the engine rpm.
4. Start the engine and slowly command the EGR open. Watch the EGR position information or the rpm. Is the EGR opening confirmed by the position information or a drop in rpm?
5. If the EGR valve moved according to the command given through the scan tool, and the EGR position indicator also moved and a corresponding drop in rpm was noticed, the system is OK, or the problem is intermittent.
6. If the EGR moved with the command and the rpm did not change, check for plugged EGR passages that would not allow exhaust gas to get to the EGR valve.
7. If the EGR valve did not move according to commands from the scan tool, then check the wiring to the EGR valve or the PCM itself.

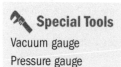

Special Tools

Vacuum gauge
Pressure gauge
Exhaust gas analyzer
Tachometer
Scanner

Using a hand-operated vacuum pump to check the operation of a vacuum-operated EGR valve:

1. Check all vacuum lines for correct routing. Ensure that they are attached securely. Replace cracked, crimped, or broken lines. Ensure that there is good manifold vacuum available from the engine (**Figure 10-23**).
2. Remove the vacuum supply hose from the **EGR valve port**.
3. Connect the vacuum pump to the port and supply 18 inches of vacuum. Observe the EGR diaphragm movement. In some applications, a mirror may be held under the EGR valve to see the diaphragm movement. When the vacuum is applied, the diaphragm should have moved. If the valve diaphragm did not move or did not hold the vacuum, replace the valve.
4. With the engine at normal operating temperature, check the vacuum supply hose to make sure there is no vacuum to the EGR valve at idle. Then plug the hose.
5. Install a tachometer.
6. On EFI engines, disconnect the idle air control (IAC) valve if equipped.

Figure 10-23 Check the engine's vacuum level before testing the EGR.

7. Slowly apply 5 to 10 inches of vacuum to the EGR valve vacuum port using a hand vacuum pump. The idle speed should drop more than 100 rpm (the engine may stall), and then return to normal (25 rpm) again when the vacuum is removed.

8. If the idle speed does not respond in this manner, remove the valve and check for carbon in the passages under the valve. Clean the passages as required or replace the EGR valve. Carbon may be cleaned from the lower end of the EGR valve with a wire brush, but do not immerse the valve in solvent and do not sandblast the valve.

9. If the EGR valve is operating properly, unplug and reconnect the EGR valve vacuum supply hose.

10. Reconnect the throttle air bypass valve solenoid if removed.

Classroom Manual
Chapter 10, page 317

Differential Pressure Feedback EGR

The **differential pressure feedback (DPFE) EGR system** is used with Ford OBD II systems to monitor the performance of the EGR systems. This system verifies the integrity of the EGR system controls and flow rate. See **Figure 10-24** for the individual components. The following is a brief description of the diagnostic routine for some of the components of the DPFE EGR system. You should always check for specific information for the vehicle you are servicing.

DPFE EGR pipes can become clogged over time.

First we will look at the DPFE sensor itself (**Figure 10-25**). The codes that set for the DPFE sensor are P1400 for low voltage from the DPFE and P1401 for high voltage (**Figure 10-26**). A low-voltage condition could occur with a short to ground in the wiring or a defective sensor or PCM. A high-voltage code could be set by a short to power at the wiring, defective sensor, or PCM.

The DPFE sensor's hoses are also checked for disconnection by commanding the EGR valve closed. If EGR flow is still indicated, the test assumes that the upstream DPFE hose is off or plugged or the EGR tube is plugged or damaged, and a P1405 code is set. If the EGR valve is commanded closed and a negative EGR flow is indicated at the DPFE sensor, then a P1406 code is set. The P1406 code is set by a damaged or disconnected downstream hose.

KOER is the abbreviation for *key on, engine running* test initiated by the technician with a scan tool.

The EGR system runs its monitor for P0401 whenever the engine is at cruise and the rate of EGR is high. The PCM checks feedback from the DPFE sensor and compares it to

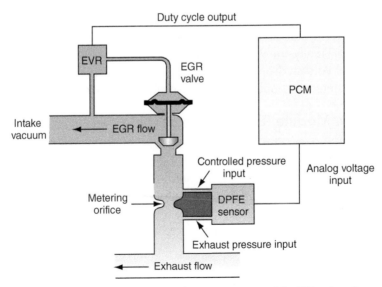

Figure 10-24 This drawing shows the exhaust gas passages of the EGR system, the metering orifice, EVR solenoid, and DPFE sensor.

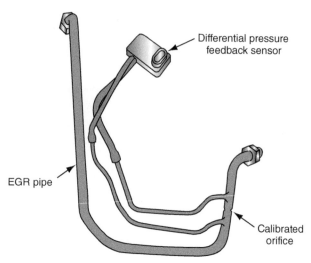

Figure 10-25 The DPFE EGR pipe with sensor, calibrated orifice, and upstream and downstream hoses.

Component	Codes
DPFE Sensor	P1400 Low Voltage from sensor P1401 High Voltage from sensor P1405 Upstream DPFE hose unplugged P1406 Downstream DPFE hose unplugged
EGR Self-Test Failure	P1408, P0401
EGR Vacuum Regulator Solenoid Circuit	P1409

Figure 10-26 Ford DPFE code definitions.

Never assume that a code indicated means a part must be replaced. Always check the circuit according to the latest diagnostic procedure.

the rate of EVR control. If the two do not agree, a P0401 is flagged in the PCM. The test for P1408 is only conducted when the KOER test is initiated, but it is similar to the P0401 test. If the KOER test was run after repairs to the EGR system, a technician could verify the repair without waiting for the PCM to conduct the test on the road. The PCM looks for an expected EGR flow at a fixed rpm. If the amount of EGR expected is not achieved, then the code is set. Finally, the P1409 code is the monitor for the **EVR valve**, which is set when the expected voltage drop across the valve is out of specification. This helps the PCM realize that there is a problem with the wiring to the valve or the valve itself.

An EGR System Module (ESM) EGR system

The ESM style EGR system (**Figure 10-27**) eliminates the DPFE restriction in the exhaust that tended to get blocked with use. The system itself works very similar to the older DPFE System, but is more reliable. The system restriction is now located in the intake manifold. The system looks at the difference between manifold pressure and the pressure at the base of the EGR valve. When the EGR is commanded open, the pressure at the base of the EGR valve rises, and the difference in pressures is sent to the PCM as a confirmation of EGR operation.

P1400 or P0405—DPFE Circuit Low
P1401 or P0406—DPFE Circuit High
P1409 or P0403—EVR circuit open or shorted

EGR System Module Components

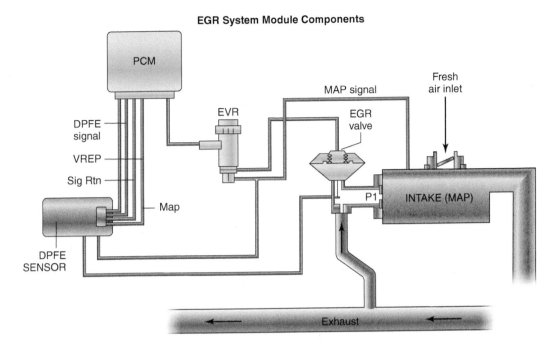

Figure 10-27 An EGR System Module (ESM) EGR system.

Digital EGR Valve Diagnosis

The **digital EGR valve** may be diagnosed with a scan tester. With the engine at normal operating temperature and the ignition switch off, connect the scan tester to the DLC. Start the engine and allow the engine to operate at idle speed. Select EGR control on the scan tester and then energize EGR solenoid number one with the scan tester. When this action is taken, the engine rpm should decrease slightly. The engine rpm should drop slightly as each additional EGR solenoid is energized with the scan tester. When the EGR valve does not operate properly, check the following items before replacing the EGR valve:

1. Check for 12 volts at the power supply wire on the EGR valve (**Figure 10-28**).
2. Check the wires between the EGR valve and the PCM.
3. Remove the EGR valve, and check for plugged passages under the valve.

Linear EGR Valve Diagnosis

The **linear EGR valve** diagnostic procedure varies depending on the vehicle make and model year. Always follow the recommended procedure in the vehicle manufacturer's service manual. The scan tester may be used to diagnose a linear EGR valve. The engine should be at normal operating temperature prior to EGR valve diagnosis. Since the linear EGR valve has a pintle position sensor (PPS), the actual pintle position may be checked on the scan tester. The pintle position should not exceed 3 percent at idle speed. The scan tester may be operated to command a specific pintle position, such as 75 percent, and this commanded position should be achieved within 2 seconds. With the engine idling, select various pintle positions and check the actual pintle position. The pintle position should always be within 10 percent of the commanded position. Refer to **Figure 10-29**. When the linear EGR valve does not operate properly:

> The digital and linear EGR valves do not use vacuum to operate.

1. Check the fuse in the 12-volt supply wire to the EGR valve.
2. Check for open circuits, grounds, and shorts in the wires connected from the EGR valve winding to the PCM.
3. Use a digital voltmeter to check for 5 volts on the reference wire to the EVP sensor.

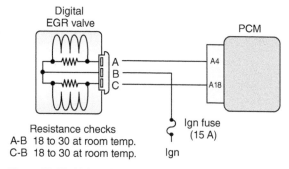

Figure 10-28 Digital EGR valve circuit.

Resistance checks
A-B 18 to 30 at room temp.
C-B 18 to 30 at room temp.

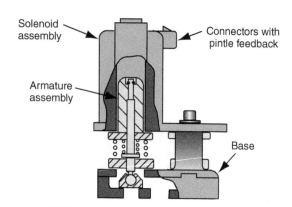

Figure 10-29 A typical linear EGR valve.

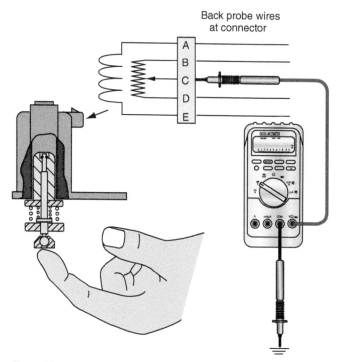

Figure 10-30 Checking the pintle position sensor.

4. Check for excessive resistance in the EVP sensor ground wire.
5. Leave the wiring harness connected to the valve, and remove the valve. Connect a digital voltmeter from the pintle position wire at the EGR valve to ground, and manually push the pintle upward (**Figure 10-30**). The voltmeter reading should change from approximately 1.0 volt to 4.5 volts.

Classroom Manual
Chapter 10, page 318

If the EGR valve did not operate properly on the scan tester and tests 1 through 5 are satisfactory, replace the valve. Remember that not all EGR valves are tested the same way. Check the appropriate service manual for specific tests.

EGR Efficiency Tests

Most testing of EGR valves involves the valve's ability to open and close at the correct time. An EGR system has one job: to control NO_X emissions. The system needs to be tested further to determine if it is doing what it was designed to do. Checking the efficiency of the system may also uncover the cause of other problems. If the exhaust passage in the valve is not the size it was designed to be, can the valve be as effective as it should? EGR efficiency testing is extremely important while diagnosing the cause of high NO_X and detonation.

Many technicians err by thinking that the EGR valve is working properly if the engine stalls or idles very rough when the EGR valve is opened. Actually, the results of this test indicate that the valve was closed and it will open. A good EGR valve opens and closes, but it also allows the correct amount of exhaust gas to enter the cylinder. EGR valves are normally closed at idle and open at approximately 2,000 rpm. This is where the EGR system should be checked. Keep in mind, anything that increases combustion temperatures can cause an increase in NO_X. Common causes of high NO_X are faulty cooling systems, incorrect ignition timing, lean mixtures, and faulty EGRs.

Classroom Manual
Chapter 10, page 319

> ⚙ **SERVICE TIP** In some EGR systems, the PCM energizes the EVR solenoid (**Figure 10-31**) at idle and low speeds. Under this condition, the solenoid shuts off vacuum to the EGR valve. When the proper input signals are available, the PCM de-energizes the EVR solenoid and allows vacuum to the EGR valve.

A five-gas exhaust analyzer can be used to check an EGR system. Allow the engine to warm up, then raise the engine speed to around 2,000 rpm. Watch the NO_X readings on the analyzer. The meter measures NO_X in parts per million (ppm). In most cases, NO_X should be below 1,000 ppm. It is normal to have some temporary increases over 1,000 ppm. However, the reading should be generally under 1,000 ppm. If the NO_X is above 1,000 ppm, the EGR system is not doing its job. The exhaust passage in the valve is probably clogged with carbon.

If only a small amount of exhaust gas is entering the cylinder, NO_X will still be formed. A restricted exhaust passage of only one-eighth inch will still cause the engine to run rough or stall at idle, but it is not enough to control combustion chamber temperatures at higher engine speeds. However, never assume the EGR passages are OK just because the engine stalls at idle when the EGR is fully opened.

ELECTRONIC EGR CONTROLS

When the EGR valve checks out and everything looks fine visually but a problem with the EGR system is evident, the EGR controls should be tested. Often, a malfunctioning electronic control will trigger a DTC. Service information will give the specific directions for

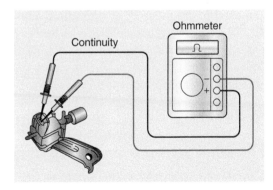

Figure 10-31 Use an ohmmeter to check the windings of an EVR solenoid.

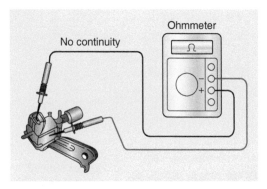

Figure 10-32 Use an ohmmeter to test an EVR solenoid for shorts to ground.

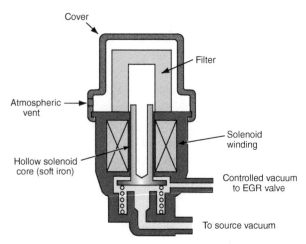

Figure 10-33 Operation of an EVR solenoid.

testing these controls. Always follow them. The tests below are given as examples of those test procedures.

EGR Vacuum Regulator (EVR) Tests

Connect a pair of ohmmeter leads to the EVR terminals to check the winding for open circuits and shorts (**Figure 10-32**). An infinity ohmmeter reading indicates an open circuit, whereas a lower-than-specified reading means the winding is shorted. Then, connect the ohmmeter leads from one of the EVR solenoid terminals to the solenoid case (**Figure 10-33**). A low ohmmeter reading indicates that the winding is shorted to ground. If the winding is not shorted, an infinity reading will be observed.

Special Tools

DMM
Scanner

> **SERVICE TIP** The same driver in a PCM may operate several outputs. For example, a driver may operate the EVR solenoid and the torque converter clutch solenoid. On General Motors computers, the drivers sense high current flow. If a solenoid winding is shorted and the **quad driver** senses high current flow, the quad driver shuts off all the outputs it controls rather than being damaged by the high current flow. When the PCM fails to operate an output or outputs, always check the resistance of the solenoid windings in the outputs before replacing the PCM. A lower-than-specified resistance in a solenoid winding indicates a shorted condition, and this problem may explain why the PCM driver stops operating the outputs.

A scan tester may be used to diagnose EGR operation. This procedure is shown in **Photo Sequence 22**. In the appropriate mode, the scan tester can command the EGR valve open and closed with the engine running. With the engine idling, the technician can open the EGR valve. This action can be shown on the scan tool screen and reflected in the idle quality of the vehicle. The engine will run roughly if the EGR valve is open at idle. This test can confirm the PCM's ability to open the EGR valve by watching the pintle position as well as the condition of the EGR passages evidenced by the idle quality.

Inspecting an EGR Position Sensor from a 2008 Chrysler

This procedure is based on an example from a 2008 Chrysler Town and Country van with a 3.3-liter engine (**Figure 10-34**). Always make sure to use the appropriate service information for the vehicle concerned. The trouble code is a P0404 EGR Position Sensor

Figure 10-34 A Chrysler EGR valve with a pintle position sensor.

Performance. The PCM needs to know the position of the EGR pintle to determine if the EGR valve is moving and is in the correct place during operation. The PCM raises and lowers the pintle to regulate the amount of EGR flow.

PHOTO SEQUENCE 22
Scan Tool Control of an EGR Valve

P22-1 Technician connects the scan tool to the vehicle.

P22-2 Enter the vehicle information into the scan tool.

P22-3 Select EGR outputs on the scan tool.

P22-4 The technician starts the car and lets the idle stabilize for few seconds.

P22-5 The technician opens the EGR valve about 10 percent. The idle should be fairly rough.

P22-6 The technician opens the EGR valve to 50 percent, the engine may not stay running, or runs very rough at idle.

P22-7 The technician has to give the vehicle some throttle to keep the engine running with the EGR at 100 percent.

P22-8 The technician is finished and disconnects the scan tool. The PCM, EGR valve, wiring and exhaust passages have all been verified as operational.

The PCM looks at two different values to determine if the position sensor is operational, the zero or closed position, and the difference between the desired (where the PCM wants the pintle to be) and actual pintle positions. The actual and desired positions may not be identical, but can vary by only a programmed amount. The possible causes are as follows:

A. Wiring circuit damage between the pintle position sensor and the ECM
B. A binding or damaged EGR valve
C. A malfunctioning computer
D. Poor ground connection

1. Confirm that the DTC is present now; if not, inspect wiring connections and terminals for damage. Test drive the vehicle under the conditions noted in the "freeze frame" for the trouble code.
2. Turn the key off and disconnect the ECM connectors. Install the breakout box (**Figure 10-35**) to test the EGR valve circuit. Start the engine and measure the voltage drop between terminals C1–C27 of the breakout box and terminal 2 of the EGR connector. The voltage drop should be less than 0.5 volt. If the voltage drop is more than 0.5 volt, the pintle position supply circuit has high resistance. If the supply circuit checks out OK, then check the EGR control circuit.

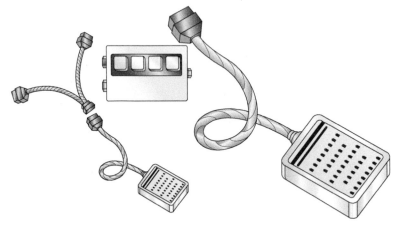

Figure 10-35 A breakout box that allows a connection to the ECM without damaging the connectors.

3. Turn the engine off. Again using the breakout box, connect the voltmeter. Start the engine and check the voltage drop between terminals C2–C22 on the breakout box and terminal 1 of the EGR assembly. The voltage drop should be below 0.5 volt. If the voltage drop is higher than 0.5 volt, then there is high resistance between C2–C22 and terminal 1.

4. Next, the diagnosis directs the technician to connect a test lamp to battery voltage and connect it to terminal 4 of the EGR connection (**Figure 10-36**). If the ground is good, then the test lamp should light brightly. If the test lamp does not light, then the ground for the EGR valve is defective.

5. After the repair, verify by driving the vehicle according to the freeze frame that was recorded when the code set, and make sure that a pending code does not reset.

CATALYTIC CONVERTER DIAGNOSIS

Detailed instructions on testing a catalytic converter are given in the chapter on exhaust systems in this book. Although a catalytic converter is definitely an emission control device, it is also an integral part of the exhaust system. What follows is a summary of the tests given in Chapter 5.

Classroom Manual
Chapter 10, page 299

A plugged converter or any exhaust restriction can cause loss of power at high speeds, stalling after starting (if totally blocked), a drop in engine vacuum as engine rpm increases, or sometimes popping or backfiring through the intake manifold.

There are many ways to test a catalytic converter (**Figure 10-37**). One of these is to simply rap the converter with a rubber mallet. If the converter rattles, it needs to be replaced and there is no need to do other testing. A rattle indicates loose catalyst substrate that will soon rattle into small pieces. This is one test and is not used to determine if the catalyst is good.

Checking for Exhaust Back Pressure

A vacuum gauge can be used to watch engine vacuum while the engine is accelerated. The vacuum gauge will drop and not recover on hard acceleration with a restricted exhaust. The best way to check for a restricted exhaust or catalyst is to insert a pressure gauge in

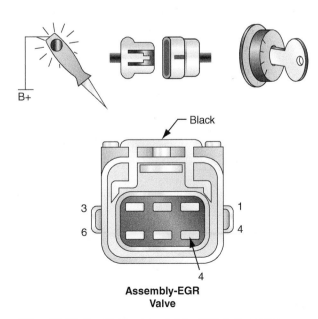

Figure 10-36 Checking for ground at the EGR pintle position.

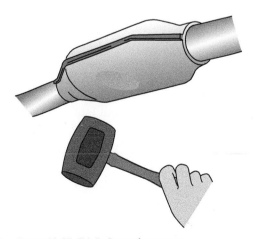

Figure 10-37 "Mallet" test of converter.

the exhaust manifold's bore for the O_2 sensor. With the gauge in place, hold the engine's speed at 2,000 rpm and watch the gauge. The desired back-pressure reading will be less than 1.25 psi. A substantial restriction will give a reading of over 2.75 psi.

Checking the Converter

Special Tools

Vacuum gauge
Pressure gauge
Pyrometer
Exhaust gas analyzer

The converter should be checked for its ability to convert CO and HC into CO_2 and water. There are three separate tests for doing this. The first method is the delta temperature test. To conduct this test, use a **handheld digital pyrometer**, or an infrared thermometer noncontact thermometer. Testing the exhaust pipe just ahead of and just behind the converter, there should be an increase of at least 100°F (or 8 percent above the inlet temperature reading) as the exhaust gases pass through the converter. If the outlet temperature is the same or lower, nothing is happening inside the converter.

The next test is called the O_2 storage test and is based on the fact that a good converter stores oxygen. Begin by disabling the air injection system. Once the analyzer and converter are warmed up, hold the engine at 2,000 rpm. Watch the readings on the exhaust analyzer. Once the numbers stop dropping, check the oxygen level on the gas analyzer. The O_2 readings should be about 0.5 to 1.0 percent. This shows the converter is using most of the available oxygen. It is important to observe the O_2 reading as soon as the CO begins to drop. If the converter fails the tests, chances are that it is working poorly or not at all.

This final converter test uses a principle that checks the converter's efficiency. Before beginning this test, make sure the converter is warmed up. Calibrate a five-gas analyzer and insert its probe into the tailpipe. Disable the ignition. Then crank the engine for 9 seconds. Watch the readings on the analyzer. The CO_2 on fuel-injected vehicles should be over 11 percent. As soon as the readings are obtained, reconnect the ignition and start the engine. Do this as quickly as possible to cool off the catalytic converter. Crank the engine while observing the HC levels, if they go over 1,500 ppm the converter is not working. Also, stop cranking once the CO_2 readings reach 10 or 11 percent. The converter is good. If the catalytic converter is bad, there will be high HC and low CO_2 at the tailpipe. Do not repeat this test more than one time without running the engine in between. If a catalytic converter is found to be bad, replace it.

Classroom Manual
Chapter 10, page 322

The OBD II system checks catalyst efficiency by comparing a pre-catalyst heated oxygen sensor with a post-catalyst heated oxygen sensor (**Figure 10-38**). The the post-catalyst sensor should show very little activity when compared to the pre-catalyst sensor. The converter, if working properly, should clean the exhaust to the point that the oxygen sensor will have very little response.

AIR INJECTION SYSTEM DIAGNOSIS AND SERVICE

Classroom Manual
Chapter 10, page 323

The air injection reaction (AIR) system injects air into the exhaust manifolds of a cold engine during warm-up to help bring the oxygen and catalytic converter up to temperature and reduce emissions.

Not all engines are equipped with an air injection system; only those that need them to meet cold-start emissions standards have them. Each system has its own test procedure—always follow the manufacturer's recommendations for testing.

OBD II systems use the air injection reactor (AIR) injection used with an electrically driven AIR pump that only supplies the exhaust manifold when the engine is cold to help eliminate HC and CO emissions until the converter becomes warm enough to oxidize them.

The secondary AIR system (**Figure 10-39**) works for only a few seconds when the engine is cold. Engines equipped with secondary AIR run for various amounts of time, depending on the manufacturer. When the AIR pump is commanded on by providing a ground for the AIR relay, the AIR pump forces air through the electronic shut-off valve and into the exhaust manifolds. When the secondary AIR is inactive, the electronic shut-off valve prevents airflow from the exhaust to the secondary AIR system.

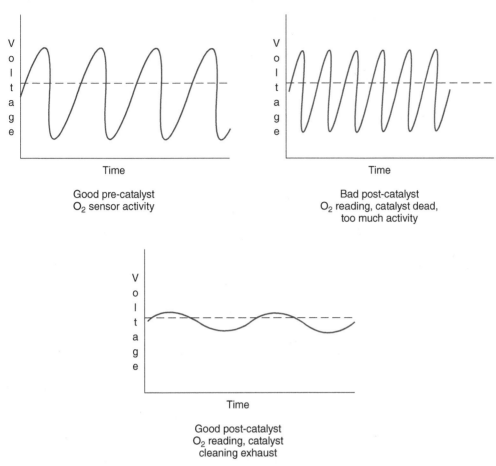

Good pre-catalyst
O_2 sensor activity

Bad post-catalyst
O_2 reading, catalyst dead,
too much activity

Good post-catalyst
O_2 reading, catalyst
cleaning exhaust

Figure 10-38 Comparison of O_2 activity with catalyst efficiency.

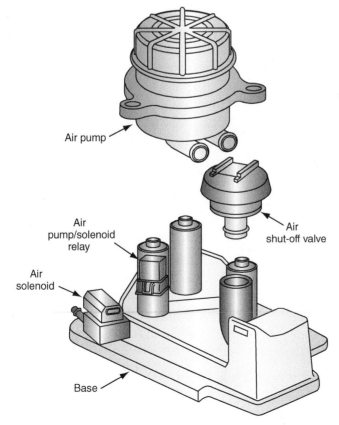

Air pump

Air
pump/solenoid
relay

Air
solenoid

Air
shut-off valve

Base

Figure 10-39 An electric air pump and air assembly (OBD II type).

Code	Possible Causes
PO411 Downstream AIR System Flow	Basic mechanical faults such as: Damaged hoses or pipes, stuck open or leaking solenoids, defective AIR pump
PO412 (AIR) Circuit Malfunction	An electrical fault in the solenoids, electric AIR pump, or circuitry involved in the control of the AIR system.

Figure 10-40 Basic OBD II codes for the AIR system.

OBD II Monitoring of the AIR System

The OBD II system monitors the **AIR system** for operation (**Figure 10-40**). The switching solenoids and relays are monitored for opens and shorts in the wiring. The AIR typically has a passive test that is run by comparing the oxygen sensor voltage shortly after start-up and comparing it to the voltage level at the O_2 sensor just before going into **closed loop**. The O_2 sensor should show very lean if the AIR pump was working. If the passive test is failed or inconclusive results are obtained, the PCM runs an intrusive test. The AIR pump is commanded on with the system in closed loop for around 3 seconds. Of course, in this situation the heated oxygen sensors go very lean. The AIR system does not have to be tested for the amount of air pushed through the system, just that the system is working.

Some codes for the AIR system are included here for the sake of discussion, and these are only some of the generic codes for the OBD II system. Be sure to use the specific service information for the vehicle concerned before beginning work.

Electric AIR Pumps

Set the scan tool display to observe the oxygen sensor(s). Start the engine, and allow it to reach operating temperature and obtain a normal idle. Enable the AIR system, and check the HO_2S voltages. If the voltages are low, the AIR pump, solenoid, and shut-off valve are working properly. If the voltages are not low, perform individual AIR system component tests as directed by the appropriate service information.

ASE-STYLE REVIEW QUESTIONS

1. While discussing catalytic converter operation,
   Technician A says a delta temperature test should be conducted.
   Technician B says a good converter will have an outlet that is 8 percent cooler than the inlet.
   Who is correct?
   A. A only
   B. B only
   C. Both A and B
   D. Neither A nor B

2. While discussing EGR valve diagnosis,
   Technician A says that if the EGR valve does not open, the engine may hesitate on acceleration.
   Technician B says that if the EGR valve does not open, the engine may detonate on acceleration.
   Who is correct?
   A. A only
   B. B only
   C. Both A and B
   D. Neither A nor B

3. While discussing EGR valve diagnosis,
   Technician A says that a defective throttle position (TP) sensor may affect the EGR valve operation.
   Technician B says that a defective engine coolant temperature (ECT) sensor may affect the EGR valve operation.
   Who is correct?
   A. A only
   B. B only
   C. Both A and B
   D. Neither A nor B

4. While discussing a linear EGR,

   Technician A says that the commanded pintle position and the actual pintle position have to be the same.

   Technician B says that the pintle position at idle cannot exceed 3 percent.

   Who is correct?

   A. A only

   B. B only

   C. Both A and B

   D. Neither A nor B

5. While discussing digital EGR valve diagnosis,

   Technician A says that a scan tester may be used to command the PCM to open each solenoid in the EGR valve.

   Technician B says that the EGR valve should open when 18 in. Hg of vacuum are supplied to the valve at idle speed.

   Who is correct?

   A. A only

   B. B only

   C. Both A and B

   D. Neither A nor B

6. While discussing EVAP valves,

   Technician A says the evaporative emissions system must be able to detect a leak as small as 0.020 inches in size.

   Technician B says that the monitor for the EVAP system works by closing the purge and vent valves and watching to see if the tank pressure increases.

   Who is correct?

   A. A only

   B. B only

   C. Both A and B

   D. Neither A nor B

7. *Technician A* says that a desired backpressure reading is 1.25 psi or less.

   Technician B says that anything less than 2.75 psi is acceptable.

   Who is correct?

   A. A only

   B. B only

   C. Both A and B

   D. Neither A nor B

8. While discussing PCV system diagnosis,

   Technician A says that an accumulation of oil in the air cleaner indicates the PCV valve is stuck open.

   Technician B says that an accumulation of oil in the air cleaner may indicate the piston rings and cylinders are worn.

   Who is correct?

   A. A only

   B. B only

   C. Both A and B

   D. Neither A nor B

9. While discussing PCV system diagnosis,

   Technician A says that a defective PCV valve may cause rough idle operation.

   Technician B says that satisfactory PCV system operation depends on a properly sealed engine.

   Who is correct?

   A. A only

   B. B only

   C. Both A and B

   D. Neither A nor B

10. While discussing knock sensor and knock sensor module diagnosis,

    Technician A says that with the engine running at 1,500 rpm, if the engine block is tapped near the knock sensor, a knock sensor signal should appear on the scan tester.

    Technician B says that with the engine running at 1,500 rpm, if a 12-volt test light is connected from a 12-volt source to the disconnected knock sensor wire, a knock sensor signal should appear on the scan tester.

    Who is correct?

    A. A only

    B. B only

    C. Both A and B

    D. Neither A nor B

ASE CHALLENGE QUESTIONS

1. IM240 and RG240 tests are basically the same *except*:

   A. They use the same test methods.

   B. They analyze the same gases.

   C. They use the same type of chassis dyno.

   D. They use constant volume sampling.

2. *Technician A* says that improper canister purge will cause an engine to run rough in hot weather.

   Technician B says that the ORVR system is part of the evaporative emissions system.

   Who is correct?

   A. A only

   B. B only

   C. Both A and B

   D. Neither A nor B

3. While discussing knock sensors,

    Technician A says that the first thing to check is the systems wiring and connections for corrosion or looseness.

    Technician B says that the knock sensor has one purpose, to control spark knock.

    Who is correct?

    A. A only

    B. B only

    C. Both A and B

    D. Neither A nor B

4. While discussing EGR systems,

    Technician A says that a too high TP sensor signal may cause the EGR to open early.

    Technician B says that DTCs from other sensors can assist in diagnosing EGR failures.

    Who is correct?

    A. A only

    B. B only

    C. Both A and B

    D. Neither A nor B

5. *Technician A* says that the computer monitors the operation of the catalytic converter by looking at the upstream and downstream oxygen sensors.

    Technician B says that a vacuum gauge installed on a manifold vacuum source will rise steadily on hard acceleration if the exhaust is restricted.

    Who is correct?

    A. A only

    B. B only

    C. Both A and B

    D. Neither A nor B

Name _____   Date _____

CHECK THE EMISSION LEVELS ON AN ENGINE

Upon completion of this job sheet, you should be able to measure the emissions levels of an engine and determine the cause of any abnormal readings.

ASE Education Foundation Correlation

This job sheet addresses the following **MAST** task: VIII. Engine Performance; B. Computerized Engine Controls Diagnosis and Repair

Task #6   Diagnose emissions or drivability concerns without stored or active diagnostic trouble codes; determine needed action. **(P-1)**

Tools and Materials

- Hand tools
- Clean engine oil
- Exhaust gas analyzer
- Spark plug gapper

Describe the vehicle being worked on:

Year _____   Make _____

Model _____   VIN _____

Engine type and size. _____   Vehicle's mileage. _____

Describe the general operating condition of the engine.

_____

_____

Procedure   **Task Completed**

1. Using a scan tool, look at the LT and ST fuel trim. Do these numbers indicate a problem with fuel control? Explain your answers below.

_____

_____

_____

2. Check timing (if possible). Timing is found to be _____

3. Verify proper idle speed.   ☐

4. Connect vacuum gauge. Make sure this is a source for manifold vacuum. Vacuum reading is: _____. What does this tell you about the engine?

_____

_____

For this engine, what are the HC and CO specs for this car? If not available, ask your instructor for a good base line reading.

_____ HC _____ CO _____ CO_2 _____ O_2 _____ NO_X

What source did you use for these specs?

_____

5. Warm up and calibrate a five-gas exhaust analyzer.

    Prior to running any tests with the engine at idle, what are your readings on the exhaust analyzer?

    _____ HC _____ CO _____ CO_2 _____ O_2 _____ NO_X

    What do these indicate?

    _____

    _____

    With the engine running at about 2,500 rpm, what are your readings on the exhaust analyzer?

    _____ HC _____ CO _____ CO_2 _____ O_2 _____ NO_X

    What do these indicate?

    _____

    _____

6. Now unplug an injector (if port fuel). How can you explain your readings? What does this tell you?

    _____

    _____

    What are your readings on the exhaust analyzer?

    _____ HC _____ CO _____ CO_2 _____ O_2 _____ NO_X

    What do these indicate?

    _____

    _____

    With the engine running at about 2,500 rpm, what are your readings on the exhaust analyzer?

    _____ HC _____ CO _____ CO_2 _____ O_2 _____ NO_X

    What do these indicate?

    _____

    _____

7. Unplug the oxygen sensor. How can you explain your readings? What does this tell you?

    _____

    _____

    What are your readings on the exhaust analyzer?

    _____ HC _____ CO _____ CO_2 _____ O_2 _____ NO_X

    What do these indicate?

    _____

    _____

    With the engine running at about 2,500 rpm, what are your readings on the exhaust analyzer?

    _____ HC _____ CO _____ CO_2 _____ O_2 _____ NO_X

    What do these indicate?

    _____

    _____

☐

8. Set airflow back to normal.

9. Plug off the PCV valve.

With the engine at idle, the vacuum reading now is: _____

What does this tell you?

_____

_____

What are your readings on the exhaust analyzer?

_____ HC _____ CO _____ CO_2 _____ O_2 _____ NO_X

What do these indicate?

_____

_____

With the engine running at about 2,500 rpm, what are your readings on the exhaust analyzer?

_____ HC _____ CO _____ CO_2 _____ O_2 _____ NO_X

What do these indicate?

_____

_____

_____

10. Return the PCV system to normal.   ☐

11. Cause a manifold vacuum leak. If fuel injected, the engine may not run with a large vacuum leak. If this happens, cause a small manifold vacuum leak.

Source of leak: _____

With the engine at idle, the vacuum reading now is: _____

What does this tell you?

_____

_____

What are your readings on the exhaust analyzer?

_____ HC _____ CO _____ CO_2 _____ O_2 _____ NO_X

What do these indicate?

_____

_____

With the engine running at about 2,500 rpm, what are your readings on the exhaust analyzer?

_____ HC _____ CO _____ CO_2 _____ O_2 _____ NO_X

What do these indicate?

_____

_____

12. Correct the vacuum leak.   ☐

13. Open the EGR valve slightly, if possible. With the engine at idle, the vacuum reading
now is: _____

What does this tell you?

_____

_____

What are your readings on the exhaust analyzer?

_____ HC _____ CO _____ CO_2 _____ O_2 _____ NO_X

What do these indicate?

_____

_____

With the engine running at about 2,500 rpm, what are your readings on the exhaust analyzer?

_____ HC _____ CO _____ CO_2 _____ O_2 _____ NO_X

What do these indicate?

_____

_____

14. Allow the EGR valve to close. ☐

15. Turn engine off. ☐

16. Take a tailpipe reading.

What are your readings on the exhaust analyzer?

_____ HC _____ CO _____ CO_2 _____ O_2 _____ NO_X

What do these indicate?

_____

_____

17. Ground a spark plug wire on one of the cylinders.

With the engine at idle, the vacuum reading now is: _____

What does this tell you?

_____

_____

What are your readings on the exhaust analyzer?

_____ HC _____ CO _____ CO_2 _____ O_2 _____ NO_X

What do these indicate?

_____

_____

With the engine running at about 2,500 rpm, what are your readings on the exhaust analyzer?

_____ HC _____ CO _____ CO_2 _____ O_2 _____ NO_X

What do these indicate?

_____

_____

18. Return the spark plug wire to normal.   ☐

19. Bypass or disconnect the electronic timing advance system (if equipped). With the engine at idle, the vacuum reading now is: _____.

What does this tell you?

_____

_____

What are your readings on the exhaust analyzer?

_____ HC _____ CO _____ CO_2 _____ O_2 _____ NO_X

What do these indicate?

_____

_____

With the engine running at about 2,500 rpm, what are your readings on the exhaust analyzer?

_____ HC _____ CO _____ CO_2 _____ O_2 _____ NO_X

What do these indicate?

_____

_____

20. Reconnect the timing control system.   ☐

21. Snap acceleration from idle speed.

Indicate the vacuum reading after the snap: _____

What does this tell you?

_____

_____

What are your readings on the exhaust analyzer prior to the snap?

_____ HC _____ CO _____ CO_2 _____ O_2 _____ NO_X

What do these indicate?

_____

_____

What are the readings on the exhaust analyzer several seconds after the snap?

_____ HC _____ CO _____ CO_2 _____ O_2 _____ NO_X

What do these indicate?

_____

_____

22. Squirt a small amount of water down the throttle plates.

    With the engine at idle, the vacuum reading now is: _____

    What does this tell you?

    _____

    _____

    What are your readings on the exhaust analyzer?

    _____ HC _____ CO _____ CO_2 _____ O_2 _____ NO_X

    What do these indicate?

    _____

    _____

    With the engine running at about 2,500 rpm, what are your readings on the exhaust analyzer?

    _____ HC _____ CO _____ CO_2 _____ O_2 _____ NO_X

    What do these indicate?

    _____

    _____

23. Insert exhaust gas analyzer probe into the oil filler tube. (*Caution*—do not immerse the probe into the oil.)

    With the engine at idle, what are your readings on the exhaust analyzer?

    _____ HC _____ CO _____ CO_2 _____ O_2 _____ NO_X

    What do these indicate?

    _____

    _____

    With the engine running at about 2,500 rpm, what are your readings on the exhaust analyzer?

    _____ HC _____ CO _____ CO_2 _____ O_2 _____ NO_X

    What do these indicate?

    _____

    _____

24. Summarize below what affects an engine's vacuum reading and the readings on an exhaust gas analyzer.

    _____

    _____

    _____

    _____

    _____

    _____

    _____

    _____

_____

_____

_____

_____

Instructor's Response

_____

_____

_____

_____

Name _____   Date _____

CHECK THE OPERATION OF A PCV SYSTEM

Upon completion of this job sheet, you should be able to check the PCV valve and hoses to determine if they are working properly.

ASE Education Foundation Correlation

This job sheet addresses the following **MLR** task: VIII. Engine Performance; D. Emissions Control Systems

Task #1   Inspect, test, and service positive crankcase ventilation (PCV) filter/breather valve, tubes, orifices, and hoses; perform needed action. **(P-2)**

This job sheet addresses the following **AST/MAST** task: VIII. Engine Performance; E. Emissions Control Systems Diagnosis and Repair

Task #1   Diagnose oil leaks, emissions, and drivability concerns caused by the positive crankcase ventilation (PCV) system; determine needed action. **(P-3)**

Task #2   Inspect, test, service, and/or replace positive crankcase ventilation (PCV) filter/breather cap, valve, tubes, orifices, and hoses; perform needed action. **(P-2)**

Tools and Materials

• Exhaust gas analyzer
• Appropriate service manual
• Oil

Describe the vehicle being worked on:

Year _____   Make _____

Model _____   VIN _____

Procedure   **Task Completed**

1. Describe the location of the PCV valve.

   _____

   _____

2. Run the engine until normal operating temperature is reached.   ☐

3. Record the CO reading from the exhaust analyzer: _____ percent

4. Remove the PCV valve from the valve or camshaft cover.   ☐

5. Record the CO reading: _____ percent

6. Explain why there was a change in CO and what service you would recommend.

   _____

   _____

   _____

7. Place your thumb over the end of the PCV valve.   ☐

8. Record the CO reading: _____ percent

9. Explain why there was a change in CO and what service you would recommend.

_____

_____

_____

10. Remove the valve from its hose and check for vacuum.   ☐

11. Record your results.

_____

_____

_____

12. Hold and shake the PCV valve.   ☐

13. Record your results.

_____

_____

14. State your conclusions about the PCV system on this engine.

_____

_____

_____

_____

Instructor's Response

_____

_____

_____

Name _____   Date _____

CHECK THE OPERATION OF AN EGR VALVE

Upon completion of this job sheet, you should be able to check the operation of an EGR valve and associated circuits.

ASE Education Foundation Correlation

This job sheet addresses the following **AST/MAST** task: VIII. Engine Performance; E. Emissions Control Systems Diagnosis and Repair

Task #3   Diagnose emissions and drivability concerns caused by the exhaust gas recirculation (EGR) system; inspect, and test, service and/or replace electrical/electronic sensors, controls, wiring, exhaust passages, vacuum/pressure controls, filters and hoses of exhaust gas recirculation (EGR) systems; determine needed action. **(P-2)**

Tools and Materials

- Hand-operated vacuum pump
- Vacuum gauge
- Exhaust gas analyzer
- Scan tool for assigned vehicle
- A vehicle with an electronic EGR EVR solenoid and solenoid

Describe the vehicle being worked on:

Year _____ Make _____

Model _____ VIN _____

Type of EGR system _____

Procedure

Task Completed

1. Inspect the hoses and connections within the EGR system. Record the results.

   _____

   _____

2. Pull the hose to the EGR EVR solenoid.   ☐

3. Run the engine and check for vacuum at the hose.   ☐

4. Turn off the engine and connect the scan tool to the DLC.   ☐

5. Insert a tee fitting into the vacuum hose and reconnect the supply line to the EVR solenoid. Connect the vacuum gauge to the tee fitting.   ☐

6. Start the engine.   ☐

7. Record the vacuum reading: _____ in. Hg

8. There should be a minimum of 15 in. Hg. Summarize your readings.

   _____

   _____

9. Actuate the solenoid with the scan tool.   ☐

10. Did the solenoid click?   ☐ Yes ☐ No

11. Did the vacuum fluctuate with the cycling of the solenoid?   ☐ Yes ☐ No

12. Disconnect the vacuum hose and backpressure hose from the EVR solenoid.   ☐

13. Disconnect the electrical connector from the solenoid.   ☐

14. Plug the EVR solenoid output port.   ☐

15. Apply 1 to 2 psi of air pressure to the backpressure port of the EVR solenoid.   ☐

16. With the vacuum pump, apply at least 12 in. Hg to the other side of the EVR solenoid. How did the EVR solenoid react?

_____

_____

What did this test tell you?

_____

_____

17. Run the engine at fast idle.   ☐

18. Insert the probe of the gas analyzer in the vehicle's tailpipe.   ☐

19. Record these exhaust levels:

_____ HC _____ CO _____ CO_2 _____ O_2 _____ NO_X

What was the injector pulse width at this time? _____ ms.

20. Remove the vacuum hose at the EGR valve and attach the vacuum pump to the valve.   ☐

21. Apply just enough vacuum to cause the engine to run rough. Keep the vacuum at that point for a few minutes and record the readings on the gas analyzer while the engine is running rough.

_____ HC _____ CO _____ CO_2 _____ O_2 _____ NO_X

What was the injector pulse width at this time? _____ ms.

22. Describe what happened to the emissions levels and explain why:

_____

_____

23. Apply the vehicle's parking brake.   ☐

24. Disconnect the vacuum pump and plug the vacuum hose to the EGR valve.   ☐

25. Firmly depress and hold the brake pedal with your left foot and place the transmission in drive.   ☐

26. Raise engine speed to about 1,800 rpm.   ☐

27. Record the readings on the gas analyzer:

_____ HC _____ CO _____ CO_2 _____ O_2 _____ NO_X

What was the injector pulse width at this time? _____ ms.

28. Return the engine to an idle speed, place the transmission into park, then shut off the engine.   ☐

29. Describe what happened to the emissions levels and explain why.

_____

_____

_____

_____

30. Reconnect the EGR valve. ☐

31. Apply the vehicle's parking brake. Start the engine. ☐

32. Firmly depress and hold the brake pedal with your left foot and place the transmission in drive. ☐

33. Raise engine speed to about 1,800 rpm. ☐

34. Record the readings on the gas analyzer:

    _____ HC _____ CO _____ CO_2 _____ O_2 _____ NO_X

    What was the injector pulse width at this time? _____ ms.

35. Return the engine to an idle speed, place the transmission into park, then shut off the engine. ☐

36. Describe what happened to the emissions levels and explain why.

    _____

    _____

    _____

    _____

    _____

    _____

37. Use the scan tool and retrieve any DTCs. Record any that were displayed.

    _____

38. Clear the codes. ☐

Instructor's Response

_____

_____

_____

_____

Name _____   Date _____

TEST A CATALYTIC CONVERTER FOR EFFICIENCY

Upon completion of this job sheet, you will be able to properly perform exhaust system back-pressure tests, diagnose emissions and drivability problems resulting from failure of catalytic converter systems, and properly inspect and test catalytic converter systems.

ASE Education Foundation Correlation

This job sheet addresses the following **AS/MAST** tasks: VIII. Engine Performance; D. Fuel, Air Induction, and Exhaust Systems Diagnosis and Repair

Task #10   Perform exhaust system back-pressure test; determine needed action. **(P-2)**
(MAST #11)

E. Emissions Control Systems Diagnosis and Repair,

Task #6   Diagnose emissions and drivability concerns caused by the catalytic
(AST #5)   converter system; determine needed action. **(P-2)(AST P-3)**

Tools and Materials

• Rubber mallet
• Exhaust gas analyzer
• Pyrometer
• A vehicle
• Pressure gauge
• Hoist
• Propane enrichment tool
• Protective clothing
• Goggles or safety glasses with side shields

Describe the vehicle being worked on:

Year _____ Make _____

Model _____ VIN _____

Engine type and size _____

Describe general operating condition.

_____

_____

Procedure

Task Completed

1. Securely raise the vehicle on a hoist.   ☐

   Make sure you have easy access to the catalytic converter and that it is not *hot*.

2. Using a rubber mallet, smack the exhaust pipe by the converter.
   Did it rattle?   ☐ Yes ☐ No

   If it did, it needs to be replaced and there is no need to do any more testing.
   A rattle indicates loose substrate, which will soon rattle into small pieces.

3. If the converter passed this test, it does not mean it is in good shape.   ☐
   It should be checked for plugging or restrictions.

4. Lower the vehicle and open the hood.   ☐

5. Remove the O_2 sensor.   ☐

6. Install a pressure gauge into the sensor's bore. On some engines, it is not easy to do.   ☐
On engines with a PFE, you can use a port to install your gauge. On other engines, you
may need to fabricate a tester from an old O_2 sensor or air check valve.

7. After the gauge is in place, start the engine and hold the engine's speed at 2,000 rpm.
Record the reading on the pressure gauge: _____ psi.

   You are looking for exhaust pressure under 1.25 psi. A very bad restriction would give
   you over 2.75 psi.

8. What does your reading tell you?

   _____

   _____

   Newer cars should have pressures well under 1.25 psi. Some older ones can be as high
   as 1.75 psi and still be good. You will notice that if you quickly rev up the engine, the
   pressure goes up. This is normal. Remember, do this test at 2,000 rpm, not with the
   throttle wide open.

9. Remove the pressure gauge, turn off the engine, allow the exhaust to cool, and then   ☐
   install the O_2 sensor.

10. If the converter passed this test, you can now check its efficiency. There are three ways   ☐
    to do this. The first way is the delta temperature test. Start the engine and allow it to
    warm up. With the engine running, carefully raise the vehicle using the hoist or lift.

11. With a pyrometer, measure and record the inlet temperature of the converter. The
    reading is _____.

12. Now measure the temperature of the converter's outlet. The reading is _____.

13. What was the percentage increase of the temperature at the outlet compared to the
    temperature of the inlet? _____

14. There should be a temperature increase of about 8 percent or 100 degrees at the
    converter's outlet. If the temperature does not increase by 8 percent, replace the
    converter. What are your conclusions based on this test?

    _____

    _____

15. Now you can do the O_2 storage test. This is based on the fact that a good converter   ☐
    stores oxygen. The following test is for closed-loop feedback systems. Non-feedback
    systems require a different procedure. Begin by disabling the air injection system.

16. Turn on your gas analyzer and allow it to warm up. Start the engine and warm the car   ☐
    up as well.

17. When everything is ready, hold the engine at 2,000 rpm. Watch the exhaust readings.
    Record the readings:

    _____HC _____CO _____ CO_2 _____ O_2 _____ NO_X

    If the converter was cold, the readings should continue to drop until the converter
    reaches light-off temperature.

Task Completed

18. When the numbers stop dropping, check the oxygen levels. Check and record the oxygen level; the reading is _____. O_2 should be about 0.5 to 1 percent. This shows the converter is using most of the available oxygen. There is one exception to this: If there is no CO left, there can be more oxygen in the exhaust. However, it still should be less than 2.5 percent. It is important that you get your O_2 reading as soon as the CO begins to drop. Otherwise, good converters will fail this test. The O_2 will go way over 1.25 after the CO starts to drop.

19. If there is too much oxygen left and no CO in the exhaust, stop the test and make sure the system has control of the air-fuel mixture. If the system is in control, use your propane enrichment tool to bring the CO level up to about 0.5 percent. Now the O_2 level should drop to zero.

20. Once you have a solid oxygen reading, snap the throttle open, and then let it drop back to idle. Check the oxygen. The reading is _____. It should not rise above 1.2 percent.

21. If the converter passes these tests, it is working properly. If the converter fails the tests, chances are that it is working poorly or not at all. The final converter test uses a principle that checks the converter as it is doing its actual job, converting CO and HC into CO_2 and water. ☐

22. Allow the converter to warm up by running the engine. ☐

23. Calibrate the gas analyzer and insert its probe into the exhaust pipe. If the vehicle has dual exhaust with a cross over, plug the side that the probe is not in. If the vehicle has a true dual exhaust system, check both sides separately. ☐

24. Turn off the engine and disable the ignition. ☐

25. Crank the engine for 9 seconds as you pump the throttle. Look at the gas analyzer and record the CO_2 reading: The CO_2 for injected cars should be over 11 percent. If you are cranking the engine and the HC goes above 1,500 ppm, stop cranking: the converter is not working. Also stop cranking once you hit your 10 or 11 percent CO_2 mark; the converter is good. If the converter is bad, you should see high HC and low CO_2 at the tailpipe. What are your conclusions from this test?

_____

_____

_____

26. Do not repeat this test more than *one* time without running the engine in between. ☐

27. Reconnect the ignition and start the engine. Do this as quickly as possible to cool off the converter. ☐

Instructor's Response

_____

_____

Name _____   Date _____

EVAPORATIVE EMISSION CONTROL SYSTEM DIAGNOSIS

Upon completion of this job sheet, you will be able to diagnose emissions and drivability problems resulting from the failure of the evaporative emissions control system, and you will also be able to inspect and test the components and the hoses of the evaporative emissions control system.

ASE Education Foundation Correlation

This job sheet addresses the following **AST** tasks: VIII. Engine Performance; E. Emissions Control Systems Diagnosis and Repair

Task #6   Inspect and test components and hoses of evaporative emissions control system (EVAP); determine needed action. **(P-1)**

Task #8   Interpret diagnostic trouble codes (DTCs) and scan tool data related to the emission control systems; determine needed action. **(P-3)**

This job sheet addresses the following **MAST** tasks: VIII. Engine Performance; E. Emissions Control Systems Diagnosis and Repair

Task #5   Diagnose emissions and drivability concerns caused by the evaporative emissions control system (EVAP); determine needed action. **(P-1)**

Task #7   Interpret diagnostic trouble codes and scan tool data related to the emissions control systems; determine needed action. **(P-3)**

Tools and Materials

- Service information
- EVAP leak tester
- Digital multimeter
- Hand-operated vacuum pump
- Scan tool
- Fuel cap tester
- EVAP tester
- Protective clothing
- Goggles or safety glasses with side shields

Describe the vehicle being worked on:

Year _____ Make _____

Model _____ VIN _____

Engine type and size _____

Procedure

Note: EVAP system diagnosis varies depending on the vehicle make and model year. Always follow the service and diagnostic procedure in the vehicle manufacturer's service information.

1. Use a scan tool and check for any EVAP-related DTCs. If an EVAP DTC is set, always correct the cause of this code before further EVAP system diagnosis. Summarize the results of your check.

_____

_____

2. Describe the EVAP system found on this vehicle.

_____

_____

3. How does this system check for leaks?

_____

_____

4. Check and describe the status of the EVAP monitor.

_____

_____

5. How many trips are necessary to run the monitor?

_____

_____

6. Observe the short-term fuel trim (STFT). How does this relate to vapor purge?

_____

_____

7. When the engine is idling, what is the status of the purge solenoid?

_____

_____

8. Keep the scan tool connected and take the vehicle for a road test that includes all required enable criteria. _Let someone else drive_ while you monitor the system from the passenger seat. What conditions does this include?

_____

_____

9. If the purge solenoid does not turn on during the road test, what is indicated?

_____

_____

10. After the road test, check all EVAP hoses for leaks, restrictions, and loose connections. Describe your findings.

_____

_____

11. Check the canister for cracks or damage. Describe what you found.

_____

_____

12. The electrical connections in the EVAP system should be checked for looseness, corroded terminals, and worn insulation. Summarize the results of your check.

_____

_____

13. Check for vacuum leaks in the EVAP system. A vacuum leak in any of the evaporative emission components or hoses can cause starting and performance problems, as can any engine vacuum leak. It can also elicit complaints of fuel odor. Summarize the results of your check.

_____

_____

14. Close the purge and intake ports at the canister and apply low air pressure (about 2.8 psi [19.6 kPa]) to the vent port. Was the canister able to hold that pressure for at least 1 minute? What does this indicate?

_____

_____

15. Measure the voltage drop across the canister purge solenoid and compare your readings to specifications. If the readings are outside specifications, replace the charcoal canister assembly.

_____

_____

16. The canister purge solenoid winding may be checked with an ohmmeter. Find the specifications in the service information and compare your measurements to them. Summarize the results of your check.

_____

_____

17. Remove the tank pressure control valve. Try to blow air through the valve with your mouth from the tank side of the valve. What happened and what does this indicate?

_____

_____

18. Connect a vacuum pump to the vacuum fitting on the pressure control valve and apply 10 in. Hg to the valve. Now try to blow air through the valve from the tank side. What happened and what does this indicate?

_____

_____

19. Allow the engine to run until it reaches normal operating temperature. Connect the purge flow tester's flow in series with the engine and evaporative canister. Zero the gauge of the tester with the engine off, and then start the engine. With the engine at idle, turn on the tester and record the purge flow rate and accumulated purge volume. What did you measure?

_____

_____

20. Gradually increase engine speed to about 2,500 rpm and record the purge flow. What did you measure?

_____

_____

21. What can you conclude from the two previous tests?

_____

_____

22. How much fuel is in the vehicle?

_____

_____

23. Use the pressure chart that accompanies the EVAP leak tester and determine how much pressure the tester should be set at. Record this amount.

_____

_____

24. Connect the tester to the EVAP service port. Where is this located?

_____

_____

25. Set the scan tool for an EVAP test. Adjust the output pressure from the tester and observe how much pressure the system can hold. What does this indicate?

_____

_____

26. What method will you use to identify the location of a leak? Briefly describe the procedure for doing this.

_____

_____

27. What equipment will you use to check the fuel cap? Briefly describe the procedure for doing this.

_____

_____

28. Remove the fuel cap and inspect the filler neck. Describe its condition.

_____

_____

29. If the fuel tank has a pressure and vacuum valve in the filler cap, check these valves for dirt contamination and damage. The cap may be washed in clean solvent. When the valves are sticking or damaged, replace the cap. Summarize the results of your check.

_____

_____

30. What are your conclusions about this EVAP system?

_____

_____

Instructor's Response

_____

_____

Name _____   Date _____

AIR INJECTION SYSTEM DIAGNOSIS AND SERVICE

Upon completion of this job sheet, you will be able to inspect and test the mechanical components of secondary air injection systems, as well as be able to inspect and test the electrical/electronically operated components and circuits of air injection systems.

ASE Education Foundation Correlation

This job sheet addresses the following **AST** task: VIII. Engine Performance; E. Emissions Control Systems Diagnosis and Repair

Task #4   Inspect and test electrical/electronically-operated components and circuits of air secondary injection systems; determine needed action. **(P-2)**

This job sheet addresses the following **MAST** tasks: VIII. Engine Performance; E. Emissions Control Systems Diagnosis and Repair

Task #4   Diagnose emissions and drivability concerns caused by the secondary air injection system; and circuits of air injection systems; inspect, test, repair, and/or replace electrical/electronically-operated components and circuits of secondary air injection systems; determine needed action. **(P-2)**

Tools and Materials
- Vehicle with an AIR system
- Exhaust gas analyzer
- Vacuum gauge
- Scan tool
- Lab scope
- Protective clothing
- Goggles or safety glasses with side shields

Describe the vehicle being worked on:

Year _____ Make _____

Model _____ VIN _____

Engine type and size _____

Describe general operating condition.

_____

_____

Procedure

Secondary Air Injection System Diagnosis

1. Check all vacuum and delivery hoses and electrical connections in the system. Summarize the results of this check.

_____

_____

_____

_____

2. Where is the air pump located?

_____

3. Describe the operation of the air pump on your assigned vehicle.

   A. When does the air pump operate?

   _____

   B. What is the purpose of the air pump?

   _____

4. How many valves are used in the air system to direct the flow of air? Describe them here.

   _____

   _____

5. What would be the consequences of the air valve diverting all the time?

   _____

6. What are the consequences of the air valve running at all times?

   _____

   _____

7. Has the monitor for the secondary air injection system run? If not, what code is stored or pending?

   _____

8. Explain when the secondary air system monitor is run by the computer.

   _____

   _____

9. Explain how the ECM knows that the air system is functional.

   _____

   _____

   _____

   _____

   _____

10. Check the hoses in the air system for evidence of burning. This would indicate a leak and could be the cause of excessive noise. Summarize the results of this check.

   _____

   _____

   _____

   _____

Check Valve Testing

1. All of the types of air injection systems have a one-way check valve. The valve opens to let air in but closes to keep exhaust from leaking out. If exhaust has been leaking from the valve, then an exhaust leak would exist, or the hose to the valve would be burned. Summarize the results of this check.

   _____

   _____

System Efficiency Test

1. Run the engine at idle and command the secondary air system on (enabled). Using a scan tool, measure the STFT. What were the readings?

_____

_____

2. Turn off the secondary air system and continue to allow the engine to idle. Again, measure and record the STFT. What were the readings?

_____

Instructor's Response

_____

_____

_____

CHAPTER 11

SERVICING COMPUTER OUTPUTS AND NETWORKS

Upon completion and review of this chapter, you should be able to:

- Diagnose problems with high-side and low-side drivers.
- Troubleshoot computer output drivers.
- Diagnose computer output actuators.
- Diagnose output drivers in the operation of fuel injectors.
- Describe computer control of the idle air control (IAC) system.
- Diagnose computer-operated exhaust gas recirculation (EGR) valves.
- Diagnose vehicle network module function.

Terms To Know

Actuators

Bidirectional control

Class 2

Computer output circuits

Customization

Electronic automatic temperature control (EATC)

Integrated diagnostic software

ISO 9141

J1850

J2534

J2534 Pass-through programming

Passive anti-theft system (PATS)

Protocols

Standard Corporate Protocol (SCP)

Vehicle communications module (VCM)

Vehicle networking

INTRODUCTION

In this chapter, we will look at diagnosing computer outputs and vehicle networks. As computers continue to become more prevalent in vehicle design, the service technician should not feel overwhelmed. Understanding computer operation is an essential skill. We actually have more information and diagnostic capability than we would have imagined a short time ago. Knowledge is very valuable in solving problems.

DIAGNOSING COMPUTER OUTPUTS

Computer output problems can be of three different sources: wiring problems, actuator problems, or computer problems. **Computer output circuits** are just as vulnerable to defects as computer inputs. Make sure you have an accurate wiring diagram, check for technical bulletins, and as always, double-check power and ground circuits. The **actuator** can, of course, be at fault, and you should be able to test the component. In the case of a computer problem, your fault might be caused by a bad input that is preventing the powertrain control module (PCM) from activating a component, or the computer itself could be defective. In this chapter we will concentrate on actuator and computer problems.

Computer output circuits are used by the computer to carry out commands by the actuators.

The most common computer-controlled devices are as follows:

- Throttle actuator control
- EGR solenoid
- Canister purge solenoid
- Fuel injectors (which are also specialized solenoids)
- Fuel pump control modules
- Fuel pump and A/C relays
- Idle speed control motor
- A/C controller
- Cooling fan module or relay
- Malfunction indicator lamp (MIL) and shift indicator
- Automatic transmission solenoids and pressure regulators

Classroom Manual
Chapter 11, page 334

Most systems allow for testing of the actuator through a scan tool. Actuators that are duty cycled by the computer are more accurately diagnosed through this method. Prior to diagnosing an actuator, make sure the engine's compression, ignition system, and intake system are in good condition. Using a scanner, serial data can be used to diagnose outputs. The displayed data should be compared against specifications to determine the condition of any actuator. Also, when an actuator is suspected as being faulty, make sure the inputs related to the control of that actuator are within normal range. Faulty inputs will cause an actuator to appear faulty. Some examples of this are as follows:

1. A throttle position (TP) sensor signal that is lower than specifications at idle speed may cause the IAC motor counts to be higher than normal at idle, resulting in a higher than desired idle speed.
2. A TP sensor signal that is higher than specifications at idle speed may cause the IAC motor counts to be lower than normal at idle, resulting in a lower than desired idle speed.
3. A low or lean-biased O_2 signal will cause the PCM to richen the mixture. This will decrease fuel economy.
4. A high or rich-biased O_2 signal will cause the PCM to lean out the mixture. This will result in engine surging and a hesitation during acceleration.
5. An engine coolant temperature (ECT) sensor signal that indicates a coolant temperature well below the actual temperature may increase injector pulse width and spark advance. It may also have a negative effect on emissions by delaying the operation of the exhaust gas recirculation (EGR), radiator cooling fan, and/or canister purge.

Actuators perform the actual work commanded by the PCM. They can be in the form of a motor, relay, switch, or solenoid.

Remember that an actuator can fail electrically or mechanically.

If the actuator is tested by other means than a scanner, always follow the manufacturer's recommended procedures. Because many actuators operate with 5 to 7 volts, never connect a jumper wire from a 12-volt source unless directed to do so by the appropriate service procedure. Some actuators are easily tested with a voltmeter by testing for input voltage to the actuator. If there is the correct amount of input voltage, check the condition of the ground. If both are in good condition, the actuator is faulty. If an ohmmeter needs to be used to measure the resistance of an actuator, disconnect it from the circuit first.

When checking components with an ohmmeter, logic can dictate good and bad readings. If the meter reads infinity, there is an open. Based on what is being measured across, an open could be good or bad. The same is true for very low resistance readings. Across some components, this would indicate a short. For example, there should not be an infinity reading across the windings of a solenoid. It should be low resistance. However, an infinity reading from one winding terminal to the case of the solenoid is normal. If the resistance is low, the winding is shorted to the case.

On late-model vehicles, more high-side drivers are being used.

Testing Actuators with a Lab Scope

Most computer-controlled circuits are ground controlled. The PCM energizes the actuator by providing the ground. On a scope trace, the on-time pulse is the downward pulse.

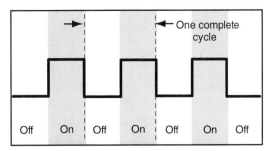

Figure 11-1 The difference between ground-controlled (low-side) and power-controlled (high-side) actuator traces.

On positive-feed circuits, where the computer is supplying the voltage to turn a circuit on, the on-time pulse is the upward pulse (**Figure 11-1**). One complete cycle is measured from one on-time pulse to the beginning of the next on-time pulse.

Actuators are electromechanical devices, meaning they are electrical devices that cause some mechanical action. Actuators can be either electrically or mechanically faulty. By observing the action of an actuator on a lab scope, you will be able to watch its electrical activity. Normally, if there is a mechanical fault, this will affect its electrical activity as well. Therefore, you have a good sense of the actuator's condition by watching it on a lab scope.

To test an actuator, you need to know what it is. Most actuators are solenoids. The computer controls the action of the solenoid by controlling the pulse width of the control signal. By watching the control signal, you can see the turning on and off of the solenoid (**Figure 11-2**). The voltage spikes are caused by the discharge of the coil in the solenoid.

Some actuators are controlled pulse width modulated signals (**Figure 11-3**). These signals show a changing pulse width. These devices are controlled by varying the pulse width and signal frequency.

Pulse width could also be called "on" time.

Both waveforms should be checked for amplitude, time, and shape. You should also observe changes to the pulse width as operating conditions change. A bad waveform will have noise, glitches, or rounded corners. You should be able to see evidence that the actuator immediately turns off and on according to the commands of the computer.

A fuel injector is actually a solenoid. The PCM's signals to an injector vary in frequency and pulse width. Frequency varies with engine speed, and the pulse width varies with fuel control. Increasing an injector's on time increases the amount of fuel delivered to the cylinders. The trace of a normally operating fuel injector is shown in **Figure 11-4**.

Practice will help you interpret good and bad waveforms.

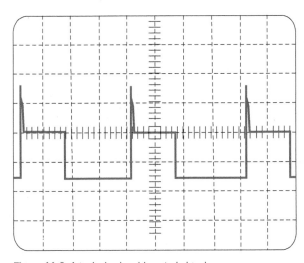

Figure 11-2 A typical solenoid control signal.

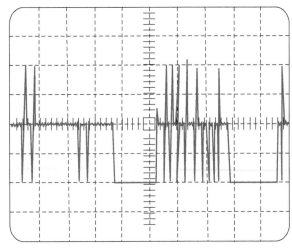

Figure 11-3 A typical pulse width modulated solenoid control signal.

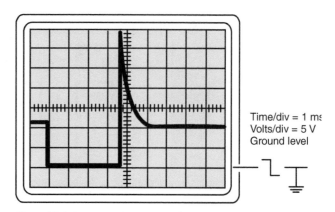

Figure 11-4 The trace of a normally operating fuel injector.

Time/div = 1 ms
Volts/div = 5 V
Ground level

PHOTO SEQUENCE 23
Bidirectional Control of the EVAP Purge and Vent Valves

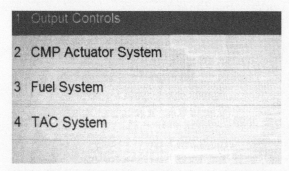

Year : 2006
Vehicle Type : Passenger Car
Product Make : (2) Pontiac
Body Style : (3) Two-Door-Convertible
Engine Type : (B) 2.4L L4 LE5
Vehicle Name : SOLTICE CONVERTIBLE
VIN : 1G2MB33B76Y100296

Is this correct?

P23-1 Connect the scan tool to the vehicle, and proceed to the opening screen. The scan tool has identified the vehicle we are testing.

1 Output Controls

2 CMP Actuator System

3 Fuel System

4 TAC System

P23-3 Next we will select "output controls" from the ECM menu, and then "EVAP Purge/Seal."

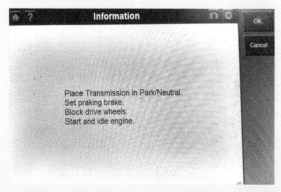

Information   OK

                              Cancel

Place Transmission in Park/Neutral.
Set praking brake.
Block drive wheels.
Start and idle engine.

P23-5 As a precaution when performing overrides, follow the on-screen prompts to prepare the vehicle for testing.

Diagnostic Trouble Codes (DTC)

2 Data Display

3 Special Functions

4 I/M System Information

5 Module ID Information

6 Module Setup

P23-2 For testing the purge and vent valves, we will go to the "Special Functions" menu from the ECM screen.

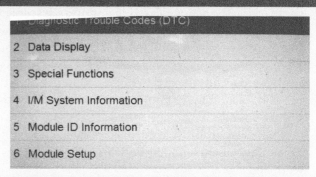

Information   OK

                              Cancel

Increase:Controls duty cycle of purge solenoid and closes vent solenoid.
Seal:Closes purge solenoid (0%) and closes vent solenoid.
Esc:Returns control to the vehicle.

P23-4 With this override, we can control the purge and vent valves to place a vacuum on the EVAP system by opening the purge valve and closing the vent valve. We can also seal the EVAP system at the purge and vent valves to check for fuel tank vapor decay. We are closing the vent valve and opening the purge valve to place a vacuum on the EVAP system.

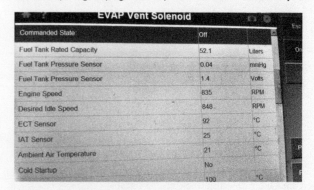

EVAP Vent Solenoid

Commanded State	Off	
Fuel Tank Rated Capacity	52.1	Liters
Fuel Tank Pressure Sensor	0.04	mmHg
Fuel Tank Pressure Sensor	1.4	Volts
Engine Speed	835	RPM
Desired Idle Speed	848	RPM
ECT Sensor	92	°C
IAT Sensor	25	°C
Ambient Air Temperature	21	°C
Cold Startup	No	
	100	°C

P23-6 The data screen before the vent valve was closed and the purge valve is at 0 percent operation. Note the tank vapor pressure.

PHOTO SEQUENCE 23 (CONTINUED)

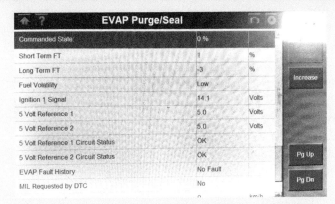

P23-7 We thought it would be interesting to check the fuel trims with the purge valve open and vent sealed, so the information was pulled up for short-term and long-term fuel trims. Short term was toggling between 1 and 2, and long term was at -3 (normal values).

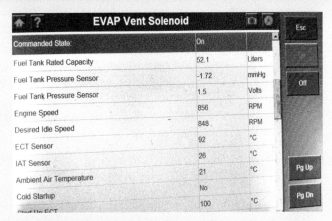

P23-8 When the vent solenoid was commanded on (closed), the fuel tank vacuum immediately started to rise.

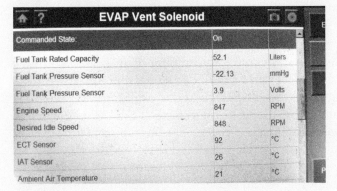

P23-9 The scan tool allowed the vacuum to build to 22.13 mmHg. (The safety in the computer program did not allow the vacuum to build much higher before automatically releasing the vacuum).

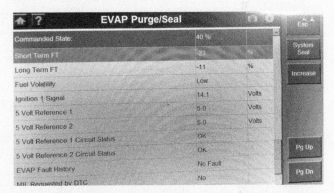

P23-10 The purge valve could be opened from 0 percent all the way to 100 percent. We checked the fuel trims with the purge valve at 40 percent. The trims show significantly rich mixture that the ECM is trying to correct. After we finished testing, we returned the vehicle to its normal state.

Testing an Actuator with a Scan Tool

The scan tool has many capabilities, but one of the best diagnostic uses of a scan tool comes from its **bidirectional control**. Bidirectional control means that the technician can tell a computer to exercise an output without the regular required input. Some bidirectional controls are not available on all scan tools. The most capable bidirectional scan tools are the manufacture-dedicated tools; it seems to take a while for some of these functions to find their way into popular scan tools used in the aftermarket. For example, we will look at bidirectional control of the EVAP purge and vent valves in **Photo Sequence 23**. Of course, output controls are different for each make and model vehicle, so do not assume that every vehicle has all these options.

Bidirectional control allows the technician to command outputs independently of the PCM.

Serial Data

"Serial data" is a term used to describe module to module or computer to computer communication. Basically scan tools "speak the language" of the computer system installed on the vehicle. Automotive computers and the modules they use speak a certain language and speed. The different languages are called **protocols**. Manufacturers often use more than one protocol, even on the same vehicle. These different protocols are the reason non-manufacturer-specific scan tools need Asian, Domestic, or European cartridges to decode data.

Serial data is the
method that
computers use to
communicate with
each other and with
the scan tool. Serial
data can help the
technician diagnose
problems on the
vehicle.

There were several types of serial data used for OBD II communication, such as Class B, ISO 9141, SCP, Class 2, and KWP2000. The communication types vary somewhat between and within manufacturers, so there were no hard and fast rules until controller area network (CAN) became the mandatory standard in 2008. Manufacturers can still use the other protocols for non-emission-related modules, such as the ABS, or entertainment.

- The earliest standard was called UART for universal asynchronous receive and transmit or Class A. UART was very slow but still used for some communications.
- Class B, which is also called **J1850** is used by most manufacturers. GM called their Class B communication **Class 2**. Ford has a special faster version of Class B that they call **Standard Corporate Protocol (SCP)**.
- **ISO 9141** is a Class B communication that is only used for a scan tool to communicate with the modules and is not used for module-to-module communication on the vehicle. ISO 9141 is usually associated with the Japanese but is also used by Ford, Chrysler, and GM.
- KWP2000 is usually associated with European manufacturers, but once again is used by GM and others as well.
- Controller area network is the protocol (language) used for all emission control–related modules after 2008. Manufacturers are still free to use different protocols for modules not related to emissions.

Idle Air Control

Classroom Manual
Chapter 11, page 336

The computer is in control of idle speed by varying the amount of air past the throttle plate with a stepper motor called the IAC (**Figure 11-5**). The stepper motor can be positioned in "steps" by the PCM. Naturally, the idle is based on coolant temperature, TPS, and other sensors depending on make and model of the vehicle. If the IAC is suspected of causing a problem or being inoperable, the valve can usually be commanded open and closed by the scan tool. This would allow the technician to check the IAC through the range of operation. If the IAC valve responds properly, then the technician would need to look for a problem with a PCM input that would cause an idle problem, a minimum air plug that had been tampered with, or a vacuum leak.

Exhaust Gas Recirculation

Vehicles with digital or linear EGR valves can be commanded open with the scan tool allowing the technician to perform a functional check of the system very quickly. Usually to check the EGR valve it is commanded open while the vehicle is at idle. This results in an immediate rough idle if the system is operating normally. Additionally, if the system

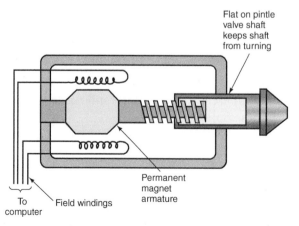

Figure 11-5 The IAC controls idle by varying the amount of airflow past the throttle plate and is controlled by PCM outputs.

uses a pintle position sensor, then the technician can see the results of the command to open the valve. The pintle position sensor tells the PCM that its command for EGR is being met by relaying the pintle location back to the PCM. If the valve does not open, then the technician would check the EGR valve wiring for response when commanded by the PCM. If the harness check showed the EGR valve was receiving commands but not opening, then the EGR valve is faulty. If the commands are not received, then the PCM or wiring is faulty. If the valve shows opening as commanded, then the valve would be removed for inspection of the passages for blockage.

COMPUTER NETWORKS

Vehicle networking has become common and will continue to be as technological advances are made. The use of many modules all connected to a common data bus allows for the use of smart modules that have diagnostic capabilities of their own. The modules have to use a common communication protocol or language. The most commonly used today is J1850. Ford calls their communication method Standard Corporate Protocol (SCP), but does not use it for all of their modules. GM calls their J1850 communication Class 2. ISO 9141 is used by Ford for some modules and in Japanese vehicles and KWP2000 for European imports. **Figure 11-6** shows a GM Class 2 network. The new CAN protocol

Classroom Manual
Chapter 11, page 340

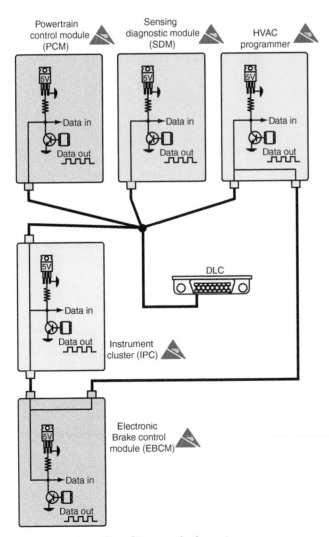

Figure 11-6 GM's (Class 2) communication system.

If network wiring is repaired, make certain to maintain the length as close as possible to the original.

was mandatory for communication beginning in the year 2008 (**Figure 11-7**). **Figure 11-8** diagrams the pins used in the data link connector (DLC). A diagram of the Ford ISO 9141 modules is shown in **Figure 11-9**, and the SCP or J1850 is shown in **Figure 11-10**. A vehicle's communication method or methods could be determined by observation of the pins in the DLC. Compare Figure 11-8 with Figure 11-9 and Figure 11-10.

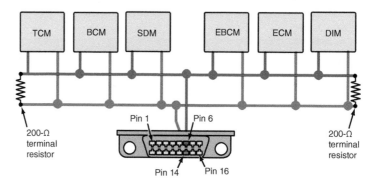

Figure 11-7 Representation of a CAN-C (high-speed) bus.

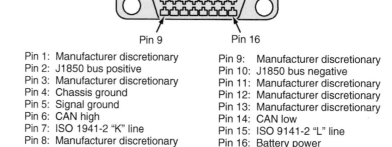

Pin 1: Manufacturer discretionary
Pin 2: J1850 bus positive
Pin 3: Manufacturer discretionary
Pin 4: Chassis ground
Pin 5: Signal ground
Pin 6: CAN high
Pin 7: ISO 1941-2 "K" line
Pin 8: Manufacturer discretionary

Pin 9: Manufacturer discretionary
Pin 10: J1850 bus negative
Pin 11: Manufacturer discretionary
Pin 12: Manufacturer discretionary
Pin 13: Manufacturer discretionary
Pin 14: CAN low
Pin 15: ISO 9141-2 "L" line
Pin 16: Battery power

Figure 11-8 Data links to the 16-pin DLC. The protocols used can be determined by the pins in the DLC.

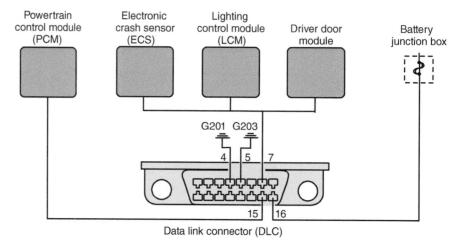

Figure 11-9 Ford ISO 9141 modules.

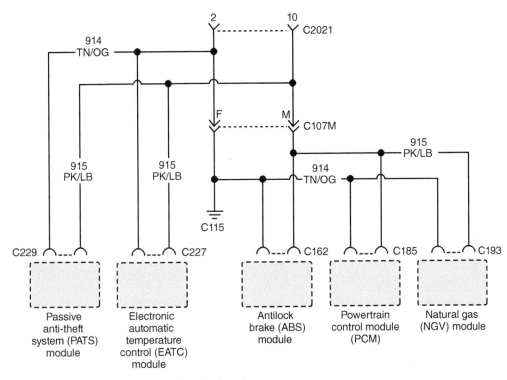

Figure 11-10 J1850 module diagram for a Ford product.

CAN Bus

A typical diagnostic procedure for CAN-C (high-speed) bus problems is in Figure 11-7. Note that this is not based on any particular vehicle. The service technician must have service information available for his or her specific vehicle.

Classroom Manual
Chapter 11, page 342

1. First, make sure to perform any preliminary diagnostic checks recommended by the manufacturer. These checks are to ensure that a very simple problem, like a loose battery cable or blown fuse, does not lead you down a long and unproductive path. Once the preliminary checks are finished, follow the chart for the specific U (networking) code.

2. Attempt to communicate with the modules on the serial data bus with a scan tool. If you can connect to some of the modules, the data bus has to be operational, at least to part of the modules used. Looking at the diagram, if connector C3 were unplugged, for instance, then the scan tool would not be able to communicate with the engine control module (ECM) and driver information module (DIM). By using a wiring diagram, the technician may be able to diagnose the problem before doing any disassembly on the vehicle. Also, each module needs three things before it can communicate on the data bus: power, ground, and serial data. Obviously, without power and ground the module cannot operate. The module also needs serial data so it can receive a "wake-up call" from the other modules. If the module is asleep, it cannot communicate on the bus!

3. The high-speed CAN system has two twisted wires; on this system, if either wire is open or shorted, communication is not possible. If you cannot connect to any of the modules, the serial data bus may be open or shorted to ground, the two wires could be shorted together, or they could be grounded inside a module.

4. Check for 2.5 volts between pin 6 and 14 pin of the DLC. If it is 0 volt, then there may be an open between the DLC and the CAN bus, an open power supply, a module internal problem, or a wiring problem that may have resulted in grounding the serial data line. If it is 12 volts, then the serial data line may be shorted to battery voltage. It is important to realize that there is not enough amperage carried in most of these systems to blow a fuse, but communication is definitely affected.

5. The technician can also check for resistance (after disconnecting the battery) across pins 6 and 14. Because the terminating resistors are in parallel, the resistance should be around 60 ohms. If the resistance were 120 ohms, then one leg of the circuit would be open. If there were little or no resistance, then the wiring could be shorted together or there could be an open circuit.

For one example of a networked module function, we will examine the function of Ford's **Passive Anti-Theft System (PATS)** from a Crown Victoria. **Figure 11-11** shows a wiring diagram for the system, which consists of a control module, transceiver, and PATS indicator.

Each key for the PATS-equipped vehicle must be digitally encoded into the PATS module. A transceiver with an antenna located in the steering column reads the key in the ignition. The PATS module and the PCM communicate with each other using sophisticated commands. Diagnosis of the system begins with ensuring the key is properly coded and not damaged. Of course, if one of the modules or PCM has a concern or there is a wiring problem, the vehicle will not start. A Vehicle Communication Module (VCM) and the Integrated Diagnostic System (IDS) are required to diagnose the system. The trouble codes for the PATS module are shown in **Figure 11-12**. Note the U codes are for network problems. It has been found that certain fast-pay keychain devices might have an adverse effect on the PATS module and that customers should hold the rest of the keys away from the ignition cylinder when starting the vehicle to help prevent interference.

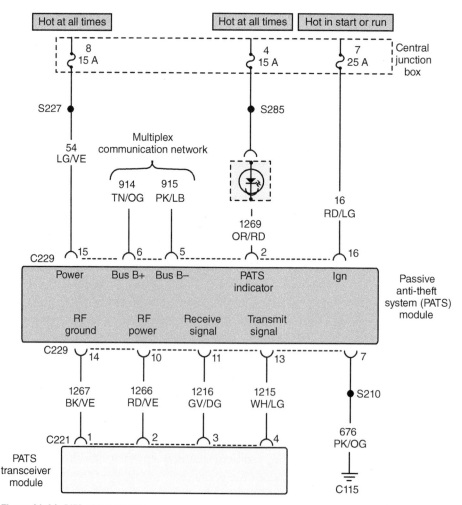

Figure 11-11 PATS wiring diagram.

PASSIVE ANTI-THEFT SYSTEM

DTC	Description
B1213	The number of programmed encoded keys is less than 2
B2103 or B1232	The antenna is not connected to the transceiver or the transceiver is defective
B1600	The key code was not received. The key is damaged or non-PATS
B1601	Encoded key—but not programmed
B1602	Only part of the key code was received
B1681	Signal not received from PATS transceiver. Possibly not connected, damaged, or wiring problem
B2139	PCM identification does not match between the PATS module and PCM
B2141	PATS and PCM did not exchange security identification
U1147	Faulty Standard Corporate Protocol link/PCM calibration incorrect
U1262	Standard Corporate Protocol message missing

Figure 11-12 PATS codes.

Electronic Automatic Temperature Control

The Ford **electronic automatic temperature control (EATC)** system is part of the SCP network. The PCM and the EATC module communicate with each other, and both these modules set codes for the A/C system. The system can perform a self-test as well as detect intermittent conditions and set them as run-time faults. The EATC can set codes during the self-test or the run-time. The PCM can also set its own codes for the EATC. The EATC self-test can be run only with the A/C off; the run-time codes are set with the system active. The PCM can set codes that will shut down the A/C system. The EATC module will not record the PCM codes; to receive all related codes, it is necessary to use the VCM. Refer to **Figure 11-13** for trouble codes from both the PCM and EATC module. A block diagram for the EATC system is shown in **Figure 11-14**.

The **EATC** is a computerized climate control system that is part of the vehicle network.

J2534 module programming is performed with a scan tool or laptop directly from the Internet.

| New Generation Star Tester Codes | Module Tests | | Description |
	Self-Test Faults	Intermittent Faults	
B1249	024	022 025	Blend door short Blend door failure
B1251	031	N/A	Open in car temperature sensor circuit
B1253	030	N/A	Grounded in car temperature sensor
B1255	041	043	Open ambient temperature sensor circuit
B1257	040	042	Ambient temperature sensor shorted to ground
B1261	050	052	Grounded sun radiation sensor
U1041	N/A	N/A	Vehicle speed data missing or invalid (SCP)
U1073	N/A	N/A	Engine coolant data missing or invalid (SCP)
U1222	N/A	N/A	Interior lamps data missing or invalid (SCP)

(PCM) Diagnostic Trouble Code Index	
DTC	Description
P1460	Internal driver malfunction in wide-open throttle cut-out of A/C
P1469	A/C cycling too often
P1474	PCM internal driver for low-speed fan failed
P1479	PCM internal driver for high-speed fan failed
P1464	A/C demand switch detected on during self-test

Figure 11-13 EATC and PCM codes related to climate control.

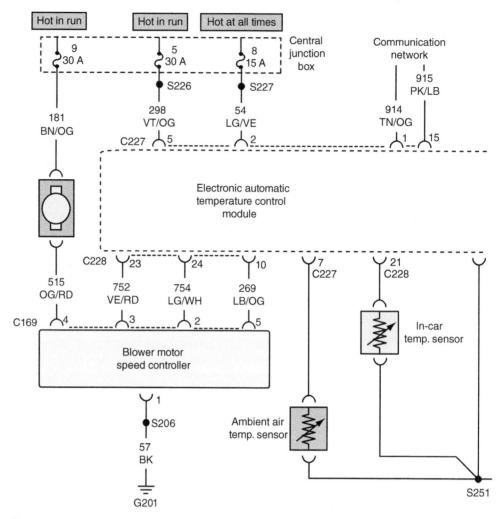

Figure 11-14 EATC electronic automatic temperature control diagram from a Ford vehicle.

Customization is possible through the use of programmable modules. GM's **J2534** pass through programming device is called the multiple diagnosis interface (MDI).

When reprogramming modules, make certain that at least 12.5 volts is available at all times. An extra battery or a very clean output battery charger is necessary for good results.

Ford calls their **J2534 pass-through program-ming** device the vehicle communications module (VCM). The VCMII was introduced in 2012. Ford calls their software subscription **integrated diagnostic software (IDS)**.

Replacing Modules

Many modules have to be reprogrammed for the vehicle if they are to be replaced, just like the PCM. Make sure you have the proper tools to replace and reprogram a module before you begin. Another factor is the **customization** of certain features that the owner has put into the systems. One example of customization is the lighting modules; some customers like the dome light to stay on after the front door is closed, some do not. In some localities, the alarm cannot chirp the horn when it is set. If the module involves tire size, the tire size may have to be entered into the module. When a module is replaced, make sure that you know how the customer had his system programmed if possible. Since it takes a scan tool to reset the customization features on some vehicles, the customer will have to make an extra trip to the shop if it is not done.

Module Reprogramming

Most reprogramming has been done from the factory scan tool, but in the last few years many vehicles have begun to use a laptop computer and **J2534 pass-through program-ming** directly from the manufacturer's website. The subscription to the website has to be purchased from the manufacturer, and each manufacturer has its own website. The laptop and vehicle module are connected through a **vehicle communications module** (VCM). This reprogramming is called J2534 reprogramming or "reflashing" as it is often referred to in the field. Ford calls their software subscription **integrated diagnostic software (IDS)**. J2534 programming is detailed in **Photo Sequence 24**.

Automatic Transmission Controls

Although this text is primarily concerned with engine performance issues, we did feel the need to mention automatic transmissions in the discussion of electronically controlled components. Automatic transmissions are some of the heaviest users of electronic actuators on a late-model vehicle. Most electronically shifted transmissions have the capability to display scan tool data and trouble codes just as the engine management system. Many transmissions can be shifted manually by commands from a scan tool or special transmission testers. This gives the technician the capability to determine whether a fault is in the transmission control system or the transmission. Additionally, the technician can see which gear ratio the computer has commanded and compare the command to the actual gear ratio. Refer to **Figure 11-15** for a chart of solenoid usage for a Ford AX4N transmission.

SHIFT SOLENOID OPERATION CHART

Transaxle range selector lever position	Powertrain Control Module Gear commanded	Eng braking	AX4N solenoids		
			SS 1	SS 2	SS 3
P/N[a]	P/N	No	Off[b]	On[b]	Off
R (Reverse)	R	Yes	Off	On	Off
	1	No	Off	On	Off
	2	No	Off	Off	Off
Overdrive	3	No	On	Off	On
	4	Yes	On	On	On
	1	No	Off	On	Off
D (Drive)	2	No	Off	Off	Off
	3	Yes	On	Off	Off
	2[c]	Yes	Off	On	Off
Manual 1	3[c]	Yes	Off	Off	Off
		Yes	On	Off	Off

a   When transmission fluid temperature is below 50° then SS 1 = Off, SS 2 = On, SS 3 = On to prevent cold creep.

b   Not contributing to powerflow.

c   When a manual pull-in occurs above calibrated speed, the transaxle will downshift from the higher gear until the vehicle speed drops below this calibrated speed.

Figure 11-15 An example of solenoid activity during different gear ranges.

PHOTO SEQUENCE 24
J2534 Pass-Through Module Programming

P24-1 Check for any applicable service bulletins for the problem before starting the repairs. There may be additional information available to help repair the problem or bulletins for replacing the concerned module.

P24-2 Before programming or replacing a module, make certain diagnosis is complete. If communication failure has occurred, the module will not be available to the scan tool. Diagnose any U-codes before replacing a module. Some modules will not program when codes are present.

P24-3 Once you have determined that the module needs to be reprogrammed or replaced and reprogrammed to the vehicle, you need to make sure that the software you have is up to date and your Internet connection is excellent. If the programming is interrupted, the module you are programming might be damaged. A subscription to the manufacturer's service information is necessary to get the required information.

P24-4 Before beginning, make sure the battery voltage is at least 12.5 volts. It is a good idea to put an extra battery on the vehicle, or a special charger that has a very clean output voltage. A normal battery charger is not acceptable; there is electrical interference from a normal charger that may disrupt the programming process.

P24-5 All doors and any electrical devices must be turned off, such as cell phone chargers, radios, any accessory current draw should be avoided. Put back-up battery on vehicle, or use special charger to maintain battery voltage. Reprogramming may take several minutes. Do not change key position in ignition. Do not open doors or windows while programming. Now proceed with programming module, per instructions from website.

P24-6 Now that programming is complete, make sure that the module does not need any additional setup, such as programming key FOBS, setting tire sizes, and the like, depending on the module you are programming.

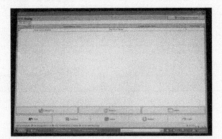

P24-7 Make sure that there are no codes set, and repair is verified before turning the vehicle back over to the customer.

CASE STUDY

A technician had diagnosed a vehicle with a defective ECM. He ordered the part from the parts department and proceeded to install and reflash the new module. After several failed attempts to program the new module, it was decided that the replacement part was defective. A new module was ordered and installed with the same results. The technician had correctly installed a battery charger during the programming process, with one mistake: it was not the special charger with a clean output. The technician did not understand that the normal battery charger has a very large amount of A/C ripple that plays havoc with computer communications. Upon understanding the procedure the technician had better luck doing pass-through programming.

ASE-STYLE REVIEW QUESTIONS

1. *Technician A* says that some scan tools have bidirectional controls to help diagnose computer problems.
   Technician B says that the best scan tools for bidirectional control are aftermarket scan tools.
   Who is correct?
   A. A only
   B. B only
   C. Both A and B
   D. Neither A nor B

2. *Technician A* says that serial data is what the scan tool uses to communicate with the PCM and modules.
   Technician B says that bidirectional control can be used to diagnose outputs.
   Who is correct?
   A. A only
   B. B only
   C. Both A and B
   D. Neither A nor B

3. *Technician A* says that the IAC valve controls the idle speed by varying the amount of air past the throttle plate.
   Technician B says that the IAC valve is usually a stepper motor.
   Who is correct?
   A. A only
   B. B only
   C. Both A and B
   D. Neither A nor B

4. *Technician A* says that the passages for the EGR might be blocked if the engine runs rough when the EGR is commanded open at idle.
   Technician B says that the pintle position sensor measures the location of the EGR pintle.
   Who is correct?
   A. A only
   B. B only
   C. Both A and B
   D. Neither A nor B

5. *Technician A* says that the PATS system can be affected by certain fast-pay modules a customer may have on his or her keychain.
   Technician B says that the transceiver and antenna are located in the steering column.
   Who is correct?
   A. A only
   B. B only
   C. Both A and B
   D. Neither A nor B

6. *Technician A* says that "U" codes are undefined codes.
   Technician B says that "U" codes are network codes.
   Who is correct?
   A. A only
   B. B only
   C. Both A and B
   D. Neither A nor B

7. *Technician A* says that Ford calls their J1850 communication language Class 2.
   Technician B says that GM calls their J1850 communication system SCP.
   Who is correct?
   A. A only
   B. B only
   C. Both A and B
   D. Neither A nor B

8. *Technician A* says that actuators perform the actual work commanded by the PCM.
   Technician B says that actuators have to be relays.
   Who is correct?
   A. A only
   B. B only
   C. Both A and B
   D. Neither A nor B

9. *Technician A* says that the EATC can set codes during a self-test for the A/C system.
   Technician B says that the EATC codes take the place of any PCM codes for the A/C system.
   Who is correct?
   A. A only
   B. B only
   C. Both A and B
   D. Neither A nor B

10. While discussing J2534 pass-through programming,
    Technician A says that before programming a module, you should make certain that the battery voltage is at least 12.5 volts.
    Technician B says that after a module is programmed, there may be additional steps to set up the module, such as setting tire size.
    Who is correct?
    A. A only
    B. B only
    C. Both A and B
    D. Neither A nor B

ASE CHALLENGE QUESTIONS

1. *Technician A* says that when a ground-controlled (low-side) driver is turned off, the voltage trace on the oscilloscope at the actuator goes high.

   Technician B says that when a driver-controlled (high-side) actuator is turned off, the oscilloscope trace at the actuator goes low.

   Who is correct?

   A. A only   C. Both A and B

   B. B only   D. Neither A nor B

2. *Technician A* says that each key for the PATS has to be coded into the PATS.

   Technician B says that the PATS and PCM share information.

   Who is correct?

   A. A only   C. Both A and B

   B. B only   D. Neither A nor B

3. *Technician A* says that most replacement network modules have to be reprogrammed.

   Technician B says that some replacement modules can be customized by programming certain features for the customer.

   Who is correct?

   A. A only   C. Both A and B

   B. B only   D. Neither A nor B

4. *Technician A* says that a linear or digital EGR valve can be commanded open with a scan tool.

   Technician B says that if the linear or digital EGR valve is commanded open at idle, the engine will run very rough or die, which verifies the operation of the EGR valve.

   Who is correct?

   A. A only   C. Both A and B

   B. B only   D. Neither A nor B

5. *Technician A* says that ISO 9141 is used only by the Japanese.

   Technician B says that SCP is used by European manufacturers.

   Who is correct?

   A. A only   C. Both A and B

   B. B only   D. Neither A nor B

Name _____   Date _____

EXHAUST GAS RECIRCULATION ACTUATION

Upon completion of this job sheet, you should be able to diagnose emissions and drivability problems caused by malfunctions in the EGR system; determine necessary action.

ASE Education Foundation Correlation

This job sheet addresses the following **AST/MAST** tasks: VIII. Engine Performance; E. Emissions Control Systems Diagnosis and Repair

Task #3   Diagnose emissions and drivability concerns caused by the exhaust gas recirculation (EGR) system; inspect, test, service and/or replace electrical/electronic sensors, controls, wiring, tubing, exhaust passages, vacuum/pressure controls, filters and hoses of exhaust gas recirculation (EGR) system; determine needed action. **(P-2)**

Tools and Materials

- A vehicle with an electronic EGR valve
- Scan tool with bidirectional capability
- Protective clothing
- Safety glasses

Describe the vehicle being worked on:

Year _____   Make _____

Model _____   VIN _____

Engine type and size _____

Procedure

Task Completed

1. Install the scan tool in the vehicle and go to the override selection. This may be called by several different names according to the scan tool manufacturer. Make sure to be familiar with your scan tool before performing this test.   ☐

2. If possible, pull up the EGR percentage on the scan tool and activate the EGR valve and look for a change in the EGR valve opening. If this is not possible, with your equipment, look for a change in engine rpm as the valve is opened.   ☐

3. If you were unable to command the EGR valve open with the scan tool, what could be a possible cause?

   _____

   _____

4. If the scan tool showed a change in EGR position, but the engine did not show the effects (rough idle), what is a possible cause?

   _____

   _____

5. How could a technician use these tests with a scan tool to save time in diagnosis?

   _____

   _____

6. List the problems encountered.

_____

_____

Instructor's Response

_____

_____

_____

Name _____   Date _____

COMPUTER CONTROL OF THE FUEL INJECTOR

ASE Education Foundation Correlation

This job sheet addresses the following **AST/MAST** task: VIII. Engine Performance; D. Fuel, Air Induction, and Exhaust Systems Diagnosis and Repair

Task #7   Inspect, test and/or replace fuel injectors. **(P-2)**

Tools and Materials

- Oscilloscope and test leads
- Fuel-injected vehicle
- Service information for vehicle being tested
- Protective clothing
- Safety glasses

Describe the vehicle being worked on:

Year _____   Make _____

Model _____   VIN _____

Engine type and size _____

Procedure

Task Completed

1. Bring vehicle into the shop, apply parking brake and wheel chocks.   ☐

2. Using electronic service information or the appropriate service manual, find the wiring   ☐
   diagram for the fuel injection circuit, in particular the injectors themselves.

3. Connect the positive lead of the oscilloscope to the negative lead of the injector. The   ☐
   negative lead should be the one activated by the computer. Attach the negative lead of
   the oscilloscope to a good ground.

4. Start the engine and adjust the oscilloscope to obtain the best view possible. *Note:*   ☐
   Refer to the manufacturer's recommendations on setting up the oscilloscope and
   triggers. *Hint:* Sometimes the best patterns are possible with the scope synchronized
   to cylinder 1. Ask your instructor for help with the setup as needed. Using the
   oscilloscope will become easier with practice.

5. Once you have established a good signal, answer the following questions:

   Looking at the pattern, see what the voltage on the ground is during the "on" time of
   the injector.

   _____ ms

   Looking at the pattern, see what the voltage on the injector ground is during the "off"
   time of the injector. How do you explain the voltage you see?

   _____ V

   What is the approximate injector on time in milliseconds?

   _____ ms

   Create a vacuum leak (large enough to force the engine to run rough). What is the
   injector "on time"? How do you explain what you see?

   _____

   _____

6. Reconnect the vacuum hose. Snap accelerate the engine from idle. Explain what happened to the injector on-time on acceleration and then as the engine slowed down.

_____

_____

7. List the problems encountered.

_____

_____

Instructor's Response

_____

_____

_____

CHAPTER 12

ON-BOARD DIAGNOSTIC (OBD II) SYSTEM DIAGNOSIS AND SERVICE

Upon completion and review of this chapter, you should be able to:

- Explain how to logically approach diagnosing a problem in an OBD II system.
- Conduct preliminary checks on an OBD II system.
- Describe how a scan tool can be used in diagnostics.
- Use a symptom chart to set up a strategic approach to troubleshooting a problem.
- Define the terms associated with OBD II diagnostics.

- Identify the cause of an illuminated malfunction indicator lamp (MIL).
- Explain the basic format of OBD II diagnostic trouble codes (DTCs).
- Monitor the activity of OBD II system components.
- Explain how to diagnose intermittent problems.
- Properly repair OBD II circuits.

 **Basic Tools**

Scan tool
Service information

Terms To Know

Drive cycle
Enable criteria
Freeze frame
Freeze-frame access mode
Induction
Long-term fuel trim (LTFT)
Mode 1
Mode 2
Mode 3

Mode 4
Mode 5
Mode 6
Mode 7
Mode 8
Mode 9
Mode 10
Modes of operation
OBD II modes of operation
Output driver

Parameter identification (PID) mode
Pending situation
Programmable read-only memory (PROM)
Quad driver
Readiness
Serial data
Short-term fuel trim (STFT)

INTRODUCTION

Regulations of **OBD II modes of operation** require that the powertrain control module (PCM) (**Figure 12-1**) monitor and perform some continuous tests on the emission control system and components. Some OBD II tests are completed at random, at specific intervals, or in response to a detected fault.

To perform the strategies and tests on the emission control system, OBD II PCMs have diagnostic management software. The many diagnostic steps and tests required of OBD II systems must be performed under specific operating conditions referred to as **enable criteria**. The PCM's software organizes and prioritizes the diagnostic routines. The software determines if the conditions for running a test are present. Then it monitors the system for each test and records the results of these tests.

The tests by the PCM are called monitors. Some monitors are performed continuously, and some only during certain conditions.

On some OBD II vehicles, the knock sensor module has to be transferred if the PCM is replaced.

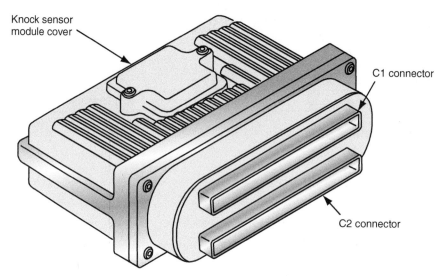

Figure 12-1 A typical OBD II PCM.

Enable criteria are the conditions that have to be met before a monitor is run by the PCM.

The PCM supplies a buffered low voltage to various sensors and switches. The input and output devices in the PCM include analog-to-digital converters, signal buffers, counters, and special drivers. The PCM controls most components with electronic switches that complete a ground circuit when turned on. These switches are arranged in groups of four and seven that are called either a **quad driver** module or **output driver** module. The *quad driver module* can independently control up to four output terminals. The *output driver module* can independently control up to seven outputs.

Classroom Manual
Chapter 12, page 360

Adaptive Learning

The PCM has a learning ability that allows the module to make corrections for minor variations in the fuel system in order to improve drivability. Whenever the battery cable is disconnected, the learning process resets. The driver may note a change in the vehicle's performance. In order to allow the ECM to relearn, drive the vehicle at part throttle with moderate acceleration. Some manufacturers have a specific idle relearn procedure that can speed the process. If available, this can be an important customer satisfaction tool.

GM calls the diagnostic function of the PCM the diagnostic executive because it is in charge of the diagnostic tests.

Electrically erasable programmable read-only memory (EEPROM) modules are soldered into the ECM. EEPROMs allow the manufacturer to update what is held in the **programmable read-only memory (PROM)** without replacing it. The ECM checks its internal circuits continuously for integrity. Using a check sum value unique to the program, it checks its EEPROM for accuracy of its data. It checks the actual values against what they are supposed to be and sets a code if they are different. Besides the hard-wired memory chip, the PCM also monitors its volatile keep-alive memory. If that has been improperly changed or deleted, the PCM sets a code. This type of code will also be set if the vehicle's battery has been disconnected.

Special Tools
Lab scope
DMM

There is a continuous self-diagnosis on certain control functions. This diagnostic capability is complemented by the diagnostic procedures contained in the service information. The language of communicating the source of the malfunction is a system of DTCs. When a malfunction is detected by the control module, a DTC will set and the MIL will illuminate.

The system monitoring diagnostic sequence is a unique segment of the software that is designed to coordinate and prioritize the diagnostic procedures as well as define the protocol for recording and displaying their results. The main diagnostic responsibilities of the PCM are as follows:

- Monitoring the diagnostic test's enabling conditions
- Requesting the MIL
- Illuminating the MIL
- Recording pending, current, and history DTCs
- Storing and erasing freeze-frame data
- Monitoring and recording test status information

The diagnostic tables and functional checks given in service information are designed to locate a faulty circuit or component through a process of logical decisions. The tables are prepared with the assumption that the vehicle functioned correctly at the time of assembly and that there are no multiple faults present.

Electrostatic Discharge Damage

In order to prevent possible electrostatic damage to the PCM, do not touch the connector pins or the soldered components on the circuit board. Electronic components used in the control systems are often designed to carry very low voltage. Electronic components are susceptible to damage caused by electrostatic discharge. Less than 100 volts of static electricity can cause damage to electronic components. There are several ways for a person to become statically charged. The most common methods of charging are by friction and by **induction**. An example of charging by friction is a person sliding across a car seat. Charging by induction occurs when a person with well-insulated shoes stands near a highly charged object and momentarily touches ground. Charges of the same polarity are drained off, leaving the person highly charged with the opposite polarity. Static charges can cause damage. Therefore, it is important to use care when handling and testing electronic components. Service information and wiring diagrams also carry the electrostatic discharge warning (**Figure 12-3**).

Aftermarket Electrical and Vacuum Equipment

In the design of a new car, there was no allowance made for add-on equipment. Therefore, do not install any add-on vacuum-operated equipment to new vehicles. Also, only connect the add-on electrically operated equipment to the vehicle's electrical system at the battery.

Caution

The PCM is designed to withstand normal current draws associated with vehicle operation. Avoid overloading any circuit. When testing for opens or shorts, do not ground any of the PCM circuits unless instructed to do so. When testing for opens or shorts, do not apply voltage to any of the control module circuits unless instructed. Test these circuits with a lab scope or digital voltmeter only (**Figure 12-2**) while the PCM connectors remain connected to the PCM.

Aftermarket equipment is most commonly referred to as add-on equipment and includes anything installed on a vehicle after it has left the factory.

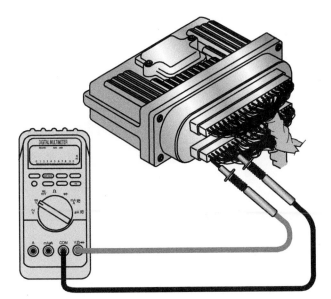

Figure 12-2 Using a digital voltmeter to check the PCM's circuit.

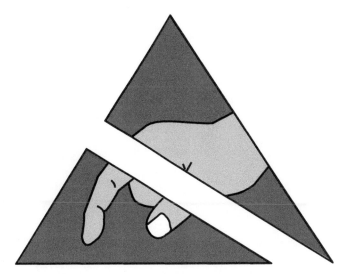

Figure 12-3 General Motors' electrostatic discharge (ESD) symbol warns technicians that a part or circuit is sensitive to static electricity.

Testing of individual components is important in diagnosing automotive systems. Although the DTCs in these systems give much more detail on the problems, the PCM does not know the exact cause of the problem. That is the technician's job.

Add-on equipment, even when installed under these guidelines, may still cause the powertrain system to malfunction. This may also include equipment not connected to the vehicle's electrical system such as portable telephones and radios. The first step in diagnosing any powertrain problem is to eliminate all aftermarket electrical equipment from the vehicle. If the problem still exists after this is done, diagnose the problem in the normal manner.

OBD II PCM

The PROM in the PCM is an EEPROM, or electrically erasable programmable read only memory chip, that can be reprogrammed without removing the PCM from the vehicle. This allows the updating of the PCM without replacing the PROM and means that a new PCM will have to be reprogrammed before it can be used in the vehicle. Some OBD II computers do have a chip that needs to be transferred to the new PCM, but this chip is a knock sensor module as shown in Figure 12-1. The procedure to reprogram a PCM is explained in Photo Sequence 25. This procedure is not meant to replace the service information from vehicle manufacturers, but it is meant to be a guide only. Always consult vehicle manufacturer information before programming a PCM or any other procedure.

PHOTO SEQUENCE 25
Reprogramming an OBD II PCM

P25-1 Technician is directed by a factory procedure or bulletin that the PCM must be reprogrammed or replaced.

P25-2 Before beginning the programming procedure, the battery voltage must be more than 12 volts. Charge the battery as needed, but never program a PCM with a battery charger on the vehicle.

P25-3 Make certain that unnecessary battery drains, such as blower motors, daytime running lights, A/C, etc., are switched off before beginning.

P25-4 Make sure the ignition switch is in the position directed by the scan tool. Do not turn the key on or off unless prompted to do so.

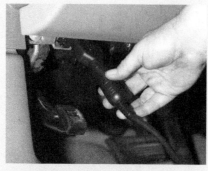

P25-5 Make certain that all connections are secure. A disconnection during programming may damage the PCM.

P25-6 Place the scan tool into the programming mode and enter the necessary information when requested.

PHOTO SEQUENCE 25 (CONTINUED)

P25-7 Make sure that the vehicle identification number (VIN) matches the VIN of the vehicle. The VIN gathered from the vehicle by the scan tool will ensure that the proper calibration is downloaded into the scan tool at the computer terminal.

P25-8 The technician takes the scan tool back to the computer terminal to download calibration information.

P25-9 After the calibration information has finished downloading to the PCM, turn off the scan tool *before* disconnecting it from the computer terminal.

P25-10 The technician reinstalls the scan tool to make certain all connections are secure, then turns the scan tool on and enters PCM programming mode.

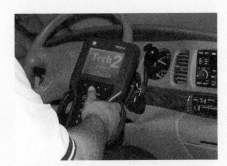

P25-11 The technician then downloads the calibration to the vehicle's PCM following appropriate service procedures prescribed by the manufacturer of the scan tool and vehicle.

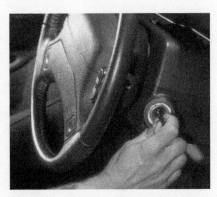

P25-12 After the download is complete, the technician should perform any necessary relearn procedures (such as idle learn) and drive the vehicle to confirm the repair.

OBD II DIAGNOSTICS

OBD II systems note the deterioration of certain components before they fail, which allows owners to bring their vehicles in at their convenience and before it is too late.

OBD II monitors the vehicle's exhaust gas recirculation and fuel systems, oxygen sensors, catalytic converter performance, and other miscellaneous components. OBD II diagnosis is best done with a strategy based on the flow charts and other information given in the service information. Before beginning to diagnose a problem, verify the customer's complaint. In order to verify the complaint, a technician must know the normal operation of the system. Compare what the vehicle is doing to what it should be doing. Also, pay close attention to the vehicle's drivability. Sometimes there is another problem or symptom that the customer is not aware of or disregarded. This symptom may be the key to properly diagnosing the system. When diagnosing a vehicle with multiple symptoms, look for a problem that may be common or related to all complaints. By identifying the circuit of seemingly unrelated codes, common power and ground circuits may be discovered. This can be an excellent clue leading to a quick diagnosis.

Visual Inspection

After verifying the complaint (**Figure 12-4**), the next step is a careful visual inspection. Make sure all of the grounds, including the battery and computer ground, have clean and tight (torqued) connections. Do a voltage drop test across all related ground circuits. The PCM and its systems cannot function correctly with a bad ground. A voltage drop of as little as 0.2 volt across a ground circuit can cause problems.

Check all vacuum lines and hoses, as well as the tightness of all attaching and mounting bolts in the induction system. Check for damaged air ducts. Check the ignition circuit, especially the secondary cables, for signs of deterioration, insulation cracks, corrosion,

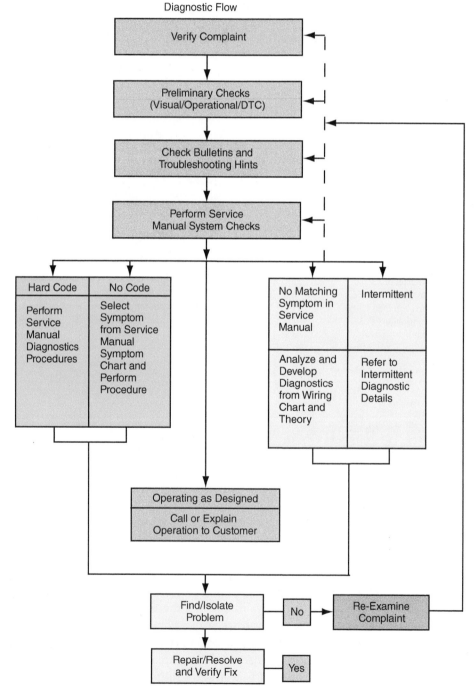

Figure 12-4 A strategy-based diagnostic tree.

and looseness. Pay attention to any unusual noises or odors. These can be the hints needed to locate the fault.

Inspect all related wiring and connections at the PCM. These may cause an intermittent fault. Any circuitry that is suspected as causing an intermittent complaint should be thoroughly checked for the following conditions: backed-out terminals, improper mating, and broken locks, improperly formed or damaged terminals, poor terminals to wiring connections, physical damage to the wiring harness, and corrosion.

⚡ WARNING **Always block the drive wheels and set the parking brake while checking the system.**

If the visual inspection did not identify the source of the customer's complaint, gather as much information as you can about the symptom. This should include a review of the vehicle's service history. Check for any technical service bulletins (TSBs) relating to the exhibited symptoms. This should include videos, newsletters, and any electronically transmitted media. Do not depend solely on the diagnostic tests run by the PCM. A particular system may not be supported by one or more DTCs. System checks verify the proper operation of the system. This will lead the technician in an organized approach to diagnostics.

When a complaint cannot be isolated or the cause found, a reevaluation is necessary. The complaint should be re-verified and could be intermittent or normal. Then check the following:

- Engine coolant temperature (ECT) sensor for initial coolant temperature reading close to ambient (then observe the rise in temperature while the engine is warming up).
- Throttle position (TP) sensor for proper sweep from 0 percent to 100 percent. (This range may vary on some vehicles.)
- Manifold absolute pressure (MAP) sensor for quick changes during changes in various engine loads.
- Check the oxygen sensor for proper rich-lean and lean-rich sweeps operation (**Figure 12-5**).
- Idle air control valve or throttle actuator control system for proper operation during warm-up, which can be a crucial step in correctly diagnosing any drivability concern.

Careful observation of these sensors during engine warm-up may reveal a slow-responding sensor or a sensor that malfunctions only within a small portion of its range.

⚡ WARNING **Vehicles are equipped with air bags, including driver- and passenger-side air bags in addition to side-curtain and rear-seat systems. Before working around or on the circuits related to the air bags (Figure 12-6), make sure you refer to the precautions given in the service information. Failure to follow these precautions can cause unexpected air bag deployment, which can cause injury and/or damage to the vehicle.**

Diagnosis should continue with preliminary system checks. Each manufacturer will recommend that certain tests be completed at this stage. The purpose of this stage in diagnostics is to make sure the basic engine and its systems are in good shape. This includes checking for vacuum leaks, checking the engine's compression, ignition system, and air-fuel system.

GM and many other manufacturers recommend that a system check be conducted before proceeding with diagnosis. The powertrain OBD system check (**Figure 12-7**) is an organized approach to identifying a problem created by an electronic engine control system malfunction. The powertrain OBD system check is the starting point for any drivability complaint diagnosis. This directs the technician to the next logical step in

Special Tools
DMM

⚠ **Caution**

Never use a test light to diagnose powertrain electrical systems unless specifically instructed by the diagnostic procedures.

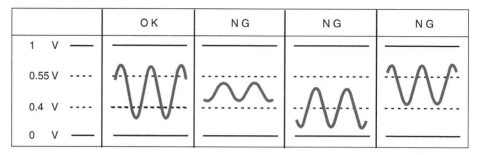

	OK	NG	NG	NG
1 V				
0.55 V				
0.4 V				
0 V				

Figure 12-5 O$_2$ sensor activity should be checked with a lab scope.

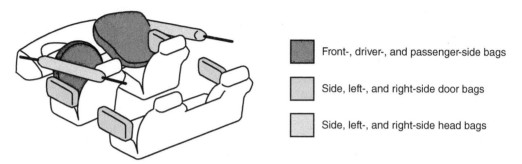

Front-, driver-, and passenger-side bags

Side, left-, and right-side door bags

Side, left-, and right-side head bags

Figure 12-6 The locations of various air bags.

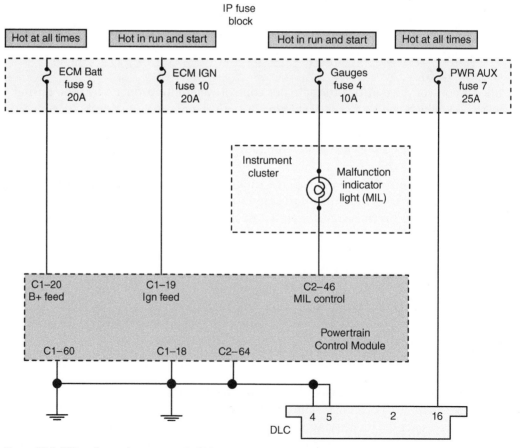

Figure 12-7 Wiring diagram for a powertrain OBD system check.

diagnosing a complaint (**Figure 12-8**). Never perform this check if a drivability complaint cannot be duplicated. Understanding the system check table and using it correctly will reduce diagnostic time and prevent the replacement of good parts.

CUSTOMER CARE As simple as it may seem, advise your customers to tighten the fuel cap at least three to four clicks when reinstalling it after filling the fuel tank. Otherwise, the OBD II system may assume a problem exists within the system resulting from a poorly sealed fuel tank and turn on the MIL. Fortunately, most new vehicles are equipped with a cap-less fuel tank that should prevent these situations from happening as often in the future.

Powertrain OBD System Check

Step	Action	Values	Yes	No
1	1. Turn the ignition switch ON with the engine lamp OFF. 2. Observe the Malfunction Indicator Lamp (MIL). Is the MIL on?	–	Go to Step 2	Go to No Malfunction Indicator Lamp
2	1. Turn the ignition switch OFF. 2. Install a scan tool. 3. Turn the ignition switch ON. Does the scan tool display PCM data?	–	Go to Step 3	Go to Data Link Connector Diagnosis
3	Using the scan tool, command the MIL OFF. Does the MIL turn OFF?	–	Go to Step 4	Go to Malfunction Indicator Lamp On Steady
4	Check for DTCs with the scan tool. Were any last test fail, history, or MIL request DTCs set?	–	Go to Step 5	Go to Step 6
5	Using the scan tool, record the freeze frame and failure records information. Is the action complete?	–	Go to applicable DTC table	–
6	Does the engine start and continue to run?	–	Go to Step 7	Go to engine cranks but does not run
7	1. Turn the ignition switch ON with the engine OFF. 2. Check the ECT and TP sensors for proper operation. 3. Start the engine. 4. Allow the running engine to reach operating temperature. 5. Allow the running engine to reach operating temperature, check the ECT, MAP, and O_2S_1 sensors and IAC valve for proper operation. 6. Compare the scan tool data with the typical values shown in the Scan Tool Data. Are the display values normal or within typical ranges?	–	Go to symptoms	Go to the applicable diagnostic section

Figure 12-8 Powertrain OBD system check.

Scan Tool Diagnosis

When the ignition is initially turned on, the malfunction indicator lamp (MIL) will momentarily flash on, then off, and remain on until the engine is running or there are no DTCs stored in memory. Now connect the scan tool. If the scanner does not automatically detect the vehicle's VIN, enter the vehicle identification information (**Figure 12-9**), then retrieve the DTCs with the scan tool. The scan tool should be used to do more than simply retrieve codes. The information that is available, whether or not a DTC is present, can reduce diagnostic time. Some of the common uses of the scan tool are identifying stored DTCs, clearing DTCs, performing output control tests, and reading serial data (**Figure 12-10**).

Bidirectional Controls

Many professional-grade scan tools are available to offer a host of bidirectional controls. Bidirectional controls provide the ability to turn actuators on and off from the scan tool itself. This is useful when checking actuators such as solenoids, relays, fan motors, and fuel injectors. There are even tests that can turn on dash warning lights, activate coolant fans, reset fuel trims, and other tasks by command of the scan tool. Many scan tools, both aftermarket and original manufacturer, have bidirectional ability. For the most coverage a manufacturer's scan tool generally offers the most controls. These features help a technician diagnose problem areas faster. For instance, if the cooling fan is inoperative, the scan tool can command the fan relay on, allowing the technician to check the circuit wiring without getting the engine to operating temperature.

Depending on the scan tool being used, OEM-enhanced data and special diagnostic routines may be available in addition to generic OBD II data. The scan tester must have the appropriate connector to fit the diagnostic link connector (DLC) on an OBD II system (**Figure 12-11**) and the proper software for the vehicle being tested. Different inserts are available for some OBDII scan tester cables depending on the age of the scan tester. Always follow the instructions in the information supplied by the scan tester manufacturer. Many similar tests, such as the key on, engine off (KOEO) and key on, engine running (KOER) tests, are performed on OBD II systems as on previous systems.

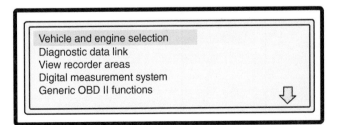

Figure 12-9 The main menu on a typical scan tool.

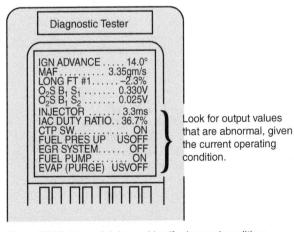

Figure 12-10 Use serial data to identify abnormal conditions.

Figure 12-11 A personality key for a scan tool.

OBD II codes are obtained only through the scan tool. OBD II DTCs, with the conditions required to set the DTCs, are shown in **Figure 12-12**. When the option DTCs are selected on the scan tool, the scan tool can be used to read those DTCs. DTCs should not be erased until the **freeze frame** has been recorded. The freeze frame provides additional assistance for technicians.

Check the DTCs with the scan tool. If multiple DTCs are stored, refer to the DTC interpretation table in the following order:

1. Network DTCs
2. PCM error DTCs
3. System voltage DTCs
4. Component-level DTCs
5. System-level DTCs

The next step in diagnostics depends on the outcome of the DTC display. If there were DTCs stored in the PCM's memory, proceed by following the designated DTC table to make an effective diagnosis and repair. If no DTCs were displayed, match the symptom to the symptoms listed in the manufacturer's symptom tables (**Figure 12-13**) and follow the diagnostic paths or suggestions to complete the repair or refer to the applicable component or system check. If there is no matching symptom, analyze the complaint, and then develop a plan for diagnostics. Use the wiring diagrams and theory of component and system operation. Refer to technical bulletins for similar cases where repair history may be available. It is possible that the customer's complaint is a normal operating condition for that vehicle. If this is suspected, compare the complaint to the drivability of a known good vehicle.

After the DTCs and the MIL have been checked, the OBD II data stream can be selected (**Figure 12-14**, see page 621). In this mode, the scan tool displays all of the data from the inputs. This information may include engine data (**Figure 12-15**, see page 622), EGR data (**Figure 12-16**, see page 624), three-way catalytic converter data (**Figure 12-17**, see page 625), oxygen sensor data (**Figure 12-18**, see page 626), and misfire data (**Figure 12-19**, see page 627). The actual available data will depend on the vehicle and the scan tool. However, OBD regulations dictate a minimum amount of data. Always refer to the appropriate service information and the scan tool's operating instructions for proper identification of normal or expected data.

Classroom Manual
Chapter 12, page 379

Negative fuel trim numbers indicate that fuel is being subtracted from the base calculation, while positive trim numbers indicate fuel is being added.

INPUTS

Component	Malfunction	J2012 DTC	Diagnostic Criteria to Set Fault	Mature Time
Sync. (cam/crank)	Rationality	P1390	Cam/Crank reference Angle>threshold	25 cts.
MAP Sensor	Rationality	P1297	Baro MAP<4 in. Hg at idle	2 sec.
	Short	P0170	Voltage signal <0.02 volt	2 sec.
	Open	P1080	Voltage signal >4.667 volts	2 sec.
Vehicle Speed Sensor	Rationality	P0500	No pulses during driving conditions	11 sec.
Coolant Temp. Sensor	Rationality	P0125	After start engine temperature <35°F	10 min.
	Short	P0117	Voltage signal <0.51 volt	3 sec.
	Open	P0118	Voltage signal >4.96 volts	3 sec.
02 Sensor Upstream	Shortened Low	P0131	Voltage stays below 0.15 volt during O_2 heater test	7 min. 3 sec.
02 Sensor Upstream	Shortened High	P0132	Voltage signal <1.22 volts	2 min.
02 Sensor Upstream	Open	P0134	Voltage stays >0.35 volt or <0.59 volt	7 min.
02 Sensor Downstream	Shortened Low	P0137	Voltage signal <1.22 volts	3 sec.
02 Sensor Downstream	Shortened High Rationality	P0138 P0139	A WOT voltage signal >0.61 volt and at decel fuel shutoff voltage signal <0.29 volt	15 sec.
Charge Temp. Sensor	Short	O0112	Voltage signal <0.51 volt	3 sec.
	Open	P0113	Voltage signal >4.96 volts	3 sec.
Throttle Position Sensor	Rationality	P0121	At idle voltage signal >4 volts OR at cruise voltage signal <0.51 volt	3 sec.
	Short	P0122	Voltage signal <0.157 volt	0.7 sec.
	Open	P0123	Voltage signal >4.706 volts	0.7 sec.
PCM	ROM sum	P0605	At power down but sum not equal to calibration	Immediate
	SPI	P0605	No communication between devices	
Cam Position Sensor	Rationality	P0340	No pulses seen	5 sec.
P/S Pressure Switch	Rationality	P0551	At vehicle speed >40 mph, high pressure	15 sec.

OUTPUTS

Component	Malfunction	J2012 DTC	Diagnostic Criteria to Set Fault	Mature Time
IAC motor	Open/Short	P0505	Voltage signal read ≠ to expected state	3 sec.
	Functionality	P1294	At idle, rpm not within ±300 of target	20 sec.
Injector 1	Open/Short	P0201	Voltage signal read ≠ to expected state	3 sec.
Injector 2	Open/Short	P1202	Voltage signal read ≠ to expected state	3 sec.
Injector 3	Open/Short	P0203	Voltage signal read ≠ to expected state	3 sec.
Injector 4	Open/Short	P0204	Voltage signal read ≠ to expected state	3 sec.
EVAP solenoid	Functionality	P0441	Shift in O_2 feedback or IAC or rpm from solenoid off to full on	25 sec.
	Open/Short	P0443	Voltage signal read ≠ to expected state	3 sec.
EGR solenoid	Open/Short	P0403	Voltage signal read ≠ to expected state	3 sec.
High-speed fan relay	Open/Short	P1489	Voltage signal read ≠ to expected state	3 sec.
Low-speed fan relay	Open/Short	P1490	Voltage signal read ≠ to expected state	3 sec.
Ignition coil 1 (DIS)	Open/Short	P0351	Max. dwell current ≠ to expected state	3 sec.
Ignition coil 2 (DIS)	Open/Short	P0352	Max. dwell current ≠ to expected state	3 sec.

Figure 12-12 OBD DTCs with the conditions required to set them.

⚠ **Caution**

Do not clear DTCs unless directed by a diagnostic procedure. Clearing DTCs will also clear valuable freeze-frame data.

One of the selections on a scan tool is the DTC data. This selection contains an important information called freeze frame, which displays the conditions that existed when a DTC was set.

Using the scan tool, record the freeze-frame information. By storing the freeze-frame data (**Figure 12-20**, see page 628) on the scan tool, an electronic copy of the data is created when the fault occurred and stored on the scan tool, which can be referred to later. Using the scan tool with a five-gas exhaust analyzer allows for a comparison of computer data with tailpipe emissions (**Figure 12-21**, see page 628). A lab scope is also extremely handy for watching the activity of sensors and actuators. Both of these testers will aid in the diagnosis of a problem.

If the PCM reacts to a misfire by shutting down the injector of a cylinder, that bad cylinder will still have an effect on the engine and emissions. The piston of that cylinder is still moving, the valves are still opening, and the spark plug is still firing. All of this leads to excessive amounts of oxygen in the exhaust stream. Most late-model vehicles will go into open loop during a cylinder misfire and shut the injector off on the misfiring cylinder as previously mentioned. Remember that the oxygen sensor reacts only to the oxygen in

⚠ **Caution**

A scan tool that displays faulty data should not be used, and the problem should be reported to the manufacturer. The use of a faulty scan tool can result in misdiagnosis and unnecessary parts replacement.

Classroom Manual
Chapter 12, page 380

Hesitation, Sag, and Stumble

Step	Action	Values	Yes	No
	Definition: Momentary lack of response as the accelerator is pushed down. Can occur at all vehicle speeds. Usually most severe when first trying to make the vehicle move, as from a stop sign. May cause engine to stall if severe enough.			
1	1. Was the Powertrain On-Board Diagnostic (OBD) System Check performed?	–	Go to Step 2	Go to Powertrain OBD System Check 2.4 L Powertrain OBD System Check 2.2 L
2	1. Perform a bulletin search. 2. If a bulletin that addresses the symptom is found, correct the condition per bulletin instructions. Was a bulletin found that addresses the symptom?	–	Go to Step 16	Go to Step 3
3	1. Perform the careful visual/physical checks as described at the beginning of the Symptoms section. Is the action complete?	–	Go to Step 4	–
4	1. Check the TP sensor for binding or sticking. Voltage should increase at a steady rate as the throttle is moved toward wide-open throttle. 2. Repair if found faulty. Was a problem found?	–	Go to Step 16	Go to Step 5
5	1. Check the MAP sensor for proper operation. Refer to MAP sensor output. Check 2.4 L or MAP sensor output. Check 2.2 L. 2. Repair if found faulty. Was a problem found?	–	Go to Step 16	Go to Step 6
6	1. Check for fouled spark plugs. 2. Replace the spark plugs if found faulty. Was a problem found?	–	Go to Step 16	Go to Step 7
7	1. Check the ignition control module for proper grounds. 2. Repair if grounds are found faulty. Was a problem found?	–	Go to Step 16	Go to Step 8

Figure 12-13 A diagnostic sequence based on a symptom. *(continued on next page)*

Step	Action	Values	Yes	No
8	1. Check for poor fuel pressure. Refer to fuel system diagnosis 2.4 L or 2.2 L. 2. If a problem is found, repair as necessary. Was a problem found?	–	Go to Step 16	Go to Step 9
9	1. Check for contaminated fuel. 2. If a problem is found, repair as necessary. Was a problem found?	–	Go to Step 16	Go to Step 10
10	1. Perform a fuel injector coil test 2.4 L or 2.2 L and fuel injector balance test 2.4 L or 2.2 L. 2. If a problem is found, repair as necessary. Was a problem found?	–	Go to Step 16	Go to Step 11
11	1. Check for proper system performance. Use table EVAP control system diagnosis 2.4 L or 2.2 L. 2. If a problem is found, repair as necessary. Was a problem found?	–	Go to Step 16	Go to Step 12
12	1. Check for proper function of the thermostat. Also check for proper heat range. 2. If a problem is found, repair as necessary. Was a problem found?	–	Go to Step 16	Go to Step 13
13	1. Check the generator for proper output voltage. The generator output voltage should be between the specified values. 2. If a problem is found, repair as necessary. Was a problem found?	11 V to 16 V	Go to Step 16	Go to Step 14
14	1. Check the intake valves for valve deposits. 2. If deposits are found, remove as necessary. Were deposits found on valves?	–	Go to Step 16	Go to Step 15
15	1. Review all diagnostic procedures within this table. 2. If all procedures have been completed and no malfunctions have been found, review/inspect the following items: *Visual/physical inspection *Scan tool data *All electrical connections within a suspected circuit and/or system. 3. If a problem is found, repair as necessary. Was a problem found?	–	Go to Step 16	Check for service bulletin
16	1. Operate vehicle within the conditions under which the original symptom was noted. Does the system now operate properly?	–	System OK	Go to step 2

Figure 12-13 A diagnostic sequence based on a symptom. *(continued)*

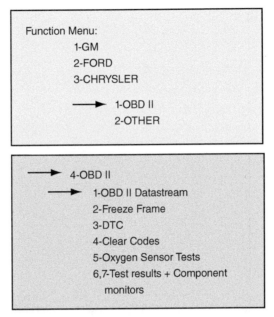

Figure 12-14 OBD II scan tool menu.

the exhaust stream, not fuel from a misfiring cylinder. When the O_2 sensor sees the excessive oxygen in the exhaust stream, it will send a very lean reading to the PCM. In turn, the PCM would tend to increase the richness of the mixture to correct for the excess oxygen coming from the misfiring cylinder. On most engines, this amount of enriching will be beyond the capacity of the system because no amount of added fuel will burn the oxygen from the dead cylinder. On older computer-controlled vehicles, the fuel that was not burned from the misfiring cylinder would go through the catalytic converter making it extremely hot, resulting in damage to the converter itself.

When a cylinder misfire is present, check to see if two cylinders are involved. If there are, look for something shared by both cylinders such as the common coil or a common vacuum leak. A misfire code should set and tell the technician which cylinder is missing.

> **Special Tools**
> Exhaust analyzer
> Lab scope

OBD II TERMINOLOGY

Drive Cycle

An OBD II **drive cycle** consists of driving a vehicle under specific conditions until all the readiness monitors have been run.

Enable Criteria

Enabling criteria defines the criteria that must be met in order for a monitor test to be completed. The enable criteria vary with the make and model of the vehicle. Therefore, refer to the service information to identify the manufacturer's requirements. If the enable criteria are not met, that particular monitor test is not completed and the condition of the tested system is unknown. An example of this is the catalyst efficiency monitor. The enable criteria to run the monitor require the vehicle to be driven on the highway at a steady cruise. If the vehicle were driven only in the city, the monitor would never run.

On-board Diagnostics

A diagnostic test is a series of steps that has a beginning and an end, the result of which is a pass or fail reported to the PCM. When a diagnostic test reports a pass result, the PCM records the following data: the diagnostic test has completed since the last ignition cycle, the diagnostic test has passed during the current ignition cycle, and the fault identified by the diagnostic test is not currently active. When a diagnostic test reports a fail result, the PCM records the following data: the diagnostic test has completed since the last ignition, the fault identified by the diagnostic test is currently active, the fault has been active during the ignition cycle, and the operating conditions at the time of the failure.

Parameter	Units Displayed	Typical Data Value
Engine Speed	rpm	± 100 rpm from desired idle
Desired Idle	rpm	PCM-commanded idle speed (varies with temperature)
ECT	°C, °F	85°C–105°C/ 185°F–220°F (varies with temperature)
IAT	°C, °F	Varies with ambient air temperature
IAC Position	Counts	10–40
TP Angle	Percent	0%
TP Sensor	Volts	0.20–0.74
Throttle at Idle	Yes/No	Yes
BARO	kPa	65–110 (Depends on engine load and barometric pressure)
MAP	kPa/volts	29–48 kPa /1–2 volts (Depends on engine load and barometric pressure)
MAF Input Frequency	Hz	1,200–3,000 (Depends on engine load and barometric pressure)
MAF	gm/s	3–6 (Depends on engine load and barometric pressure)
Injection Pulse Width	milliseconds	Varies with engine load
Air Fuel Ratio	Ratio	14.2–14.7
Fuel Pump	On/Off	On
VTD Fuel Disable	Active/Inactive	Inactive
ECT	°C, °F	85°C–105°C/ 185°F–220°F (varies with temperature)
Engine Run Time	Hr: Min: Sec	Depends on time since startup
Loop Status	Open/Closed	Closed
Fuel Trim Learn	Disabled/Enabled	Enabled
HO$_2$S Bn1 Sen. 1	Not Ready/Ready	Ready
HO$_2$S Bn2 Sen. 1	Not Ready/Ready	Ready
HO$_2$S Bn1 Sen. 1	Millivolts	0–1,000, constantly varying
HO$_2$S Bn2 Sen. 1	Millivolts	0–1,000, constantly varying

Figure 12-15 Engine data. *(continued on next page)*

Parameter	Units Displayed	Typical Data Value
Rich/Lean Bn 1	Rich/Lean	Constantly changing
Rich/Lean Bn 2	Rich/Lean	Constantly changing
HO$_2$S Bn 1 Sen. 2	Millivolts	0–1,000, constantly varying
HO$_2$S Bn 1 Sen. 3	Millivolts	0–1,000, varying
Short-Term FT Bn 1	Percent	+10%–10%
Long-Term FT Bn 1	Percent	+10%–10%
Short-Term FT Bn 2	Percent	+10%–10%
Long-Term FT Bn 1	Percent	+10%–10%
Fuel Trim Cell	Cell #	0
Engine Load	Percent	2%–5%
Desired ERG Pos.	Percent	0%
Actual ERG Pos.	Percent	0%
Vehicle Speed	MPH, Km/h	0
Engine Speed	rpm	100 rpm from desired idle
TCC command	Engaged/Disengaged	Disengaged
Current Gear	1/2/3/4	1
Ignition 1	Volts	13 (varies)
Commanded GEN	On/Off	On
Cruise Engaged	Yes/No	No
Cruise Inhibited	Yes/No	Yes
Engine Load	Percent	2%–5% (varies)
TWC Protection	Active/Inactive	Inactive
Hot Open Loop	Active/Inactive	Inactive
ECT	°C, °F	85°C–105°C/ 185°F–220°F (varies with temperature)
Decel Fuel Mode	Active/Inactive	Inactive
Injector Pulse Width	milliseconds	Varies with engine load
Power Enrichment	Active/Inactive	Inactive
TP Angle	Percent	0%

Figure 12-15 Engine data. *(continued)*

Pending Situation

Under certain circumstances, the PCM will postpone and not run a particular monitor if the MIL is on and a DTC stored. This creates a **pending situation** in that a test will not run or be completed until the fault is corrected or goes undetected for the required number of driving cycles. Pending situations occur when a sensor is malfunctioning but its signal is important to the testing of another component or system. An example of this is the oxygen sensor. If the O$_2$ sensor signal is out of acceptable range or if it is not responding according to the guidelines set in the PCM, a DTC will be stored. However, the O$_2$ signal is needed to measure the efficiency of the catalytic converter. This prevents the catalyst efficiency monitor from being conducted and the test is pending.

A **pending situation** prevents a monitor from being performed due to an uncorrected fault.

Parameter	Units Displayed	Typical Data Value
Desired EGR Position	Percent	0%
Actual EGR Position	Percent	0%
EGR Test Count	Counts	0
EGR Feedback	Volts	0.14–1.0 volt
EGR Duty Cycle	Percent	0%
EGR Pos. Error	Percent	0%–9%
EGR close Valve Pintle Position	Volts	0.14–1.0 volt
Engine Speed	rpm	± 100 rpm from desired idle
Desired Idle	rpm	PCM commanded idle speed
ECT	°C, °F	85°C–105°C / 85°F–220°F
IAT	°C, °F	Varies to ambient
IAC Position	Counts	10–40
TP Angle	Percent	0%
Baro	kPa	65–110
MAP	kPa/volts	29–48 kPa, 1–2 volts
Loop Status	Open/Closed	Closed
Fuel Trim Learn	Disabled/Enabled	Enabled
Rich/Lean Bn1	Rich/Lean	Constantly changing
Rich/Lean Bn2	Rich/Lean	Constantly changing
Fuel Trim Cell	Cell #	0
Engine Load	Percent	2%–5% (varies)
Hot Open Loop	Active/Inactive	Inactive
Ignition 1	Volts	13 (varies)
Knock Retard	Degrees	0%
KS Activity	Yes / No	No
Current Gear	1/2/3/4	1
Trans. Range	P/N, Reverse, Low, Drive 2, Drive 3 & 4	P/N
Commanded TCC	Engaged/Disengaged	Disengaged
Trans Hot Mode	Active/Inactive	Inactive
Engine Load	Percent	2%–5% (varies)
TWC Protection	Active/Inactive	Inactive
Decel Fuel Mode	Active/Inactive	Inactive
Power Enrichment	Active/Inactive	Inactive

Figure 12-16 EGR data.

Serial Data

Serial data is the PCM information sent to the scanner and other control modules.

Serial data is a term that refers to a method of data transmission. Serial data is transmitted, one bit at a time, over a single wire. The data travels through a data bus where it is made available to the PCM and other associated control modules. The computer talks to the modules on the network and the scan tool using serial data. Serial data is vital to most diagnostic routines on the computer network.

Similar Conditions

This term defines the operating conditions that must be present during a second trip to set a code when there is a defect and to erase a code after a repair has been made. For the conditions to be similar, these conditions must be present: the load conditions must be

Parameter	Units Displayed	Typical Data Value
TWC Monitor Test Counter	Counts	0
TWC Diagnostic	Enabled/disabled	Enabled
Engine Speed	rpm	± 100 rpm from desired idle
ECT	°C, °F	85°C–105°C/185°F–220°F (varies with temperature)
IAT	°C, °F	Varies with ambient temperature
IAC Position	Counts	10–40
TP Angle	Percent	0%
TP Sensor	Volts	0.20–0.74
Throttle at Idle	Yes/No	Yes
Engine Speed	rpm	±100 rpm from desired idle
MAF	gm/s	3–6 (Depends on engine load and barometric pressure)
Engine Load	Percent	2%–5% (varies)
Air-Fuel Ratio	Ratio	14.2–14.7
ECT	°C, °F	85°C–105°C/185°F–220°F (varies with temperature)
Engine Run Time	Hr: Min: Sec	Depends on time since startup
Loop Status	Open/Closed	Closed
Fuel Trim Learn	Disabled/Enabled	Enabled
HO_2S Bn1 Sen. 2	Millivolts	0–1,000, constantly varying
HO_2S Bn1 Sen. 3	Millivolts	0–1,000, varying
Vehicle Speed	mph, km/h	0
MIL	On/Off	Off
Engine Load	Percent	2%–5% (varies)
TWC Protection	Active/inactive	Inactive
Hot Open Loop	Active/inactive	Inactive
ECT	°C, °F	85°C–105°C/185°F–220°F (varies with temperature)
Decel Fuel Mode	Active/inactive	Inactive
Inj. Pulse Width	Milliseconds	Varies with eng. load
Power Enrichment	Active/inactive	Inactive
TP Angle	Percent	0%

Figure 12-17 Three-way catalytic converter data.

within 10 percent of the vehicle load present when the diagnostic test reported the malfunction, the engine speed must be within 375 rpm of the engine speed present when the DTC was set, and the engine coolant temperature must have been in the same range present when the PCM turned the MIL on.

Trip

The ability for a diagnostic monitor to run depends on whether or not a trip has been completed. A trip for a particular monitor is defined as a key on and key off cycle in which all the enabling criteria for a given monitor have been met. The requirements for trips

Parameter	Units Displayed	Typical Data Value
Start-up IAT	°C, °F	Depends on intake air temperature at time of startup
Start-up ECT	°C, °F	Depends on engine coolant temperature at time of startup
HO$_2$S Bn1 Sen. 1	Millivolts	0–1,000, constantly varying
HO$_2$S Bn2 Sen. 1	Millivolts	0–1,000, constantly varying
HO$_2$S Bn1 Sen.2	Millivolts	0–1,000, constantly varying
HO$_2$S Bn 1 Sen. 3	Millivolts	0–1,000, varying
HO$_2$S X Counts Bn1	Counts	Varies
HO$_2$S X Counts Bn2	Counts	Varies
HO$_2$S Warm-Up Time Bn1 Sen. 1	Hr: Min: Sec	Depends on startup intake air temperature, startup engine coolant temperature, and time to HO$_2$S activity.
HO$_2$S Warm Up Bn2 Sen 1	Hr: Min: Sec	Depends on startup intake air temperature, startup engine coolant temperature, and time to HO$_2$S activity.
HO$_2$S Warm Up Bn1 Sen 2	Hr: Min: Sec	Depends on startup intake air temperature, startup engine coolant temperature, and time to HO$_2$S activity.
HO$_2$S Warm Up Bn1 Sen 3	Hr: Min: Sec	Depends on startup intake air temperature, startup engine coolant temperature, and time to HO$_2$S activity.
Engine Speed	rpm	±100 rpm from desired idle
ECT	°C, °F	85°C–105°C/185°F–220°F (varies with temperature)
MAF	gm/s	3–6 (Depends on engine load and barometric pressure).
EVAP Purge PWM	Percent	0%–25% (varies)
TP Angle	Percent	0%
Ignition 1	Volts	13 (varies)

Figure 12-18 Heated oxygen sensor data.

vary as they may involve items of an unrelated nature such as driving style, length of trip, ambient temperature, and so on. Some monitors run only once per trip, such as the catalyst monitor. Others run continuously, such as the misfire and fuel system monitor systems. If the proper enabling conditions are not met during that ignition cycle, the tests may be incomplete or the monitor may not have run.

Two Trip Monitors

A maturing code (pending code) is a type B code that has not been repeated nor has it caused the MIL to illuminate.

The first time the OBD II monitor detects a fault during any drive cycle, it sets a pending code in the memory of the PCM. Before a pending code becomes a DTC and turns on the MIL, the fault must be repeated under similar conditions. A pending code can remain in memory for some time while it waits for the conditions to repeat themselves. When the same conditions exist, the PCM checks the fault. If the fault is still present, a DTC is stored and the MIL is illuminated. Two trip monitors are those monitors that require a second trip to set a code. Some misfire and catalyst faults and the comprehensive component monitor can cause the PCM to turn the MIL on after only one trip.

Parameter	Units Displayed	Typical Data Value
Misfire Current #1	Counts	0
Misfire History #1	Counts	0
Misfire Current #2	Counts	0
Misfire History #2	Counts	0
Misfire Current #3	Counts	0
Misfire History #3	Counts	0
Misfire Current #4	Counts	0
Misfire History #4	Counts	0
Misfire Current #5	Counts	0
Misfire History #5	Counts	0
Misfire Current #6	Counts	0
Misfire History #6	Counts	0
Misfire Failures Since First Fail	Counts	0
Misfire Passes Since First fail	Counts	0
Total Misfire Current Count	Counts	0
ECT	°C, °F	85°C–105°C/185°F–220°F (varies with temperature)
Hot Open Loop	Active/Inactive	Inactive
Loop Status	Open/Closed	Closed
Fuel Trim Learn	Disabled/Enabled	Enabled
HO2S Bn 1 Sen. 1	Not Ready/Ready	Ready
HO2S Bn 2 Sen. 1	Not Ready/Ready	Ready
Rich/Lean Sen. 1	Rich/Lean	Constantly changing
Rich/Lean Sen. 2	Rich/Lean	Constantly changing
HO2S X Counts Bn 1	Counts	Varies
HO2S X Counts Bn 2	Counts	Varies
Short-Term FT Bn 1	Percent	–10%–10%
Long-Term FT Bn 1	Percent	–10%–10%
Short-Term FT Bn 2	Percent	–10%–10%
Long-Term FT Bn 2	Percent	–10%–10%
Fuel Trim Cell	Cell #	0
Engine Load	Percent	2%–5% (varies)
Engine Speed	rpm	±100 rpm from desired idle
Ignition Mode	IC/Bypass	IC
Spark	Degrees	16
Knock Retard	Degrees	0
Desired ERG Position	Percent	0%
Actual ERG Position	Percent	0%
Vehicle Speed	mph, km/h	0
Current Gear	1/2/3/4	1
Engine Speed	rpm	±100 rpm from desired idle
TCC Enable	On/Off	Off
IAC Position	Counts	10–40
Current Gear	1/2/3/4	1
Engine Load	Percent	2%–5%
TWC Protection	Active/Inactive	Inactive
Decel Fuel Mode	Active/Inactive	Inactive
Power Enrichment	Active/Inactive	Inactive
A/C Request	Yes/No	No (Yes with A/C "ON")
Commanded A/C	On/Off	Off (On with A/C compressor engaged)

Figure 12-19 Misfire data.

At Idle/Upper Radiator/Closed Throttle/Park or Neutral Closed Loop/Acc. OFF		
Scan Tool Parameter	**Units Displayed**	**Typical Data Value**
DTC Freeze Frame	#	DTC #
Scan Tool Parameter	Units Displayed	Typical Data Value
Air-Fuel Ratio	Ratio	Varies
Calc. Air Flow	g/S	Varies
ECT	°C–°F	Varies
BARO	kPa, V	Varies
Base PWM Cyl. 1	ms	Varies
MAP	kPa,V	Varies
Short-Term FT	Counts, %	Varies
Long-Term FT	Counts, %	Varies

At Idle/Upper Radiator/Closed Throttle/Park or Neutral Closed Loop/Acc. OFF		
Scan Tool Parameter	**Units Displayed**	**Typical Data Value**
Loop Status	Open/Closed	Varies
mph	mph and km/h	Varies
Load (2.4-L Engine)	%	15–21% (16–29% @2,500 rpm)
Load (2.2-L Engine)	%	22–26% (19–25% @2,500 rpm)
TP Angle	%	Varies
Engine Speed	rpm	Varies

Figure 12-20 Freeze-frame and failure records data.

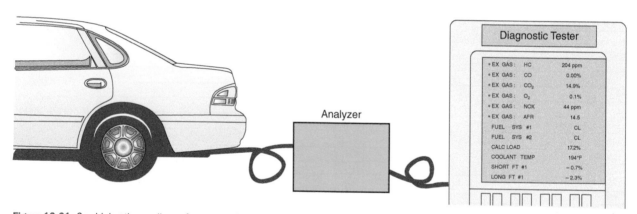

Figure 12-21 Combining the readings of a scan tool with those of an exhaust gas analyzer are very valuable in diagnostics.

Warm-up Cycle

Classroom Manual
Chapter 12, page 374

A warm-up cycle consists of engine startup and vehicle operation such that the coolant temperature has risen greater than 40°F from startup temperature and reached at least 160°F. If this condition is not met during the ignition cycle, the diagnostic monitors may not run.

MALFUNCTION INDICATOR LAMP

The malfunction indicator lamp (MIL) is on the instrument panel. The MIL informs the driver that a fault that affects the vehicle's emission levels has occurred. The owner should take the vehicle in for service as soon as possible. The MIL will illuminate if a component or system that has an impact on vehicle emissions indicates a malfunction or fails to pass an emissions-related diagnostic test (**Figure 12-22**). The MIL will stay illuminated until

Components Intended to Illuminate MIL
Automatic Transmission Temperature Sensor
Engine Coolant Temperature (ECT) Sensor
Evaporative Emission Canister Purge (EVAP Canister Purge)
Evaporative Emission Purge Vacuum Switch
Idle Air Control (IAC) Coil
Intake Air Temperature (IAT) Sensor
Ignition Control (IC) System
Ignition Sensor (Cam Sync, Diag)
Ignition Sensor High Resolution 7x
Knock Sensor (KS)
Manifold Absolute Pressure (MAP) Sensor
Mass Air Flow (MAF) Sensor
Throttle Position (TP) Sensor A, B
Transmission Range (TR) Mode Pressure Switch
Transmission Shift Solenoid A
Transmission Shift Solenoid B
Transmission TCC Enable Solenoid
Transmission 3/4 Shift Solenoid
Transmission Torque Converter Clutch (TCC) Control Solenoid
Transmission Turbine Sensor (HI/LO)
Transmission Vehicle Speed Sensor (HI/LO)
Transmission Vehicle Speed Sensor (HI/LO)

Figure 12-22 A list of components that will cause the MIL to turn on when the PCM detects they are defective or are malfunctioning.

the system or component passes the same test for three consecutive trips with no emissions-related faults. After making the repair, technicians may need to take the vehicle for three trips to ensure the MIL does not illuminate again.

As a bulb and system check, the MIL comes on with the ignition switch on and the engine off. On most vehicles, the MIL will go off after a few seconds even if the engine is not started. When the MIL remains on while the engine is running or a malfunction is suspected due to a drivability or emissions problem, perform an OBD system check.

If the vehicle is experiencing a malfunction that may cause damage to the catalytic converter, the MIL will flash once per second. If the MIL flashes, this indicates a catalyst damaging misfire, and the driver needs to get the vehicle to the service department as soon as possible. If the driver reduces speed or load and the MIL stops flashing, a code is set and the MIL stays on. This means that the misfire that presented potential problems to the converter has passed with the changing operating conditions.

On some systems, the MIL may also light if the gas cap is off or is loose. This is due to the evaporative leak in the fuel vapor system and will be identified with an evaporative system DTC when the EVAP monitor runs. Some vehicles will illuminate a "check fuel cap" lamp if a large EVAP leak is detected.

MIL History Codes

The PCM must acknowledge when all the emissions-related diagnostic tests have reported a pass or fail condition since the last ignition cycle. Each diagnostic test is separated into four types: when it is a emissions-related test that turns on the MIL the first time the PCM reports a fault, this is a type A DTC; when it is a emissions-related test that turns on the MIL if the fault is active for two consecutive driving cycles, this is a type B DTC; when it is non-emissions-related, like a code for the cruise control, and does not turn on the MIL but will turn on the service light, this is a type C code; and when it is non-emissions-related but does not turn on the MIL or the service light, this can be for various types of trouble codes, such as a cruise control fault. A U code is a communications network problem code.

A type-B DTC requires two failures to illuminate the MIL. It sets a pending code on the first failure that can be read on most scan tools.

Given the comprehensive nature of these diagnostic tests and readiness monitors, many state I/M programs are implementing a simple readiness code check for OBD II vehicles. If all the monitors have run and passed, the vehicle does not have to be run on a dyno, or have the exhaust emissions checked. The readiness code would reveal any conditions affecting emissions by failing the monitor for a particular system.

Not all vehicles have a service light; all vehicles do have an MIL. The PCM has the option of turning the MIL off when the diagnostic test that caused the MIL to light passes for three consecutive trips. In the case of misfire or fuel trim malfunctions, there are additional requirements, as follows:

- The load conditions must be within 10 percent of the vehicle load present when the diagnostic test reported the malfunction.
- The engine speed must be within 375 rpm of the engine speed present when the DTC was set.
- The engine coolant temperature must have been in the same range present when the PCM turned on the MIL.

When the PCM requests the service light to be turned on, a history DTC is also recorded for the diagnostic test. The provision for clearing a history DTC for any diagnostic test requires 40 subsequent warm-up cycles during which no diagnostic tests have reported a fail, a battery disconnect, or a scan tool clear info command.

Unique to the misfire diagnostic, the PCM has the capability of alerting the driver of potentially damaging levels of misfire. If a misfire condition exists that could potentially damage the catalytic converter, the PCM will command the MIL to flash at a rate of once per second during the times that the catalyst-damaging misfire condition is present. The PCM may also shut down injectors for the misfiring cylinders to prevent raw fuel from entering the converter.

With OBD II, there are multiple oxygen sensors. One sensor is located before the catalyst and detects rich-lean swings in the exhaust stream. Another oxygen sensor is located after the catalytic converter. It also looks at the rich-lean swings of the exhaust stream. If the catalytic converter is working properly, there should be no swings! Large swings mean that the converter is not working properly (**Figure 12-23**). The PCM responds by setting a fault code. If the downstream O_2 sensor signal is the same as the upstream sensor, this means the catalytic converter is not functioning properly.

Each time a fuel trim malfunction is detected, the engine load, engine speed, and engine coolant temperature are recorded. When the ignition is turned off, the last reported set of conditions remains stored. During subsequent ignition cycles, the stored conditions are used as a reference for similar conditions. If a fuel trim malfunction occurs during two consecutive trips, the PCM treats the failure as a regular type B code and turns on the MIL. However, if a fuel trim malfunction occurs on two nonconsecutive trips, the stored conditions are compared with the current conditions. The MIL will illuminate if the engine load, rpm, and coolant temperature are in the same range as the first failure.

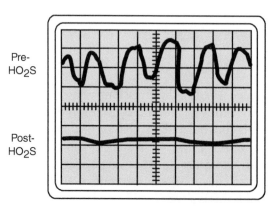

Figure 12-23 O_2 sensor signals from before and after a good converter.

In the case of an intermittent fault, the MIL may illuminate and turn off after three trips. However, the corresponding DTC will be stored in memory. When unexpected DTCs appear, check for an intermittent malfunction (**Photo Sequence 26**).

PHOTO SEQUENCE 26
Comparing O$_2$ Signals

P26-1 A two-channel oscilloscope can read both front and rear O$_2$ sensors simultaneously. The technician will compare the two signals.

P26-2 The technician connects the scope to the O$_2$ sensor using a jumper harness to avoid piercing the wire. Both sensors will be connected in the same way. The red lead of the scope will be connected to the sensor output wire and the black to ground.

P26-3 Turn the key on and look for about 0.5 volt on both channels.

P26-4 The engine is started and allowed to warm up. Take note of the cold O$_2$ sensor activity.

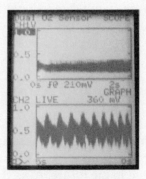

P26-5 After the engine is warmed up, raise the engine speed to 2,000 rpm and watch the patterns.

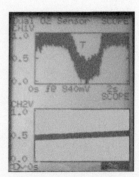

P26-6 Note the difference in the upstream and downstream sensors. Compare the upstream and downstream average voltage.

P26-7 If the throttle is raised quickly, the rear O$_2$ sensor may pick up on the additional fuel flow and show a small bit of activity for a short time.

FREEZE FRAME

Government regulations require that engine operating conditions be captured whenever the MIL is illuminated, or a pending code is recorded. The data captured is called freeze-frame data. Whenever the MIL is illuminated, the corresponding record of operating conditions is recorded to the freeze-frame buffer. Each time a diagnostic test reports a failure, the current engine operating conditions are recorded in the freeze-frame buffer. A subsequent failure will update the recorded operating conditions. The following operating conditions for the diagnostic test that failed typically include the following parameters:

- Air-fuel ratio
- Airflow rate
- Fuel trim
- Engine speed
- Engine load
- Engine coolant temperature
- Vehicle speed
- TP angle
- MAP/BARO
- Injector base pulse width
- Loop status

Freeze-frame data can only be overwritten with data associated with a misfire or fuel trim malfunction. Data from these faults take precedence over data associated with any other fault. The freeze-frame data will not be erased unless the associated history DTC is cleared. Technicians should use freeze frame to help duplicate intermittent conditions.

GM vehicles have the ability to store several "failure records" instead of only one freeze frame.

Classroom Manual
Chapter 12, page 379

DIAGNOSTIC TROUBLE CODES

OBD II standards call for standardized diagnostics and DTCs. The standardized diagnostic codes give somewhat detailed descriptions of the faults detected by the PCM (**Figure 12-24**). In the standardized list, there are gaps in the numbering that will allow for codes to be added in the future. Also, the list of codes includes some equipment not included on all engines. This means some of the codes are not available on some engines. The importance of the standardized codes is simply that every car make and model will use the same code to define a fault in generic OBD II mode. In enhanced mode, the manufacturer decides what the code definition is for a particular fault. P0XXX and P2XXX code are generic codes, and all manufacturers will use the same numbering. A P1XXX code is a manufacturer-specific code that can have different definitions. P3XXX can be either generic or manufacturer specific. Additionally, generic codes apply only to emissions-related faults; other codes, such as transmission codes, may not be available with a generic scan tool.

Classroom Manual
Chapter 12, page 377

All DTCs are displayed as a five-character alphanumeric code in which each character has a specific meaning. The first character is the prefix letter that indicates the area of the fault.

| B = body | C = chassis | P = powertrain | U = network |

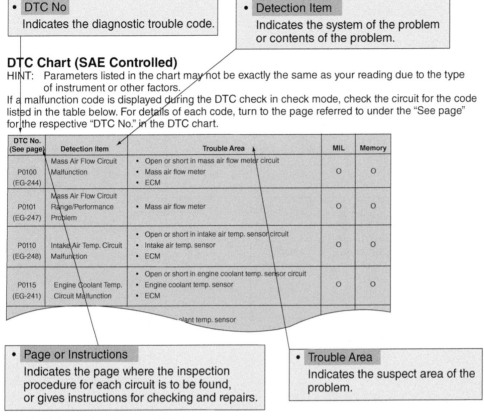

Figure 12-24 Explanation of a typical DTC chart.

The second character is a number that indicates whether the DTC that follows is a standard, common, or manufacturer-specific code. A 0 indicates that it is a generic or common code, and a 1 means the code is manufacturer specific. The latter means the same code may indicate something different on another vehicle make or model.

The third character indicates the subsystem of the area that the fault lies in. Engine performance problems will set a DTC that begins with a P. The third character defines where in the powertrain the fault resides. The following list includes the possible third characters and their associated subgroups:

0 = entire system	5 = idle speed control
1 = air-fuel control	6 = PCM and its inputs and outputs
2 = air-fuel control	7 = transmission
3 = ignition system	8 = non-OBD II system powertrain components
4 = auxiliary emissions controls	

The fourth and fifth characters identify the specific detected fault. These characters not only indicate the component or circuit, but also give a description of the type of fault. As an example of this, consider DTC P0108. This is a generic powertrain DTC. The fault is in the air-fuel system. The last two characters indicate that the condition that triggered the fault is that the input voltage to the MAP sensor is above the

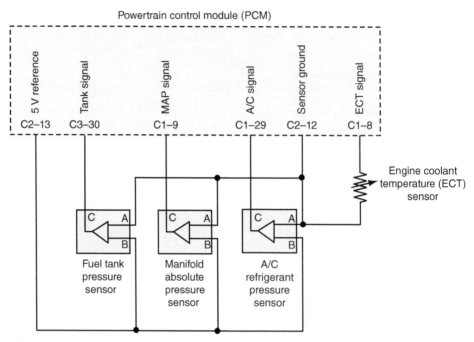

Figure 12-25 Circuit pertaining to PO 108 DTC.

acceptable standard (**Figure 12-25**). With this sort of information, a technician knows what to check (**Figure 12-26**).

OBD II mandates that all DTCs be stored according to a priority. DTCs with a higher priority take precedence over a DTC with a lower priority. The higher priority codes are set the first time the fault occurs and turn on the MIL immediately. These are type A codes. The next level of code priorities are those faults that set a code the first time they are detected, but the MIL is not lit until they occur a second time. These codes are called type B codes. The lower-level codes are faults that relate to non-emissions systems and are called type C and D codes.

DATA LINK CONNECTOR

Classroom Manual
Chapter 12, page 376

It is possible to determine which protocols a vehicle uses by looking at the terminals of the DLC.

Prior to the advent of OBD II, there were several DLCs in various locations inside and under the hood. Each manufacturer also had a specific way of connecting a scan tool. Since 1996, the connectors are all standardized as shown in **Figure 12-27**. Additionally, any scan tool can communicate with any OBD II vehicle in the generic mode (global OBD II). Most of the pin assignments are also mandated, with a few left to the individual manufacturers. This common connector and pin assignment have led to the streamlining of scan tool use. Most scan tools used J11350 or ISO 9141 at the onset of OBD II, but all vehicles made since 2008 use the controller area network (CAN) as the standard communication language (or protocol) to communicate with the scan tool and the emission-related module network. The DLC is also a door for the scan tool to communicate with many other modules in the network (**Figure 12-28**).

DIAGNOSTIC SOFTWARE

The diagnostic management software in the PCM is designed to organize and prioritize all of the monitor tests and their procedures to record and display DTCs and freeze-frame information and control operation of the MIL. All OBD II systems have the same basic

DTC PO108 MAP sensor circuit high voltage

Step	Action	Values	Yes	No
1	Was the Powertrain On-Board Diagnostic (OBD) System Check performed?	–	Go to Step 2	Go to Powertrain OBD System Check 2.4 L Powertrain OBD System Check 2.2 L
2	1. Install a scan tool 2. Engine at idle. Does the scan tool display the MAP voltage specified?	4.0V	Go to Step 3	Go to Step 4
3	1. Turn the ignition switch OFF. 2. Disconnect the MAP sensor electrical connector. 3. Turn the ignition switch ON. Does the scan tool display the MAP voltage specified?	1.0V	Go to Step 5	Go to Step 6
4	1. Turn the ignition switch ON, with the engine OFF, review freeze frame data. 2. Operate the vehicle within the freeze-frame conditions. Does the scan tool display the MAP voltage specified?	4.0V	Go to Step 3	Go to diagnostic aids
5	Probe the MAP sensor signal ground circuit with a test light connected to battery voltage. Does the test light illuminate?	–	Go to Step 7	Go to Step 11
6	Check the MAP sensor signal circuit for a short to voltage and repair as necessary. Was a repair necessary?	–	Go to Step 14	Go to Step 12
7	With a DVM connected to ground, probe the 5-volt reference circuit. Does the DVM display the specified voltage?	5V	Go to Step 8	Go to Step 9
8	Check the MAP sensor vacuum source for being plugged or leaking. Was a problem found?	–	Go to Step 10	Go to Step 13
9	Check the 5-volt reference circuit for a short to voltage and repair as necessary.	–	Go to Step 14	Go to Step 12
10	Repair the vacuum hose as necessary. Is the action complete?	–	Go to Step 14	–
11	Check for an open in the MAP sensor ground circuit and repair as necessary. Was a repair necessary?	–	Go to Step 14	Go to Step 12
12	Replace the PCM. Is the action complete?	–	Go to Step 14	–
13	Replace the MAP sensor. Is the action complete?	–	Go to Step 14	–
14	1. Using the scan tool, clear the DTCs. 2. Start the engine and idle at normal temperature. 3. Operate the vehicle within the conditions for setting the DTC. Does the scan tool indicate that the diagnostic ran and passed?	–	Go to Step 15	Go to Step 2
15	Check for any additional DTCs. Are any DTCs displayed that have not been diagnosed?	–	Go to specific DTC charts	System OK

Figure 12-26 Diagnostic chart for a DTC–PO108.

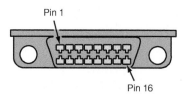

Pin 1	Manufacturer's discretion
Pin 2	J1850 Bus + line on 2 wire systems, or single wire (class 2)
Pin 3	Manufacturer's discretion
Pin 4	Chassis ground pin
Pin 5	Signal ground pin
Pin 6	CAN High
Pin 7	K line for International Standards Organization (ISO) application
Pin 8	Manufacturer's descretion
Pin 9	Manufacturer's desretion
Pin 10	J1850 Bus-line for J1850-2 wire applications
Pin 11	Manufacturer's discretion
Pin 12	Manufacturer's discretion
Pin 13	Signal Ground
Pin 14	CAN low
Pin 15	L line for International Standards Organization (ISO) application
Pin 16	Battery power from vehicle unswitched (4 AMP MAX)

Figure 12-27 The scan tool connects through a common DLC for OBD II.

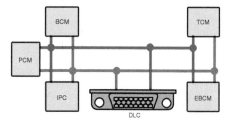

Figure 12-28 Many modules on the network connect to the scan tool through the DLC.

test modes. These test modes are accessible through an OBD II scan tool. The following list of diagnostic modes is available in the generic OBD II selection. If you are using a manufacturer-specific scan tool, such as the GDS 2, Ford IDS, or an aftermarket scan tool such as Snap-on Solus or Modis with "enhanced" (factory) information, you will not see these modes unless the generic or global OBD selection is made from the scan tool menu (**Figure 12-29**). More information is available in the "enhanced" scan tools functions; in generic mode, only the mandated parameters are in the data list, and the only trouble codes that are defined are the POXXX and the P2XXX codes. P1XXX codes are the "manufacturer-specific" codes. These codes will be listed, but no definition will be given. To find the definition of the manufacturer-specific codes, you will have to use a look-up chart or an "enhanced" scan tool. Generic mode is also much slower than the enhanced mode.

Modes of Operation

The PCM is programmed for specific **modes of operation** and communication with the scan tool. **Mode 1** is the **parameter identification (PID) mode**. This mode allows access to data values, input and output information, calculated values, and system status. As stated before, a generic scan tool has access to a smaller number of PIDs than a manufacturer-specific one. Mode 1 is also the location of the monitor **readiness** information. All

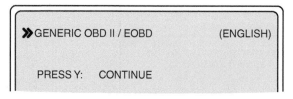

Figure 12-29 OBD II generic function is chosen from the scan tool screen.

READINESS MONITORS

ID: $	
MISFIRE	TEST COMPLETE
FUEL SYSTEM	TEST COMPLETE
COMPONENTS	TEST COMPLETE
HEATED CATALYST	TEST COMPLETE
EVAPORATIVE SYSTEM	TEST COMPLETE
SEC. AIR SYSTEM	TEST COMPLETE
A/C SYSTEM REFRIG.	TEST COMPLETE
O_2 SENSOR	TEST COMPLETE

Figure 12-30 Readiness monitors.

FREEZE FRAME DATA

ID: $	10
P0306 CYLINDER 6 MISFIRE DETECTED	
ENGINE SPEED (1/MIN)	706
ABSOLUTE THROTTLE POSITION (%)	0.4
FUEL SYSTEM 1	OPEN LOOP
FUEL SYSTEM 2	NOT USED
ENGINE COOLANT TEMPERATURE (F°)	165
AIR FLOW RATE (g/s)	5.70
INTAKE MAP (inHg)	13.0
SHORT-TERM FUEL TRIM BANK 1 (%)	–0.7
LONG-TERM FUEL TRIM BANK 1 (%)	3.9
VEHICLE SPEED (mph)	0
CALCULATED LOAD VALUE (%)	3.5

Figure 12-31 Freeze-frame data as displayed on a scan tool.

of the available monitors are listed, as well as their current status. If a particular monitor is not used, it is labeled as "not supported" as shown in **Figure 12-30**. This readiness information is currently used in some state IM240 tests in lieu of a traditional tailpipe testing. As you recall, readiness monitors will not reset until a complete drive cycle has run. If there are problems that prevent a monitor from passing to completion, then the vehicle will fail the IM240 test.

Mode 2 is the **freeze-frame access mode**, which stores specific emissions-related data at the time that a DTC was set (**Figure 12-31**). Having freeze-frame information is a good way for the technician to duplicate the conditions needed to set the code. A technician can also use this to verify a repair, along with some other information we will discuss later in this chapter. Only one freeze frame has been mandated by OBD II, so if additional codes set later, they will not have an associated freeze frame. The only trouble codes that can overwrite a stored freeze frame is a misfire or fuel trim code. For this reason, some manufacturers provide more than one freeze frame that can be accessed by an enhanced scan tool. Also remember that freeze frames are cleared when a code is cleared, so you might want to write down or print the information. Also, freeze frame is not a snapshot; a snapshot is generally a "movie" of information before and after a particular event. Snapshot is a function of the scan tool, not the PCM.

Mode 3 is a request from the scan tool to display DTCs. The information is transmitted from the PCM to the scan tool. The DTC, its descriptive text, or both will be displayed.

Mode 4 is the PCM reset mode that allows the scan tool to clear all emissions-related diagnostic information from its memory. This includes freeze frames and the readiness information. The readiness information will not pass till a drive cycle is completed. Do not send a customer to the IM station immediately after clearing codes without making sure all the monitors have passed.

Mode 5 is the oxygen sensor-monitoring test. This mode contains the information on oxygen sensor testing performed by the PCM. This mode gives the oxygen sensor fault limits and the actual oxygen sensor outputs during the test for all the oxygen sensors on the vehicle. It is important to note that some controller area network systems reports oxygen and AF sensor information in Mode 5.

Mode 6 contains the test results and values for non-continuously monitored systems. These results can be used to see which components failed and may help in the diagnosis of intermittent faults by checking to see if a sensor may be near the limit, and possibly crossing over the line at times, and then going back into specification.

The PCM has modes of operation that have been programmed for specific purposes of communication with the scan tool.

Parameter identification mode allows the technician to read information from the scan tool.

The readiness information shows the technician if a particular monitor has run. If it is unsuccessful, then a pending or current code will set (depending on the area that failed).

Freeze frame is a picture of the data from the PCM when the conditions for setting a code were present. This information allows the technician to operate the vehicle under the same conditions to duplicate the problem.

Test ID	Component ID	Description of Test	Units
$04	$15	Bank one sensor one amplitude and threshold	To obtain voltage, convert hexadecimal number to decimal and then multiply by 0.00098.
$04	$16	Bank two sensor one voltage amplitude and voltage threshold	To obtain voltage, convert hexadecimal number to decimal and then multiply by 0.00098.
Monitor ID	**Test ID**	**Description for CAN**	
$01	$17	HO$_2$S$_{11}$ voltage amplitude and voltage threshold	Some vehicles with CAN may read voltage directly.

Figure 12-32 A very small section of a typical chart to decipher modes 5 and 6 data from the manufacturer. GM, Ford, Toyota, and Honda test mode Information are free on the Internet; many others charge a subscription fee. (*Note:* This is not an actual chart; use the specific chart for your vehicle.)

Modes 5 and 6 Information Use. In the generic mode, there is a wealth of information available if you have the manufacturer-specific information to interpret the various tests. This information is critical to understanding both modes 5 and 6 with a generic scan tool. For example, in **Figure 12-32**, Test ID $04, component $15 for Bank one sensor one heated oxygen sensor could be looked at in the generic function before and after you replaced a sensor to verify the repair. Of course, you could do another drive cycle and look for a pending code, but mode 5 might be faster. If you waited for the MIL to come back on, it would take two complete trips for the oxygen sensor monitor. There are some pros and cons to using this data. It can be difficult and even expensive to obtain. The manufacturer may not give you the conversion factors at the website, so actual values are not obtainable. All manufacturers have their own way of displaying the data, and no two manufacturers are the same (the introduction of CAN should help, because of interest in using mode 6 is increasing). At this time, there is quite a bit of inconsistency between different manufacturers and the usefulness of modes 5 and 6 for a typical technician. Sometimes reading the information with an enhanced scan tool is placed in an easier-to-use format for the technician (especially in a manufacturer-specific tool). The key to a beginning technician is to use the tools available, practice using modes 5 and 6, across different manufacturer's vehicles; if it is useful, you have another tool to help you in diagnosis.

The scientific calculator function on a personal computer can also convert hexadecimal to decimal.

Mode 7 is used by the PCM to store *pending codes*, which are codes that have not yet illuminated the MIL, such as a type B code (two trips) that has failed only once. A technician can use this information to verify repairs by performing a trip for the specific code once instead of twice. Some scan tools on some models of vehicles can be "forced" into performing a monitor without meeting the enable criteria, generally using a manufacturer-specific scan tool.

Mode 8 is used for PCM output control. Using mode 8, a technician can command various actuators to switch on and off. An example would be the EGR, idle air control, transmission solenoids, A/C relay, purge solenoids, vent solenoids, and so on. This output control can be valuable for a technician, enabling checks of various systems without having to wait for conditions to be right to activate them normally, and usually in the shop instead of on a road test.

Mode 9 allows the technician to read the VIN, as well as the calibration number of the PCM directly through the scan tool. The proper VIN is essential, because each vehicle PCM is loaded with the software for its particular option package. Say, for instance, that someone took a PCM from one vehicle and switched it to another vehicle. That vehicle may come into your shop with a "strange" drivability problem, because of an incorrect calibration. The calibration ID is also important because if an updated version is released due to a bulletin or recall, the technician needs to know if the update is necessary.

Mode 10 was brought into existence around the 2010 model year and is listed as permanent codes. Permanent codes are not truly permanent, but they can only be cleared by the PCM. Even if the vehicle passes its readiness monitors, the code will stay in the PCM memory until the computer decides the situation that set the code has been fixed. This prevents someone from "tricking" the PCM into setting the readiness monitors as OK and passing the emission test.

ADAPTIVE FUEL CONTROL STRATEGY

OBD II's **short-term fuel trim (short-term FT)** and **long-term fuel trim (long-term FT)** strategies monitor the oxygen sensor signal. The information gathered by the PCM is used to make adjustments to the fuel control calculations. The adaptive fuel control strategy allows for changes in the amount of fuel delivered to the cylinders according to the operating conditions.

During open loop, the PCM changes pulse width without any feedback from the O_2 sensor, and the short-term adaptive memory value is one. The one represents a 0 percent change. Once the engine warms up, the PCM moves into closed loop and begins to recognize the signals from the O_2. The system remains in closed loop until the engine stops unless the throttle is fully opened or the engine's temperature drops below a specified limit. In both of these cases, the system goes into open loop.

The PCM controls the pulse width of the injectors in response to the amount of oxygen in the exhaust. When the system is in open loop, the injectors operate at a fixed-base pulse width, reinforcing the airflow calculations of a MAF sensor or load calculations of a MAP sensor. During closed-loop operation, the pulse width is either lengthened or shortened. It is positively or negatively corrected to ensure the proper air-fuel mixture for the operating conditions. As the voltage from the oxygen sensor increases in response to a rich mixture, the short-term FT decreases, which means the pulse width is shortened. Decreases in short-term FT are indicated on a scan tool as a percentage. For example, the short-term adaptive value of −25 percent means the pulse width was shortened by 25 percent. A short-term adaptive value of +25 means the pulse width was lengthened by 25 percent.

Once the engine reaches a specified temperature (normally 180°F), the PCM begins to update the long-term FT. The adaptive setting is based on engine speed and the short-term FT. If the short-term FT moves 3 percent and stays there for a period of time, the PCM adjusts the long-term FT. The long-term FT becomes a new but temporary base. In other words, the long-term FT changes the length of the pulse width that is being changed by the short-term FT. The long-term FT works to bring the short-term FT close to 0 percent correction.

If a lean condition exists because of a vacuum leak or restricted fuel injectors, the long-term FT will have a plus (+) number on the scan tool. If the injectors leak or the fuel pressure regulator is faulty, there will be a rich condition. The rich condition will be evident by minus (−) numbers in the long-term adaptive on the scan tool.

Should the exhaust from a vehicle indicate a rich condition, the O_2 sensor signal prompts the computer to subtract fuel. The reverse condition is an engine running lean in which the PCM adds fuel.

The **short-term FT** and **long-term FT** refer to the PCM's temporary or permanent fuel control correction as part of its adaptive memory.

On an OBD II vehicle, zero is the midpoint of the fuel strategy when the computer is in closed loop. Ford's fuel cells are illustrated as a percentage: Numbers without a minus sign indicate fuel is being added, and numbers with a minus sign indicate fuel is being subtracted. The constant change or crossing above and below the zero line indicates proper system operation. If the short-term FT readings are constantly on either side of the zero line, the engine is not operating efficiently.

The GM long-term FT strategy is displayed on the scan tool in the same way as their short-term FT. Long-term FT represents the PCM learning driver habits, engine variables, and road conditions. If the number displayed on the scan tool is above 0 percent, the computer has learned to compensate for a lean exhaust. If the number is below 0 percent, the computer has learned to compensate for a rich exhaust.

The OBD II long-term FT strategy is displayed as a percentage on the scan tool in the same way as the short-term FT. Long-term FT is the computer learning the driver's habits, engine variables, and road conditions. Numbers without a minus sign indicate the computer has compensated for a lean exhaust. Numbers with a minus sign indicate the PCM has compensated for a rich exhaust. As the long-term FT learns to compensate for an exhaust that is rich or lean, the short-term FT returns to a value that crosses the zero reference point. If the engine's condition is too far toward either the lean or rich side, the long-term FT will not compensate and a DTC will be set.

Some intermittent fuel-related drivability problems may be diagnosed by making a recording of the computer's data stream. Before you begin recording with a scan tool, test your understanding of long-term FT and short-term FT.

To diagnose this system, connect the scan tool, start the engine, and pull up the PCM's data stream display. Observe the PID called short-term FT and note the value. With the engine running, pull off a large vacuum hose and watch for the short-term FT value to rise above zero to as high as 33 percent. Now look at the long-term FT value, which should have risen above zero, to learn the engine condition. Be aware that a smaller vacuum leak must be made to watch the short-term FT return to zero as the result of long-term FT compensation. Since these systems use two oxygen sensors, a pair of short-term FT PIDs will appear on the scan tool. Use short-term FT to isolate the side of the engine that is running rich or lean by making two separate recordings. For example, let's say the engine bucks and jerks at highway speeds. After performing a self-test, we know whether the drivability problem did or did not set a fuel-related code. If the fuel injector's wiring or connector opened for a moment causing the engine to run lean, the recording corresponding to that injector bank would reveal the problem.

Classroom Manual
Chapter 12, page 377

OBD II MONITOR TEST RESULTS

All OBD II scan tools include a readiness function showing all of the monitoring sequences on the vehicle and the status of each, complete or incomplete. If vehicle travel time, operating conditions, or other parameters were insufficient for a monitoring sequence to complete a test, the scanner will indicate which monitoring sequence is not yet complete.

The specific set of driving conditions that will set the requirements for all OBD II monitoring sequences follows:

1. Start the engine. Do not turn the engine off for the remainder of the drive cycle.
2. Drive the vehicle for at least 4 minutes.
3. Continue to drive until the engine's temperature reaches 180°F or more.
4. Idle the engine for 45 seconds.
5. Open the throttle to about 25 percent to accelerate from a standstill to 45 mph in about 10 seconds.

6. Drive between 30 and 40 mph with a steady throttle for at least 1 minute.
7. Drive for at least 4 minutes at speeds between 20 and 45 mph. If during this phase you must slow down below 20 mph, repeat this phase of the drive cycle. During this time, do not fully open the throttle.
8. Decelerate and idle for at least 10 seconds.
9. Accelerate to 55 mph with about half throttle.
10. Cruise at a constant speed of 40 to 65 mph for at least 80 seconds with a steady throttle.
11. Decelerate and allow the engine to idle.
12. Using the scan tool, check for diagnostic system readiness and retrieve any DTCs.

OBD II readiness status indicates if the various monitors in the OBD II system have completed or not. This mode does not indicate whether the emissions levels were excessive or not during each monitor.

When most monitor tests are run and a system or component fails a test, a pending code is set. When the fault is detected a second time, a DTC is set and the MIL is lit. The DTC will define the fault but will not give the exact location of the fault. The PCM does not know that. It only knows the description of the problem. A technician must use the available information (DTC and freeze frame) from the scan tool and the DTC charts in the service information to locate the source of the problem.

Once the MIL has been turned on for misfire or a fuel system fault, the vehicle must go through three consecutive drive cycles, including the operating conditions similar to those that were present when the code was set.

It is possible that a DTC for a monitored circuit may not be entered into the PCM memory even though a malfunction has occurred. This may happen when the monitoring criteria has not been met.

> The comprehensive component monitor is used for components that do not have a dedicated monitor. These include the TP sensor, MAP, MAF, IAT, and so on.
>
> **Classroom Manual**
> Chapter 12, page 372

Comprehensive Component Monitor

The comprehensive component monitor is used to continuously monitor components that do not have their own monitor, such as the TP sensor, MAF sensor, MAP sensor, and the like. The PCM's input and output signals are constantly being monitored by the diagnostic management software. These signals are observed to detect opens and shorts in the circuits. Some of the input and output devices are also monitored for their effectiveness and functionality or defects in components. The PCM is programmed to expect certain things to happen when it commands a device to do something. Perhaps the simplest example of this is idle speed. The PCM has control over the engine's idle speed. This is accomplished in many different ways, one of which is through the throttle actuator control. The PCM sends voltage signals to the motor that changes the idle speed. When the PCM commands the motor to extend, the engine's speed should increase. The PCM watches engine speed and continues to adjust the motor until the desired speed is reached. If the PCM commands the motor to extend and the engine's speed does not increase, the PCM knows the motor is not doing what it is supposed to be doing. As a result, the PCM sets a DTC and illuminates the MIL.

> The comprehensive component monitor runs at all times.

INTERMITTENT FAULTS

An intermittent fault is a fault that is not always present. This type of fault may not activate the MIL or cause a DTC to be set. Therefore, intermittent problems can be difficult to diagnose. To use the DTC charts and test procedures in a service information, the fault must be present. Prior to testing for the source of an intermittent problem, make sure the preliminary diagnostic checks are completed first.

⚙ SERVICE TIP On certain vehicles, the oxygen sensor may cool down after only a short period of operation at idle. This will cause the system to go into open loop. To restore closed-loop operation, run the engine at part throttle several minutes and accelerate from idle to part throttle a few times.

By studying the system and the relationship of each component to another, you should be able to create a list of suspect sources of the intermittent problem. To help identify the cause of an intermittent problem, follow these steps:

1. Observe the history DTCs, the DTC modes, pending DTCs, and the freeze-frame data. Operate the vehicle under the same conditions as recorded in the freeze frame.
2. Check for bulletins for similar cases where repair history may be available. Combine your knowledge and understanding of the system with the service information that is available.
3. Evaluate the symptoms and conditions described by the customer.
4. Use a check sheet or other method to identify the circuit or electrical system component that may have the problem.
5. Follow the suggestions for intermittent diagnosis found in service information.
6. Visually inspect the suspected circuit or system.
7. Use the data capturing capabilities of the scan tool.
8. Test the circuit's wiring for shorts, opens, and high resistance. This should be done with a DMM in a typical manner unless instructed differently in the service manual.

Most intermittent problems are caused by faulty electrical connections or wiring. Refer to a wiring diagram for each of the suspected circuits or components. This will help identify all of the connections and components in that circuit. The entire electrical system of the suspected circuit should be carefully and thoroughly inspected. Check for the following problems:

- Burnt or damaged wire insulation
- Damaged terminals at the connectors
- Corrosion at the connectors
- Loose connectors
- Wire terminals loose in the connector
- Disconnected or loose ground wire or strap

To locate the source of the problem, a voltmeter can be connected to the suspected circuit and the wiring harness wiggled (**Figure 12-33**). As a guideline for what voltage should be expected in the circuit, refer to the reference value table in the service manual. If the voltage reading changes with the wiggles, the problem is in that circuit. The vehicle can also be taken for a test drive with the voltmeter connected. If the voltmeter readings become abnormal with the changing operating conditions, the circuit being observed is probably the circuit with the defect.

The vehicle can also be taken for a test drive with the scan tool connected to it.

Scan tools have several features that make intermittent problem identification easier. The scan tool can be used to monitor the activity of an input or circuit while it is being driven. This allows for the observation of that circuit's response to changing operating conditions. The snapshot feature stores engine conditions and operating parameters or when the PCM sets a DTC. If the snapshot can be taken when the intermittent problem is occurring, the problem will be easier to diagnose.

⚠ Caution

If you drive with a scan tool, take someone with you either to operate the tool or to drive. Do not drive and try to operate a scan tool at the same time; in fact, in many states this is against the law!

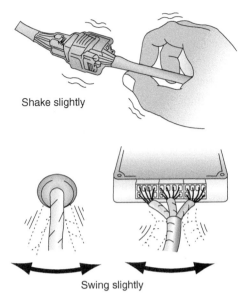

Shake slightly

Swing slightly

Figure 12-33 The wiggle test can be used to locate intermittent problems.

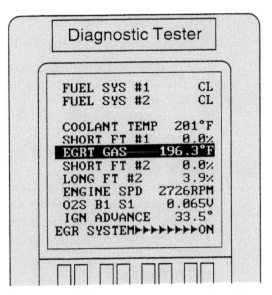

Figure 12-34 OBD II scan tools allow you to observe the results of an active test and to watch the effects of an actuator.

With an OBD II scan tool, actuators can be activated and their functionality tested. The consequences of the change in output operation can be monitored (**Figure 12-34**). Also, the activity of the outputs can be monitored as they respond to changes in sensor signals to the PCM. This can be helpful in locating an intermittent problem.

When an actuator is activated, watch the response on the scan tool. Listen for the clicking of the relay that controls that output. If no clicking is heard, measure the voltage at the relay's control circuit. There should be a change of more than 4 volts when the output is activated.

To monitor how the PCM and an output responds to a change in sensor signal, use the scan tool. Select the mode that relates to the suspected circuit and view and record the scan data for that circuit. Compare the reading to specifications, then create a condition that would cause the related inputs to change. Observe the scan data to see if the change was appropriate.

REPAIRING THE SYSTEM

After isolating the cause, repairs should be made. Then the system should be rechecked to validate proper operation and to verify that the symptom has been corrected. This may involve road testing the vehicle in order to verify that the complaint has been resolved. Test drive the vehicle, and make sure that the monitor runs for the component you replaced without setting a pending code.

Parts Replacement

The components in an OBD II system are designed to perform within specifications for 100,000 miles. The parts in these systems may not be interchangeable with parts from previous systems even though they have the same exterior appearance. For example, if an electronic ignition (EI) module from one ignition system might use a different advance program than a similar part does inaccurate crankshaft position (CKP) sensor signals are sent to the PCM, which activates the engine misfire monitor and illuminates the MIL.

OBD II Circuit Repairs

When servicing or repairing OBD II circuits, the following guidelines are important:

- Do not connect aftermarket accessories into an OBD II circuit.
- Do not move or alter grounds from their manufactured locations.
- Only repair OBD II circuits in accordance with their manufactured configuration.
- Always replace a relay in an OBD II circuit with the same replacement part. Damaged relays should be thrown away, not repaired.
- After repair of connectors or connector terminals, make sure to achieve proper terminal retention.
- Make sure all connector locks are in good condition and are in place.
- Completely clean the contact areas of the ground wire before fastening to the chassis.
- When installing an electrical ground fastener, be sure to apply the specified torque.
- After repair of connectors, make sure to reinstall the connector seals.

HO$_2$S Repair

If the HO$_2$S wiring, connector, or terminal is damaged, the entire oxygen sensor assembly should be replaced. Do not attempt to repair the wiring, connector, or terminals. For this sensor to work properly, it must have a clean air reference. This clean air reference is obtained by way of the oxygen sensor signal and heater wires. Any attempt to repair the wires, connectors, or terminals could result in the obstruction of the air reference and degraded oxygen sensor performance.

The following guidelines should be followed when servicing the heated oxygen sensor:

- Do not apply contact cleaner or other cleaning materials to the sensor or wiring harness connectors. These materials may get into the sensor, causing poor performance.
- The sensor pigtail and harness wires must not be damaged in such a way that the wires inside are exposed. This could provide a path for foreign materials to enter the sensor and cause performance problems.
- Neither the sensor nor the vehicle lead wires should be bent sharply or kinked. Sharp bends, kinks, and so on could block the reference air path through the lead wire.
- Do not remove or defeat the oxygen sensor ground wire. Vehicles that utilize the ground wire sensor may rely on this ground as the only ground contact to the sensor. Removal of the ground wire will cause poor engine performance.
- To prevent damage due to water intrusion, be sure the peripheral seal remains intact on the vehicle harness.

Special Tools

Crimp and seal splice sleeves
Crimping tool
Electrician's tape

Wire Repair

When a damaged wire is found during the visual inspection or if it is the cause of a problem, it should be replaced or repaired. Proper repair to the wires and/or connectors is critical to the proper operation of an OBD II system. Improper repairs can cause excessive resistance immediately after the repair or in the future. Always follow the manufacturer's guidelines for repairing wire and connectors.

Most often, manufacturers recommend that damaged sensor pigtails should be replaced rather than repaired. Since the pigtail comes with the sensor, replacing the pigtail means replacing the sensor. To repair a wire, cut out the damaged part, then strip the insulation off the ends of the wire and splice them together. Normally, splice clips or solder are the recommended ways to join two wire ends together. After the wire has been spliced, tightly wrap electrical tape around the splice. Make sure the tape covers the entire splice. The tape should not be flagged; rather, it should be tightly rolled onto the spliced area (**Figure 12-35**). If the spliced wire does not belong in some conduit or other harness covering, tape the wire again. Wrap the tape to cover the entire first splice (**Figure 12-36**).

> ⚙ SERVICE TIP Never replace a computer until the ground wires and voltage supply wires to the computer are checked and proven to be in satisfactory condition. High resistance in computer ground wires may cause unusual problems. It is also important to make sure that all computer-controlled actuators are at least 20 ohms in resistance (excepting injectors) to prevent excessive current flow through the PCM.

Splice Sleeve Color	Wire Gauge
Salmon	18, 20
Blue	14, 16
Yellow	10, 12

Crimp and seal splice sleeves may be used on all types of insulation, except those designated by the manufacturer, which normally includes coaxial cable used for networking. They are also used where there are special requirements such as moisture sealing. There are different splice sleeves and different sizes of wire. Always use the proper splice sleeve. The splice sleeves are color coded according to applicable wire size:

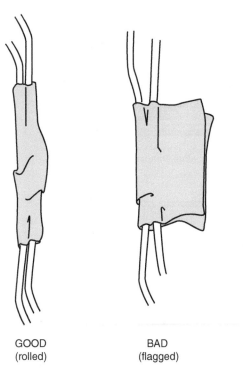

GOOD
(rolled)

BAD
(flagged)

TAPE AGAIN
(if needed)

Figure 12-35 A properly and improperly taped splice.

Figure 12-36 The appearance of the second taping on a splice.

To install a splice sleeve, strip away about 5/16 inch of the insulation on both wire ends. Check the stripped wire for nicks or cut strands. If the wire is damaged, cut the damaged section and restrip. Select the proper splice sleeve according to wire size. Using the splice crimp tool, position the splice sleeve in the proper color nest of the crimp tool (**Figure 12-37**). The nests of the crimp tool are color coded and should be matched to the crimp sleeve color as follows:

Splice Sleeve Color	Crimp Tool Nest Color
Salmon	Red
Blue	Blue
Yellow	Yellow

Place the splice sleeve in the nest so that the crimp falls midway between the end of the barrel and the stop. The sleeve has a stop in the middle of the barrel to prevent the wire from going further (**Figure 12-38**). Lightly squeeze the crimper handles to hold the splice sleeve firmly in the nest. Insert the wire into the sleeve until it hits the barrel stop, and squeeze the crimper handles until the handles open when released. The crimper handles will not open until the proper amount of pressure is applied to the crimp sleeve. Repeat this same procedure to the other wire end.

Apply heat where the barrel is crimped. Gradually move the heat to the open end of the tubing, shrinking the tubing completely as the heat moves along the insulation.

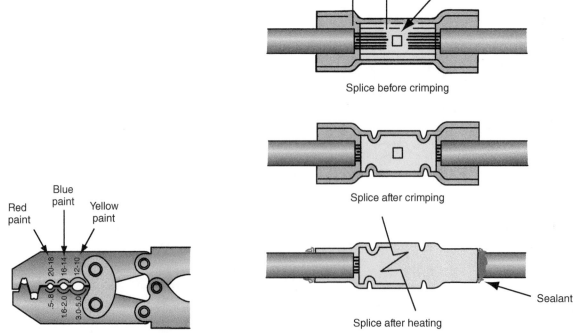

Figure 12-37 A hand crimping tool.

Figure 12-38 Installation sequence for a crimp and seal splice sleeve.

A small amount of sealant will come out of the end of the tubing when sufficient shrinking is achieved.

Connector Repair

Follow these steps to repair push-to-seat (**Figure 12-39**) and pull-to-seat (**Figure 12-40**) connectors:

1. Remove any external locks used to hold the connector's halves together.
2. Remove any external locks used to keep the terminals from backing out of the connector.
3. Open any secondary locks on the connector. Secondary locks aid in the retention of the terminals in the connector, and they are normally molded to the connector shell.
4. Separate the connector halves and back out the seals.
5. Grasp the lead and push the terminal to its forwardmost position. Hold the lead in this position.
6. Locate the terminal lock in the connector canal.
7. Insert a pick directly into the canal.
8. Depress the locking tang to unseat the terminal. On push-to-seat connectors, gently pull on the lead to remove the terminal through the back of the connector. On pull-to-seat connectors, gently push the lead through the front of the connector.

GM refers to these retaining locks as Connector Position Assurance (CPA) locks.

GM refers to these terminal locks as Terminal Position Assurance (TPA) locks.

 WARNING **Never force a terminal out of a connector.**

9. Inspect the terminal and connector for damage or corrosion. Repair or replace the terminal as required.
10. Reform the lock tang and reseat the terminal in the connector. Apply grease to the terminal if it was originally coated with grease.
11. Install the terminal and connector locks, close the secondary locks, and join the connector halves together.

To repair Weather Pack® connectors (**Figure 12-41**):

1. Separate the connector halves.
2. Open the secondary lock.
3. Grasp the lead and push the terminal to its forwardmost position. Hold the lead in that position.

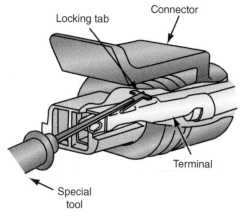

Figure 12-39 Typical push-to-seat connector and terminal.

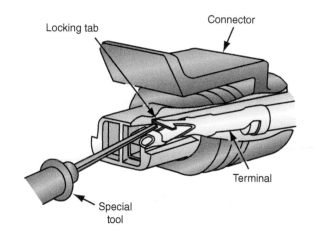

Figure 12-40 Typical pull-to-seat connector and terminal.

1. Open secondary lock on connector.

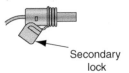

Secondary
lock

2. Remove terminals using special tool.

3. Cut wire immediately behind cable seal.

4. Repair as shown below.

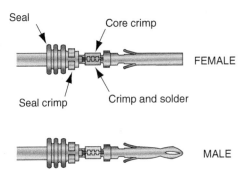

Seal

Core crimp

FEMALE

Seal crimp

Crimp and solder

MALE

Figure 12-41 Typical Weather Pack connector and terminal.

4. Insert the Weather Pack terminal removal tool into the front of the connector cavity until it rests on the cavity shoulder.
5. Gently pull on the lead to remove the terminal through the back of the connector.
6. Inspect the terminal and connector for damage. Repair or replace as required.
7. Reform the lock tang and reseat the terminal in the connector.
8. Close the secondary locks and join the connector halves.

Terminal Repair

The following procedure can be used to repair most types of terminal connectors, including pull-to-seat, push-to-seat, and Weather Pack terminals (**Figure 12-42**).

1. Cut off the terminal between the core and the insulation crimp and remove any seals.
2. Apply the correct seal according to the wire size and slide it far enough on the wire to allow for insulation stripping.
3. Strip the insulation the amount required for the terminal.
4. Align the seal with the end of the wire's insulation.
5. Position the stripped wire into the terminal.
6. Hand crimp the terminal's core wings.
7. Hand crimp the insulation wings. If the terminal is from a Weather Pack connector, make sure the seal is under the insulation wing before crimping.

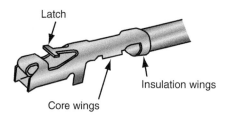

Typical pull-to-seat terminal

Typical push-to-seat terminal

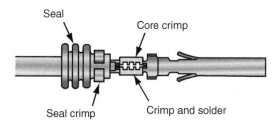

Typical Weather Pack terminal

Figure 12-42 Typical connector terminals.

VERIFYING VEHICLE REPAIR

Verification of repair is more comprehensive for vehicles with OBD II system diagnostics than earlier vehicles. After a repair, a technician should perform the following steps:

1. Review the fail records and/or the freeze-frame data for the DTC that was diagnosed. Record the fail records or freeze-frame data.
2. Clear the DTCs.
3. Operate the vehicle within the conditions noted in the fail records or the freeze-frame data.
4. Monitor the DTC status information for the specific DTC that has been diagnosed until the diagnostic test associated with that DTC runs. Some OEM scan tools can tell the PCM to run a monitor without waiting for enable criteria. Check for pending codes.

Check the PCM memory with the scan tool after driving and ensure that the monitor for the system repaired has been run successfully by the PCM. This means there are no pending codes in the PCM and the vehicle has been repaired.

AUTHOR'S NOTE Any time a vehicle is repaired it is very important to verify that the problem is fixed. If a trouble code is repaired, always drive the vehicle until the monitor for the system has been run and no pending codes have been set in memory.

Clearing DTCs

In order to clear DTCs, use the diagnostic scan tool to clear DTCs or clear info function. When clearing DTCs, follow the instructions supplied by the tool manufacturer.

CASE STUDY

A customer brought in his OBD II–equipped car into the dealership complaining of poor fuel economy. The customer stated that the gas mileage has been declining since the car was new. This is not normal. Gas mileage normally improves slightly as the engine is broken in. The customer had no other complaints. Because this is a very difficult problem to verify, the technician began the diagnostic process with a visual inspection and found nothing out of the ordinary. The car's MIL was not lit.

He then connected a scan tool to the DLC and reviewed the data. Comparing the input data being displayed to the normal range of values listed in the service information, he discovered that the upstream O_2 sensor was biased rich. That meant the PCM was seeing a lean condition and adding fuel to correct for this problem. To verify this, the technician watched the fuel trim. Sure enough, the long-term FT had moved to add more fuel. Normally this is caused by a vacuum leak or restricted fuel injectors. The latter probable cause seemed unlikely since the O_2 sensor showed that added fuel was being delivered. Therefore, the technician began to look for a possible cause of a vacuum leak.

The process continued for quite some time as he checked all of the vacuum hoses and components. Nothing appeared to be leaking. As he leaned over to check something in the back of the engine, he heard a slight "Pffft." The noise had somewhat of a rhythm to it, and he focused his attention to it. It did not sound like a vacuum leak. As he increased the engine's speed, the noise pulses became closer together and soon became a constant noise. The noise appeared to be coming from the lower part of the engine.

Using a stethoscope, he was able to identify the source of the noise. It appeared that the gasket joining the exhaust manifold to the exhaust pipe was not seated properly. To verify this, he raised the car on a hoist and took a look. He found that the retaining bolts were loose. He took a quick look at the gasket and found it to be in reasonable shape, then tightened the bolts to specifications.

Not sure that the noise or the exhaust leak was related to the problem but suspected that it could be, he connected a lab scope to the oxygen sensor and watched its activity. The sensor's signal no longer showed a bias. It appears that the leak in the exhaust was pulling air into the exhaust between each pulse. This was adding oxygen to the exhaust stream causing the computer to think the mixture was lean.

ASE-STYLE REVIEW QUESTIONS

1. While discussing the adaptive learning of a PCM, *Technician A* says that when the battery is disconnected from the vehicle, the learning process resets. *Technician B* says that the vehicle must be driven under part throttle with moderate acceleration for the PCM to relearn the system.
   Who is correct?
   A. A only
   B. B only
   C. Both A and B
   D. Neither A nor B

2. *Technician A* says that P0 codes with definitions can be accessed through a generic scan tool. *Technician B* says that P1 codes with definitions are only available on an enhanced or manufacturer-specific scan tool.

   Who is correct?
   A. A only
   B. B only
   C. Both A and B
   D. Neither A nor B

3. *Technician A* says that if the PCM is correcting for a lean condition, the long-term FT will indicate a negative number. *Technician B* says that a rich condition causes the long-term FT to show a positive number.
   Who is correct?
   A. A only
   B. B only
   C. Both A and B
   D. Neither A nor B

4. While discussing the MIL on OBD II systems,
Technician A says that the MIL will flash if the PCM detects a fault that would damage the catalytic converter.
Technician B says that whenever the PCM has detected a fault, it will set a pending code but not illuminate the MIL on the first failure if it is a type B code.
Who is correct?

A. A only   C. Both A and B
B. B only   D. Neither A nor B

5. While discussing diagnostic procedures,
Technician A says that after preliminary system checks have been made, the DTCs should be cleared from the memory of the PCM.
Technician B says that the data stream can be helpful when there are no DTCs but there is a fault in the system.
Who is correct?

A. A only   C. Both A and B
B. B only   D. Neither A nor B

6. While discussing freeze-frame data,
Technician A says that this feature displays the conditions that existed when the PCM set a DTC.
Technician B says that the freeze-frame feature stores data even when a fault does not turn on the MIL.
Who is correct?

A. A only   C. Both A and B
B. B only   D. Neither A nor B

7. *Technician A* says that the enable criteria must be met before the PCM completes a monitor test.
Technician B says that a drive cycle includes operating the vehicle under specific conditions so that a monitor test can be completed.
Who is correct?

A. A only   C. Both A and B
B. B only   D. Neither A nor B

8. *Technician A* says that freeze frames should always be recorded before erasing a DTC.
Technician B says that the vehicle should be driven in the same conditions indicated on the freeze frame to duplicate the conditions when the code was set to verify the repair.
Who is correct?

A. A only   C. Both A and B
B. B only   D. Neither A nor B

9. *Technician A* says that freeze-frame data is accessed through mode 2 with a generic scan tool.
Technician B says that some manufacturers have more than the mandated one freeze frame.
Who is correct?

A. A only   C. Both A and B
B. B only   D. Neither A nor B

10. While discussing adaptive fuel control,
Technician A says that if there is a lean condition, the short-term FT will show a minus value on the scan tool.
Technician B says that the PCM will extend the injector pulse width if there is excess oxygen in the exhaust.
Who is correct?

A. A only   C. Both A and B
B. B only   D. Neither A nor B

ASE CHALLENGE QUESTIONS

1. A vehicle surges at cruise and has an erratic idle. The first operation the technician should perform after verifying the customer's complaint is to:

A. check service history.
B. check the fuel pressure.
C. perform a visual inspection.
D. disconnect any add-on equipment and repeat the road test.

2. *Technician A* says that mode 5 contains the tests for non-continuously monitored systems.
Technician B says that mode 3 contains the oxygen sensor monitoring tests.
Who is correct?

A. A only   C. Both A and B
B. B only   D. Neither A nor B

3. An HO$_2$S DTC is retrieved and the fuel trim is incorrect.

   Technician A says that this can be the result of a blocked referenced air path.

   Technician B says that the sensor's wiring connectors may have been exposed to an engine cleaning agent.

   Who is correct?

   A. A only

   B. B only

   C. Both A and B

   D. Neither A nor B

4. *Technician A* says that the comprehensive component monitor is used to check devices that do not have their own monitor.

   Technician B says that the comprehensive component monitor also checks the wiring for short and opens.

   Who is correct?

   A. A only

   B. B only

   C. Both A and B

   D. Neither A nor B

5. *Technician A* says that a ground wire fastener must be torqued to specifications.

   Technician B says that the contact areas of the ground wire must be completely cleaned before fastening to the chassis.

   Who is correct?

   A. A only

   B. B only

   C. Both A and B

   D. Neither A nor B

Name _____   Date _____

CONDUCT A DIAGNOSTIC CHECK ON AN ENGINE EQUIPPED WITH OBD II

Upon completion of this job sheet, you should be able to conduct a system inspection and retrieve codes from the PCM of an OBD II–equipped engine.

ASE Education Foundation Correlation

This job sheet addresses the following **MLR** tasks: VIII. Engine Performance;
B. Computerized Controls

Task #1	Retrieve and record diagnostic trouble codes (DTC), OBD monitor status, and freeze frame data; clear codes when applicable. **(P-1)**
Task #2	Describe the use of of OBD monitors for repair verification. **(P-1)**

This job sheet addresses the following **AST** tasks: VIII. Engine Performance;
B. Computerized Controls Diagnosis and Repair

Task #1	Retrieve and record diagnostic trouble codes (DTC), OBD monitor status, and freeze frame data; clear codes when applicable. **(P-1)**
Task #4	Describe the use of OBD monitors for repair verification. **(P-1)**

E. Emissions Control Systems Diagnosis and Repair

Task #8	Interpret diagnostic trouble code and scan tool data related to the emissions control systems; determine needed action. **(P-3)**

This job sheet addresses the following **MAST** tasks: VIII. Engine Performance;
B. Computerized Controls Diagnosis and Repair

Task #1	Retrieve and record diagnostic trouble codes (DTC), OBD monitor status, and freeze frame data; clear codes when applicable. **(P-1)**
Task #4	Describe the use of OBD monitors for repair verification. **(P-1)**
Task #5	Diagnose the causes of emissions or drivability concerns with stored or active diagnostic trouble codes (DTC); obtain, graph, and interpret scan tool data. **(P-1)**

E. Emissions Control Systems Diagnosis and Repair,

Task #7	Interpret diagnostic trouble codes (DTCs) and scan tool data related to the emissions control systems; determine needed action. **(P-1)**

Tools and Materials
- A vehicle equipped with OBD II
- Service information
- Scan tool

Describe the vehicle being worked on:

Year _____ Make _____

Model _____ VIN _____

Engine type and size _____

Procedure   **Task Completed**

1. Check all vehicle grounds, including the battery and computer ground, for clean and tight connections. Comments:

   _____

   _____

2. Perform a voltage drop test across all related ground circuits. State where you tested and what your findings were:

   _____

   _____

3. Check all vacuum lines and hoses, as well as the tightness of all attaching and mounting bolts in the induction system. Comments:

   _____

   _____

4. Check for damaged air ducts. Comments:

   _____

   _____

5. Check the ignition circuit, especially the secondary cables, for signs of deterioration, insulation cracks, corrosion, and looseness. Comments:

   _____

   _____

6. Are there any unusual noises or odors? If there are, describe them and tell what may be causing them.

   _____

   _____

7. Inspect all related wiring and connections at the PCM. Comments:

   _____

   _____

8. Gather all pertinent information about the vehicle and the customer's complaint. This should include detailed information about the symptom from the customer, a review of the vehicle's service history, published TSBs, and the information in the service information.   ☐

9. Are there any vacuum leaks?   ☐ Yes   ☐ No

10. Is the engine's compression normal?   ☐ Yes   ☐ No

11. Is the ignition system operating normally?   ☐ Yes   ☐ No

12. Are there any obvious problems with the air-fuel system?   ☐ Yes   ☐ No

13. Your conclusions from the above:

   _____

   _____

Task Completed

14. Check the operation of the MIL by turning the ignition on. Describe what happened and what this means.

_____

_____

15. Connect the scan tool to the DLC. ☐

16. Enter the vehicle identification information into the scan tool. ☐

17. Retrieve the DTCs with the scan tool. ☐

18. List all codes retrieved by the scan tool.

_____

_____

_____

_____

19. Work through the menu on the scan tool until its screen reads "DATA LIST." Then, monitor the activity of the system and record any unusual activity. Describe your findings here.

_____

_____

_____

_____

20. Review any freeze-frame data that may be related to the DTC or abnormal serial data. Are you able to identify the conditions present when the problem occurs?

_____

_____

_____

_____

_____

21. What are your service recommendations?

_____

_____

_____

_____

_____

22. After all codes have been recorded and repairs are made or planned, the DTCs should be cleared. This is done to verify that the service corrected the problem. Codes can be cleared with the scan tool. Make sure the scan tool is connected to the DLC. What did you find?

_____

_____

_____

_____

_____

23. After the vehicle has been repaired, it is important to drive the vehicle until all the OBD II monitors have been run and completed. List the OBD II monitors that have not been run by the ECM at this time (before driving).

_____

_____

24. Describe the reason that it is important to make sure that all monitors have been run and completed before returning the vehicle to the customer.

_____

_____

_____

_____

_____

Instructor's Response

_____

_____

_____

Name _____   Date _____

MONITOR THE ADAPTIVE FUEL STRATEGY ON AN OBD II–EQUIPPED ENGINE

Upon completion of this job sheet, you should be able to monitor the short-term and long-term fuel strategies of an OBD II system and determine if the trim is indicative of an abnormal condition.

ASE Education Foundation Correlation

This job sheet addresses the following **MAST** tasks: VIII. Engine Performance; B. Computerized Engine Controls Diagnosis and Repair

Task #6   Diagnose emissions or drivability concerns without stored or active diagnostic trouble codes; determine needed action. **(P-1)**

Tools and Materials

- A vehicle equipped with OBD II
- Scan tool

Describe the vehicle being worked on:

Year _____ Make _____

Model _____ VIN _____

Procedure

Task Completed

1. Connect the scan tool.   ☐

2. Start the engine.   ☐

3. Pull up the PCM's data stream display.   ☐

4. Observe the PID called short-term FT and note and record the value.

  _____

5. With the engine running, pull off a large vacuum hose and watch the short-term FT value. What did it do?

  _____

6. Look at the long-term FT value. What did it do?

  _____

7. State your conclusions from this test.

  _____

8. Reconnect the large vacuum hose.   ☐

9. Observe the short-term FT and long-term FT values. Record what happened.

  _____

10. What was happening to the fuel trim, and why did it happen?

  _____

Instructor's Response

_____

_____

_____

CHAPTER 13

DIAGNOSING RELATED SYSTEMS

Upon completion and review of this chapter, you should be able to:

- Diagnose related systems for drivability problems.
- Name the common problems that occur with clutches that can directly affect drivability.
- Describe how to accurately perform a road test analysis.
- Explain when and how pressure tests are performed.
- Recognize conditions that adversely affect the performance of drums, shoes, linings, and related hardware.

- Describe how problems in the hydraulic brake system may seem like a drivability problem.
- Explain the importance of correct wheel alignment angles.
- Describe why certain factors affect tire performance, including inflation pressure, tire rotation, and tread wear.

🔧 Basic Tools

Basic mechanic's tool set

Appropriate service manual

Terms To Know

Clutch slippage
Four-wheel alignment

INTRODUCTION

The purpose of this chapter is to make you aware of the relationship that the engine has with the rest of the vehicle. This understanding is especially important when diagnosing drivability problems. Whenever there is a customer complaint that relates to how the car operates, always check the systems (**Figure 13-1**) that could cause the problem before diving into the engine and its systems. Often, diagnosis of these systems is much easier and less time consuming than those in the engine systems.

There is no attempt made in this chapter to cover all of the possible problems that will affect the way a vehicle operates. What is covered are the basic systems and components that are common causes of drivability problems. Further, the repair techniques for these problems are not covered; those should be covered in courses and textbooks that focus on these systems.

A drivability complaint often centers on asking the vehicle to do something it was not designed to do. Carrying extremely heavy objects will put any vehicle to a test, especially if it is not powered by a larger engine or if the transmission's and final drive gear ratios are not designed for the heavy load. When this is the complaint, the owner should be advised not to use the vehicle with those kinds of loads. Heavy loads will cause poor acceleration, a general lack of power, and poor fuel economy. They can also create a very unsafe condition because the vehicle will have a hard time stopping and will not handle correctly.

Look at the complete vehicle

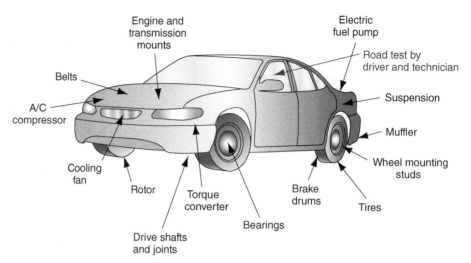

Figure 13-1 Possible sources of drivability symptoms.

CUSTOMER CARE As a technician, it is necessary to become familiar with all aspects and systems of the vehicles being serviced. Always be aware of or inspect for other problems while the vehicle is being serviced. Regardless of the service being performed, take a look at other items such as abnormal tire wear, fluid or oil leaks, and fluid condition. Observe how the engine was running when it was driven into the service bay, or listen for any abnormal brake noise. In other words, just because the vehicle is being serviced for perhaps only an oil and filter change, do not ignore other services that may be necessary even though they are unknown to the customer.

CLUTCHES

Classroom Manual
Chapter 13, page 387

Clutches are used on vehicles with manual transmissions or transaxles. The clutch is used to mechanically connect the engine's flywheel to the transmission or transaxle input shaft (**Figure 13-2**). It does this through the use of a special friction plate that is splined to the input shaft. When the clutch is engaged, the friction plate contacts the flywheel and transfers power through the plate to the input shaft.

A bucking or jerking sensation can be felt if the clutch slips or if its splines to the transmission shaft are worn. Slipping can be caused by a contaminated clutch disc or incorrect clutch pedal free play (**Figure 13-3**). An unbalanced clutch assembly may cause the engine to run rough. This same problem can be caused by a loose flywheel or pressure plate. If the clutch does not release, the engine may be hard to start or perhaps will not crank at all unless the transmission is in neutral.

Slippage

Clutch slippage occurs because of damage, wear, or contamination of the friction material on the clutch. Slippage is usually most evident taking off from a stop, or in higher gears.

Clutch slippage is a condition in which the engine overspeeds without generating any increase in torque to the driving wheels. It occurs when the clutch disc is not gripped firmly between the flywheel and the pressure plate. Instead, the clutch disc slips between these driving members. Slippage can occur during initial acceleration or subsequent shifts but is usually most noticeable in higher gears.

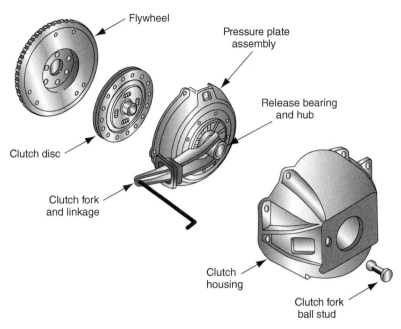

Figure 13-2 Parts of a clutch assembly.

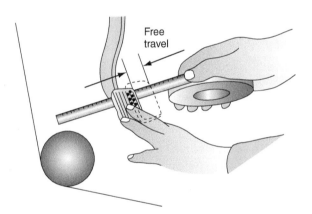

Figure 13-3 Checking clutch pedal free-play.

One way to check for slippage is by driving the vehicle. Normal acceleration from a stop and several gear changes indicate whether the clutch is slipping.

Slippage can also be checked in the shop. Check the service manual for correct procedures. A general procedure for checking clutch slippage follows. Be sure to follow the safety precautions stated earlier.

With the parking brake on, disengage the clutch. Shift the transmission into third gear and increase the engine speed to about 2,000 rpm. Slowly release the clutch pedal until the clutch engages. The engine should stall immediately.

If it does not stall within a few seconds, the clutch is slipping. Safely raise the vehicle, and check the clutch linkage for binding or broken or bent parts. If no linkage problems are found, the transmission and the clutch assembly must be removed so that the clutch parts can be examined.

Clutch slippage can be caused by an oil-soaked or worn disc facing, warped pressure plate, weak diaphragm spring, or the release bearing contacting and applying pressure to the release levers.

Drag and Binding

If the clutch disc is not completely released when the clutch pedal is fully depressed, clutch drag occurs. Clutch drag causes gear clash, especially when shifting into reverse. It can also cause hard starting because the engine attempts to turn the transmission input shaft.

To check for clutch drag, start the engine, depress the clutch pedal completely, and shift the transmission into first gear. Do not release the clutch. Next, shift the transmission into neutral and wait for 5 seconds before attempting to shift smoothly into reverse.

It should take no more than 5 seconds for the clutch disc, input shaft, and transmission gears to come to a complete stop after disengagement. This period, called the clutch spin-down time, is normal and should not be mistaken for clutch drag.

If the shift into reverse causes gear clash, raise the vehicle safely and check the clutch linkage for binding or broken or bent parts. If no problems are found in the linkage, the transmission and clutch assembly must be removed so that the clutch parts can be examined.

Clutch drag can occur as a result of a warped disc or pressure plate, a loose disc facing, a defective release lever, or incorrect clutch pedal adjustment that results in excessive pedal play.

Binding can result when the splines in the clutch disc hub or on the transmission input shaft are damaged or when there are problems with the release levers.

Vibration

Clutch vibrations, unlike pedal pulsations, can be felt throughout the vehicle and can occur at any clutch pedal position. These vibrations usually occur at normal engine operating speeds (more than 1,500 rpm).

There are other possible sources of vibration that should be checked before disassembling the clutch to inspect it. Check the engine mounts and the crankshaft damper pulley. Look for any indication that engine parts are rubbing against the body or frame.

Accessories can also be a source of vibration. To check them, remove the drive belts one at a time. Set the transmission in neutral and securely set the emergency brake. Start the engine and check for vibrations. Do not run the engine for more than 1 minute with the belts removed.

If the source of vibration is not discovered through these checks, the clutch parts should be examined. Be sure to check for loose flywheel bolts, excessive flywheel runout, and pressure plate cover balance problems.

MANUAL TRANSMISSIONS

The purpose of the transmission or transaxle is to use gears of various sizes to give the engine a mechanical advantage over the driving wheels. During normal operating conditions, power from the engine is transferred through the engaged clutch to the input shaft of the transmission or transaxle. Gears in the transmission or transaxle housing alter the torque and speed of this power input before passing it on to other components in the powertrain. Without the mechanical advantage the gearing provides, an engine can generate only limited torque at low speeds. Without sufficient torque, moving a vehicle from a standing start would be impossible.

Classroom Manual
Chapter 13, page 390

Unfortunately, nearly all drivability complaints caused by a manual transmission are actually the fault of the driver. The driver has simply selected the wrong gear for the conditions. Shifting quickly through the gears will cause the engine to seem sluggish. Without the benefit of gear reduction, the engine has a very hard time overcoming a load.

AUTOMATIC TRANSMISSIONS

In the past, all automatic transmissions were controlled by hydraulics. However, systems now feature computer-controlled operation of the torque converter and transmission. Based on input data supplied by electronic sensors and switches, the computer sets the torque converter's operating mode, controls the transmission's shifting sequence (**Figure 13-4**), and, in most cases, regulates transmission oil pressure.

A transmission that shifts too early or too late will affect fuel consumption and overall performance. When the transmission shifts too early, the available power to the wheels decreases, causing poor acceleration. When the gears change later than normal, the customer may complain about noise or may notice that fuel consumption has increased.

Classroom Manual
Chapter 13, page 395

Slippage can normally be traced to a leaking hydraulic circuit, low fluid level, or misadjusted band.

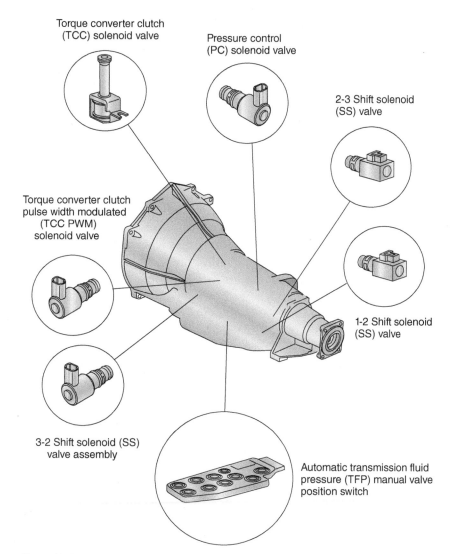

Figure 13-4 An electronic automatic transmission.

| Range | | Gear ratio | Clutch | | Low and reverse brake | Lockup | Band servo | | One way clutch | Parking pawl |
			High-reverse clutch (front)	Forward clutch (rear)			Operation	Release		
Park										X
Reverse		2.364	X		X					
Drive	D1	2.826		X					X	
	D2	1.543		X			X			
	D3	1.000	X	X		X	X	X		
2	2₁	2.865		X					X	
	2₂	1.543		X			X			
1	1₁	2.826		X	X				X	
	1₂	1.543		X			X			

Figure 13-5 A band and clutch application chart can be helpful when diagnosing transmission problems.

⚙️ **SERVICE TIP** A range reference chart is shown in **Figure 13-5**. The chart is useful in diagnosing problems with transmissions based on clutch and band application. For instance, the chart shows the high-reverse clutch on in both reverse and third gear. If this transmission had neither third gear nor reverse, then the high-reverse clutch is probably inoperative. Always use a range reference chart along with the road test to help accurately diagnose transmission problems.

Shift quality is also important to good drivability. If the transmission slips, there will be a lack of power, poor acceleration, and poor fuel economy. Transmission slipping is normally caused by bad bands or clutches. Sometimes when there is slippage, the engine will seem to lurch or buck once the gear is finally fully engaged. There is a power loss and a sudden engagement of the gear that causes a surge of power.

The engine can seem to run rough if the transmission—automatic or manual—mounts are loose or faulty. This is also true of bad engine mounts. The normal vibration of an engine is dampened by the mounts. If the mounts are bad, no or little dampening takes place.

Automatic transmission diagnosis and service is a highly specialized field of automotive work. Entire books cannot cover all the variables offered by the different manufacturers. But half of all automatic transmission problems result from improper fluid levels, fluid leaks, or electrical malfunctions.

Typical automatic transmission diagnostic steps may include the use of manufacturers' diagnostic guides, input from the customer, fluid and linkage checks, electrical checks, stall testing, road testing, and pressure testing.

Diagnostic sheets, such as the one shown in **Figure 13-6**, are extremely helpful in accurately recording all useful information during testing and troubleshooting.

Road testing is exceedingly important when troubleshooting automatic transaxle and transmission problems. The objective of the road test is to confirm the customer's complaint. It is always wise to have the customer accompany the diagnostician on a road test to confirm the malfunction. The road test allows for the opportunity to check the transaxle or transmission operation for slipping, harshness, incorrect upshift speeds, and incorrect downshifts.

Date ____/____/____

TRANSMISSION/TRANSAXLE-CONCERN CHECK SHEET
(*information required for technical assistance)

*VIN _____ * Mileage_____ R.O.# ____

*Model year _____ *Vehicle Model_____ *Engine ____

* Trans. Model _____ *Trans. Serial _____

S
E
R
V
I
C
E

A
D
V
I
S
O
R

CUSTOMER'S CONCERN
_____
_____
_____
_____

Check the items that describe the concern:

WHAT:	WHEN:	OCCURS:	USUALLY NOTICED:
_no power	_vehicle warm	_always	_idling
_shifting	_vehicle cold	_intermittent	_accelerating
_slips	_always	_seldom	_coasting
_noise	_not sure	_first time	_braking
_shudder			at ____MPH

T
E
C
H
N
I
C
I
A
N

Preliminary Check Procedures   NOTE FINDINGS

Inspect
*fluid level and condition   _____
*engine performance-vacuum & EMC codes _____
*TV cable and/or modulator vacuum   _____
*manual linkage adjustment
road test to verify concern   *Simulate the conditions
                                    under which the customer's
                                    concern was observed

* PROPOSED OR COMPLETED REPAIRS
_____
_____
_____
_____

TRANS. TEMPERATURE   ____HOT   ____COLD

*VACUUM READINGS AT MODULATOR
(180C, 250C, 350C, 400, 410-T4)
Readings at Modulator_____IN. HG.(Engine at Hot Idle, Trans. in Drive)
Check for Vacuum Response During Accelerator Movement

P
R
E
S
S
U
R
E

T
E
S
T

P		MINIMUM		MAXIMUM	
R					
N		SPEC.	ACTUAL	SPEC.	ACTUAL
D		(from manual)		(from manual)	
O					
2					
1					

*FINDINGS BASED ON ROAD TEST

R
O
A
D

T
E
S
T

Check items on road test that describe customer's comments about:

Garage Shift	Upshifts	Downshifts	Torque Converter
Feel	_early	_busyness	Clutch
_engine stops	_harsh	_harsh	_busyness
_harsh	_delayed	_delayed	_harsh
_delayed	_slips	_slips	_no release
_no drive	_no upshift	_no downshift	_shudder
			_early apply
			_no apply
			_late apply

Concerns Occur When/During:

Check	Gear	Range	During
		P-N	_1-2 upshift
	1st		_2-3 upshift
	2nd	D	_3-4 upshift
	3rd		_4-3 downshift
	4th		_3-2 downshift
	1st		_2-1 downshift
	2nd	D	
	3rd		
	1st	2	Throttle Position
	2nd		_light
	1st	1	_medium
	Reverse	R	_heavy
			_W.O.T.

Noise

Type	When Noticed
_buzz	_always
_whine	_load sensitive
_clunk	_steering sensitive
	at _____MPH
	in _____gear

Pitch	Level
_low	_light
_medium	_medium
_high	_heavy

Figure 13-6 Typical transmission diagnostic checklist.

Torque Converters

An automatic transmission eliminates the use of a mechanical clutch and shift lever. In place of a clutch, it uses a fluid coupling called a torque converter to transfer power from the engine's flywheel to the transmission input shaft. The torque converter allows for smooth transfer of power at all engine speeds.

A malfunctioning torque converter may provide less torque multiplication and a customer's complaint of a lack of power. An unbalanced or loose torque converter and/or flexplate will cause a vibration, which will make the engine seem to run rough or have a misfire (**Figure 13-7**).

Classroom Manual
Chapter 13, page 392

Lockup Torque Converters

A lockup torque converter eliminates the 10 percent slip that takes place between the impeller and turbine at the coupling stage of operation. The engagement of a clutch between the engine crankshaft and the turbine assembly has the advantage of improving fuel economy and reducing torque converter operational heat and engine speed.

The lockup torque converter clutch assembly is controlled by the PCM. When the computer receives electronic signals from the different sensors confirming the requirements for lockup have been met, lockup clutch engagement begins. These sensors include an engine coolant sensor, vehicle speed sensor, engine vacuum sensor, and throttle position sensor.

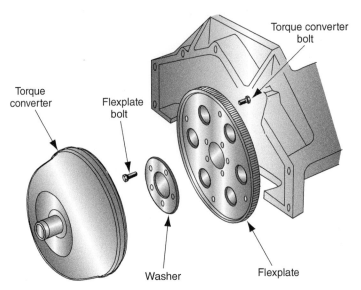

Figure 13-7 The torque converter mounted to a flexplate that is bolted to the engine's crankshaft.

If the converter clutch does not lock during cruising speeds, the customer will notice a decrease in fuel economy. Check the lockup circuit and the converter control circuit to determine why the clutch is not locking.

If the clutch locks too soon or remains locked throughout all engine speeds, many different complaints are possible. There may be a lack of power and poor acceleration. This is because while locked, the torque converter does not provide any torque multiplication. If the clutch is locked while the engine is idling, the engine will stall. If this situation exists, the engine may also stall during deceleration, especially when the wheels stop turning. Since the engine is mechanically connected to the drivetrain when the clutch is locked, if the wheels cannot turn, or if the engine's power is too low to make them turn, the engine will stop running.

If the converter locks too soon, there may be a surge or bucking as the clutch locks. The additional load of the drivetrain on the engine causes it to slow down until it produces enough power to overcome the load.

Diagnosing Transmission Fluid

If the fluid level is low and off the crosshatch section of the dipstick, the problem could be external fluid leaks. Low fluid levels cause a variety of problems. The condition allows air to be drawn into the pump inlet circuit. Aeration of the fluid causes slow pressure buildup and low fluid pressures, contributing to slipping automatic shifts.

High fluid levels cause the rotating planetary gears and parts to churn up the fluid and admit air developing foam. This condition is very similar to low fluid level. Aerated fluid can cause overheating; fluid oxidation, which contributes to varnish buildup; and interference with normal spool valve, clutch, and servo operation.

Uncontaminated automatic transmission fluid (ATF) is pinkish or red in color. A dark brownish or blackish color or a burnt odor indicates overheating. If the fluid has overheated, the ATF and the filter must be changed and the transmission inspected. A milky color can indicate that engine coolant is leaking into the transmission cooler in the radiator outlet tank. Bubbles on the dipstick indicate the presence of air. The bubbles are usually caused by a high-pressure leak.

Wipe the dipstick on absorbent white paper. Look at the fluid stain. Dark particles in the fluid indicate band or clutch material. Silvery metal particles indicate excessive wear on metal parts or housings. Varnish or gum deposits on a dipstick indicate the need to change the ATF and the transmission filter.

Some late-model transmissions are sealed and require special procedures to check the transmission fluid level.

Overheated fluid has been described as smelling like burnt popcorn. It is also very dark and often contains particles of clutch material.

Electronically Shifted Automatic Transmissions

In modern automatic transmissions, the modulator has been replaced by the MAP sensor, the throttle valve cable by the throttle position (TP) sensor, and the governor by the vehicle speed sensor (VSS). Most transmissions have input and output speed sensors that the PCM uses to determine the actual gear ratio and compare it to the desired gear ratio. Some transmissions have line pressure controlled by pulse width modulated valves, and line pressure is determined by the PCM. Since these sensors also govern engine performance, a fault in these sensors can lead to complaints in both the engine and transmission. For instance, a faulty throttle position (TP) sensor can cause erratic torque converter clutch operation, high or low line pressure, erratic or fixed shift points, and improper engine operation, all of which could be interpreted as a major transmission failure (**Figure 13-8**). The technician must also be capable of determining if a transmission fault is electrical or hydraulic. Many transmissions have been replaced or disassembled only to find that the problem was electrical. Solid electrical diagnosis is a must on these systems. Some transmissions can be shifted by the scan tool or special diagnostic tools to aid in determining the fault (see **Photo Sequence 27**). For instance, if the transmission can be externally shifted by the use of a break-out box, then obviously the transmission is not at fault.

Components	Pass thru pins	Resistance at 20°C	Resistance at 100°C	Resistance to ground (case)
1-2 shift solenoid valve	A, E	19–24	24–31	Greater than 250M
2-3 shift solenoid valve	B, E	19–24	24–31	Greater than 250M
TCC solenoid valve	T, E	21–24	26–33	Greater than 250M
TCC PWM solenoid valve	U, E	10–11	13–15	Greater than 250M
3-2 shift solenoid valve assembly	S, E	20–24	29–32	Greater than 250M
Pressure control solenoid valve	C, D	3–5	4–7	Greater than 250M
Transmission fluid temp. (TFT) sensor	M, L	3088–3942	159–198	Greater than 10M
Vehicle speed sensor	A, B Vss conn	1420 @ 25°C	2140 @ 150°C	Greater than 10M

IMPORTANT: The resistance of this device is necessarily temperature dependent and will therefore vary far more than that of any other device. Refer to transmission fluid temp (TFT) sensor specifications.

Figure 13-8 Service manual page showing resistance values of a particular transmission.

P27-1 A vehicle is brought into the shop with a complaint of irregular transmission shifting and the MIL illuminated.

P27-2 After a preliminary quick check (fluid levels and fuses) and a test drive that confirmed the transmission was not shifting, the technician installs a scan tool to read the trouble codes stored.

P27-3 The trouble code is P0751 "Shift Solenoid A Performance" (stuck off).

P27-4 The technician gathers service information on the shift solenoid action and DTC diagnostic information for the vehicle. The A solenoid is normally closed and is opened when activated.

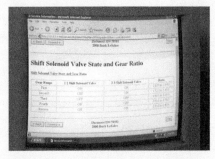

P27-5 The technician finds that according to the chart, if shift solenoid A is not able to turn on, the vehicle would start from first, be unable to shift into second or third, and finally shift into fourth. (The technician should not attempt to achieve fourth during a road test due to engine rpm.)

P27-6 As directed in the diagnostic chart for P0751, the technician moves the shifter through its range D1, D2, D3, D4, N, R, and Park while watching the scan tool "TR Switch" for a matching reading. The gear switch was determined to be OK.

P27-7 The technician raises the drive wheels off the ground and, with the scan tool, commands the first, second, third, and fourth gears. Only first and fourth gears were present regardless of the commanded range.

P27-8 With the engine off and key on, the technician used the scan tool to turn the solenoid on and off while listening with a stethoscope for the clicking sound of the solenoid as it activates. No click is heard.

P27-9 The technician checks for power and ground to the shift solenoid at the harness connector end when commanded by the scan tool. There is power to the connector, but the PCM cannot or does not ground the solenoid when commanded.

P27-10 On close inspection, the technician finds that the terminal for the A solenoid is deformed. The technician replaces the terminal.

P27-11 The technician clears the codes and test drives the vehicle to verify the repair.

Computer and Solenoid Valve Tests

Refer to the manufacturer's service manual procedures for specific transmission computer tests. Be aware that improper test procedures can damage or destroy computer units immediately. Specific tests for input and output signals are made. If input signals are correct and output signals are incorrect, the computer unit must be replaced.

Solenoid valves (**Figure 13-9**) are tested by removing the electrical connector. Then the resistance of the solenoid coil can be checked between the connector and ground. Another way of testing a solenoid is to apply current with a small jumper wire to the disconnected solenoid valve. Correct operation can be heard or felt when the solenoid moves sharply. Refer to the manufacturer's service manuals and service bulletins for specific tests.

The transmission solenoids can often be activated by a scan tool. This allows the technician to check solenoid operation by shifting the transmission with the scan tool.

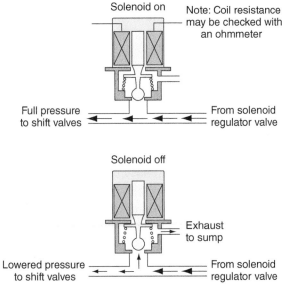

Figure 13-9 A converter clutch solenoid is checked the same way as any other solenoid.

Classroom Manual
Chapter 13, page 403

Drive shafts are used on rear-wheel-drive vehicles and four-wheel-drive vehicles. They connect the output shaft of the transmission with the gearing in the rear axle housing on rear-wheel-drive vehicles. They are also used to connect the output shaft to the front and rear drive axles on a four-wheel-drive vehicle.

A drive shaft consists of a hollow drive or propeller shaft that is connected to the transmission and drive axle differential by universal joints. These U-joints allow the drive shaft to move up and down with the rear suspension, preventing damage to the shaft. Drive shaft problems can set up a vibration that may appear as a rough running engine, especially at cruising or steady vehicle speeds.

Diagnosis of Drive Shaft and U-Joint Problems

A failed U-joint or damaged drive shaft can exhibit a variety of symptoms. A clunk that is heard when the transmission is shifted into gear is the most obvious. You can also encounter unusual noise, roughness, or vibration.

To help differentiate a potential drivetrain problem from other common sources of noise or vibration, it is important to note the speed and driving conditions at which the problem occurs. As a general guide, a worn U-joint is most noticeable during acceleration or deceleration and is less speed sensitive than an unbalanced tire (commonly occurring in the 30 to 60 mph range) or a bad wheel bearing (more noticeable at higher speeds). Unfortunately, it is often very difficult to accurately pinpoint drivetrain problems with only a road test. Therefore, expand the undercar investigation by putting the vehicle up on the lift where it is possible to achieve a good view of what is going on underneath.

Differential

Classroom Manual
Chapter 13, page 391

All vehicles use a differential to provide an additional gear reduction (torque increase) above and beyond what the transmission or transaxle gearing can produce. This is known as the final drive gear.

In a transmission-equipped vehicle, the differential gearing is located in the rear axle housing. In a transaxle, the final reduction is produced by the final drive gears housed in the transaxle case.

The final drive gears consist of the pinion shaft pinion gear and the large differential ring gear. The fact that the driving pinion gear is much smaller than the driven ring gear leaves no doubt that there is substantial gear reduction and torque multiplication in the final drive gears.

Differentials are seldom the cause of drivability complaints unless the owner or previous owner has changed the ratio of the gear-set. The new ratio may not offer the performance or fuel economy that the vehicle was designed to provide. If this is suspected, check the ratio of the final drive. If it is not a standard ratio, advise the customer. Also, check the accuracy of the speedometer. If the gear-set was installed without installing the appropriate speedometer drive gear, the speedometer will read wrong. This alone can make the customer believe the vehicle is using too much fuel or is not accelerating quick enough (depending on the ratio installed). Engineers determine the ideal final drive ratio by considering the engine's power, the desired vehicle speed range, and the typical load on the vehicle.

The differential produces a final gear reduction ratio of around 3:1 in a car or light truck.

A drivability complaint that can be caused by a final drive gear-set is one of poor fuel economy. If the bearings in the final drive or transmission are bad, the shafts and gears will have a difficult time rotating. Extra engine power will be needed to keep them rotating. This wastes fuel.

HEATING AND AIR CONDITIONING

An automotive air-conditioning (A/C) system is a closed pressurized system. It consists of a compressor, condenser, receiver/dryer or accumulator, an expansion valve or orifice tube, and an evaporator. A considerable amount of power from the engine is required to turn the compressor. Power loss when the air conditioner is running is normal. However, too much power loss suggests further diagnosis of the compressor and/or compressor drive (**Figure 13-10**).

Classroom Manual
Chapter 13, page 406

Older vehicles were more prone to idle fluctuation as the A/C compressor and cooling fans operated and was considered normal operation. Although the operation of the A/C is still a heavy load for some engines, modern vehicles generally use "anticipatory" systems. Anticipatory means that since the PCM has control of both the idle speed and the air conditioning it "anticipates" the load of adding air conditioning. The idle can be adjusted and alternator output increased by the PCM to minimize the effect of a cycling air-conditioning compressor before it comes on.

A normal condition may end up as a customer complaint—that is, a reduction in fuel economy. The load of the compressor uses power from the engine. The lost power requires a larger throttle opening to move the vehicle at a desired speed. This complaint is more common during very hot weather and when the vehicle is used in stop-and-go traffic.

Even though using an air conditioner will consume more fuel, it is more efficient than opening the windows. This is due to the added aerodynamic drag on the vehicle.

Engine Cooling Fans

With the advent of transverse-mounted engines, electrical cooling fans found their way under the hood of the modern automobile. Today, most new cars are equipped with an electric cooling fan.

Air-conditioned vehicles, especially those with small engines, usually have additional circuitry to ensure that the cooling fan comes on when the compressor cycles on. Airflow through the condenser must be present for A/C to function correctly. With the condenser mounted in front of the radiator, the logical method of ensuring airflow is to turn the cooling fan on as the A/C compressor cycles on.

On some engines, the turning on of the electric cooling fan will cause the engine to slow down. This is especially true if the air-conditioning compressor was engaged at the same time. This is a normal condition and cannot be corrected.

Most cooling fans do not run above 35 mph.

Airflow is actually reduced with the fan operating above 35 mph.

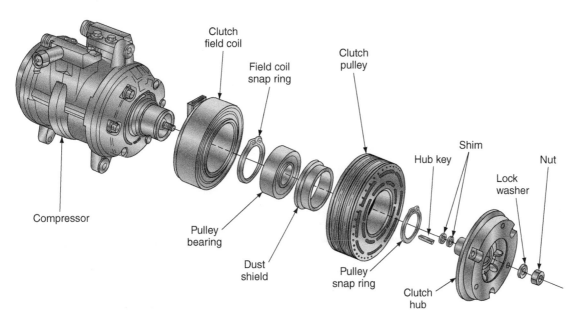

Figure 13-10 The air-conditioning compressor clutch and drive assembly.

Although most newer vehicles have an electric cooling fan, some new vehicles have a cooling fan driven by the engine. These fans can cause a rough running engine if the fan is not balanced properly, if its drive hub is damaged, or if it is loose on its mounting. A vibration may also result from a bad drive belt. Of course, anything that is out of round or imbalanced and rotates with the engine will cause the engine to vibrate or run rough.

CRUISE CONTROL SYSTEMS

Cruise or speed control systems are designed to allow the driver to maintain a constant speed without having to apply continual foot pressure on the accelerator pedal. Selected cruise speeds are easily maintained and can be easily changed. Several override systems also allow the vehicle to be accelerated, slowed, or stopped. Because of the constant changes and improvements in technology, each cruise control system may be considerably different.

Most modern cruise control systems are a function of the throttle actuator control (TAC) system. The same motor that operates the throttle normally can also operate the cruise control. Any problems with the TAC should set diagnostic trouble codes (DTCs) for the condition.

BRAKE SYSTEMS

The *caliper* of a disc brake assembly is a housing containing the pistons and related seals, springs, and boots, as well as the cylinders and fluid passages necessary to force the friction linings or pads against the rotor. The caliper resembles a hand in the way it wraps around the edge of the rotor. It is attached to the steering knuckle. Some models employ light spring pressure to keep the pads close against the rotor. In other caliper designs, this is achieved by a unique type of seal that allows the piston to be pushed out the necessary amount, then retracts it just enough to pull the pad off the rotor (**Figure 13-11**).

Unlike shoes in a drum brake, the pads act perpendicular to the rotation of the disc when the brakes are applied. This effect is different from that produced in a brake drum, where frictional drag actually pulls the shoe into the drum. Disc brakes are said to be non-energized, and so require more force to achieve the same braking effort. For this reason, they are ordinarily used in conjunction with a power brake unit.

Some cruise controls can actually slow the vehicle if a slower moving vehicle is detected ahead.

Classroom Manual
Chapter 13, page 406

Disc brakes are released by the action of the caliper seal. Drum brakes are released by return springs.

Disc brakes require more pressure to apply because the surface area is smaller and disc brakes are not self-energizing.

Classroom Manual
Chapter 13, page 409

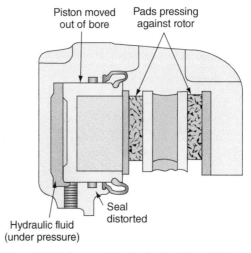

Figure 13-11 Cross section of a brake caliper. Notice the seal around the piston. When it relaxes, it helps pull the piston back into the bore.

Failure of the brakes to release is often caused by a tight or misaligned connection between the power unit and the brake linkage. Broken pistons, diaphragms, bellows, or return springs can also cause this problem.

To help pinpoint the problem, loosen the connection between the master cylinder and the brake booster. If the brakes release, the problem is caused by internal binding in the unit. If the brakes do not release, look for a crimped or restricted brake line or similar problem in the hydraulic system.

Proper adjustment of the master cylinder pushrod is necessary to ensure proper operation of the power brake system (**Figure 13-12**). A pushrod that is too long causes the master cylinder piston to close off the replenishing port, preventing hydraulic pressure from being released and resulting in brake drag. A pushrod that is too short causes excessive brake pedal travel and causes groaning noises to come from the booster when the brakes are applied. A properly adjusted pushrod that remains assembled to the booster with which it was matched during production should not require service adjustment. However, if the booster, master cylinder, or pushrod is replaced, the pushrod might require adjustment.

Inspection and Service

Road testing allows the brake technician to evaluate brake performance under actual driving conditions. Whenever practical, perform the road test before beginning any work on the brake system. In every case, road test the vehicle after any brake work to make sure the brake system is working safely and properly.

A vehicle's brakes are only as good as the tires.

⚡ WARNING **Before test driving any car, first check the fluid level in the master cylinder. Depress the brake pedal to be sure there is adequate pedal reserve. Make a series of low-speed stops to be sure that the brakes are safe for road testing. Always make a preliminary inspection of the brake system in the shop before taking the vehicle on the road.**

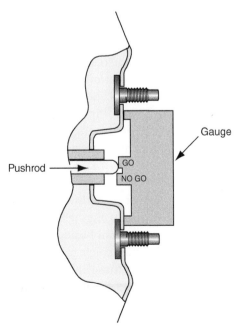

Figure 13-12 A method for correctly adjusting the master cylinder's pushrod.

Brakes should be road tested on a dry, clean, reasonably smooth, and level roadway. A true test of brake performance cannot be made if the roadway is wet, greasy, or covered with loose dirt. Not all tires grip the road equally. Testing is also adversely affected if the roadway is crowned so as to throw the weight of the vehicle toward the wheels on one side or if the roadway is so rough that wheels tend to bounce.

Test brakes at different speeds with both light and heavy pedal pressure. Avoid locking the wheels and sliding the tires on the roadway. There are external conditions that affect brake road test performance. Tires having unequal contact and grip on the road cause unequal braking. Tires must be equally inflated, and the tread pattern of right and left tires must be approximately equal. When the vehicle has unequal loading, the most heavily loaded wheels require more braking power than others. A heavily loaded vehicle requires more braking effort. A loose front-wheel bearing permits the drum and wheel to tilt and have spotty contact with the brake linings, causing erratic brake action. Misalignment of the front end causes the brakes to pull to one side. Also, a loose front-wheel bearing could permit the disc to tilt and have spotty contact with brake shoe linings, causing pulsations when the brakes are applied. Faulty shock absorbers that do not prevent the car from bouncing on quick stops can give the erroneous impression that the brakes are too severe.

Cylinder binding can be caused by rust deposits, swollen cups due to fluid contamination, or by a cup wedged into an excessive piston clearance. If the clearance between the pistons and the bore wall exceeds allowable values, a condition called heel drag might exist. It can result in rapid cup wear and can cause the piston to retract very slowly when the brakes are released. Brake caliper binding can be caused by sticking caliper slides.

Classroom Manual
Chapter 13, page 412

Parking brakes can also be a cause of poor performance and bad fuel economy. If the parking brakes do not release, the rear brakes will drag. Even if the brakes are only slightly dragging, drivability will be affected.

SUSPENSION AND STEERING SYSTEMS

Wheel alignment allows the wheels to roll without scuffing, dragging, or slipping on different types of road conditions. This gives greater safety in driving, easier steering, longer tire life, reduction in fuel consumption, and less strain on the parts that make up the front end of the vehicle.

Proper alignment of both the front and the rear wheels ensures easy steering, comfortable ride, long tire life, and reduced road vibrations.

There is a multitude of angles and specifications that the automotive manufacturers must consider when designing a car. The multiple functions of the suspension system complicate things a great deal for design engineers. They must take into account more than basic geometry. Durability, maintenance, tire wear, available space, and production cost are all critical elements. Most elements contain a degree of compromise in order to satisfy the minimum requirements of each.

Most technicians do not need to be concerned with all of this. All they need to do is restore the vehicle to the condition the design engineer specified. To do this, the technician must be totally familiar with the purpose of basic alignment angles.

The alignment angles are designed in the vehicle to properly locate the vehicle's weight on moving parts and to facilitate steering. If these angles are incorrect, the vehicle is misaligned. Before making adjustments, conduct the following pre-alignment checks.

Begin the alignment with a road test. While driving the car, check to see that the steering wheel is straight. Feel for vibration in the steering wheel as well as in the floor or seats. Notice any pulling or abnormal handling problems such as hard steering, tire squeal while cornering, or mechanical pops or clunks. This helps find problems that must be corrected before proceeding with the alignment.

Conduct a visual inspection. This includes tire wear and mismatched tire sizes or types. Look for the results of collision damage and towing damage. It should also include a ride height measurement. Every car is designed to ride at a specific curb height. Curb height specifications and the specific measuring points are given in service manuals.

With the vehicle raised, inspect all steering components such as control arm bushings, upper strut mounts, pitman arm, idler arm, center link, tie-rod ends, ball joints, and shock absorbers. Check the CV joints (if equipped) for looseness, popping sounds, binding, and broken boots. Damaged components must be repaired before adjusting alignment angles.

Caster is designed to provide steering stability. The caster angle for each wheel on an axle should be equal. Unequal caster angles cause the vehicle to steer toward the side with less caster. Too much negative caster can cause the vehicle to have sensitive steering at high speeds. The vehicle might wander as a result of negative caster. Caster is not a wear angle.

Caster adjustments are not possible on some strut suspension systems. Where they are provided, they can be made at the top mount of the strut assembly.

Camber angle changes, through the travel of the suspension system, are controlled by pivots. Camber is affected by worn or loose ball joints, control arm bushings, and wheel bearings. Anything that changes chassis height also affects camber, such as heavy loading. Camber is adjustable on most vehicles (**Figure 13-13**). Some manufacturers prefer to include a camber adjustment at the spindle assembly. Camber adjustments are also provided on some strut suspension systems at the top mounting of the strut. Very little adjustment of camber (or caster) is required on strut suspensions if the tower and lower control arm positions are in their proper place. If serious camber error has occurred and the suspension mounting positions have not been damaged, it is an indication of bent suspension parts. In this case, diagnostic angle and dimensional checks should be made on the suspension parts. Damaged parts should be replaced.

Incorrect toe increases the rolling resistance of the tires. As the vehicle moves straight, the tires are dragged on their sides with excessive toe in and toe out. If this is suspected, check the tires for feathering.

Ride height is also called the curb or trim height.

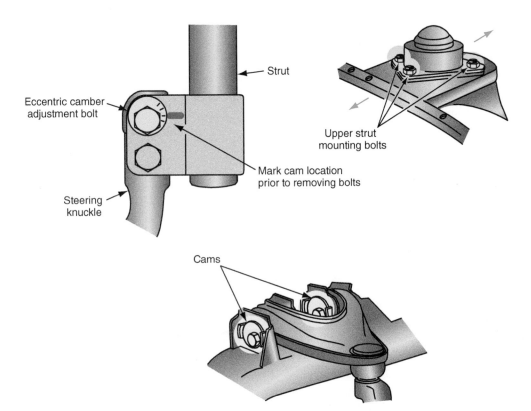

Figure 13-13 Different suspension designs and the various locations and methods for adjusting camber and caster.

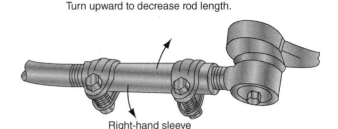

Turn downward to decrease rod length.
Turn upward to increase rod length.

Turn downward to increase rod length.
Turn upward to decrease rod length.

Left-hand sleeve

Right-hand sleeve

Figure 13-14 Toe is adjusted at the tie-rod sleeves.

Toe adjustments are made at the tie rod (**Figure 13-14**). They must be made evenly on both sides of the car. If the toe settings are not equal, the car may tend to pull due to the steering wheel being off-center. This pull condition is especially common with power-assisted rack-and-pinion gears. It can be corrected by making the toe adjustments equal on both sides of the car. Toe is the last adjustment made in an alignment.

If rear toe does not parallel the vehicle centerline, a thrust direction to the left or right is created. This difference of rear toe from the geometric centerline is called the thrust angle. The vehicle tends to travel in the direction of the thrust line rather than straight ahead.

To correct this problem, begin by setting individual rear-wheel toe equally in reference to the geometric centerline. Four-wheel-alignment machines check individual toe on each wheel. Once the rear wheels are in alignment with the geometric centerline, set the individual front toe in reference to the thrust angle. Following this procedure assures that the steering wheel is straight ahead for straight-ahead travel. If you set the front toe to the vehicle geometric centerline ignoring the rear toe angle, a cocked steering wheel results.

Checking the steering axis inclination (SAI) angle can help locate various problems that affect wheel alignment. For example, an SAI angle or SAI angle that varies from side to side may indicate an out-of-position upper strut tower, a bowed lower control arm, or a shifted center cross member.

On a short-long arm suspension, SAI is the angle between true vertical and a line drawn from the upper ball joint through the lower ball joint. In strut-equipped vehicles, this line is drawn through the center of the strut's upper mount down through the center of the lower ball joint.

When the camber angle is added to the SAI angle, the sum of the two is called the included angle. Comparing SAI included and camber angles can also help identify damaged or worn components. For example, if the SAI reading is correct but the camber and included angles are less than specifications, the steering knuckle or strut tower may be bent.

When a car has a steering problem, the first diagnostic check should be a visual inspection of the entire vehicle for anything obvious: bent wheels, misalignment of the cradle, and so on. If there is nothing obviously wrong with the car, make a series of diagnostic checks without disassembling the vehicle. One of the most useful checks that can be made with a minimum of equipment is a jounce-rebound check.

This jounce-rebound check determines if there is misalignment in the rack-and-pinion gear. For a quick check, unlock the steering wheel and see if it moves during the jounce or rebound. For a more careful check, use a pointer and a piece of chalk. Use the chalk to make a reference mark on the tire tread and place the pointer on the same line as the chalk mark. Jounce and rebound the suspension system a few times while someone watches the chalk mark and the pointer. If the mark on the wheel moves unequally in and out on both sides of the car, chances are there is a steering arm or gear out of alignment. If the mark does not move or moves equally in and out on both sides of the car, the steering arm and gear are probably all right. Each wheel or side should be checked.

Like front camber, rear camber affects both tire wear and handling. The ideal situation is to have zero running camber on all four wheels to keep the tread in full contact with the road for optimum traction and handling.

⚠ **Caution**

Never jack up or lift a front-wheel-drive vehicle on its rear axle. The weight of the vehicle may cause the axle to bend and result in misalignment of the rear wheels. Always lift the vehicle at the recommended lifting points.

Camber is not a static angle. It changes as the suspension moves up and down. Camber also changes as the vehicle is loaded and the suspension sags under the weight.

Besides wearing the tires unevenly across the tread, uneven side-to-side camber (as when one wheel leans in and the other does not) creates a steering pull just like it does when the camber readings on the front wheels do not match. It is like leaning on a bicycle. A vehicle always pulls toward a wheel with the most positive camber. If the mismatch is at the rear wheels, the rear axle pulls toward the side with the greatest amount of positive camber. If the rear axle pulls to the right, the front of the car drifts to the left resulting in a steering pull even though the front wheels may be perfectly aligned.

Rear toe, like front toe, is a critical tire rear angle. If toed in or toed out, the rear tires scuff just like the front ones. Either condition can also contribute to steering instability as well as reduced braking effectiveness. (Keep this in mind with antilock brake systems.)

If rear toe is not within specifications, it affects tire wear and steering stability just as much as front toe. A total toe reading that is within specifications does not necessarily mean the wheels are properly aligned—especially when it comes to rear toe measurements. If one rear wheel is toed in while the other is toed out by an equal amount, total toe would be within specifications. However, the vehicle would have a steering pull because the rear wheels would not be parallel to center.

Remember, the ideal situation is to have all four wheels at zero running toe when the car is traveling down the road. This is especially true with antilock brakes where improper toe can affect brakes. Such a condition can affect brake balance when braking on slick or wet surfaces, causing the antilock brakes to cycle on and off to prevent a skid. Without antilock brakes, this condition may upset traction enough to cause an uncontrollable skid.

If rear toe cannot be easily changed, the next best alternative is to align the front wheels to the rear axle thrust line rather than the vehicle centerline. Doing this puts the steering wheel back on center and eliminates the steering pull—but it does not eliminate dog tracking.

Four-Wheel Alignment

The primary objective of **four-wheel alignment** (or total wheel alignment, as it is frequently called), whether front or rear drive, solid axle, or independent rear suspension, is to align all four wheels so the vehicle drives and tracks straight with the steering wheel centered. To accomplish this, the wheels must be parallel to one another and perpendicular to a common centerline.

A **four-wheel alignment** involves adjusting all four wheels instead of only the front two.

Total toe for all four wheels must be determined and rear toe adjusted where possible to bring the rear axle or wheels into square with the chassis. The front toe setting can then be adjusted to compensate for any rear alignment deviation that might persist.

Four-wheel alignment also includes checking and adjusting rear-wheel camber as well as toe and performing all the traditional checks of front camber, toe, caster, toe out on turns, and steering axis inclination. The most important thing a four-wheel alignment job tells the technician is whether or not the rear axle or rear wheels are square with respect to the front wheels and chassis.

WHEELS AND TIRES

The only contact a vehicle has with the road is through its tires and wheels. Tires are filled with air and made of rubber and other materials to give them strength. The air in the tires cushions the ride of the vehicle. Wheels are made of metal and are bolted to the axles or spindles. Wheels hold the tires in place. Wheels and tires come in many different sizes. Their sizes must be matched to each other and to the automobile. Different wheel widths with the same sized tires will cause pulling and will change the circumference of the tires.

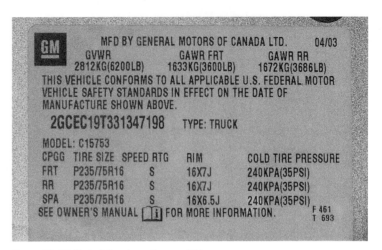

Figure 13-15 Typical tire placard on a vehicle.

Tires should be inflated to the recommended amount of pressure. The ideal amount of pressure is listed on the identification decal of the vehicle (**Figure 13-15**) and is given in the vehicle owner's manual. Underinflated tires have more rolling resistance and, therefore, will cause fuel economy, as well as available power to decrease. Tires that are only a little low will still have a negative effect on fuel economy. Check the pressure in the tires and look at the tires for outside wear. If the tires are severely worn, tell the customer that they should be replaced and explain why.

If replacement tires that are installed on a vehicle are not the same size as original equipment, then speedometer calibration, ABS braking activation, and engine performance may suffer. This happens because the tire diameter affects the overall gear ratio depending on its circumference. The PCM can be recalibrated to correct for the difference in tire size by "re-flashing." This corrects the calculations needed for proper operation.

Classroom Manual
Chapter 13, page 416

All new vehicles are now equipped with tire pressure monitoring systems.

CASE STUDY

A regular customer brings a recently purchased 2008 Mustang GT into the shop. The used car was bought just 2 months ago. The customer's main complaint is that the car gets extremely poor gas mileage. The customer states that the car runs well and has plenty of power, but it seems to use much more fuel than he had anticipated it would.

The technician takes the car out for a test drive. The car runs very well. In fact, the car accelerates much quicker than the technician experienced in similar cars. Knowing the customer, the technician knows the complaint is legitimate. Therefore, she begins to think of the possible causes for poor fuel mileage. Because the engine runs well, it is unlikely that an engine problem is the cause unless the engine has been modified. Another possible cause is a fuel leak. After a detailed visual inspection, the technician finds no evidence of modifications or fuel leaks.

The technician decides to road test the car again. This time she notices a problem with the speedometer. It seems to be reading much higher speeds than the car is actually traveling. To verify this she drives the car onto the freeway and observes the odometer and the mileage markers along the side of the road. The speedometer is in error. It records speeds that are higher than actual speeds and measures a mile well before completing an actual mile.

Odometer error was causing the owner to perceive that the gas mileage was very poor. But why is the odometer in error? Another careful visual inspection is conducted. This time attention is paid to the transmission and the rear axle. Sealing compound is present around the outside of the differential housing. This indicates that the unit has been disassembled. She rotates the rear wheels while observing the drive shaft. Doing so, she is able to closely determine the final drive gear ratio. Comparing her findings to the specifications listed in the service manual, she finds that the gear ratio is much higher numerically than the standard gear ratio. She concludes that the gears had been changed for better performance and that is what caused the error in the speedometer.

ASE-STYLE REVIEW QUESTIONS

1. While discussing manual transmissions,

   Technician A says that the clutch disc is splined to the transmission's input shaft.

   Technician B says that clutch dragging is the result of the clutch fully releasing.

   Who is correct?

   A. A only
   B. B only
   C. Both A and B
   D. Neither A nor B

2. *Technician A* says that if the fluid in an automatic transmission looks milky, it has been contaminated with engine oil.

   Technician B says that overheated fluid smells like burnt popcorn.

   Who is correct?

   A. A only
   B. B only
   C. Both A and B
   D. Neither A nor B

3. *Technician A* says that the drive shaft consists of a hollow drive shaft connected to the transmission and differential by U-joints.

   Technician B says that failed U-joints can be diagnosed by a clunk when putting the vehicle into gear.

   Who is correct?

   A. A only
   B. B only
   C. Both A and B
   D. Neither A nor B

4. *Technician A* says that the differential provides a final gear reduction (torque increase) above what the transmission produces.

   Technician B says that the the differential produces an overdrive gear ratio to help save fuel.

   Who is correct?

   A. A only
   B. B only
   C. Both A and B
   D. Neither A nor B

5. The air-conditioning system can cause a heavy load on a small engine.

   Technician A says that late-model systems are "anticipatory" and set idle speed up just before the A/C system engages.

   Technician B advises his customers to "roll the windows down" to save fuel while driving on the highway.

   Who is correct?

   A. A only
   B. B only
   C. Both A and B
   D. Neither A nor B

6. *Technician A* says that the disc brake caliper contains the pistons and related seals.

   Technician B says that disc brakes take less force to stop the vehicle than drum brakes.

   Who is correct?

   A. A only
   B. B only
   C. Both A and B
   D. Neither A nor B

7. While discussing vehicle brakes,

   Technician A says that a master cylinder pushrod that is too long can cause the brakes to drag.

   Technician B says that a master cylinder pushrod that is too short can cause extended pedal travel.

   Who is correct?

   A. A only
   B. B only
   C. Both A and B
   D. Neither A nor B

8. *Technician A* says that too much positive caster on the front wheels can cause tire wear.

   Technician B says that camber is the difference of the rear toe from centerline.

   Who is correct?

   A. A only
   B. B only
   C. Both A and B
   D. Neither A nor B

9. *Technician A* says that automatic transmissions have input and output shaft speed sensors.

   Technician B says that the PCM can determine the actual gear ratio to the desired gear ratio of the transmission by using these sensors.

   Who is correct?

   A. A only
   B. B only
   C. Both A and B
   D. Neither A nor B

10. *Technician A* says that the some power loss is normal when the A/C is operating.

    Technician B says that too much power loss suggests that further diagnosis is necessary.

    Who is correct?

    A. A only
    B. B only
    C. Both A and B
    D. Neither A nor B

ASE CHALLENGE QUESTIONS

1. *Technician A* says that improper tire sizes can cause the speedometer to be inaccurate.
   Technician B says that the PCM can be recalibrated to correct this condition.
   Who is correct?
   A. A only
   B. B only
   C. Both A and B
   D. Neither A nor B

2. While discussing a vibration problem,
   Technician A says as a general rule, a U-joint problem is most noticeable in the 30 to 60 mph range.
   Technician B says that tire balance problems are most noticeable on acceleration and deceleration.
   Who is correct?
   A. A only
   B. B only
   C. Both A and B
   D. Neither A nor B

3. *Technician A* says that a four-wheel alignment tells the technician whether the rear axle or wheels are square with the front wheels.
   Technician B says that total toe for all four wheels must be determined and rear toe adjusted where possible to bring the rear axle or wheels into square with the chassis.
   Who is correct?
   A. A only
   B. B only
   C. Both A and B
   D. Neither A nor B

4. While discussing electronically controlled transmissions,
   Technician A says that line pressure is controlled by a pulse width modulated solenoid.
   Technician B says that the solenoid is controlled by the PCM.
   Who is correct?
   A. A only
   B. B only
   C. Both A and B
   D. Neither A nor B

5. While discussing wheel alignment angles,
   Technician A says that camber angle changes while driving.
   Technician B says that the camber angle also changes when the vehicle is loaded and sags under the weight.
   Who is correct?
   A. A only
   B. B only
   C. Both A and B
   D. Neither A nor B

Name _____   Date _____

ROAD TEST A VEHICLE TO CHECK THE OPERATION OF THE AUTOMATIC TRANSMISSION

Upon completion of this job sheet, you should be able to determine if the cause of a drivability complaint is the transmission or torque converter.

ASE Education Foundation Correlation

This job sheet addresses the following **AST/MAST** task: VIII. Engine Performance; B. Computerized Controls Diagnosis and Repair

Task #2   Access and use service information to perform step-by-step (troubleshooting) diagnosis. **(P-1)**

Tools and Materials

- A vehicle with an automatic transmission
- Service manual
- Clean shop rag
- Pad of paper and pencil

Describe the vehicle being worked on:

Year _____ Make _____

Model _____ VIN _____

Model and type of transmission _____

Procedure

Task Completed

1. Park the vehicle on a level surface.   ☐

2. Wipe all dirt off of the protective disc and the dipstick handle (if applicable).   ☐

3. Start the engine and allow it to reach operating temperature.   ☐

4. Remove the dipstick and wipe it clean with a lint-free cloth or paper towel.   ☐

5. Reinsert the dipstick, remove it again, and note and record the reading.

   _____

6. Describe the condition of the fluid (color, condition, and smell):

   _____

7. What is indicated by the fluid's condition?

   _____

   _____

8. Find and duplicate from a service manual the chart that shows the band and clutch   ☐
   application for different gear selector positions. Using these charts will greatly simplify
   your diagnosis of automatic transmission problems. It is also wise to have a notebook
   or piece of paper to jot down notes about the operation of the transmission.

9. Inspect the transmission for signs of fluid leakage. Comments:

   _____

   _____

10. Drive the vehicle at normal speeds to warm the engine and transmission. Describe the behavior of the transmission and torque converter.

_____

_____

11. Place the shift selector into the drive position and allow the transmission to shift through all of its normal shifts. Describe the operation of the transmission and torque converter.

_____

_____

12. Check for proper operation in all forward ranges, especially the 1-2, 2-3, 3-4 upshifts and converter lockup during light throttle operation. Describe the operation.

_____

_____

13. Force the transmission to "kickdown," and record the quality of this shift and the speed at which it downshifts.

_____

_____

14. Manually cause the transmission to downshift. How did it react?

_____

_____

15. Record any vibrations or noises that occur during the test drive.

_____

_____

16. Park the vehicle and move the shifter into each gear range. Pay attention to shift quality as each range (including park) is selected. Record the results:

_____

_____

17. Compare your notes with the specifications and shifting chart. What are your conclusions about the transmission and torque converter? Could the transmission and/or torque converter be the cause of a drivability complaint?

_____

_____

_____

Instructor's Response

_____

_____

_____

Name _____   Date _____

INSPECTING DRIVE BELT

Upon completion of this job sheet, you should be able to visually inspect a serpentine drive belt and check its tightness.

ASE Education Foundation Correlation

This job sheet addresses the following **AST/MAST** task: VIII. Engine Performance; A. General: Engine Diagnosis

Task #3   Diagnose abnormal engine noises or vibration concerns; determine needed action. **(P-3)**

Tools and Materials

- Vehicle with a serpentine belt
- Service information for the above vehicle

Describe the vehicle being worked on:

Year _____ Make _____

Model _____ VIN _____

Procedure

1. Carefully inspect the belt and describe the general condition.

   _____

   _____

   _____

   _____

2. Check the tension of the belt. Belt tension and wear are generally assessed by looking at a wear indicator on the belt tensioner. (See Service Information.)

   You found _____

3. Based on the above, what are your recommendations?

   _____

   _____

4. Describe the procedure for checking or adjusting the tension of the belt.

   _____

   _____

   _____

   _____

   _____

   _____

   _____

   _____

5. Based on the above, what are your recommendations?

_____

_____

Instructor's Response

_____

_____

1. *Technician A* says that hydrocarbon (HC) emissions can be caused by a rich mixture. *Technician B* says that carbon monoxide levels are highest when the mixture is lean. Who is correct?
   A. A only
   B. B only
   C. Both A and B
   D. Neither A nor B

2. *Technician A* says that engine detonation may be caused by low octane fuel. *Technician B* says that detonation can be caused by engine overheating. Who is correct?
   A. A only
   B. B only
   C. Both A and B
   D. Neither A nor B

3. While discussing the PCV system, *Technician A* says that a PCV valve that is stuck open can cause rough idle and stalling. *Technician B* says that the PCV valve relieves crankcase pressure caused by piston ring blow by. Who is correct?
   A. A only
   B. B only
   C. Both A and B
   D. Neither A nor B

4. An EI-equipped vehicle will not start. *Technician A* says to check for correct fuel pressure. *Technician B* says to check for spark at more than one plug wire during diagnosis. Who is correct?
   A. A only
   B. B only
   C. Both A and B
   D. Neither A nor B

5. While discussing electronic fuel injection systems, *Technician A* says that a hard-to-start engine may have a leaking one-way check valve in the fuel pump. *Technician B* says that a hesitation when accelerating from idle may be caused by dirty injectors. Who is correct?
   A. A only
   B. B only
   C. Both A and B
   D. Neither A nor B

6. *Technician A* says that OBDII AIR systems do not supply air to the catalytic converter. *Technician B* says that OBD II secondary air systems are all belt driven. Who is correct?
   A. A only
   B. B only
   C. Both A and B
   D. Neither A nor B

7. A vehicle has a consistent slow- or no-crank condition. This may be caused by.
   A. a slipping AC generator drive belt.
   B. a defective AC generator.
   C. a malfunctioning voltage regulator.
   D. any of the above.

8. A fuel-injected vehicle comes in with a miss on acceleration. *Technician A* says that there could be an ignition problem. *Technician B* says that the intake to MAF duct could be opening up at a tear on acceleration. Who is correct?
   A. A only
   B. B only
   C. Both A and B
   D. Neither A nor B

9. *Technician A* says that static electricity generated by clothing in contact with vehicle upholstery must be discharged before beginning work on electronic devices. *Technician B* says that the best control for static electricity is the wearing of a waist or wrist grounding strap. Who is correct?
   A. A only
   B. B only
   C. Both A and B
   D. Neither A nor B

10. *Technician A* says that a camshaft installed one tooth retarded can be compensated by adjusting the ignition's base timing. *Technician B* says to use the starter to crank the engine with the timing belt removed to determine if the engine is interference type or free-wheeling. Who is correct?
    A. A only
    B. B only
    C. Both A and B
    D. Neither A nor B

11. A fuel-injected engine runs rough and sluggish on acceleration. When checking the fuel pressure, the technician notices that the fuel pressure is at specifications during idle, but falls off on acceleration. *Technician A* says that the fuel pump could be faulty. *Technician B* says that the fuel filter could be clogged. Who is correct?

    A. A only
    B. B only
    C. Both A and B
    D. Neither A nor B

12. Cylinder leakage testing is being discussed. *Technician A* says that air escaping through the PCV system indicates leaking exhaust valve. *Technician B* says that bubbles in the cooling systems during this test indicates a blown head gasket between the tested cylinder and an adjacent cylinder. Who is correct?

    A. A only
    B. B only
    C. Both A and B
    D. Neither A nor B

13. A test light connected between the ignition coil's positive terminal and ground illuminates but does not flash during cranking. This condition is.

    A. caused by a grounded positive contact on the ignition module.
    B. a normal condition.
    C. indicative of a grounded ignition pick-up coil.
    D. caused by an open coil primary winding.

14. While diagnosing a poorly performing engine with no stored DTCs, *Technician A* says that the first diagnostic step is to follow the manufacturer's "diagnose by symptom" routine. *Technician B* says that gas analyzers and scopes can be the best diagnostic method in this instance. Who is correct?

    A. A only
    B. B only
    C. Both A and B
    D. Neither A nor B

15. *Technician A* says that the installation of a magnetic crankshaft position sensor may require the use of a special tool to set the air gap. *Technician B* says that some crankshaft sensors are not adjustable. Who is correct?

    A. A only
    B. B only
    C. Both A and B
    D. Neither A nor B

16. While discussing the AF ratio sensor, *Technician A* says that the AF ratio sensor is more accurate than the conventional oxygen sensor. *Technician B* says that the AF ratio sensor does not use a heater. Who is correct?

    A. A only
    B. B only
    C. Both A and B
    D. Neither A nor B

17. While discussing the catalytic converter, *Technician A* says that the converter oxidizes HC and CO. *Technician B* says that it also oxidizes NO_x. Who is correct?

    A. A only
    B. B only
    C. Both A and B
    D. Neither A nor B

18. A vehicle seems to have a key-off battery drain. *Technician A* says that a test lamp connected in series with the battery negative terminal can be used to determine if there is a draw. *Technician B* says that an ammeter connected in the same manner will also determine if the draw is excessive. Who is correct?

    A. A only
    B. B only
    C. Both A and B
    D. Neither A nor B

19. A vehicle has high NO_x emissions. *Technician A* says that this could be caused by carbon accumulation in the combustion chambers. *Technician B* says that this could be caused by a thermostat that is stuck open. Who is correct?

    A. A only
    B. B only
    C. Both A and B
    D. Neither A nor B

20. While discussing the spray pattern of a fuel injector, *Technician A* says that an open or empty space in the cone indicates a dirty injector. *Technician B* says that it is normal to have small fuel drips in the center of the cone during injector firing. Who is correct?

    A. A only
    B. B only
    C. Both A and B
    D. Neither A nor B

21. *Technician A* says that a low steady engine vacuum reading may indicate late ignition timing. *Technician B* says that a burned exhaust valve will cause a low steady vacuum reading. Who is correct?

    A. A only
    B. B only
    C. Both A and B
    D. Neither A nor B

22. A cylinder power balance test is being performed. *Technician A* says that if there is no rpm drop on any cylinder, then each cylinder is producing the same amount of power. *Technician B* says that a cylinder showing a larger rpm drop than the others means less power is being produced by that cylinder. Who is correct?

    A. A only
    B. B only
    C. Both A and B
    D. Neither A nor B

23. While discussing EI ignition systems, *Technician A* says that waste spark systems do not need a cam sensor to operate the ignition system. *Technician B* says that coil-over-plug systems do not need a cam sensor to operate the ignition system. Who is correct?

    A. A only
    B. B only
    C. Both A and B
    D. Neither A nor B

24. *Technician A* says that formed plastic fuel lines may be replaced with flexible fuel line and clamps. *Technician B* says that a fuel tank vent valve is part of the EVAP system. Who is correct?

    A. A only
    B. B only
    C. Both A and B
    D. Neither A nor B

25. While discussing turbochargers, *Technician A* says that an exhaust leak between the converter and muffler may cause decreased boost pressure. *Technician B* says that a binding waste-gate linkage may produce higher-than-specified boost pressure. Who is correct?

    A. A only
    B. B only
    C. Both A and B
    D. Neither A nor B

26. An EGR system is being tested. *Technician A* says that using a scan tool to command the EGR valve open and noting engine operation is a quick method to confirm or eliminate the valve as the malfunctioning component. *Technician B* says that this method can be used to test a linear EGR valve. Who is correct?

    A. A only
    B. B only
    C. Both A and B
    D. Neither A nor B

27. While discussing the secondary air injection, *Technician A* says that air directed to the exhaust manifold during idling at normal operating temperatures may be a normal condition. *Technician B* says that air directed to the manifold during operation at normal temperatures will cause the O_2 sensor voltage signals to be low. Who is correct?

    A. A only
    B. B only
    C. Both A and B
    D. Neither A nor B

28. The starter motor drags during cranking. *Technician A* says that corrosion on the battery cables may cause this problem. *Technician B* says that low voltage in the ignition switch to starter relay (solenoid) circuit can be the cause. Who is correct?

    A. A only
    B. B only
    C. Both A and B
    D. Neither A nor B

29. A vehicle comes into the shop with a code set for a large evaporative emission system leak. *Technician A* says that the first thing to check for is a loose or missing fuel cap. *Technician B* says that the next step is to do a visual inspection of the evaporative emission hoses. Who is correct?

    A. A only
    B. B only
    C. Both A and B
    D. Neither A nor B

30. While discussing OBD II systems, *Technician A* says that not all emissions-related trouble codes set on the first failure. *Technician B* says that these codes are called "B" type codes. Who is correct?

    A. A only
    B. B only
    C. Both A and B
    D. Neither A nor B

31. *Technician A* says that the actual catalyst efficiency monitor is the O_2 sensor at the outlet of the catalytic converter.
    Technician B says that a vacuum gauge can be used to check for a suspected clogged converter.
    Who is correct?
    A. A only
    B. B only
    C. Both A and B
    D. Neither A nor B

32. *Technician A* says that lean mixtures cause NO_x emissions to increase.
    Technician B says that rich mixtures cause high NO_x emissions.
    Who is correct?
    A. A only
    B. B only
    C. Both A and B
    D. Neither A nor B

33. A technician investigating a drivability complaint sees that long-term fuel trims are showing positive 20 percent.
    Technician A says that an excessively restricted air filter may cause this condition.
    Technician B says that a vacuum leak may be the cause.
    Who is correct?
    A. A only
    B. B only
    C. Both A and B
    D. Neither A nor B

34. While discussing the TP sensor,
    Technician A says that the voltage level increases as the throttle opens.
    Technician B says that the ECM also looks at the speed of the throttle opening for fuel delivery calculations.
    Who is correct?
    A. A only
    B. B only
    C. Both A and B
    D. Neither A nor B

35. A fully electronic EGR system is not functioning properly. The *least likely* cause is.
    A. that the PCM is not cycling the solenoids.
    B. a loss of vacuum at the valve's port.
    C. low voltage to the valve.
    D. that there is no reference voltage to the EVP sensor.

36. *Technician A* says that a type A code turns on the MIL with the first fault.
    Technician B says that a flashing MIL indicates a catalyst damaging misfire.

Who is correct?
    A. A only
    B. B only
    C. Both A and B
    D. Neither A nor B

37. A digital auto-ranging multimeter connected between a power feed conductor and ground is registering up and down the mVDC scale.
    Technician A says five-gas this could indicate that there is no voltage on that conductor.
    Technician B says five-gas this could indicate a ground in the conductor prior to the meter connection.
    Who is correct?
    A. A only
    B. B only
    C. Both A and B
    D. Neither A nor B

38. While discussing computer systems grounds, *Technician A* says to use an ohmmeter to check the resistance to ground.
    Technician B says that it is better to check for a voltage drop to ground with the system operating.
    Who is correct?
    A. A only
    B. B only
    C. Both A and B
    D. Neither A nor B

39. No visible leaks are found on an engine that requires periodic topping off of the cooling system.
    Technician A says that a stuck open vacuum valve in the radiator cap may be the cause.
    Technician B says that an infrared gas analyzer can be used to check for combustion chamber to cooling system leaks.
    Who is correct?
    A. A only
    B. B only
    C. Both A and B
    D. Neither A nor B

40. While discussing oxygen sensors,
    Technician A says that the O_2 sensor can be killed by coolant entering the exhaust.
    Technician B says that the oxygen sensor can be checked with an oscilloscope.
    Who is correct?
    A. A only
    B. B only
    C. Both A and B
    D. Neither A nor B

41. The AC generator output measured at the battery is 11 VDC at 5 amps. The output is over 13 VDC when measured at the AC generator output terminal.

Technician A says that a bad wire to the terminal connection may be the cause.
Technician B says that a faulty voltage regulator could be the cause.
Who is correct?

A. A only
B. B only
C. Both A and B
D. Neither A nor B

42. Below is the rough data obtained from an exhaust test using a five-gas analyzer.

Emission	Idle result	Cruise result
O_2	Low	Low
CO_2	Low	Low
CO	High	Higher
HC	High	Higher
NO_x	Low	Low

Technician A says that a leaking injector can cause these readings.
Technician B says that high fuel pressure could be the cause.
Who is correct?

A. A only
B. B only
C. Both A and B
D. Neither A nor B

43. *Technician A* says that most vehicles made in the last 10 years do not have a fuel return line.
Technician B says that many vehicles made in the last 10 years have a "lifetime" fuel filter.
Who is correct?

A. A only
B. B only
C. Both A and B
D. Neither A nor B

44. *Technician A* says that a forward-biased diode will block current flow.
Technician B says a reverse-biased diode will allow current flow.
Who is correct?

A. A only
B. B only
C. Both A and B
D. Neither A nor B

45. While discussing automotive computer networks, *Technician A* says that some computer networks use a gateway module to talk to a scan tool.

Technician B says that all vehicles built after 2006 were CAN network only.
Who is correct?

A. A only
B. B only
C. Both A and B
D. Neither A nor B

46. *Technician A* says that a direct injection system uses a lower fuel pressure than port fuel injection systems.
Technician B says that port fuel injection systems have a mechanical high-pressure fuel pump driven by the camshaft.
Who is correct?

A. A only
B. B only
C. Both A and B
D. Neither A nor B

47. An engine's performance gets worse as the engine speed increases.
Technician A says that an engine vacuum test may reveal the system causing the problem.
Technician B says that an extended overly rich fuel mixture may cause the cruise vacuum to be lower than the idle vacuum.
Who is correct?

A. A only
B. B only
C. Both A and B
D. Neither A nor B

48. Which exhaust emission is an efficiency indicator?

A. Oxygen
B. Carbon dioxide
C. Carbon monoxide
D. Oxides of nitrogen

49. While discussing the MAP sensor,
Technician A says that the MAP sensor measures changes in manifold pressure.
Technician B says that the MAP sensor is a voltage-generating sensor.
Who is correct?

A. A only
B. B only
C. Both A and B
D. Neither A nor B

50. Which of the following failures would be covered by the comprehensive component monitor?

A. EGR valve not opening

B. MAF sensor failure

C. Catalytic converter not storing O_2

D. Engine misfire

APPENDIX B
METRIC CONVERSIONS

To convert these	To these,	Multiply by:
TEMPERATURE		
Centigrade degrees	Fahrenheit degrees	1.8 then + 32
Fahrenheit degrees	Centigrade degrees	0.556 after −32
LENGTH		
Millimeters	Inches	0.03937
Inches	Millimeters	25.4
Meters	Feet	3.28084
Feet	Meters	0.3048
Kilometers	Miles	0.62137
Miles	Kilometers	1.60935
AREA		
Square centimeters	Square inches	0.155
Square inches	Square centimeters	6.45159
VOLUME		
Cubic centimeters	Cubic inches	0.06103
Cubic inches	Cubic centimeters	16.38703
Cubic centimeters	Liters	0.001
Liters	Cubic centimeters	1000
Liters	Cubic inches	61.025
Cubic inches	Liters	0.01639
Liters	Quarts	1.05672
Quarts	Liters	0.94633
Liters	Pints	2.11344
Pints	Liters	0.47317
Liters	Ounces	33.81497
Ounces	Liters	0.02957

To convert these	To these,	Multiply by:
WEIGHT		
Grams	Ounces	0.03527
Ounces	Grams	28.34953
Kilograms	Pounds	2.20462
Pounds	Kilograms	0.45359
WORK		
Centimeter-kilograms	Inch-pounds	0.8676
Inch-pounds	Centimeter-kilograms	1.15262
Meter kilograms	Foot-pounds	7.23301
Foot-pounds	Newton-meters	1.3558
PRESSURE		
Kilograms/square centimeter	Pounds/square inch	14.22334
Pounds/square inch	Kilograms/square centimeter	0.07031
Bar	Pounds/square inch	14.504
Pounds/square inch	Bar	0.06895

AES Automotive Electronics Services
3849 N. Fine Ave. 102
Fresno, CA 93727
559-292-7851
www.aeswave.com/

AllData Corp.
9412 Big Horn Blvd.
Elk Grove, CA 95758
800-829-8727
www.alldata.com

Autoland Scientech
1001 Cypress Creek Rd., Suite 101
Cedar Park, TX 78613
877-833-4728
www.autolandscientechusa.com

etool cart.com
866-251-4267
www.etoolcart.com

Ferret Instruments
1310 Higgins Dr.
Cheboygan, MI 49721
213-627-5664
www.ferretinstruments.com

Fluke Corp.
Box 9090
Everett, WA 98206
800-44-FLUKE
www.fluke.com

Hickok
10514 DuPont Ave.
Cleveland, OH 44105
800-342-5080
www.hickok-inc.com

Kleen Tec
Box 91
1212 Sykes St.
Albert Lea, MN 56007
800-435-5336
www.kleentec.com

Mac Tools
4635 Hilton Corp. Dr.
Columbus, OH 43232
800-MAC-TOOLS
www.mactools.com

Matco Tools
4403 Allen Rd.
Stow, OH 44224
800-433-7098
www.matcotools.com

Mitchell Repair Information Co.
14145 Danielson St.
Poway, CA 92064
858-391-5000
www.mitchell1.com/

OTC Tools
655 Eisenhower Dr.
Owatonna, MN 55060
800-533-6127
www.otctools.com

SK Hand Tool Corp.
9500 W. 55th St., Suite B
McCook, IL 60525
800-822-5575
www.skhandtool.com

Snap-on Diagnostics
2801 80th St.
Kenosha, WI 53143
877-762-7664
www1.snapon.com/diagnostics

Snap-on Tools Co.
2801 80th St.
Kenosha, WI 53143
262-656-5200
www.store.snapon.com

SPX Service Solutions
28635 Mound Rd.
Warren, MI 48092
586-574-2332
Fax: 1-800-578-7375
www.servicesolutions.spx.com/

Note: **Terms are highlighted in bold**, followed by Spanish translation in color.

14.7:1 air-fuel ratio The preferred air to fuel ratio for efficient engine operation 14.7 parts of air to 1 part of fuel.

Relación de aire a combustible 14.7:1 La relación preferida de 14.7 partes de aire a 1 parte de combustible que provee la operación más eficiente del motor.

Abrasive cleaning Any friction method used to clean components.

Limpieza abrasiva Cualquier metodo de fricción que se usa para limpiar los componentes.

Acceleration simulation mode (ASM) A test that incorporates steady-state and transient testing.

Modo de simulación de aceleración (ASM) Una prueba que incorpora las pruebas de los régimenes permanentes y transitorios.

Actuation test mode A scan tester mode used to cycle the relays and actuators in a computer system.

Modo de prueba de activación Instrumento de pruebas de exploración utilizado para ciclar los relés y los accionadores en una computadora. Actuator Electrical devices that a computer uses to carry out mechanical actions, usually a relay solenoid or lamp.

Actuator Electrical devices that a computer uses to carry out mechanical actions, usually a relay solenoid or lamp.

Actuador Los dispositivos eléctricos que utiliza una computadora para realizar las acciones mecánicas, por lo generral un soleniode de relé o una lámpara.

Adjusting pads, mechanical lifters Metal discs that are available in various thicknesses and are positioned in the end of the mechanical lifter to adjust valve clearance.

Cojines de ajuste, elevadores mecánicos Discos metálicos disponibles en diferentes espesores que se colocan en el extremo del elevador mecánico para ajustar el espacio libre de la válvula.

Advance-type timing light A timing light that is capable of checking the degrees of spark advance.

Luz de ensayo de regulación del encendido tipo avance Luz de ensayo de regulación del encendido capaz de verificar la cantidad del avance de la chispa.

AF sensor The air-fuel ratio sensor is more accurate and has a broader operating range than the conventional O_2 sensor.

Sensor de aire y combustible El sensor de relación de aire y combustible es más preciso y tiene un rango de funcionamiento más amplio que el sensor de O_2 convencional.

After top dead center (ATDC) Any measurement after top dead center.

Después de punto muerto superior (ATDC) Cualquier medida después del punto muerto superior.

AIR bypass (AIRB) solenoid A computer-controlled solenoid that directs air to the atmosphere or to the AIR diverter solenoid.

Solenoide de paso AIR Solenoide controlado por computadora que conduce el aire hacia la atmósfera o hacia el solenoide derivador AIR.

AIR diverter (AIRD) solenoid A computer-controlled solenoid in the secondary air injection system that directs air upstream or downstream.

Solenoide derivador AIRD Solenoide controlado por computadora en el sistema de inyección secundaria de aire que conduce el aire hacia arriba o hacia abajo.

Air-fuel ratio The proportion of air to fuel that an engine is burning. 14.7:1 is ideal.

Relación de aire-combustible La proporción de aire al combustible que quema en un motor. El 14.7:1 es ideal.

Air-fuel (A/F) ratio sensor A sensor that measures oxygen in the exhaust stream similar to an oxygen sensor but with the capability of reporting a much more precise air-fuel ratio.

Sensor de relación aire/combustible (A/F) Sensor que mide el oxígeno en la corriente de escape. Es similar a un sensor de oxígeno, pero tiene la capacidad de informar la relación de aire/combustible con mucha más precisión.

Air-operated vacuum pump A vacuum pump operated by air pressure that may be used to pump liquids such as gasoline.

Bomba de vacío accionada hidráulicamente Bomba de vacío accionada por la presión del aire que puede utilizarse para bombear líquidos, como por ejemplo la gasolina.

AIR system An air injection system.

Sistema de aire Un sistema de inyección de aire.

Alternating current (AC) An electric current that reverses its direction at regularly recurring intervals.

Corriente Alterna Una corriente eléctrica que reserva su dirección a intervalos de repetición regular.

American wire gauge (AWG) A system that designates wire sizes established by the SAE.

Calibre de alambre Americana (AWG) Un sistema que usa la SAE para establecer los tamaños del alambre.

Ammeter A device used to measure electrical current.

Amperímetro Un dispositivo para medir la corriente eléctrica.

Amplitude The difference between the highest and lowest voltage in a waveform signal.

Amplitud La diferencia entre el voltaje más alto y el voltaje más bajo en una señal en forma de onda.

Analog meter A meter with a moveable pointer and a meter scale.

Medidor analógico Medidor provisto de un indicador móvil y una escala métrica.

API American Petroleum Institute. An organization that sets standards for petroleum-based products, such as engine oil.

API Instituto Americano del petróleo. Una organización que establece estándares para productos basados en petróleo.

ASE blue seal of excellence An individual or a shop that employs technicians certified by ASE in automotive repair may display the ASE seal on their uniforms and place of business.

Sello azul de excelencia de ASE Cuando un individuo o un taller emplea a técnicos certificados por ASE en la reparación de automóviles, puede mostrar en los uniformes o el establecimiento el sello de ASE.

ASE technician certification Certification of automotive technicians in various classifications by the National Institute for Automotive Service Excellence (ASE).

Certificación de mecánico de la ASE Certificación de mecánico de automóviles en áreas diferentes de especialización otorgada por el Instituto Nacional para la Excelencia en la Reparación de Automóviles (ASE).

Atomization The process of breaking up a liquid into small particles or droplets.

Atomización El proceso de romper un líquido adentro de partículas pequeñas o gotas.

Automatic shutdown (ASD) relay A computer-operated relay that supplies voltage to the fuel pump, coil primary, and other components on Chrysler fuel-injected engines.

Relé de parada automática Relé accionado por computadora que les suministra tensión a la bomba del combustible, al bobinado primario, y a otros componentes en motores de inyección de combustible fabricados por la Chrysler.

Average responding Meters that show the average voltage peak.

Respuesta promedio Los medidores que enseñan la picadura promedio de voltaje.

"Banjo" fitting A round fitting through which a bolt with a passage drilled through it is used to attach the fitting to a housing. Looking at the fitting and line, it resembles a banjo.

Conexión banjo Una conexión redondo por la cual un perno perforado por un taladro se usa para conectar la conexión con un cárter. La conexión con la linea parecen un banjo.

Barometric (Baro) pressure sensor A sensor that sends a signal to the computer in relation to barometric pressure.

Sensor de la presión barométrica Sensor que le envía una señal a la computadora referente a la presión barométrica.

Baud rate The rate at which a PCM is able to transfer and receive data. Baud rate is measured in bits per second.

Velocidad Baud La velocidad a la cual el PCM (módulo de control de la potencia del motor) es capaz de transferir y recibir data. La velocidad Baud es medida en mordidas por segundo.

Before top dead center (BTDC) Any measurement, usually in degrees, prior to the top dead center point.

Antes de punto muerto superior (ATDC) Cualquier medida, suele ser en grados, antes del punto muerto superior.

Belt tension How tightly the belt is pulled, which has to be correct in order to prevent belt or component damage or belt slippage.

Tensión de la correa Fuerza con que se tira de la correa; debe ser correcta para evitar que se la correa se deslice o que se dañen la correa o los componentes.

Belt tension gauge A gauge designed to measure belt tension.

Calibrador de tensión de la correa de transmisión Calibrador diseñado para medir la tensión de una correa de transmisión.

Bidirectional control Command of a function usually performed by the PCM via a scan tool (i.e., shifting of an electronic transmission) used for diagnostics.

Control bidireccional Mando de una función que suele controlar el PCM por medio de una herramienta exploradora (por ejemplo, de cambio de marcha de una transmisioón electrónica) que se usa para los diagnósticos.

Blow-by Compression and exhaust gases that blow by the piston rings and enter into the engine's crankcase.

Fuga en la cámara de la combustión Compresión y gases de escape que se escapan atraves de los anillos del pistón y entran adentro de la cacerola del aceite.

Blowgun A device attached to the end of an air hose to control and direct airflow while cleaning components.

Soplete Dispositivo fijado en el extremo de una manguera de aire para controlar y conducir el flujo de aire mientras se lleva a cabo la limpieza de los componentes.

Boost pressure The amount of intake manifold pressure created by a turbocharger or supercharger.

Presión de sobrealimentación Cantidad de presión en el colector de aspiración producida por un turbocompresor o un compresor.

Bore The diameter of a cylinder in the engine block.

Calibrado del cilíndro El diámetro de un cilindro en el bloque motor.

Breakout box A terminal box that is designed to be connected in series at Ford PCM terminals to provide access to these terminals for test purposes.

Caja de desenroscadura Caja de borne diseñada para conectarse en serie a los bornes del módulo del control del tren transmisor de potencia de la Ford, con el objetivo de facilitar el acceso a dichos bornes para propósitos de prueba.

Burn time The length of the spark line while the spark plug is firing, measured in milliseconds.

Duración del encendido Espacio de tiempo que la línea de chispas de la bujía permanece encendida, medido en milisegundos.

Cam (camshaft) A straight shaft with lobes machined along its length to act on a valve lifter to open intake and exhaust valves.

Árbol de levas Un árbol recto que tiene lóbulos maquinados por su lengitud que actúan un levantador de válvulas para abrir las válvulas de entrada y salida.

Canister purge solenoid A computer-operated solenoid connected in the evaporative emission control system.

Solenoide de purga de bote Solenoide accionado por computadora conectado en el sistema de control de emisiones de evaporación.

Canister-type pressurized injector cleaning container A container filled with unleaded gasoline and injector cleaner and is pressurized during the manufacturing process or by the shop air supply.

Recipiente de limpieza del inyector presionizado tipo bote Recipiente lleno de gasolina sin plomo y limpiador de inyectores,

presionizado durante el proceso de fabricación o mediante el suministro de aire en el taller mecánico.

Carbon dioxide (CO₂) A gas formed as a by-product of the combustion process.

Bióxido de carbono (CO₂) Gas que es un producto derivado del proceso de combustión.

Carbon monoxide A gas formed as a by-product of the combustion process in the engine cylinders. This gas is very dangerous or deadly to the human body in high concentrations.

Monóxido de carbono Gas que es un producto derivado del proceso de combustión en los cilindros del motor. Este gas es muy peligroso y en altas concentraciones podría ocasionar la muerte.

Carbon tracking The formation of carbonized dust between distributor cap terminals.

Rastro de carbon La formación de polvo de carbon entre los terminales de la tapa del distribuidor.

Cetane A rating used to classify diesel fuel, refers to the volatility of the fuel.

Cetano Una relación usada para clasificar el combustible diesel, se refiere a la votálidad del combustible.

Charcoal canister A container of activated charcoal where gasoline vapors are stored by the evaporative emissions system until they can be burnt in the engine.

Bote de carbon Un contenedor de carbón activado en donde se almacenan los vapores de gasolina del sistema de emision hasta que se pueden quemar en el motor.

Chassis ground The ground of the circuit.

Tierra del armazón La tierra de un circuito.

Chemical cleaning Removing dirt or buildup from parts using a variety of solvents and other chemicals.

Limpieza por proceso químico Remover la suciedad o la aumentación de las partes usando una variedad de disolventes u otros productos químicos.

Class 2 Communication protocol that toggles a voltage from zero to seven volts. The signal can be sent in long or short pulse widths. Class II is medium speed at 10,400 bits per second.

Clase 2 Un protocolo de comunicación que escoja un voltaje de cero a siete voltíos. El señal se puede mandar en anchuras de pulso largos o cortos. El clase II es de velocidad mediana en 10,400 bits por segundo.

Closed loop A computer operating mode in which the computer uses the oxygen sensor signal to help control the air-fuel ratio.

Bucle cerrado Modo de funcionamiento de una computadora en el que se utiliza la señal del sensor de oxígeno para ayudar a controlar la relación de aire y combustible.

Closed-loop system A closed-loop system uses input from a sensor about a condition; the computer then uses this input to correct or adjust a condition according to the computer's programming. The term, closed loop, as applied to automotive systems, generally refers to the fuel control system, but can apply in several situations. The exhaust oxygen, or AF sensor, outputs a signal that corresponds to the air-fuel mixture. If the mixture appears to be lean or rich, the computer can inject more or less fuel, and then recheck the signal from the sensor, which then adjusts the mixture as needed. This is basically a continuous check and adjust process.

Sistema de ciclo cerrado Los sistemas de ciclo cerrado utilizan la entrada que envía un sensor sobre una condición; la computadora utiliza esa entrada para corregir o ajustar la condición, según su programación. En los sistemas automotores, el término generalmente hace referencia al sistema de control de combustible, pero puede aplicarse a varias situaciones.El sensor de oxígeno del escape, o sensor AF, emite una señal respecto a la mezcla de combustible y aire. La computadora puede inyectar más o menos combustible, según la mezcla sea rica o pobre, y luego vuelve a verificar la señal del sensor, que ajustará la mezcla según sea necesario. Es, esencialmente, un proceso continuo de verificación y ajuste.

Clutch slippage Occurs when the clutch disc does not firmly grip connect the flywheel and pressure plate.

Deslizamiento del embrague Ocurre cuando el disco del embrague no une el volante y la placa de presión con firmeza.

CNG Compressed Natural Gas. An alternative fuel source for engines.

CNG Gas Natural Comprimido. Una fuente de combustible alternativo para los motores.

Cold fouling The result of an excessively rich air-fuel mixture characterized by a layer of dry, fluffy black carbon deposits on the tip of the plug.

Ensuciamiento en frío El resultado de una mezcla de aire y combustible demasiado rico que se caracteriza por una capa de depósitos vellosos secos y negros en la punta de la bujía.

Combustion chamber The area in the cylinder head above the piston where fuel burning takes place.

Cámara de combustión El área en la cabeza del cilindro arriba del piston en donde se quema el combustible.

Compact discs (CD-ROM) Electronic data systems that can store large amounts of information.

Disco compacto (Cd-ROM) Los sistemas de datos electrónicos que pueden almacenar grandes cantidades de la información.

Comprehensive tests A complete series of battery, starting, charging, ignition, and fuel system tests performed by an engine analyzer.

Pruebas comprensivas Serie completa de pruebas realizadas en los sistemas de la batería, del arranque, de la carga, del encendido, y del combustible con un analizador de motores.

Compression fitting A common type of mating surfaces in fittings.

Montaje de presión Un tipo común de superficies parejos en los montajes.

Compression gauge A gauge used to test engine compression.

Manómetro de compresión Calibrador utilizado para revisar la compresión de un motor.

Compression ratio The amount the air-fuel mixture is compressed in the cylinder during the compression stroke. It is expressed in terms of volume, such as 8:1.

Relación de compresión La proporción de aire y mezcla de combustible durante el ciclo de compresión en el cilindro. Se expresa en términos de volumen, como 8:1.

Compression sense ignition A waste spark ignition which determines which cylinder is on compression by measuring the amount of secondary ignition voltage thereby eliminating the need for a cam sensor. The cylinder on compression has a higher spark voltage than the companion cylinder on the exhaust stroke.

Relación de compresión La cantidad de la mezcla del aire y combustible se comprimen en el cilíndro durante la carrera de compresión. Se expresa en términos de volúmen, tal como el 8:1.

Computed timing test A computer system test mode on Ford products that checks spark advance supplied by the computer.

Prueba de regulación del avance calculado Modo de prueba en una computadora de productos fabricados por la Ford que verifica el avance de la chispa suministrado por la computadora.

Computer output circuits Electrical circuits from the PCM that are used to activate the computers actuators.

Circuitos de salida de la computatdora Los circuitos eléctricos del PCM que se usan para activar los actuadores de la computadora.

Constant volume sampling (CVS) Ensures a constant volume of air is sampled by the emission analyzer.

Muestreo de volumen constante (CVS) Asegura que un volumen constante de aire se muestra por el analizador de emisión.

Continuity tester A self-powered test light.

Probador de continuidad Un foco de probador que tiene su propria energía.

Continuous self-test A computer system test mode on Ford products that provides a method of checking defective wiring connections.

Prueba automática continua Modo de prueba en una computadora de productos fabricados por la Ford que proporciona un método de verificar conexiones defectuosas del alambrado.

Coolant hydrometer A tester designed to measure coolant specific gravity and determine the amount of antifreeze in the coolant.

Hidrómetro de refrigerante Instrumento de prueba diseñado para medir la gravedad específica del refrigerante y determinar la cantidad de anticongelante en el refrigerante.

Cooling system pressure tester A tester used to test cooling system leaks and radiator pressure caps.

Instrumento de prueba de la presión del sistema de enfriamiento Instrumento de prueba utilizado para revisar fugas en el sistema de enfriamiento y en las tapas de presión del radiador.

Corona effect A glow around the spark plug cables indicating that the cable should be replaced.

Efecto de corona Una incandescencia alrededor de los cables de las bujías indicando que el cable debe ser cambiado.

Corrosive Materials that burn the skin or dissolve metals or other materials.

Material corrosivo Material que daña la piel o disuelve metales u otros tipos de materiales.

CPP sensor Gives the PCM the location of the clutch pedal for engine load purposes.

Sensor de posición del pedal del embrague El sensor de posición del pedal del embrague (CPP, por sus siglas en inglés) le informa al módulo de control del tren de potencia (PCM, por sus siglas en inglés) la ubicación del pedal del embrague a efectos de la carga del motor.

Crude oil Oil as it is taken from the ground, before it has been refined.

Aceite en crudo El aceite en su estado de salir de la tierra, antes de que se ha refinado.

Current limiting A feature that saturates the coil with high current for 1 second.

Limitador de corriente Una característica que satura la bobina en una alta corriente por un segundo.

Current ramp Graphing current on an oscilloscope; usually utilizing current probes.

Rampa de corriente El corriente gráfico en un osciloscópio, normalmente utilizando las sondas de corriente.

Custom tests A series of tests programmed by the technician and performed by an engine analyzer.

Pruebas de diseño específico Serie de pruebas programadas por el mecánico y realizadas por un analizador de motores.

Customization Refers to the ability of the customer to customize certain accessories such as door locks, power sliding doors, etc., to act according to a set routine (such as all doors lock when the vehicle is put into gear).

Personalización Refiere a la habilidad del cliente a personalizar ciertos acesorios tal como los cierres de puerta, las puertas corredizas de poder, etc., a funcionar según una rutina predeterminada (como todas las puertas se cierran cuando el vehículo se pone en marcha).

Cycle One complete set of changes in a recurring signal.

Ciclo Un juego de cambio completo en una señal recurrente.

Cylinder leakage The amount of air or volume lost from a sealed cylinder measured in percent.

Fuga del cilíndro La cantidad del aire o del volúmen que se pierde de un cilíndro sellado que se mide en porcentaje.

Cylinder leakage tester A tester designed to measure the amount of air leaking from the combustion chamber past the piston rings or valves.

Instrumento de prueba de la fuga del cilindro Instrumento de prueba diseñado para medir la cantidad de aire que se escapa desde la cámara de combustión y que sobrepasa los anillos de pistón o las válvulas.

Data link connector (DLC) A computer system connector to which the computer supplies data for diagnostic purposes.

Conector de enlace de datos Conector de computadora al que ésta suministra datos para propósitos diagnósticos.

Delta pressure feedback EGR sensor (DPFE) A sensor that uses a pressure difference between two passages to verify the operation of the EGR valve.

Sensor de retroalimentación EGR de presión Delta (DPFE) Un sensor que utiliza una diferencia de presión entre dos pasajes para verificar la operación de la válvula EGR.

Detonation Abnormal combustion. Refers to the ignition of the air-fuel mixture inside the combustion chamber prior to the firing of a spark plug.

Detonación Combustión anormal. Se refiere a la ignición de la mezcla de aire/combustible adentro de la cámara de combustión antes de que la chispa de la bujía ocurra.

Diagnostic trouble code (DTC) A code retained in a computer memory representing a fault in a specific area of the computer system.

Códigos indicadores de fallas para propósitos diagnósticos Código almacenado en la memoria de una computadora que representa una falla en un área específica de la computadora.

Diesel engine An internal combustion engine that utilizes diesel fuel.

Motor de Diesel Un motor de combustión interna que utiliza el combustible diesel.

Diesel particulates Small carbon particles in diesel exhaust.

Partículas de diesel Pequeñas partículas de carbón presentes en el escape de un motor diesel.

Differential pressure feedback (DPFE) EGR system A method for monitoring the opening of an EGR valve by using a pressure differential on both sides of a restriction in the EGR exhaust feed tube.

Sistema EGR de retroalimentación de presión diferencial (DPFE) Método para monitorear la apertura de la válvula de recirculación de gases del escape (EGR) utilizando presión diferencial a ambos lados de una obstrucción en el tubo de entrada de EGR.

Digital EGR valve An EGR valve that contains a computer-operated solenoid or solenoids.

Válvula EGR digital Una válvula EGR que contiene un solenoide o solenoides accionados por computadora.

Digital meter A meter with a digital display.

Medidor digital Medidor con lectura digital.

Digital multimeter (DMM) The same as a multimeter, but uses a digital readout and can usually be used to test a computerized vehicle.

Multímetro digital (DMM) Lo mismo que un multímetro pero tiene una lectural digital y se puede usar para efectuar una prueba en un vehículo computerizado.

Digital storage oscilloscope (DSO) A device that converts voltage signals to digital information and stores it in memory.

Osciloscopio de almacen digital (DSO) Un dispositivo que convierta las señales de voltaje en la información digital y la deposita en la memoria.

Direct current (DC) An electrical current that flows in one direction.

Corriente Directa Una corriente eléctrica que fluye solamente en una dirección.

Direct fuel injection or DFI Injects fuel directly into the combustion chamber.

Inyección directa de combustible La inyección directa de combustible (DFI, por sus siglas en inglés) inyecta combustible directamente en la cámara de combustión.

Direct injection A fuel injection system in which fuel is injected directly into the combustion chamber instead of the intake manifold.

Inyección directa Sistema de inyección de combustible en el cual el combustible se inyecta directamente en la cámara de combustión en lugar de en el múltiple de admisión.

Displacement The total volume of the cylinders in an engine, usually expressed in cubic centimeters or liters.

Desplazamiento El volúmen total de los cilíndros en un motor, expresado normalmente en centímetros cúbicos o litros.

Display pattern Cylinder patterns displayed from left to right on an oscilloscope.

Diagrama de presentación Las diagramas de los cilíndros que se presentan de la izquierda a la derecha en un osciloscopio.

Distributor ignition A mechanically operated ignition system using a distributor that is indexed to the camshaft.

Encendido con distribuidor Sistema de encendido mecánico que utiliza un distribuidor indizado con el árbol de levas.

Distributor ignition (DI) system SAE J1930 terminology for any ignition system with a distributor.

Sistema de encendido con distribuidor Término utilizado por la SAE J1930 para referirse a cualquier sistema de encendido que tenga un distribuidor.

Double flare Refers to the type of mating surfaces in fittings.

Avellanado doble Hace referencia al tipo de superficie de contacto de los accesorios.

Double-flare fitting A tubing fitting made with a special tool that has an anvil and a cone. The process is performed in two steps: first, the anvil begins to fold over the end of the tubing, then the cone is used to finish the flare by folding the tubing back on itself doubling the thickness and creating two sealing surfaces.

Montaje de doble abocinado Un montaje de tubo hecho con una herramienta especial que consiste de un yunque y un cono. El proceso se efectúa en dos pasos: primero el yunque pliega de la extremidad del tubo por encima de si mismo, luego se usa el cono para acabar el abocinado plegando el tubo así doblando el espesor para crear dos superficies de estanqueidad.

Downstream Usually refers to the oxygen sensor on the downstream side of the catalytic converter.

Corriente abajo Generalmente hace referencia al sensor de oxígeno que se encuentra en el lado inferior del conversor catalítico.

Downstream air Air injected into the catalytic converter.

Aire conducido hacia abajo Aire inyectado dentro del convertidor catalítico.

Drive cycle A specified set of driving conditions. The drive cycle is important to adaptive strategies of a computer. The proper drive cycle is also required of OBD II systems to complete monitor tests on certain systems.

Ciclo de ejecución Un juego especifico de condiciones de manejar. Estas condiciones son importantes para la estrategia adaptiva de la computadora.El ciclo de manejo apropiado es también requerido para el sistema OBD II (diagnóstico abordo del vehículo II) para competir con las pruebas del monitor en algunos sistemas.

Drive lugs Protrusions at the end of the shaft that seat in mating grooves in the end of the camshaft.

Lenguetas de mando Parte saliente en la extremidad de un árbol que se asientan en las ranuras apareadas en la extremidad del árbol de levas.

Dual overhead camshafts (DOHC) An engine with two camshafts per cylinder head, one each for intake and exhaust.

Árboles de levas elevados duales (DOHC) Un motor con dos árboles de levas por cada cabeza del cilíndro, un para cada entrada y salida.

Duty cycle On-time to off-time ratio, as measured in a percentage of pulse width or degrees of dwell.

Ciclo de duración La relación de apagado y encendido, según es medido en porcentaje de la amplitud del pulso o grados Dwell (tiempo en que los puntos están cerrados medidos en grados).

Dwell The amount of time the current is flowing through a circuit. Most often, this term is applied to ignition systems.

Tiempo en que los puntos están cerrados medidos en grados La cantidad de tiempo que la corriente está fluyendo atraves de un circuito. Más común, este termino es aplicado a sistemas de ignición.

Dwell meter A meter used to indicate the on time of a coil primary in degrees of crankshaft rotation.

Medidor de parada Un medidor que se usa para indicar en grados que la bobina primaria de la rotación del cigueñal esta en tiempo.

Dwell section On an ignition waveform, the time the ignition coil is energized.

Sección Dwell En la onda de encendido, tiempo durante el cual se energiza la bobina de encendido.

Dyno Dynamometer, a device that simulates road-load conditions in the shop.

Dino Dinomómetro Un dispositivo que simula las condiciones de la pista dentro del taller.

Efficiency The measure of the relationship between the amount of energy put into an engine and the amount of energy available from the engine.

Eficiencia La medida de la relación entre la cantidad de energía que se aplica a un motor y la cantidad de energía que es disponible del motor.

EGR control solenoid (EGRC) *See* EGR vent solenoid (EGRV).

Solenoide de control EGR (EGRC) Véase EGR vent solenoid (EGRV).

EGR pressure transducer (EPT) A vacuum switching device operated by exhaust pressure that opens and closes the vacuum passage to the EGR valve.

Transconductor de presión EGR Dispositivo de conmutación de vacío accionado por la presión del escape que abre y cierra el paso del vacío a la válvula EGR.

EGR vacuum regulator (EVR) solenoid A solenoid that is cycled by the computer to provide a specific vacuum to the EGR valve.

Solenoide regulador de vacío EGR Solenoide ciclado por la computadora para proporcionarle un vacío específico a la válvula EGR.

EVR valve A valve that introduces spent exhaust gas into the intake manifold to cool the combustion chamber and help prevent NO_x production during combustion.

Válvula EVR Válvula que introduce los gases de escape utilizados en el múltiple de admisión para enfriar la cámara de combustión y ayudar a prevenir la producción de óxidos de nitrógeno (NO_x) durante la combustión.

EGR valve port The port where exhaust gasses are fed into the intake manifold by the EGR.

Puerto de válvula EGR Puerto a través del cual la válvula EGR introduce gases de escape en el múltiple de admisión.

EGR vent solenoid (EGRV) A solenoid that normally cycles on and off frequently when EGR flow is controlled.

Solenoide de respirado (EGRV) Un solenoide que suele ciclarse prendiendose y apagandose frecuentemente cuando se controla el flujo del EGR.

Electromagnetic interference (EMI) Appears as AC noise on an electrical signal.

Interferencia electromagnética (EMI) Aparece como ruido de CA en una señal eléctrica.

Electromotive force (EMF) The force that exists between a positive and a negative point within an electrical circuit.

Fuerza electromotiva (EMF) La fuerza que existe entre un punto positivo y negative dentro de un circuito eléctrico.

Electronic automatic transmission control (EATC) A transmission that is shifted by electronic actuators via computer control.

Control electrónico de la transmisión automática (EATC) Una transmisión que cambia de marcha por medio de los actuadores electrónicos controlado por computadora.

Electronic distributorless ignition system (EDIS) A type of ignition system produced by Ford Motor Company that uses a high data rate signal for crankshaft inputs.

Sistema de ignición electrónica sin distribuidor (EDIS) Un tipo de sistema de ignición producido por la companía FordMotor que utiliza una señal de información de alta velocidad para las entradas del cigueñal.

Electronic fuel injection (EFI) A generic term applied to various types of fuel injection systems.

Inyección electrónica de combustible Término general aplicado a varios sistemas de inyección de combustible.

Electronic ignition (EI) system SAE J1930 terminology for any ignition system without a distributor.

Sistema de encendido electrónico Término utilizado por la SAE J1930 para referirse a cualquier sistema de encendido que no tenga distribuidor.

Electronic vacuum regulator valve (EVRV) Used to activate the EGR valve by controlling the vacuum available via a solenoid.

Válvula electrónica reguladora del vacío (EVRV) Se usa para activar la válvula EGR por CA controlando el vacío disponible por medio de un solenoide.

Enable criteria The specific operating conditions for diagnostic tests.

Criterios de capacidad Las condiciones específicas de operacion para las pruebas diagnósticas.

Energy conserving oils Engine oils designed to reduce friction and therefore save fuel.

Acietes que conservan energía Los aceites de motor diseñados para disminuir la fricción así preservando el combustible.

Engine analyzer A tester designed to test engine systems such as battery, starter, charging, ignition, and fuel plus engine condition.

Analizador de motores Instrumento de prueba diseñado para revisar sistemas de motores, como por ejemplo los de la batería, del arranque, de la carga, del encendido, y del combustible además de la condición del motor.

Engine coolant temperature (ECT) sensor A sensor that sends a voltage signal to the computer in relation to coolant temperature.

Sensor de la temperatura del refrigerante del motor Sensor que le envía una señal de tensión a la computadora referente a la temperatura del refrigerante.

Engine lift A hydraulically operated piece of equipment used to lift the engine from the chassis.

Elevador de motores Equipo accionado hidráulicamente que se utiliza para levantar el motor del chasis.

Engine off natural vacuum (EONV) An EVAP monitoring system that runs pressure and vacuum tests based on the natural volatility of the fuel as the fuel warms up and cools off after driving.

Vacío normal de motor apagado (EEONV) Un sistema monitor del EVAP que efectúa las pruebas de presión y vacío basado en la volatilidad normal del combustible al calentarse y enfriarse el combustible después de estar en marcha.

Enhanced EVAP An EVAP system used on OBD II in which the PCM conducts several tests to determine if the system is operational and there are no leaks.

EVAP intensificado Un sistema EVAP que se usa en OBD II en el cual el PCM ecfectúa varias pruebas para determinar si funciona el sistema y no hay fugas.

Environmental Protection Agency (EPA) The federal agency that enforces laws to prevent activity that may be harmful to the environment.

Agencia de Protección del Medio Ambiente (EPA) Una agencia federal que impone los leyes para prevenir cualquier actividad que podría ser dañosa al medio ambiente.

Evaporative (EVAP) system A system that collects fuel vapors from the fuel tank and directs them into the intake manifold rather than allowing them to escape to the atmosphere.

Sistema de evaporación Sistema que acumula los vapores del combustible que escapan del tanque del combustible y los conduce hacia el colector de aspiración en vez de permitir que los mismos se escapen hacia la atmósfera.

Exhaust gas analyzer A tester that measures carbon monoxide, carbon dioxide, hydrocarbons, and oxygen in the engine exhaust.

Analizador del gas del escape Instrumento de prueba que mide el monóxido de carbono, el bióxido de carbono, los hidrocarburos, y el oxígeno en el escape del motor.

Exhaust gas recirculation (EGR) valve A valve that circulates a specific amount of exhaust gas into the intake manifold to reduce NO_x emissions.

Válvula de recirculación del gas del escape Válvula que hace circular una cantidad específica del gas del escape hacia el colector de aspiración para disminuir emisiones de óxidos de nitrógeno.

Exhaust gas recirculation valve position (EVP) sensor A sensor that sends a voltage signal to the computer in relation to EGR valve position.

Sensor de la posición de la válvula de recirculación del gas del escape Sensor que le envía una señal de tensión a la computadora referente a la posición de la válvula EGR.

Exhaust gas temperature sensor A sensor that sends a voltage signal to the computer in relation to exhaust temperature.

Sensor de la temperatura del gas del escape Sensor que le envía una señal de tensión a la computadora referente a la temperatura del escape.

Expansion tank The volume in a fuel tank that cannot be filled with fuel in order to allow for expansion in warm weather.

Tanque de expansión El volumen en un tanque de combustible en el que no se puede llenar con combustible para permitir su expansión en una clima cálida.

Federal Test Procedure A transient-speed mass sampling emissions test conducted on a loaded dynamometer. This is the test used by manufacturers to certify vehicles before they can be sold.

Procedimiento de la Prueba Federal Una prueba de la velocidad transitoria para la prueba de la masa en un dinamómetro cargado. Esta es la prueba usada por los fabricantes para certificar los vehículos antes de que ellos puedan ser vendidos.

Feedback A term used to describe a PCM's ability to control the activity of an actuator in response to input from sensors.

Retroalimentación Un termino usado para describir la habilidad del PCM (módulo de control de la potencia del motor) de controlar la actividad de un actuador en respuesta a la información de los sensores de entra.

Feeler gauge Metal strips with a specific thickness for measuring clearances between components.

Calibrador de espesores Láminas metálicas de un espesor específico para medir espacios libres entre componentes.

Firing order The order in which the individual cylinders in an engine are fired.

Orden de encendido El orden en el cual los cilíndros individuales de un motor se encienden.

Firing voltage The voltage needed to start the spark.

Voltaje de encendido El voltaje requerido para prender la bujía.

Five-gas emissions analyzer An analyzer designed to test carbon monoxide, carbon dioxide, hydrocarbons, oxides of nitrogen, and oxygen in the exhaust.

Analizador de cuatro tipos de emisiones Analizador diseñado para revisar el monóxido de carbono, el bióxido de carbono, los hidrocarburos, y el oxígeno en el escape.

Flash code diagnosis Reading computer system diagnostic trouble codes (DTCs) from the flashes of the malfunction indicator light (MIL).

Diagnosis con código de destello La lectura de códigos indicativos de fallas para propósitos diagnósticos de una computadora mediante los destellos de la luz indicadora de funcionamiento defectuoso.

Flat-rate manual A book that helps determine the length of time a particular task or job will take.

Manual de precios uniformes Un libro que asista en determinar la cantidad del tiempo que va tomar un trabajo o una tarea específica.

Floor jack A hydraulically operated device mounted on casters used to raise one end or corner of the chassis.

Gato de pie Dispositivo accionado hidráulicamente montado en rolletes y utilizado para levantar un extremo o una esquina del chasis.

Four-wheel alignment A process where the front and rear wheel angles are adjusted to the vehicle's thrust angle.

Alineación de cuatro ruedas Un proceso en el cual los ángulos de las ruedas delanteras y traseras se ajusten al ángulo de empuje del vehículo.

Freeze frame A feature of some scan tools and a requirement of OBD II. This feature takes a snapshot of the operating conditions present when the PCM sets a diagnostic trouble code.

Marco congelado Una característica de algunas herramientas exploratorias y un equipo de OBD II (diagnóstico abordo del vehículo II). Esta característica toma una foto de las condiciones de operaciones presente cuando el PCM (módulo de control de la potencia del motor) establece un código de diagnostico de problema.

Freeze-frame access mode Freeze frame is a picture of the data from the ECM when the conditions for setting a code were present. This information allows the technician to operate the vehicle under the same conditions to duplicate the problem.

Modo de acceso con imagen congelada La imagen congelada refleja los datos del módulo de control del motor (ECM, por sus siglas en inglés) en el momento en el que estaban dadas las condiciones para establecer un código. Esta información permite al técnico hacer funcionar el vehículo en esas mismas condiciones para replicar el problema.

Frequency The number of complete cycles that occur in a specific period of time.

Frecuencia El número de ciclos completos que ocurren en un periodo especifico de tiempo.

Fuel control Engine running with the PCM in control of fuel mixture based on inputs from the various sensors instead of programmed values (such as during cold start).

Control de combustible El motor en marcha con el PCM en control de la mezcla del combustible basado en las entradas de varios sensores en vez de los valores programados (tal como durante un encendido frío).

Fuel cut rpm The rpm range in which the computer stops operating the injectors during deceleration.

Detención de combustible según las rpm Margen de revoluciones al que la computadora detiene el funcionamiento de los inyectores durante la desaceleración.

Fuel pressure test port A threaded port on the fuel rail to which a pressure gauge may be connected to test fuel pressure.

Lumbrera de prueba de la presión del combustible Lumbrera fileteada que se encuentra en el carril del combustible a la que puede conectársele un calibrador de presión para revisar la presión del combustible.

Fuel pump volume The amount of fuel the pump delivers in a specific time period.

Volumen de la bomba del combustible Cantidad de combustible que la bomba envía dentro de un espacio de tiempo específico.

Fuel tank level sensor The input that the PCM uses to determine the fuel level in the fuel tank important for EVAP leak detection sensing.

Sensor de nivel del combustible del tanque La entrada que usa el PCM para determinar el nivel del combustible en el tanque de combustible que es importante en la detección por sensor de las fugas del EVAP.

Fuel tank pressure control valve The valve designed to control fuel tank pressure while the vehicle is sitting still. If tank pressure builds too high the valve opens and lets the vapor into the canister.

Válvula de control de presión del tanque de combustible La válvula diseñada a controlar la presión del tanque de combustible mientras que el vehículo esta imóbil. Si la presión del tanque sube demasiado abre la válvula y permite que entra el vapor al bote.

Fuel tank pressure sensor Informs the PCM of the pressure inside the fuel tank for EVAP system tests.

Sensor de presión del tanque de combustible Informa al PCM de la presión dentro del tanque de combustible en las pruebas EVAP.

Fuel tank purging Removing fuel vapors from the fuel tank.

Purga del tanque del combustible La remoción de vapores de combustible y de material extraño del tanque del combustible.

Gap bridging Occurs when carbon deposits accumulate in the combustion chamber due to stop-and-go driving.

Puesta en cortocircuito de los electrodos Ocurre cuuando los depósitos de carbon se acumulan en la cámara de combustión debido a la marcha con mucha ida-y-parada.

Glazing A glaze over the insulator from combustion chamber deposits.

Satinado Un lustro en un aislador que viene de los depósitos de la cámara de combustión.

Glitch A momentary disruption in an electrical circuit.

Falla Una interrupción momentánea en un circuito eléctrico.

Glow plug An electrically heated point inside the diesel combustion chamber to aid cold starting.

Bujía de incandescencia Un punto calentado electrónicamente dentro de la cámara de combustión diesel para asistir en el arranque frío.

Graphite oil An oil with a graphite base that may be used for special lubricating requirements such as door locks.

Aceite de grafito Aceite con una base de grafito que puede utilizarse para necesidades de lubrificación especiales, como por ejemplo en cerraduras de puertas.

Hall-effect pickup A pickup containing a Hall element and a permanent magnet with a rotating blade between these components.

Captación de efecto Hall Captación que contiene un elemento Hall y un imán permanente entre los cuales está colocada una aleta giratoria.

Handheld digital pyrometer A tester for measuring component temperature.

Pirómetro digital de mano Instrumento de prueba para medir la temperatura de un componente.

Hand press A hand-operated device for pressing precision-fit components.

Prensa de mano Dispositivo accionado manualmente para prensar componentes con un fuerte ajuste de precisión.

Hazard communication standard The part of the right-to-know laws that requires employers to train employees about hazardous materials they may encounter on the job.

Norma de comunicación de peligros La parte de los leyes derechode-saber que requiere que los patrones entrenan a los empleados con respeto a los materiales peligroso que puedan encontrar en el trabajo.

Heated resistor-type MAF sensor A MAF sensor that uses a heated resistor to sense air intake volume and temperature and sends a voltage signal to the computer in relation to the total volume of intake air.

Sensor MAF tipo resistor térmico Sensor MAF que utiliza un resistor térmico para advertir el volumen y la temperatura del aire aspirado, y que le envía una señal de tensión a la computadora referente al volumen total de aire aspirado.

Hertz A unit of measurement for counting the number of times an electrical cycle repeats every second. One hertz is one pulse per second.

Hertz Una unidad de medida para contar los números de veces que un ciclo eléctrico se repite cada segundo. Un hertz es un pulso cada segundo.

High input impedance For our purposes, impedance is best defined as operating resistance.

Alta impedancia de entrada A nuestros efectos, la impedancia se define como la resistencia operativa.

High-pressure pump On a direct fuel injection (DFI) system, a mechanical fuel pump that is driven by the camshaft.

Bomba de alta presión En los sistemas de inyección directa de combustible (DFI), bomba de combustible mecánica accionada por el árbol de levas.

History code Is a code that set but the condition is not present at the time the scan tool is connected because the condition is intermittent.

Código histórico Es un código que se establece aunque la condición no esté presente al conectar la herramienta de exploración por tratarse de una condición intermitente.

Horsepower The force required to move 550 pounds one foot in one second.

Potencia *Es la fuerza necesaria para mover 550 libras, un pie (250 kg, 30 cm), en un segundo.*

Hot wire-type MAF sensor A MAF sensor that uses a heated wire to sense air intake volume and temperature and sends a voltage signal to the computer in relation to the total volume of intake air.

Sensor MAF tipo térmico *Sensor MAF que utiliza un alambre térmico para advertir el volumen y la temperatura del aire aspirado, y que le envía una señal de tensión a la computadora referente al volumen total de aire aspirado.*

Hybrid Electric Vehicles (HEVs) Vehicles that use both an electric motor and an internal combustion engine.

Vehículo eléctrico híbrido (HEVs) *Los vehículos que utilisan un motor eléctrico y tambien un motor de combustión interna.*

Hydraulic press A hydraulically operated device for pressing precision-fit components.

Prensa hidráulica *Dispositivo accionado hidráulicamente que se utiliza para prensar componentes con un fuerte ajuste de precisión.*

Hydraulic valve lifters Round, cylindrical, metal components used to open the valves. These components are operated by oil pressure to maintain zero clearance between the valve stem and rocker arm.

Desmontaválvulas hidráulicas *Componentes metálicos, cilíndricos y redondos utilizados para abrir las válvulas. Estos componentes se accionan mediante la presión del aceite para mantener cero espacio libre entre el vástago de válvula y el balancín.*

Hydrocarbons (HC) Leftover fuel from the combustion process.

Hidrocarburos *El combustible restante después del proceso de combustión.*

Hydrometer A tester designed to measure the specific gravity of a liquid.

Hidrómetro *Instrumento de prueba diseñado para medir la gravedad específica de un líquido.*

Hyperlink An electronic link that provides a short cut to another document or media directly by clicking on a word in the text that is usually set apart in a different color and underlined, such as *engine performance.*

Hipervínculo *Una conexión electronica que provee una ruta directa a otro documento o pagina directamente al hacer clic en una palabra del texto que suele ser distincto en otro color y subrayado, tal como engine performance.*

Idle air control bypass air (IAC BPA) motor An IAC motor that controls idle speed by regulating the amount of air bypassing the throttle.

Motor para el control de la marcha lenta con el paso de aire *Un motor IAC que controla la velocidad de la marcha lenta regulando la cantidad de aire que se desvía de la mariposa.*

Idle air control bypass air (IAC BPA) valve A valve operated by the IAC BPA motor that regulates the air bypassing the throttle to control idle speed.

Válvula para el control de la marcha lenta con el paso de aire *Válvula accionada por el motor IAC BPA que regula el aire que se desvía de la mariposa para controlar la velocidad de la marcha lenta.*

Idle air control (IAC) motor A computer-controlled motor that controls idle speed under all conditions.

Motor para el control de la marcha lenta con aire *Motor controlado por computadora que controla la velocidad de la marcha lenta bajo cualquier condición del funcionamiento del motor.*

Ignition crossfiring Ignition firing between distributor cap terminals or spark plug wires.

Encendido por inducción *Encendido entre los bornes de la tapa del distribuidor o los alambres de las bujías.*

Ignition module tester An electronic tester designed to test ignition modules.

Instrumento de prueba del módulo del encendido *Instrumento de prueba electrónico diseñado para revisar módulos del encendido.*

Ignition systems An electrical system that delivers spark and controls timing.

Sistema de encendido *Sistema eléctrico que suministra chispa y controla la sincronización.*

Induction The spontaneous creation of an electrical current in a conductor as the conductor passes through a magnetic field or a magnetic field passes across the conductor.

Inducción *La creación espontanea de una corriente eléctrica en un conductor según el conductor atraviesa un campo magnético o un campo magnético atraviesa el conductor.*

Initial vacuum pull down A step in the EVAP leak detection cycle where a vacuum is pulled on the EVAP system and then checked for the rate of decay. (*See also* weak vacuum test.)

Rebajada inicial del vacío *Un paso en el ciclo de detección de fugas del EVAP en el cual se cree un vacío en el sistema EVAP y luego se revisa para ver la cantidad de degradación. (véase tambien weak vacuum test.)*

Injector balance tester A tester designed to test port injectors.

Instrumento de prueba del equilibrio del inyector *Instrumento de prueba diseñado para revisar inyectores de lumbreras.*

Inspection and maintenance (I/M) testing Emission inspection and maintenance programs that are usually administered by various states.

Pruebas de inspección y mantenimiento *Programas de inspección y mantenimiento de emisiones que normalmente administran diferentes estados.*

Intake air temperature sensor A sensor that sends a signal to the computer in relation to intake air temperature.

Sensor de temperatura del aire de admisión *Sensor que envía una señal a la computadora sobre la temperatura del aire de admisión.*

Integrated diagnostic software (IDS) Ford calls their software subscription integrated diagnostic software (IDS).

Software de diagnóstico integrado *Ford llama software de diagnóstico integrado (IDS, por sus siglas en inglés) a su software por suscripción.*

Intermediate section The section of the scope trace pattern that shows the remaining coil voltage as it dissipates or drops to zero.

Sección intermedio *La sección de la diagrama de una muestra de un osciloscopio que enseña lo que queda del voltaje de la bobina en el momento que dispersa o baja hasta el cero.*

International Lubricant Standardization and Approval Committee (ILSAC) An organization that rates the quality of a lubricating oil and applies the API Service Symbol to those oils which pass requirements of Japanese and American automobile manufacturers.

Comisión internacional de regularización y aprobación de lubricantes (ILSAC) *Un organización que evalúa la calidad de un aceite lubricante y aplica el signo de Servicio API a aquellos aceites que supera los requerimientos de los fabricantes de automóviles japoneses y americanos.*

International System (SI) A system of weights and measures in which each unit may be divided by 10.

Sistema internacional *Sistema de pesos y medidas en el que cada unidad puede dividirse entre 10.*

Ion sensing ignition An ignition system which uses the spark plug as a combustion sensor by placing a voltage across the spark plug electrode immediately after the spark plug firing event.

Encendido por detección de iones *Sistema de encendido que emplea una bujía para detectar la combustión a través de un voltaje en el electrodo de la bujía, inmediatamente después del evento de encendido de la bujía.*

ISO 9141 Specifies the requirements for one of the digital languages used between the PCM and scan tools per the International Standards Organization.

ISO 9141 *Especifica los requerimientos de una de las idiomas digitales que se usan entre el PCM y las herramientas exploradoras según la Organización de Regularización Internacional.*

J1850 Specifies the requirements for a communication standard used as a digital language between the PCM, individual modules, and scan tools per SAE. Also known as Class B and SCP.

J1850 *Especifica los requerimientos para una comunicación regularizado que se usa como idioma digita entre el PCM, los módulos individuales y las herramientas exploradores según el SAE. Tambien se llaman Clase B y SCP.*

Jack stand A metal stand used to support one corner of the chassis.

Soporte de gato Soporte de metal utilizado para apoyar una esquina del chasis.

Key on, engine off (KOEO) test A computer system test mode on Ford products that displays diagnostic trouble codes (DTCs) with the key on and the engine stopped.

Prueba con la llave en la posición de encendido y el motor apagado Modo de prueba en una computadora de productos fabricados por la Ford que muestra códigos indicadores de fallas para propósitos diagnósticos cuando la llave está en la posición de encendido y el motor está apagado.

Key on, engine running (KOER) test A computer system test mode on Ford products that displays diagnostic trouble codes (DTCs) with the engine running.

Prueba con la llave en la posición de encendido y el motor encendido Modo de prueba en una computadora de productos fabricados por la Ford que muestra códigos indicadores de fallas para propósitos diagnósticos cuando el motor está encendido.

Kilopascal (kPa) The metric version of psi.

Kilopascal (kPa) La version métrica de libras por pulgada cuadrada.

Kilovolts (kV) Thousands of volts.

Kilovoltios (kV) Miles de voltios.

Knock sensor (KS) A sensor that sends a voltage signal to the computer in relation to engine detonation.

Sensor de golpeteo Sensor que le envía una señal de tensión a la computadora referente a la detonación del motor.

Knock sensor module An electronic module that changes the analog knock sensor signal to a digital signal and sends it to the PCM.

Módulo del sensor de golpeteo Módulo electrónico que convierte la señal analógica del sensor de golpeteo en una señal digital y se la envía al módulo del control del tren transmisor de potencia.

Leak detection pump (LDP) A pump used on some vehicles to pressurize the EVAP system which is used in conjunction with a pressure sensor to detect leaks.

Bomba detector de fugas (LDP) Una bomba usado en algunos vehículos para presurizar el sistema EVAP que se usa de acuerdo con un sensor de presión para detectar las fugas.

Lift A device used to raise a vehicle.

Elevador Dispositivo utilizado para levantar un vehículo.

Lift pump or low-pressure pump A pump inside the fuel tank and a main pump located outside the fuel tank.

Bomba de elevación o bomba de baja presión Bomba situada dentro del tanque de combustible y bomba principal situada fuera del tanque de combustible.

Linear Refers to a straight line. The linear Hall-effect TP sensor produces a voltage that moves in a straight line from closed to wide-open throttle.

Lineal Hace referencia a una línea recta. El sensor lineal de ángulo de apertura del acelerador (TP) de efecto Hall genera un voltaje que produce un movimiento en línea recta desde el punto en que el acelerador está cerrado hasta el punto en que está completamente abierto.

Linear EGR valve An EGR valve containing an electric solenoid that is pulsed on and off by the computer to provide a precise EGR flow.

Válvula EGR lineal Válvula EGR con un solenoide eléctrico que la computadora enciende y apaga para proporcionar un flujo exacto de EGR.

Lobes Elliptical projections on the camshaft that contact the valve lifter in order to open and close the valve lifters.

Lóbulos Las proyecciones elípticas en el árbol de levas que hacen contacto con la levanta-válvulas para abrir y cerrar las levanta-válvulas.

Logic probe A test light with three different colored light emitting diodes (LEDs) indicating voltage levels.

Explorador lógico Un foco de prueba que tiene diodos emisores de luz (LED) con tres luces de colores distinctos indicando los niveles de voltaje.

Long-term adaptive fuel trim Long-term fuel injector pulse width compensation determined by the PCM according to operating conditions. This fuel trim is set to maintain minimum emissions output and is the base point for short-term fuel trim.

Restricción del combustible por tiempo largo Compensación amplia de los pulsos del inyector de combustible determinado por el PCM (módulo de control de la potencia del motor) de acuerdo con las condiciones de operaciones. Esta restricción de combustible es calibrada para mantener las emisiones a un mínimo y es el punto de base para la restricción del combustible por tiempo corto.

Long-term fuel trim (LTFT) The long-term trends in fuel control. Positive numbers mean that fuel is being added, negative numbers mean fuel is being subtracted from the long-term fuel calculation.

Eficiencia de combusitible de largo plazo (LTFT) La tendencia de control de combustible en largo plazo. Los números positivos quieren decir que se agrega el combustible, los números negativos reflejan que se substrae el combustible de la calculación de largo plazo.

Low-amp probe A probe to measure small amounts of current with an amp clamp.

Explorador de bajo ampério Un explorador para medir las pequeñas cantidades de corriente con una grapa de ampérios.

Magnetic-base thermometer A thermometer that may be retained to metal components with a magnetic base.

Termómetro con base magnética Termómetro que puede sujetarse a componentes metálicos por medio de una base magnética.

Magnetic probe-type digital tachometer A digital tachometer that reads engine rpm and uses a magnetic probe pickup.

Tacómetro digital tipo sonda magnética Tacómetro digital que lee las rpm del motor y utiliza una captación de sonda magnética.

Magnetic probe-type digital timing meter A digital reading that displays crankshaft degrees and uses a magnetic-type pickup probe mounted in the magnetic timing probe receptacle.

Medidor de regulación digital tipo sonda magnética Lectura digital que muestra los grados del cigüeñal y utiliza una sonda de captación tipo magnético montada en el receptáculo de la sonda de regulación magnética.

Magnetic sensor A sensor that produces a voltage signal from a rotating element near a winding and a permanent magnet. This voltage signal is often used for ignition triggering.

Sensor magnético Sensor que produce una señal de tensión desde un elemento giratorio cerca de un devanado y de un imán permanente. Esta señal de tensión se utiliza con frecuencia para arrancar el motor.

Malfunction indicator light (MIL) A light in the instrument panel that is illuminated by the PCM if certain defects occur in the computer system.

Luz indicadora de funcionamiento defectuoso Luz en el panel de instrumentos que el módulo del control del tren transmisor de potencia ilumina si ocurren ciertas fallas en la computadora.

Manifold absolute pressure (MAP) sensor An input sensor that sends a signal to the computer in relation to intake manifold vacuum.

Sensor de la presión absoluta del colector Sensor de entrada que le envía una señal a la computadora referente al vacío del colector de aspiración.

Mass The measurement of an object's inertia.

Masa La medida de la inercia de un objeto.

Mass airflow (MAF) sensor A sensor that monitors incoming airflow, including speed, temperature, and pressure.

Sensor de flujo de aire en masa (MAF) Un sensor que amonesta el flujo del aire de entrada, incluyendo la velocidad, la temperatura y la presión.

Material safety data sheet (MSDS) An information sheet that specifies the dangers of human contact with substances used in any business that employees may be exposed to on the job. They must be available to all employees.

Hojas de datos de seguridad de materias (MSDA) Un página de información que especifica los peligros del contacto entre los seres humanos con las sustancias que se usan en cualquier empleo que se pueden exponer los empleados en el trabajo. Deben ser disponibles a cada empleado.

Mechanical valve lifters, or solid tappets Round, cylindrical, metal components mounted between the camshaft lobes and the pushrods to open the valves.

Desmontaválvulas mecánicas, o alzaválvulas sólidas Componentes metálicos, cilíndricos y redondos montados entre los lóbulos del árbol de levas y las varillas de empuje para abrir las válvulas.

Meter impedance The total internal electrical resistance in a meter.

Impedancia de un medidor La resistencia eléctrica interna total en un medidor.

Metering The process of controlling the flow of something. Metering of gasoline is controlled by the time it is allowed to flow, by the size of the opening it is flowing through, or by the pressure causing it to flow.

Regulación El proceso de controlar el flujo de algo. La regulación de la gasolina es controlada por el tiempo que es permitida que fluya, por el tamaño de la apertura que está fluyendo, o por la presión que la está causando que fluya.

Mild hybrid A hybrid vehicle that has an internal combustion engine and electric motor. The internal combustion engine runs to charge the batteries for the electric motor. The electric motor only runs when the small internal combustion motor needs assistance during acceleration or hill pulling.

Híbrido moderado Un vehículo híbrido que tiene un motor de combustión interno y un motor eléctrico. El motor de combustión interno esta en marcha para cargar las baterías para el motor eléctrico. El motor eléctrico sólo funciona cuando el pequeño motor interno de combustión requiere asistencia durante la aceleración o en jalar una carga en una colina.

Mode 1 Parameter identification (PID) mode.

Modo 1 Modo de identificación de parámetros (PID).

Mode 2 Freeze-frame data access mode.

Modo 2 Modo de acceso a los datos existentes en un momento dado ("datos congelados").

Mode 3 Permits scan tools to obtain stored DTCs.

Modo 3 Permite que las herramientas de exploración obtengan los códigos de problemas de diagnóstico (DTC) almacenados.

Mode 4 Allows the scan tool to clear all emission-related diagnostic information from its memory.

Modo 4 Permite que la herramienta de exploración borre de la memoria toda la información de diagnóstico relacionada con las emisiones.

Mode 5 Oxygen sensor monitoring test.

Modo 5 Prueba de monitoreo del sensor de oxígeno.

Mode 6 On-board test of non-continuously monitored systems.

Modo 6 Prueba de a bordo para los sistemas sin monitoreo continuo.

Mode 7 Monitoring of test results for continuously monitored systems.

Modo 7 Monitoreo de los resultados de las pruebas en los sistemas con monitoreo continuo.

Mode 8 A request for control of an on-board system test or component.

Modo 8 Solicitud para controlar un componente o una prueba del sistema de a bordo.

Mode 9 Allows the PCM to identify the vehicle's identification number (VIN), and the PCM's calibration ID and calibration verification

Modo 9 Permite que el PCM determine el número de identificación del vehículo (VIN), así como la verificación de calibración y la id. de calibración del PCM.

Mode 10 Codes that can only be erased by the PCM/ECM.

Modo 10 Códigos que solo pueden ser borrados por el PCM/ECM.

Modes of operation The ECM has modes of operation that have been programmed for specific purposes of communication with the scan tool.

Modos de funcionamiento El módulo de control del motor (ECM, por sus siglas en inglés) tiene diversos modos de funcionamiento que se han programado específicamente para la comunicación con la herramienta de exploración.

Muffler chisel A chisel that is designed for cutting muffler inlet and outlet pipes.

Cincel para silenciadores Cincel diseñado para cortar los tubos de entrada y salida del silenciador.

Multimeter A handheld tool capable of measuring volts, amps, and ohms.

Multímetro Una herramienta de mano que es capaz de medir los voltíos, los amperios y los ohmios.

National Institute for Automotive Service Excellence (ASE) An organization responsible for certification of automotive technicians in the United States.

Instituto Nacional para la Excelencia en la Reparación de Automóviles Organización que tiene a su cargo la certificación de mecánicos de automóviles en los Estados Unidos.

Neutral A transmission gear selection that disengages the transmission from the drive wheels.

Neutro Selección del sistema de transmisión que desconecta la transmisión de las ruedas motrices.

Neutral/drive switch (NDS) A switch that sends a signal to the computer in relation to gear selector position.

Conmutador de mando neutral Conmutador que le envía una señal a la computadora referente a la posición del selector de velocidades.

New Generation Star Tester (NGS) A proprietary scan tool used by Ford dealer technicians.

Probador Estrella de Nueva Generación (NGS) Una herramienta exploradora propietario usado por los técnicos de sucursales Ford.

Noid light Used to determine if a fuel injector is receiving the proper voltage pulse.

Luz noid Usado para determinar si un inyector de combustible recibe el impulso apropiado de voltaje.

Non sinusoidal A voltage trace that is not uniform in size and shape. Usually a DC voltage in automotive terms.

No sinusoidal Un razgo de voltaje que no es uniforme en tamaño ni en forma. Suele referirse en términos automativos de voltaje DC.

OBD II The second generation of on-board diagnostics.

OBD II La segunda generación de los diagnósticos a bordo.

Occupational Safety and Health Act (OSHA) A set of rules and guidelines implemented to protect employees from a dangerous work environment.

Acta de seguridad y salud en el lugar de trabajo (OSHA) Un grupo de reglas y guías implementados para proteger los empleados de los peligros en un medio ambiente del trabajo.

Octane A rating used to classify gasoline, refers to the volatility of the fuel.

Octano Una clasificación usada para clasificar la gasolina, se refiere a que tan volátil es el combustible.

Offset A misfire that results in one and one half times the emission output standard.

Desviado Un fallo del encendido que resulta en un escape que excede lo indicado por la ley de emisiones por uno y medio.

Ohmmeter A tool that measures resistance in an electrical circuit.

Ohmiómetro Una herramienta que mide la resistencia en un circuito eléctrico.

Oil pressure gauge A gauge used to test engine oil pressure.

Manómetro de la presión del aceite Calibrador utilizado para revisar la presión del aceite del motor.

Oil pump A pump used to pump oil through the engine for cooling and lubrication.

Bomba de aceite Bomba que se utiliza para impulsar el aceite a través del motor para enfriamiento y lubricación.

Oil pump pickup The screen in the oil pan that supplies engine oil to the oil pump.

Recobro de la bomba de aceite La rejilla en el cárter de aceite que suministrs el aceite a la bomba de aceite.

On-board refueling vapor recovery (ORVR) An EVAP system that is designed to eliminate HC emissions during refueling by storing them in a large canister and burning them during engine operation.

Recobro de vapor de reaprovisionamiento de combustible (ORVR) Un sistema de EVAP que es diseñado a eliminar las emisiones de HC durante el reaprovisionamiento al almacenarlos en un gran bote y quemarlos durante la operación del motor.

Open loop A computer operating mode in which the computer controls the air-fuel ratio and ignores the oxygen sensor signal.

Bucle abierto Modo de funcionamiento de una computadora en el que se controla la relación de aire y combustible y se pasa por alto la señal del sensor de oxígeno.

Optical-type pickup A pickup that contains a photo diode and a light emitting diode with a slotted plate between these components.

Captación tipo óptico Captación que contiene un fotodiodo y un diodo emisor de luz entre los cuales está colocada una placa ranurada.

Oscilloscope A cathode ray tube (CRT) that displays voltage waveforms from the ignition system.

Osciloscopio Tubo de rayos catódicos que muestra formas de onda de tensión provenientes del sistema de encendido.

Output driver A transistor in the output area of a control device that is used to turn various output devices off and on.

Ejecutador de salida Un transistor en el área de salida de un dispositivo de control, que es usado para apagar y prender varios dispositivos de ejecución.

Output state test A computer system test mode on Ford products that turns the relays and actuators on and off.

Prueba del estado de producción Modo de prueba en una computadora de productos fabricados por la Ford que enciende y apaga los relés y los accionadores.

Overhead camshaft (OHC) An engine designed with the camshaft acting directly on the top of the valve spring.

Árbol de levas en cabeza (OHC) Un motor diseñado con el árbol de levas accionando directamente arriba del resorte de la válvula.

Overhead valve (OHV) An engine with the camshaft mounted in the block. Pushrods are utilized to open and close the valves.

Válvula en cabeza (OHV) Un motor que tiene el árbol de levas montado en el bloque. Las varillas empujadoras se usan para abrir y cerrar las válvulas.

Overheating Takes place from using the wrong heat range spark plug, timing, or an overheating engine.

Sobrecalentamiento Se produce cuando se emplea una sincronización o una bujía con un rango térmico incorrecto, o un motor sobrecalentado.

Oxygen (O₂) A gaseous element that is present in air.

Oxígeno (O₂) Elemento gaseoso presente en el aire.

Oxygen (O₂) sensor A sensor mounted in the exhaust system that sends a voltage signal to the computer in relation to the amount of oxygen in the exhaust stream.

Sensor de oxígeno (O₂) Sensor montado en el sistema de escape que le envía una señal de tensión a la computadora referente a la cantidad de oxígeno en el caudal del escape.

Parade pattern The pattern that starts with the spark line of the number one cylinder and ends with the firing line of the number one cylinder.

Diagrama de desfile La diagrama que comienza con la línea de encendido cilíndro número uno y termina con la línea de encendido del cilíndro número uno.

Parallel hybrid A hybrid vehicle that uses both the electric motor and the internal combustion engine to power the drive wheels. The electric motor in the parallel system is small and uses principally the internal combustion engine except during times of low load, such as in town driving or freeway cruising.

Híbrido paralelo Un vehículo híbrido que usa un motor eléctrico y tambien un motor de combustión interno para propulsar las ruedas motrices. El motor eléctrico en el sistema paralelo es pequeño y usa principalmente el motor de combustión internos salvo cuando hay carga baja, tal como la marcha en la ciudad o en la carretera.

Parameter identification mode (PID) A mode that is accessed through an OBD II compliant scanner that provides manufacturer-specific or common vehicle information.

Modo de identificación de parámetro (PID) Un modo al cual hay acceso por medio de un explorador sumiso OBD II que provee la información específica de fabricador o común del vehículo.

Parasitic drain Battery drain on a computer-equipped vehicle.

Descarga parásita La descarga de la batería en un vehículo equipado con una computadora.

Park A transmission gear selection that disengages the transmission, and locks the output shaft so that the vehicle will not roll.

Estacionamiento Selección del sistema de transmisión que desconecta la transmisión y bloquea el eje de salida para que el vehículo no ruede.

Park/neutral switch A switch connected in the starter solenoid circuit that prevents starter operation except in park or neutral.

Conmutador PARK/neutral Conmutador conectado en el circuito del solenoide del arranque que evita el funcionamiento del arranque si el selector de velocidades no se encuentra en las posiciones PARK o NEUTRAL.

Particulates Very small pieces of matter, usually carbon, that can be harmful.

Partículos Pedacitos muy pequeños de la material, normalmente de carbon, que pueden ser peligrosos.

Parts per million (ppm) The volume of a gas such as hydrocarbons in ppm in relation to one million parts of the total volume of exhaust gas.

Partes por millón (ppm) Volumen de un gas, como por ejemplo los hidrocarburos, en partes por millón de acuerdo a un millón de partes del volumen total del gas del escape.

Passive anti-theft system (PATS) An anti-theft system that requires no action from the driver to activate.

Sistema pasivo anti-robo (PATS) Un sistema de anti-robo que no requiere acción del conductar para activarse.

Peak and hold injector circuits Circuits that incorporate unique circuitry that applies higher amperage to initially open the injector, then lower amperage for the remainder of the pulse width.

Circuitos de inyector picadura y retención Los circuitos que incorporan la circuitería que aplica un amperaje más alto para abrir el inyector desde un principio, y luego el amperaje más bajo por lo que queda de la duración del impulso.

Pending situation A circumstance that prevents a test from being performed due to an uncorrected fault.

Situación pendiente Una circunstancia que previene que se efectúa una prueba debido a un fallo no corregido.

Photoelectric tachometer A tachometer that contains an internal light source and a photoelectric cell. This meter senses rpm from reflective tape attached to a rotating component.

Tacómetro fotoeléctrico Tacómetro que contiene una fuente interna de luz y una célula fotoeléctrica. Este medidor advierte las rpm mediante una cinta reflectora adherida a un componente giratorio.

Pinging noise A shop term for engine detonation that sounds like a rattling noise in the engine cylinders.

Sonido agudo Término utilizado en el taller mecánico para referirse a la detonación del motor cuyo ruido se asemeja a un estrépito en los cilindros de un motor.

Pipe expander A tool designed to expand exhaust system pipes.

Expansor de tubo Herramienta diseñada para expandir los tubos del sistema de escape.

Pneumatic tools Tools such as impact wrenches that operate with controlled compressed air.

Herramientas neumáticas Las herramientas tal como las llaves neumáticas que operan con el aire comprimido controlado.

Polyurethane air cleaner cover A circular polyurethane ring mounted over the air cleaner element to improve cleaning capabilities.

Cubierta de poliuretano del filtro de aire Anillo circular de poliuretano montado sobre el elemento del filtro de aire para facilitar la limpieza.

Port EGR valve An EGR valve operated by ported vacuum from above the throttle.

Válvula EGR lumbrera Válvula EGR accionada por un vacío con lumbreras desde la parte superior de la mariposa.

Port fuel injection (PFI) A fuel injection system with an injector positioned in each intake port.

Inyección de combustible de lumbrera Sistema de inyección de combustible que tiene un inyector colocado en cada una de las lumbreras de aspiración.

Positive crankcase ventilation (PCV) valve A valve that delivers crankcase vapors into the intake manifold rather than allowing them to escape to the atmosphere.

Válvula de ventilación positiva del cárter Válvula que conduce los vapores del cárter hacia el colector de aspiración en vez de permitir que los mismos se escapen hacia la atmósfera.

Pounds per square inch (psi) A measurement of pressure.

Libras por pulgada cuadrada (psi) Una medida de la presión.

Power balance tester A tester designed to stop each cylinder from firing for a brief time and record the rpm decrease.

Instrumento de prueba del equilibro de la potencia Instrumento de prueba diseñado para detener el encendido de cada cilindro por un breve espacio de tiempo y registrar el descenso de las rpm.

Power tools Tools that are operated by outside power sources.

Herramienta de motor Las herramientas que se operan por medio de un suministro de potencia exterior.

Power train control module (PCM) SAE J1930 terminology for an engine control computer.

Módulo del control del tren transmisor de potencia Término utilizado por la SAE J1930 para referirse a una computadora para el control del motor.

Prelubrication Lubrication of components such as turbocharger bearings prior to starting the engine.

Prelubrificación Lubrificación de componentes, como por ejemplo los cojinetes del turbocompresor, antes del arranque del motor.

Pressure gauge A tool used to measure a gas, a liquid, or air in pounds per square inch.

Manómetro de presión Una herramienta que se usa para medir un gas, un líquido, o el aire en libras por pulgada cuadrada.

Pressurized injector cleaning container A small, pressurized container filled with unleaded gasoline and injector cleaner for cleaning injectors with the engine running.

Recipiente presionizado para la limpieza del inyector Pequeño recipiente presionizado lleno de gasolina sin plomo y limpiador de inyectores para limpiar los inyectores cuando el motor está encendido.

Programmable read only memory (PROM) A computer chip containing some of the computer program. This chip is removable in some computers.

Memoria de solo lectura programable (PROM) Pastilla de memoria que contiene una parte del programa de la computadora. Esta pastilla es desmontable en algunas computadoras.

Protocol The language or method used by a PCM to communicate to other computers.

Protocolo El lenguaje o método usado por una PCM (módulo de control de la potencia del motor) para comunicarse con las otras computadoras.

Pulse A voltage signal that increases from a constant value and then decreases back to its original value.

Pulso Una señal de voltaje que incrementa desde un valor constante y después disminuye de regreso a su valor original.

Pulse modulated A circuit that maintains average voltage levels by pulsing the voltage on and off.

Pulso modulado Un circuito que mantiene niveles promedios de voltaje pulsando en una frecuencia de voltaje de apagado a encendido.

Pulse rate The number of pulses that take place over a specific period of time.

Velocidad del puso Los números de pulso que toman lugar sobre un periodo de tiempo especifico.

Pulse width The duration from the beginning to the end of a signal's on time or off time.

Amplitud del pulso La duración desde el principio al final de una señal prendida y apagada.

Pulse width modulated injector circuit A system that uses a high current to open the injector, then lowers the current to hold the injector open by pulsing the ground circuit.

Circuito de inyector con modulación por ancho de pulso Sistema que utiliza una corriente elevada para abrir el inyector y luego disminuye la corriente para mantenerlo abierto mediante pulsos al circuito a tierra.

Pulsed secondary air injection system A system that uses negative pressure pulses in the exhaust to move air into the exhaust system.

Sistema de inyección secundaria de aire por impulsos Sistema que utiliza impulsos de la presión negativa en el escape para conducir el aire hacia el sistema de escape.

Purge valve A valve that is opened electrically by the PCM to allow the EVAP canister to be purged of stored vapors and burned in the engine when conditions dictate.

Válvula de purga Una válvula que es abierto electrónicamente por el PCM para permitir la purga del bote de EVAP de los vapores que se han almacenado y quemado en el motor cuando permiten las condiciones.

Pyrometer An electronic device that measures heat.

Pirómetro Un dispositivo electrónico que mide el calor.

Quad driver A group of transistors in a computer that controls specific outputs.

Excitador cuádruple Grupo de transistores en una computadora que controla salidas específicas.

Quick-disconnect fuel line fittings Fuel line fittings that may be disconnected without using a wrench.

Conexiones de la línea del combustible de desmontaje rápido Conexiones de la línea del combustible que se pueden desmontar sin la utilización de una llave de tuerca.

Radiator shroud A circular component positioned around the cooling fan to concentrate the airflow through the radiator.

Bóveda del radiador Componente circular que rodea el ventilador de enfriamiento para concentrar el flujo de aire a través del radiador.

Radio frequency interference (RFI) Electrical noise that may come from the ignition system.

Interferencia de frecuencia radio (RFI) El ruido eléctrico que puede provenir del sistema del encendido.

Raster pattern A pattern that stacks voltage patterns one above the other.

Diagrama raster Una diagrama que sobrepone las diagramas de voltaje uno arriba del otro.

Reactive The tendency of a chemical to combine with another substance, including human flesh. This can be violent in nature and can cause severe injury.

Reactividad Tendencia de una sustancia química a combinarse con otra sustancia, incluida la carne humana. Puede ser de naturaleza violenta y causar lesiones graves.

Readiness Information shows the technician if a particular monitor has ran successfully or not.

Disponibilidad La información de disponibilidad le indica al técnico si un monitor ha funcionado correctamente o no.

Recovery tank A reservoir that stores coolant expelled from the radiator by expansion as the coolant warms up. When the radiator cools, the coolant is drawn back into the radiator.

Tanque de recobro Un suministro que almacena el líquido refrigerante expulsado del radiador por la expansión que ocurre al calentarse el refrigerante. Cuando se enfría el radiador el líquido refrigerante regresa al radiador.

Reference pickup A pickup assembly that is often used for ignition triggering.

Captación de referencia Conjunto de captación que se utiliza con frecuencia para el arranque del encendido.

Reference voltage A constant voltage supplied from the computer to some of the input sensors.

Tensión de referencia Tensión constante que le suministra la computadora a algunos de los sensores de entrada.

Refractometer Is used to measure the coolant's freezing level.

Refractómetro Se utiliza para medir el nivel congelación del refrigerante.

Repair grade (RG-240) An economical version of the standard I/M 240 test.

Grado de reparación (RG-240) Una version económica de la prueba de convención I/M240.

Resource Conservation and Recovery Act (RCRA) A law that basically states hazardous material users are responsible for hazardous materials from the time they are produced until they are properly disposed.

Acta de Conservación y Recobro de los Recursos (RCRA) Un ley que afirma basicamente que los que usan las materias tóxicas son responsables de las materias tóxicas del momento que éstos se producen hasta que se disponen de una manera responsable.

Revolutions per minute (rpm) drop The amount of rpm decrease when a cylinder stops firing for a brief time.

Descenso de las revoluciones por minuto (rpm) Cantidad que descienden las rpm cuando un cilindro detiene el encendido por un breve espacio de tiempo.

Right-to-know law A law that states employees should be informed about the materials that they are using on the job.

Ley de Derecho en saber Un ley que afirma que los empleados deben ser informado de las propriedades de la materiales que usan en el trabajos.

Ring ridge A ridge near the top of the cylinder created by wear in the ring travel area of the cylinder.

Reborde del anillo Reborde cerca de la parte superior del cilindro ocasionado por un desgaste en el área de carrera del anillo del cilindro.

Root mean square (RMS) Meters that convert the AC signal to DC voltage signal.

Raíz de la media del los cuadrados (RMS) Los metros que conviertan los señales AC (corriente alterna a un señal de voltaje DC (corriente continua).

Rotary engine An engine that uses a rotary motion instead of a reciprocating motion. The rotary engine uses a rotor that forms combustion chambers as it turns.

Motor rotatorio Un motor que usa un movimiento rotario en vez de un movimiento reciprocativo. El motor rotario usa un rotor que forma las cámaras de combustión al girar.

Safety glasses Eyewear that protects eyes from foreign material.

Lentes de seguridad Los lentes de seguridad son anteojos que protejan los ojos de materia foránea.

Scan tester A tester designed to test automotive computer systems.

Instrumento de pruebas de exploración Instrumento de prueba diseñado para revisar computadoras de automóviles.

Scan tool May be called a scanner or a scan tester.

Herramienta de exploración También se denomina escáner o probador de escaneo.

Schrader valve A threaded valve on the fuel rail to which a pressure gauge may be connected to test fuel pressure.

Válvula Schrader Válvula fileteada que se encuentra en el carril del combustible a la que puede conectársele un calibrador de presión para revisar la presión del combustible.

Secondary air injection (AIR) system A system that injects air into the exhaust system from a belt-driven pump.

Sistema de inyección secundaria de aire Sistema que inyecta aire dentro del sistema de escape desde una bomba accionada por correa.

Self-powered test light A test light powered by an internal battery.

Luz de prueba propulsada automáticamente Luz de prueba propulsada por una batería interna.

Self-test input wire A diagnostic wire located near the diagnostic link connector (DLC) on Ford vehicles.

Alambre de entrada de prueba automática Alambre diagnóstico ubicado cerca del conector de enlace diagnóstico en vehículos fabricados por Ford.

Sequential fuel injection (SFI) A fuel injection system in which the injectors are individually grounded into the computer.

Inyección de combustible en ordenamiento Sistema de inyección de combustible en el que los inyectores se ponen individualmente a tierra en la computadora.

Serial data The PCM information sent to the scanner and other control modules.

Información en serie La información PCM enviado al explorador y a otros módulos de control.

Series hybrid Uses a small internal combustion engine to charge batteries when they become discharged. The electric motor provides power to the drive wheels. The series hybrid uses a complex controller to provide seamless acceleration from the electric motor.

Híbrido en serie Usa un pequeño motor de combustión interno para cargar las baterías cuando éstas se discargan. El motor eléctrico provee la fuerza a las ruedas motrices. El híbrido en serie usa un controlador complejo para proveer la aceleración sín interrupción del motor eléctrico.

Shimmy The rapid side-to-side vibration of a wheel/tire assembly.

Abaniqueo La vibración lateral rápida de una asamble de rueda/neumático.

Shop layout The design of an automotive repair shop.

Arreglo del taller de reparación Diseño de un taller de reparación de automóviles.

Short-term adaptive fuel trim Short-term fuel injector pulse width compensation determined by the PCM according to operating conditions. This fuel trim is set to minimize emissions output and represents minor adjustments to the long-term fuel trim strategy.

Tiempo corto para la adaptación de la restricción del combustible Compensación determinada del pulso corto para la amplitud del inyector de combustible de acuerdo con el PCM (módulo de control de la potencia del motor) para las condiciones de operaciones. Esta restricción del combustible es establecida para minimizar las emisiones y representan pequeños ajustes a la restricción del combustible por tiempo largo.

Silicone grease A heat-dissipating grease placed on components such as ignition modules.

Grasa de silicón Grasa para disipar el calor utilizada en componentes, como por ejemplo módulos del encendido.

Sinusoidal Equal rise and fall from positive and negative.

Sinusoidal La subida y caída del positivo y negativo son iguales.

Slitting tool A special chisel designed for slitting exhaust system pipes.

Herramienta de hender Cincel especial diseñado para hendir los tubos del sistema de escape.

Small leak test An EVAP test run after the vehicle passes the large leak test. The small leak test can detect a hole as small as 0.020".

Prueba de fugas pequeñas Una prueba de EVAP que se efectúa después de que el vehículo haya pasado por la prueba de fugas grandes. La prueba de fugas pequeñas puede detectar un hoyo tan pequeno que de 0.020".

Snap shot testing The process of freezing computer data into the scan tester memory during a road test and reading this data later.

Prueba instantánea Proceso de capturar datos de la computadora en la memoria del instrumento de pruebas de exploración durante una prueba en carretera y leer dichos datos más tarde.

Spark duration A measurement in milliseconds used to determine function of the spark plugs.

Duración de la chispa del encendido Una medida en milisegundos que se usa para determinar la función del las bujías de chispa.

Spark line The display of the voltage level and time of the sparks duration across the spark plug electrodes.

Línea de la bujía La presentación del nivel del voltaje y el tiempo de duración de la chispa através de los electrodos de la bujía.

Specific gravity The weight of a liquid in relation to the weight of an equal volume of water.

Gravedad específica El peso de un líquido de acuerdo al peso de un volumen igual de agua.

Splash fouling Loosened deposits found in the combustion chamber due to misfiring.

Ensuciamiento por salpicadura Los depósitos desalojados que se encuentran en la cámara de combustión debido a un fallo del encendido.

Standard corporate protocol (SCP) The term used by Ford to describe their J1850 module communication language.

Protocolo establecido de corporación (SCP) El término usado por Ford para describir su idioma de comunicación módulo J1850.

Steam cleaning Uses hot water vapor mixed with chemical cleaning agents to clean dirt from an object.

Limpieza al vapor Utiliza vapor de agua caliente mezclado con agentes químicos de limpieza para quitar la suciedad de un objeto.

Stethoscope A tool used to amplify sound and locate abnormal noises.

Estetoscopio Herramienta utilizada para amplificar el sonido y localizar ruidos anormales.

Stoichiometry An air-fuel mixture of 14.7:1 that promotes low emissions.

Stoichiometria Una mezcla de aire-combustible de 14.7:1 que promueba las bajas emisiones.

Strategy A plan. Typically refers to the programs of a PCM that insure low emissions levels.

Estrategia Un plan. Típicamente se refiere a los programas del PCM (módulo de control de la potencia del motor) que aseguran niveles bajos de emisiones.

Sulfuric acid A corrosive acid mixed with water and used in automotive batteries.

Ácido sulfúrico Ácido sumamente corrosivo mezclado con agua y utilizado en las baterías de automóviles.

Superimposed pattern The pattern that displays all patterns one on top of the other using the full width of the screen.

Diagrama superpuesta La diagrama que muestra todas las diagramas una sobre otra usando todo lo ancho de la pantalla.

Switch test A computer system test mode that tests the switch input signals to the computer.

Prueba de conmutación Modo de prueba de una computadora que revisa las señales de entrada de conmutación hechas a la computadora.

Synchronizer (SYNC) pickup A pickup assembly that produces a voltage signal for ignition triggering or injector sequencing.

Captación sincronizadora Conjunto de captación que produce una señal de tensión para el arranque del encendido o para el ordenamiento del inyector.

Tachometer A device used to measure the revolutions per minute of an engine.

Taquímetro Un dispositivo que se usa para medir las revoluciones por minuto de un motor.

Tachometer (TACH) terminal The negative primary coil terminal.

Borne del tacómetro Borne negativo de la bobina primaria.

Temperature-responding switches Switches that respond to temperature.

Interruptores accionados por temperatura Interruptores que reaccionan a la temperatura.

Temperature switch A mechanical switch operated by coolant or metal temperature.

Conmutador de temperatura Conmutador mecánico accionado por la temperatura del refrigerante o del metal.

Test spark plug A spark plug with the electrodes removed so that it requires a much higher firing voltage for testing such components as the ignition coil.

Bujía de prueba Bujía que a consecuencia de habérsele removido los electrodos necesitará mayor tensión de encendido para revisar componentes, como por ejemplo la bobina del encendido.

Thermal cleaning The process of cleaning using extremely high temperatures.

Limpieza termal El proceso de limpieza usando las temperaturas extremadamente altas.

Thermal vacuum switch (TVS) A vacuum switching device operated by heat applied to a thermo-wax element.

Conmutador de vacío térmico Dispositivo de conmutación de vacío accionado por el calor aplicado a un elemento de termocera.

Thermal vacuum valve (TVV) A valve that is opened and closed by a thermo-wax element mounted in the cooling system.

Válvula térmica de vacío Válvula de vacío que un elemento de termocera montado en el sistema de enfriamiento abre y cierra.

Thermostat tester A tester designed to measure thermostat opening temperature.

Instrumento de prueba del termostato Instrumento de prueba diseñado para medir la temperatura inicial del termostato.

Throttle body injection (TBI) A fuel injection system with the injector or injectors mounted above the throttle.

Inyección del cuerpo de la mariposa Sistema de inyección de combustible en el que el inyector, o los inyectores, están montados sobre la mariposa.

Throttle position (TP) sensor A sensor mounted on the throttle shaft that sends a voltage signal to the computer in relation to throttle opening.

Sensor de la posición de la mariposa Sensor montado sobre el árbol de la mariposa que le envía una señal de tensión a la computadora referente a la apertura de la mariposa.

Tier 2 emissions Strict emission control standards made mandatory in 2009 that place vehicles in emission categories called bins, depending on the levels of pollutants they emit.

Conmutador de la posición de la mariposa Conmutador que le advierte a la computadora si la mariposa se encuentra en la posición de la mariposa.

Tier 3 emissions Emissions and fuel mileage standards that are based on the footprint of the vehicle.

'dEmisiones de Nivel 3 Normas de emisiones y de combustible que se basan en la huella del vehículo.

Timing connector A wiring connector that must be disconnected while checking basic ignition timing on fuel-injected engines.

Conector de regulación Conector del alambrado que debe desconectarse mientras se verifica la regulación básica del encendido en motores de inyección de combustible.

Timing meter A device that shows timing degrees and engine rpm.

Medidor de tiempo Un dispositivo que muestra los grados del tiempo y las rpm del motor.

Top dead center (TDC) The highest in the cylinder a piston travels.

Punto muerto superior (TDC) Lo más alto que llega un piston en un cilíndro.

Toxic Materials are toxic if they leach one or more of eight heavy metals in concentrations greater than 100 times the primary drinking water standard.

Tóxico Los materiales son tóxicos si dejan filtrar uno o varios de entre ocho metales pesados en concentraciones que superan cien veces las normas básicas para el agua potable.

Trace *See* waveform.

Rastro Véase waveform.

Turbulence burning Is evidenced by the wearing away of the insulator on one side.

Combustión turbulenta Se manifiesta cuando el aislador se desgasta en un costado.

United States customary (USC) A system of weights and measures.

Sistema usual estadounidense (USC) Sistema de pesos y medidas.

Vacuum Any air pressure less than atmospheric pressure.

Vacío Cualquier presión de aire menos de lo atmoférico.

Vacuum pressure gauge A gauge designed to measure vacuum and pressure.

Manómetro de la presión del vacío Calibrador diseñado para medir el vacío y la presión.

Valve overlap The few degrees of crankshaft rotation when both valves are open and the piston is near TDC on the exhaust stroke.

Solape de la válvula Los pocos grados que gira el cigüeñal cuando ambas válvulas están abiertas y el pistón se encuentra cerca del punto muerto superior durante la carrera de escape.

Valve stem installed height The distance between the top of the valve retainer and the valve spring seat on the cylinder head.

Altura instalada del vástago de la válvula Distancia entre la parte superior del retenedor de la válvula y el asiento del resorte de la válvula en la culata del cilindro.

Vane-type airflow sensor A sensor containing a pivoted vane that moves a pointer on a variable resistor. This resistor sends a voltage signal to the computer in relation to the total volume of intake air.

Sensor tipo paleta Sensor con una paleta articulada que mueve un indicador en un resistor variable. Este resistor le envía una señal de tensión a la computadora referente al volumen total de aire aspirado.

Vaporization The process in which a liquid changes to a gas or vapor.

Vaporización El proceso en el cual un líquido cambia de un gas a un vapor.

Variable rate sensor signal (VRS) Provides the ignition module with engine position and rpm information.

Señal del sensor de regimen variable (VRS) Provee la información de la posición del motor y las rpm al módulo del encendido.

Variable valve timing Allows the PCM to fine tune the camshaft profiles to meet engine operating conditions.

Sincronización variable de válvulas Permite que el módulo de control del tren de potencia (PCM, por sus siglas en inglés) afine los perfiles del árbol de levas para ajustarlos a las condiciones de funcionamiento del motor.

Vehicle communications module (VCM) or J2534 pass-through Ford calls their J2534 pass-through programming device the vehicle communications module (VCM). The VCMII was introduced in 2012.

Módulo de comunicaciones vehicular o dispositivo de paso J2534 Ford llama módulo de comunicación vehicular (VCM, por sus siglas en inglés) al dispositivo de programación de paso J2534. El VCMII se introdujo en 2012.

Vehicle networking The use of several modules to perform tasks on a vehicle that are all interconnected and using a common language.

Intercomunicación del vehículo El uso de varios módulos para efectuar los deberes del vehículo que son interconectados y que usan una idioma común.

Vehicle speed sensor (VSS) A sensor that is usually mounted in the transmission and sends a voltage signal to the computer in relation to engine speed.

Sensor de la velocidad del vehículo Sensor que normalmente se monta en la transmisión y que le envía una señal de tensión a la computadora referente a la velocidad del motor.

Vent valve An EVAP valve opened to allow fresh air through the charcoal canister. It is closed when the PCM needs to measure pressure or vacuum changes.

Válvula de ventilación Una válvula de EVAP que abre para permitir entrar aire fresca entrar al bote de carbón. Se cierra cuando el PCM necesita medir la presión o cuando cambia el vacío.

Viscosity A term used to describe a liquid's ability to flow.

Viscocidad Un termino usado para describir la habilidad de un líquido en fluir.

Viscous-drive fan clutch A cooling fan drive clutch that drives the fan at higher speed when the temperature increases.

Embrague de mando viscoso del ventilador Embrague de mando del ventilador de enfriamiento que acciona el ventilador para que gire más rápido a temperaturas más altas.

Volt-amp tester A tester designed to test volts and amps in such circuits as battery, starter, and charging.

Instrumento de prueba de voltios y amperios Instrumento de prueba diseñado para revisar los voltios y los amperios en circuitos como por ejemplo, de la batería, del arranque y de la carga.

volt/ampere tester (VAT) Used to test batteries, starting systems, and charging systems.

Tester o multímetro de VA El multímetro de VA (VAT, por sus siglas en inglés) se utiliza para probar baterías, sistemas de arranque y sistemas de carga.

Voltmeter A tool used to measure electrical pressure.

Voltímetro Una herramienta que se usa para medir la presión eléctrica.

Wankel engine *See* rotary engine.

Motor Wankel Véase motor rotario.

Waste-gate stroke The amount of turbocharger waste-gate diaphragm and rod movement.

Carrera de la compuerta de desagüe Cantidad de movimiento del diafragma y de la varilla de la compuerta de desagüe del turbocompresor.

Waveform The pattern on an oscilloscope.

Forma de onda La diagrama en un osciloscopio.

Weak vacuum test This will establish if there is a large system leak. Once the enable criteria are set, the PCM will close the vent valve and open the purge valve. This would of course cause a vacuum to be indicated on the fuel tank pressure sensor if the system were sealed. If the vehicle should fail to pull a vacuum of up to 7 in. Hg, then the MIL would illuminate after the second consecutive failure under similar conditions and set a code PO455, gross evaporative system leak.

Prueba de vacío débil Ésto establece si hay una fuga grande en el sistema. Una vez que se establece el criterio autorizado, el PCM cerrará la válvula de ventilación y abrirá la válvula de purga. Ésto por supuesto causaría que se indicara un vacío en el sensor del tanque de combustible si estuviera sellado el sistema. Si el vehículo fallara en sostener un vacío llegando a 7 in. Hg, entonces el MIL iluminaría después del segundo fallo consecutivo bajo condiciones parecidas y pondría un código de PO455, fuga masiva del sistema evaporativo.

Wet compression test A cylinder compression test completed with a small amount of oil in the cylinder.

Prueba húmeda de compresión Prueba de la compresión de un cilindro llevada a cabo con una pequeña cantidad de aceite en el cilindro.

Wet fouling A term describing the tip of the spark plug being drowned in excess oil.

Ensuciamiento en húmedo Un término describiendo que la punta de la bujía de chispas se ha sumerjido en un exceso de aceite.

Wide range air/fuel ratio sensor A sensor that measures the air/fuel ratio by sending different voltages and amperages to the PCM.

Sensor de relación aire de alto alcance/combustible Sensor que mide la relación entre el aire y el combustible enviando distintos voltajes y amperes al PCM.

Wiggle test A test in which the PCM "sees" and indicates a problem with wires or connections while the technician wiggles or moves the wiring.

Prueba de meneo Una prueba en la cual el PCM "vea" e indica un problema con las alambres o las conexiones mientras que el ténico menea o mueva los alambres.

Workplace hazardous materials information system (WHMIS) The Canadian version of MSDS sheets.

Sistema de información de materiales peligrosos del trabajo (WHMIS) La versión canadiense de las hojas MSDS.

Zener diode A special type of diode that allows current to flow in the desired direction only when certain conditions are met.

Diodo Zener Un tipo especial de diodo que permite que la corriente fluya en la dirección deseada solamente cuando ciertas condiciones son cumplidas.

INDEX

Note: page references with *f* notation refer to a figure on that page.